International Conference “Ecological Paradigms of Sustainable Development: Political, Economic and Technological Dimension of Biosphere Problems” (EPSD 2021)

E3S Web of Conferences Volume 311

St. Petersburg, Russia
30 - 31 August 2021

Editor:

Stanislav Germanovich Eremeev

ISBN: 978-1-7138-3895-1

Printed from e-media with permission by:

Curran Associates, Inc.
57 Morehouse Lane
Red Hook, NY 12571

CURRAN ASSOCIATES INC.
proceedings.com

Some format issues inherent in the e-media version may also appear in this print version.

For additional information, please contact EDP Sciences – Web of Conferences at the address below.

EDP Sciences – Web of Conferences
17, Avenue du Hoggar
Parc d'Activité de Courtabœuf
BP 112
F-91944 Les Ulis Cedex A
France

Phone: +33 (0) 1 69 18 75 75
Fax: +33 (0) 1 69 28 84 91

contact-edps@webofconferences.org

Additional copies of this publication are available from:

Curran Associates, Inc.
57 Morehouse Lane
Red Hook, NY 12571 USA
Phone: 845-758-0400
Fax: 845-758-2633
Email: curran@proceedings.com
Web: www.proceedings.com

TABLE OF CONTENTS

PREFACE

International Scientific and Practical Conference

"Ecological Paradigms of Sustainable Development: Political, Economic and Technological Dimension of Biosphere Problems"

(EPSD 2021)

We are pleased to present this collection of scholarly articles of the International Scientific and Practical Conference " Ecological Paradigms of Sustainable Development: Political, Economic and Technological Dimension of Biosphere Problems" (EPSD 2021). The conference was held in Leningrad State University named after A.S. Pushkin, Saint Petersburg, Pushkin, Russian Federation, on August 30-31, 2021.

The key sections of the Conference were:

- Environmental and social responsibility
- "Green universities" and the formation of environmental culture
- Environmental policy instruments
- Environmental justice
- Reflections on environmental risks
- Energy saving and energy efficiency improvement
- Technologies in regional governing planning
- Ecological technologies of smart cities: moving to sustainable development
- Policy of human resource management in context of sustainable development
- Rational use of natural resources
- Food biotechnologies for human health

The main scope of the conference was to create a discussion space for the participants to consider the strategic tasks of sustainable development of ecosystems, the main political economic and social processes in connection with environmental problems, contribute to the research of the new trends of innovative ecology and environmental security, as well as to work out new approaches and strategies in the field of biosphere management and protection of national environmental interests.

The Conference has received 257 manuscripts, of which 82 submissions have been accepted by our reviewers (well-known international experts in the field of sustainable development). Full papers submitted to the conference were peer-reviewed by two - tree reviewers per manuscript before being selected for publication. In case when two experts had disagreements, the paper was sent to another two reviewers for evaluation.

We would like to express our high appreciation to the organizers, members of the program committees and reviewers of EPSD 2021 for their hard work. Without their participation, it would be impossible to hold this meeting successfully and ensure high quality of papers of the conference proceedings. We also would like to express our gratitude to the unbelievable number of participants for contributing their research results to the Conference. Special acknowledgements go to E3S Web of Conferences team. We also hope to meet again in the upcoming conferences.

edp sciences

E3S Web of Conferences

Statement of Peer review

In submitting conference proceedings to E3S Web of Conferences, I certify to the Publisher that I adhere to the **Policy on Publishing Integrity** of the journal in order to safeguard good scientific practice in publishing.

1. All articles have been subjected to peer review administered by the proceedings editors.
2. Reviews have been conducted by expert referees, who have been requested to provide unbiased and constructive comments aimed, whenever possible, at improving the work.
3. Proceedings editors have taken all reasonable steps to ensure the quality of the materials they publish and their decision to accept or reject a paper for publication has been based only on the merits of the work and the relevance to the journal.

Title, date and place of the conference

EPSD 2021 - International Conference "Ecological Paradigms of Sustainable Development: Political, Economic and Technological Dimension of Biosphere Problems"
Saint Petersburg, Russia
August 30 - 31, 2021

Proceedings editor(s):

Eremeev Stanislav Germanovich

Date and editor's signature

12/10/2021

Immersive and cultural transformation of the system of additional education as a tool for the development of environmental and social responsibility of the society

Roman Oparin[1,2,*], *Elena* Arbuzova[3,4,5], and *Sergey* Nazarov[6]

[1]Moscow State Regional University, Russia
[2]Novosibirsk State Pedagogical University, Russia
[3]Omsk Academy of Humanities, Russia
[4]Omsk State University named after F.M. Dostoevsky, Russia
[5]Moscow State Regional University, Russia
[6]Omsk State Pedagogical University, Russia

Abstract. The article scientifically substantiates the cultural and creative concept of additional environmental education, as a methodological basis for the formation and development of environmental and social responsibility, which involves the development of four levels of value orientations of students: reproductive, heuristic, creative. The concept is aimed at the formation of the individual activity of students in solving environmental problems on the basis of social, scientific and aesthetic axiological systems that determine the content of ecological-biotechnical, ecological-research, ecological-aesthetic activities. The main result of the implementation of this concept is the ability of students to plan their activities for the study, preservation and creative transformation of the environment independently, using traditional and immersive technologies. Scientists have designed a cultural and creative system of additional environmental education for students using immersive learning technologies, which is considered as a tool for the development of environmental and social responsibility of society. The article presents the results of an experimental test of the effectiveness of the cultural and creative system of additional environmental education using traditional and immersive learning technologies, which has shown the high effectiveness of the influence of innovative forms, means and methods of additional environmental education in the formation of environmental value orientations of students.

1 Introduction

In the context of global environmental problems and the inevitably impending environmental crisis on the planet, to ensure the survival of mankind, both cardinal political, technological, science-intensive solutions and innovative changes in the field of socio-economic and

* Corresponding author: 89236613134@inbox.ru

cultural policy, especially in the field of education, are required. In this context, the determining factor is the rethinking of the traditional value and worldview attitudes that determine the attitude to nature, the degree of its transformation, the formation of a balanced lifestyle of a person, his ecologically balanced needs. According to the Concept and Goals of Sustainable Development adopted by the world community, by 2030 the priority areas of education are the formation of a new world outlook, lifestyle, culture of sustainable development, based on the modern scientific picture of the world, the values of environmentally sustainable development, the ability to identify in the surrounding world interconnections of natural, economic and social processes, global, local and personal, past, present and future; as well as experience of practical actions in the interests of sustainable development of the local community.

Continuous environmental education acts as a platform for the implementation of education for sustainable development in many countries of the world, as in our country. In fact, this is a new social order for the environmental education system at all its levels, including in the system of additional education. The field of additional education carries a huge potential for the formation of environmental and social responsibility, since it provides students with an independent choice of an educational trajectory, integrating humanitarian, natural science and practice-oriented approaches to environmental education. This approach to additional environmental education does not “cancel”, but complements classical education in the field of bio ecology and geo ecology, significantly changing the ideas about the goals and methods of ecologization of the educational process, the content of the environmental competence of teachers, the role of general education in its formation.

An important direction in the development of additional environmental education is its information and digital orientation. A new generation of students (generation Z, children of the 2000s [1]) lives in a digital environment, which is formed by immersive technologies, including such educationally significant ones as immersive technologies [2]: 3D visualization technologies, virtual, augmented, mixed reality, 360° video, virtual laboratories, cryolized hypertexts, etc. The relevance and significance of the use of immersive technologies in all areas of education is reflected in the new UNESCO recommendations ("UNESCO ICT Competency Framework for Teachers. VERSION 3", 2018) [3], corresponding to "The 2030 Agenda for Sustainable Development" adopted by the UN General Assembly [4].

Summarizing the above approaches to the study of the problem of environmental education, it should be noted that there have been no special studies devoted to the design of an innovative system of additional environmental education as an integrating factor of the general education system in terms of the formation of environmental and social responsibility of society. This urgent problem, in our opinion, requires further research, especially in terms of conceptualization and development of methodological and technological aspects. Without solving this problem, the continuity of the formation of an ecological culture of general additional, university and postgraduate education is violated. The educational process takes place without a sufficiently complete and deep substantiation of the methodological, theoretical and technological foundations of environmental education.

The relevance of the research topic determined the scientific problem of the research, which is the search for an answer to the question: "What should the system of additional environmental education be, which would effectively form the environmental and social responsibility of society?"

The basic idea of the research is to form students' environmental and social responsibility in the system of additional environmental education based on the cultural concept with the use of immersive technologies.

The purpose of the study is to increase the level of environmental and social responsibility of students through the use of a cultural-creative system of additional environmental education in the educational process with the use of immersive learning technologies.

2 Materials and Methods

The main idea of the Cultural concept, analyzed in the context of our article, is the cultural and cultural edifying role of the student's personality. Environmental education within the framework of this concept is understood not only as a reproducing activity, ensuring the maintenance of the achieved level in social and ecological development, but also as the creativity of the student's personality, aimed not only at the assignment (de-objectification) of cultural values (at the level of environmental value orientations), their preservation (maintenance and restoration of natural and cultural heritage), but also the creation of something new in culture (artistic and creative transformation of the environment).

The main theoretical positions, on which the cultural concept is built, include four main components.

1. The activity of the object of pedagogical influence, aimed at the study, preservation and artistic and creative transformation of the surrounding socio-natural environment. An important guideline for such activities is the system of environmental value orientations, acquired in the procedures of empathy, active involvement, personal participation in solving environmental problems, on the scale of their region - in particular, and our planet - as a whole.

2. The activity of the subject of pedagogical influence sets the algorithms for an ecological-upbringing dialogue with an object in the process of mastering ecological value orientations and diagnosing the quality of this mastery. The activity of the subject allows the progressive development of environmental and social responsibility to be deployed in the spatial vector of environmental education.

3. Environmental and social responsibility and environmental value orientations as an indicator of their formation are the main, system-forming component of the spiritual and material space of the concept.

4. Diagnostics at each age stage of environmental education has a common vector of consistent correction of the development of environmental value orientations of the individual, building in the mind of the pupil a holistic picture of nature and man in it. At each level of diagnostics, its own intermediate goals, problem-content and psychological-pedagogical determinants are distinguished, which makes it possible to build horizontal interdisciplinary connections, to model in the interaction of forms and means of cultural activity the axiological space to be mastered.

- The interaction of the activities of a teacher, student and diagnostic tools in the field of formation and development of environmental value orientations is carried out in the axiological space of the concept. The axiological space includes a system of ecological value orientations, which we differentiate into eight groups: vital and biological; ecological and social; ecological-patriotic; ecological and ethical; ecological and aesthetic; environmental research; ethno-ecological and pedagogical.
- The implementation of the cultural and creative concept of environmental education in the vector of the formation of environmental value orientations presupposes an awareness of the multilevel nature of cultural and educational activities. Taking into account the stages of the formation of environmental and social responsibility (ID Zverev [5], IT Suravegina [6]) as well as the psychological and physiological characteristics of this group (IV Shapovalenko [7], VA Yasvin [8]) we have identified four levels of development of ecological value orientations, each of which determines the content of forms and means of ecological education.
- The adaptive level is the starting one and is characterized by an unstable attitude of students to the development of value orientations that are within the vital biological axiological group. The system of ecological value orientations, for the most part, is absent.

- The reproductive level is manifested in the motivation to assimilate the spiritual ecological values of one's ethnic group and the ways of their interpretation, based on the algorithmic model proposed by the teacher. Students maintain constant contact with the teacher and have self-control over their activities and their effectiveness. At the same time, the degree of independence of students is rather low.
- The heuristic level is characterized by greater purposefulness, stability, awareness of the development of environmental moral and patriotic values. Guided by them, the student continues to seek and discover new ways of environmentally sound solutions. It increases the independence and the creative activity of the students, their ability to plan individual operations when solving environmental problems at the local level.
- The creative level is characterized by a high degree of effectiveness of the individual activities of students in solving environmental problems on the basis of social, scientific and aesthetic axiological systems that determine the content of ecological-biotechnical (environmental), ecological-research (scientific), ecological-aesthetic (associated with the creative transformation of the environment) activities. The structure of the student's personality harmoniously combines scientific and practical interests in the field of ecology, a high level of reflection and creative independence. The main characteristic of achieving this level of development of value orientations is the ability to independently plan their activities for the study, preservation and creative transformation of the environment.

Thus, we consider the cultural-creative concept as a system of ideas, views, attitudes, which should serve as a starting theoretical base, a strategic guideline for a specific search for the necessary model of modern environmental education in the context of the development of environmental and social responsibility of society.

3 Results and Discussion

The main task of the first series of experiments was to prove that the system of additional environmental education developed on the basis of the formulated conceptual provisions creates conditions for the holistic and effective formation of environmental value orientations of students.

The first series of experiments showed a fairly high efficiency of the influence of innovative forms of cultural and educational activities (environmental and cultural training, environmental movements, interactive games; environmental and cultural excursions, ethno-environmental expositions, environmental and cultural clubs) on the formation of environmental value orientations. So, in the experimental groups of students at the reproductive, heuristic and creative stages of the development of ecological value orientations, the curves of the levels of formation of their main determinants are, on average, 14 - 25% higher than in the control groups.

The second series of experiments showed a fairly high efficiency of the means of organizational and methodological support (on the example of the special course "Altai Zapovedny") in the formation of ecological value orientations of schoolchildren in the context of national and cultural values. The high and above average levels of development of ecological value orientations, taking into account the regional aspect, in the experimental group are from 39.3 to 41.2%, while in the control group the similar levels are from 34.5 to 35.6%.

The third series of experiments showed that the most effective forms and methods for the development of ecological value orientations of schoolchildren are practical cultural and educational activities of an ecological orientation (cultural and creative practices that are revealed through the ecological-local history, ecological-biotechnical and ecological-aesthetic directions). Thus, the level of formation of the cultural and creative determinant of the ecological value orientations of "eco-researchers" is on average 21.2% - than that of

 https://doi.org/10.1051/e3sconf/202131101001

senior pupils and students of the control group. The level of formation of the cultural determinant of the ecological value orientations of “eco-designers” is, on average, 25.9% higher than in the control.

The reliability of the experimental results was determined on the basis of the X^2 (chi-square) goodness-of-fit criterion at a significance level of 0.05. On the basis of this criterion, the hypotheses put forward about the statistical significance of the observed differences in the control and experimental groups at the initial and final stages of training are tested. The calculation results show that Tnabl ≥ Tkrit, which confirms the reliability of the results of our research.

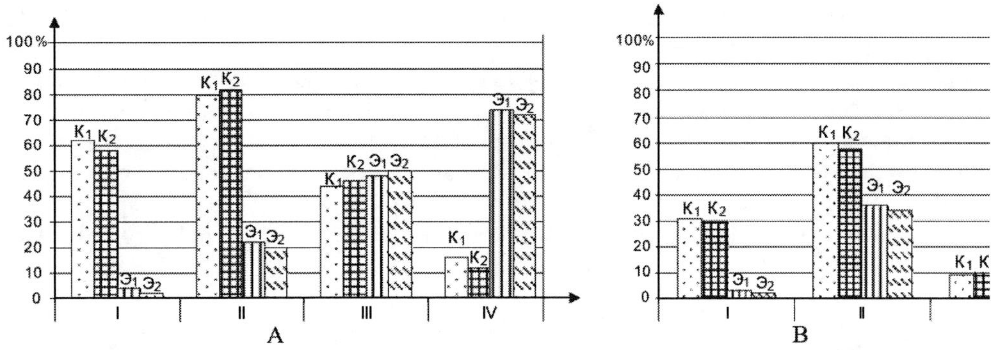

Fig. 1. Formation of environmental and social responsibility of students of the first (A) and second (B) series of experiments. C1 - control group No. 1; C2 - control group No. 2; E1 - experimental. group No. 1; E2 - experimental group No. 2.

4 Conclusion

In the course of the study, the following results, reflecting its novelty, theoretical and practical significance were obtained.

1. The cultural and creative concept of additional environmental education has been scientifically substantiated as a methodological basis for the formation and development of environmental and social responsibility using a complex of immersive technologies. The main result of the implementation of this concept is the ability to plan their activities for the study, preservation and creative transformation of the environment independently, using traditional and immersive technologies.

2. A cultural-creative system of additional environmental education for students has been designed using immersive learning technologies, the conditions for the effective functioning of which are: cultural-creative orientation (providing for the ordering of levels and types of additional environmental education due to the idea of environmental cultural creativity), dynamism (ensuring the formation and development of environmental value orientations through a set of innovative forms and means of interaction between a teacher and a student in a natural and immersive environment), predictiveness (determining the possibility of assessing indicators of the formation of environmental value orientations using the technology of axiological diagnostics of the results of environmental education using immersive technologies).

3. A regionally-oriented software and methodological complex has been created, including a package of software and normative documentation, a set of management tools for the implementation of environmental education, scientific and methodological recommendations for its organization, a set of methods regulating the current and final diagnostics. The effectiveness of the relationship of forms, means and methods within the

framework of the implementation of the regional program and methodological complex was monitored in all the listed structures using a system of competence trainings, the results of which were used as the basis for the experimental part of the work [9].

4. Experimental and experimental testing of the effectiveness of the cultural and creative system of additional environmental education using traditional and immersive learning technologies was carried out, which showed the high efficiency of the influence of innovative forms, means and methods of additional environmental education in the formation of environmental value orientations of students. The most effective forms and methods are the inclusion of students in practical cultural and educational activities of environmental orientation (interactive exhibitions, virtual polygons, interactive games and AR-quests) [10].

So, in the course of the study, it was confirmed that the cultural concept and the system of organizational and methodological support create favorable conditions for the implementation of additional environmental education for students. Experimental testing of its effectiveness has confirmed the advisability of choosing the forms and means of environmental education, showing a high level of development by students of environmental value orientations, a wide range of use of skills in the field of ecology and nature protection.

The authors do not consider this study to be exhaustive; research prospects can be associated with the further development of immersive technologies in the theory and practice of environmental education.

References

1. V.I. Blinov, Draft didactic concept for digital vocational education and training, 72 (2019)
2. I.V. Robert, Pedagogical informatics, **3**, 141 (2020)
3. UNESCO ICT Competency Framework for Teachers. VERSION 3, (2018), gcedclearinghouse.org
4. The 2030 Agenda for Sustainable Development. Official site of UNESCO (Russian version), https://www.un.org
5. I.D. Zverev, Formation of ecological education in Primorsk Territory. Those trends and perspectives. Materials for the conference, 16 (1998)
6. I.T. Suravegina, Theory and practice of the formation of a responsible attitude of students to nature in the process of teaching biology: avtoef. dis., 215 (1996)
7. I.V. Shapovalenko, Developmental psychology and developmental psychology, 349 (2014)
8. S.D. Deryabo, Methods of diagnostics and correction of attitude to nature, 147 (1995)
9. R. Oparin, E3S Web of Conferences 210, 22033, Annual scientific conference "Innovative technologies in science and education" (2020), https://itno.donstu.ru/
10. R. Oparin, SAHD 2021 - 5th International Scientific and Practical Conference 2021 "Modern Science: Problems and Development Prospects" (Social and Humanitarian Directions), 101 (03018) (2021)

Environmental safety of the air transport: the problems and the solutions

E. E. Rumyantseva[1,*], *A.V.* Gubenko[2], and *A. R.* Pankratova[2]

[1]Economic Policy and Business Center, Moscow, Russia
[2]Federal State Budgetary Educational Institution of Higher Education «Saint-Petersburg State University of Civil Aviation», St. Petersburg, Russia

Abstract. This article presents the results of an interdisciplinary study of problems and ways to solve them in the field of ensuring the environmental safety of air transport in Russia and in the world, carried out by its authors. The environmental safety of air transport in Russia is provided in Russian legislation to a greater extent formally, but based on a rather deeply developed concept of "safety", covering all areas of life. Wherein, there is an acute lack of collection and publication of quantitative indicators on which development of this research topic shall be based, in the adopted regulatory legal acts there is inconsistency and lack of development is felt. The authors focus on the need for scientific research in Russia and preparation of regulatory legal acts at a new qualitative level as overcoming Russia's position as a catching-up country in this and a number of other areas, presenting new promising areas as a result of their scientometric work, including personal responsibility of passengers for reducing carbon emissions from their air travel.

1 Introduction

Being an important part of the transport sector, air transport provides convenient transportation and significant economic benefits, but it is also a major environmental pollutant. The greatest attention of the international community is paid to the problem of carbon emissions from air transport. In 2009, the International Air Transport Association (IATA) has launched an initiative to reduce carbon emissions by 50 % by 2050 to 2005 levels. In 2016, the International Civil Aviation Organization (ICAO) drew attention to this initiative, indicating that without mitigation measures, carbon emissions from air travel will reach 2.6 billion tonnes by 2050, which is 22 % of the global total and will become almost 11 times more than the current level of pollution (about 2 %) [1]. ICAO recently adopted a "carbon neutral growth from 2020" resolution requiring CO_2 emissions from international aviation not to exceed 2019–2020 levels [2].

The progress in ensuring the environmental safety of air transport is made up of the efforts of many countries around the world. Due to the fact that Russia is tacitly guided by the position of the catching-up country on this issue, which is also manifested in the approaches to settlement of territorial environmental problems [3] and the country's

* Corresponding author: e.p.centre@mail.ru

orientation towards use of foreign-made air transport, it seems relevant to draw the attention of Russian researchers and authorities to these problems in order to raise the quality of their scientific developments and regulatory legal acts to a new level.

2 Materials and methods

Systematically, the topic of environmental safety of air transport is part of the safety concept currently applied in the Russian legislation and also one of the aspects of an integrated approach to ensuring the safety of air transport. The nationwide theoretical concept of safety, shared by Russian science, is sufficiently high-tech reflected in the Federal Law "On Safety", and considering the specifics of industry management, for example, in the Concept of the Federal Target Program "Ensuring the Flight Safety of Aircraft of the State Aviation of the Russian Federation in 2010-2014".

In Article 1 of the Federal Law "On Safety", environmental safety is part of the general concept of "safety" along with its other parts, the study of which is based on application of the following five principles of their provision that are common for all types of safety, namely:

observance and protection of human and civil rights and freedoms;

legality, i.e. legal certainty and control over their implementation;

consistency and complexity of application of political, organizational, socio-economic, informational, legal and other safety measures;

priority of planned preventive safety measures in comparison with emergency rescue;

interaction of federal bodies of state power, bodies of state power of the constituent entities of the Russian Federation, other state bodies with public associations, international organizations and citizens in order to ensure safety [4].

The extent to which these principles are implemented in the practice of ensuring the environmental safety of air transport is the subject of research in more than one scientific work, including the analysis of information sources for which there is an obvious deficit. However, even in the text of the Law itself, the concept of "environmental safety" is used only once, and its specificity, accordingly, is not disclosed. This determines the advisability of disclosing the essence of the concept and the mechanism for ensuring "environmental safety" in other regulatory legal acts, including the legal regulation of the activities of Russian air transport.

In the Concept of the federal target program "Ensuring the Flight Safety of Aircraft of the State Aviation of the Russian Federation in 2010-2014":

a fairly clear definition of the categories of "flight safety" and "hazardous factors" subdivided into internal and external ones (i.e., the leading role of the need to establish causal relationships and use of factor analysis in studies of air transport safety) is given;

the most common causes of accidents related to the actions or inaction of aviation personnel and/or the state (behavior) of the aircraft or flight support facilities were named;

the feasibility of creating a unified flight safety management system based on application of the standards and recommendations of the International Civil Aviation Organization was determined;

the most important role of quality and timeliness of collection, processing and analysis of complete and reliable information on the functioning of the air transport system that is part of the continuous monitoring of the effectiveness of the unified safety management system as a whole and its links is substantiated;

the unified safety management system also emphasizes the important role of aircraft accident investigations, information on which shall be concentrated within the framework of a single accident investigation center for the state aviation designed to identify and eliminate hazards even before it becomes necessary to consider them as the causes of aviation incidents.

Wherein, it was noted in this Concept that at the time of its development and adoption, achievement of such a level of safety within the framework of the main activities of the bodies responsible for development and functioning the state aviation was not ensured [5]. That is, as of 2009, at the official level, a significant problem of the insufficient level of ensuring the safety of flights of state aviation aircraft was recognized. However, at the same time, the concept of "environmental safety" has never been used in this document, which indicates insufficient attention to this problem and incompleteness of creation of the system of state and public control over implementation of regulatory legal acts, since information about the Concept of the Federal Target Program "Ensuring the Flight Safety of Aircraft of the State Aviation of the Russian Federation in 2010-2014" with establishment of the maximum (forecast) amount of financing of the Program from the federal budget in the amount of 11 billion rubles, were officially published, and the indicators of its implementation, including data on cancellation or prolongation, estimates its effectiveness is absent.

Wherein, in the State Program for Ensuring the Safety of Civil Aviation Aircraft, adopted a year earlier, one of the tasks of "achieving an acceptable level of flight safety" is "to ensure a systematic approach to identifying hazard sources and controlling risk factors to minimize the number of human casualties, as well as the size of material, environmental and social damage". It is clear that the environmental safety of flights is not put in the first place in comparison with the preservation of human lives, but it is nevertheless considered in this document along with property and social damage. However, within defining safety management, along with minimizing human casualties, property damage, financial ones are added to the social and environmental consequences. That is, even one state document contains inconsistencies and contradictions between conceptually important formulations of related parts – definition of safety management, which (according to the requirements of science) shall include the most important features of the concept, and one of the tasks, the wording of which shall not coincide with the definition. At the same time, in this document, the adjective "ecological", except for the above two fragments, is not used anywhere else, and, accordingly, the mechanism for minimizing environmental damage is not considered, which is one of the reasons that it is still substantially incompletely implemented in practice [6].

The report of the Federal Air Transport Agency "On the Results of the Work of the Federal Air Transport Agency in 2019, the Main Tasks for 2020 and for the Medium Term" does not reflect the problem of the environmental safety of air transport in any way, even purely formally and fluently, which indicates an underestimation of significance of this complex problems for Russia in the current activities of the federal authority responsible for development of this industry [7].

These and other regulatory legal acts that do not consider the environmental safety of air transport as the most important direction of state policy in compliance with achievements of the world and all-Russian theory of effective management – from goal-setting to assessing their implementation with the transition from it to the next stage of goal-setting, at the same time significantly emphasize attention to the main methods that shall be used not only in fundamental scientific research, but also in current analytical activities - systemic and integrated approaches, causal and factor analysis, fundamental principles of safety assurance, including in relation to the environmental safety of air transport.

3 Results and discussion

Despite the significant shortcomings while reflecting the problems of ensuring the environmental safety of air transport in the regulatory framework, which, according to researchers [for example, 8], cannot be considered complete, and the search for ways of sustainable development of air transport in Russia has not yet been completed [9], the Transport strategy of the Russian Federation for the period up to 2030, among the tasks aimed at achieving goal 6 "Reducing the negative impact of the transport system on the environment" sets "increasing environmental requirements for aircraft (in terms of noise and emissions)" and also "introduction of new environmentally friendly clean substances for aviation technologies, reduction of use of harmful substances, development of technologies for their utilization, implementation of research programs in the field of aviation biofuels". Appendix 3 to this Strategy provides a change (in relation to the level of 2011) in the number of quantitative indicators for achieving goal 6 "Reducing the negative impact of the transport system on the environment" for air transport in two ways:

on an innovative option

first level indicators

1. the volume of CO_2 emissions per one reduced t-km – up to 66 %;

2. the volume of emissions of air pollutants per reduced t-km – up to 60 %;

second level indicators

3. average specific fuel/electricity consumption per one reduced t-km – up to 66 %;

4. the share of air transport organizations that have implemented environmental management systems for environmental quality management and environmental safety in transport in the total number of air transport organizations – an increase from less than 1 % in 2011 to 100 % in 2030;

according to the basic (conservative) option

first level indicators

1. the volume of CO_2 emissions per one reduced t-km – reduction up to 80 %;

2. the volume of emissions of air pollutants per reduced ton-km – reduction of up to 70 %;

second level indicators

3. average specific fuel/electricity consumption per one reduced t-km – reduction up to 80 %;

4. the share of air transport organizations that have implemented environmental management systems for environmental quality management and environmental safety in transport in the total number of air transport organizations – increase from less than 1 % in 2011 to 95 % in 2030 [10].

As follows from the data provided, the two options considered in the Transport Strategy of the Russian Federation have significant differences and, of course, lead to different levels of environmental safety of air transport, which, in our opinion, are seriously underestimated – indicators of environmental safety of air transport could be much higher in the context of a closer integration of the Russian government and science.

Of course, not only the planned indicators provided in the two options are of interest, but also their implementation. Reports on implementation of the Transport Strategy of the Russian Federation for 2015-2020 have been published on the website of the Ministry of Transport of the Russian Federation, which provide some selected indicators for achievement of goal 6. For example, the Report for 2020 provides some (not all) indicators for rail and road transport only. It is specified that these indicators are provided by official statistical observation. There are no data to check the planned indicators for the environmental safety of air transport in the Report for 2020, that is, monitoring of their implementation has not started and is not reflected in any way [11]. The problem of the published information reliability is also important [12].

Our analysis shows that the Russian legislation has not yet implemented a systematic and integrated approaches to ensuring the environmental safety of air transport, and research developments in this area have not been fully considered. For example, an international group of researchers offers to develop transport projects based on the methodology of balancing economic interests and negative environmental and social consequences [13]. Spanish researchers provided a dynamic analysis of changes in CO_2 emissions in the European Union as a result of renewal of the aircraft fleet [14], which, in our opinion, may be required for monitoring the implementation of the Transport Strategy of the Russian Federation for this first level indicator. E.Yu. Starkov and N.I. Nikolaikin focus attention on the problems of emergency volley environmental impact at aviation catastrophes by offering a range of measures to reduce the negative impact on the environment of this part of the environmental safety of the air transport system, along with the problem of CO_2 emissions and thereby negative impact to climate change [15]. Australian researcher M. Mkono draws attention to the environmental movement of citizens who refuse to fly on airplanes that emerged in 2018 in Sweden (including publicly, attracting media attention) due to implementation of their civil liability for the negative impact of air travel on the environment [16]. M. Mkono and a number of other foreign researchers believe [17] that the participants in this movement have a distorted perception of the impact of air transport on the environment, and, in general, condemn intervention of society in specific issues of industry development. Other researchers, opposing the negative attitude to the movement of flight refusals due to climate change, on the contrary, conclude that the increased awareness of citizens as one of the stakeholders in reducing toxic air emissions from air travel and their impact on formation of long-term climate policy [18]. These publications represent a completely new direction of research in Russia, initiating development of an analysis of the interaction of the state, airlines and passengers in resolving the most complex problems associated with ensuring the environmental safety of air transport in terms of accelerating their scientific developments and solutions. According to the American researcher K. Button, environmental damage from air transport is often associated with serious government failures [19]. In this regard, the positive role of public institutions, influencing adoption of more elaborated government decisions on the basis of an open discussion, seems obvious. This conclusion is also confirmed by a group of American researchers who identified 10 criteria for prioritizing stakeholders when making decisions in the field of environmental management [20], which is also relevant for analyzing the decision-making system to ensure the environmental safety of air transport. Wherein, the above critical publications do not disclose the history of formation of the movement of refusal of a number of active citizens with a sense of personal responsibility from flights, which was preceded by experiments on introducing voluntary monetary compensation for carbon emissions for passengers of some airlines: some airlines have introduced voluntary quotas on carbon emissions, which can be purchased by passengers, giving them the opportunity to pay offsets for carbon emissions arising from their personal air travel. From December 2010 to January 2011, 1,339 Taiwanese air passengers were polled at Taoyuan International Airport on their willingness to pay for carbon offset [21].

In Australia, voluntary carbon offset programs for the aviation sector were part of the Clean Energy Future Plan and a carbon tax from July 1, 2012 to July 17, 2014, which was canceled due to severe political changes [22]. According to the survey conducted in Malaysia, 71 % of households were willing to financially participate in the carbon reduction program to improve the environment and protect the climate [23]. According to the survey of 1,228 Italians, they are willing to pay compensation for their carbon footprint in the environment from use of air transport, and this willingness ranges from 12 to 38 euros per tonne and from 14 to 66 euros per flight [2]. ICAO developed a method of calculating carbon dioxide emissions as a result of air travel for use in compensation programs and, moreover, provided the possibility for any person wishing to calculate online calculator carbon CO_2 emissions per passenger. According to these calculations, carbon emissions within Moscow-Anapa flight per passenger are 116.9 kg [24], which makes one think not only of the world's governments loaded with various issues, but also of citizens outside the system of government and, of course, contributes to combining the efforts of different stakeholders.

4 Conclusions

Despite the fact that for several decades the topic of environmental safety of air transport has been reflected in international and Russian legislation and is included in educational standards, and has been developing in applied terms, not enough attention has yet been paid to it in Russia. In our opinion, the environmental safety of air transport shall be provided in modern Russian legislation as one of the constituent parts of the concept of "safety of air transport" (in its definition) with the appropriate disclosure of each of the parts in accordance with the theoretical concept of safety and environmental management standards adopted in Russia and in the world. In the field of science, it is of interest for Russian researchers to involve in development of new scientific directions supported by researchers from different countries of the world, including the discussion of details of problems and measures to limit the negative impact of air transport on the environment, the procedure for interaction of interested parties, including passengers, and etc.

References

1. R. Qiu, S. Hou, Z. Meng, Transport Policy, **107**, 1 (2021)
2. L. Rotaris, M. Giansoldati, M. Scorrano, Transportation Research Part A: Policy and Practice, **142**, 71 (2020)
3. E. E. Rumyantseva, A. G. Bezdudnaya, M. G. Treiman, Problems of Modern Economics, **2(70)**, 195 (2019)
4. Federal Law of December 28, 2010 No. 390-FZ (as amended on November 09, 2020) "On Safety", The original text was published in the Corpus of Legislation of the Russian Federation, January 03, 2011, No. 1, Art. 2
5. The concept of the federal target program "Ensuring the Flight Safety of Aircraft of the State Aviation of the Russian Federation in 2010-2014" approved by Order of the Government of the Russian Federation of April 22.04.2009 No. 554-r. Corpus of Legislation of the Russian Federation, May 11, 2009, No. 19, Art. 2354
6. State program for ensuring the safety of flights of civil aviation. Approved by Order of the Government of the Russian Federation of May 06, 2008 No. 641-r. Corpus of Legislation of the Russian Federation, May 19, 2008, No. 20, Art. 2373
7. Report of the Federal Air Transport Agency "On the Results of the Work of the Federal Air Transport Agency in 2019, the Main Tasks for 2020 and for the Medium Term", https://favt.gov.ru/
8. P. N. Ishmuratov, Administrative Law and Process, **6**, 58 (2014)
9. A. V. Gubenko, Yu. I. Rastova, A. R. Pankratova, Scientific Magazine of National Research University of Information Technologies, Mechanics and Optics, **2**, 82 (2019)

10. Transport strategy of the Russian Federation approved by Order of the Government of the Russian Federation of November 22, 2008 No. 1734-r. Corpus of Legislation of the Russian Federation, December 15, No. 50, Art. 5977
11. Report on implementation of the Transport Strategy of the Russian Federation for the period up to 2030 for 2020. Published on August 20, 2021. Official website of the Ministry of Transport of Russia, https://mintrans.gov.ru/
12. E. E. Rumyantseva, Modern management: problems and prospects. Collection of articles on the results of the XVI international scientific and practical conference (St. Petersburg, 2021), 81
13. K.-J. Wu, M.-L. Tseng, M. H.,Ali, B.Xue, A. S. F. Chiu, M. Fujii, M. Xu, S. Lan, M. Ren and Y. Bin, Environmental Impact Assessment Review, **88** (2021)
14. F. Amizadeh, G. Alonso, A. Benito, G. Morales-Alonso, J. of Air Transport Management, **55**, 9 (2016)
15. E. Yu. Starkov, N. I. Nikolaykin, XXI century: the results of the past and the problems of the present plus, **2**, 13 (2016)
16. M. Mkono, J. of Tourism Futures, **6(3)**, 223 (2020)
17. P. Chiambaretto, E. Mayenc, H. Chappert, J. Engsig, A.-S. Fernandez, F. Le Roy, J. of Air Transport Management, **93** (2021)
18. S. Gössling, A. Humpe, T. Bausch, J. of Cleaner Production, **266** (2020)
19. K. Button, J. of Air Transport Management, **26**, 1 (2013)
20. L. M. Sharpe, M. C. Harwell, C. A. Jackson, J. of Environmental Management, **282**, 15 (2021)
21. J.-L. Lu, Z. Yi Shon, Transportation Research Part D: Transport and Environment, **17(2)**, 124 (2012)
22. A. S. Choi, S. Gössling, B. W. Ritchie, Transportation Research Part D: Transport and Environment, **62**, 225 (2018)
23. M. M. Masud, Quasem A. Al-Amin, R. Akhtar, F. Kari, R. Afroz, M. S. Rahman, M. Rahman, J. of Cleaner Production, **89**, 41 (2015)
24. ICAO Carbon Emissions Calculator, Official site of ICAO, https://www.icao.int/

E3S Web of Conferences **311**, 01003 (2021) https://doi.org/10.1051/e3sconf/202131101003
EPSD 2021

International Tourism and Sustainable Development Challenges in Southern and Eastern Africa

Anna A. Suchkova*

Institute of Asian and African Studies, Lomonosov Moscow State University, 125009, Mohovaya 11, Moscow, Russia

Abstract. The article discusses the role of international tourism and tourist companies in Eastern and Southern Africa in addressing sustainable development challenges such as hunger eradication, transition to affordable clean energy, climate change and ecosystem conservation. Facts and figures show that African companies that are developing their businesses within the framework of sustainable tourism can make a major contribution through their daily procedures to the challenges of biodiversity conservation, combating poaching and reforestation. The article investigates positive trends in business practices of African companies that operate in a globally competitive environment, such as the switch to solar energy in their tourist destinations. The main lines of action to mitigate the negative environmental and social consequences of the decline in international tourism due to the Covid-19 pandemic are highlighted.

1 Introduction

In the XXI century the rapid growth of tourism in a number of Eastern and Southern African countries has resulted in tourism industries playing a significant role in their economies, contributing considerably to their GDP and employment. In Namibia, Rwanda and Tanzania, tourism accounted for 15.5%, 11.4 % and 10.7% of their respective GDP in 2019 compared to 10.4%, which represented the share of tourism in the global GDP. In Ethiopia, Zambia, Kenya and Botswana, this rate was slightly lower but still showed a noteworthy score of 6.8% to 9.6% [1]. All of the countries mentioned above, as well as South Africa, Uganda and Zimbabwe, have adopted national tourism development strategies where tourism is identified as a driving force of economic growth.

One of the main competitive advantages of African countries in the global tourism market is their exceptional biodiversity and the opportunity to travel in comfort across the wilderness. A variety of activities are offered to foreign and local tourists in national parks and private reserves where tourist camps and lodges are built. Tourism infrastructure development projects have generated mixed reactions from environmentalists and conservationists, who see them as a potential threat to the ecosystems of national parks. Local communities living near parks and reserves that are actively promoting tourism can also

*Corresponding author: onemoresuchkova@gmail.com

express discontent with the losses they might incur due to the changes in their traditional business environment. Thus, the balance of costs and benefits for individual territories is not always positive. To counterbalance the negative effects that accompany the expansion of tourism-related businesses and to reconcile the people's interests with nature conservation, sustainable tourism concepts were developed.

The Covid-19 pandemic caused the biggest crisis in tourism history. From January to May 2021, the number of tourists visiting Africa decreased by 81% compared to 2019 [2]. However, this suspension has provided an opportunity to reassess the role of tourism not only from an economic point of view but also as a tool for implementing sustainable development plans.

This article discusses the role of tourist companies in Eastern and Southern Africa in addressing hunger, improving the health and well-being of local communities, energy conservation and use of clean energy, and ecosystem conservation in general.

2 Sources and Methods

For the purpose of writing this article, data from the reports of international organizations were analyzed and systematized, such as World Tourism Organization, World Tourism and Travel Council, World Wildlife Fund, Food and Agriculture Organization of the United Nations (FAO), African Union, and others. Publications by tourism professionals, investigations into the illegal trade in wildlife, websites of travel companies in Tanzania, Kenya and Botswana, surveys of travel companies' employees in Tanzania, South Africa and Zimbabwe were also used while working on this article.

3 Results and Discussion

3.1 Climate change, hunger and the role of international tourism

Climate change, population growth and armed conflict have resulted in 250 million people in Africa (19.1 % of the population) suffering from undernourishment. Africa is the second-largest region in the world in absolute terms and the largest in proportion of the starving population. Malnutrition is particularly acute in Eastern Africa, which has the highest absolute number of undernourished population in sub-Saharan Africa: according to FAO in 2019, it accounted for half of the total undernourished population in the subcontinent, i.e. 117.9 million out of 234.7 million people. The situation in Southern Africa is much better, but according to the FAO forecasts, the number of starving people in the region will double by 2030 [3].

The complexity of hunger and malnutrition challenges requires a range of coordinated interventions ranging from international and government programs to targeted projects at the village level aimed at adapting to climate change, improving agricultural productivity and reducing postharvest losses. To meet these challengers, tourist companies in Eastern and Southern Africa work in two different ways. First, they work with local communities to implement agricultural quality standards that would enable farmers to become food suppliers for the tourist companies, improve the marketability of their farms and increase income. To illustrate how difficult this could be, we will consider the case of Greystoke Mahale – a lodge located on the shore of Lake Tanganyika (Tanzania). In local villages, agriculture is cultivated on small farms, and its products are used for personal consumption. Poor seed quality, rudimentary production technologies, lack of harvest storage facilities, poor roads or complete lack of them – all of the above made it virtually impossible for a five-star lodge to buy food locally. It took considerable effort and time on the part of lodge managers to turn subsistence farming into a supplier of vegetables. In 2019, according to Greystoke Mahale managers, they began working with local poultry farmers. They also set up a fund to raise money to build a road to one of the villages. Without downgrading the merits of numerous projects by international and governmental organizations aimed at educating farmers to improve agricultural productivity and marketing, we would like to emphasize the important advantage of the tourist companies' strategy, which guarantees a market for farmers and provides clear incentives for them to adopt new technics.

The Covid-19 pandemic, which reduced the number of tourists traveling to Africa by more than a half, has undermined the ability of farmers to sell agricultural products and handicrafts to tourist camps, reduced their cash income and aggravated the problem of hunger. The proportion of the undernourished population has increased in all countries, including the relatively prosperous Southern African region. In Botswana and Namibia, this rate was 17% and 22%, respectively, in September 2021. In tourism countries in East Africa, these figures are somewhat lower, but the absolute number of starving people is estimated at millions: in Kenya – 8 million people, in Tanzania – 5 million [4]. In this context, direct assistance to people suffering from hunger and malnutrition becomes more and more prominent. This is the second major effort undertaken by responsible tourist companies in their fight against hunger. In Tanzania, Nomad company provides 1,500 schoolchildren in three primary schools with hot meals every school day [5]. Most parents in villages around Ruaha National Park, where the Ruaha Carnivore project is being implemented, cannot afford to pay for their children's lunches at school. As a result, children often did not attend school or dropped out early. Therefore, such projects not only help to combat malnutrition, but also support another important goal of sustainable development: affordable education.

3.2 Forest loss and rehabilitation

The existing economic models in African countries, especially in rural areas, result in a conflict between the challenges of preserving the environment and eradicating hunger. Low agricultural productivity and high losses (up to a third) of food during storage and transportation provide a strong incentive to expand arable land, including at the expense of forests. Involvement in illegal timber extraction and its further sale is becoming a source of survival for many low-income families. The loss of forest area is also due to legal commercial logging, burning of wood for charcoal and its subsequent use for cooking and heating. The challenges of survival conceal the fact that forest loss in the short term will lead to serious deterioration of their lifestyle – drying out of rivers, lack of drinking water, droughts and bad harvest are the direct results of the destruction of forests. Eco-activists are working with local communities in a number of countries to raise awareness of the need for forest conservation and rehabilitation. However, these efforts will be of little use unless they are supported by measures to increase agricultural productivity, improve access to alternative energy sources, restrain illegal logging, and manage forest rehabilitation.

In the territories granted on concession to safari operators, the protection of forests and bushes from illegal logging becomes the direct responsibility of the concessionaire, and its financing is partly covered by the revenues from tourist activities. For example, the Phinda Forest Lodge, located in the private Phinda Resource Reserve in Kwazulu Natal Province of South Africa, organizes mass weeding to remove invasive plants to preserve the unique forest. In Kenya, Serena company, which owns a network of hotels and camps throughout the country, developed its own reforestation program. The program includes planting in company-owned land endemic drought-resistant tree species, such as *Acacia*, *Mellia*, *Croton* and *Commiphora*. Every hotel guest gets the opportunity to plant «his» or "her" tree. Recognizing the importance of forests in combating desertification and mitigating negative impacts of climate change, the company has also been planting forests in local communities' land, involving in such activities neighboring schools and accompanying these events with educational presentations [6]. The efforts of tourist companies to maintain forests and support their rehabilitation are also important for the conservation of various animal species, which is essential for the development of tourism in African countries.

3.3 Green energy initiatives by tourist companies

The problem of energy shortage in Africa is so acute that even the pandemic has not removed it from the agenda. This is well attested for by the publication of the joint report of the African Development Bank (ADB) and the African Union Development Agency "New Partnership for African Development", dedicated to the African energy masterplan, which appeared on the websites of these organizations on 23 October 2020. According to ADB, "over 640 million Africans have no access to energy, corresponding to an electricity access rate for African countries at just over 40 percent, the lowest in the world. Per capita consumption of energy in sub-Saharan Africa (excluding South Africa) is 180 kWh, compared to 13,000 kWh per capita in the United States and 6,500 kWh in Europe» [7].

The challenges of climate change and the fight against carbon emissions provide an incentive to address energy deficiency by shifting to clean energy. However, most African families, especially those living in rural areas, cannot afford to buy solar panels or biogas devices. Therefore, all known projects on the continent are carried out at the expense of international organizations such as ADB, the European Union, WWF, etc. The scale of the task does not imply quick solutions, but there are some encouraging examples to date. In Tanzania, 350 households in two pilot villages received access to solar energy. Biogas devices replaced 30% of wood and coal stoves. Similar programs are being implemented in Madagascar and Uganda. The Government of Madagascar has decided on a national program for the training of solar engineers. It is planned that 750 women will be trained under this program, who will then service 630,000 households that will be provided with solar energy facilities [8].

For tourist companies that have facilities in national parks and reserves, electricity supply is a serious problem. The majority of tourist camps use diesel generators. And the severity of the problem lies not only in the high cost of fuel and its transportation to remote locations, but also in the accusations coming from the environmentalists who object to the use of diesel generators in nature reserves and parks. Tourist companies are under even greater pressure from potential clients, who want not only to enjoy every moment of being exposed to a unique natural environment, but through their participation as tourists hope to maintain and preserve this uniqueness. Thus, renewable energy and eco-efficiency are turning into marketing strategies that companies in Southern and Eastern Africa use to sell their tourism products.

There are several eco-efficiency strategies that are currently being used in tourism businesses in Eastern and Southern Africa. The first is to replace diesel generators with solar panels. According to the Ecotourism Society of Kenya, of the 87 camps and hotels ranked as the most environmentally friendly, 10 camps have switched to solar power. At Tortilis camp a power plant consisting of 110 solar panels and equipped with inverters and batteries is controlled by a digital system that even calculates the amount of Carbon (IV) Oxide emissions avoided. A further 45 camps and lodges from the ranking list use solar energy in combination with other types of sources such as diesel fuel, liquefied gas, charcoal briquettes, and firewood [9]. A similar shift from diesel to solar energy is observed in Botswana. For example, out of 22 camps owned by Okavango Wilderness Safaris and located in the Okavango delta, 10 are completely solar-powered. Switching from diesel to solar energy had a positive economic impact on the company. The company's report states that the complete conversion of five camps to solar energy has resulted in cost savings of almost $400,000. However, more than 80% of tourist destinations in the country still use fossil fuels [10]. The same situation is in Ethiopia, which in 2018 became the world leader in tourism growth. Surveys show just isolated cases of using solar energy for power supply of camps and lodges in the country [11].

The second strategy for improving the environmental performance of the tourism business is to apply various energy-saving practices, including use of energy balances, energy-saving lamps, elimination of sockets in guest tents and houses, etc. Jack's Camp, which is located in the Kalahari Desert and recreates the atmosphere of a XIXth century camp, abandoned electricity altogether, replaced it with oil lamps, and provided access to it only in public areas. The third eco-efficiency strategy is to manage the resources used, including control over the quality of diesel fuel and the disposal of spent fuel.

The importance of the environmental responsibility issues in marketing strategies of African tourist companies competing for global markets is evidenced by the fact that on their English-language sites, these companies inform their potential clients about the company's energy efficiency and energy conservation achievements. For example, a Kenyan company Serena has announced that it has equipped its Kilaguni Serena Safari Lodge in Tsavo National Park with a power plant consisting of 960 solar panels, covering 95% of the camp's electricity needs [6]. Asilia company, which manages tourist facilities in four East African countries (Tanzania, Kenya, Rwanda and Uganda), in its 2019/2020 report informed interested readers that during that period it was able to offset CO2 emissions by 1,425 tons, which is equivalent to the removal of some 118 medium-sized vehicles from traffic for a year [12].

3.4 Biodiversity conservation, tackling poaching and illegal wildlife trade

The development of tourism in Africa is directly linked to the wildlife conservation. The decline in animal populace is due not only to climate change, population growth, economic pressure on the environment, but also to poaching and illegal trade in African animals. Existing estimates of losses in the tourism industry due to poaching can only give an approximate picture. According to these estimates, elephant poaching costs the continent $25 million per year. The impact of poaching is particularly evident in Eastern and Southern Africa, where national and private parks provide the bulk of tourism revenue. According to David Sheldrick Wildlife Trust, in 2014 Africa's tourism industry, local communities, and economies together have lost more than $44.5 million to poaching alone. A survey of 216 protected areas in 2016 showed that there is a consistent inverse correlation between the number of tourists and their expenditures on the one hand and the elephant population on the other. Decline in the rhino population has the same negative effect on tourism. According to Ecologic Institute, in 2006-2014 rhino poaching resulted in a total annual loss to tourism revenue in the South African Republic, Namibia, Kenya and Zimbabwe of between €205.76 million and €230.76 million. Moreover, according to the Institute's estimates the extinction of one species from the "Big Five" (elephant, rhino, buffalo, lion, leopard) reduces ecotourism by 20% [13].

There are three kinds of poaching. First is killing animals for food. It is carried out by residents of villages located near protected areas. The second category of poaching is illegal commercial hunting. Finally, the third and most dangerous group of poachers are poachers who are members of international criminal alliances that organize the killing, transportation and sale of rare animals (such as pangolins) and animal parts (rhino horn and ivory). The third category of poachers is the most dangerous one – consequences of their activities are not limited to economic losses only.

Poachers kill not only animals, but rangers who guard them. The number of people killed in clashes with poachers vary but they are all in the hundreds [14]. Paramilitary guards sent to protect animals from poaching raise tensions in local communities, which see the military as a threat to their normal way of life. On the other hand, the residents of protected areas are under constant pressure from criminal groups to join in their criminal activities. These groups are extremely difficult to oppose. They are very well equipped: they have not only weapons, but also night vision devices, helicopters and small aircraft at their disposal. Such resources are not available to the police and environmental agencies involved in tackling poaching. Investigations carried out by the authorities of African countries and international organizations show that organized crime comprises representatives of different countries, and that there is a «division of labor» among them. Hunters are either local residents or foreign recruits. For example, employees of national parks and private reserves in South Africa reported that most poachers were Mozambique nationals. Criminal managers bribe police, customs and national parks officers. A recent trial held a few years ago in South Africa, revealed that leaders of one of the criminal groups were white people – lawyers, attorneys and even veterinarians [15].

Until recently, the main markets for ivory and rhino horns were in China, Vietnam and Arab countries. At the end of 2017 China imposed a ban on trade in ivory and ivory products. Factories producing these products were closed. Legal markets have moved to neighboring countries. And while a black market continues to operate in China, the impact of measures undertaken by the Chinese authorities remains to be significant. A consumer survey conducted in China in April 2021 showed that demand for ivory declined by 43% after the ban on trade was introduced [16]. As a result, the price of elephant tusks in Southern and Eastern Africa has fallen dramatically and have ceased to compensate for the high risks associated with organized poaching of elephants.

Attempts to solve the problem of illegal trade in rhino horns by means of bans have so far failed. In China and Vietnam it is still used in traditional medicine as a cure for a range of diseases, including cancer. Although numerous studies have shown no positive effect of rhino horn powder, surveys of Chinese consumers show that almost half of them buy rhino horn for medical purposes. Another 12% buy it as a luxury item and as protection against inflation (the price of a kilo of horn in 2012-13 was $60,000) and more than 20% pursue both objectives [17]. However, a quarter of the respondents were unaware of the ban on the horn trade, and most of them did not consider the information on the catastrophic decline in the number of rhinos to be important for their purchasing preferences. The failure to protect rhinos by altering demand, which worked in the case of ivory, has intensified the search for other means to tackle rhino poaching. Measures being considered include legalization of trade in cut-off horns of rhinos, grown on farms or in private reserves, and production of synthetic horns.

For tourist companies, the fight against all types of poaching is an important part of their daily activities and a significant addition to their expenses. Patrolling the territories, removing snares and traps, building fences, cutting off rhino horns to prevent their killing, and educating the local population – to name just a few of its components. In souvenir shops located at tourist camps the product range is strictly monitored to ensure that no product made from an endangered plant or animal is sold there. Recently, there has been a tendency to label the products of these shops with tags containing information on environmentally friendly production methods and the target use of the proceeds of their sale. These targets generally include reforestation, protection of individual species and support to local communities. Such information is also part of the companies' marketing strategies to attract a special group of high-income tourists who appreciate conservation efforts and are willing to spend money for a noble cause.

The Covid-19 pandemic has aggravated the poaching problem. As a result of the pandemic, there was practically no tourists and visitors entering the national parks and private reserves, and that greatly facilitated poachers' access to the closed areas. The loss of income by local residents due to the closure of tourist camps has forced people to engage in illegal hunting which previously they did not dare to do. Disturbing information comes from all over the world. In Ethiopia at the beginning of the summer 2020 at least 6 elephants were killed in a single day, compared to 10 elephants killed in all 2019, which became the «largest slaughter» in the country ever to be recorder. It turned out difficult to apprehend the perpetrators because local residents living in the Mago National Park area were armed and unwilling to cooperate with the officials [18]. In the North of Serengeti National Park in Tanzania in 2020, the rangers rescued thousands of animals, removing 9,000 snares and saving 135 animals from traps [19].

4 Conclusion

Tourist companies in Eastern and Southern Africa, committed to sustainable and responsible tourism, can contribute to solving many critical environmental and social problems. Being engaged in the international tourism market competition, they are under pressure from consumers, competitors and the environmental community to move towards a more environmentally friendly tourism. The need to maintain their main competitive advantage - unique biodiversity and beautiful landscapes – compels them to invest their resources, time and energy in conservation of endangered species. In search for solutions to prevent conflicts with local communities, which often and not without reason view tourism activity close to their habitat as undesirable due to the loss of resource base and change in their natural living conditions, tourist companies came up with numerous economic and social initiatives to compensate for these losses. First of all, these initiatives include commitments to hire and buy locally, to provide funds for scholarships, medical treatment and food to members of local communities, etc. Originally driven by business reasoning, these initiatives presently become the best practices and eventually develop into social responsibility standards of sustainable tourism companies. The associations of tourist companies that exist in all countries of the region contribute to the dissemination of these standards.

The Covid-19 pandemic, which has dramatically halted international tourism to Africa, led to the increase in the number of hungry and undernourished people, including those who directly or indirectly depended on tourism. Another unfortunate consequence of the decline in tourism was the increase in poaching in the deserted parks and reserves.

References

1. Economic Input Reports, Country, Region Data, https://wttc.org
2. International Travel Largely On Hold Despite Uptick In May, https://www.unwto.org
3. The State of Food Security and Nutrition in the World 2020, http://www.fao.org
4. World Food Programme. HungerMap, https://hungermap.wfp.org
5. Nomad Conservation Partners and Communities, https://www.gofundme.com
6. Serena News. Insights and Inspirations in the World of Serena Hotels, **8** (2018) https://www.ethosmarketing.co.uk
7. Africa's energy masterplan takes shape as African Development Bank and AUDA-NEPAD release key report, https://www.afdb.org
8. Conservation: The People's Perspective. WWF AfricaFY17, 15
9. Ecotourism Kenya, https://ecotourismkenya.org
10. J. E. Mbaiwa et al., Int. J. Tourism Policy, **8,** 1 (2018) https://www.inderscienceonline.com

11. Eco-Lodges and Tourist Infrastructure Development in and Around Abijata Shalla Lakes National Park; From the Perspective of Evaluating their Sustainability, Journal of Tourism, Hospitality and Sports, www.iiste.org
12. An International Peer-reviewed Journal, **45** (2019) https://www.researchgate.net
13. Positive Impact 2019/20, https://www.asiliaafrica.com
14. Tackling Illegal Wildlife Trade in Africa: Economic Incentives and Approaches, https://www.chathamhouse.org
15. The Price of Protecting Rhinos, https://www.theatlantic.com
16. Millions seized in rhino horn smuggling group, https://www.gov.za
17. Demand for Elephant Ivory in China Drops to Lowest Level Since National Ban, https://www.worldwildlife.org
18. An Undercover Investigation on Rhino Horn Trafficking in China and Vietnam, https://elephantleague.org
19. Six elephants killed in one day by poachers in Ethiopia, https://www.theguardian.com
20. Nomad Conservation Partners and Communities, https://www.gofundme.com

Philosophy of ecologized economics

Nikolay Privalov[1], and *Elena* Fursova[1]

[1]Emperor Alexander I St. Petersburg State Transport University, Saint Petersburg, Russia

Abstract. The ecological crisis goes a long way back and has been brewing for centuries. The main factors of human alienation from nature: technical progress; suppression of pagan culture, that used to be tied to nature, world religions; spread of atheism; consolidation of the positivism paradigm in scientific methodology; triumph of the market economy model. As a result, humans were pulled out of their natural environment. They live and work by rhythms and rules contradicting natural laws. The result is the growing global crisis of industrial civilization.

1 Introduction

Industrial society has exacerbated environmental problems to their limit. Back in 1992, Professor of Georgetown University, USA, Dr. Murray Feshbach, published the book *Ecocide in the USSR: Health and Nature Under Siege*. There, he made an attempt to analyze data on the ecology and health of the population. The book's overall result was inauspicious: the population of Russia is threatened with environmentally induced degradation and extinction [1].

Evident is the "alienation of nature" as one of the types of alienation, described by K. Marx. The essence of the theory of alienation is that the external environment, as well as artificially created nature, are included in the production process, humanized, transformed by human labor. At the same time, the laws of nature are transformed from the prerequisites and conditions of production into hostile, destructive forces. What is important is methodology of the question, and above all the root causes of the problem. They can be found both in technical progress and in the areas external to the economy – politics, spirituality, etc. Having outlined the solution to the main issues, we can finally focus on a specific necessary mechanism.

2 Literature overview

There are quite a lot of research papers on the problems of interaction between ecology and economics. As a whole, articles indexed by the Web of Science in recent years can be divided as follows:

1. General approach – environmental pollution [2], green economy and relationship with politics [3], impact of globalization [4], at the level of methodology and global economics theories [5];
2. On more specific issues: interaction of the economy and the environment with a positive (also in terms of economics) effect [6-7], water systems [8], agriculture [9], COVID-

19 pandemic [10], architecture and design [11], territorial aspect [12], international cooperation [13].

The largest group of papers is concerned with interdisciplinary scientific relations [14], the problem of culture and morality for solving environmental issues [15].

The papers indexed in the e-library over the last 6-7 years can also be divided as follows:

1. field-specific – industrial agriculture [16], legal issues [17], civil engineering [18], subsoil use [19], waste reduction [20], pedagogy [21];
2. within the framework of the traditional market approach to the economy [22];
3. on state regulation of problems, in particular, on financial instruments [23];
4. on improving sustainability and efficiency of natural resource use in general [24];
5. quite a few are engaged with the issue of innovation and the latest technology;
6. some authors write from the standpoint of sustainable development [26];
7. less (although more often compared to previous decades) is spoken about the need for a systematic approach [27] and building a new type of ecologized economics in view of the crisis of the old, industrial, consumeristic type of mass production and consumption [28];
8. very few speak directly about the problem of culture and morality for solving the environmental issue [29].

Thus, we observe a shortage in comprehensive study of the problem covering not only technologies as factors of environmental risks, but also industrial (institutional, social) relations, linking the problem with the industrial production crisis and the transition to a different type of social reproduction. A deeper analysis will bring the researcher to a philosophical level of understanding, and the predicted tendencies and recommendations for managing society, which are of a systemic nature, must inevitably take into account social psychology, rooted in the depths of mentality and mass culture.

3 Materials and Methods. Analysis and Results

Mainly, the systems theory, method of scientific abstraction, formal logic, dialectical-theoretical methods, method of empirical analysis, and other economics methods were applied. We used the concept of the equilibrium paradigm as the methodological basis.

The main results of researching.

1. As the initial principle of constructing the concept, following the example of classical economists (A. Smith, K. Marx, etc.), we took the anthropological principle, i.e. the initial ideology should proceed from general ideas about human nature. As we have argued earlier, a human is a social, spiritual, and biological being, since they have characteristics generated by society, nature, and spiritual forces [30. P.101]. Hence the main statement of the equilibrium paradigm: it is necessary to restore the lost equilibrium of man with nature, society, and God [30. P.85; 31. P. 373]. Equilibrium here is viewed as maintaining the quality of the system, i.e. its founding elements and connections between them. This paradigm's necessity and essence stems from the deepening global crisis, i.e. systemic crisis of industrial Christian civilization, covering the economy, ecology, social relations, and spirituality. The Positivism paradigm prevailing in modern science was created in the 1830s-40s by Auguste Comte and is reducing scientific knowledge only to accurate, observable, accumulated, and verifiable information. It has exhausted its capabilities, since it is an obstacle to obtaining new knowledge not exactly meeting the described criteria. Therefore, the separation of science from religion, culture, and morality led to a rapid growth of technical progress, while creating global risks throughout the whole world, including above all others the environmental ones.

Therefore, the topic of this paper is identifying the root causes of the loss of the former balance between man and nature, not only as an environment or a source of resources, but as an ecological niche or habitat.

2. The first, initial, reason for the alienation of nature is inextricably linked to the technical progress. Hunting and agricultural types of economy were built into the natural environment. A human was afraid (maybe excessively) to go beyond its limits, the conscience was syncretic, believing in mystics and magic. Eventually, the artificially created environment (especially industrial technology and related sciences), for no less than 100 years, freed humans from their direct dependence on the natural climatic environment and created a tangible foundation for leaving it. However, given that many processes tend to cycle, it is possible to assume that a dialectical return to nature may occur at a new spiral of progress.

Technological progress, especially accelerated by the industrial production, has generated a lot of negative side effects (negative externalities). Humans have been long living in the artificial environment, from the moment about 20 thousand years ago, during the final period of the Younger Dryas glaciation, when they left their caves and began building their first artificial houses, usually dugouts or semi-dugouts. And while to do so for thousands of years natural materials were used – primarily wood and stone, in the twentieth century they were replaced by a plethora of artificial building materials using a variety of chemicals. The result is well-known – the spread of cancer, allergies, and other diseases caused by radioactive and other harmful elements.

According to the research done by biologists, physicists, and medical doctors, a discrepancy between the internal and external electromagnetic environment of a person is possible. Experiments with water show the negative impact on its energy by various electromagnetic devices – plasma TVs, computers, cell phones, microwave ovens, etc. [32]. This can especially harm the health of children whose cranial bones have not yet developed sufficient thickness. This issue is gradually being resolved. In particular, the Federal Service for Surveillance on Consumer Rights Protection and Human Wellbeing adopted new sanitary rules in the field of child education and upbringing, stating that minors are now prohibited from using smartphones for the purposes of education [33].

3. The mass junk food industry creates addiction to certain ersatz foods, the constant consumption of which can lead, at least, to poisoning, or, worse, to dangerous diseases such as oncology (coffee, sweets, hot dogs, etc.). Food, medicine, clothing, household items, building materials for the sake of cost reduction and convenient mass production contain more and more artificial, chemical-based components, which also potentially contradicts human nature. Thus, according to TV presenter Dr. A.L. Myasnikov, today 200 million people on Earth are diagnosed with prediabetes. This is the initial stage of type 2 diabetes, which develops in adults due to the consumption of unhealthy food. Since the late 70s, the food industry began to produce left-handed glucose as sweeter and therefore applicable for the confectionery industry. However, our body is used to the other, right-handed, glucose. As a result, blood sugar does not enter the cells. Cells are hungry for energy – and blood sugar decomposes into acetone and poisons the body. In order not to die from poisoning, you need to follow a strict diet and take certain pills. Otherwise, you'll develop real diabetes. This is an example of the market being an economy without conscience.

4. There is a separation from natural cycles, violation of daily and other life cycles that have a cosmic origin. The use of electricity has practically erased the shift between the times of day, which, in addition to purely technical progress, has greatly increased human capabilities by creating industries with a continuous cycle (metallurgy, chemistry, transport, medicine, all security services, partially trade, entertainment industry, etc.). However, separating a person from natural rhythms is manifested in the discrepancy between their individual life cycle and natural cycles, e.g. daily, seasonal, etc. All this, especially paired with unhealthy pastime, leads to the risk of various health problems. Due to the lack of sleep, a person can make grave mistakes, even the ones leading to severe casualties.

5. For a long time, humans were engaging into unhealthy habits that animals do not have – smoking, alcohol, drugs, overeating. Throughout the history of mankind, from time to time

the cult of physical pleasure intensifies, tearing humans away from reality, making them dependent on their vices and bound. Obesity leads to excessive strain on the heart, and the abuse of alcoholic beverages and tobacco leads to liver cirrhosis, lung disfunction, and other diseases.

These are only several of the most striking negative consequences of technological progress and spreading urbanization generated by the industrial production system and artificial (mainly urban) habitat.

6. The spread of modern world religions also contributed to the alienation of man from nature, since their followers did their best to destroy pagan cults. Pagan religions are, first of all, the worship of nature, where humans are inextricably linked with all of animate and inanimate life. Actually, all life in paganism is considered to have spirit and a person is supposed to treat it accordingly. This means numerous prohibitions (taboos), including many environmental ones. Christianity, and later Islam, proclaimed man to be created in the image and likeness of God, and therefore, the master over other earthly creatures. The next step from here is to proclaim man the ruler of nature. In fairness, it should be pointed out that Islam retains some environmental provisions. For example, during the Hajj (pilgrimage), you cannot pick the grass, and killing animals is only allowed in self-defense. Christianity, on the other hand, ignored the issue of environmental conservation, and various magical actions in relation to animals, for example, during a hunt, are denied as superstitious. At the same time, paganism returned the favor, since Christianity absorbed elements of local pre-Christian cults. For example, the Orthodox menologium significantly altered the Christian calendar. Hence, the holidays with agricultural subtext such as Apple Feast of the Saviour, Saviour of the Honey Feast Day, etc. This suggests that it is impossible to completely detach a person from nature, and a person keeps striving for it in various ways.

7. The assertion of atheism in the general mainstream of the development of positivist science finally torn man away from nature, which was proclaimed to be "not a temple, but a workshop". Positivism, as a science freed from the influence of morality, has put into the hands of man a weapon dangerous to the entire world.

8. The unrestrained development of the market, especially in the twentieth century, exacerbated the ecological situation, spiritual degradation of society to their limits, there is a sharp increase in risks to the physical health of people.

Thus, according to the main conclusions drawn by J. Forrester and D. Meadows, in the period 2030–2050 there will actually be a world catastrophe caused by the lack of natural resources and the rapidly growing population. Even if we assume that the world's population will consume resources at a rate of 4 times less than that of 1970, it will still reach its maximum number by 2030, and then, as a result of hunger, disease, mutual extermination, will die out by five-sixths of the reached 10.9 billion people. [34. P.24].

4 Conclusions

Thus, based on long-term studies of economic, political, social, and cultural processes of recent decades, we came to the following conclusions – recommendations for contemporary politicians:

1. The use of enlightenment, propaganda, and education, taking into account that public consciousness, especially ideologies and religious beliefs (including such secular forms of religion as atheism, and liberal popular culture), are conservative. This requires a general sobering up, i.e. liberation of the psyche and consciousness from various substances. State support for culture, especially traditional culture, is needed. The introduction of moral censorship is advisable.

2. The problem of depletion of natural resources is solved by a more complete, "deep", integrated extraction and recycling of available resources. There is considerable experience,

for example, in more complete oil production by the organization "Gazpromneft-Development". An obligatory element in modern circumstances is energy saving and loss reduction in heat, water, and other resources. Another direction is more efficient use of traditional types of energy and development of non-traditional types, complex recycling of household waste.

3. It is necessary to study the existing experience of the past, for example, the traditional economy, which for hundreds of thousands of years has provided people with food and other resources, without upsetting the ecological balance.

4. However, the development of new technologies is not enough. A new philosophy and a complex methodology of interaction with the environment is required, that was started by the Russian cosmism philosophers (Chizhevsky, Tsiolkovsky, Vernadsky, etc.), who created the concepts of sustainable development, biosphere, noosphere.

In conclusion, let us designate new possible *principles of human interaction with the natural environment within the Ecologized Economics model*:

- accepting human nature as social, spiritual, and biological, as opposed to being only viewed as a social being. Through accepting their inextricable ties with the environment, the perception of a human should change both as an object of management and as a goal of the existence of society. At the same time, this imposes new responsibilities in the world due to humans' unique position among other beings;

- developing directions of state policy and social regulation on these problems in accordance with the human nature. For example, it is necessary to recognize that socializing a person by means of ordinary "enlightenment" is not enough. People accept a lot "in good faith". This is specific to the everyday thinking, which often prevails in behavior. This is the basis of the concepts of "routine", "stereotypes" in institutional economics, management, marketing, the concept of "bounded rationality" by the Nobel laureate H. Simon, etc. Therefore, a deeper consideration of psychology is required, for example, *personality types* (according to C. Jung), generated not by society, but by genetics (meaning, partly by the natural environment);

- applying to the social sciences (economics and other human studies) the interdisciplinary connections, for example, with biology. Even A. Marshall recognized the influence on the neoclassical theory of biology he created (for example, Darwin's theory of evolution). However, the account of new data is necessary – on behalf of genetics, zoopsychology, cybernetics, and other systemic theories. Thus, at one time, the Theory of general equilibrium by L. von Bertalanffy and the Theory of tectology by A. Bogdanov were created mainly based on medical data.

All of this allows us to consider a person in a more systematic, complex way, socialize them accordingly and manage them more efficiently, benefitting both society and the person. This concept can be viewed as a new dialectical stage largely denying the concept of "industrial man" (as the embodiment of *Homo economicus*).

References

1. A. I. Pereskokov, Bulletin of the Ural Ecological Fund. Financial crisis as a catalyst for solving environmental problems, **3(44)** (1999)
2. R. Irwin, Educational philosophy and theory, Jul, 834 (2020)
3. J. M. Kramer, Journal of economic issues, Apr, 416 (2021)
4. F. Bran, M. Iovitu, Dumitru-Alexandru Bodislav, 20th International symposium - The Environment and the industry, Sep, 318 (SIMI 2017)
5. M. Treacy, Revista colombiana de sociologia, Jul-Dec, 241 (2020)

6. P. Jones, *Design for a sustainable culture: perspectives, practices and education*, 38 (2017)
7. S. A Inkpen, C. T. DesRoches. Ethics policy & environment, **23(2)**, 203 (2020)
8. C. Tyler DesRoches, S. A. Inkpen, T.L. *Green, Contemporary philosophy and social science: an interdisciplinary dialogue,* 39 (2019)
9. S. M. Gardner, S.J. Ramsden, R.S. Hails, *Agricultural Resilience: Perspectives from Ecology and Economics* (2019)
10. Q. Guo, D.C. Lee, AMBIO, Jul (2020)
11. M. Hackel, A. Hackel, ICSBE-2018, 03001 (2019)
12. E. Miller, *Canadian study in population,* SI, 84 (2018)
13. M. V. Shatokhin, A.L. Nikiforov-Nikishin, A.I. Romanenko, R.I. Lazareva, N.A. Golovacheva, V.A. Klimov, *Education excellence and innovation management through vision 2020*, 6048 (2019)
14. H. Thoren, S. Stalhammar, *Ecology and society,* 44 (2018)
15. J. McMullen, Twenty-first-century ecology lessons from seventeenth-century Japan: climate change, deforestation, and moral regeneration, Oct (Japan forum, 2020)
16. I. M. Donnik, B.A. Voronin, Agrarian Bulletin of the Urals, **1(143)**, 77 (2016)
17. S. A. Bogolyubov, Journal of Russian law, **4(268)**, 105 (2019)
18. N. V. Karpova, Yu.A. Komarets, Modern trends in economics and management: a new look: collection of materials of the XLVI International Scientific and Practical Conference, Novosibirsk, 110 (2017)
19. O. V. Kosolapov, M.N. Ignatiev, News of higher educational institutions, Mining Journal, **2**, 79 (2019)
20. I. V. Gordin, Economic science of modern Russia, **3(90),** 63 (2020)
21. V. S. Shilova, *Socio-ecological education of students (in the context of the activity approach)* (2020)
22. I. D. Mylov, Financial Economics, **9,** 77 (2019)
23. A. A. Salamatov, Professional project: ideas, technologies, results, **1(26),** 31 (2017)
24. E. R. Mukhina, A.V. Marinina, Science diary, **10(10),** 27 (2017)
25. 25 S. M. Sukhorukova, A.M. Burnt, Biosphere economy: theory and practice, **2(20),** 5 (2020)
26. I. U. Shakhgiraev, Z. M. Makoev, Bulletin of the complex scientific research institute, H.I. Ibragimov RAS, **1(1),** 126 (2020)
27. P. A. Kokovin, O.B. Mezenin, Bulletin of the Bashkir State Agrarian University, **4(36),** 123 (2015)
28. A. A. Malyshev, N.N. Solodkov, N.A. Korobkov, Stolypinsky Bulletin, **2, 4** (2020) https://stolypin-vestnik.ru/
29. E. Yu. Zakharova, I.A. Shchetkin, Historical, philosophical, political and legal sciences, cultural studies and art history. Questions of theory and practice, **1-2(27),** 88 (2013)
30. N. G. Privalov, *The third way of Russia. New Hope in the 21st Century* (2012)
31. N. G. Privalov, *Economics of the Non-Profit Sector: Textbook* (2020)
32. Film "The Great Secret of Water", Russia (2006)
33. News mail.ru, https://news.mail.ru/society/
34. J. Ferrester, World dynamics, 24 (1978)

Real-virtual platforms for organizing events on environmental education of students of technical universities

O. V. Semaeva[1,*], *V. N.* Novikova[2], *R. F.* Azmetova[3], *M. E.* Ivanova[4], and *N. N.* Pivkina[1]

[1]National Research University Moscow Power Engineering Institute, Moscow, Russian Federation
[2]Moscow State Institute of International Relations (University) of the Ministry of Foreign Affairs of the Russian Federation (MGIMO), Moscow, Russian Federation
[3]Moscow State University of Technology and Management. named after K.G. Razumovsky (First Cossack University), Moscow, Russian Federation
[4]Russian University of Economics named after G.V. Plekhanov, Moscow, Russian Federation

Abstract. The problem of environmental conservation is currently one of the most acute for the entire world community, and therefore one of the strategic directions for the development of modern higher technical education in Russia is to increase the level of environmental education of university students. The article discusses the ways of organizing educational work in technical universities using real-virtual platforms, provides a theoretical justification for the need to improve the effectiveness of methods of environmental education of university students, and identifies the specifics of organizing discussion clubs using real-virtual platforms. The paper presents the results of an empirical study of students ' participation in an experiment, during which they discussed environmental problems on the site of a real-virtual discussion club. The results of the study prove the high efficiency of conducting environmental educational activities on real-virtual sites.

1 Introduction

The twenty-first century has become a turning point in the life of human society. Introduction of high technologies in modern enterprises, production and use of innovative materials, new methods of processing and production of food products, unfortunately, become not only a factor in improving the living standards of the world's population, they often lead to irreversible consequences for the environment and, as a consequence, human health. It is no coincidence that the issues of preserving the environment today are on a par with the most pressing problems of the entire world community. The Russian government is also making every effort to address these problems. According to the environmental safety of the Russian Federation for the period up to 2025, the country's environmental safety is an integral part of the national safety. Among the goals and objectives aimed at implementing the state policy in the field of environmental safety are such as:

* Corresponding author: semaewa.olga@yandex.ru

1) activation of fundamental and applied scientific research in the field of environmental protection and nature management, including environmentally friendly technologies;

2) development of the system of environmental education and awareness, professional development of personnel in the field of environmental safety. [1]

These problems can be solved only through the competent organization of environmental education and education of the young generation of Russians at all stages of the educational process. However, technical universities play an especially important role in this process, because they train specialists involved in the design and development of machines and technologies, the functioning of which has a direct impact on the natural environment.

In addition, today's students tomorrow will become responsible officials and heads of enterprises, on whose decisions the rational use of the natural resources of our country will depend.

Meanwhile, it is impossible not to agree with B. Nevzorov that modern entrepreneurs and government officials often make decisions that contradict the principles of nature conservation under the influence of the free market laws [2].

Environmental education as an element of general education is associated with the mastery by students of the scientific foundations of interaction between nature and society. Its goal is formation of a system of knowledge, views and beliefs aimed at fostering the moral responsibility of the individual for the state of the environment, awareness of the need for constant care for it in all types of activities [3].

Today, scientists agree that arrangement of environmental education in technical universities requires an integrated approach that provides not only for the acquisition of scientific knowledge in the field of rational environmental management, it shall be aimed at formation of moral convictions in the need to preserve the country's natural resources [4, 5, 6, 7, 8]. Environmental education shall permeate all training courses taught at technical universities, students shall be well aware not only of environmental problems in various regions of Russia, but also of the causes of these problems [2, 4, 7, 8]. It is necessary to teach students to evaluate their projects and developments in terms of their possible consequences for the environment.

Researchers identify the following factors in the dynamics of development of "environmental responsibility of the individual":

– "alienation" and "estrangement" of man from nature;

– the growth of scientific and technological progress that changed the strength of interaction and the nature of relationships in the "man-society-nature" system that led to its instability and the danger of the global ecological crisis;

– the emergence of environmental ethics, the main criterion of which is a high level of development of environmental responsibility of the individual. It is it that becomes the basis for formation of a new ecological ecocentric consciousness that will provide a strategy for transition of civilization to the path of sustainable development;

- determination by the world pedagogical education system of the need to form students' environmental responsibility that shall become the goal and expected result of environmental education both in secondary and higher education [9].

Educational work in technical universities shall also be built considering the need to improve the environmental literacy of students. It is important to arrange various events aimed at increasing the environmental literacy of students. This can be visits to specialized enterprises, participation of students in preparation of projects for restoration of natural objects affected by the irrational activities of industrial enterprises, discussion clubs, where each of the students will have the opportunity to express their point of view on existing environmental problems. In other words, the entire life of the student community must, in one way or another, come into contact with the solution of environmental problems in specific regions of the country.

2 Materials and Methods

Modern information technologies make it possible to use not only real, but also virtual platforms for arranging student discussion clubs. We conducted an experiment on the basis of Moscow Power Engineering Institute in order to find out how effective real-virtual platforms are for arranging discussions among students on environmental topics.

The empirical basis of the study was the results of a questionnaire survey of students conducted within the framework of a real-virtual discussion club "Raising the Environmental Literacy".

Real-virtual discussion platforms imply holding events both online and offline, which is very convenient for participants and organizers, since this form of holding does not require significant financial and time costs. The experiment was attended by 1st year students of technical areas of training in the amount of 58 people. The experiment was carried out in three stages. At the first stage, an offline oral conversation was held with the participants about the fact that after receiving a diploma of higher education, they will have every chance to become the very people on whom decisions will depend considering the problem of preserving natural resources. At the second stage, they were asked to answer the questions of an online questionnaire in order to find out how aware they are of the serious environmental problems that humanity is facing today. At the third stage, an offline conversation was held with them again, during which they were provided with the results of the survey and they had the opportunity to express their own opinion about the sources and possible solutions to the problems raised.

3 Results and Discussion

The results of the survey analysis showed that the majority of first-year students of Moscow Power Engineering Institute are aware of the most pressing environmental problems that is associated with the successful promotion of environmental values among them at school. The results of the survey are provided in the table below.

Table 1. The results of the questionnaire survey of the participants of the real-virtual discussion club "Raising the Environmental Literacy

1. Are you familiar with the law of irreversibility of human-biosphere interaction? Renewable natural resources become non-renewable in the case of a profound change in the environment, significant overexploitation, reaching total destruction or extreme depletion, and therefore exceeding the possibilities of their restoration. (https://professia-uc.ru/f/101_osnovnye_ponyatiya_ekologii.pdf)	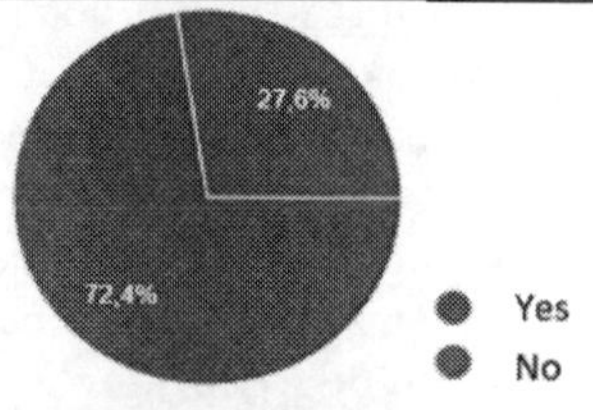 58 answers
2. Are you familiar with the law of the indispensability of the biosphere (V.I. Vernadsky, D.P. Marsh, E. Reclus)? There is no reason to hope for construction of artificial communities that stabilize the environment with the same degree of accuracy as natural communities. Therefore, reduction of natural biota in a volume exceeding the threshold value deprives the environment of sustainability, which cannot be restored through creation of treatment facilities and transition to waste-free production. (https://studme.org/260227/ekologiya/osnovnye_zakony_pravila_printsipy_ekologii)	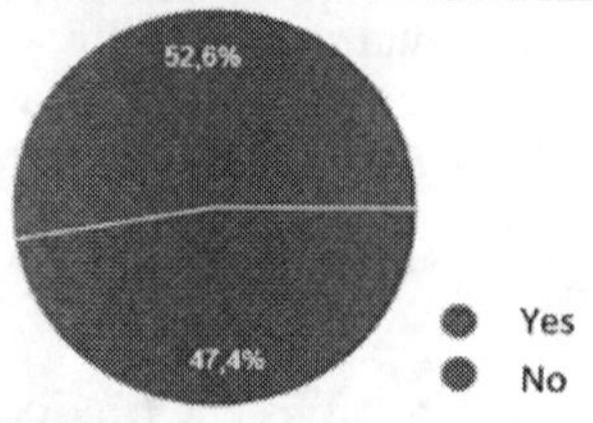 57 answers

Table 1. Continued

3. Do you know the feedback law of interaction in the human-biosphere system (P. Dansero, 1957)? Any change in the environment caused by **human** economic activity, "back" and has undesirable consequences, affecting the economy, social life and **health.** (https://studme.org/260227/ekologiya/osnovnye_zakony_pravila_printsipy_ekologii)	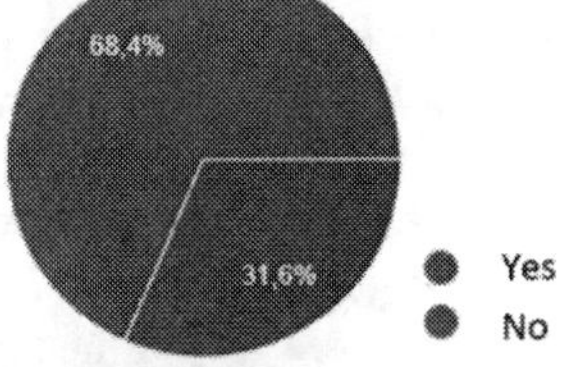 57 answers
4. Did you know that in order to keep one tree from being cut down, you have to find 80 kg of recycled paper? (https://sdelatiz.ru/skol-ko-nuzhno-derev-yev-chtoby-sdelat-knigu/)	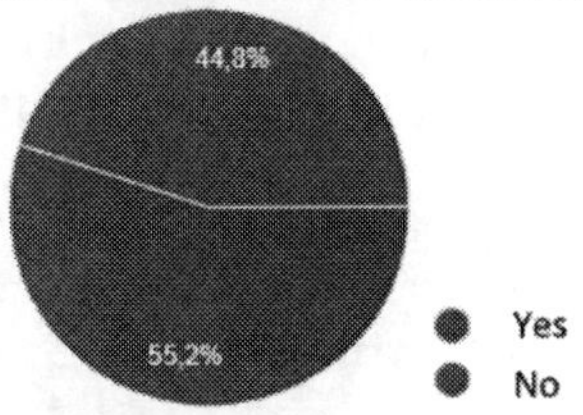 58 answers
5. Did you know that 24 trees need to be cut to produce one ton of ordinary office paper? (https://sdelatiz.ru/skol-ko-nuzhno-derev-yev-chtoby-sdelat-knigu/)	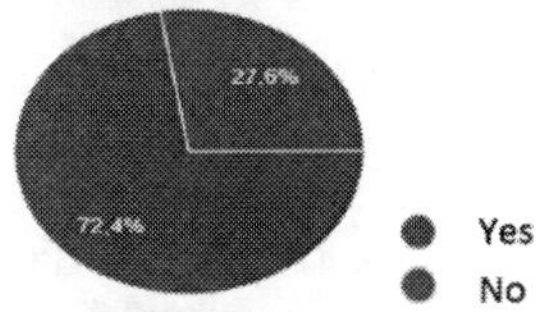 58 answers
6. Did you know that it takes 55 liters of fresh water to make a white cotton T-shirt (https://www.inopressa.ru/article/06Dec2019/bild/tshirt.html)	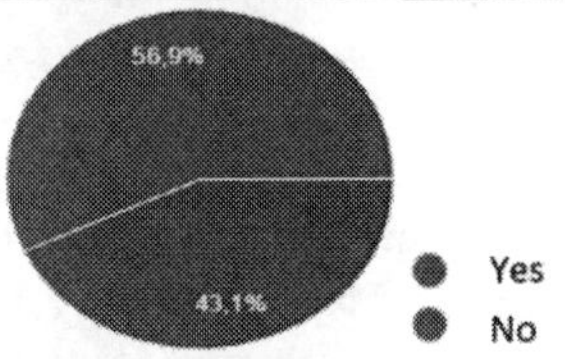 58 answers
7. Did you know that the average decomposition time of plastic products created using different technologies ranges from 400 to 700 years? Do plastic bags that people use on a daily basis take 100 to 200 years to decompose in nature? (https://ru.wikipedia.org/wiki/Пластиковое_загрязнение#:~:text=Среднее%20время%20разложения%20пластмассовых%20изделий%2C,прочности%20и%20долговечности%20пластиковых%20изделий)	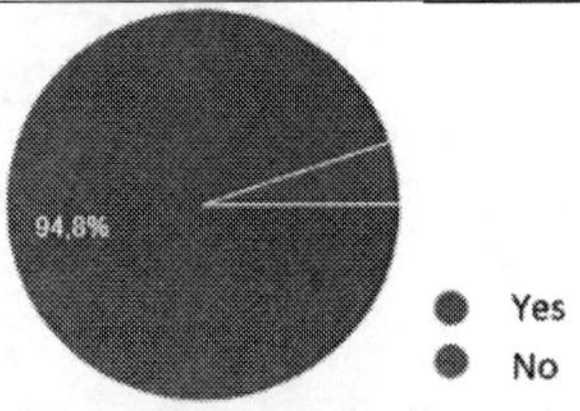 58 answers
8. You are aware of the existence of a large garbage patch in the Pacific Ocean, rough estimates of the area of which vary from 700 thousand to 1.5 million km² and more. May there be more than a hundred million tons of garbage on this site? https://www.independent.co.uk/climate-change/news/the-world-s-rubbish-dump-a-tip-that-stretches-from-hawaii-to-japan-778016.html	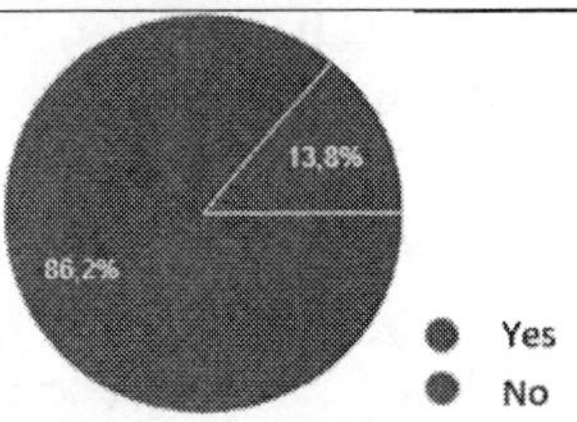 58 answers

Wherein, less than 50 % of survey participants were familiar with the laws of the Biosphere: only 47.4 % of freshmen answered positively to the question whether they knew the law of the indispensability of the biosphere (V.I. Vernadsky, D.P. Marsh, E. Reclus),

31.6 % of the respondents were familiar with the feedback law of interaction in the human-biosphere system (P. Dansero, 1957), but the law of the irreversibility of human-biosphere interaction was known to 72.4 % of the respondents.

With regard to specific environmental problems, such as pollution with plastic waste, conservation of water resources, here the participants in the experiment showed great awareness: 82.2 % of students knew about the existence of a large garbage patch in the Pacific Ocean, 94.8 % are aware that the average decomposition time plastic products created using different technologies range from 400 to 700 years, 72.4 % know how many trees are required to produce one ton of office paper, but only 43.1 % of respondents knew how much water is required to produce one cotton T-shirt.

Summarizing these results, we can confidently assert that for about half of the participants in the discussion club, the information provided in the questionnaire turned out to be new, which means a high educational potential of arranging such discussion platforms, and if we consider that all this information was provided to the trainees in a new context for them that they will have to make production decisions in the future considering the requirements of respect for nature, then we can talk about the high efficiency of the event. This is also supported by the active participation of freshmen in offline discussions.

4 Conclusions

The topic of environmental education of university students, who in the future will become managers who make important government decisions related to use of natural resources of our country, has acquired national importance today. In this regard, the task of teachers involved in arranging the process of teaching and educating students is to search for new modern pedagogical tools for interacting with students in order to increase the level of environmental education and environmental literacy among them. Real-virtual platforms for arranging educational activities aimed at increasing the environmental literacy of students of technical universities are an effective pedagogical tool. An experiment carried out among students of Moscow Power Engineering Institute proved that about 50 % of students who come to the university from school are not aware of the basic laws of the Biosphere and not all of them are familiar with even the most well-known global environmental problems. Wherein, they are happy to take part in discussions arranged on real-virtual platforms dedicated to environmental problems. An important task for teachers is to convey to their consciousness that in the future they will be responsible for making important environmental decisions. Real-virtual discussion clubs are undoubtedly an effective platform for formation of the ecological culture of future managers.

References

1. The strategy of ecological safety of the Russian Federation for the period up to 2025, https://www.garant.ru/
2. B. P. Nevzorov, T. B. Nevzorov, Bulletin of Kemerovo State University. Series: Humanities and Social Sciences, **1**, 18-27 (2017)
3. Pedagogical Dictionary, http://niv.ru/
4. L. Popova, Higher education in Russia, **10** (2012)
5. Yu. N. Nazarova, Formation of environmental responsibility of university students as a moral basis for environmental competence (in the classroom on floristry)
6. E. N. Kamenskaya, Engineering Bulletin of the Don, **4** (2015)
7. N. N. Moiseev, The problems of philosophy, **3**, 3-28 (1991)

8. V. N. Dopshak, Bulletin of the Kuzbass State Technical University, **2** (2009)
9. V. M. Makarov, A. P. Iosus, E. A. Donskaya, *Ecological foundations of nature management textbook*, 156 (Volgograd: IUNL VolgGTU, 2014)

Psychological factors of environmental responsibility of student youth

Elena Mititsina*, *Nadezhda* Parfenova, *Anna* Bizaeva, *Ekaterina* Kovalevskaya, and *Tatiana* Guchkova

Pskov State University, Psychology and Child Development Support Chair, 180000 Pskov Lenin Squire 2, Russia

Abstract. The study addresses the current social problem of psychological determinants of environmentally responsible behavior. Environmental responsibility acts as a mediating link between the cognitive, values, and emotional components of ecological consciousness and the predisposition of young people to pro-ecological actions. The empirical material was collected using a questionnaire method followed by factor and cluster analysis. It was found that the motivation of environmental activity is most associated with the social and territorial identity of young people, in the least with the declared ecological values. Identity with interest groups determines the readiness to volunteer in environmental organizations. A sense of connection with large social communities increases the desire of young people for socio-political activity. Three equal in number groups of students were identified, who attribute responsibility for the natural environment to one of three factors: personal behavior, environmental organizations of various scales, as well as economic conditions, and state policy.

1 Introduction

The scientific community and public figures are unanimous in the opinion that the manifestations of various environmental problems recently are primarily determined by psychological factors. Even though ecological technologies are intensively developing, the overhaul of the consciousness of modern man in the field of safe nature management should be of paramount importance. Activities promoting a careful attitude towards nature should be addressed primarily to young people as the most receptive, socially active age category with intensively developing worldview positions.

Some concepts actively discussed in the scientific literature reflect the content of the personality consciousness regarding the attitude to the surrounding natural environment and ensure environmental friendliness of behavior. In ecological psychology, such concepts include categories of different scales, ranging from relatively simple phenomena such as environmental values [1,2] and attitudes [3,4] to complex integral phenomena - ecological consciousness [5-8], ecological culture [9,10].

* Corresponding author: pskovprof@mail.ru

Among these categories, the concept of environmental responsibility occupies a particular position, since it can act as a mediating link between the emotional attitude to nature and environmental protection activity [3,11,12]. At the philosophical level, ecological responsibility is a moral imperative of human existence, which is the internal regulator of his relationship in the "man-biosphere" system, based on awareness of his duty and prompting self-government in the field of sound nature management [13]. Pedagogical practice approaches environmental responsibility as the principal goal of environmental education. They consider it a moral and legal personal quality, which, through awareness of its involvement in socio-ecological processes, contributes to the ability to predict the consequences of its behavior in the natural environment and regulate it [14,15]. According to a well-known methodologist of Russian environmental psychology V.I. Panov [16], environmental responsibility is a component of environmental consciousness. It has a practice-oriented character, ensuring the unity of consciousness and activity. It is viewed at the transcendental level when a person acts as a subject of self-development in the system "Humanity-Nature".

Accepting environmental responsibility as a prerequisite for pro-ecological behavior, we set the goal of this study to identify the relationship of cognitive, value, emotional components of ecological consciousness and predisposition to environmental actions. Research in the field of environmental psychology suggested the following components of ecological consciousness as factors of environmental responsibility: ecological perceptions reflecting the role of the natural environment in human life [3,6,17]; environmental values [18,19]; concern about the impact of environmental threats [20,21]; the identity of the individual with the socio-natural environment [22,23].

In our work, we attempt to answer two research questions:

1) What is the nature of the relationship between these phenomena, both among themselves and to carry out environmental activity among young people.

2) To what extent are the selected components of environmental responsibility expressed in students.

2 Materials and methods

The method of obtaining empirical data was a questionnaire method. The questionnaire aimed at identifying conditions affecting the responsible behavior of young people with concern for the natural environment. The study involved 254 students of humanities and natural sciences of Pskov State University aged 18 to 23 years.

The respondents were addressed with 92 questions grouped into 4 assessment scales that underwent a psychometric check procedure. The 4-point Likert scale assessed the degree of agreement of the respondents with the answer options.

The generalized content of the assessment scales is represented by the following characteristics:

The scale "Ecological values and ideas about nature" reflects vital values, values of environmental safety, the aesthetic value of the natural environment, a set of ideas about nature as a complex fragile system and a source of development of science, art, and human personality.

The scale "Identity with nature and social groups" characterizes a sense of connection with social groups of various scales: identity with the immediate environment (family, study group), identity with interest groups and with the territory (leisure, religious and other associations, place of residence), identity with large social groups (humanity, the population of the country, etc.).

The scale "Concern about threats to environmental safety" includes three blocks of environmental threats. These are the environmental problems of the globe (for example, the greenhouse effect, the decline of biological diversity), natural disasters and energy problems, pollution, and waste.

The scale "Behavior aimed at nature protection" allows identifying the respondents' ideas about the subjects that should ensure environmental protection at the level of state policy, large organizations, volunteer associations, as well as about the actions that they are personally ready to take in this direction.

The obtained data of the questionnaire were subjected to factor analysis using the method of main components (Varimax rotation). Several factors were allocated whose eigenvalues are more than one, the significance of the factor weights of variables is more than 0.4, the accumulated percentages of variance on the integral index were above 50%. Cluster analysis helped to characterize the levels of severity of these components in young people. As a result, groups of students with similar manifestations of environmental responsibility components were described.

3 Results and discussion

Factor analysis made it possible to reveal the structural and substantive characteristics of students' environmental responsibility and to identify complexes of indicators that reflect the relationship of students' attitude to the natural environment and trends in their ecological behavior. As a result of the analysis of empirical data, five factors were obtained.

Factor 1 "Environmental values, concerns about pollution, economic and political environmental regulation" (13.4% variance) includes variables: pollution and waste (0.732); the impact of economic structures and public policies on the protection of nature (0.599); environmental values (0.541); the value of the natural environment as opposed to the values of civilization (0.513).

Environmental values take a significant place in the structure of this factor, which include vital values (life, health), awareness of responsibility for the well-being of future generations, the importance of an environmentally safe environment. At the same time, there is a dichotomy between the understanding of nature as the universal human habitat and the values that determine the consumer attitude to nature. These include, for example, the pursuit of material enrichment and the use of natural resources for technological purposes. Thus, young people realize that the needs of modern civilized society harm the preservation of the natural environment and lead to its irreversible changes. Concern about pollution of all habitats takes a special place among other ecological problems in this factor. Probably mass media's active attention to the issue facilitates this concern in students.

At the same time, with normative values, this factor did not include a single indicator that would reflect the pro-ecological behavior of young people at the individual level. The main contribution to the organization of nature conservation events is associated with macro factors - state policy, officials, the media, the administration of industrial enterprises. This situation reflects some contradiction when the awareness of the importance of nature protection does not yet reach the level of personal involvement in environmental measures.

Factor 2 "Activity of public environmental organizations" (12.3% variance) includes variables: the aesthetic value of nature as opposed to its industrial use (0.746); the impact of environmental organizations, scientists, and volunteers on nature (0.680); identity with interest groups and territory (0.612); identity with the immediate environment (-0.487).

In the composition of this factor, in our opinion, the most interesting is the negative relationship between two types of identity, which allows us to distinguish two forms of personality orientation. One of them reflects the concentration of the individual on the life of the family and study groups, which are the immediate sphere of youth activity, the other - indicates the expansion of the interests of students, their orientation to groups of like-minded people, and connection with the territory of residence. This orientation determines confidence in the efficient activities of scientific communities, regional and international public organizations in environmental protection. One can assume that such an attitude will create youth readiness for volunteer activities. This factor also emphasizes the aesthetic value of the natural environment, which is opposed to its pragmatic use.

Factor 3 "Perceptions of nature as a fragile system and concern for global environmental problems" (11.5% of the variance) includes variables: nature as a complex system and a source of development of science, art, and human personality (0.738); environmental problems of the globe (0.671).

The factor reflects a complex of diverse ideas about nature. On the one hand, according to respondents, nature gives impetus to the development of science and art, is a source of recreation, inspiration, and the formation of a man's personality. On the other hand, students believe that nature is a sensitive system, the imbalance in which poses a danger to humanity. These perceptions lead to a high level of concern about global environmental problems. However, this concern does not lead to behavioral activity for nature conservation. The respondents probably perceive the planetary scale of environmental issues (climate warming, destruction of the ozone layer, etc.) as something abstract that does not directly affect their lives.

Factor 4 "Individual pro-environmental behavior" (11.4% variance) includes variables: daily pro-environmental behavior and its promotion among loved ones (-0.571) and concern about natural disasters and the energy problem (0.695).

The negative relationship between indicators of individual environmental behavior and concerns about environmental threats in the form of natural disasters and lack of resources constitute the contents of Factor 4. At the same time, this relationship may indicate that everyday environmentally friendly behavior helps to reduce the level of anxiety from environmental insecurity.

Factor 5 "Students' own social and political activity on nature protection" (10.3% of variance) includes variables: students' socio-political activity on nature protection (0.779) and identity with large social groups (0.693).

Factor 5 is similar to factor 2, as it also reflects the relationship between indicators of identity and social activity. However, in factor 2 identity with small groups is related to participation in broad environmental organizations. Identity in factor 5 is associated with increased political activity, media activity, and involvement in ecological movements with large groups (humanity, population of a territory). Variables in factor 5 also indicate readiness for environmental protection as a volunteer.

The undertaken cluster analysis characterized the level of selected factors among students. Noteworthy that the subjects were divided into three main groups equal in number (32.3% of students each) depending on which methods of protecting nature they consider the most effective. Representatives of the first cluster can be designated as "Aware of individual responsibility" since the most pronounced indicator of this group is readiness for daily nature-saving behavior. At the same time, their faith in the impact of regional organizations and international movements on the environmental situation is not that strong. At an average level, this group of students demonstrates their ecological values, anxiety about global ecological issues, assessing the influence of economic and political factors.

The second cluster included students focused on social activity in environmental behavior. In this group, the highest level of the indicator "Activity of public environmental organizations" is observed among all groups. These students identify more with like-minded communities and with the territory of residence and are ready to participate in the work of public associations. The third cluster is composed of students who show passivity in the field of nature conservation. Although they have a fairly high level of environmental values and concern about world ecological problems, they demonstrate a weak desire to participate in pro-environmental activities as a part of environmental organizations and individuals. At the same time, these students attribute responsibility in the protection of nature to economic factors and public policy. A small number of respondents (3.1%) entered the fourth cluster with a very high level of pro-environmental activity in its various forms and increased anxiety about the impact of environmental threats.

4 Conclusions

The undertaken study of environmental responsibility revealed the following trends characteristic of students of Pskov State University:

1. The development of environmental values is not directly related to the increase in personal responsibility in ecological activities. Environmental values and perceptions of nature as a fragile system requiring protection remain declarative and do not lead to action.

2. To the greatest extent, readiness for environmental activity has correlations with various forms of identity. Moreover, the higher the identity of students with large social communities, the higher their desire for socio-political activity. Students who actively participate in social groups united by common interests show readiness for volunteer activities as a part of public environmental associations.

3. Active day-to-day conservation activities reduce global environmental concerns.

4. The opinion of students regarding who is primarily responsible for the natural environment was divided equally, taking into account three factors: each person personally in his daily life, environmental organizations of various scales, political and economic decisions at the state level.

In general, students' awareness of environmental problems and ways to overcome them prevails over their involvement in environmental issues. However, the findings can be accepted only by taking into account some limitations. They require clarification of the results of the study on large samples of respondents covering different age, professional, regional groups.

References

1. P. W. Schultz, V. V. Gouveia, L. D. Cameron, G. Tankha, P. Schmuck, M. Franěk, Journal of Cross-Cultural Psychology, **36(4)**, 457 (2005)
2. J. I. De Groot, L. Steg, Environment and Behavior, **40**, 330 (2008)
3. F. G. Kaiser, M. Ranney, T. Hartig, P.A. Bowler, European psychologist, **4(2)**, 59 (1999)
4. M. Wright, B. Klyn, Journal of Empirical Generalisations in Marketing Science, **4**, 42 (1998)
5. V. I. Panov, *Ecological psychology: Experience in constructing methodology* (2004)
6. V. A. Yasvin, *Psychology of attitude to nature* (2000)
7. S. D. Deryabo, *Ecological psychology: diagnostics of ecological consciousness* (1999)
8. I. A. Shmeleva, Psychological Journal, **32(5)**, 5 (2011)
9. S. A. Bortnikova, Bulletin of the International Academy of Sciences (Russian section), **2**, 113 (2011)
10. E. G. Vinogradova, Economic and Social Research, **4(28)**, 70 (2020)
11. X. Zhang, N. Ye, *Proceedings of international symposium on green management and local governments responsibility*, 51 (2017)
12. A. Paco, R. G. Rodrigues, International Journal of consumer studies, **40(4)**, 466 (2016)

13. E. V. Ryabova, Humanities, Social-economic and Social Sciences, **5**, 269 (2012)
14. O. V. Petunin, A. A. Mzhelskaya, Innovations in Education, **5**, 58 (2005)
15. V. L Krainik, Zh. Zh. Sergazina, World of Science, Culture, Education, **3(70)**, 203 (2018)
16. V. I. Panov E. V. Lidskaya, Vestnik MSU named after M.A. Sholokhov. Socioecological technologies, **1**, 38 (2012)
17. S. Jovanovic, O. Miljkovic, O. L. Zivkovic, D. Sabic, D. Gataric, M. Dzinovic, Journal of environmental protection and ecology, **18(3),** 1223 (2017)
18. S. Jovanovic, L. Živković, A. Sladjana, G. Dragica, Procedia - Social and Behavioral Sciences, **171**, 317 (2015)
19. G. Liobikiene, R. Juknys, Journal of cleaner production, **112(4)**, 3413 (2016)
20. P. Janmaimool, S. Chudech, Sustainability, **12(4)**, 1610 (2020)
21. A. Ferrer-i-Carbonell, J. M. Gowdy, Rensselaer Working Papers in Economics, **0503**, 67 (2005)
22. S. Claiton, B.D. Irkhin, S.K. Nartova-Bochaver**,** Psychology. Journal of the Higher School of Economics, **16(1)**, 85 (2019)
23. E. A. Sorokoumova, E. I. Cherdymova, Psychological Science and Education, **26(1)**, 102 (2021)

Features of design solutions in the design of a special-purpose building of urban infrastructure

Larisa Filimonova, *Elena* Matys, and *Nadezhda* Skvortsova

Industrial University of Tyumen, 625000, Lunacharskiy Str., 2, Tyumen, Russia

Abstract. The article discusses approaches to the formation of the optimal combination of design solutions in the construction of a building for funeral services. The projected facility fulfills a triple purpose: social, communal and industrial. The industrial purpose is due to the peculiarities of the assembled technological equipment and its maintenance. These are the parameters of the crematorium construction project. The contradiction in the purpose of the building can be traced in the architectural solution, the decision on the choice of engineering communications and the transport and logistics infrastructure of the environment of the object. What the authors pay attention to in their publication. The classical task of substantiating a design solution is the formation of a single information space of a capital construction object, taking into account information reflecting all stages of its life cycle. The principal distinguishing feature of the research results is the prepared set of key characteristics of the formation of project documentation, taking into account the functional parameters of the specific purpose of the projected capital construction facility, the relevance of which is undeniable both from an environmental point of view and from the standpoint of socializing society and preventing religious fanaticism.

1 Introduction

According to the results of an investigation by a Russian blogger, "about two million people die in Russia every year, which determines a number of factors that hinder the civilized development of society and regions. For example, the ideological function in the Russian practice of the sphere of ritual services is sometimes performed by criminalized structures. According to the blogger's investigations, "the turnover of the funeral industry only officially amounts to about 60 billion rubles a year, the size of its shadow sector, according to official sources, can reach 250 billion"[1]. There fore this cache stream does not fall under the Russian taxation and budgeting system. The cemetery plays an equally important role in the ecological potential of the region, since the traditional cemetery is a potential source of negative impact on the human environment. As a result of the decomposition of biological waste, bacteria, organic compounds, nitrogen compounds, chlorides, hydrocarbons, heavy metals. Enter the soil, rocks of the aeration zone and groundwater. «Before the ceremony, the body of the deceased is treated with formalin, which slows down the decomposition process, which is a toxic substance» [2].

2 Literature Review

According to the following sources [2,3,4,5] and the report of the Tyumengeomonitoring, the current state of groundwater requires the abandonment of the traditional option of waste disposal. The report [5] was prepared on the basis of long-term data obtained in the process of conducting state monitoring of groundwater at the basic state observation network of the federal level, polygons for state monitoring of groundwater, created at the expense of the regional budget, as well as objects of local monitoring of groundwater. In general, over the year, the number of sites on which local contamination of groundwater was recorded increased from 15 to 17, which allows us to conclude that their condition has worsened.

According to the provisions of the following regulatory documents [6-12] during the design and construction of the crematorium, it is necessary to solve simultaneously two main tasks:

1. Formation of an orderly system that will provide a given passage of the service personnel.

2. Creation of the necessary environment for conducting farewell ceremonies, if possible, while maintaining the spiritual, ideological, psychological and moral halo for most categories of clients.

In addition to the basic regulatory documents for the development of project documentation, the Urban Planning Code of the Russian Federation and the Decree of the Government of the Russian Federation No. 87 are presented in table 1, a set of documents required for the development of our project of the building of the crime scene is presented.

Table 1. The normative legal support of the project.

The normative Legal act	Name
GOST R 53999-2010	Household services. Crematorium services. General specifications 2011
SNiP 2.07.01-89*	Urban planning. Planning and development of urban and rural settlements 2011
SanPiN 2.1.2882-11	Hygienic requirements for the placement, arrangement and maintenance of cemeteries, buildings and structures for funeral purposes 2011
MDC 31-5.2000	Recommendations for the design of objects of ritual significance 2000
SanPiN 2.2.1/2.2.1200-03	Sanitary protection zones and sanitary classification of enterprises, structures and other objects

The main provisions for the placement of funeral services should be highlighted:

1. Land plots for the construction of crematoria are recommended to be located near existing utilities and roads.

2. The location of the crematoria should be in separate areas, near the motorway.

3. The site for the construction of the crematorium should not be located near objects with a high noise regime (airports), as well as near objects and areas with unpleasant odors.

4. If the crematorium is intended only for the cremation of bodies, then it is recommended to place it outside the city, on a separate site.

5. Bus stops and parking lots should be located at a distance of no more than 150 m from the crematorium.

6. The sanitary protection zone from the crematorium to residential buildings must be at least 1000 meters.

7. The zone of moral protection from the crematorium, columbarium buildings, cemeteries with burials after cremations, to residential buildings must be at least 100 meters.

3 Results of approbation

The result of the conducted bibliographic research (search, critical and clarifying) [6-19] should be recognized as the substantiation of the author's approach to the choice of a constructive, space-planning solution when formulating a working hypothesis.

Features of cremation architecture. Cremation architecture - is the main part of the infernal architecture (in translation "inferno" - the afterlife), which has its starting point from the Egyptian pyramids. Today, in cremation architecture, the main role is played by the ethical setting on the intensity of the influence of the image as opposed to its artistic expression, which by its nature is always subject to certain trends and trends, and therefore is not acceptable in buildings for mourning purposes. The purpose of cremation architecture is to reflect the inner content associated with grief. It should not suppress the mourning processes taking place in crematoria, and at the same time, it should not impose artistic ideas and images on one architect, who has his own associations with grief and mourning. In this connection, the authors of the article made an emphasis on figurative neutrality when forming a design solution.

The subject of this study was the demographic processes taking place in the region, as well as the theoretical and regulatory framework that determine the value and purpose of the construction of a specific-purpose facility, and the process for the disposal of biological waste that meets the principles of social and environmental efficiency.

Throughout Russia and foreign countries, where the cost of land is very high, the ratio of the number of cremations to the total number of deaths reaches 91%. Based on the long-term practice of the nearest crematorium, which is located in Yekaterinburg (about 330 km from Tyumen), it was revealed that 60% of the total number of deaths are cremated. According to the data of the same crematorium, 12,518 cremations were carried out in 2014, of which about 3,500 services were provided to residents of Tyumen and the South of the Tyumen region, which is 21-22% of the total mortality statistics for 2014 in the South of the Tyumen region.

Of course, at the present stage of development of society and respect for the ecosystem, an important factor is the presence of a crematorium and a columbarium in the settlement, as well as the methods of burial accepted by society and, especially, religion. Despite its technological effectiveness and rationality, cremation has not yet become widespread - both in Russia and here in Tyumen.

Sanitary protection zones and sanitary classification of enterprises, buildings and structures for crematoria are at least 1000 meters. The key provision of the working hypothesis when deciding on the choice of a land plot for the construction of a building is the remoteness from the settlement. Land plots for the construction of crematoria are chosen by the introduction of engineering communications and transport infrastructure. At the same time, the sanitary protection zone from the crematorium to residential buildings should be at least 1000 meters. The composition of the premises of the building is taken in accordance with MDS 31-5.2000 "Recommendations for the design of ritual facilities". Space-planning solution. The cremation building is classified as a production facility. The building is one-story with dimensions in the axes of 48.0 x 18.0 m. The height of the building is 7.4 m. The following functional groups of premises are provided for in the building: Premises for receiving the deceased (vestibule, reception and registration room, room for refrigeration chambers, loading room); rooms for cremation of the dead (cremation room, room for processing cremated remains, rooms for storing urns, room for storing and upholstery of coffins); service (storage room and washing trolleys, rest and meal room, bathroom for staff, medical office, administration). Technical premises (boiler room, switchboard, ventilation chamber, compressor room).

The main elements of the building are the entrance, lobby and funeral hall. The funeral hall according to MDS 31-5.2000 "Recommendations for the design of ritual facilities" is accepted according to the project, with a capacity of 20-30 people. When designing this chain, the rule of "Funnels" was laid, when, in addition to perceiving information of direct influence, the participants in the ceremony perceive the information codes contained in the architectural style of the building, as well as its design (fig.1, fig.2). We propose the following architectural technique as a basis for designing a crematorium. It consists in the ability to perceive in the most detail various parts of the furnishings (interior), such as carved wooden doors, garlands of stones. The epicenter of the congestion of participants - the lobby, should adapt people, reduce their prevailing emotional state, distract the participants in the direction of the philosophical-contemplative direction. The process of ritual procession to the funeral hall is the beginning of the farewell ceremony and should create the strongest effect of the impression of solemnity. The funeral hall should provide a final impression of the entire ritual building.

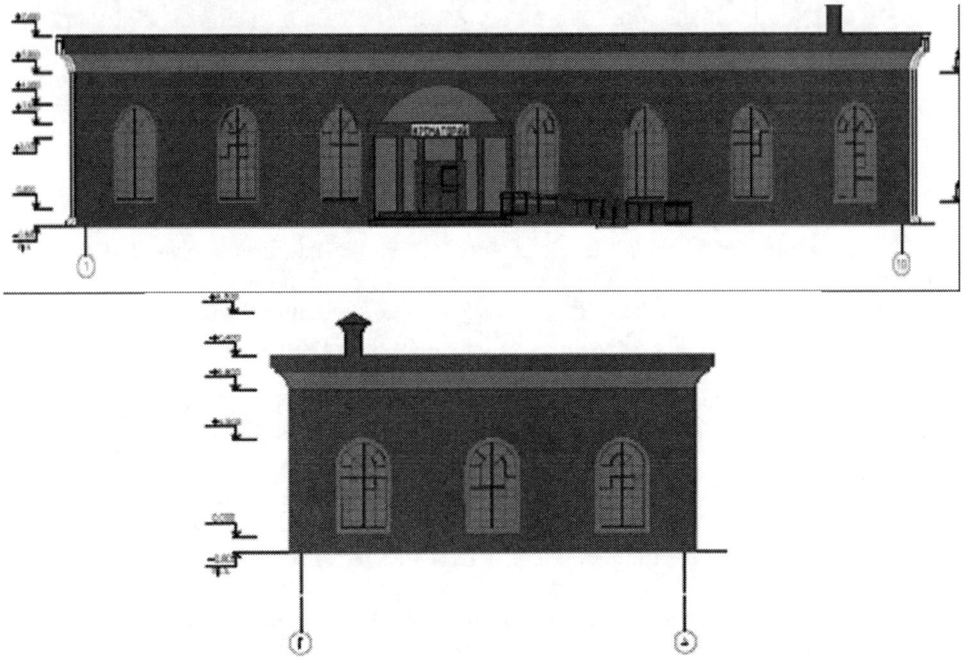

Fig. 1. Facades of the building.

Constructive solutions of the projected building: Foundations. A type: Columnar with monolithic grillages for columns. Concrete class: 20. Frost resistance grade: F100. Waterproof grade: W4. Columns. Cross section: Rectangular. Dimensions (edit) 400 * 600. Column step: Regular according to the main center lines. Concrete class: 20. Frost resistance grade: F100. Waterproof grade: W4. Armature: A600 GOST 5781-82. Partitions. Material: Ceramic brick GOST 530-2012. Masonry mark: M50. Solution grade: M50. Thickness: 88, 120mm. Overlapping and covering. Reinforced concrete. Device method: Monolithic. Concrete class B30. Frost resistance grade: F100. Waterproof grade: W4. Armature: A600 GOST 5781-82 *, Bp-1200. Roof. A type: Flat. Sediment removal method: Internal drain. Outdoor decoration. Finishing type: Facade ceramic tiles in accordance with GOST 13996-93. Interior decoration. Finishing type: Walls and partitions are made and faced according to the "Knauf Tigi" system, followed by painting with water-based paints. Floors. A type: Ceramic tile.

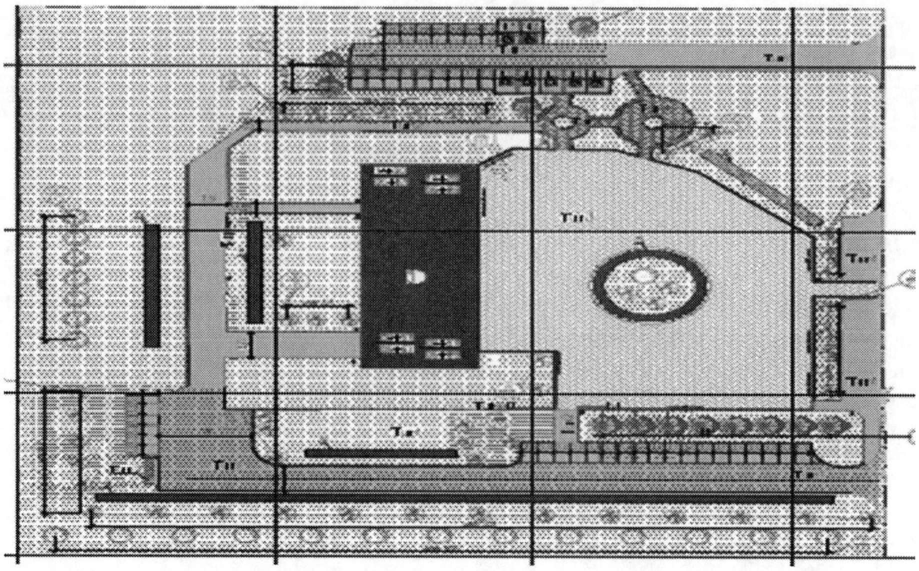

Fig. 2. General plan of the projected building.

Calculation of the throughput of the crematorium.

It is recommended to install two crematoria in single-hall crematoria, and three in two-hall crematoria. The throughput of a crematorium (one crematorium, type 1) is determined on average at the rate of 1 hour 15 minutes per one cremation.In crematoria (type 2), the number of incinerators is determined by calculation. Cremation ovens must undergo an environmental and sanitary-hygienic examination, as well as have a sanitary-epidemiological conclusion. The entire crematorium building must be equipped with supply and exhaust ventilation, and the use of air conditioners (systems with air recirculation) is also not allowed.

The throughput of crematoria is set by calculation depending on the number of clients served. The capacity of the building for the funeral rites is determined by the number of possible cremations. The time of the ritual is taken from 35 to 40 minutes, the break between rituals is from 10 to 15 minutes. The time of one cremation is from 40 minutes to 1 hour, depending on the type of cremation oven. When calculating the capacity of an enterprise, the following are taken into account:

1. The mortality rate.
2. Estimated (on assignment) percentage of cremations of the total number of deaths.
3. The coefficient of unevenness of mortality (from the conditions of the season), equal to 1.33-1.5, where the maximum falls on the spring-autumn period.

Calculation of the crematorium throughput according to MDS 31-5.2000 "Recommendations for the design of ritual facilities":

The number of funerals per year is calculated using the formula (1):

$$Nf = Np \times Mr \rightarrow \max, \tag{1}$$

Nf - number of funerals per year, persons; Np- number of population served, people; Mr - mortality rate.

The number of deaths in January - December 2019 in the Tyumen region without districts was 6,598 people (100.7% compared to the corresponding period of 2018) [14, 15]. The population as of 12.2019 in the Tyumen region is 1 million 537,42 thousand people

[14,15]. The mortality rate for 2019 was 10.8 thousand people. The mortality rate was 0.007. The number of funerals per year will be 1537.42 * 0.007 = 10.76.

Let's determine the maximum number of cremations per day:

Of the total number of funerals per year, the percentage of cremations determined by the task should be taken into account, which will determine the maximum number of cremations per day according to the formula (2):

$$MAX_{crj}^{day} = \frac{Nf \times K_{cu}}{Dw} \times K_{cr} \rightarrow \max, \quad (2)$$

$$MAX_{crj}^{day} = \frac{10761 \times 2{,}1}{300} \times 0{,}3 = 22{,}6 \approx 23 \rightarrow \max,$$

MAX cr - the maximum possible number of cremations per day (capacity of the crematorium); *Nf* - is the number of funerals per year for a given city; *K cr* - cremation rate, (In Russia it is 20%, during a pandemic from 30-40%); *K cu* - coefficient of unevenness (1.35-2.5); *Dw* - the number of working days of the crematorium in a year (taken equal to 300 days, taking into account weekends, holidays and sanitary days).

Calculation of the number of ovens.

The required number of ritual halls and cremation furnaces (chambers) is determined by the capacity of the crematorium, the throughput of one ritual hall and the time of one cremation in a furnace of a certain design. When determining the number of cremation ovens, it is necessary to provide for backup ovens (at least one).

The required number of furnaces is determined by the formula (3):

$$F = \frac{T \times MAX_{cr}}{480} = \frac{40 \times 23}{480} = 1.74 \approx 2, \quad (3)$$

where F - is the required number of furnaces; T - Burning time with continuous cremation, min (1 cremation 25-50 min, on average 40 min); *MAX cr* - the number of cremations per day.

For the crematorium, ovens will be consumed, in the amount of 2 pieces. The maximum capacity of the crematorium for continuous cremation will be 23 cremations per day.

Based on the comparison of a wide range of furnace options presented on the Russian and foreign markets, it was decided to use TABO VL-A.

Epidemiological, hygienic and infrastructural requirements for sanitary cleaning of areas adjacent to the building and the infrastructure of the project are also important in substantiating the design solution.. To implement the operational stage of the project, you need:

Modern water supply system. Whenever possible, we recommend starting an autonomous system (for example, a well) with recycling. Multi-stage purification with multiple use of industrial water for economic purposes will reduce the operating costs of the project not only for the building's water supply, but also for water disposal (sewerage). All drains from the crematorium containing toxic elements must be deeply cleaned, filtered, disinfected, disinfected to acceptable, regulated treatment and re-launched in the water supply system without being released into the general sewage system.

Internal networks of cold and hot water supply are made of PP-R PN 20 polypropylene pipes. The source of hot water supply is a universal gas boiler.External fire extinguishing is provided from 2 fire tanks with a volume of 60 m3 each. Water consumption for external fire extinguishing is determined according CP 8.12120.2009 (rev. No. 1) and is 10 l / s.

Modern heating system. According to the project of the building according to the project and the use of modern technological (gas) equipment for cremation, it is planned to launch an autonomous closed heating system. Allows you to approach the reduction of environmental risks, since the object is not of a level in this area of this energy during the operation of cremation furnaces.

Modern power supply system. The project should provide, in addition to traditional sources of electricity, present in the vicinity of the location of the building. Electricity is supplied through two inputs from existing external networks. The voltage at the inputs is 380/220 V. Also, alternative sources, for example, when performing roofing work, provide for the installation of solar panels.

Modern ventilation system. The building is designed in accordance with the normative requirements MDS 31-5.2000 "Recommendations for the design of ritual facilities", supply and exhaust ventilation with mechanical and natural induction. A single supply ventilation system is provided for the premises: ritual halls and vestibules, cremation hall, bathrooms, refrigeration chambers. Exhaust ventilation is provided from the premises of the ritual halls with the provision of measures to prevent overturning of the draft. All air ducts to the ventilation chambers are covered with perlite plaster 20 mm to achieve the fire resistance limit of 0.5 hours. Exhaust ventilation is provided from the bathroom premises through ventilation ducts with natural motivation. Ventilation and ventilation are provided by means of ventilation shafts, opening windows, transoms or vents.

To exclude the penetration of unpleasant odors and noise into the ritual hall of the crematorium, it is not allowed to combine the room for the preparation of the bodies of the deceased before being placed in the coffins with the hall of farewell by the elevator shaft. To ensure safety, an automatic shading of the lift shaft opening should be provided after lowering the lift platform.

As part of the existing foreign practice, we propose to include a design solution for the implementation of BIM (Building Information Modeling) technology on the blockchain platform with adaptation to the goals and objectives of the projected building. This technology is able to provide the results of the analysis of the causes of emergencies (their manifestations) in the building not only at the request of the interested party, but also in the current mode, which allows minimizing human casualties in emergency cases. The authors of the article give recommendations on how to launch BIM in order to minimize risks and uncertainties for both the owner of the building and residents of nearby territories. The proposed technology takes into account fire-prevention measures in the project in accordance with the requirements of SNiP 2.01.01-97 "Fire safety of buildings and structures." The fire resistance of the building is II; accordingly, all structures are accepted.

The finishing materials used in the premises, as well as the space-planning and structural solutions have been selected to prevent the spread of fire.

4 Conclusion

The principal distinctive feature of the research results is the formulated key approaches to the formation of an optimal design solution in the development of a multipurpose building project. The authors of the article are based on the results of monitoring the existing import and "modest" domestic practice of developing and launching projects for the construction of crematoria. Thus, on the basis of the above functional capabilities of the designed building for a specific purpose, the relevance of adapting the traditional approach to design through the prism of innovative digital technologies is revealed. Digital BIM technology is able to provide up-to-date results of diagnostics of a building, technological equipment and analysis of the causes of emergency situations at the facility, not only at the request of an interested person, but also during operation, which allows minimizing human casualties in emergency situations.

The crematorium building, according to the authors of the article, must meet a variety of requests, which corresponds to modern trends and features of the development and purpose of the construction industry and housing and communal services:

1. Modern water supply system. Launch of an autonomous system (well) with recycling. Multi-stage purification with multiple use of industrial water for economic purposes will reduce the operating costs of the project not only for the building's water supply, but also for water disposal (sewerage). All drains from the crematorium containing toxic elements must be deeply cleaned, filtered, disinfected, disinfected to an acceptable, regulated concentration and re-launched into the water supply system without being released into the general sewage system.

2. Modern heating system. According to the purpose of the building according to the project and the use of modern technological (gas) equipment for cremation, it is advisable to launch an autonomous closed heating system. This approach helps to reduce environmental risks, since the facility will not release thermal energy into the environment during the operation of cremation furnaces.

3. Modern power supply system. The project provides, in addition to traditional sources of electricity, which are present in the vicinity of the location of the building, alternative sources. When performing roofing works, installation of solar panels is provided.

4. The ergonomics of the building is oriented towards the existing foreign practice. The project provides for the launch of BIM (Building Information Modeling) technology on the blockchain platform with adaptation to the goals and objectives of the projected building. This technology is able to provide the results of the analysis of the causes of emergency situations (their manifestations) in the building, not only at the request of the interested party, but also in the current mode, which allows minimizing human casualties in emergency cases. The authors of the article give recommendations on how to launch BIM in order to minimize risks and uncertainties for both the owner of the building and residents of nearby territories.

The proposed technology takes into account fire-prevention measures in the project in accordance with the requirements of SNiP 2.01.01-97 "Fire safety of buildings and structures." The fire resistance of the building is II; accordingly, all structures are accepted.

5. Based on the comparison of the options presented, it was decided to use the TABO VL-A furnace for the project.

This publication can serve as a basis for replicating the design solution and bringing the design documentation into the status of a reuse project.

The results of the generalization of design solutions allow us to minimize labor costs at the pre-investment stage of the implementation of a socially significant project, the relevance of which is undeniable. The authors of the article presented the rationale, conclusions and recommendations for the development and launch of a project for the construction of a building for the provision of ritual services for the cremation of organic material.

The authors of the article presented the rationale, conclusions and recommendations for the development and launch of a construction project for a building for the provision of funeral services in order to minimize risks and uncertainties for all project participants and residents of nearby territories.

References

1. I. Golunov, Coffin, cemetery, hundreds of billions of rubles (2018) https://meduza.io

2. Administration of the Tyumen region: official site, Groundwater Contamination of the Tyumen Region: Report on the environmental situation in the Tyumen region (2020) https://admtyumen.ru
3. Scientists: Avoiding Coffins and Cremation Will Save Wildlife (2019) https://www.vesti.ru
4. Which is more environmentally friendly: cremation or burial (2019) https://www.popmech.ru
5. Administration of the Tyumen region: official site, Report of the Department of Subsoil Use and Ecology of the Tyumen Region (2020) https://pandia.ru
6. Ministry of Construction, Housing and Utilities of the Russian Federation, http://www.minstroyrf.ru
7. SanPiN 2.1.1279-03. Hygienic requirements for the placement, arrangement and maintenance of cemeteries, buildings and structures for funeral purposes, https://files.stroyinf.ru
8. MDS 31-5.2000 Recommendations for the design of objects of ritual significance, https://files.stroyinf.ru
9. SanPiN 2.2.1 / 2.2.1200-03 Sanitary protection zones and sanitary classification of enterprises, structures and other facilities, https://files.stroyinf.ru
10. SNiP 23-03-2003: SP 51.13330.2010 Protection against noise, 69 (2011)
11. CP 60.13330.2010 Heating, ventilation and air conditioning, 57 (2012)
12. SanPiN 2.1.1279-03 Hygienic requirements for the location, arrangement and maintenance of cemeteries, buildings and structures for funeral purposes (2020) https://files.stroyinf.ru
13. N. A. Surnacheva, *Features of cremation architecture* (2019) https://moluch.ru
14. Analytical report of the Federal Antimonopoly Service on the analysis of the funeral services market (in the context of municipal districts and urban districts (2020) http://www.ritual-grk.ru
15. Report on the analysis of the funeral services market, https://www.mos-ritual.ru
16. Ritual portal: official site, https://ritual-rus.ru
17. Cemetery "Memorial", Tyumen: official site, https://ritual-rus.ru
18. Buka cemetery. Tyumen: official site, http://izgotovleniepamyatnikov.ru
19. E. A. Yumaev, Journal of Economic Theory, **2**, 181 (2017)
20. S. Logunov, Draft strategy for the development of the construction industry (2019) https://zen.yandex.ru

E3S Web of Conferences **311**, 01008 (2021) https://doi.org/10.1051/e3sconf/202131101008
EPSD 2021

Sustainable development: new socio-environmental trends

Yulia Sergeenko[1,*], *Elena* Abanina[1], *Oksana* Galkina[1], *Yulia* Nikitenko [2], and *Olga* Averyanova[3]

[1]Saratov State Law Academy, Volskaya St., 1, 410056 Saratov, Russia.
[2]Leningrad State University named after Alexander Pushkin, St. Petersburg sh., 10., Pushkin, 196605, St. Petersburg, Russia
[3]St. Petersburg State University of Economics, 21 Sadovaya street, 191023 Saint-Peterburg, Russia

Abstract. The policy of each state is aimed at preserving the life and health of people. Longevity of people depends on environmental and legal protection guaranteed and respected by the adopted normative legal acts of states, as well as interstate agreements. Sustainable development is based on three areas: environmental, economic, and social. In the case of accidents (industrial injuries, environmental emergencies, and etc.), non-observance and violation of the rights of people and employees of enterprises in the environmental sphere of activity is observed. Therefore, it became necessary to solve the problem in the direction of social and economic responsibility. The article authors have developed exceptional offers for implementation and maintenance of these areas, both in environmental law and in labor law (social protection). By applying and following these offers, we will try to preserve and protect the ecology of the world countries, thereby implementing social and labor protection of people.

1 Introduction

The world around our planet is a priceless universe created for people. The way we treat it depends on the quality and duration of life on earth, preservation of the environment (birds, animals, insects, the green world of the planet). With the arrival of developed industrial complexes, plants, factories in the countries, changes also occur in the surrounding world, or rather, human intervention in nature is a violation of the animal and green worlds. This affects negatively the human himself/herself, his/her health, social guarantees within performance of labor functions. Various measures are being taken to correct this situation in the environment and protect humanity, for example, "regional standards of social servicing of the population" [1], "increasing public awareness of disaster impacts and risks" [2], "integrated management on land resources" [3], "the exploitation of forest heritage as an important economic resource" [4], "mechanisms for protecting the land" [5], "advanced training of enterprise managers in order to ensure environmental safety" [6], "the transition to sustainable development in the context of ecology" [7], "the ecological safety of

* Corresponding author: iulya.sergeenko@yandex.ru

forestry" [8], and etc. However, in order to restore fully the balance of environmental and social protection in emergency situations in the surrounding world, it is necessary to restore and implement fair responsibility in the field of sustainable development.

2 Materials and Methods

While studying and researching environmental and social responsibility, the article authors participated in conferences and discussions that focus on environmental and labor responsibility. This is the VI International Research Conference "Advanced Digital, Information and Communication Technologies and Innovations in Education, Industry, Public Health, Modern Socio-economic, Humanitarian and Interdisciplinary Research - 2021", the International Scientific and Practical Conference "Sustainable Development in Agriculture, Environmental Safety and Energy Efficiency"(EESTE2021), II International Scientific and Practical Conference "Information Technologies and Intelligent Decision Making Systems" (ITIDMS-II-2021), and etc.

The following works have been studied:

- "Risk Analysis and Protection Strategy of Coastal Urban Construction Project under Extreme Typhoon Climate -- A Case Study of Xiamen City" [9];
- "Environmental Education and Management: Exclusive, Inclusive, and Indigenous Doctrines" [10];
- "The Assessment of the Feasibility of Using Degraded Lands as a Resource in the Production of Environmentally Friendly Food for Vulnerable Populations" [11];
- "The social costs of health- and climate-related on-road vehicle emissions in the continental United States from 2008 to 2017" [12].

Within the study of offers for preservation and protection of the ecology of the world countries, implementation of social and labor protection of people, the following methods were used:

- Empirical method. It consists in observing the ongoing emergency situations of a global scale affecting the protection of people and consequences after violation of environmental safety affecting responsibility;
- Analytical method. It promotes in-depth study of environmental and social responsibility, within performance of work duties, both within performance of work and in emergency situations;
- Analysis. The study was carried out on violations associated with accidents in production, which entailed environmental safety at the enterprise and residents living in the adjacent territories of these facilities.

The materials used facilitated the process of research by the authors of the issue on responsibility in case of disasters occurring not only for persons working at enterprises, but causing harm to residents inhabiting the area.

3 Results and Discussion

In the 20th and 21st centuries, emergencies, environmental and man-made disasters, natural disasters resulting in the death of people, the disability of people, housing not suitable for further living, climatic changes on the globe, the death of forests, air pollution occurred and are occurring in the territories of countries, loss of large-scale territories for development and human population. The legislation provides for criminal liability for violation of environmental protection rules within performance of works. This is a fine, forced labor, prohibition to hold certain positions or imprisonment.

 https://doi.org/10.1051/e3sconf/202131101008

The accident at Chernobyl nuclear power plant, the accident at Mayak enterprise, Fukushima-1 is not a complete list of disasters that have occurred in the countries. The consequences of these emergencies are dire. To comply with the rights of residents of territories where climatic and environmental violations have occurred, strict control by state bodies is required for implementation of those guarantees that are provided for by law (improvement of housing conditions, compensation for utilities). At the federal level, in accordance with Law of the Russian Federation of May 15, 1991, No. 1244-1 (as amended on May 26, 2021) "On Social Protection of Citizens Exposed to Radiation as a Result of the Chernobyl Disaster" (as amended and supplemented, entry into force dated September 01, 2021) "provision of those in need of improvement of living conditions with living space, compensation of expenses for payment of residential premises and utilities in the amount of 50 percent - rent and (or) payment for maintenance of residential premises, payment for overhaul of common property in an apartment building, payments for cold water, hot water, electricity consumed in maintenance of common property, utility bills, payment of fuel costs are guaranteed" [13], Resolution of the Government of the Russian Federation of August 21, 2001 No. 607 (ed. of July 10, 2020) "On the Procedure for Payment of Monthly Monetary Compensation in Reimbursement of Harm Caused to the Health of Citizens in Connection with Radiation Exposure Due to the Chernobyl Disaster or Work to Eliminate the Consequences of the Disaster at Chernobyl Nuclear Power Plant" [14], at the level of subjects in accordance with Order of the Ministry of Labor and Social Protection of the Saratov Region dated July 28, 2021 No. 862 "On Amendments to Order of the Ministry dated December 7, 2012 No. 1185" for victims at Chernobyl Nuclear Power Plant, a state service is provided to provide citizens with compensation for harm caused to health and compensation for health improvement, compensation to families for the loss of a breadwinner [15].

Scientists and specialists in various fields are still investigating the dimensions and consequences of disasters, trying to develop measures for elimination, analyze, and identify the harm caused to both people and the environment. This direction requires closer attention from countries and specialized organizations for preservation of health and environment.

Of course, the heads of states are making efforts to eliminate them, but the issue of protecting human health, protecting and preserving the ecological space remains acute, in addition, there is no strict regulation of the responsibility of persons who are responsible for an emergency to the population of the affected region. In practice, bringing to responsibility of those responsible for the accidents occurred is not fully observed.

For the social protection of employees of enterprises and the population in the territory of which the facility is located, it is necessary to assign responsibility in the field of social protection to the head and (or) managers of shops associated with performance of especially dangerous duties, if it is proved that the superior governing body is guilty, then to make it responsible providing for this in the employment agreement (contract). Both environmental responsibility and social responsibility, in spite of the time that passes after the accident, must be assigned to the perpetrators and must be enforced as a sustainable development of the present time.

For workers injured in accidents and persons living in these territories, the legislation establishes benefits, allowances, pensions for material assistance and physical support of the body. However, the consequences of the accidents affected not only the participants and residents themselves, but also the children born within the accident and after it. Therefore, complex work of the countries of our planet is required. It is necessary to provide data collection into an unified system on the consequences of accidents that affected human health and the earth's ecosystem. When developing the unified database, we will together identify ways to deal with the consequences of emergencies and measures to prevent diseases and prevent pathologies in people affected by disasters. At the same time, when employing people who have been affected by radiation, it is necessary to provide them with specially protected conditions within performance of their work duties, thereby we will be able to provide them with timely assistance and prolong their quality of life.

4 Conclusion

"Nature is critical to our survival: nature supplies us with oxygen, regulates our weather conditions, pollinates our crops, produces food, feed and fiber for us. But she is under increasing negative influence. Human activity has changed almost 75 % of the Earth's surface, displacing the animal world and nature in a very scanty corner of the planet" [16].

In order to eliminate such consequences, it is necessary:

1. to impose responsibility on the perpetrators (heads of enterprises) in case of accidents and establish control over its implementation, both environmental and social, regardless of the time elapsed after the disaster.

2. Creation of the unified database at the international level to identify the consequences of accidents and develop offers for combating the results of emergencies and measures to prevent diseases and prevent pathologies in people affected by disasters.

3. When concluding an employment contract with persons affected by radiation, it became necessary to prepare special labor protection for them within performance of their duties, so we will gradually change the quality of life of the workers injured.

As a result of the study, the article authors identified offers for the sustainable development of ecological and social protection of the population in unfavorable ecological territories. By steadily developing them, observing and implementing them, a high and persistent protection of people and the environment will be established. This will discipline workers responsible for the quality of works at hazardous enterprises and eliminate the loss of lives of people, reduce disability, and preserve the ecology of the globe.

References

1. G. I. Zhurukhin, et al. IOP Conf. Ser.: Earth Environ. Sci., 666, 062145 (2021)
2. N. Khotimah, et al. IOP Conf. Ser.: Earth Environ. Sci., 271, 012003 (2019)
3. Yixiao Kuang, et al. IOP Conf. Ser.: Earth Environ. Sci., 702, 012052 (2021)
4. Omar Mekhlouf, IOP Conf. Ser.: Earth Environ. Sci., 779, 012131 (2021)
5. E. V. Provodina, et al. IOP Conf. Ser.: Earth Environ. Sci., 723, 042058 (2021)
6. E. N. Abanina, et al. IOP Conf. Ser.: Mater. Sci. Eng., 582, 012032 (2019)
7. E. N. Abanina, et al. IOP Conf. Ser.: Earth Environ. Sci., 543, 012022 (2020)
8. Yu. S. Sergeenko, et al. IOP Conf. Ser.: Earth Environ. Sci., 723, 022100 (2021)
9. Jianxun Deng, et al. IOP Conf. Ser.: Earth Environ. Sci., 510, 062002 (2020)
10. A. I. Akhmetova, et al. IOP Conf. Ser.: Earth Environ. Sci., 459, 052086 (2020)
11. V. F. Stukach, et al. IOP Conf. Ser.: Earth Environ. Sci., 459, 052012 (2020)
12. Sarah E Zelasky, Jonathan J Buonocore, Environ. Res. Lett. 16 065009 (2021)
13. Law of the Russian Federation of May 15, 1991 No. 1244-1 (as amended on May 26, 2021) "On Social Protection of Citizens Exposed to Radiation as a Result of the Chernobyl Disaster" (as amended and supplemented, entry into force dated September 01, 2021)

14. Resolution of the Government of the Russian Federation of August 21, 2001 No. 607 (ed. of July 10, 2020) "On the Procedure for Payment of Monthly Monetary Compensation in Reimbursement of Harm Caused to the Health of Citizens in Connection with Radiation Exposure Due to the Chernobyl Disaster or Work to Eliminate the Consequences of the Disaster at Chernobyl Nuclear Power Plant"
15. Order of the Ministry of Labor and Social Protection of the Saratov Region dated July 28, 2021 No. 862 "On Amendments to Order of the Ministry dated December 7, 2012 No. 1185" (together with "The administrative regulations for provision of state services by the Ministry of Labor and Social Protection of the Saratov Region to provide citizens with compensation for harm caused to health as a result of the Chernobyl disaster, compensation for health improvement, as well as compensation to families for the loss of a breadwinner")
16. Goal 15: To protect, restore and promote sustainable use of terrestrial ecosystems, sustainably manage forests, combat desertification, halt and reverse land degradation and halt biodiversity loss, https://www.un.org/sustainabledevelopment/ru/biodiversity/

Infrastructure road construction environmental risks minimization

Tatyana Sukhadolets[1,*], Elena Stupnikova[2], and Olga Olenina[2]

[1]Analytical Center Social Infrastructure, 115088, Moscow, Russian Federation
[2]Institute of Economics and Finance, Russian University of Transport (MIIT), 127994, Moscow, Russian Federation,

Abstract. Measures to reduce environmental risks within the infrastructure construction are mainly reduced to recommendations on use of renewable energy sources, and use of vehicles with hybrid and electric motors. The article shows that the predominant factor in reducing environmental risks is investment in fixed assets for environmental protection. The analysis performed using OLS confirmed that the degree of P-value dependence in all three constructed models was obtained for the Investments variable.

1 Introduction

Control over the state of atmospheric air and its pollution is increasing every year, however, almost all the largest agglomerations of the Russian Federation continue to face a number of environmental problems. At the same time, if earlier the sources of air pollution were industrial enterprises, nowadays, the sources of pollution are mobile sources, first of all, the road transport. So, it turns out that the more developed the agglomeration, the greater the need to improve the environmental friendliness of transport, reduce emissions through environmental protection measures and environmental costs.

In addition, Eurasian integration demanded construction of an international transport bridge Europe - Western China that is a transcontinental project similar to the Silk Road, which has already linked Europe and China for several millennia. According to the project, Europe - Western China road will have a length of 8,500 km, and it will be possible to drive along it in just 10 days. In addition, central bypass roads have already been built or are being built around large agglomerations - these are the Central Ring Road around Moscow, the ring road around St. Petersburg, the ChKAD - bypass around Chelyabinsk, the ring roads of Magnitogorsk, Saratov, Rostov-on-Don, Omsk, Voronezh, and etc. As a result, to accelerate movement of population and goods, not only regional, intercountry, but also intercontinental roads are being built.

Currently, an important factor affecting the environment is the fact that in the last decade all Russian regions have increased the volume of goods delivery via e-commerce. All this led to an increase in intercontinental and intracity movement by road. The flows of couriers significantly load highways, which will require additional implementation of road infrastructure projects in the future to improve movement and logistics. And in this case, the

[*] Corresponding author: sutv@yandex.ru

challenge was the COVID-19 pandemic, which has already changed the way of life of many people in all Russian regions in the area of displacement. This becomes the reason for formation of new requirements for assessment of environmental risks, infrastructure road construction, especially in huge metropolitan areas.

2. Literature Review

The work of Lukashevich [1] shows that a significant proportion of air pollution occurs in the road transport complex. The negative consequences of global pollution are manifested primarily in megalopolises, where there is an active road construction and an increase in the number of vehicles. Transport and environmental risks are becoming even more relevant due to development of transport corridors and the lack of multiple fixation of environmental problems, Zhanbirov [2]. The main pollution of roadside areas is determined by six heavy metals: lead, zinc, nickel, copper, chromium and cadmium, Kosinova & Fonova [3], Dunn [4]. These particles, which are formed from movement of cars, form an aerosol cloud up to two meters high with a maximum concentration depending on the density of the traffic flow, Coffin at all [5]. Therefore, road infrastructure companies must apply the latest environmental technologies to reduce the number of indicators of the impact of environmental risk on the environment, An [6]. Fuel is the source of both nuclear lead particles in the form of soot and battery decomposition products. Zinc oxide, chromium is used in production of automobile tires, paints and varnishes. Metals in the form of nickel, copper, cadmium are found in nickel-plated parts of batteries, cables, and alloys. Unfortunately, emissions are reduced very weakly (Fig. 1). According to Savko [7] , these particles have a density from 3,000 kg/m3 of shell particles to 9,000 kg/m3 of nuclear particles.

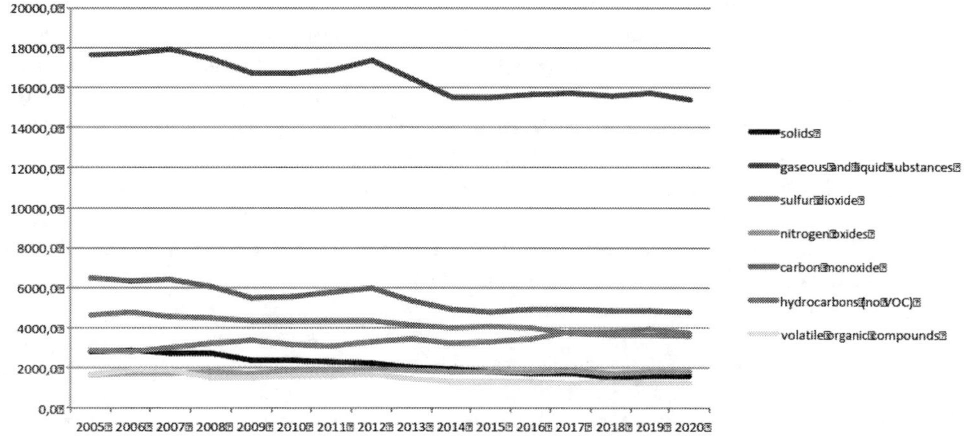

Fig. 1. Emissions of the most common air pollutants from use of road infrastructure (thousand tons).

3 Materials and methods

The study was carried out in two stages. At the first stage, data were collected for 15 years, from 2005 to 2020, for variables (Table 1):

1. Waste from production and consumption in transport.
2. Highways.
3. Investments in fixed assets for atmospheric air protection.
4. Expenses for environmental protection.

5. Emissions of solids and their dynamics are assessed.

Table 1. Variable notation used.

Waste from production and consumption, transport, mln tons	Waste
Automobile roads, ths km	Autoroads
Investments in fixed capital for atmospheric air protection, mln rubles	Investments
Environment protection expenditures, mln. rubles	Expenses
Emissions of solids, ths tons	Emissions

The dynamics of changes in emissions into the atmosphere and the length of roads were assessed (Fig. 2).

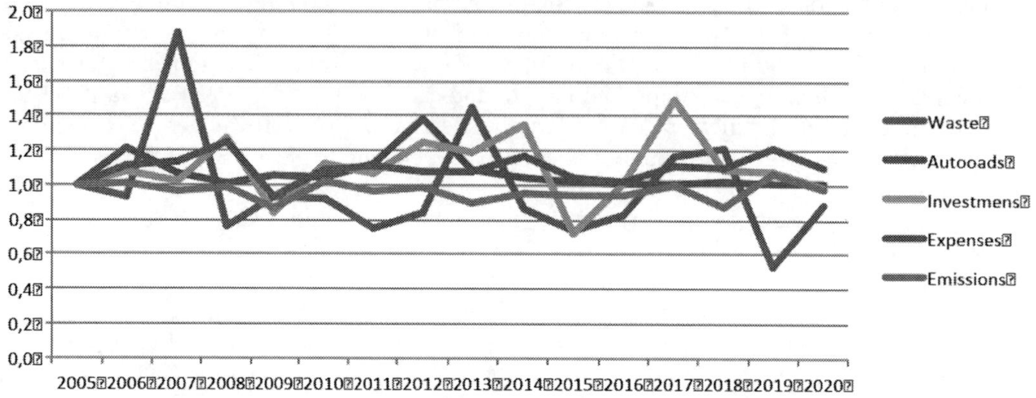

Fig. 2. Dynamics of changes in air emissions, investments and the length of roads.

At the second stage, an assessment of the causal relationship (using regression analysis (OLS)) was carried out for: a) the amount of emissions of solid substances from production and consumption waste, the length of roads, investments in fixed assets for environmental protection and environmental protection costs; b) consumption waste from the length of roads, investments in fixed assets for environmental protection, environmental protection costs, the amount of solid emissions; c) the amount of emissions of solid substances from the length of roads, investments in fixed assets for environmental protection.

The following conditions were met: the mathematical expectation of random errors is zero, factors and random errors were estimated. Factors exogenousness was verified using Schwartz, Akaike and Hennan-Quinn criteria.
Calculation formula:

$$\sum_i e_i^2 = \sum_i (y_i - f_i(x))^2 \rightarrow \min_x \quad (1)$$

e_i^2 - a proximity measure for the sum of squares of the variables under consideration;
x is a set of variables;
$f_i(x)$, where i=1.....m, m > n.

4 Results and Discussion

With Emissions dependent, the most significant factor was the Expenses variable (0.71), Autoroads (0.24). The highest P-value is obtained for the Investments variable (0.62), Table 2.

Table 2. Model 1. OLS, observations 1-16 are used. Dependent variable: Emissions.

	Coefficient	Std. error	t-statistic	P-value	
Waste	−0,0269948	0,0470547	−0,5737	0,5768	
Autooads	0,248074	0,122319	2,028	0,0653	*
Investmens	−0,0493472	0,0982575	−0,5022	0,6246	
Expenses	0,712652	0,149078	4,780	0,0004	***

In the case of Waste dependent, the Expenses variable (0.94) and Emissions variable (0.98) turned out to be the most significant factor. The highest P-value is obtained for the Investments variable (0.75), Table 3.

Table 3. Model 2: OLS, observations 1-16 are used. Dependent variable: Waste.

	Coefficient	Std. error	t-statistic	P-value	
Autooads	0,665946	0,126605	5,260	0,0001	***
Investmens	0,226994	0,122050	1,860	0,0840	*

In the case of Emissions dependent, the most significant factor was the Autoroads variable (0.66). The highest P-value is obtained for the Investments variable (0.08), Table 4.

Table 4. Model 3: OLS, observations 1-16 are used. Dependent variable: Emissions.

	Coefficient	Std. error	t-statistic	P-value	
Autooads	0,665946	0,126605	5,260	0,0001	***
Investmens	0,226994	0,122050	1,860	0,0840	*

Therefore, the constructed regression models 1-3 show that Emissions and Waste are more dependent on investments in fixed assets for atmospheric air protection.

5 Conclusion

Construction of road infrastructure and further use of highways is associated with emissions of heavy metals and gases into the atmosphere. The negative consequences of such negative environmental risks can be minimized to a greater extent due to investments in fixed assets for atmospheric air protection. With Emissions dependent (Waste, Autoroads, Investments, Expenses variables), the highest P-value was obtained for the Investments variable (0.62). With Waste dependent (Autoroads, Investments, Expenses, Emissions variables), the highest P-value was obtained for the Investments variable (0.75). With Emissions dependent (Autoroads, Investments variables), the highest P-value was obtained for the Investments variable (0.08).

Therefore, our assessment confirms the priority importance of minimizing environmental risks while increasing financing of capital investments in fixed assets for environmental protection. Investments in fixed assets aimed at environmental protection and rational use of natural resources include investments in fixed assets aimed at environmental protection, carried out at the expense of all sources of financing, both within newly constructed enterprises and at existing enterprises [8]. Such investments that will help preserve the environment and reduce risks include: the cost of construction, reconstruction (including expansion and modernization) of facilities, purchase of machinery, equipment, vehicles. It also makes sense to stimulate economically the use of clean fuels, renewable energy sources, vehicles with hybrid and electric motors, materials and technologies that reduce environmental risks and negative impact on the environment.

References

1. O. D. Lukashevich, Chemistry. Ecology. Urban studies. **1**, 142 (2020)
2. Zh. G. Zhanbirov, Science, new technologies and innovations of Kyrgyzstan, **2**, 3 (2021)
3. I. I. Kosinova & Fonova, Engineering-geological and geoecological problems of urban agglomerations, **1**, 418 (2015)
4. F. L. Dunn, The Arboviruses: epidemiology and ecology, **27**, 281 (2020)
5. A. W. Coffin, D. S. Ouren, N.D, at al, Issues In Ecology, **27**, https://www.esa.org
6. D. Anh, Innovation and Investment, **4**, 224 (2021)
7. A.D. Savko, Bulletin of Voronezh State University. Geological series **1**, 47 (2003)
8. Rosstat order of July, **383** (2020)

Environmental and social instruments of public policy in the transition to sustainable development

Elena Abanina[1,*], *Yulia* Sergeenko[1], *Dmitry* Petrov[1], *Yulia* Litvinova[2], and *Natalia* Bobrakova[2]

[1]Saratov State Law Academy, Saratov, Russia
[2]Leningrad State University named after A.S. Pushkin, Pushkin, Russia

Abstract. The article examines the problems of the formation of state policy in the context of the transition to sustainable development in terms of the search for new tools for its implementation. The authors analyzed modern factors influencing the formation of environmental and social policy in the field of achieving sustainable development goals. The factors that influence the process of achieving sustainable development goals in environmental policy are identified: environmental, organizational, legal and political. Also noted are the factors of a legal and political nature that affect governmental social policy. in the field of achieving sustainable development. New environmental and social instruments of modern governmental policy in the context of the transition to sustainable development have been developed: environmentalization and socialization of economic activity. Key words: sustainable development, ecologization of economic activity, socialization of economic activity.

1 Introduction

A little over a year has passed since the publication of data on the process and progress in achieving the Sustainable Development Goals (hereinafter also SDGs) prepared by UN experts for each of the 17 goals [1]. This means that society can take stock of the validity of the experts' conclusions and the effectiveness of the measures introduced to improve the process of achieving the SDGs. Achieving the Sustainable Development Goals requires transforming financial, economic and political systems to guarantee human rights for all. Taking into account that the experts and the Secretary-General of the United Nations found the pace of progress towards achieving the set goals insufficient, the authors of this article seem to be able to contribute to the overall process by proposing new tools to improve the situation in certain areas.

The transition to sustainable development should ensure a long-term balanced solution to the problems of socio-economic development, preservation of a favorable environment and natural resource potential, and meeting the needs of the present and future generations of people. Thus, the concept of sustainable development links together the environment,

* Corresponding author: elena-abanina@yandex.ru

society and the economy. However, taking into account the current state of the economic, environmental and social spheres, one can draw attention to the fact that the economy does not experience such a load as the environmental and social spheres. In this regard, there is a need to study modern problems affecting the formation of environmental and social policy in the field of achieving sustainable development goals in terms of maintaining their balance.

The aim of this work is to develop environmental and social instruments of modern public policy in the context of the transition to sustainable development.

2 Materials and methods

In the study of environmental and social instruments of modern state policy in the context of the transition to sustainable development, we proceeded from the understanding of sustainable development as such a development in which the needs of present generations are met by creating a sustainable "green" economy, rational sustainable use of natural resources, and at the same time the needs future generations are not compromised [2].

We put forward a hypothesis that, given the limited time frame for achieving the SDGs, it is necessary to talk not about individual instruments of state policy in achieving the set goals in certain areas, but about the development of such instruments that allow achieving goals while maintaining a balance of all interests. At the moment, in the conditions of insufficient development of the environmental and social sphere, there is a need to develop environmental and social instruments of state policy in the context of the transition to sustainable development, while excluding their priority over each other.

To develop environmental and social instruments of governmental policy in the context of the transition to sustainable development, the following methods were used:

the dialectical method (when considering the categorical series of the institute “sustainable development”, accompanying categories “ecological component of sustainable development”, “social component of sustainable development”) ”;

methods of analysis and synthesis (when identifying factors influencing the formation of environmental and social policy in the field of achieving sustainable development goals);

systemic and structural method (for the development of measures (methods) of greening and socialization of economic activity as instruments of state policy).

3 Results and discussion

Sustainable development is fully ensured through the implementation of all envisaged measures in its three components in total.

The economic component of sustainable development consists in the optimal use of natural resources, including their extraction, procurement, collection, processing while maintaining their ecological value.

The ecological component of sustainable development should fully guarantee the safety of all natural resources and ensured by their ecological value. For example, a natural feature such as a forest prevents land degradation and desertification, reduces the risk of floods, forests are necessary for the functioning of main catchments and affect the rainfall regime; maintaining clean air, reducing the risk of landslides, avalanches, droughts, dust and sand storms and other natural disasters; biodiversity conservation (forests and woodlands contain more than 60,000 tree species and, according to UN experts, forests are home to 80 percent of all terrestrial biological species). Under the conditions of global climate change, the role of Russian forests as a carbon sink is increasing. The absorption of carbon by Russian forests in recent years has been estimated at between 200 and 400, and even 600 million tons per year.

The social component of sustainable development is ensured by taking into account all possible interests of organizations and groups of people in the process of using all natural resources, their protection, protection and reproduction of renewable resources. For example, UN experts estimate 1.6 billion people, or 25 percent of the world's population, depend on natural resources for livelihoods, livelihoods, employment and income-generating activities.

Having studied the current state of the economic, environmental and social component, we drew attention to the fact that in terms of the economic sphere, despite the economic crisis associated with the pandemic, the economy of natural resources does not experience such a burden as the environmental and social spheres. In this regard, we have studied modern factors influencing the formation of environmental and social policy in the field of achieving sustainable development goals.

1. Environmental policy:

Environmental factors:

— violation of biological diversity (due to the destruction of rare animals and plants);

— insufficient efficiency of restoration, protection and protection of natural resources.

Organizational factors:

— insufficient accuracy of accounting for natural resources;

— low relevance of information about natural resources;

— excessive administrative barriers;

— poorly developed infrastructure, including road one;

— low level of technical, scientific and personnel support for many natural resource industries.

Legal and political factors:

— imperfection of the legal framework in the natural resource complex and related industries;

— a high level of shadow turnover of natural resources (animals, forests, minerals);

— high latency of environmental violations.

2. Social policy.

The prevailing number of cases related to the violation of the social rights of the population are associated with factors of a legal and political nature:

— lack of legal regulation of socially significant issues (lack of order and form of participation of the local population in the preparation of decisions, the implementation of which may have an impact on nature during its use);

— insufficient regulation of the use of natural resources by the local population for their own needs;

— the lack of regulations on the preservation of natural areas that are of particular importance to the local population (for example, recreation sites, local reservoirs for fishing, places for collecting medicinal herbs, berries, mushrooms; hunting places);

— high differentiation of the level of income of different strata and groups of the population.

These circumstances indicate that in the context of the intensification of the process to achieve sustainable development goals in the environmental and social spheres, new, adequate response measures are needed.

In our opinion, given the limited time frame for achieving the SDGs - 2030, it makes sense to talk not just about recommendatory, "soft" measures, but to introduce clear restrictive, prohibitive, mandatory requirements. At the moment, in the conditions of insufficient development of the ecological and social sphere, we propose such measures as the greening and socialization of economic (economic) activities.

Greening economic activity means implementing the idea of “green” activities, that is, “green” economy = “green” industry = development of “green” technologies = compliance with “green” standards.

The measures (methods) of greening economic activity can include:

— improvement of existing production without capital replacement and introduction of new technologies (for example, reconstruction of systems for trapping pollutants at industrial enterprises [3, 4, 5]; introduction of resource-saving and waste-free technologies and the latest circulating water supply systems at enterprises, etc.) [6];

— improvement of measures allowing the rational use of natural resources [7, 8];

— legal measures (establishment and consolidation of standards and plans for the gradual reduction of pollution from industrial facilities to levels corresponding to the best environmentally friendly world technologies, the introduction of digital technologies into production) [9, 10];

— financial measures (attraction of budgetary funds to finance environmental protection activities, etc.).

Socialization of economic activity means not just the introduction of short-term social measures to relieve social tensions that arise both in connection with economic reasons (decrease in income) and in balance with the ecological sphere, namely, in the use of the natural environment to the detriment of the local population, but the construction of long-term policy taking into account the specified circumstances.

The measures (methods) of socialization of economic activity can include:

— clear consolidation in the legislation of the rights of the local population to the natural environment and privileges from its use;

— an increase in the degree of involvement of natural resources in social forms of use (mass involvement of the local population in work in the natural resource industry);

— improving the state of affairs in the field of labor protection of workers in the natural resource industry; increase in the level of wages - equalizing the level of wages in cities and rural areas [11];

— availability of social infrastructure (medical, educational and other institutions) in remote natural resource regions.

We believe that such measures will make it possible to more quickly achieve the goals of sustainable development in the environmental and social spheres precisely due to the building of a balance of environmental and social interests.

4 Conclusions

Features of the current stage in achieving sustainable development goals are in the implementation of all planned activities in the context of the global environmental crisis and social tension.

As a result of the study, the authors of the article proposed tools for enhancing the process of achieving sustainable development goals in the form of "greening" and "socialization" of economic activity. In our opinion, this will make it possible to solve in a balanced way the current problems of social and ecological development and preserve a favorable environment and natural resource potential.

References

1. The-Sustainable-Development-Goals-Report-2020, https://unstats.un.org/
2. E. Abanina, Yu. S. Sergeenko, Iu. Litvinova, Yu. M. Nikitenko, N. Bobrakova, SHS Web of Conferences, 93 (2021)
3. V. L. Abramov, F. A. Kodirov, A. A. Gibadullin, V. N. Nezamaikin, O. I. Borisov and N. V. Lapenkova, J. Phys: Conf. Ser., **1515** (2020)
4. V. N. Zakharov, V. Y. Linnik, Y. N. Linnik, A. B. Zhabin, Mining Informational and Analytical Bulletin, **5** (2019)
5. A. V. Nikonorova, P. V. Stroev, D. E. Morkovkin, O. N. Bykova, N. I. Isaichikova, A. A. Kvak and O. O. Skryabin, J. of Interdisciplinary Research, **9** (2020)
6. A. Caverzan, M. L. Tornaghi, P. Negro, IOP Conf. Ser: Mater. Sci. Eng., 442 (2018)
7. A. A. Gibadullin, et al, IOP Conf. Ser.: Earth Environ. Sci., 421 (2020)
8. L. M. Musina, M. U. Neucheva, IOP Conf. Ser.: Earth Environ. Sci., 107 (2018)
9. K. A. Artamonova, et al, J. Phys.: Conf. Ser., **1515** (2020)
10. D. E. Morkovkin, et al, J. Phys.: Conf. Ser. **1399** (2019)
11. A. Shilovtsev, N. Sorokina, K. Stozhko, J. L. L. Garcia, E3S Web of Conferences, 222 (2020)

The role of the ecological culture of the teacher for the sustainable development of the state

N. A. Shaidenko[1,*] , *S. N.* Kipurova[1], *E. V.* Shelispanskaya[2], *A. V.* Sergeeva[2], and *N. Yu.* Zubenko[3]

[1] Institute for Advanced Studies and Professional Retraining of Educators of the Tula Region, Center for Strategic Education planning the development of education, expertise and scientific consulting, Tula, Russia
[2] Tula State Pedagogical University named after L.N. Tolstoy, Tula, Russia
[3]School №. 1080, Moscow, Russia

Abstract. The concept of sustainable development (SD) was born, developed and is actively evolving in many national and international strategies, official reports, modern scientific research in various directions. The context of sustainable development is present in the official and scientific chronicle of the development of society, summarized in documents and studies. However, in pedagogical works, this tendency is not covered in such detail. The idea of SD is aimed at systematic, multidimensional development, at the well-being of countries, peoples, an individual, at eliminating inequalities: demographic, social and political stability and neutralizing factors dangerous for the development of nature. The basic element of SD in the 21st century is higher education, which was declared at the Paris World Conference on Higher Education in 2009. The evolution of education in a society of sustainable development contributes to a change in the ecological culture of an individual and groups of people based on a value attitude towards nature, safety, health, well-being, unity and cultural diversity. The article is based on the idea of the leading importance of the formed ecological culture of the teacher for the sustainable development of the state. The developed foundations for the formation of environmental culture in the context of sustainable development of university students imply the definition of its goals and conditions, clarification of the structure and relationships between the components, patterns and principles, methods and forms of organization of training, criteria and indicators for assessing the development of components of environmental culture.

1 Introduction

The concept of the country's transition to sustainable development was approved by the President of Russia in 1996. The concept is designed to ensure a balance in solving social and economic problems and problems of preserving the natural resource potential of the

* Corresponding author: nashaidenko@gmail.com

environment, which is necessary to fully meet the needs of the population. In 2015, the UN 2030 Agenda for Sustainable Development was approved.

The main strategic and program documents (doctrines, programs, concepts) of Russia contain activities that involve the implementation of sustainable development goals. "The analysis conducted by the Analytical Center for the Government of the Russian Federation in the spring of 2020, 12 national projects and the Comprehensive Plan for the Modernization and Expansion of the Backbone Infrastructure directly or indirectly affected 107 of the 169 tasks identified in the UN document" [1].

On July 21, 2020, by decree of the President of Russia, the national development goals until 2030 were approved, taking into account which national projects are now being adjusted in the direction of preserving the health and well-being of the population, solving demographic problems in connection with the improvement of the environment. A comfortable and safe environment for life is one of the national goals.

The COVID-19 pandemic and a number of global environmental disasters required the country's leadership to take urgent measures to improve the state of the environment and production technologies in the country, to join efforts with the world community.

On July 19, 2021, a meeting of the Council for Strategic Development and National Projects under the President of the country was devoted to this issue.

President Vladimir Putin introduced a new position of the special representative of the President of the Russian Federation for relations with international organizations to achieve the goals of sustainable development. A.B. Chubais, who believes that a common task with other states can become: "green" topics and sustainable development" [2]. He is convinced that Russia has a big start on sustainable development "thanks to the government, business and civil activists" and that Russia needs to develop a national action plan for sustainable development.

Unfortunately, there is a transfer to the technocratic side of sustainable development and a decrease in the attention of the humanitarian component [3].

The Citizenship Review on the implementation of the Sustainable Development Goals (SDGs) was compiled by the Coalition for Sustainable Development of the Country (CSDC).

"Education for sustainable development (ESD) is mainly of an educational nature (education about, not for, sustainable development) in the form of separate activities, most often environmental. The system of lifelong ESD is not built, there is no regulatory base, the general institutional approach is practically not applied. The global context is poorly represented, which increases the cultural gap between Russia and the world" [4]. In the same document, among the recommendations to the state, the following was given: "Development of the Concept of Education for Sustainable Development and the National Action Plan for 2021-2030. Approval of the state program "Education for Sustainable Development" for 2021-2030 with appropriate funding. Widespread popularization of the concept of sustainable development and SDGs in formal, non-formal (compulsory) and informal education (enlightenment). Preparation and implementation of educational modules on sustainable development for all levels of education and refresher courses for teachers".

The purpose of the study is to substantiate the processes in Russian higher education, which bring it to the world level of training modern teaching staff based on the model of creative, advanced and interdisciplinary education with an orientation towards the tasks of sustainable development of the state. This is an attempt to construct education as a social activity aimed at building a society for sustainable development and the development of a new person through the formation of his ecological culture.

2 Materials and Methods

The development of the problem was carried out at the scientific-theoretical, methodological and experimental-practical levels in a complex system of knowledge, including: the theory of education and pedagogy, psychology of thinking, cultural studies and sociology, ecology and the theory of innovative processes, public policy.

The combination of the principles and methods of scientific knowledge of social phenomena used made it possible to study in dynamics the processes and phenomena that take place in higher pedagogical education that determine the socio-political development of the state and society in dynamics, as well as to identify their influence on the formation of the outlook of a future teacher in the interests of sustainable development of the Russian Federation.

The methodological basis of the study was the following methodological approaches:

– the dialectical approach was used to identify and study contradictions in the processes of changing the functions of the higher pedagogical school in the formation of the worldview of future teachers in the interests of sustainable development of the state in specific historical periods;
– the deductive approach is used to study social phenomena and processes in determining the interdependencies of the development of higher education and sustainable development of the state;
– the inductive approach is used in the transition from specific facts of the transformation of higher education and regulatory documents to characterizing general trends, identifying the objective essence of sustainable development of the state;
– the sociocultural approach made it possible to take into account the changes in society, state and regional characteristics when building the pedagogical process in higher education, aimed at social education, the formation of the ecological culture of students.

The comparative-historical, problem-chronological, retrospective and systemic approaches were used.

The main methods and forms of organizing scientific knowledge were also typologization of the main ideas of sustainable development of the state from the standpoint of the influence of the higher pedagogical school on it.

At the theoretical level, methods of analysis of sociological, political, historical, pedagogical, psychological, etc. were used. research on the problem of sustainable development, higher pedagogical school, environmental culture.

Socio-political analysis was used to study the influence of the socio-political and economic situation on the development of higher education and its influence on the sustainable development of the state, including the formation of the ecological culture of students.

Psychological and pedagogical analysis was applied to assess the effectiveness of the implementation of state educational policy, as well as to study the quality of professional activities of university teachers in the formation of the ecological culture of students in the interests of sustainable development of the state.

Scientific, theoretical and methodological work on the project due to its interdisciplinarity required the use of methods from different fields of knowledge, including: methods of structural and functional analysis, theory of knowledge, didactic analysis of cognitive situations and pedagogical experiment, theory of creative spaces, social anthropology, theory social mobility, social modeling, theory of innovation processes.

Among the empirical methods, methods of included pedagogical observation, polls, testing, conversations, analysis of the results of the formation of ecological culture were used.

Thus, the solution of the tasks set was carried out on the basis of a comprehensive study of general scientific and specific disciplinary approaches and methods.

3 Results and Discussion

"Scientists consider the quality of man's relationship to nature, people, and himself as indicators of the development of his ecological culture. The concept of ecological culture is concretized at each stage of personality development by certain indicators, which is due to the psychological characteristics of age" [5-8].

"A person who has mastered ecological culture subordinates all types of his activities to the requirements of rational nature management, takes care of improving the environment, does not allow its destruction and pollution. Therefore, he needs to master scientific knowledge, master moral value orientations in relation to nature, as well as develop practical skills and abilities to preserve favorable conditions, the natural environment" [9].

The study is based on the original scientific concept that a state needs a new type of social and individual culture of the population for its sustainable development, which can become an ecological culture.

Despite the fact that in scientific research and official documents there is a requirement for the formation of ecological culture, the idea of the conditionality of ecological culture by a new level of development of society and that, in turn, a society of sustainable development can be created only by people with an ecological culture is rarely considered.

The theoretical foundations of the sustainable development of the Russian state were considered in the space of influence on this process of pedagogical education, in particular, the formation of students' ecological culture.

We did not set ourselves the task of conducting a detailed analysis of existing approaches to the structure of ecological culture.

A theoretical analysis of the study of the structure of ecological culture in the works of philosophers, sociologists, psychologists and teachers gave us the opportunity, based on the tasks of a society for sustainable development, to propose the structure of the ecological culture of future teachers in the context of sustainable development, the components of which, interacting with each other, form an integral system:

- motivational and value component;
- cognitive component;
- emotional-volitional component;
- activity-practical component.

Table 1. Characteristics of the levels of formation of the worldview

Motivational value criterion		
High level	**Average level**	**Low level**
1. Always shows a steady interest in socio-ecological and other problems of sustainable development. 2. Realizes the personal need to raise the level of knowledge about the ways to achieve sustainable development of Russian society. 3. The motivation for value-oriented environmental activities is at a high level.	1. Shows interest in socio-ecological and other problems of sustainable development. 2. Has a personal need to raise the level of knowledge about ways to achieve sustainable development of Russian society; 3. At the middle level is the motivation of value-oriented environmental activities.	1. Shows no interest in socio-ecological and other problems of sustainable development. 2. There is no personal need to raise the level of knowledge about ways to achieve sustainable development of Russian society. 3. The motivation for value-oriented environmental activities is at a low level.

Table 1. Continued

Cognitive criterion		
1. The volume, stability and reliability of knowledge about SD and the possibilities of achieving sustainable development of society are high. 2. Knowledge has passed into stable attitudes and beliefs. 3. Owns the methods of making socially significant and environmentally oriented decisions in professional activities, in the formation of the ecological culture of schoolchildren.	1. Sufficient volume, strength of socially significant and environmentally oriented knowledge about sustainable development of society. 2. environmental attitudes and beliefs are not sustainable. 3. Shows interest in ways of making socially significant and environmentally friendly decisions in professional activities.	1. Lack of knowledge and understanding of the meaning of the values of sustainable development of society, ignores explanations from others. 2. Occasional manifestations of environmental views and beliefs. 3. Does not show interest in ways of making socially significant and environmentally oriented decisions in professional activity.
Emotional-volitional criterion		
1. Shows a positive emotional attitude towards nature, environmental and other social issues. 2. Formed readiness for personal and professional self-development of ecological culture. 3. Has a strong-willed readiness to solve environmental problems in the interests of sustainable development of society.	1. Shows neutral emotional attitude to nature, environmental and social problems at the regional and global levels; 2. The readiness for personal and professional self-development of ecological culture is insufficiently formed. 3. Knows how to solve environmental problems for sustainable development.	1. Emotional attitude to nature, economic, environmental and social problems is not formed. 2. The readiness for personal and professional self-development of ecological culture has not been formed. 3. Does not know how to solve environmental problems in the interests of sustainable development. .
Activity-practical criterion		
1. Has experience in environmental activities to change the natural world based on knowledge about sustainable development and ways to achieve sustainable development of Russian society. 2. Knows how to design and implement the ecological culture of the idea of sustainable development in professional and socially significant activities, including the competence of the development of ecological culture. 3. Actively participates in practical, including value-oriented socially significant environmental activities.	1. Does not fully have experience in environmental activities to change the social and natural world on the basis of knowledge about sustainable development and ways to achieve sustainable development of Russian society. 2. Knows how to design, but does not show interest in the embodiment of ecological culture, ideas of sustainable development in professional and socially significant activities. 3. Owns the methods of practical, including value-oriented socially significant environmental activities.	1. Does not have experience in environmental activities to change the social and natural world on the basis of knowledge about sustainable development and ways to achieve sustainable development of Russian society. 2. Does not know how to design and does not show interest in the embodiment of ecological culture, ideas of sustainable development in professional and socially significant activities. 3. Does not know how to practice environmental, including value-oriented socially significant environmental activities.

Table 1. Continued

Reflexive-evaluative criterion		
1. Owns reflexive methods of assessing his attitudes towards the natural world and himself as a part of it. 2. Able to assess their readiness to carry out targeted work to create a society of sustainable development. 3. Ready to assess the level of development of the components of their ecological culture in schoolchildren	1. Knows reflexive techniques, ways of assessing his attitudes towards the natural world and himself as a part of it. 2. Sometimes he assesses his readiness to carry out targeted work to create a society of sustainable development. 3. Insufficiently ready to assess the level of development of the components of ecological culture.	1. Does not possess reflexive techniques, the assessment of his attitudes towards the natural world and himself as a part of it. 2. Does not assess its own readiness to carry out targeted work to create a sustainable development society. 3. Not ready to assess the level of development of the components of ecological culture.

We used general approaches to the essence of the ecological culture of university students from the point of view of the bearer of this culture - the student. Sustainable development cannot be imagined without work on the ecological culture of modern students, which are actively involved in building a new Russian society. It is customary to view the student body as a distributor of a new ideology and culture. That is why young people are the object and subject of national and state interests.

The developed foundations for the formation of environmental culture in the context of sustainable development of university students imply the definition of goals and conditions for formation, clarification of the structure and relationships between the components, patterns and principles, methods and forms of organization of training, criteria and indicators for assessing the development of components of environmental culture.

The goals of the formation of the ecological culture of students in the context of sustainable development in the educational space of a pedagogical university can be expressed through the performance of the following functions by culture: social, educational and creative, upbringing, developing.

The principles of shaping the ecological culture of future teachers in the context of sustainable development are:

- co-evolutionary, which is based on the organization of interaction between man and the natural environment, which in turn influences the formation of the components of ecological culture;
- humanistic, based on the priority of an ecophilic attitude to the natural world, on a positive perception of the World and Nature by a person;
- conformity to nature, based on the correct understanding of the processes occurring in nature;
- cultural conformity.

In general, the direction of the entire system for the formation of ecological culture is set by the following guidelines:

- the main provisions of the humanistic paradigm of training and education;
- anthropocentric approach to teacher training;
- the continuity of the development process of the personality of the future teacher;
- multicultural approach to education;
- the openness of the educational space of the university of social reality by means of pedagogical influence on the subjects of society;
- dialogical nature of interaction with the outside world.

4 Conclusions

As a result of the research:

- the range of criteria and indicators of the effectiveness of the transformation of the systems of higher pedagogical education in Russia has been expanded, aimed at the sustainable co-development of the ecological, economic and social spheres of the states;
- the laws, principles, content, forms, methods and means of transforming the higher education of future teachers in the interests of the sustainable development of the Russian state have been determined;
- the specificity of ecological culture is shown, which consists in the development and assimilation by students of universal human ecological values and norms of behavior, the development of a social ideal in the interests of sustainable development of the state;
- the effectiveness of interdisciplinary and axiological approaches to building the content of education for the ecological culture of future teachers has been proved;
- the tendencies of the formation of the ecological culture of students in the context of the idea of sustainable development of the state are determined;
- substantiations are given for the choice of interactive technologies of the educational process, which activate the development of the ecological culture of students, which will increase the efficiency of the transformation of the systems of training students - future teachers who can contribute to the sustainable development of the Russian state.

The following use of the research results is assumed:

- formation of proposals for state policy in the field of education based on a comparative analysis of modern models of scientific and educational systems and development trends;
- introduction of the developed and tested method of modern research education into the educational process of Russian universities;
- replication of the created scientific and educational model of the innovative educational complex, which ensures the formation of the ecological culture of future teachers;
- development of the state system of full-time and distance learning and tutoring for promising young teachers;
- creation of scientific and methodological manuals for the system of networked scientific and educational training of promising young teachers with an ecological culture in the innovative environment of modern universities.

References

1. T. Radchenko, K. Bannikova, Rossiyskaya Gazeta, **211** (8265) (2020)
2. A. Chubais, The Russian government has leaped forward in matters of sustainable development, Interview TASS 02/12/2021 (2021)
3. N. Shaidenko, Personality education, **2**, 56-60 (2017)
4. 2020–2030: Decade of Action for the SDGs in Russia. Challenges and solutions, 142 (2020)
5. E. Boyarov, S. Abramova, D. GershinkovaSiberian Journal of Life Sciences and Agriculture, **13(1)**, 111-133 (2021)

6. L. I. Cvetkova, G. I. Kopina, S.V. Makarova, T. N. Baryshnikova, Water and ecology: problems and solutions, **2(70)**, 79-89 (2017).
7. V. Golovanov, T. Bykovskaya, N. Panteleeva, S. Fedorova, and A. Tretyakov, Perspectives for Science and Education, **1(49)**, 299-317 (2021)
8. V. Mezinov, M. Zakharova, I. Karpacheva, Prospects for Science and Education, **1(49)**, 113-126 (2021)
9. E. Cherdymova, News of the Samara Scientific Center of the Russian Academy of Sciences, 3-1 (2010)
10. N. Shaidenko, N. Zubenko, New psychology of professional work of a teacher: from an unstable reality to sustainable development, 81-85 (2021).

Monitoring of water bodies as a technology for sustainable development

Natalia Lazareva[1*]

[1]International Banking Institute, Nevsky prospect, 60, 191011 St. Petersburg, Russia

Abstract. To achieve the goals of sustainable development, environmental monitoring of water bodies is considered as a modern technology for sustainable development in the field of clean water and a tool for the rational use of water resources. The works of the participants of the Baltic scientific and engineering competition on environmental monitoring, assessment of the ecological state of water bodies are analyzed. Biological purification of water bodies is not unlimited. When the intensity of the intake of pollutants exceeds the sanitary capabilities of water bodies, the ecosystem is destroyed. A set of measures for the environmental safety of river water is outlined. Reducing pollution, refusal from unauthorized dumping of waste, untreated wastewater are considered as means of improving water quality. Based on the review of competitive projects, priority measures were proposed to ensure the sustainable development of territories, including the conduct of massive information and educational projects, the implementation of internal control of the cleanliness of coastal zones of water bodies, and continuous environmental monitoring of water bodies.

1 Introduction

Environmental monitoring is a tool for achieving sustainable development goals. The availability and high quality of drinking water determine the health and life expectancy of the population of the territory. Surface sources are the main sources of water supply along with underground sources. Despite the fact that water resources are renewable, their rational use is a priority of our time. Human activities should be reasonable and aimed at preserving aquatic ecosystems, which ensures the preservation of environmental safety. The ecological monitoring of water bodies carried out for the conservation of water resources is aimed at the rational use of water resources. Indicators of achieving the goals of sustainable development of territories in the field of clean water are a significant proportion:

1. the population that uses the water supply, observing safety requirements
2. safely treated waste water;
3. reservoirs of good water quality;
4. transboundary water basins with water use cooperation;
5. administrative bodies involved in water resources management.

To assess the achievement of the goals, the dynamics of changes in the water use efficiency and the area of ecosystems associated with water are also analyzed; the level of

* Corresponding author: Lazarevan69@mail.ru

water load is calculated: the percentage of fresh water intake to the available fresh water resources; the degree of implementation of integrated water resources management is determined (from 0 to 100). The main measures for the rational use of water resources have demanded directions for practical implementation (Table 1).

Table 1. Rational use of water resources

Measures	Directions for the practical implementation of measures
Preservation of natural habitat	a decrease in human consumption of water; preservation of the natural aquatic environment important for local flora and fauna; increase in total watercourse
Renewability	ensuring the availability of water resources for future generations; non-exceeding the natural coefficient of water substitution with fresh water intake
enhancing environmental liability	application of measures of responsibility to enterprises; introduction of measures of responsibility of the population for the condition and protection from pollution and depletion of water resources
Restoration of water bodies	restoration and conservation of water bodies; ensuring environmentally friendly living conditions for population; dynamics of growth of the specific gravity of safely treated wastewater
Institutional regulation	legislative enforcement of the rational use of water resources; professional regulation of the rational use of water resources with methodological support
Development of the water management complex	improvement of drinking water preparation technology; modernization of wastewater treatment technology; reconstruction and new construction of water supply facilities; introduction of new technologies for water purification.

In February-April 2021, the Baltic Science and Engineering Competition vk.com/balticsef was held in St. Petersburg, where the projects aimed at studying the state of water and environmental monitoring of water bodies were presented:

–Environmental monitoring of the lake Kustarevsky Demsky administrative district of Ufa (I. Zyablitsky);

–Assessment of the ecological state of the Belaya River in the city of Meleuz by zoobenthos for the period 2019-2020. (K. Valeeva);

–Study of biodiversity of the State Natural Reserve of Regional Significance "Lake Lenevo" (P. Prokasheva).

The participants of the competition are Russian schoolchildren, which confirms their growing scientific acceleration. The author of the article was a member of the jury, which resulted in the need to review the problems and prospects of environmental monitoring of water bodies.

The problems raised by the participants of the competition are especially relevant in the context of ensuring sustainable development of territories while achieving the global goals of *SDGs* , in particular clean water and sanitation. And that is the main thing, since in the modern conditions of hyper digital development, a reassessment of relations is necessary. The vector of further development should be a reasonable attitude to the environment of everyone, careful attitude to water resources, excluding their barbaric consumer ruinous use. The garbage left near water bodies after a mass visit by humans makes us not to talk again about the right to clean water and an ecologically clean environment, but to outline the priority measures to ensure the sustainable development of territories with clean water. These measures are the subject of research in this article.

Environmental monitoring of the lake of the Kustarevsky Demskiy administrative district of Ufa, the "EcoPost" project was carried out within the framework of the federal project “Environmental Patrol in the Republic of Bashkortostan”, which was started in 2008 by the pupils of the "Rostok" Economic and Social Center. The Demsky administrative district of Ufa includes only two water bodies, one of which is Kustarevskoye Lake, therefore, with an increasing anthropogenic influence due to the intensification of agriculture, recreational use in the territories of national and natural parks, the rational use of the lake ecosystem requires continuous environmental monitoring.

"The symmetry of goals provides a clear framework for developing a comprehensive policy to ensure universal accessibility, quality, efficiency in the use of natural resources, stimulate cooperation and technology transfer" [1].

The introduction of new technologies "can only be achieved in harmony with the established behavioral attitudes of residents" [2].

2 Materials and Methods

Aquatic plants remove impurities from water. The latter enter water bodies more often with surface runoff. The environmental monitoring of Kustarevskoye Lake by Zyablitsky showed that lake reeds and narrow-leaved cattails absorb phosphorus and nitrogen from wastewater, accelerate the oxidation of oil products in the reservoir. The length of Kustarevskoe Lake is more than 4 km with a depth of up to 2.5 m. During this monitoring, special attention was paid to saprobity, as the degree of pollution of watercourses with organic substances, which made it possible to distinguish between pure polysaprobic, β, α-mesosaprobic with an average degree of pollution and dirty polysaprobic watercourses. It was the dynamics of saprobity in macrophytes and invertebrates that was the subject of the study of the ecological monitoring of Lake Kustarevskoe, which showed the following:

- The conditional division of the reservoir into four parts and the laying in each part of the test site for the morphological and geobotanical description of the reservoir made it possible to classify it as a higher aquatic vegetation, calculate the saprobity index for each part, and showed that at the place of the fuel oil spill, the lake became α - mesosaprobic;
- aquatic invertebrates live in water bodies of a certain degree of pollution, organisms of a medium degree of saprobity - dragonfly larvae and molluscs appeared in the lake at the end of August;
- determination of the parameters of atmospheric air and water, carried out using digital sensors, showed the presence of household waste in the beach area;
- on a sunny day, carbon dioxide is predominantly more in the green zone, since the process of photosynthesis is actively going on, compared to the railway zone, while the oxygen content is higher than normal; the content of hydrocarbons was estimated by the presence of an iridescent film on the surface of the water, which was absent by the end of August;
- analysis of the floristic characteristics of coastal and aquatic vegetation showed the disappearance of some macrophytes.

Valeeva's assessment of the ecological state of the Belaya River in the city of Meleuz in terms of zoobenthos for the period 2019-2020 showed that river water pollution was negatively affected by:

- nonpoint discharges, liquid waste;
- wash-offs from agricultural facilities and settlements;
- pollution by industrial hub, oil production facilities;
- industrial and domestic wastewater;
- deposition of pollutants from the atmosphere.

This influence resulted in the quality of the baseline water being assessed as "very polluted". The zoobenthos of the river in the area of the Meleuz town is insufficiently studied. The conducted studies of the chemical composition of water in the absence of biological monitoring of the water body served as the basis for assessing the pollution of the water body. Valeeva investigated that the composition of bottom sediments depends on the speed of the river flow, the nature of the terrain. The bottom dwellers have an important influence on the process of biological self-purification of water bodies. Carried out by Valeeva collection, processing and analysis of zoobenthos from six points showed that the water quality of the river Belaya in the section downstream of the town of Meleuz was affected by waste water from the city water utility. Zoobenthos is very diverse in October. As the water pollution increases, the taxonomic structure of the biocenosis becomes simpler. The disappearance of different groups of animals from the reservoir is characteristic as the pollution grows.

In the study of the Prokasheva area of the natural reserve of regional significance "Lake Lenevo", local monitoring and a route method were used, which made it possible to note changes in the number and species diversity of vegetation, as well as to compile a list of lichen and fungi species. Similar to the conclusions of other participants in the competition, Prokasheva identified the main sources of water pollution: industrial and domestic wastewater. It is she who considers bottom sediments of reservoirs as an indicator for identifying the intensity and composition of technogenic pollution, which made it possible to evaluate the effectiveness of using the RU170251 floating filter nozzle for cleaning the tested wastewater of a particular sanatorium. The methods of analysis, comparison, experiment used by her made it possible to draw the following conclusion:

– floating filtering nozzle RU170251 allows with high efficiency and productivity to purify wastewater from mechanical impurities, bio-utilize organic matter, eliminate specific odors, reduce chemical indicators;

– after the closure of the sanatorium due to covid period , the catalase activity of coastal bottom sediments is at the level of a medium-rich degree of soil enrichment in all sampling sites.

A global problem is the shortage of clean water. Internal control of sustainable development of territories is aimed at solving this problem while achieving a similar goal. Improving water quality by reducing industrial and household pollution, eliminating unauthorized waste disposal, reducing the proportion of untreated wastewater and increasing the scale of recycling, safe multiple reuse of wastewater is a priority for the sustainable development of each territory.

3 Results and Discussion

An obvious massive tool for internal control of sustainable development of territories is environmental monitoring of water bodies, which makes it possible to assess the safety of river water.

Environmental safety of river water can be achieved through a set of measures:

– expansion of construction of treatment facilities;

– decrease in the concentration of pollutants in water [3];

– remote monitoring of SBR wastewater treatment based on MCGS [4];

– determination of the degradation characteristics of the main pollutants of river mouths [5];

– an inventory of emissions from ships based on data from the automatic identification system (AIS), scheme of internal emission control areas (DECA) [6];

– elimination of bad living habits of residents, gaps in state supervision [7];

– improvement of the technology of efficient wastewater treatment of production facilities [8].

4 Conclusions

Based on the review of the competition projects, the following priority measures have been developed to ensure the sustainable development of territories:

1. conducting mass educational work through social networks, mass lecture courses on ensuring the rational use of water and respect for water resources;
2. implementation of continuous monitoring of the cleanliness of coastal areas of water bodies;
3. creation of tourist and recreational zones for the development of educational tourism;
4. continuous environmental monitoring of water bodies;
5. educational online excursions www.youtube.com/watch?v=BjrcGzG35SM for mass viewing.

Acknowledgements

The article includes an overview of the competitive works of the participants of the XVII Baltic Science and Engineering Competition. The author expresses gratitude to the organizers and participants of the Competition for the research that has confirmed the importance of monitoring water bodies for ensuring sustainable development of territories.

References

1. D. Sklarew, J. Sklarew, *Integrated Water-Energy Policy for Sustainable Development* (2018)
2. Z. Mingaleva, et al. *Waste Management in Green and Smart Cities: A Case Study of Russia* (2020)
3. LI Jiatong, et al., IOP Conf. Cep .: Earth Environ. Sci., 651, 042040 (2021)
4. Hu Zhihao, et al IOP Conf. Ser.: Earth Environ. Sci., 651, 042049 (2021)
5. He Chengqi, et al. IOP Conf. Cep .: Earth Environ. Sci., 651, 042048 (2021)
6. Qin Cuihong, et al IOP Conf. Ser.: Earth Environ. Sci., 651, 042054 (2021)
7. Yu Xialei, et al IOP Conf. Ser.: Earth Environ. Sci., 651, 042056 (2021)
8. Ni Tingting, et al., 2021 IOP Conf. Cep .: Earth Environ. Sci., 651, 042065 (2021)

Eco-labeling is a useful piece of informational metabolism in the youth environment

E. I. Khabarova*, *S. V.* Nikitina, and *M. V.* Strebkova

MIREA - Russian Technological University (M.V. Lomonosov Institute of Fine Chemical Technologies), Moscow,119571, Russia

Abstract. Environmental awareness of consumers and the popularity of a healthy lifestyle are the trends of our time both in the world and in Russia. The article discusses the ecolabelling sign as a means of offline information about the environmental characteristics of products and an important criterion when choosing and buying various goods. In our country, there are only prerequisites for an approach to greening business and positioning "environmentally friendly" products. However, it is vital that everyone understands the eco-label and distinguishes the certificate of declared quality from the marketing ploy. The paper touches upon the issues of ecological education and formation of the population ecological culture. To clarify the importance of eco-labeling in the formation of an environmentally oriented lifestyle among young people, a sociological research method was used, suitable for clarifying any aspects of social relations and capable of providing operational primary sociological information. The results of the survey of the 1st and 2nd year students of the Moscow State University of Fine Chemical Technologies named after M.V. Lomonosov in 2019/2020 academic year, concerning their interest in the environmental characteristics of the purchased goods, are provided.

1 Introduction

It is known that at present we are dealing with many changes in all spheres of life, perceiving them as a significant acceleration of the pace of life. Wherein, we are required to be very flexible in our actions. The main thing is that these changes are perceived consciously, in accordance with the real course of events. There can be an adequate response to what is happening only under these conditions [1]. Wherein, the exchange of information with the environment, i.e. the so-called informational metabolism [2,3], is our main vital function, like breathing.

The term "information metabolism" was offered by the Polish psychiatrist and philosopher Anton Kempinski, who compared the biological term of metabolism (a set of processes in the body, as a result of which the incoming food, water and air are converted into substances necessary for the body for life) with the process of perception, accumulation, assimilation, processing, storage and transmission of information by the human psyche. According to his theory, external information signals are likened to food, being transformed

* Corresponding author: khabarova@mitht.ru

and perceived by certain channels of the human psyche, and also (depending on the quantity and quality of this information) significantly affecting the mental health of each person.

All information we perceive affects our thinking and behavior. Everything that we see, hear and feel penetrates into us and affects our life. Now in the information age, the quality of the information that we receive is extremely important. Even if we are dealing with reliable, not falsified information coming to us from the environment, we need a lot of work to analyze information, process it and select the most beneficial adaptive behavior [1].

Most of the information that is widely available today for every citizen for private use is used to create a variety of benefits. The only question is whether we know how to handle this information in such a way that it benefits us, not harm.

According to international studies, there are two important for us among ten trends in demand that are relevant today. The first one is environmental awareness of consumers, the second is the popularization of a healthy lifestyle. In Russia, it is still not easy to implement the idea of the need to keep track of what goods we use, and to think not only about ease of use, but also about how this or that product was produced and how it shall be disposed of. Wherein, the demand for eco-friendly products has grown significantly in recent years: now buyers are ready to pay about 30 % more for products, if they belong to the "eco" category, the population begins to simply care more about their own health, and in fact, at the same time, about preserving environment. Eco-labels and other symbols on products are intended to inform the user about the environmental effects of products, production processes or safety for consumer health [4].

The eco-label is awarded only to those manufacturers who have successfully passed the examination and, accordingly, have proven the environmental safety and high quality of their products. At this stage, mainly large world-class companies can meet international environmental safety criteria, for which reducing the negative impact on the environment becomes a real competitive advantage.

Due to above, the goal of environmental education and upbringing is not to preserve the world in its original form, but to find forms of harmonious co-evolution of Man and Nature, which would ensure the joint development of the biosphere and human population, which is an integral part of the biosphere [5].

Therefore, ecological education is formation among the general population of the consistency of all types of human activity that are associated with the knowledge, development and transformation of nature. Wherein, educational and upbringing tasks are not limited only to formation of a sense of respect for nature, but also consist in deepening and expanding environmental knowledge.

The biosphere consciousness and worldview can provide the basis for reasonable cooperation between man and the natural environment.

Environmental consciousness (an indispensable component of environmental culture) is understood as the result of the logical processing of knowledge, which turns this knowledge into beliefs, value orientations and attitudes of people, as well as stimulating the cognitive activity of citizens in order to improve the natural environment.

Wherein, the ecological worldview is considered as a multidimensional integral component of the intellectual and spiritual culture of the individual, as a subjective, systemic, multidimensional experience of the individual, which ensures his/her creative self-realization in understanding and resolving environmental problems [6].

That is why for our society, where the number of problems requiring careful sociological study is growing, the public engagement of sociology is becoming actual. The public and the authorities often do not notice many sociological conclusions about modern challenges and threats (including environmentally oriented ones), as well as the offered ways and recommendations for overcoming them considering the methods of forecasting and constructing options for situation development. In this regard, it is necessary to update communication channels through the media considering not only paper, but also electronic media, the Internet and other discussion platforms [7, 8].

When comparing three methods by which sociological study is usually carried out (sociological survey, the method of observation and the study of documentary information in all forms), preference was given not to the theoretical method of analysis, but to empirical study, which is aimed at obtaining general information about human actions in all kinds of arising situations [9].

In order to reveal what young people think about certain positions, it is possible to collect the information necessary for study by conducting a sociological survey (one of the best and most reliable study methods) based on the principle of randomness with involvement of a large number of people. Additional advantages of this method are cost-effectiveness, efficiency, mobility, simplicity and direct contact with the respondent of the received information array. These data will concern the collective, group, public and, of course, the purely individual opinion of each person participating in the experiment. Today, using a sociological survey, it is possible to obtain more than 90 % of sociological data [10].

Via the efforts of the All-Russian Center for the Study of Public Opinion (VTs5I9OM) and other organizations, public opinion polls are conducted regularly in our country. It was found that 78 % of the respondents believe that such polls help the authorities to learn about the mood in society, and 52 % report that the results of the polls are useful to scientists who study social processes [11].

Document "Agenda of the 21 century" adopted at the UN Conference on Sustainable Development (1992) (Chapter 4, para. 21) states that "Governments, in cooperation with the industrial sector ... need to encourage the expansion of information programs that provide for introduction of environmental labeling of goods and dissemination of information about the environmental characteristics of the products sold so that buyers can make informed choices about certain products" [12].

Accordingly, the ecological component of sustainable development shall orient society towards maintaining the sustainability of ecosystems, protecting the environment and adopting an environmentally oriented lifestyle by citizens, accompanied not only by a clear understanding of environmental problems and threats, but also by rejection of the consumer-utilitarian attitude to the environment [13].

Unfortunately, it is too early to talk about the wide distribution and great popularity of ecolabels in Russia. However, in fairness, it shall be noted that already now in our country there are many legal acts that affect the need to improve the quality of product production: regulations in the field of consumer protection, environmental protection, standardization and advertising, as well as state standards and the conformity mark of Gosstandart of the Russian Federation [14].

2 Materials and Methods

RTU MIREA teachers are also trying to introduce the useful concept of eco-labeling into the everyday life of student buyers. Socio-statistical diagnostics also helps to obtain information about the impact of environmentally oriented transformations on young people.

Supporting the development of "a targeted system of general and continuous environmental education and upbringing that shall be based on the principles of consistency, continuity, integrity, complexity and interdisciplinarity" [15], we considered it expedient to conduct regular sociological surveys in the form of written questionnaires among junior students of RTU MIREA and high school students in the Western District of Moscow regarding awareness of environmental labeling signs [16,17].

In the context of our study, we wanted to convince young people of the need to balance environmental issues and their primary needs as consumers of the goods and services provided to them.

3 Results and Discussions

The results of an anonymous questionnaire survey (self-filling of questionnaire forms) of 229 1st year students and 222 2nd year students for 2019/2020 full-time education at the Moscow State University of Fine Chemical Technologies named after M.V. Lomonosov are shown in Fig. 1-10. The total number of respondents was 451 people.

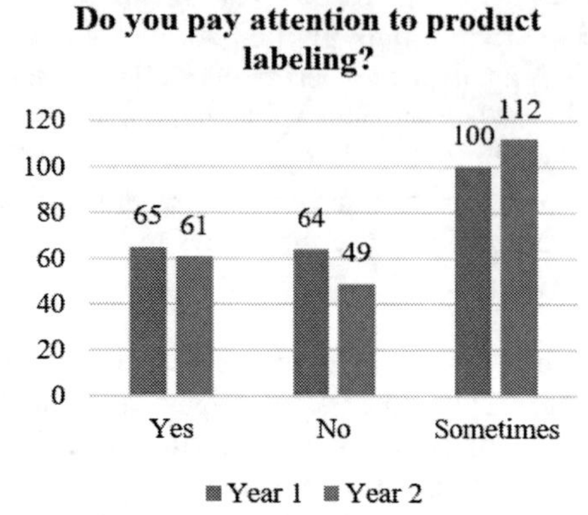

Fig. 1. Answers to question No. 1

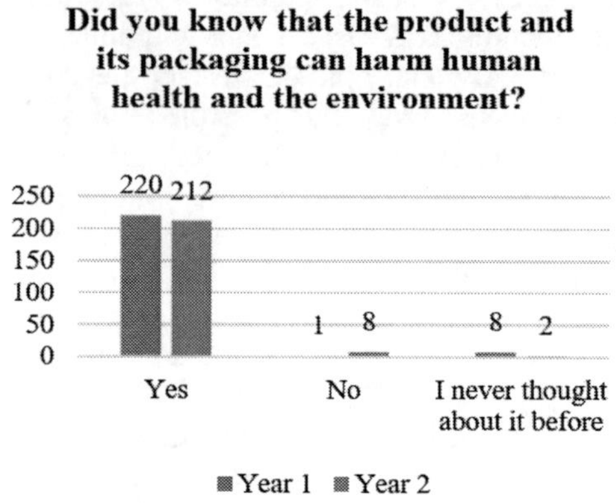

Fig. 2. Answers to question No. 2

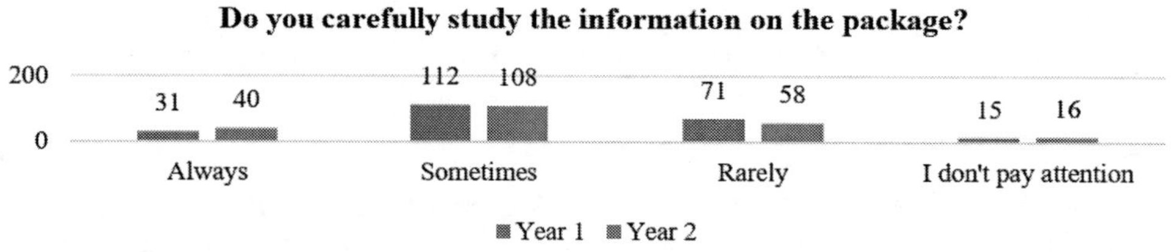

Fig. 3. Answers to question No. 3

Fig. 4. Answers to question No. 4

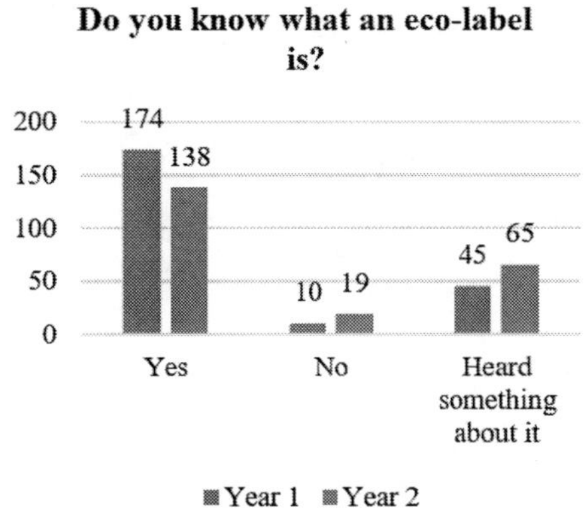

Fig. 5. Answers to question No. 5

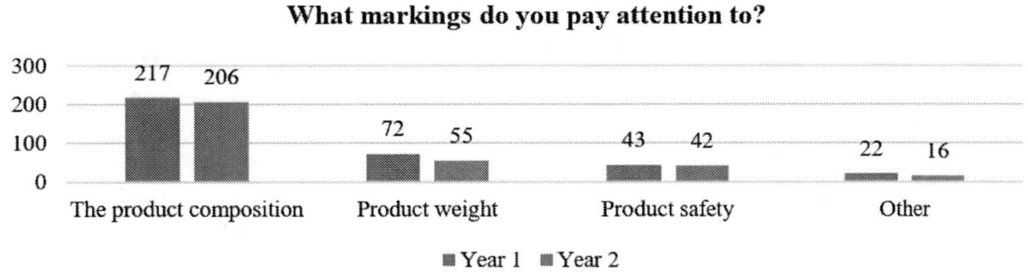

Fig. 6. Answers to question No. 6

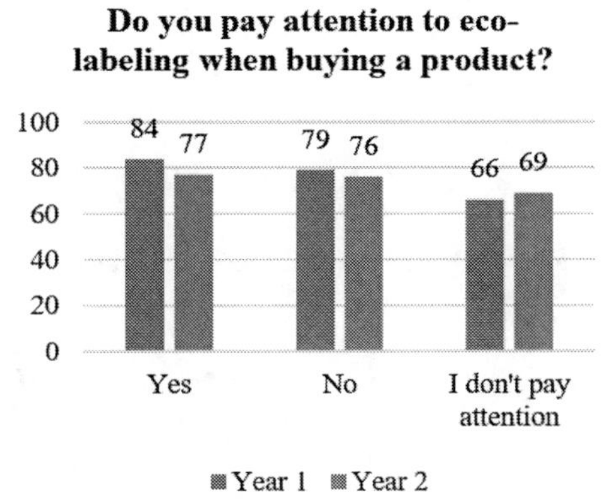

Fig. 7. Answers to question No. 7

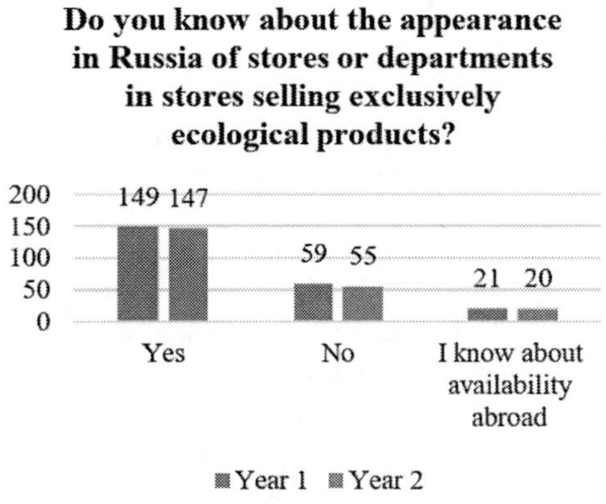

Fig. 8. Answers to question No. 8

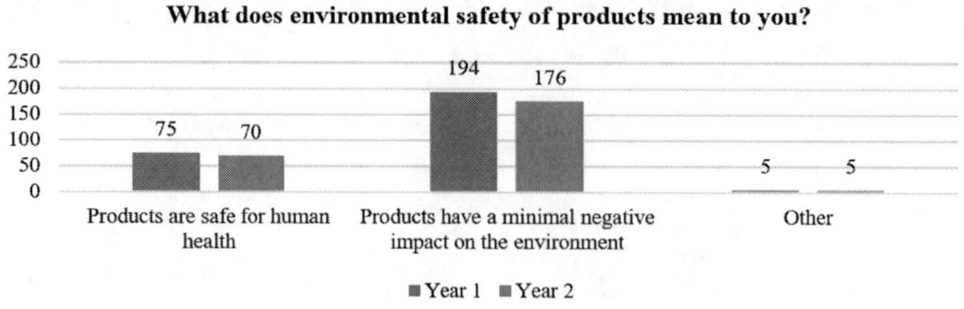

Fig. 9. Answers to question No. 9

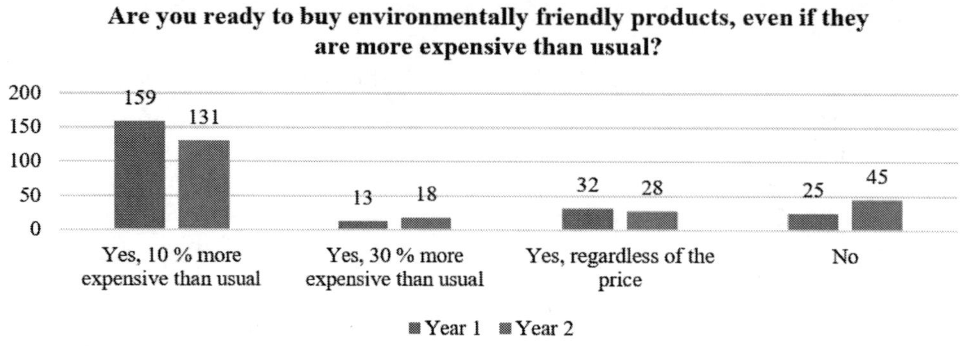

Fig. 10. Answers to question No. 10

4 Conclusions

From the results of the questionnaire, it can be seen that:

1) all answers of the 1st and 2nd year students are generally comparable to each other (in percentage);
2) half of the respondents are not attentive to the labels containing information about the purchased goods, and only occasionally study them;
3) the same 50 % practically do not pay attention to such labeling of goods on labels as warning notices intended to ensure the safety of the consumer and the environment, and are guided, most likely, only by the basic consumer properties (weight characteristic, country and manufacturer, term suitability, methods of transportation and storage, etc.);
4) wherein, the majority of respondents (almost 100 %) are really aware that any product and its packaging can harm human health (35 % think) and the environment (70 % think);
5) 80 % of students were directly familiar with the word "ecolabelling", although in life they ignore goods with specific signs on the packaging: only 40 % admitted that they pay attention to ecolabels when buying a product.
6) About 80 % know or have heard of appearance in Russia of stores or departments in stores selling exclusively ecological products.
7) Interestingly that 70 % of the respondents are prone to buy environmentally friendly products 10 % more expensive than usual, and 10 % - even regardless of the price.

Although in this study the results of a sociological survey represent only a vector for measuring certain values, the conversation is about formation of adequate environmental values and types of everyday behavior that correspond to the goals and principles of sustainable development of society, i.e. harmonious relationship with nature.

Interviewing junior students has 2 goals:

1) focusing the attention of students, as consumers of products and services, on presence of labels with ecolabels on goods, as a result of which young people are armed with accessible information that allows them to make a choice in favor of high-quality products considering the minimum negative impact on the environment in the process of their production;
2) formation of a forward movement in any trajectory of teaching today's students, as tomorrow's entrepreneurs and leaders, on coordination of used and implemented technologies with identification of the focus of the future target activities of today's students and tomorrow's entrepreneurs on coordination of the interests of society and nature from the standpoint of introduction and use of technologies friendly to the ecosphere.

References

1. B. Krieger, Slavyane, **2(11)**, 32-35 (2018)
2. A. Kempinsky, *Existential Psychiatry*, 320 (Sovershenstvo, 1998)
3. A. Kempinsky, *Melancholy*, 405 (Nauka, 2002)
4. A. G. Ishkov, V. A. Grachev, *Popular Ecology*, 196 (NEF named after V.I. Vernadsky, 2014)
5. N. N. Moiseev, *Human ecology through the eyes of a mathematician*, 251 (Molodaya Gvadiya Publishing House, 1988)
6. N. M. Mamedov, S.N. Glazachev, Environmental education: concepts and technologies: Collection of scientific papers, Volgograd, 16-26 (1996)
7. L. N. Vdovichenko, Sociological studies, **2**, 134-140 (2014)
8. L. N. Vdovichenko, Sociological Studies, **7**, 25-29 (2015)
9. Sociological research, Research agency ZOOM MARKET, https://www.mazm.ru/
10. E. Potapova, Sociological survey: an example. The results of a sociological survey, Project FB.ru, https://fb.ru/
11. I. Zakharov, 10 most interesting opinion polls in Russia 2018-2019, Site BASETOP, https://basetop.ru/
12. Agenda for the XXI century, http://www.un.org/
13. I. A. Sosunova, L. G. Titarenko, Use and protection of natural resources in Russia, **4(142)**, 99-102 (2015)
14. A. N. Kazantseva, O. I. Malikova, Innovations, **4(210)**, 71-75 (2016)
15. S. N. Glazachev, E. A. Kogay, *Environmental culture and education*, 34 (Horizont, 1999)
16. L. L. Budaeva, S. V. Nikitina, E. I. Khabarova, Understanding ecolabels that reflect relevant environmental information in a consumer-accessible form, from college students. Materials of the VI All-Russian scientific-practical. conference "Environment and Health. Hygiene and ecology of urbanized areas", Moscow, September 13-14, 99-103 (2016)

17. E. I. Khabarova, S. V. Nikitina, L. L. Budaeva, M. V. Strebkova, A. M. Pogorely. Assistance in introducing environmental culture through familiarity with eco-labels. // Life Safety: Problems and Solutions, 2018, Collection of articles based on the materials of the II Int. scientific and practical. conferences. Under the general ed. of S.F. Sukhanovoy, Kurgan: publishing house of KGSKhA named after T.S. Maltsev, 2018, pp. 220-223.

The role of Eco-tourism in sustainable development: case study eco-tourism challenges in Iran

Ismael Rezaeinejad*, and *Amol* Khaniwadekar

Peoples' Friendship University of Russia (RUDN University), Miklukho-Maklaya Street, 6, 117198, Moscow, Russian Federation

Abstract. Sustainable development is a new concept formed after the industrial revolution and the problems created concerning industrialization of cities and technological development of cities and linking different economic, cultural, political, and social development and exceptional attention to ecological considerations. Sustainable development has a comprehensive, integrated and human-oriented approach. It includes valuable concepts in preserving, promoting, and improving health and ecological integrity in the long term and providing human needs. Without destroying the next generation's ability and achieving economic dynamism, a livable environment and social equality are critical dimensions. The importance of conservation of natural resources in the urban environment is as significant as one of the essential areas of development of communities and, at the same time, a criterion for improving the quality of life. The consequences of urban development and the complexity of environmental problems have the inevitable existence of green spaces and their development. Due to the role of urban management in achieving sustainable urban development, this research tries to explain the concept of urban management, its objectives, and its components and provide effective urban management to achieve sustainable development. Eco-tourism is mainly based on resources that the natural environment is available to tourists that use it and appropriate conditions and requires protection of natural resources, which can be realized through evaluation of capability and capacity of the natural environment to attract tourists. The unique geography and diversity of natural phenomena caused Iran to be recognized as the fifth most crucial natural diversity globally, but planning for using these conditions is still not provided. In addition, the country's eco-tourism is faced with social, economic, and environmental problems, and for planning in the field of overcoming this challenge, we first need to recognize these challenges.

1. Introduction

In the current era, the acceleration of modern urban life and culture led to decreased social interaction between citizens and neglect of the importance of public spaces for these social

*Corresponding author: 1032205254@pfur.ru

interactions. Therefore, the importance of urban green space in life and its sustainability and its physical and social impacts in the urban system is impossible, so the existence of green space in cities and its appropriate distribution is one of the fundamental issues in urban planning. one of the other issues related to the importance of green space use in cities is multifunctional. The natural, aesthetic, social of this land use makes it essential to deal with this issue in the urban area. Population increase and its concentration at the regional level directly affect the quality of the environment [5]. Population growth, poverty, and environmental degradation in developing countries have created a vicious circle. The vicious circle of quality of life has affected people's quality of life and has made the efforts of the third world countries to achieve sustainable development. One way to achieve sustainable development is to have welfare and comfort in the life of society [3]. Today, all governments and authorities of society are providing welfare and peace to the people. The welfare and comfort of life are divided into material and spiritual parts. That spiritual comfort is one of the most critical factors in providing welfare and well-being in crowded spaces of urban green spaces, which has a significant impact on the welfare of citizens. Today, urban planning emphasizes maintaining the relationship between the urban environment and green space and nature [7].

In recent years, eco-tourism has grown significantly in different policy communities and public institutions and sustainable development and created a wide range of challenges between extremist supporters and irrational usage of natural capacity with conservation perspectives sustainable use of these natural resources. However, in most developing countries, little attention has been given to global warming, environmental protection, or commercial activities' carbon effects [8].

Increasing population growth and limited capacity of natural resources include the use of divine gifts for future generations, which is, directly and indirectly, related to nature and is the primary motivation of the visit and the use of natural attractions and indigenous culture without any disruption [2]. The most important incentive to encourage tourists to visit natural attractions today is tourists' curiosity in recognizing and studying rare plant and animal species and natural resources that increase their deeper appreciation of nature, which ultimately stimulates the sense of protection of natural environments [1]. Adventure goals in meeting phenomena and specific natural effects associated with sport, recreation, and hedonic and compelling attractions in health care and physical health, such as hot water springs, are also the other incentives known as Tourist factors. The problems of urbanization caused by migration and population in cities, climate change and ecosystems change due to environmental factors, the development of technology, the rise of income, duplication of living conditions in the city forced people to take refuge in nature to relieve the fatigue. Sustainable tourism will respond to tourists and the next generation's current needs and balance environmental, economic, social, and cultural dimensions. That is, sustainable tourism is looking for socio-economic, social, and cultural dimensions. Meanwhile, Iran has unique opportunities that can be used in order to develop eco-tourism. This study reveals the importance and challenges of eco-tourism in Iran.

2. Materials and Method

2.1 Research method

The present paper is conducted in an analytical-descriptive method. The gathering method is library study, including books, magazines, papers, and electronic resources. This paper explores challenges in sustainable eco-tourism in Iran to introduce eco-tourism and tourism activities in the sustainable development of eco-tourism and natural resources.

2.2 Theoretical bases

2.2.1 Sustainable development

The term sustainable development was used for the first time in 1970 in the Declaration on the Environment and Development. The Global Commission of sustainable development will meet the needs of the present generation without compromising the ability of future generations to meet its needs. In other words, radical changes in exploiting the resources, investments, and development of technology are defined as the potential for existing and future resources to meet the needs of the human being [11].

Sustainable development is a process in which natural, social, physical, economic, financial, trade, energy, agriculture, energy, industry, and all other policies are designed that cause development that can be socially and economically sustainable. In other words, sustainable development means managing and conserving natural resources and directing economic and institutional changes so that the present and future needs are regularly met and sustainable [4].

2.2.2 Sustainable development theory

The concept of sustainable development theory has emerged in the form of environmental quality in the early 1980. this theory is presented to support natural resources. Sustainable development meets the needs of the present era without compromising the supply of future needs. Sustainable development is a development that is based on human beings and causes human happiness and growth. So, such development ensures sustainable growth and development in all economic, cultural, and social fields. The primary rationale for sustainable development is to improve the standard level of people's lives and the incredible well-being of those who enjoy the benefits of society while avoiding unnecessary costs. Therefore, the main issue of sustainable development is how to achieve continuous development, regard nature and human beings without injustice, and for all human generations. This new attitude is also necessary for urban planning and urban development policy [6].

2.2.3 Sustainable Urban Development

The theoretical foundations of land conservation planning are sustainable urban development pattern, balanced density, land use efficiency; attention to the network is continuous from the green spaces of the district and city [3]. This theory suggests a combination of nature-oriented and sustainable development and a strong-based, technology-based, and robust-based structuralism. The most important definition of sustainable development at the Rio summit is as follows: " the development that meets the current needs of the human without compromising the needs of the future generations, and to be considered in the environment and future generations." Although many definitions of sustainable development have been introduced yet, the axis of all definitions is the next generation and attention to the future. In all definitions, attention to the environment and conservation of the environment is considered too.

The principles of sustainable development can be summarized as follows:

1. Attention to the use of renewable sources, such as solar and wind energy.
2. Using less of non - renewable and polluting energies, such as fossil fuels.
3. Pay attention to future generations
4. Attention to the environment and its pollution reduction and attention to environmental cycles.

A sustainable city is a city that has the lowest consumption of renewable energies, minimum waste, and minimal impact on the environment and can sustain its current function in the coming decades.

2.2.4 Sustainable Tourism and Eco-tourism

Sustainable tourism is ecologically safe, economically applicable, socially acceptable, and economically feasible.

Eco-tourism is a form of tourism involving responsible travel to natural areas, conserving the environment, and improving the well-being of the local people. Its purpose may be to educate the traveler, provide funds for ecological conservation, directly benefit local communities' economic development and political empowerment, or foster respect for different cultures and human rights. Since the 1980s, environmentalists have considered eco-tourism critical, so future generations may experience destinations relatively untouched by the human intervention [7].

Generally, eco-tourism deals with interaction with biotic components of the natural environment. Eco-tourism focuses on socially responsible travel, personal growth, and environmental sustainability [11]. Eco-tourism typically involves travel to destinations where flora, fauna, and cultural heritage are the primary attractions. Eco-tourism is intended to offer tourists an insight into the impact of human beings on the environment and to foster a greater appreciation of our natural habitats.

Responsible eco-tourism programs include those that minimize the negative aspects of conventional tourism on the environment and enhance the cultural integrity of local people. Therefore, besides evaluating environmental and cultural factors, an integral part of eco-tourism is promoting recycling, energy efficiency, water conservation, and creating economic opportunities for local communities. For these reasons, eco-tourism often appeals to advocates of environmental and social responsibility [10].

3. Result and Discussion

3.1 Relationship between sustainable eco-tourism development with Environmental Protection and Natural Resources

Indeed, sustainable eco-tourism development depends on maintaining the environment and optimum use of available facilities. The impact of eco-tourism on the environment is not unique to participate in leisure activities. However, they need facilities and equipment for their time in the region and have their leisure role. Therefore, planners and designers of tourist facilities must develop comprehensive and detailed plans to prevent and decrease the adverse effects of eco-tourism activities in the environment. On the other hand, it reflects the positive and negative impacts on the environment. Without attention and neglect, this matter has severe risks to wildlife and plant species, and other natural objects and may cause ecological changes in the environment [2].

Undoubtedly, attention to national parks and wildlife and many of their functions such as preserving and supporting watersheds and ecosystems, improving environmental conditions in surrounding areas, and preserving the value of natural heritage in most countries, especially the tourist countries, promoted eco-tourism too.

Environmental protection and optimal use of existing natural attractions are costly today, especially for developing countries and the third world. Therefore, these countries should promote eco-tourism in their national park and the areas that have natural attractions. However, such a move will show people's willingness and effort to find calmness, solitude, and contact with nature, and the opportunity to familiarize them with their natural heritage and others. Eco-tourism activities in national parks or any other protected area can be a financial mechanism for regulating these areas and the first means to protect them.

These activities are possible if the type and level of tourism management are favorable and proportionate to the carrying capacity of these areas [5]. Another point is that this attitude will arouse public awareness about the importance of protecting and protecting the environment. the experiences gained from the tourist countries by applying scientific and knowledgeable management in this regard indicate that the obtained income in most cases is necessary for the management of national parks and eco-tourism and sometimes more than necessary. It is evident that achieving sustainable tourism development due to the increasing population growth and the need for their people to play their leisure role, the tourist activity is one of the most complex forms, is possible when host countries think of mixed tourism [9].

Respecting the environment and maintaining it is a culture, and culture is aware of its personality as a thinking being and its relationship with others and the natural environment. Therefore, the care of the environment is a necessary part of a cultural policy. It requires the best training methods to take systematic and rational exploitation of natural resources and data. It is also necessary to create a balance between recreational satisfaction and the protective requirements of natural attractions. The search for such equilibrium must determine the concept and philosophy of meeting the natural and environmental attractions and eco-tourism.

3.2 Analysis of eco-tourism development in Iran

Due to the diversity of its natural environment and the abundance of geography and diverse geographical attractions such as waterfalls, straits, rivers, mountains and mountains, temperate climate, and thousands of flora and aquatic plants, Iran can receive many domestic and foreign tourists annually. In other words, such as the infrastructure for eco-tourism is to be created for this country, it will undoubtedly create jobs for residents and economic prosperity.

Culture can help accelerate the positive process of eco-tourism developments; what can be done not only on the media level of the public, but at the textbook level, including the benefits of mass travel in the form of tours, the use of expertise, and company information, prevent waste of time, greater security, and even the cost of travel . all of this needs to be benefited from the atmosphere of friendly social interaction with humans, both thinking and taste as well as learning and influencing generations. In Iran, it is as if the cost of traveling with a tour is above a personal journey. Either way, it should be done cheap. However, it can be seen that people spend more money on foreign trips. One of the reasons for the difference is that Iranian people do not pay enough to see the sights and values of their communities and do not know them. As long as Iranian people do not know the value of their country's attractions, their balance of tourism statistics will be negative. The correction of these perspectives requires deep culture and curriculum.

The other challenge is the awareness of host communities about the value of its natural, cultural, and historical attractions. They need to look at these attractions like an asset. Otherwise, they wipe them out with their own hands, trash the trees, pollute the fountains, and destroy rivers, write souvenirs, and land on the old fort wall, all of this means taping into eco-tourism and sustainable development.

Organizing training classes, encouraging and rewarding popular information in the form of defined projects, providing facilities to the people of host communities to provide facilities for drawing and lodging, and including beautiful solutions for the gradual of environmental protection, eco-tourism resources, and cultural and historical values of societies and finally sustainable development.

On the other hand, policymaking in the field of introducing natural attractions and going to nature without focusing on responsible tourism standards, getting away from the correct method of presence in nature by using expertise and information of authorized and specialist companies in nature, cause environmental pollution, impact on eco-tourism resources and even tourism industry.

Lack of comprehensive planning in the country and lack of systematic advertising are critical challenges. Eco-tourism in Iran, with all the extent and diversity, has not yet been identified and compiled, and because of these shortcomings, it cannot be promoted. As a result, there is no ad, or if we have, the organization is not targeted and usually leads to the influx of people to these areas. Suppose the people of the region are trained. In that case, they will design and implement new and rational ways of protecting the environment and the natural and cultural attractions of the region and prevent injury to the region.

Generally, the main challenges of eco-tourism in this country are:

3.2.1 Economic challenges

Due to the lack of detection of eco-tourism regions and not being infrastructure for that. Among other economic challenges in eco-tourism, the following are:

- Lack of quality of life: a broader definition of human well-being should be offered in a way that goes beyond economic welfare
- The non - compliance of economic, environmental integration: economic decisions should be adopted according to their impacts on the environment
- The lack of guidance and economic plans in the eco-tourism industry
- The deficiencies and failure of infrastructure services such as the roads

3.2.2 Social challenges

Among other challenges, could refer to cultural, social, and cultural challenges that the program of modernization and the culture of all sides profoundly impact. Collective travel culture in the form of the tour is one of the indicators of development by taking advantage of the expertise of host companies that have not yet been institutionalized in Iran. Host communities ' awareness of the value of their natural, cultural, and historical attractions is fundamental. Lack of macro planning in the country and lack of systematic advertising are challenges facing eco-tourism in Iran. Eco-tourism routes in Iran have not yet been identified and developed with all their extent and diversity. Due to the lack of sufficient knowledge, it is not even possible to promote it or even do it; advertising is not organized and purposeful and usually leads to the intrusion of public policy into these areas. Another example of the challenging problem of eco-tourism is the following:

- Lack of commitment between generations: At present, decisions have to be made according to the results on the environmental impact of future generations
- Social justice: All people have the right to have the environment in which they grow and flourish
- Lack of involvement: Institutions should be restructured to make all sounds in the decision-making process
- Lack of security in some desert areas in Iran
- The lack of proper management in the eco-tourist regions
- Lack of ease in obtaining visas to enter the country
- lack of Hotels and resort centers in the eco-tourist regions

- The cultural divide between the host community and the visitor
- Lack of understanding of the actual value of the country's natural resources by the people (host community)

3.2.3 Environmental challenges

Some of the most critical challenges of eco-tourism in Iran can be pointed out as environmental challenges. Tourism has caused irreparable damage to the Iranian environment by tourism so far, including the excessive exploitation of coastal areas and forest areas of north province by tourists, which have already dealt a deadly blow to the environment of Northern provinces.

Among other challenges in this regard are natural protection, protecting resources, and protecting the animal and animal world.

Therefore, eco-tourism in Iran seems to have a little attention that has the weakness points and threats mentioned in table (1).

Table 1. The most significant weaknesses and threats of eco-tourism in Iran.

Threats	weaknesses.
showing the negative image from Iran In international media	Lack of comprehensive strategic programs
Tension, unrest, insecurity in the Middle East	unsuitable transportation system to visit specific areas
competition among world countries to attract tourists	weakness and lack of hotels and suitable residences for tourists in eco-tourism area
The global economic downturn, which has led to a decline in large numbers of tourists	Lack of skill and human resources in eco-tourism major
environmental degradation, drought, and natural disasters	weakness in marketing and introduction of Iranian ecosystems
Non - promotion and exclusion of Iran except some attractive areas by Westerners	weakness health care in some destination areas
	Inadequate or non - access to technologies such as the Internet in some destination areas
	lack of attention to eco-tourism attractions and lack of responsibility for the environment
	Weakness and lack of social security, especially in some remote areas
	high inflation rate and economic problems in the country
	Cultural problems

3.2.4 Opportunities of Iran eco-tourism

As a four-season country with pristine and beautiful nature in different parts, Iran has a lot to offer nature lovers and eco-tourists. Another example of the opportunity of eco-tourism in Iran is the following:

- Iran will be on the path of the Silk Road and the link between the world of the West and East Africa as one of the most eco-tourist countries in the world
- Inland waters are recognized as important bird areas with more than 50 inland lakes and 22 internationally important wetlands
- Coastal areas: Caspian Sea (630 km) and the Persian Gulf and Oman Sea (1880 km) and a multitude of islands

- 80 registered protected areas under the jurisdiction of the Department of Environment
- Many caves, springs, waterfalls, rivers, and geological structures
- the hottest deserts in the world (Lut Desert)
- unique animal and plant species
- the first phase of development of the National Nature-based Tourism Strategy, 2000 naturally attractive sites were identified

4. Conclusions

Eco-tourism can have different and significant effects in the region. These effects can be social, cultural, economic, and environmental. Planners and policymakers also use the eco-tourism industry as the central pillar of sustainable development. According to the extent of sustainable development of the term, this term is described in different ways. The principle of transformation of relations between the system is social, economic, and natural processes. Considering the importance of the eco-tourism industry and cultural, social, and political factors, natural environment factors play an important role in tourism development and attract tourists. Tourists use a geographical space. This space has a physical and natural structure, including environmental factors, geology, topography, flora, fauna, and some of the factors caused by human beings' activity.

The arrival of tourists to each region has led to employment at the macro level that provides the context for economic, social, and cultural development. Today, the eco-tourism industry has found much development globally. Many countries have improved their status considerably and reduce many of their problems such as unemployment, low-income level, and shortage of foreign exchange income by a city with many eco-tourism potentials.

The results show that Iran can be one of the main axes of eco-tourism. Considering the current situation and identifying the strengths and opportunities of tourism development in the region and on the other hand, the lack of suitable communication channels between cities and, lack of facilities for residential, service and welfare facilities for passengers, weakness of advertising and lack of people willing to invest are main obstacles of eco-tourism development. The development and reconstruction of roads between the provinces, rebuilding roads leading to attractions, creating service facilities, utilities and residential areas for tourist entry, and using well-trained people as tour guides can be grounds for eco-tourism development.

Due to the variety of attractions and natural landscapes, Iran is one of the eco-tourism poles. Eco-tourism in terms of economic causes prosperity and construction of rural areas. Planning in eco-tourism is considered the most crucial part of eco-tourism and undoubtedly is one way to achieve sustainable development and eco-tourism.

Eco-tourism can reduce threats through monitoring, local practices, educated people, environmental education, and protected areas. Eco-tourism needs comprehensive planning and management as well as understanding the potential.

As one of the essential tools among all tourism industry sectors, eco-tourism is growing and challenges economic, social, and environmental issues that can be used with proper planning to achieve its primary goal, sustainable development.

References

1. M. Altinay, K. Hussain, International Journal of Contemporary Hospitality Management (2005)

2. S. Blangy, H. Mehta, Journal for Nature Conservation, **14(3-4),** 233 (2006)
3. D. B. Carson, A.J. Taylor, The Rangeland Journal, **30(1),** 77 (2008)
4. J. Davenport, J.L.Davenport, Estuarine, coastal and shelf science, **67(1-2),** 280 (2006)
5. D. Harrison, S. Schipani, *Lao Tourism and Poverty Alleviation: Community-Based Tourism and the Private Sector. In Pro-poor Tourism: Who Benefits?,* 84 (2007)
6. I.G. Malkina-Pykh, Integrated assessment models and response function models: pros and cons for sustainable development indices design. Ecological indicators, **2(1-2),** 93 (2002)
7. G.P. Nyaupane, B. Thapa, Journal of Ecotourism, **3(1),** 20-45 (2004)
8. I. Rezaeinejad, A. Khaniwadekar, K.S. Rana, SHS Web of Conferences, **114,** 01028 (2021)
9. P. Tremblay, Desert tourism scoping study. Desert Knowledge CRC (2006)
10. D.B. Weaver, L.J. Lawton, Tourism management, **28(5),** 1168 (2007)
11. M. Whitehead, Urban Studies, **40(7),** 1183 (2003)

Ecological communications of modern students (experience of the Moscow automotive-road state technical university)

I. S. Voznesensky*, *T. A.* Mironova, and *L. O.* Ternovaya

Moscow Automobile and Highway State Technical University (MADI), Moscow, Russia

Abstract. The article presents the results of the analysis of the process of expansion and diversity of communication channels in the university environment, which is relevant from the standpoint of achieving the goals of sustainable development in the field of education. It is shown that one of the most accessible and open communication channels, in which modern students are actively involved, is the channel of environmental communications. They have many manifestations, which corresponds to the most exceptionally wide ecological space, reflecting both individual and collective interests, features of a narrow professional attitude to nature and its civilizational understanding. A significant part of the disciplines studied in higher education contributes to the formation of students' environmental competencies. The specificity of the ecological approach to education and upbringing contributes to involvement of students in activities for the nature protection. The regularities of environmental communications functioning can be traced on the example of environmental education carried out at the Moscow Automobile Road State Technical University (MADI).

1 Introduction

Scientists' attention to environmental problems became clearly visible when this environment began to show its unfriendly attitude towards humans. This process was triggered by accelerated industrialization. It began in the second half of the 19th century, when people from the countryside moved en masse to the sprawling cities. They got a job working for industrial enterprises, increasing the pressure on nature in the megalopolises formed at that time. Just then researchers began to show an increased interest in the issues of the relationship between man and nature. First of all, these were natural scientists, who felt and understood the fragility of the environment more subtly and better than others. The very concept of "ecology" arose, based on ancient Greek terms: οἶκος — dwelling, dwelling place and λόγος — learning. The space was designated not only for scientific discourse, but also for social activity aimed at preserving nature. This required both development of the newest science — ecology, and allocation of environmental communication channels that could be filled with nature-saving meanings that attract caring people.

* Corresponding author: iv@sociomadi.ru

Since any environmental issue is characterized by two parameters, the first of which is its transboundary nature, all-embracing, commonality for the entire planet, even despite the locality of any action, and the second is the impact not only on living people, but also on future generations, to involve young people in solving problems of nature protection is of special significance. This directs them towards a more conscious movement towards their own future and taking responsibility for it already in the present. For the full participation of young people in environmental initiatives, they must have the necessary competencies and be ready for the difficulties that arise in the process of environmental communications. Students receive the fundamentals of environmental knowledge from disciplines that are directly related to conservation issues. However, it is important that such information be supplemented with the skills to formulate an environmental problem, propose ways to solve it, as well as develop environmental communication skills even during the period of university studies, in order to demonstrate this knowledge, skills and abilities in subsequent work. The article presents some developments in this direction, accumulated by teachers of the Moscow Automobile and Road Construction State Technical University (MADI).

2 Materials and methods

The main source of the research presented in this article was the pedagogical practice at MADI, where environmental issues are widely introduced into the content of various training courses based on use of an integrative approach and use of transdisciplinary methods. To comprehend the experience gained while reading various academic disciplines, a systematic and comparative analysis of the results of students' mastering of environmental topics was used. Namely, a comparison of the methods and means of forming the environmental competence of students studying at universities in different countries was made, which showed that, despite the difference in training programs and the specifics of the educational and general socio-cultural environment, there are more common characteristics in the approaches to developing the foundations of environmental competence among students than differences. This allows a wider use of foreign experience in attracting students to environmental projects in the academic programs of Russian higher educational institutions. [12]

It also summarized the mutual influence of theory and practice in implementation of international strategies for the environmental education development, as part of the UNESCO strategy in the field of education for sustainable development (ESD). To identify the effectiveness of pedagogical technologies used to expand the ecological communications of students, the process of forming their ecological culture was modeled. Also, the diagnostics of the components of environmental communications was carried out, during which the methods of observation, questioning and graphical presentation of the students' environmental projects were used.

As a study innovation, the toolkit of the spiral dynamics concept was tested, which helps to analyze the human value world, organically inscribing environmental values into it, as well as to build the logic of the development of a value system in which environmental values are promoted to priority positions. [3, 4]

3 Results and discussion

To involve modern student youth in environmental communications, first of all, it is necessary to understand the nature of these actions, which over a relatively short period, covering a little more than a hundred years, from a narrow understanding of the dissemination of knowledge about the relationship between man and nature, have turned into an exceptionally wide field of interactions between individuals, groups, organizations, social movements on various aspects of environmental issues, and even opened up the possibility of connecting virtual participants to these activities.

The starting point for formation of ideas about environmental communications can be considered the decisions of the third International Botanical Congress (eng. *International Botanical Congress*), which was held in Brussels in 1910. Its participants identified three subsections of ecology: autoecology (ancient Greek αὐτός — itself), synecology (from Greek συν — together) and demecology (from ancient Greek δῆμος — people). The first direction studies the interaction of an individual organism or species with the environment, in particular, it studies life cycles and behavior as a way of adapting to the environment. The second is interaction of populations of individuals of the same species within a given population and with its environment. The third direction studies the features of the functioning of communities and their interaction with biotic and abiotic factors. In fact, these branches of ecology shall be considered as the first channels of environmental communication identified by specialists.

Obviously, allocation of these subsections of ecology reflected not only the picture of biological science at the beginning of the 20th century, but also the fairly stable ideas that had developed by that time about the need to develop this area within oneself, to understand its specifics and highlight the basic positions. Under the influence of the changes associated with the scientific and technological revolution, the need to expand such channels and connect the subject area of ecology with other sciences and, in general, with the changing world outlook of a person was revealed. As a result, ecology has turned not into a narrow section of biology, but, on the contrary, a complex of sciences that study the functional relationships between organisms, which include humans and human society as a whole, and their environment, including the circulation of substances and energy flows that make life itself possible.

In this complex of sciences we see: bioecology, geoecology, geoethics, hydroecology, technoecology, landscape ecology, ethnoecology, medical ecology, social ecology, ecolinguistics, chemical ecology, radioecology, human ecology, antecology (ecology of flowering and pollination), information ecology and others branches of ecological knowledge. They also turn out to be communication channels both for communication within the actual ecological scientific space, and with adjacent areas, which are becoming more and more.

Insofar as the liberal model of management has become dominant in the promotion of global flows, the meanings of this management have penetrated into the understanding of the conditions for the emergence and definition of the framework of environmental communications. We see a clearer formulation of this framework already in the 21st century. This is undoubtedly due to the fact that environmental risks and challenges have jeopardized the continued existence of mankind. Environmental communications began to be identified as a separate scientific and at the same time practical problem when research on environmental ethics and ecolinguistics appeared. In the same period, experts noted the start of what was called "globalization", which allows us to say that environmental communications, on the one hand, opened with the beginning of globalization and, on the other hand, gave it a new impetus. [5, 6, 7]

According to R.J. Cox, former president of the *Sierra Club* that teaches rhetoric and communication courses at the University of North Carolina, the concept of environmental communication is based on three basic principles:

1. human communication is a form of a symbolic act;

2. beliefs, attitudes, behavior of people regarding nature and environmental problems are established by communication;

3. the public sphere is a field for functioning and development of eco-communicative discourse. [8]

Cox identifies the following seven main areas of study and practice of environmental communication: environmental rhetoric and discourse; media and environmental journalism; public participation in environmental decision-making; social marketing and outreach campaigns; environmental cooperation and conflict resolution; risk communication; representations of nature in popular culture and green marketing. [8, p. 18]

Despite the fact that among the noted areas we do not find the field of education, it is present in environmental communications not only as an initial platform for development of environmental knowledge and the foundations of environmental behavior of students, but also as a space in which environmental rhetoric unfolds for a young person. In this regard, we can mention: Felix Finkbeiner, a German schoolchild from Bavaria, who initiated the planting of a million trees, for which he was awarded the *Sea of Excellence-Avard prize* for the best projects and ideas "with a view to the future"; Swedish schoolgirl, eco-activist Greta Thunberg, who promotes the version of the inevitable existential crisis of mankind due to the effects of climate change, which in 2019 was included in the list of the 100 most influential people in the world according to Time magazine and recognized by the same publication as "person of the year"; her young opponent from Germany, Naomi Seibt, who opposes environmental alarms.

More and more opportunities are opening up for young people to participate in environmental decision-making, namely, through participation in the eco-volunteer movement. In Moscow, young eco-volunteers in 2020 took part in the Clean Park action, the *Zero Waste* festival, a volunteer night at the Sborka eco-center, and also joined the celebration of the World Day of Cleanliness on September 18 and other events. Students within the framework of "Moscow cedar trees" project help workers of Moscow cedar mini-nurseries, and then plant small cedar trees in the territory of educational institutions of the Russian capital.

The fact that environmental communications actually permeate the educational process at all levels of education is drawn attention in UNESCO documents. Namely, the UNESCO Strategy "Education for Sustainable Development" (ESD) has been developed and is being implemented, according to which education itself has come to be understood as the most important factor in social change. The UNESCO World Conference on Education for Sustainable Development, held on November 10 - 12, 2014 in Aichi Prefecture, Nagoya City (Japan), completed the United Nations Decade of Education for Sustainable Development (2005 - 2014). The outcome of this conference was the Aichi Nagoya Declaration on ESD and the Implementation Plan for the Global Program of Action on ESD.

The above-mentioned features of environmental communications allow, firstly, to talk about their breadth, the possibility of including not only teachers and students in the communication process at the university, but also representatives of local communities, enterprises and public organizations. Secondly, ecological communications can be considered as a tool of an integral approach in education and a means of developing interdisciplinary connections, due to the fact that almost every discipline can see its ecological background. Thirdly, thanks to participation of students in environmental initiatives, as well as a wider immersion in environmental communications, they become interested not only in protecting the environment, maintaining environmental safety, but also in the socio-political component of this process. Namely, they can understand how to defend environmental rights in parliamentary-type structures, take part in local public hearings on environmental issues, and join youth parliaments. It is known that environmental volunteering leads many participants in these actions to understand the importance of political advocacy of this issue within any political party or social movement, primarily environmental.

Undoubtedly, the educational process creates the greatest opportunities for development of environmental communications in higher education. In recent years, new areas of training have emerged in higher education, reflecting the need of the modern world, and even more of the world of the future, for environmentally qualified specialists. Environmental topics are also firmly embedded in the content of many disciplines. For example, the Moscow Automobile and Highway State Technical University has an Energy and Ecology Faculty (EEF), where there is a direction of training "Technosphere Safety" in the bachelor's profile "Environmental Engineering " and curricula for the training of masters "Engineering Road Transport Complex" and "Sustainable Urban Transport". Corresponding member of the Russian Academy of Sciences (RAS) V.N. Lukanin is the initiator and head of the state scientific and technical program "Environmentally Friendly Car".

Most of the disciplines taught by the teachers of the Department of Sociology and Management of MADI include environmental topics in their content. Classes in these subjects contribute to formation of environmental competencies of students. [9, 10] As an example, we can refer to the problems of the course "Corporate Social Responsibility", which analyzes the principles of environmental management and the prospects for transition to a "green economy". [11] Students' attention is drawn to the model of environmental, social and corporate governance (eng. *Environmental, Social, and Corporate Governance, ESG*), which is a unity of management approaches aimed at maximizing involvement of company organizations in not only solving environmental, social and management problems that meet their commercial interests, but also the needs of local communities, as well as the environment, generally. In the classroom, students analyze this problem using examples of the functioning of the honey lines of the world's largest auto concerns (*Rolls-Royce, Bentley, Porsche*), which, having taken care of a sharp decline in the world's bee population, have created their own apiaries. Students also analyze the experience of the most successful startups (*BeeHero* and *Edete* [Israel], *Pollenity* [Bulgaria]), which, by developing and implementing various projects, help to combat the decline in the global population of bees, but also show examples of a responsible attitude to nature.

It is known that, despite the increased environmental responsibility of enterprises, until now it has not been possible to completely prevent the risks of man-made accidents causing enormous harm to nature. Therefore, in order to improve the environmental competence of students, and to understand the features of environmental communications in critical situations associated with natural disasters and man-made disasters, in a business game, students present projects of Russian companies related to elimination of the consequences of such emergencies. For example, students submit a report on the actions of the Russian mining and metallurgical company PJSC Norilsk Nickel to eliminate a diesel leak that occurred in May 2020 when a diesel fuel tank was depressurized at TPP-3 in Kayerkan (Norilsk region). This leak of oil products has created a threat to the ecosystem of the Arctic Ocean. Experience of Rosatom's actions to prevent an environmental disaster in Usolye-Sibirskoye city in the Irkutsk region, where one of the largest enterprises in Siberia, Usoliekhimprom, operated and after its closure left the environment poisoned by toxic chemicals is also interesting. Rosatom is the builder of enterprises for disposal of industrial waste throughout Russia, and an ecotechnopark is planned in the territory of Usoliekhimprom, which will also be involved in the waste processing process. The last example is most interesting from the point of view of the formation of students' environmental communications in that it makes it possible to see how such communications can be built not only horizontally between organizations and territories, but also vertically.

Based on the materials related to this environmental problem, it is possible to compare how such communications were carried out in the past, are being carried out now and can be organized in the future, namely, with the greater use of information technologies. The attention of students in the course "Information resources and technologies in economics and management" is drawn to the gigantic possibilities of such communications.

In "Fundamentals of Personnel Safety" course the most important part is also the problem of environmental safety. This topic is raised in various forms of educational work, lectures, case studies, educational quizzes and business games. In the latter, namely, GIS technologies are used that help students to map objects mentioned in game plots, which also expands the environmental competence of students - future road engineers [12].

Experience of the teachers of Russian higher education allows us to assert that there are practically no academic subjects, the content of which would not allow use of training sessions to advance through the channels of environmental communications. Huge opportunities for such work open up when studying foreign languages. Presence of such opportunities is explained by combination of environmental and linguistic communications in the educational process. Students highlight the study of literature in foreign languages on environmental culture and the study of foreign languages itself as significant factors that play a decisive role in formation of their individual environmental culture, considering these areas of environmental communication as a component of intercultural communication. [13, p. 47 - 74]

Currently, an increasing number of university students can be attracted to participate in implementation of international environmental projects, and absolutely everyone who wants to — in local environmental actions in their higher educational institution and in the local community. Ecological tourism is becoming diverse in geography and rich in topics, where students are also involved. For example, from September 16 to September 27, 2013 in the Krasnodar Territory, events were held within the framework of the Student Ecological Tourism project, in which 32 students from the Russian State Social University (RSSU) from Moscow, Klin, Dedovsk, Maikop and Sochi took part. The organizers of this project were the Scientific and Pedagogical Center "Family, Gender, Demography" and the National Public Committee "Russian Family". The tourist direction of environmental communications can also be considered as such a means of military-patriotic education, which helps to form a sense of belonging to the great feats of ancestors, to develop students' interest in search activities.

4 Conclusion

A feature of environmental communications is their transboundary nature. Since the problems of nature are of a global nature, they can only be solved by universal efforts. Therefore, there is an international component in environmental communications. Thanks to the environmental information obtained during classroom work and self-study, the student does not just get acquainted with the environmental traditions of various countries, but with the help of this knowledge they build bridges to the culture and history of states and peoples.

Therefore, environmental education, which is a part of environmental education, opens up space for creation in higher educational institutions of the widest network of environmental communications, through which the student, already in the university period of his/her life, receives an education that meets modern needs, but also acquires the necessary skills to include not only in social life. This shall enable future specialists to solve complex environmental problems, which is an important prerequisite for immersing young people in wider public communications.

References

1. O. Özdemir, H.U. Journal of Education, **43**, 373 (2012)
2. I. Ugulu, M. Sakin, S. Baslar, Int. l J. of Educational Science, **5(4)**, 415 (2013)
3. A. M. Butters, Approaching Religion, **5(2)**, 67 (2015)
4. N. Reitter, J. of Conscious Evolution, **11(5)** (2018)
5. S. Comfort, Y. E. Park, Environmental Communication, **12(7)**, 862 (2018)
6. M. Katz-Kimchi, Environmental Communication, **9**, 367 (2015)
7. M. Katz-Kimchi, B. Goodwin, Environmental Communication, **9(3)**, 367 (2015)
8. R. J. Cox, *Environmental Communication and the Public Sphere* (2010)
9. L. Berényi, Regional formation and Development Studies, **8(3)**, 15 (2014)
10. M. Mulder, J. Gulikers, H. Biemans, R. Wesselink, J. of European Industrial Training, **33(8/9)**, 755 (2009)
11. O. Perfilova, Y. Alizade, (2011), Procedia-Social and Behavioral Sciences, **15**, 2293 (2011)
12. S. Hryshchenko, V. Morkun, Metallurgical and Mining Industry, **4**, 139 (2015)
13. L. M. Andryukhina, N. Yu. Fadeeva, J. De Negri, Education and Science, **19(8)**, 47 (2017)

Overcoming risks in the formation of environmental culture of students

Olga Shefer*, *Svetlana* Kraineva, *Tatyana* Lebedeva, *Vera* Elagina, and *Elena* Erentraut

South Ural State Humanitarian Pedagogical University, 454080 Chelyabinsk, Russia

Abstract. The article presents the result of a study on overcoming risks in the formation of ecological culture among undergraduate students of a pedagogical university. Their readiness and ability to lead an ecological lifestyle and understanding of the essence of sustainable development and environmental policy of Russia, as well as their organization of their professional activities in the formation of the ecological worldview of schoolchildren. It is high-quality education and ecological consciousness that are the basis for sustainable personal and social development, which affects the formation of the ecological culture of all members of society. Therefore, ecological culture is the basis of activities to overcome complex environmental problems (both global and regional) and prevent their occurrence in the future. The study of ways and methods of overcoming risks in the formation of ecological culture is extremely important for Russia. Russia is striving to overcome the negative environmental consequences and to raise people with their professional activities who implement the ideas of sustainable development in life. One of the ways identified by the authors of the study is the organization of work of undergraduate students with online platforms: Stepik, Skillbox, GeekBrains, Coursera, Udemy. Open education is represented by electronic resources functioning at the pedagogical university that form an ecological culture. The article presents the results of a survey to identify the state of environmental culture among students of a pedagogical university and students of schools, leading to the conclusion that it is necessary to involve the ideas of "green" universities in the practice of teaching the younger generation. This will allow us to overcome risks in the formation of ecological culture in a new way in modern conditions of pedagogical universities. It will also contribute to improving the quality of life of the population in the conditions of sustainable development of society. The authors of the article associate the overcoming of most risks in the formation of environmental culture with the creation of conditions for improving the environmental training of a university student. To increase students ' motivation to study by means of solving research environmental-oriented tasks of a professionally applied orientation, the introduction of mathematical methods in solving environmental problems. Students develop mini-projects on environmental topics, prepare course and final qualifying works with environmental content, develop the content of educational practices. This allows you to simulate the professional activity of a teacher

*Corresponding author: shefer-olga@yandex.ru

on the formation of ecological culture among students of schools, the participation of students in the All-Russian ecological quest «Share with us».

1 Introduction

Globalization, the growing power of transnational corporations and the competition of economic markets, the mobility of capital and human resources, the expansion of social and environmental risks are the main challenges in the first quarter of the XXI century [1]. The increase in environmental risks by the beginning of the XXI century was associated with a systematic aggravation of environmental problems (environmental pollution, increased consumption of non-renewable resources, the formation of a planetary anthropogeocenosis, etc.), which allowed us to state not their localization, but the transition to a new stable state – the global environmental crisis. Overcoming the latter is connected with the policy in the field of socio-economic development carried out in the leading countries.

It is known that for the first time the problem of protecting the social and natural environment at the planetary level was considered by the Club of Rome in the early 70s of the twentieth century. The problems of the ecological state of the world are reflected in the works of many domestic and foreign scientists:Y. Burgess, R. Park (the beginning of the direction "social ecology" as a subject area of knowledge), W. Catton, R. Dunlap (paradigmatic understanding of the relationship between humanity and the biosphere), A. Vil, D. Huber, A. Mola, G. Spaargen ("ecological modernization" as an approach), A. T. Bolotov, N. N. Zlatovratsky, G. I. Uspensky (socio-ecological knowledge, culture of the population's attitude to nature as a methodology), V. O. Klyuchevsky, L. M. Mechnikov, N. A. Severtsev (understanding the role of nature as a social background and the potential for the formation of the Russian people and their mentality), P. A. Sorokin (justification of the triad "society-nature-values" and the study of the spiritual and value component of culture as a problem) , etc.

According to the long-term policy in the field of socio-economic development of the Russian Federation, in particular in the field of youth policy, the state pursues the goal of «creating conditions for successful socialization and effective self-realization of young people, developing the potential of young people and using it in the interests of innovative development of the country», including overcoming environmental risks [2]. Among the means to achieve this goal, we can distinguish the educational sphere: where training and education at all levels of education [3; 4; 5].

It is in the educational sphere of Russian universities that environmental programs and projects are implemented. Environmental education in the Russian Federation allows students and graduates of universities to assess the environmental risks that arise in the course of their professional activities and the negative impact of society on the environment, develop appropriate competencies and instill an environmental culture [6].

2 Materials and methods

The term "ecological culture" was defined by L. N. Kogan, a Soviet cultural critic, who noted that «it is culture that allows us to talk about sustainable development, as not about a utopia, but the real prospect of the existence of humanity on Earth» [7]. L. N. Kogan in his works draws attention to the fact that ecological culture is a new type of culture that involves the search for new, optimal ways of interaction between man and nature, without a consumer attitude to the latter.

All components of ecological culture (Fig. 1) determine the conformity of social activity and the laws of natural integrity (in particular, the correspondence of human activity and such quality of the natural environment as its Planetary habitability).

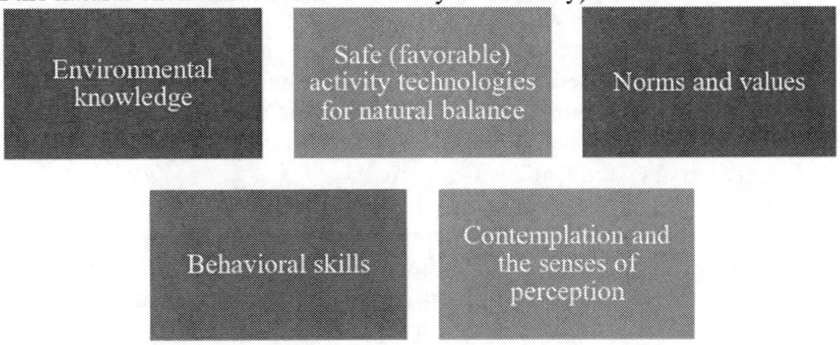

Fig. 1. Components of ecological culture

According to experts, ecological culture reproduces the review that nature is not only a source of resources, natural minerals, but also a place of our habitation and development [8; 9]. In these works, it is proved that the ecological culture of a person is fundamental for the awareness and understanding of problems. Which are caused by human interaction with the environment, that is, environmental problems, as well as in their effective solution.

The concept of ecological culture is extensive, it extends to the entire spectrum of interaction of our civilization and an individual with the environment. It also permeates the entire structure of the personality of the lman. The formation and development of ecological culture is not only the most important problem of the humanitarian sphere, but also the main factor in the development of a student's personality. What is reflected in the dissertation research (S. N. Glazachev, E. V. Girusov, S. D. Deryabo, A. P. Zakhlebny, I. D. Zverev, V. A. Ignatova, D. N. Kavtaradze, B. T. Likhachev, N. M. Mammadov, I. N. Ponomareva, V. P. Saleeva, I. T. Suravegina, G. A. Yagodin, V. A. Yasvin et al.). However, the issues related to the formation of environmental culture among the younger generation, the search for methods and forms of environmental education at the university, are still open. The correction of this situation is connected with the need to solve a number of tasks. This is ensuring the dissemination of modern environmental knowledge, instilling the need for measures to improve the state of the environment, attracting students to identify and understand the environmental problems of their region.

The result of environmental education in universities is the formation of a basic level of moral responsibility for the state of the environment in the process of economic activity. And also the development of a way of attitude to nature, manifested in certain actions of the individual [10]. Environmental education consists in the accumulation of a certain amount of knowledge about the relationship between society and nature. This is necessary for any person, no matter who he works or what he does. Environmental education is the assimilation of norms of behavior in the natural environment and the formation of a civil position and moral conviction in one's attitude to nature [11]. It is based on the principles of sustainable development and a «green» economy.

To overcome the problems in the formation of students in various areas of training internal readiness to overcome environmental hazards [12] and the need to protect the environment in the context of sustainable development and a «green» economy, the Association of Green Universities was established at Russian universities. The Association unites 117 universities across the country (http://https://greenuniversity.ru//).The Association of «green» Universities of Russia is part of the federal partner program "Green Universities of Russia", which is implemented by the All-Russian green Movement EKA and the

Foundation for Support of Youth Initiatives «ERA» (http://xn--b1afaaheyr0d3de.xn--p1ai/). "ERA" is an association of student teams, the purpose of which is the introduction of specific environmental measures and practices on the basis of universities and professional educational institutions in Russia, the exchange of experience and training in methods to reduce the «ecological footprint» of the university. The first participants of the Association were 25 student teams-winners of the All-Russian environmental quest "Share with us" (www.pco.vuzekokvest.Russian Federation), which was implemented by the ERA Foundation with the support of The Coca-Cola Foundation.

The Association of Green Universities promotes:

- Environmental education of schoolchildren and students. Most of the tasks of the All-Russian quests contain activities aimed at forming an ecological culture, a healthy lifestyle and eco-education of children, schoolchildren and students.
- Development of the environmental student movement in Russia. The Association carries out the ideas of ecological sustainable development within the framework of All-Russian quests, attracting teachers as experts and scientific supervisors. Thematic mass events, such as round tables, conferences, exhibitions, festivals, fairs and many others, help to attract students to topical environmental problems and form a team of active volunteers who want to unite, join the Association and change the ecological footprint of humanity on the planet Earth and near-Earth outer space.
- Attracting teachers to participate in the development of the environmental movement in educational institutions. The Association holds online scientific conferences dedicated to topical environmental problems and education in the context of sustainable development and the «green» economy.
- Development of international student interaction and cooperation of environmentally oriented student and youth associations. The Association is engaged in the development of international cooperation, actively participates in various international events, internships and interacts with universities in Europe and the CIS countries.

In order to interest university students in environmental conservation issues, to involve them in responsible consumption of the planet's resources, to help in the implementation of student environmental initiatives, the university teacher, according to V. A. Simonenkova, V. S. Simonenkov, S. R. Gilasieva [13], it is necessary to be competent in the field of environmental education and be able to present the necessary information to the university community qualitatively. In 2019, a project was launched for teachers and employees of higher educational institutions – the School of mentors of "green" universities. The aim of the project is to help participants gain the knowledge and competencies necessary for the implementation of environmental practices and initiatives in universities, and to share tools for productive interaction with students and exchange of experience with colleagues.

The purpose of the program «Green Universities of Russia» is the formation of ecological thinking in the university community. The program includes:

The educational part is the formation of an ecological culture and skills of an ecological life among students and employees of universities.

The practical part is the introduction of specific «green» practices in universities: responsible waste management, measures for water conservation and energy efficiency, landscaping, eco-transport, responsible procurement, eco-education.

The Association of «green» universities, like all educational institutions, switched to remote mode in the realities of 2020, mastered zoom and other Internet platforms, participated in a huge number of online meetings «while at home». This contributed to the integration of information technologies with other technologies, including environmental ones, in order to achieve greater efficiency and quality of professional activity, including education [14; 15].

Taking into account the requirements of the Federal Law "On Education in the Russian Federation", federal state educational standards, each educational institution must provide students with access to an electronic library system, electronic educational resources. As an electronic library system, our university uses the EBS "IPRbooks", which focuses on both fundamental and popular scientific, methodological literature on environmental education. Electronic educational resources are one of the components of the electronic information and educational environment of the university, allowing students to refer to the regulatory and legal documentation regulating the organization of the educational process of the university, data on the structural organization of the university, news, to the content of working programs of disciplines, methodological content of course courses from the personal account. Within the framework of the development of an individual educational trajectory, improving professional competencies [16; 17], distance learning organizations students can independently master the disciplines of the main professional educational program based on the use of online platforms: Stepik, Skillbox, GeekBrains, Coursera, Udemy, Open Education, etc.

The Open Education platform contains a huge number of courses and programs from the leading universities of the country (Lomonosov Moscow State University, MEPhI, Polytech, St. Petersburg State University, UrFU, NUST MISiS b, SPbSETU LETI) related to ecology: Radiation Ecology, Ecology and the environment, Human Ecology, Ecology, Basic concepts of biology and Ecology, System Dynamics of Sustainable Development (System Ecology), Introduction to Biology and Ecology, Nuclear Physics, Biophysics, Concepts of Modern natural sciences, etc.

The list of courses presented on the Coursera platform is shown in Figure 2.

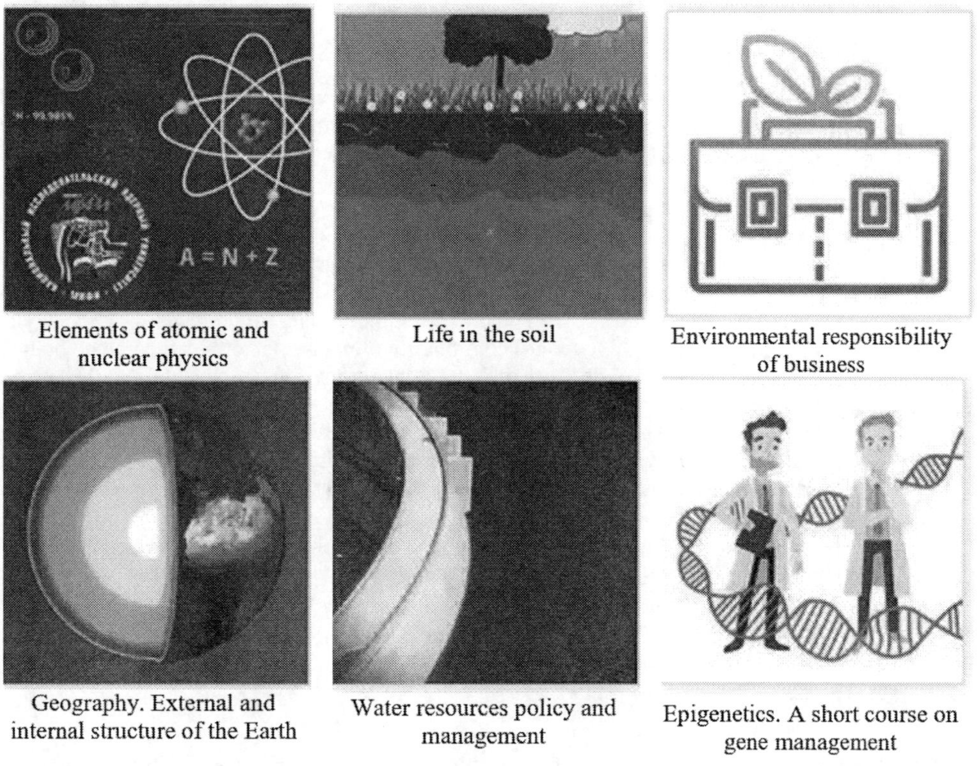

Fig. 2. The list of courses on ecology from Coursera

The analysis of courses for students that contribute to the formation of an ecological culture is presented in Table 1.

Table 1. Electronic resources forming ecological culture

Platform	Course Name	Features
Stepik	Radiation ecology	The purpose and objective of the online course is to create conditions for studying the theoretical foundations and practical applications of radiation ecology of natural and agricultural ecosystems.
	Ecology and the environment	The purpose and objective of the online course is: 1) formation of an objective understanding of the basic concepts of ecology and environmental protection problems; 2) formation of knowledge that forms the basis of autecology, demecology, synecology, about the main environmental problems in the modern world, about the structure, composition and mineral resources of the lithosphere of the planet Earth, about environmental problems of their development, about the principles of rational use and protection of natural waters, about the types of anthropogenic impact, including chemical and radiation pollution of the environment, the main pollutants and the dangers of these types of impact; 3) study of basic concepts about the ecological functions of the soil cover and the basic laws of its formation, 4) study of the basic concepts of the structure, functions and basic laws of the formation and evolution of the biosphere; 5) getting ideas about the current state of the regulatory framework in the field of environmental protection, rational use of natural resources and ensuring environmental safety, the role and tasks of environmental monitoring in ensuring a favorable human habitat and preserving the quality of the environment, about the features of the environmental situation in various regions, the main environmental problems and ways to solve them.
	Human ecology	The purpose and objective of the online course is to form an objective understanding of the basic concepts of ecology and environmental protection problems among students through the development of the discipline «Human Ecology», consisting of four modules: «Human ecology as a new interdisciplinary scientific direction», «Ecological physiology», «The influence of natural factors on the human body», «The influence of anthropogenic environmental factors on the human body».
Udemy	Introduction to the technology Blockchain	The purpose and objective of the online course is to form an objective idea of ecology in students from the point of view of the culture of energy consumption by mastering the topic «Energy consumption and ecology».
	Ten strong goals	The purpose and objective of the online course is to form students ' objective understanding of ecology from the point of view of culture in achieving the goal of ecology by mastering the topic «Ecology of the goal».

3 Results

Many scientists associate the deterioration of the environmental situation, which is especially relevant for such an industrial city as Chelyabinsk, with a low level of environmental culture of society as a whole. This is confirmed by the results of our research conducted on the basis of the South Ural State Humanitarian and Pedagogical University and educational partner organizations of general education (table 2). 150 people took part in this study using the methods proposed by S. S. Kashlev and S. L. Glazachev [18]. 4th-year students – future teachers of physics, mathematics, computer science, primary school teachers and secondary school teachers, show a low level of environmental knowledge, insufficient understanding of existing environmental problems at the level of the city, region, region, country. Their knowledge is fragmentary and limited to the level of individual disciplines. Therefore, unfortunately, they are not motivated to environmental protection activities, because the younger generation has formed only a consumer attitude to the environment, there are no systemic mechanisms for overcoming negative impacts on environmental safety, sustainable development, which reflects negative trends in the general context of the development of society. Only a small part of the students of secondary educational institutions (30 %) understand the environmental issues of the country, region, region, city, due to participation in mandatory project activities.

Table 2. Analysis of the answers to the questions of the questionnaire «Ecological culture»

N0	**Question**	**Survey participants**	**The answers, %**		
			Yes	**No**	**I find it difficult to answer**
1	Are you concerned about the environmental situation in the country, the world?	students	32	29	39
		schoolchildren	41	12	47
2	Do you think about environmental problems in everyday life?	students	69	13	18
		schoolchildren	57	19	24
3	Can you influence the solution of environmental problems in your city (or locality)?	students	21	68	11
		schoolchildren			
4	Do you think it is right and necessary to involve young people in the work of cleaning up the natural environment?	students	22	57	21
		schoolchildren	33	48	19
5	Do you think it is possible to improve the environmental situation in the country through joint efforts of citizens?	students	82	17	1
		schoolchildren	86	12	2

A special feature of the study is not only the choice of one of the three answers for respondents (yes, no, I find it difficult to answer), but also the ability for respondents to provide an explanation for each question in a detailed form. The analysis of the detailed answers shows that when disclosing the second question, the respondents named such problems of environmental problems of the city (or locality): air pollution by emissions from factories, car exhaust gases (92%), water and solid waste pollution (62%). 74 % believe that organizations responsible for environmental control and protection can solve these problems, 69% – city residents, 62 % – local authorities. 1 person (a student of the school) offered his own answer (2%) - joint efforts of local authorities and organizations responsible for environmental control and protection.

When answering the question «Can you influence the solution of environmental problems in your city (or locality)?», unfortunately, many respondents lead a passive lifestyle. At the same time, 81% suggest not to litter themselves, 69% – to fight environmental pollution by enterprises, transport, 47% – to inform others (relatives, friends) about the harmfulness of urban pollution. 4% believe that it is impossible to keep the city clean, since it is a city of various enterprises (JSC «Chelyabinsk Electrometallurgical Combine», PJSC «Chelyabinsk Metallurgical Combine», JSC «Chelyabinsk Zinc Plant», JSC «Ferroalloy», LLC Chelyabinsk Paint and Varnish Plant «Fest Pro»).Thus, the respondents understand the importance of maintaining an ecological lifestyle, which can be based regardless of the motivation-helping the environment or improving their own health.

Despite the opinions expressed on the previous question, the following results were obtained for the fourth question:

- 57% take care of plants, animals, feed birds in winter, build feeders, etc.; 26% - attend subbotniks and engage in ecological tourism;
- 8% participate in environmental movements;
- 7% organize separate garbage collection at home and hand over waste by type to reception points. At the same time, the percentage of respondents with the above answers is higher among school students than among students;
- 11% do not take any part in this.

By asking the last question, we wanted to find out measures to improve the environmental situation of our country. Thus, 82% of all students and 90% of schoolchildren are acutely concerned about global environmental problems, 18% and 10% are not concerned at all, respectively. Among the global problems known to them, the respondents identified: the quality of drinking water, deforestation, air pollution and destruction of the ozone layer, garbage accumulation, viruses and a pandemic.

The results of the survey cause concern that there is reason to think about the lack of readiness of the younger generation to take full responsibility for the environmental future of their country.

The level of development of ecological culture is becoming one of the most important indicators of the maturity and progressiveness of the social system in modern conditions [19; 20; 21].

4 Conclusion

The environmental policy in our country is reflected in the Law «On Environmental Protection» [22], which, in particular, states: «In order to improve the ecological culture of society and professional training of specialists, a system of universal, integrated and continuous environmental education and education is established, covering the entire process of preschool, school education and education, primary vocational and university education, improving their qualifications...». It was this law that became the starting point for overcoming risks in the formation of an ecological culture of students at all levels of education. Further movement in the direction of overcoming these risks is associated with federal state educational standards for all levels of general education – the mandatory implementation of a student project, the topics of which are directly related to the problems of the ecology of the native land. As evidenced by the data of our study. Fourth-year students studying at the school were not involved in mandatory project activities. And when studying subject disciplines in junior courses at the university, little attention was paid to the formation of ecological culture. This is one of the main risks for the generation of twenty-year-olds in the formation of their ecological culture.

Overcoming risks in the formation of environmental culture among future teachers of physics, mathematics, computer science, primary school teachers is achieved by using them in the educational process:

- research environmental-oriented tasks of professional and applied orientation;
- mathematical methods (processing of results, modeling) when solving environmental problems;
- mini-projects on environmental topics (development, registration and protection);
- topics with environmental content for course and final qualifying works;
- educational practices that allow simulating the professional activity of a teacher to form an ecological culture among students of schools;

–the experience of students 'participation in the All-Russian ecological quest "Share with us";

- the possibilities of electronic resources that form an ecological culture;
- opportunities of online platforms: Stepik, Skillbox, GeekBrains, Coursera, Udemy, Open education, etc.;
- ideas of sustainable development of society.

The full implementation of the above-described directions in the practice of university education increases the possibility of forming an environmentally oriented lifestyle among the younger generation, which acts as one of the components of the high quality of life of the population and as a consequence of the sustainable development of society.

References

1. B. Talukder, R. Matthew, G. W. vanLoon, M. J. Bunch, K. W. Hipel, J. Orbinski, Current Opinion in Environmental Sustainability, **50**, 98-108 (2021)
2. Free Online "ConsultantPlus Legal Reference System".Decree of the Government of the Russian Federation No. 1662-r of 17.11.2008 (ed. of 28.09.2018) "On the Concept of long-term socio-economic development of the Russian Federation for the period up to 2020", http://www.consultant.ru/
3. E. O. Kazakova, S. P. Kulikov, S. V. Novikov, Humanities, socio-economic and social sciences, **11**, 35-39 (2017)
4. S. P. Kulikov, S. V. Novikov, N. V. Prosvirina, A. E. Sorokin, Regional problems of economic transformation, **10**, 76-83 (2018)
5. A. P. Anisimov, Ethics in Science and Environmental Politics, **19**, 13-19 (2019)
6. V. A. Semykin, Bulletin of the Kursk State Agricultural Academy, **8**, 157-159 (2015)
7. L. N. Kogan, Personality. Culture. Society. Selected works 1961-1987 (Yekaterinburg: CPI "Mask") 323 (2009).
8. A. V. Ivashchenko, A. V. Gagarin, S. A. Stepanov, Bulletin of the Moscow State University for the Humanities named after M. Sholokhov. Social and Environmental Technologies, **1**, 58-67 (2012)
9. J. Papenfuss, E. Merritt, Sustainability (Switzerland), **11(14)**, 3880 (2019)
10. D. G. Silantiev, E. I. Sharova, M. N. Kurochkina, F. M. Ordokova, G. A. Kyarova, International Journal of Educational Sciences, **31**, 1-3, 61-66 (2020).
11. E. A. Masasina, M. A. Timokhina, Improving the educational process in a changing environment. Kurgan: Kurgan State Agricultural Academy named after T. S. Maltsev (Lesnikovo), 103-106 (2021)
12. O. R. Shefer, T. N. Lebedeva, M. V. Goryunova, Espacios, **39(52)**, 14 (2018)
13. V. A. Simonenkova, V. S. Simonenkov, S. R. Gilasieva, Environmental Problems of Large River Basins, **818**, 170740 (2021)
14. Y. Grishaeva, A. Gagarin, I. Spirin, O. Matantseva, T. Mitrofanova, K. Vishnevskaya, 1st International Conference on Technology Enhanced Learning in Higher Education, TELE 2021, 9482753, 268-273 (2021)
15. A. Chakir, M. Chergui, J. F. Andry, Environment, Development and Sustainability, **23(6)**, 8857-8871 (2021)
16. I. Balandina, T. Lebedeva, A. Milyutina, T. Moskvitina, O. Shefer, G. Shiganova, L. Yuzdova, ICERI2020 (13th annual International Conference of Education, Research and Innovation), 9644–9649 (2020)

17. S. V. Kraineva, O. R. Shefer, Scientific and Technical Information Processing, **44(2)**, 94-98 (2017)
18. S. S. Kashlev, S. L. Glazachev, *Pedagogical diagnostics of ecological culture of students*, 94 (Moscow: Horizon, 2000)
19. I. A. Voedilova, Astrakhan Bulletin of Environmental Education, **1**, 90-92 (2012)
20. G. E. Ivanovna, E. Yu. Ivanova, K. I. Sergeevna, K. A. Vladlenovich, International Journal of Civil Engineering and Technology, **9(13)**, 1211-1222 (2018)
21. O. Shefer, S. Kraineva, T. Lebedeva, Ural Environmental Science Forum "Sustainable Development of Industrial Region" (UESF-2021), E3S Web Conf, 258, 10004 (2021)
22. FreeOnlineSKB "Kontur".Russian Federation. Federal Law on Environmental Protection No. 342-FZ of 02.07.2021, https://normativ.kontur.ru/

Formation of the foundations of ecological culture among preschoolers by means of art

O. P. Radynova[1,*], *I. V.* Gruzdova[2], and *I. A.* Korsakova[1]

[1]Moscow State Institute of Music named after A.G. Schnittke, Moscow, Russia
[2]Togliatti State University, Togliatti, Russia

Abstract. The article reveals the problem of forming the foundations of ecological culture in preschool children by means of musical art. Formation of the foundations of ecological culture occurs not only on the basis of accumulation of natural history knowledge, in preschool childhood this process proceeds more efficiently if it is based on sensory cognition of the beauty and harmony of nature, the motives of a respectful attitude towards nature crystallize in the experience of emotional and value experiences. The article authors consider the mechanisms of formation of a child's ecological culture in the process of perceiving the artistic and figurative content of music. Sensual experience as the basis of cognition in art makes it possible to form a system of value experiences, including those associated with the child's attitude to animate and inanimate nature. Emotional responsiveness to the beauty of music, works of art reflecting images of nature is associated with responsiveness in life. The need to protect, preserve and increase the wealth of native nature can be brought up in the experience of communicating with music, if the child experiences the content of the musical image as a personally significant event.

1 Introduction

The problem of ecology on a global scale today attracts the attention of scientists from different countries. The search and development of the latest technologies are underway to protect the planet from constant destructive tendencies as a result of the phenomena of technological progress and the inconsistency of development of the culture of the human personal factor for humane communication with nature.

The definition of ecological culture is provided in the concept of its formation that was developed by B.T. Likhachev, N.S. Dezhnikova, I.V. Tsvetkova and a number of other scientists. In this concept, ecological culture is understood as a system of special knowledge, as well as assimilation of methods of activity focused on compatibility with nature. According to the concept, ecological culture is characterized by emotional responsiveness, a sense of responsibility for the environment, it is based on the desire to live in harmony with it, on a person's strict observance of legal laws [1].

* Corresponding author: olgaradynova@mail.ru

It is very valuable that this definition includes both knowledge, activity-humane components, and emotional-evaluative (emotional responsiveness), and strong-willed (personal quality of responsibility for preservation of the natural environment).

Many studies have been carried out in this "coordinate system", including in the system of vocational education [2].

2 Materials and Methods

In preschool age, the child's connection is established with the spheres of life that are leading for the personality: with the world of people and nature. The fundamentals of ecological culture are brought up in preschool childhood as a result of accumulation of positive experience of communication between a child and nature, an emotional assessment of its beauty and a variety of natural phenomena, formation of a conscious emotional-evaluative, value attitude towards the natural world and the animal world.

In modern standards of preschool education (as well as in the standards of elementary school), among the tasks of the educational system, the following are named: "Ensuring the continuity of goals, objectives and content of education implemented in the framework of educational programs at various levels"; "Development of a general culture of personality" [3].

Development of the foundations of ecological culture in preschool age is one of the most important tasks of upbringing a general culture of the individual and involves formation of moral and aesthetic consciousness (aesthetic emotions, interests, ideas about beauty in nature, art and in life and moral ideas and concepts about behavior, actions, good and evil).

The preschooler receives the first ecological representations from fairy tales, works of folklore (nursery rhymes, jokes, chants), observations of natural phenomena, listening to music, drawing, modeling, moulding, making crafts from natural material, as well as in the process of creative and didactic games. Integration of the impressions received by the child from different sources most fully reveals to the preschooler the ecological picture of the world in the field of available knowledge about nature and respect for it. Wherein, the experience of a humane value attitude towards nature, obtained at the stage of preschool childhood, can, under certain conditions, become basic in formation of the ecological culture components. This is due to the fact that vivid emotional and value experiences received in childhood have a profound impact on the psyche and personal development of the person [4].

In the study of topical problems of preschool pedagogy, works devoted to environmental education are provided in a rather diverse way. The concepts of "the basis of the preschooler ecological culture", "ecological literacy" are considered, the effective role of play in the formation of spiritual and moral consciousness, an emotionally-colored caring attitude of a child to nature is proved.

Formation of the ecological culture of preschool children is considered as the dominant goal of upbringing in S.N. Nikolaeva "Young Ecologist", N.A. Ryzhova "Our Home is Nature", T.A. Koptseva "Nature and the Artist". The culturological approach implemented in the author's programs of ecological education of preschoolers presupposes not so much the assimilation of knowledge about animals and plants by children, but development of the ability to see the beauty and harmony of the rich and diverse world of nature, take care of all life on Earth, take care of the preservation of nature, observe ecological rules of behavior in nature.

The ecological culture of a preschool child is interpreted by N.A. Ryzhova [5] from the standpoint of the moral aspect of personality development that implies the education of ideas about the intrinsic value of nature, an emotional positive attitude towards it, as well as development of primary skills for environmentally competent and safe behavior in nature and in everyday life.

In the theory of preschool education, a search is carried out for effective pedagogical technologies, conditions and methods that contribute to formation of the foundations of ecological culture in preschool children. It is proved that creation of a developing subject-spatial environment, accumulation of observation experience, communication with nature, the enrichment of environmental knowledge, aesthetic and ethical feelings and experiences, interaction with parents contribute to this process (N.F. Vinogradova [6], I.D. Zverev [7], I.I. Komarov [8], N.N. Kondratyev [9], P.G. Samorukova [10], S.N. Nikolaeva [11], N.A. Ryzhov [12] and others).

Consideration by researchers of the general structure of a personality's worldview based on knowledge also presupposes the formation of a value attitude towards the world (including the natural world) in the process of artistic education. Here, the role of the emotional background is played by a variety of works of art used as a means of bringing natural science knowledge to life. The deep foundations of art are little used as a subject of research in the science of education and in the practice of working with preschool children in the field of forming a child's ecological culture.

Images of art are included in the programs of environmental education of children, but technologies that implement artistic and practical activities in introducing preschoolers to eco-cultural values in reliance on sensory, artistic and imaginative perception and experience of art, formation of a positive emotional-evaluative, value conscious attitude and transferring it to the natural world. The opinion of scientists is quite widespread that it is precisely environmental education that aims at formation of an environmental culture, considered as a harmonious, scientifically grounded interaction between men and the environment. In this case, environmental education is a system of scientific knowledge about human interaction with the environment, about the norms and rules of behavior in a socio-natural environment, as well as a system of practical skills and abilities to preserve favorable environmental conditions for life (A.N. Zakhlebny, E.N. Dzyatkovskaya [13], D.S. Ermakov, I.D. Zverev, T.I. Suravegina [14] and et. al).

The possibilities of art (music, fine arts, poetry, musical theater) in formation of the foundations of the ecological culture of preschool children have not been sufficiently studied. Works of art are included in the programs, but outside the deep immersion in their content. Music plays the role of only an emotional background (N.G. Kuprina).

In this regard, the dissertation of N.G. Kuprina "Ecological and Aesthetic Approach in the Artistic Education of Children" [15] that examines the pedagogical concept of artistic education of children on the basis of the ecological and aesthetic approach. The work is carried out in reliance on philosophical, culturological, art criticism theories of culture and considers the ecological and aesthetic approach as formation of a person's attitude to nature as a spiritual and aesthetic value.

The dissertation shows the effectiveness of the method of emotional-activity experience by children of ecological and aesthetic values in polyartic play activity, emphasizes the role of artistic education of children in the formation of a person's value relationship to nature.

In another work, dissertation of V. G. Kezin "Formation of the ecological culture of the personality of younger adolescents in the pedagogical process of the school by means of musical art" [16], the problem of the formation of ecological culture is associated with the possibilities of music in this process. However, it was carried out in a different age category – the adolescent period of childhood and considers the possibilities of using music in the lessons of natural history, biology and geography - forms of arranging the cognitive activity that are not used in preschool childhood. The experimental part of the study presents lessons that include both traditional didactic methods and innovative ones. For example, the method of "artistic representation of natural objects" from environmental pedagogy

(S.D. Deryabo, V.A. Yasvin), "the method of thinking about music" (D.B. Kabalevsky, L.V. Goryunova), the method of "comprehending the personal meaning of a musical work" (A. Pelichauskas) from musical pedagogy. Conclusions of the author of the thesis are that musical art combined with the comprehension of scientific knowledge of natural history, influencing the development of abstract theoretical thinking, can form a holistic worldview, a holistic personality and its ecological culture.

3 Results and Discussion

Returning to the preschool age of children, we emphasize that the feeling of beauty, the foundations of musical, artistic, moral-aesthetic, ecological, general culture, are brought up in a child from early childhood, to a great extent - by means of art. The images of nature in these types of art are presented through the feelings of the composer, artist, poet, who inspired him/her to create artistic images that express the experiences that have arisen from the aesthetic perception of their beauty in life. These artistic images were created in the process of creativity of talented people at different times, in different countries, and have come down to us as values of artistic culture, affecting our perception through the prism of their consciousness, sharpening the expressive features of the artistic language of the created images, enhancing their impact on our impressions.

L.S. Vygotsky repeatedly in his works draws attention to "the issue of the practical life action of art." Emphasizing the idea that art at all times was considered as a part and a means of education, the scientist argued that it acts as "the organization of our behavior for the future, a forward orientation" [17]. L.S. Vygotsky emphasized that perception of works of art contributes to the personal development of the child - empathy, thinking and speech, imagination and creativity, and his/her cultural development.

Different types of the art have their own means of expressing content, their own capabilities for the personal development of the child. In music, these are sounds, melodies, harmonies, which in different ways express the change of images and moods. When emphasizing the power of the influence of music on development of the emotional sphere of the child, L.S. Vygotsky wrote that "the expansion and deepening of feeling, its creative restructuring, is the psychological basis of the art of music" [18].

To carry out emotional-evaluative creative activity, a child needs a "dictionary of emotions", with the help of which he/she can express his/her attitude, helping to recognize mood, feelings, their change and variety in the content of the work. This dictionary is practically absent in children, but it is important to form it, based on the use of these words-images when characterizing the content of music, fine art, poetry, natural phenomena, accompanying the explanation of their meanings with examples from life, performance of fragments of musical compositions contrasting in content.

The well-known Russian psychologist, B.M. Teplov, noted that music can depict natural phenomena, specific images of reality, but at the same time it always expresses "feelings, emotions, moods", which is its content [19]. For example, one of the parts of the sixth "Pastoral" symphony by L.V. Beethoven is called "The Thunderstorm". Bass drums sound, timpani roar and flutes whistle depicting the flashing of lightning and thunder are heard, but this music (and this is important!) conveyed feelings of anxiety, anxiety, consonant with the picture of bad weather.

If the music made an impression on the child, he/she emotionally reacts to it (shows attention, concentrates, or, conversely, "makes" the rhythm of the music, involuntarily moves, barely noticeable fine motor movements or real movements are observed) [20]. These emotional responses are very important for experiencing and understanding musical imagery. L.S. Vygotsky wrote that "Art is the work of thought, but of a very special emotional thinking" [21].

Repetitive emotional reactions to music that convey images of nature, animals and birds, fairy-tale images and their awareness contribute to manifestations of persistent interest, the birth of an aesthetic attitude to music and the natural world, testifying to development of the foundations of musical, as well as ecological culture. If a child examines pictures with interest, depicting natural phenomena, the animal world, he/she accumulates experience in the value artistic perception of painting, and, at the same time, the experience of ecological representations formed on the perception of artistic images of art; first of all, emotional responsiveness and assessments appear, which are gradually interpreted as a positive emotional and evaluative attitude to the world of ecology and to the art of painting that is expressed in the need for drawing, creative independent manifestations are born, which indicates formation of the foundations of artistic and ecological culture.

The artistic perception of poetic images of nature, images of animals in fairy tales develops in a similar way, especially if they are sounded by classical music (medeclomation, musical fairy tale games). These impressions are also aimed at formation of artistic and ecological culture in children, formation of moral and aesthetic consciousness, love and interest in works of art and real objects of the natural world.

Analysis of philosophical, cultural, psychological, art history, pedagogical research (M.S. Kagan, L.S. Vygotsky, N.A. Vetlugina, B.T. Likhachev, E.V. Nazaikinsky, B.M. Teplov) makes it possible to single out the following components as mechanisms for formation of the foundations of musical, artistic, moral-aesthetic, ecological, general culture, in their close interrelationships in preschool children:

- accumulation of experience of aesthetic perception of nature, cultural values and its transfer to life practice;
- positive emotional assessment, emotional responsiveness;
- awareness of their impressions, manifestation of an emotional, evaluative, value attitude;
- manifestation of creative activity, creative imagination in various types of artistic and aesthetic activities;
- transfer of accumulated ideas into life practice.

If a child empathizes with a work of art, feels the beauty of nature, he/she is able to express his/her impressions in words, creative musical, artistic activity, accumulates the experience of emotional responsiveness to beauty, making it his/her cultural heritage in various fields of culture (musical - artistic - moral and aesthetic - ecological - general).

O.P. Radynova carried out an experimental study based on the Musical Masterpieces program she created [22], in which the developed method of developing the musical perception of 6-7 years old children was tested through the characteristics of mood changes in the work using figurative speech, the "dictionary of emotions". Methods of contrasting comparisons, assimilation to the character of music (verbal, motor, polyartistic, intonational, instrumental), the explanation of new word-images using examples from the life of children, methods of dialogue, collective creativity, play methods - a musical fairy tale game, and etc.) were used.

It has been proved that distinguishing the change of feelings and moods expressed in a work leads to an understanding of its *meaning* - the "sensory program of a musical work", expands figurative speech, a vocabulary of emotions and enriches children's ideas about human feelings that exist in real life, manifestations of emotional responsiveness to phenomena of nature, ecological views.

The program is aimed at formation of a value concept of the world of nature and animals, which is carried out using emotionally colored, different in content, contrasting in moods works of musical classics.

The Musical Masterpieces program presents the themes "Nature and Music", "Music Tells About Animals and Birds", "A Tale in Music", "Moods, Feelings in Music".

Since culture is "appropriated" by a person, if cognitive activity is based on value content and is associated with creative activity and communication (M.S. Kagan), a system of attitudes was necessary to implement this approach:

- an *axiological* attitude aimed at development of emotional responsiveness, aesthetic emotions in the process of perceiving the values of artistic culture;
- a *cognitive* attitude, aimed at comprehending, distinguishing and understanding the beauty of artistic images of nature in art and their connection with life;
- a *creative* attitude aimed at expressing emotionally conscious impressions in creative activity - a figurative word, musical and rhythmic improvisations, composing their own melodies in the character of a value standard, orchestration, musical theatrical activities, drawing images of nature, and etc.

The Russkoe Slovo Publishing House has published the Program of O.P. Radynova "A child in the world of beauty" [23] and the methodological manual of the authors - O.P. Radynova, N.V. Barysheva and Yu.V. Panova "We Listen and Draw Music": four albums for children 3-7 years old, in which the theme "Nature" is widely represented [24]. The albums contain reproductions of paintings by great artists, as well as musical fragments of classical works, and poems that are close to them in emotional content. A short conversation about these works and the task of the child to draw his/her picture of nature under the impression of high standards of art is offered.

L.S. Vygotsky emphasized the most important regularity - the influence of the emotional factor on the combining fantasy, which psychologists call the law of the general emotional sign. "The essence of this law boils down to the fact that impressions or images having a common emotional sign, i.e. those that produce a similar emotional impact on us tend to unite with each other" [25]. This law is directly related to the possibility of integrating arts based on the proximity of emotional content, as well as the transfer of vivid artistic impressions to life phenomena.

Experimental study has shown that children began to transfer the experience of perceiving music to life situations gradually, their ideas about human feelings that exist in real life expanded, the "dictionary of emotions" was replenished, and a persistent interest in music and visual activity appeared. Children drew elements of scenery, costumes, in preparation for matinees, expressively pronounced the lines of their characters in musical fairy tales.

Polyartistic assimilation was also used, with the leading role of music. Children drew musical images, played musical fairy tale games, voiced by musical images similar in character to the characters of the fairy tale, musical and theatrical activities were used at festive matinees.

L.S. Vygotsky wrote: "If music does not directly dictate the actions that should follow it, then all the same depends on its main action, on the direction that it gives to psychological catharsis and what forces it will give to life, what it will release, and what will push back into the depths" [26].

It is music, by virtue of the specificity of its emotional content, that has become the core of musical and artistic activity, the integration of arts, in various forms of arranging children's activities - in the classroom, cognitive entertainment, festive matinees.

Here is an example: in works about nature, for example, the song by E. Tilicheyeva "Birch" ("The beauty of the birch tree glows silver") to the words of P. Voronko, it is sung that the goats "began to bend the birch, and the birch went into tears". Tears of a birch tree are amazingly conveyed in the music by slow drops in a high register. This artistic image is humanized and evokes vivid empathy reactions in children. There is a transfer of emotions from the artistic image into life – the emergence of feelings and ideas of the child that the birch is alive, that it must be protected.

This example illustrates the possibilities of art influence on formation of the ecological culture foundations in children, the connection between life and art.

The conducted study limits the possibilities of studying the mechanisms of transferring the emotional-value experience accumulated in the process of perceiving works of art into the activity of mastering ecological behavior in nature. However, the data obtained in the study of music perception by children make it possible to expand the boundaries of study into development of value and activity components in the structure of the ecological culture of preschool children.

4 Conclusions

The philosophical concept of "appropriation" of cultural values by a person is considered in detail in the works of M.S. Kagan. According to the philosopher, the future of mankind does not depend on the level of its education and application technology, but on those "value attitudes that determine the direction of the practical application of knowledge" [27]. The issues of early ecological education are especially relevant at the present stage of human society development. Formation of ecological culture as a value attitude of a person to nature is developed in preschool age, when emotional and value experiences have a personal meaning. Musical art enriches the experience of feeling, the images of nature reflected in music, evoke in children emotional responsiveness to the beauty of consonances that convey fragility and strength, the uniqueness of the harmony of sounds and colors of nature and its phenomena, cause the need to preserve and increase the natural wealth of their native land. Developing in preschool age, the emotional-value content in the structure of the child's ecological culture sets the vector in formation of all its components in the future personality development.

References

1. B. T. Likhachev, Pedagogy, **2**, 19-23 (1993)
2. N. A. Chikildina, *Formation of the ecological culture of schoolchildren in the conditions of the educational and industrial complex*, 190 (2004)
3. Federal State Standards for Preschool Education, 3 (2014)
4. A. V. Zaporozhets, *Emotional development of a preschooler*, 8-29 (1985)
5. N. A. Ryzhova, *Our home is nature: a program of environmental education for preschoolers*, 20 (1998)
6. N. F. Vinogradova, O. A. Rydze, *Implementation of the requirements of the standard in the study of the subject "The world around": from substantive actions to universal ones* (Pedagogical University "September First", 2014)
7. I. D. Zverev, *Ecology in Schooling*, 96 (Knowledge, 1980)
8. T. N. Melnikov, A. S. Moskvina, I. I. Komarova, A. L. Tretyakov, *Ecological education of preschool children*, 80 (MGOU, 2018)
9. N. N. Kondratyeva, *Environmental Education Program for Children "We": The ABC of Ecology*, 140 (1997)
10. P. G. Samorukova, *How to acquaint preschoolers with nature,* 207 ("Education", 1983)
11. S. N. Nikolaeva, *Methods of ecological education of preschoolers*, 184 (Academy, 2001)

12. N. A. Ryzhova, The OMEP Ireland Journal of Early Childhood Studies, **10**, 51–69 (2016)
13. A. N. Zakhlebny, E. N. Dzyatkovskaya, Environmental education: before school, at school, outside of school, **3**, 3-8 (2007)
14. D. S. Ermakov, I. D. Zverev, I. T. Suravegina, Learning to solve environmental problems, 112 (Moscow: Shkolnaya Pressa, 2006)
15. N. G. Kuprina, *Ecological and aesthetic approach in the artistic education of children*, 361 (2008)
16. V. G. Kezin, *Formation of the ecological culture of the personality of younger adolescents in the pedagogical process of the school by means of musical art*, 252 (1998)
17. L. S. Vygotsky, *Psychology of art*, 313-314 (Azbuka, 2016)
18. L. S. Vygotsky, *Imagination and creativity in childhood*, 15 (Soyuz, 1997)
19. B. M. Teplov, *Psychology of musical abilities*, 6 (Nauka, 2003)
20. B. M. Teplov, *Psychology of musical abilities*, 216-217 (Nauka, 2003)
21. L. S. Vygotsky, *Psychology of art*, 61 (Azbuka, 2016)
22. O. P. Radynova, The program "Musical Masterpieces" (a series of books): "Moods, Feelings in Music", "Song, Dance, March", "Music about Animals and Birds", "Nature and Music", "A Tale in Music", "Musical Instruments", M., Shopping Center "Sphere" (2019)
23. O. P. Radynova, "A child in the world of beauty", The program of artistic and aesthetic development of preschoolers (2021)
24. O. P. Radynova, N. V. Barysheva, Yu. V. Panova, *We listen and draw music* (Russkoye Slovo, 2020)
25. L. S. Vygotsky, *Imagination and creativity in childhood*, 13 (SPb.: Soyuz, 1997)
26. L. S. Vygotsky, *Psychology of art*, 313 (Azbuka, 2016)
27. M. S. Kagan, Selected works in seven volumes, **3**, 672 (2010)

Features of the design of social posters of the XXI century and their role in the formation of ecological culture

Elena Vasilenko[1], *Pavel* Vasilenko[1,*], *Valentina* Pallotta[2], *Natalia* Barsukova[3], and Tatyana Sichkar[4]

[1]Federal State Budgetary Educational Institution of Higher Education " Moscow State University of Technology and Management named after K. G. Razumovsky (First Cossack University)"
[2]Moscow international University, Leningradsky prospect, 17, Moscow, Russian Federation
[3]National Design Institute, Russian Federation, Moscow
[4]Institute of World Civilisations, Moscow, Russia

Abstract. The article considers the poster design as a means of broadcasting and its role in the formation of ecological culture. The impact of the development of technological progress on the birth of completely new forms of the poster, which are not limited to one paper medium. The value of a social poster is in its content aspect, conceptual solution, well-chosen palette, composition. The language of graphic design changes along with the situation in society, becomes extremely figurative, exaggerated, metaphorical, symbolism takes the place of verbatim. The main task of the design of a social poster focused on nature conservation is to contribute to the formation of an ecological culture, starting with personal motivation for self-preservation and preservation of the environment, where the knowledge and skills of both each person individually and entire human communities. The purpose of the social poster is to orient the viewer to the commitment to the highest social values - such as humanism, love for people, for life, and nature. The poster is an iconic carrier of information, simple and understandable, which does not require intellectual efforts from the viewer and educates him both morally and aesthetically, influences a person's emotions, his attitude to the environment.

1 Introduction

Today, the topics of environmental responsibility and sustainable development of society are becoming more and more relevant at the state level around the world. Environmental pollution and eco-friendly business and lifestyle are increasingly relevant and urgent for humanity.

In the XXI century, some issues are most acute, and one of them is large — scale environmental problems that lead to serious crises in different parts of the world. The society faces the task of switching to a mode of conscious consumption, saving resources, caring for nature, and for this it is necessary to seriously transform our social and economic

* Corresponding author: pashavasilenko@mail.ru

models of existence. Open to change, able to think critically and ready to act, activists of the environmental movement are already leaders in promoting the culture of careful and rational use of natural resources.

They introduce separate waste collection for further processing, organize environmental clubs and conduct eco-educational work, involve people of different ages and different social strata of society in public activities, learn to build a socially and environmentally responsible business taking into account the achievements of modern science. Greening can become and is becoming one of the indicators of effective life-activity of any society. The use of energy-efficient solutions, economical water supply, rational waste management in the future leads to cost reduction and careful attitude to natural resources. The activity in the implementation of "green" initiatives and projects gives significant advantages. One of the spheres in the formation of ecological culture is mass agitation, which depends on the visualization of socially significant concepts and images for society.

2 Materials and methods

Conducting a comparative analysis and comparison of the features of the design of a social poster, which is aimed at the formation of ecological culture, the trends in the change of poster art and their causes are noted.

Technological progress has contributed to the transformation of poster forms, which are not limited to one paper medium. The analysis of the appearance, creation and development of the poster design in the historical chronological sequence allows us to note its trend of change based on the trends and realities of the time. The latest trends in creating a graphic poster allow you to move from an abstract figurative image to a more concrete one. A comparison of visualization methods used in a social poster in different periods allows us to identify the stages of its development, which are characterized by the influence of political, ideological, cultural and other factors that characterize the era. When studying the current state in the field of graphic poster design, reconstruction of the past, namely the genesis-the emergence of stages of historical heritage, it is possible to identify a forecast of future trends for further development.

3 Results and discussion

The means for the formation of eco-culture are social videos, films, posters. The poster is the most concretizing, capacious, accessible and mass means of communication. The design of modern social posters plays an important role in the educational and cultural sphere of the environmental movement. Using all the possibilities of technological progress and a modern artistic language focused on the style preferences of the XXI century.

Currently, a poster means a type of large-format printed graphics that combines visual and font elements and performs an advertising, propaganda, educational or instructive-methodical function. His predecessors are considered to be propaganda engravings: for example, "flying leaflets" of the Reformation and peasant wars in Germany of the XVI century. The birth of the poster in its current form is associated with the invention of lithography in the late 1700s. The printing process in this way was quite long and expensive, but the artist Jules Cheret (1836-1932), considered one of the founders of the modern poster, after studying the English technology of color lithography, simplifies the technological process itself, which contributes to an increase in the size of circulations and the availability of production.

Throughout its history, the poster has been a reflection of reality, and its content and form have been determined by specific goals and dominant styles in art. So, the development of the poster fell on the modern era. The personalities who made the greatest contribution in this period are Jules Cheret, Henri de Toulouse-Lautrec, Alphonse Mucha. In Russia, these were artists who were part of the World of Arts association, especially Mikhail Vrubel. The poster of the Art Nouveau era was a full-fledged work of art made by masters of their craft, the main purpose of which was advertising [1]. Everything was advertised: from social events, exhibitions and entertainment venues to books, cigarettes and other products. This could be both the use of a pleasant image that catches the viewer's eye (most often – a female one, as in A. Mukha), and a verbatim image of the advertised object. The" fathers " of the advertising poster also outlined its main principles: catchiness (for which bright colors and accents were often used), conciseness, the ability to perceive information "on the go". At the same time, traditional pictorial and graphic means were used, which actually made a full-fledged illustration of the poster. Such posters had no purpose to morally educate society, it was only necessary to correctly submit the goods.

Since the beginning of the XX century, propaganda posters have become popular. The impetus for this was given by the First World War and the October Revolution that followed it, when political propaganda became the main task. The content is no longer a product, but an idea. The genre of the propaganda poster revealed itself most fully and vividly during the Great Patriotic War [2]. The elegance of modernity was replaced by loud slogans calling for a fight, a passionate image of the defender of the Motherland and the image of the enemy opposing him. The posters of this period were extremely straightforward and filled with emotions, their purpose is to raise morale and cause strong feelings. The first such poster was the famous poster of Irakli Toidze "The Motherland is calling!", released a few days after the start of the war and became a kind of personification of the unity of the country.

A social poster in the history of Soviet art calls for certain ideological things. This happened against the background of the domination of socialist realism, which sought to show the country rich, healthy, bursting with energy. The heroes of these posters are ideal citizens of their country, working people, happy families. Much attention was paid to the promotion of sports, a healthy lifestyle, ecology-the preservation of nature, motherhood, parasitism and drunkenness were strongly condemned [3, 4].

The Soviet environmental poster of the 1980s is not just beautiful pictures, they call for a conscious attitude to nature: "Stop, read, think, do what depends on you!". Posters of the 1980s are the heyday of social advertising of the Soviet state. There is no censorship, there are no strict and categorical artistic advice, there are no prohibited topics. An artist can create as he sees fit in order to reach the viewer, to convey what he wants to say. Social advertising calling for the protection of nature appeared in our country already in the sixties, although the topic of ecology was not as popular as a political poster, but it had its own established language, its own understandable images, traditional plots. Among the posters of the Soviet period concerning the topic of ecology, there are also unusual ones - for example, the poster "Alcohol is a pain of nature" by Nikolai Ivanovich Litvinenko-a combination of two relevant topics at once: anti-alcohol propaganda in the framework of the 1985-1990 campaign and environmental protection, the work of Yakov Markovich Zlatopolsky "Less and less of the surrounding nature, more and more of the environment". Posters of graphic artist Mikhail Vsevolodovich Gromyko (in 1989, Uzbekistan) — " Is there a gift like spring water in the world? A drop is like a pearl, reflecting the heat of the sun." These paired posters tell the viewer about important things. We see how much depends on the skill of the artist who creates a bright and memorable image.

Vitaly Yevgenyevich Wolf, a well-known poster artist, spoke about the importance of poster art, that an image is important in a poster, the significance of a metaphor. A metaphor can only be expressed through an image. The main purpose of the poster is to stop the gaze of a random viewer, to interest, to capture his attention, to make him look, think, to cultivate certain positive qualities and attitudes in him. This fully applies to the environmental poster. But it is not enough just to catch and attract attention: it is important to show each person what he can do for himself and his children, his city, his country, his planet.

Technological progress has contributed to the birth of completely new forms of the poster, which are not limited to one paper medium [5]. But despite this, the value of the poster primarily lies in the content aspect, in the message and its readability, which work together with a well-chosen palette and composition. Using the rich experience of poster artists, contemporaries are guided by the global environmental problems of our time, trying to convey their thoughts in a new, understandable language. The role of social posters, namely posters that form the ecological culture of society, resonates in the modern world, where there is a global transition to a "green" economy. Reducing energy consumption, limiting the use of raw materials and water consumption, using highly efficient technologies, as well as reducing greenhouse gas emissions, preventing environmental pollution, reducing all types of waste in order to protect and restore ecosystems and biodiversity. To ensure the correct transition to a green economy, it is necessary to educate a certain eco-culture, to train not only influential politicians, but also all segments of the population.

The formation of "green" skills, ecological culture depends on the desire to learn throughout life, personal motivation for self-preservation and preservation of the environment. The knowledge and skills of both each individual and entire human communities, their commitment to the highest social values - such as humanism, love for people, life, and nature – will help preserve the ecology of the planet. To do this, it is necessary, first of all, to change the public consciousness of people. It is the consciousness of people that should be ready for changes. It will not be enough to train only the younger generation or only employees of a certain sphere, it is necessary to educate and teach "green" skills to all citizens [6. 7].

Since the birth of the poster and until the middle of the XX century, with all its variety of subjects and messages, it remained uncomplicated, straightforward, clearly expressing a certain idea through clearly readable pictorial images. At the moment, we can say that the traditional look of the poster has undergone a significant transformation [8]. The reason for this is a variety of factors. From the technical side, this is a large selection of tools without any restrictions: from traditional materials and mixed techniques to photographs, collaging, the use of computer and electronic technologies. Another significant reason is the change in values and the allocation of the most acute social problems of the XXI century, which previously receded into the background. The graphic language changes along with the situation in society, becomes extremely figurative, exaggerated, metaphorical, symbolism takes the place of verbatim. If earlier the aesthetic (setting the choice of materials and artistic techniques) and social (dictating the semantic content) components in the poster were balanced, now more and more often the main emphasis is placed on the latter, and questions of aesthetics either obey it or are discarded altogether [9, 10]. This is due, firstly, to a change in the attitude to art as such. Modern art has long not gravitated towards the sublime, since it is no longer focused on the same creator, but on a simple person, "communication" with whom is sometimes reduced to primitive things and simplified silhouettes. This applies most closely to the design of the poster, since it is the kind of art that a person regularly encounters in everyday life and his task is to be understandable. Secondly, the reason lies in the very appearance of the urban environment [11].

Fig. 1. Posters of students of the Moscow State Technical University named after K. G. Razumovsky (Anastasia Glukhova, Daria Chuvilina).

4 Conclusion

The poster has always been and remains the most common way to convey information by minimal means. However, with the development of industrial society and mass communication media, the city turned out to be oversaturated with information: the streets were filled with bright signs, advertising billboards and banners, posters. Given the dynamic rhythm of urban life, it has become more difficult for a traditional poster to attract the attention of a passing viewer, and therefore most modern posters are simple forms (and sometimes even their absence with an emphasis on text), made in a flashy limited palette. Here there is a fine line between a symbol that is understandable to everyone, and the complete erasure of the image, its abstractness and ambiguity for the sake of visibility. If the recipient who receives such a message now needs to think about its message, despite the simplicity of the forms and a cursory glance is not enough, the question arises about the effectiveness of this form of dialogue. Here, the problem of copying and self-copying arises: the simpler forms are used, and the design becomes typical, the more difficult it is to visualize information so that it attracts attention and at the same time causes the necessary response from others [12]. The poster as an iconic carrier of socially significant information should not give way to tasteless advertising, which does not require any intellectual effort from the viewer and does not educate him both morally and aesthetically. Art must have a direct impact on a person's emotions [13, 14].

To preserve life on our planet, it is necessary, first of all, to change the public consciousness of people. It is the consciousness of people that should be ready for changes.

The role of the modern social poster is undeniable, its task is to participate in the process of educating and teaching the ecological culture of society in all segments of the world's population, of all ages, using visualization tools.

References

1. S. N. Begidova, P. G. Vasilenko, Bulletin of the Adyghe State University. Series 3: Pedagogy and Psycholog, **3** (2016)
2. L. K. Vartanova, *The space of dialogues: fine art and design* (2017)

3. A. M. Glukhova, T. E. Belyakova, Science: the view of the young. Collection of materials of the scientific conference (2020)
4. E. V. Demkina, E. V. Lopasova, Formation of compositional thinking among students of fine arts in the system of additional education: state, problems and prospects (2016)
5. M. V. Domasev, Color, color management, color calculations and measurements (2009)
6. A. V. Kotyshov, Education. The science. Culture. GSU, Electrical Insulator (2015)
7. M. A. Smirnova Pedagogical education at the junction of epochs: innovations and traditions in the field of educational technologies (2017)
8. E. V. Vasilenko, Modern science, **9-1** (2019)
9. I. V. Alekseeva, N. I. Barsukova, V. I. Pallotta, N. A. Skovorodnikova, European Journal of Modern Education, **6** (2017)
10. M. A. Smirnova, E. V. Vasilenko, International Scientific and Practical Conference "Actual Problems of Linguistics and Didactics: an interdisciplinary approach in the Humanities and Social Sciences" (CILDIAH-2019), (2019)
11. G. G. Sunaeva, S. G. Sunaeva, T. E. Belyakova, K. N. Chevalier, SHS Web of Conferences. International scientific and practical conference "Theory and practice of project management in education: horizons and risks" (2020)
12. E. V. Fomina, N.I. Barsukova, Intercession, **44**, 4 (2019)
13. E. Yu. Ivanova, T. E. Belyakova, L. V. Novikova, V. V. Koreshkov, Espacios, **40**, 17 (2019)
14. T. Pushkareva, D. Agaltsova, E. Ivanova, T. Belyakova, Economic and social development (2018)

Influence of personal value orientations on the successful professional development undergraduates of a Pedagogical University

N. V. Chekaleva[*], *S. N.* Shirobokov, and *V. V.* Lorents

Omsk State Pedagogical University, Omsk, Russian Federation

Abstract. The article examines the features of the implementation of the value orientations of the individual when receiving pedagogical education in modern conditions, studies the influence of stable value orientations on the successful professional development of a master's student at a pedagogical university. The results of the study of personal value orientations are presented by the authors on the basis of diagnostics of the real structure of personal value orientations according to the method of S.S. Bubnova. Level determination of value orientations development of Omsk State Pedagogical University undergraduates made it possible to identify the motivational field of students, the level of development of self-organization, the direction of building long-term professional goals, the development of reflexive skills and time management skills. The article also formulates the directions for improving the training of pedagogical personnel, taking into account the value orientations of undergraduates.

1 Introduction

One of the current trends in modern education is its internationalization and continuity, diversification and multi-levelness. As noted by A.V. Kiryakova, the urgency of the problem of personality orientation in the world of values is due to the social situation of the development of modern society, cardinal changes in the socio-economic and social life of the country [1]. The development of the Bologna process, the organization of the Common Educational Space, the introduction of standardization of education led to the creation of conditions for the development and formation of a successful competitive specialist, ready for constant self-improvement and advanced training, which determines his demand in the labor market. Modern education is "education throughout life", with constant self-learning and self-development, building a professional Akme (the pinnacle of professional excellence). Today, it is especially important to build an individual educational trajectory (training route) in order to improve qualifications, develop professional experience and exchange the best pedagogical practices. Modern higher education provides an opportunity

[*] Corresponding author: nchekaleva@yandex.ru

for a bachelor student to continue their studies in Master's degree programme, and then in postgraduate and doctoral studies.

A master's degree is the most important component of career planning. It allows each person to have a solid foundation for further successful professional development and career growth.

In our study, we identified the features of the implementation of value orientations of the personality of a master's student in real life conditions. Our conclusions and recommendations can have a positive impact on improving the training of teaching staff and making management decisions. Domestic and foreign researchers note that obtaining a master's degree and improving professional qualifications are not always accompanied by promotion and career and an increase in wages, which is possibly associated with the level of development of the motivational field and reflexive skills of the individual. In order to propose concrete practical steps and determine directions for overcoming this contradiction, it is necessary to investigate the level of development of value orientations of undergraduates and evaluate their role in building the career of teaching employees studying in a magistracy. At the stage of obtaining a master's decree, the uniform and harmonious development of the individual has a significant impact on the implementation of the value orientations of the individual and the specialist's efficiency. What determines the practical significance and relevance of the study.

2 Materials and Methods

The methodological basis of the research was the system-activity, personal and competence-based approach.

As empirical research methods, the method of questionnaire, including pedagogical observation, was used. 42 internal undergraduates of Omsk State Pedagogical University passed the survey

State Pedagogical University, including its branch. In this study, methodology "Diagnostics of the real structure of value orientations of the personality of S.S. Bubnova ",was used , which is designed to study the implementation of personal value orientations in real life conditions [2].

The methodology contains 66 closed-ended questions aimed at studying the implementation of the value orientations of the individual. The severity of each of the polystructural value orientations of the individual was determined using the key. Accordingly, the number of positive answers in all eleven value columns is counted. Based on the results of processing individual data, a graphical profile was built, reflecting the severity of each value. The following groups of personal value orientations were investigated:

I- Pleasant pastime, recreation;
II -High material well-being;
III - Search and enjoyment of the beauty;
IV -Help and mercy towards other people;
V - Love;
VI - Perception of the new things in the world, nature, human;
VII- High social status and people management;
VIII - Social esteem and respect of people and influence on wider public;
IX -Social activity to achieve positive changes in society;
X - Communication.

When building a model of the structure of value orientations, we focus on the level of the value system. The high level of values includes spiritual, social and material, while all isolated cases of manifestation of certain values in the form of personal characteristics are ranked as a private level.

M. Rokich distinguishes two levels of the hierarchy: values - goals, or the ultimate goals of existence, and values - means, or ways of personality behavior. Since the means help to achieve the set goals, we consider the diagnostic results as the motivational structure of students, their potential and targets in building a professional pedagogical career.

In this case, it is important to take into account four hierarchical levels of personal value orientations:

I . A high level of values that are generalized, abstract: nonmaterial, social, material;

II . Components of three types of abstract values. For the nonmaterial sphere - cognitive, aesthetic, humanistic and others. In the system of social values, the values of social respect, social achievements and social activity, etc. are highlighted.

I and II level of values refer to the values-ideals, allowing to build the life and professional path of the respondent.

III . Values manifested in life as personality traits (sociability, curiosity, activity).

IV . The most typical ways of a person's behavior as a result of the implementation of activities through the prism of life attitudes, real value orientations.

Levels III and IV refer to values-means that help to realize life and professional goals.

These groups of values allow you to build an individual trajectory of learning and self-development, taking into account personal motivation, value attitudes and career prospects of the student.

In accordance with the professional standard of the teacher and the requirements of the Federal State Educational Standard of the new generation, pedagogical activity is effective and of high quality provided that the teacher has developed functional literacy, the content of which is social adaptation, the readiness to carry out their professional duties in the face of changes, the ability for professional growth and effective solution of professional problems in non-standard situations.

3 Results

Based on the findings of the study, the following results were obtained. The respondents in their answers often refer to the lack of time for a pleasant pastime, recreation and social activity, which is directly related to time management and has a significant impact on the development of personality abilities not only during the learning period, but throughout life.

This approach allows students to rationally organize a pleasant pastime and recreation. In addition, the correct distribution of time by undergraduates, based on the planned goals based on introspection, has a significant impact on social activity in order to achieve positive changes in society.

Moreover, in accordance with individual priorities, personal improvement also depends on a reasonably formed balance between physical, spiritual, professional, social, mental, emotional development, and in no case from an absolute disregard for one factor or another. The results of the survey are presented in Fig. 1.

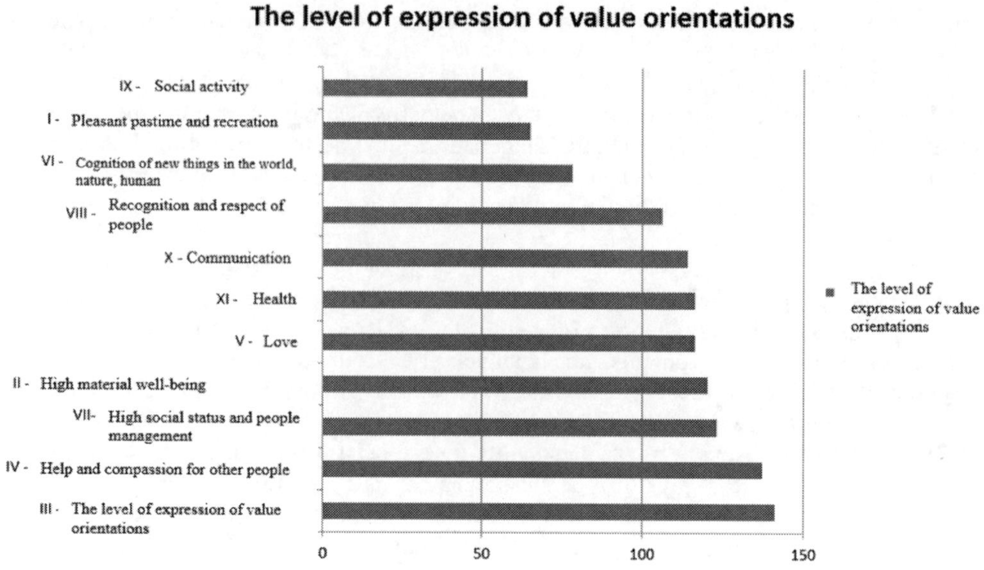

Fig. 1. The level of expression of value orientations.

It should be emphasized that a high level of expression of such value orientations as the search for and enjoyment of the beautiful, help and mercy to other people, high social status and people management, high material well-being, was recorded in the answers of many undergraduates. At the intermediate level, undergraduates have the following important value orientations: love, health, communication, recognition and respect of people, influence on others, as well as learning new things in the world, nature, and human. The study showed that among undergraduates there is a low level of expression of value orientations regarding pleasant time spending, recreation and social activity in order to achieve positive changes in society, which is confirmed by a large number of students.

The study of the opinion of active undergraduates of the Pedagogical University gives us a detailed picture of the state of the modern personality. To what extent do undergraduates perceive this stage of their development? The characteristics of time management show us, directly or indirectly, to what extent the learner is able to control his own development and balance his life. Do they understand short and long term goals and plans in the field of education? How do they manage their time

study and recreation? How much time are they wasting during the day?

The conclusions and forecasts for the future obtained on the basis of the study contain important and relevant information and recommendations not only for individuals of a particular group, but also on the whole functioning of the system that regulates various areas, especially education and labor.

In the context of significant changes in the labor market and in the higher education sector in Russia, the discourse of employment opportunities for undergraduates is becoming dominant. Organizations of higher education are strongly encouraged to ensure the release of in-demand personnel capable of achieving their goals in rapidly changing situations due to the mastery of methods for solving a large class of professional problems, and the graduates themselves are strongly encouraged to constantly develop their personal skills, qualities and experience in order to compete in the graduates' labor market.

Our article offers a critical assessment of job opportunities. In contrast to the assumption of equal conditions for the realization of value orientations, the educational space of the university significantly affects the available opportunities. The scholarly debate and discourse on the use of potential, with its emphasis on individual responsibility and disregard for social inequality, has potentially detrimental consequences for these graduates.

In the context of the development of knowledge-based competitiveness of countries, where human capital theories provide a platform on which education policy is formulated, it is perceived by governments as a way to enhance national growth and prosperity. Such convictions have led to increased pressure on higher education organizations, and the topic of employment of graduates of the magistracy currently occupies an important place in modern socio-economic conditions.

Based on the data collected during the survey of undergraduates, we emphasize the close connection between the expectations of undergraduates from being in a university with the transformation of society. Scientists note that the experience of graduates in the process of finding a job as part of the transition from university to work is being transformed on the basis of global changes in higher education and in the labor market.

Foreign research confirms that modern graduates expect more from higher education than their predecessors in the past, and that different groups of graduates receive different levels of opportunities, depending, for example, on gender, ethnicity and social class, which makes it difficult for undergraduates to realize the value orientations of the individual during the training period. Nevertheless, the comprehension of some of the results of the study of the value orientations of the student's personality also makes it possible to analyze the context and realities that the university has been going through recently.

In recent decades, the higher education system has undergone major transformations. The constantly changing context of graduates' employment leads to a critical study of the discourse of academic mobility of subjects of the educational process in the global educational space. This discourse relates to the experience of graduates, their perception of the labor market and the transformation of universities, based on the principles of the availability of master's decree education in the world. It is therefore not surprising that higher education institutions have been urged to focus on the third mission.

The knowledge economy, along with globalization, are unprecedented challenges that greatly contribute to the redesign and expansion of university missions. The dynamics of knowledge production is changing, as is the attitude of society towards expectations and values.

Indeed, in recent years, there has been increasing pressure on universities to move from teaching and research to a third mission. These universities become engines that contribute to the social, economic and cultural development of the regions in which they operate by transferring knowledge and technology to society. At the same time, the academic community is faced with the problem of the need to demonstrate both its responsibility and effective training of personnel through the implementation of strategic management. In other words, the third mission is a complex and evolving phenomenon that has been articulated in the public policies of many countries over the past few decades.

We assume that the presentation of conditions for the implementation of value orientations, including within the framework of federal projects, is also of key importance for undergraduates. For example, the national project "Education" is aimed at providing opportunities for self-realization of citizens [3]. Moreover, the federal project "Social Elevators for Everyone" is aimed at creating opportunities for citizens for professional and career growth through the formation and development of a system of professional competitions, and the federal project "Social Activity" allows creating conditions for the development and support of volunteerism as a key element of social responsibility developed civil society [4, 5] .

4 Conclusions

Thus, according to the leading role of universities, it is necessary to strive making master's decree training with a clear social impact, leading to the realization of personal value orientations in modern conditions of higher education. Higher psychological and pedagogical education, according to the Federal State Educational Standard, should train new type specialists by means of competence-based and system-activity approaches. In summary, the training of competitive personnel provides for the formation of professional and pedagogical reflective and critical thinking, the development of educational and creative potential, as well as the ability for continuous personal growth and professional development.

References

1. A.V. Kiryakova, *The theory of personality orientation in the world of values* (Orenburg, 1996)
2. S.S. Bubnova, *Methods for diagnosing the individual structure of personal value orientations, Methods of psychological diagnostics*, 2 (Moscow: Publishing house of the Institute of Psychology of the Russian Academy of Sciences, 1994)
3. Ministry of Education of the Russian Federation. National project "Education", https://edu.gov.ru/
4. Ministry of Education of the Russian Federation. Federal project "Social lifts for everyone", https://edu.gov.ru/
5. Ministry of Education of the Russian Federation. Federal project "Social Activity", https://edu.gov.ru/

E3S Web of Conferences **311**, 02010 (2021) https://doi.org/10.1051/e3sconf/202131102010
EPSD 2021

Regional specific features of architectural space ecologization for preschool education in the Urals

Natalia Lamekhova*
Ural State University of Architecture and Art, Russia

Abstract. The article is concerned with the study of methods on architectural space ecologization for preschool education in Ural region of Russia. The main goal of introducing architectural methods is to educate the child in the correct attitude toward nature, to form a basic reference point in the education of preschool children by architectural means.

1 Introduction

The lack of communication with nature – a burning issue of today. Children began to spend less time with their families, and more – in front of TV, computer and telephone screen. Any child knows more about their childhood idols from the cartoon series than about nature. At the same time, nature-deficit disorder is the price one pays for detachment. As a result, it manifests itself in deterioration of the senses, attention problems, an increase in physical and mental illnesses, including obesity. Any child needs nature for the development of feelings and creative function.

2 Materials and Methods

Landscaping in modern architectural practice is currently suffering major changes and acts as a structural element more than as an element that performs aesthetic functions. This circumstance proves the growing role of landscaping *as a structural component* in architectural environment formation, exerting a positive effect not only on the psychological state of a person, but also on physical well-being [1 – 5]. Particular attention should be paid to objects for preschool education, demonstrating experience and attitudes toward nature, as a kind of standard of environmental culture for the younger generation. The improvement of ecological component of objects of this kind will be the growth of qualitative characteristics by means of including various methods for architectural space ecologization.

*Corresponding author: lamekhova@mail.ru

2.1 Features of environmental culture formation

Childhood – one of the most important periods in the life of any person. At this age, a person is most susceptible to the perception of information in life, being an "open" system, absorbing absolutely everything that surrounds him. At the same time, translating the right approach to nature is at odds with upbringing and education. And to a large extent, this circumstance depends on the conditions where preschool children are – in preschool educational institutions. The architectural environment of existing buildings for preschool education is largely outdated, and new approaches require comprehensive study and forward-looking solutions.

The problem of natural deficiency is very acute today, which requires a comprehensive approach and high professionalism from specialists in various fields [6-8]. This fact, increases the speed of decision-making, the development of tools that would help to solve a number of problems at the level of preservation of two components of the architectural space, such as *nature conservation* and *health conservation*. In this regard, the architectural environment for preschool education is to be considered as a kind of model, which includes these qualities, and the model, translating a system of environmental values, resulting from interaction of positive-creative, positive-creative principles of detail children, aimed at harmonizing interests and the correct interaction of "man-nature".

The functions of the *health* -saving space are based on taking into account the development of:

- child's physiological characteristics (these include: age needs and physiological capabilities). For example, the need to learn to walk on a small inclined surface (for children under 2 years) or to roll down an inclined surface (for children under 3-4 years), to master running on an inclined surface (for children from 5-7 years), etc;
- child's psycho-emotional characteristics (need for children in protective function of space – presence of "quiet corners", "quiet zones", the need for social contact – the presence of zones of active games, areas for competitions), etc.;

The process of forming a correct worldview in preschool children, pursuing, spiritual formation of the personality, based on the subjective attitude to nature in preschool children is possible with the inclusion in the educational process of three channels of knowledge of the surrounding world – cognitive (mastering knowledge about the world and man and related activities), perceptive (involvement in the educational process feelings and emotions) and practical (the inclusion of the personality in socially significant, nature-like, creative activities) [9, p. 284].

The functions of an *environmentally friendly* space are based on taking into account the development of feelings:

- unity with the environment, admiration for it and respect for the processes on which it is built;
- community with other people and living beings;
- responsibility for the made choice;
- developed imagination and creativity;
- continuity between past, present and future;
- ability to find needed information;
- ability to predict and assess [9, p.291].

Features of teaching and educational activities. The task of education, upbringing, and socialization of preschool children is currently one of the highest priorities associated with the modernization of preschool education. This implies the need to improve the architectural environment for preschool education with regard to the environmental approach, within the framework of which children can form the basics of environmental culture – a conscious-right attitude towards phenomena and objects of animate and inanimate nature.

All of the above arguments prove the importance of essence of the formation of sustainable educational space from perspective of environmentally-oriented methods introduction by incorporating different forms of landscaping.

2.2 Climatic features of Ural region

The features of the Ural natural and climatic conditions are due to the meridional direction of the mountain range, which prevents the movement of prevailing westerly air masses, and the large territory length from north to south. Hence the climatic differences between the western and eastern slopes of the Ural Mountains and the Ural plains (climate continentality, sunshine duration, precipitation).

Winter winds are characterized by high stability, their frequency ranges from 40 to 70%. In January, south-western, western and southern winds prevail in the Urals. Protecting facilities from their effects is a serious problem due to the fact that protection from prevailing winds often conflicts with the insolation requirements. Therefore, protection from cold winds can be achieved through the use of natural landscape elements (forest, park, mountains, high river banks, artificially created slopes).

The main goal of space ecologization for preschool education in Ural region is the development of sensitivity to nature, the natural environment and its features in preschool children.

In this regard, preschool institutions with their territory should primarily become zones of ecological safety, which would contribute to the development of successful education in kindergarten, as well as demonstrate a kind of model, which in generalized form reflected the relationship of living and nonliving nature. Thus, the architectural environment of preschool educational institution should be a standard of ecological culture, taking into account the modern requirements of the educational process.

2.3 Ecologization of architectural space for preschool education

From perspective of the development and sustainable growth of architectural space that translates the principles of environmentally oriented architectural, environment can be the following:

2.3.1 Method of territorial defense

The inclusion of *method of territorial protection,* which implies, at the level of site formation, the creation of a *screen* by installing artificial barriers in the form of climbing walls or elements of play and physical education sites. Also, the building volume itself can serve as protective functions (due to plasticity and shape of individual parts).

Territory screening can also be done through the use of natural site forms, the difference in ground level marks, topography and zoning of play activities. The use of natural forms in the site improvement also plays a positive role, because it demonstrates the ability to make do with small and accessible means.

It is also worth mentioning that the monotonous environment affects the person much more, and children need to be protected, so the impact of visual characteristics are in the first place. In nature, there are no right angles, rectangular planes and so on, it is worth striving for smoother forms, rounded corners, demonstrating unity with nature. It is worth trying to recreate them on the site as game elements. The existing imbalance between man-made and natural has a negative impact on the lives of people in general, not to mention children. The development of environmentally oriented architectural environment for preschool education should go within the framework of inclusion of site protective functions, the building and the internal component of the architectural space.

Artificially recreated architectural volumes with elements of pitched roofs, slopes can be landscaped, or laid out of different building materials (stone, wood, the combined use of several types of materials), which somehow can be part of landscaping for the purpose of all kinds of outdoor events. Buildings-hills, elevated objects are structures that actively use the site relief. In the process of geoplastics, the site natural reliefs can be restored, and new forms with protective functions can be made, forming a protected space of architectural environment for preschool education.

2.3.2 Method of complementing the natural landscape

Or otherwise the use of *naturalness method*, which involves the organic inclusion in the structure of existing landscapes objects for construction, without destroying their integrity and reasonably correcting [10]. The combination of architectural forms, based on natural analogues, the use of smooth shapes in combination with unified elements – provide an excellent opportunity to effectively solve many problems. These include: preserving and supplementing natural landscapes, increasing the artistic qualities of architectural volume of the building, returning the plots used for construction to recreational resource base, as well as providing microclimate parameters with minimal energy consumption.

2.3.3 Method of incorporating green components into the internal building frame

This *method allows to include in the building structure, an artificially created natural environment in the form of "phyto-modules"*, which allows to form within the building specialized zones with different microclimatic parameters, some of which can serve as a kind of buffer between the main building volume and the aggressive environment. "Phyto-modules" contribute to enrichment of the internal environment of architectural space, creating additional conditions for contact with nature for preschool children.

2.3.4 Method of inclusion of natural component of walking area into internal structure of architectural space

This method can be carried out by *combining indoor and outdoor spaces by bringing the walking areas closer to the entrances* allows to be closer to nature, which is akin to the parameters of the homestead plot and the living unit of an individual house.

In this sense of creating a comfortable protective environment in Ural region is greatly influenced by the placement of the promenade in relation to the building for preschool education. Optimal conditions that activate the protective function of the architectural environment is the maximum proximity of walkway to group unit entrance. In this case, the internal group unit space merges with the external one, bypassing the gap laid by the public territory.

So, the input element plays the role of a communication node – a distributor, as well as a play module in the form of a slide with a rope descent. The idea of bringing the walking area closer to exit from the group will help children to strengthen the sense of ownership and belonging of the given object to a certain group, bring natural elements closer to group units.

2.3.5 Method of inclusion on multi-component lighting by natural light

The method of *combining different forms of lighting* involves the inclusion of various architectural combinations in the composition of the building, as well as the inclusion of *overhead* and *overhead-side lighting through the window openings*, reflective systems to light recreational areas, creating light halls, atrium spaces in exhibition areas, recreation, waiting areas (corridors and halls), in group rooms – this allows a child to contact the outside environment, time of day, weather conditions. To develop a sense of community with nature, and at the same time protection, calmness, peace, creating an opportunity to fill the inner world with new emotions baby.

2.3.6 Method of inclusion of an ecological path in the structure of the complex

The creation of ecological trails – a fairly new approach to the organization of architectural space for preschool education. At the same time, this phenomenon for the most part falls on the teachers' shoulders, and does not in any way affect the features of the building's architecture formation. At the same time, the use of this method could form the basis of building volume composition, providing an opportunity to reveal the natural environment potential, creating many angles to perceive the environment; become part of the creative function of children, bringing the natural components closer to themselves, actively participating in shaping the world around them, not excluding humans as creators

3. Conclusions

Considering the problem of ecologization of architectural space for preschool education in the Ural region this study contributes to a qualitative change in the architectural environment for preschool education in the application of a particular method, pursuing the main goal – to demonstrate the right attitude of children to nature. Moreover, given that this typology has already been introduced in most countries in one way or another.

The above methods for including a component of an environmentally friendly architectural environment are not limited to the above set of measures. This is a field of activity that constantly needs new suggestions from different specialists. But it explicitly identified areas to be followed and developed, to introduce them into the practice of architectural design. This provides a tremendous incentive to keep moving forward and putting the grain that has been lost today into the architecture of preschool buildings. Otherwise, children will not get what they sincerely deserve from nature. It is worth remembering that environmental content is a crucial factor in this process. A rich, broad environment will constantly present the possibility of alternative choices for creativity. An uninhibited, impersonal environment limits the opportunities for healthy growth and development, both as an individual and as a group.

The task of specialists-architects to help children, to love nature, bringing up in the appropriate environment, demonstrating a model of unparalleled union "man and nature".

References

1. E. A. Dorozhkina, IOP Conf. Series: Materials Science and Engineering, **451** (2018)
2. O. Finaeva, IOP Conf. Series: Materials Science and Engineering, **451** (2018)
3. E. A. Dorozhkina, IOP Conf. Series: Materials Science and Engineering, **451** 2018
4. I. Yu. Shelekhov, V. V. Pozhidaev, T. I. Shishelova, IOP Conf. Series: Materials Science and Engineering, **451** (2018)

5. A. Giyazov, B. I. Giyazov, Yu. G. Baratov, Science and Innovation, **1** (2017)
6. A. N. Gushchin, M. N. Divakova, N. S. Mironova, IOP Conf. Series: Materials Science and Engineering, **451** (2018)
7. E. Lyaskovskaya, G. Khalilova and K. Shmykova, IOP Conf. Series: Materials Science and Engineering, **451** (2018)
8. N. A. Saprykina, Ya. A. Saprykin, IOP Conf. Series: Materials Science and Engineering, **451** (2018)
9. S. A. Ivanov, N. L. Abramova, Scientific dialogue, **9(57),** 283 (2016)
10. A. V. Merenkov, N. S. Akchurina, T. M. Matveeva, IOP Conf. Series: Materials Science and Engineering, **687** (2019)

China's green insurance system and functions

Sergey A. Belozyorov, and *Xiaohe* Xie

St.-Petersburg State University, 199034, Universitetskaya Emb., 7–9, Saint Petersburg, Russian

Abstract. The article aims at the relationship between China's green insurance, social and economic development, and environmental protection. It analyzes China's green insurance development measures and operating mechanism. According to the China Statistical Yearbook, based on data from 2011-2019, research variables are selected for empirical analysis and qualitative analysis. It is concluded that the green insurance industry is conducive to sustainable and sustainable development, reduces industrial transformation risks, improves environmental protection capabilities, and improves residents' living standards. China's experience in developing the green insurance industry provides a useful reference for the sustainable development and development of green finance in developing countries and regions.

1 Introduction

Green insurance provides insurance products that support environmental improvement, respond to climate change, and promote resource conservation and efficient use. It also provides environmental risk management services. China is at the forefront of green finance initiatives globally with most beneficiaries being developing countries with a stable political environment with significantly high levels of creditrisk.[1]

From the perspective of the industrial chain, China's insurance industry has gradually enriched green insurance products and services, providing risk protection and risk management services for green energy, green transportation, green buildings, green technology, climate governance, and forest carbon sinks. With the continuous growth of insurance investment in China's green industry projects, an insurance fund operation model with Chinese characteristics has been formed. According to statistics, from 2018 to the end of 2020, the green investment stock of China's insurance funds has increased from 3954 billion rubles to 5,615 billion rubles, a growth rate of 19.17%.[2]

"China's Fourteenth Five-Year Plan for National Economic and Social Development and the Outline of 2035 Vision Goals" and carbon emission targets force its industries to face the need for green transformation and are in the process of transformation and development. The risk of the process requires the insurance industry to undertake corresponding social responsibilities, which meets the market demand for social and economic development. [3]From the perspective of the sustainable development of the insurance industry, the dramatic increase in insurance risk events caused by global climate change, rising claims costs, and challenges in underwriting and pricing have led the insurance industry to actively seek solutions that take into account both availability.

Green insurance is an important part of green finance. China has improved its legal system to guide the development of green insurance and build a comprehensive green insurance system.[4] The establishment of an environmental pollution liability insurance system should promote cooperation between insurance companies and other financial institutions to jointly promote the development of environmental pollution liability insurance.[5]

Establishing a unique green financial system and actively promoting and developing green insurance will help improve the corporate risk management system and help the insurance industry to carry out supply-side structural reforms. Build a sound environmental protection system The credit system and the green financial system can accelerate the green transformation of the economy. [6] The functions of timely compensation, reduction of management costs, risk control and other functions of green insurance meet the needs of China's environmental protection. The establishment of a green insurance system requires legitimacy Corresponding policies and measures shall be adopted in terms of system models, insurance coverage and types of insurance, and the design of supporting mechanisms. [7]Studying the impact of financing constraints on green innovation and exploring the financing constraints of enterprises on green innovation are important topics for the development of Chinese enterprises.

The green insurance mechanism provides property and personal insurance products and services for the needs of energy-saving and environmentally-friendly enterprises, and provides comprehensive risk protection for their production and operation. It can disperse risks in the process of energy conservation and environmental protection technology and equipment research and development, promote the development of energy conservation and environmental protection technology, improve energy conservation and emission reduction, accelerate the construction of a green production system, and form a new economic growth point. By improving the green insurance supply system, the insurance industry will give full play to its functions of economic compensation, financing and risk management, increase the "green supply" of the whole society, and promote supply-side structural reforms.

Based on the green insurance development model of developed countries, China has formulated relevant administrative regulations and local regulations on the basis of environmental protection law and insurance law, and established a legal guarantee for the development of green insurance in the form of legislation. In the form of legislation, the areas of insurance, the methods of insurance, and incentive and restraint mechanisms are clarified. At the same time, a strict environmental damage compensation system is implemented to clarify the responsibilities that enterprises should bear for harming the environment, and to increase the cost of damage to the ecological environment. An effective accountability mechanism forces companies to purchase environmental pollution liability insurance to diversify and resolve environmental risks. In addition, the Chinese government supports the establishment of specialized insurance companies, takes the lead in formulating insurance policies, and supervises implementation, etc., and actively exerts the government's guiding and supporting role.

With policy support, we will accelerate the pace of insurance product R&D and innovation, green insurance products and services, increase the effective supply of green insurance, and guide insurance funds to support the development of green industries. Promote the insurance industry's in-depth participation in environmental risk management, in-depth participation in the entire environmental risk management process of insured companies, and comprehensively improve the company's environmental risk management level.

In order to develop the green insurance industry, an innovation-driven development strategy has been implemented to accelerate innovation in the insurance industry's institutional mechanisms, products and services, and business models. The development of the digital economy makes full use of insurance technologies such as the mobile Internet, big data, and cloud computing to improve the level of operation and management, and improve the efficiency of operation and management of green insurance companies. Insurtech has changed the business model and process of insurance companies, the insurance industry has undergone a deep reorganization, and has reshaped the value chain of the insurance industry. The insurance industry is realizing process digitization, experience intelligence, content customization, and online channels.

The green insurance strategy has also differentiated business models, such as professional operations by customer groups, insurance + technology, and insurance + ecology. The development of insurance business attaches importance to improving the quality and efficiency of development, strengthening the management of asset-liability matching, and taking the road of sustainable development.

China has established a credit system for the insurance industry, and is connected to platforms such as bank credit reporting systems to realize information sharing, form an effective trustworthy incentive mechanism and untrustworthy punishment mechanism, and optimize the credit environment for the green and sustainable development of the insurance industry.

2 Data, descriptive statistics and empirical strategy

Developed countries have a relatively high level of economic development and relatively low environmental pressure, so the development of green insurance is mainly to deal with climate change. Unlike developed countries, green insurance in developing countries must take into account environmental pollution control and climate change. The main direction of China's development of green insurance is to support environmental improvement, address climate change, and promote resource conservation and efficient use. China introduce both green insurance and insurance subsidy.[8]

For the purpose of the research, we put forward a hypothesis: the insurance industry promotes China's economic growth and industrial upgrading, improves the quality of life of residents, and improves China's environment at the same time. And use the data from the China Statistical Yearbook and the Environmental Bulletin to do a trend analysis, and then obtain a qualitative analysis of the long-term impact of the development of green insurance on China's social and economic development.

The empirical analysis in this paper is based on the panel data variable indicators from 2011 to 2019 from the "China Statistical Yearbook" and "Chinese Ministry of Environmental Protection Bulletin." In order to study the correlation between insurance, economic growth and environmental protection capabilities, five representative variables were selected: X1- green insurance compensation funds, X2- industrial pollution control funds, X3- environmental protection investment funds, X4-GDP, X5-per capita GDP as research variable. Variable description and data are shown in Table 1. We use insurance compensation funds to measure the contribution of the insurance industry to economic development and environmental protection. GDP is used as a regional economic development indicator, per capita GDP is used to measure the living standards of residents, and industrial pollution control funds and environmental protection investment funds are used as environmental protection construction capabilities.

Table 1. Research variables and data.

year	X1(RUB bln)	X2(RUB bln)	X3(RUB bln)	X4(RUB bln)	X(Yuan)
2011	3929	444.36	7114.03	483392.8	36277
2012	4716	500.45	8253.46	537329	39771
2013	6213	849.66	9037.2	588141.2	43497
2014	7216	997.7	9575.5	644380.2	46912
2015	8674	777.3	8806.3	685571.2	49922
2016	10515.68	819	9216.8	742694.1	53783
2017	11181	681.5	9538.95	830945.7	59592
2018	12297	621.3	9539	915243.5	65534
2019	12893.97	615.2	9800	983751.2	70328

Note: National Bureau of Statistics of China, Ministry of Environmental Protection

Because the variable numbers are large and the units are not uniform, we first perform logarithmic processing on the variables and then perform subsequent analysis. From the results of the analysis in Table 2, it can be seen that the overall quality of the data is good, and the skewness and kurtosis of the variables are within a reasonable range, and the next step can be analyzed.

Table 2. Descriptive analysis results of variables.

Variables	Obs	Mean	Std. Dev.	Min	Max	p1	p99	Skew.	Kurt.
x1	9	8.987	.428	8.276	9.465	8.276	9.465	-.47	1.853
x2	9	6.523	.26	6.097	6.905	6.097	6.905	-.243	2.064
x3	9	9.099	.101	8.87	9.19	8.87	9.19	-1.419	3.984
x4	9	13.451	.241	13.089	13.799	13.089	13.799	.002	1.831
x5	9	10.832	.225	10.499	11.161	10.499	11.161	.036	1.834

3 Results and Discussion

Table 3. Correlation analysis results.

Variables	x1	x2	x3	x4	x5
(1) x1	0.118314				
(2) x2	0.043062	0.067781			
(3) x3	0.036926	0.017397	0.010176		
(4) x4	0.10065	0.014807	0.020013	0.058079	
(5) x5	0.093575	0.013286	0.018515	0.054099	0.050407

From the correlation analysis results in Table 3, it can be seen that there is a significant positive correlation between the variable insurance claims X1 and the variables, indicating that the development of green insurance can have a positive impact on the growth of GDP and per capita GDP, and can promote economic growth and improve residents. Average income. At the same time, the development of the insurance industry has increased industrial pollution control funds and environmental protection investment funds, and improved industrial pollution control capabilities and environmental protection capabilities. The empirical results support the hypothesis.

Table 4. Matrix of correlations.

Variables	(1)	(2)	(3)	(4)	(5)
(1) x1	1.000				
(2) x2	0.386	1.000			
(3) x3	0.855	0.662	1.000		
(4) x4	0.976	0.236	0.823	1.000	
(5) x5	0.974	0.227	0.817	1.000	1.000

The correlation covariance matrix in Table 4 shows the significance of the correlation between insurance industry compensation funds and other variables. It can be seen that the development of the insurance industry has a significant relationship with social and economic development, environmental pollution control and environmental protection, and the results support the previous hypothesis.

Table 5. Pairwise correlations.

Variables	(1)	(2)	(3)	(4)	(5)
(1) x1	1.000				
(2) x2	0.386	1.000			
	(0.973)				
(3) x3	0.855	0.662	1.000		
	(0.032)	(0.413)			
(4) x4	0.976	0.236	0.823	1.000	
	(0.000)	(1.000)	(0.062)		
(5) x5	0.974	0.227	0.817	1.000	1.000
	(0.000)	(1.000)	(0.069)	(0.000)	

Through the double correlation significance test in Table 5, it can be found that x1, x2, and x3 have passed the significance test, indicating that the development of green insurance has a greater contribution to the governance of industrial pollution and the improvement of environmental protection capabilities than to economic growth. The development of green insurance is conducive to the green transformation of Chinese industrial enterprises and plays an important role in the upgrading and transformation of Chinese industries.

In order to meet regulatory requirements and market demands, Chinese insurance companies have successively developed environmental pollution liability insurance, new energy insurance, forest and grassland insurance, wildlife accident insurance, chemical disposal insurance, green construction insurance, and other green insurances. Green insurance also includes the green use of insurance funds. Environmental pollution liability insurance is a typical green insurance.[9]

Accordingly, to integrity green financial system, governments, financial institutions, enterprises, and consumers need sparticipation and cooperation among each other.[10] China's green insurance mechanism is constructed under a government-guided and market-led model. Covers risk management, compensation for environmental accidents and compensation for environmental restoration. A green insurance operating mechanism with insurance companies as the core organization has been established. In the form of commercial insurance, the government, enterprises, and individual stakeholders are connected. After an environmental risk event occurs, green insurance can provide compensation for enterprises and individuals and diversify risks. Findings indicate that the mere presence of a market for insurance increases, overall, investment in profitable but risky inputs. [11]At the same time, it reduces the burden of the government on pollution control and environmental protection. In order to maintain sustainable development and profit, insurance companies will actively monitor the environmental risk behavior of the insured, which helps to detect and prevent the occurrence of environmental risk accidents in advance, there by reducing insurance compensation and increasing the benefits of green insurance.

Green insurance enables companies to obtain risk prevention and ex post compensation. Such an insurance mechanism can disperse the risks faced by enterprises in the process of transformation and innovation, especially the environmental risks of heavy industry enterprises. By providing guarantees for the quality and safety liability risks of innovative products and technologies in the green industry, providing compensation commitments and effective risk control services, enhancing the credit of green environmental protection behaviors, and providing funds for the development of environmentally friendly enterprises. At the same time, after the company has purchased environmental pollution liability insurance, once an environmental pollution accident occurs, the insurance company will provide economic compensation to the victim in accordance with the insurance contract, so that the environmental rights and interests of the victim are properly protected and avoid damage to the victim. A long-term and complicated situation. And high environmental risk companies have insured environmental pollution liability insurance, which is equivalent to the government and environmental protection departments finding a risk consultant and pollution accident handler, and the pressure on the government and environmental protection departments will also be relieved to a certain extent. This type of insurance operation is also conducive to safeguarding the public's environmental rights and interests and social equity.

Despite the rapid development of green insurance in China, there are still problems such as small premiums, low coverage, and weak corporate insurance and insurance institutions' weak willingness to conduct business. According to statistics, the percentage of all industrial companies insuring environmental liability insurance is less than 5%, and the premiums for environmental liability insurance are only 1% of all premiums charged. Insurance companies face challenges such as information asymmetry and lack of environmental risk analysis capabilities. In terms of green insurance business, due to the large investment in product innovation, small customer base, high risk, low profit and other reasons, small and medium-sized insurance companies are not very active in developing green insurance business. Although these problems are improving due to the development of digital economy and insurance technology, they still obviously restrict the development of green insurance in China.

4 Conclusion

The development of China's green insurance has made up for the shortcomings of government and corporate functions, and has changed the previous development model of relying on the sacrifice of the environment to obtain economic growth. At the same time, it can provide financial support for environmental pollution control, improve environmental protection capabilities. China is poised to be a global leader in green finance policy implementation and regulators need to accelerate the formulation of green insurance products and enhance the capacity of financial institutions to offer green insurance. While minimizing the systemic risk fintech poses, policy makers should encourage fintechs to actively participate in environmental protection initiatives that promote green insurance. [13]

China committed to being carbon neutral by 2060, Either past successes or future ambitions cannot be achieved without policy support. Intensive efforts have been made by the Chinese government to cope with environmental degradation and pollution. In addition to comprehensive regulations and administrative enforcements, green finance policies have been explicitly introduced to support green development.[14]

China's 13th Five Year Plan proposed the creation of agreen financial system that encourages the private sector to play a more active role insustainable development.[15]On the basis of improving green insurance laws and regulations, the Chinese government has introduced systematic support policies, studied and formulated special development plans, implementation rules, and work plans, and improved preferential policies in terms of premium subsidies and green credit linkage. A coordination mechanism for relevant stakeholders in environmental protection, land, finance, agriculture, and insurance supervision has been formed, which has promoted the development of the green insurance industry from the national perspective.[16]

Establish a unified green insurance database to collect and manage national environmental pollution, environmental damage, energy conservation and emission reduction, pollution damage compensation expenditures and other data to form a shared database to provide data support for insurance product research and development, rate determination, underwriting, claims, services, etc. , Reshaping the green insurance industry chain and value chain.[17]

Realize the innovation of green insurance products through the development of financial technology, increase product coverage, and expand the scope of green insurance protection. With the help of financial technology and digital technology, insurance companies have designed scenario-based green insurance, launched green insurance with multiple risk protection functions, and developed the needs of the insured and the green insurance market.

The China Banking and Insurance Regulatory Commission has formulated the project standards for green insurance, which is conducive to the development of various green insurances such as catastrophe insurance, research and development liability insurance, green enterprise loan guarantee insurance, green construction insurance, and green enterprise important equipment insurance. [18]It encourages insurance funds to explore new paths for green investment, enhances green insurance's participation in green project investment and construction, and better meets the financial needs of green project investment and construction.[19]

Green insurance helps local governments, enterprises and households face losses from natural disasters caused by climate change, smoothes financial fluctuations, and enhances risk resilience. Green insurance provides insurance applicants with compensation commitments and effective risk control services through risk compensation integrated services, technology and other resource elements, which is conducive to the sustainable development of environmental protection companies. These measures can be used as a reference for developing countries to develop green finance and achieve sustainable development.

References

1. H. Niczyporuk, J. Urpelainen, J. Cleaner Pro. **281,** 1 (2021)
2. Y.Y. Zheng, Economic Daily, **8,** 5(2021), http://www.ce.cn.
3. Y.Q. Chen. B. Zhi, J. Miao, Construction Sci. Tech., **379,** 51(2019)
4. J.Y. Chen, Southern Finance, **9,** 14 (2016), http://www.nfjr.gd.cn
5. S.Q. Wang, J. Pang. B.Tianjin U. **3,** 30(2019)
6. M. Zheng. B. Liaoning U. **1,** 66 (2021)
7. L.H. Chen, J. Puyang Voc. Tech. Col. **2,** 13 (2020)
8. W.G. Chan, P.Y. Nie, D.H. Peng, Z.H. Li. J. Cleaner Pro. **148,** 112 (2017)
9. J.J. Wang, X.P. Jia, Resources,Con.& Rec. **170,** 1 (2021)
10. H.Cui, R.Wang, H.Wang, J. Cleaner Pro. **269,** 14 (2020)
11. C. Falco, V. Rotondi, D. Kong, V. Spelta, Ecological Eco., **188,** 1 (2021)
12. S. An, B. Li, D. Song, X. European. J. Ope.Res., **1,** 126 (2020)
13. T. Muganyi, L.N. Yan, H.P. Sun. Environmental Sci. Eco., **7,** 5 (2021), http://creativecommons.org
14. C.H. Yu, X.Q. Wu, D.Y. Zhang, S. Chen, J.S. Zhao. Energy Pol., **153,** 1(2021)
15. J.Wang, X. Chen, X. Li, J. Yu, R. Zhong. P-B Finance J., **60,** 1(2020)
16. M.S. Zhang, Y. Lian, H. Zhao, X.B.Chun. Energy Str. Rev., **30,** 1(2020)
17. X.H. Xie, B. Tomsk S. U. Eco., **54,** 235 (2021)

18. Y.Shen, Z. W. Su, M. Y. Malik, M. Umar, Z. Khan, M. Khan. Total Env., **755,** 1(2021)
19. H. Xu, Q. Mei, F. Shahzad, S.Liu, X. Long, J. Zhang, Sustainability, **12,** 21(2020)

https://doi.org/10.1051/e3sconf/202131103002

Compensation ecogenic harm: experience of legal regulation

Yulia Sergeevna Cherepantseva[1,*], and *Elena Mikhailovna* Ruzaeva[2]

[1]Department of Labor and Social Security Law of the Orenburg Institute (branch) of the O.E. Kutafina (Moscow State Law Academy), Orenburg, Russia
[2]Department of Civil Law and Procedure, Orenburg State University (OSU), Orenburg, Russia

Abstract. The article is devoted to the elucidation of the problems of compensation for environmental damage caused to human life and health by the adverse effects of the environment caused by radiation exposure. Based on the fact that the radioactive factor is the primary cause causing ecogenic harm, a definition of such harm is given, its content and characteristics are disclosed, the provisions of the current legislation that enshrine the main legal means to compensate for the harm caused to the victims are analyzed.

1 Introduction

A healthy and supportive environment is one of the main conditions for a safe human existence. The problem of environmental pollution is most clearly identified before mankind due the rapid development of industry and development of new technologies. Wherein, spheres were opened that were completely unknown to people, and the consequences of their use were only assumed. One of them was radiation. Being both a blessing and a disaster, radiation has confidently entered modern life, becoming an integral part of it.

It is impossible not to notice that nature is characterized by natural pollution by radiation. It comes from the earth's crust, which contains a certain amount of radioisotopes. Radiation hits the earth from space. However, the increasing anthropogenic impact associated with man-made disasters, testing of nuclear weapons, operation of atomic and nuclear enterprises and their radioactive waste has become the main factor in the negative impact on the environment. The number of territories where the dose rate from the radiation of technogenic radionuclides is dozens of times higher than the background that existed in the pre-atomic period is steadily growing [1]. Unfortunately, we have to admit that of all known types, radiation pollution remains one of the most dangerous and causes a social risk to the life and health of the population. According to the conclusions of the UN Scientific Committee on the Effects of Atomic Radiation (2000), the pathologies caused by radiation after the accident at the Chernobyl nuclear power plant are thyroid cancer, blood diseases, anomalies in the development of children in parents exposed to the factors of a radiation accident. In the remote period after the accident, the pathology of the digestive system

* Corresponding author: chus1973@mail.ru

ranks third in the structure of primary morbidity, second only to diseases of the organs of the cardiovascular and respiratory systems [2]. The reason for this is the penetrating effects of radiation. As a result of the release of radioactive substances, damage to the territory, objects and all living organisms is caused at a considerable distance.

Consideration of the issues of compensation for environmental damage has become the subject of study of legal science in connection with development of environmental legislation as a result of major radiation accidents that have occurred. However, for a long time, the definition of environmental harm caused to a person, including as a result of radiation exposure, its volumes and methods of compensation, was not included in the sphere of legal study, remaining the prerogative of economics, medicine and sociology. In this connection, this article is an attempt to fill the gap, substantiate the legal structure of environmental harm and assess the formed methods of compensation for such harm from the standpoint of a critical analysis of the current legislation in the area under consideration. As a result, the main positions are determined that ensure protection of violated rights, among which the primary is the right to life and health protection.

2 Materials and Methods

The issues of environmental pollution and the problems of radiation disasters from the point of view of their social component, in scientific and practical terms, have been studied since the middle of the last century. Over the years of fruitful scientific study, considerable theoretical, methodological and empirical material has been accumulated. However, its entire resource is related either to the study of the catastrophe itself as a phenomenon (natural or technogenic), or its components, which determine the direction of social processes, leaving outside the legal impact on the relevant social relations in order to regulate them. The fact that it is in the law that the methods and means of influencing the consciousness and behavior of people are determined, determined the choice of study methods.

The methodological basis of the study presented in the article is formed by the dialectical method of understanding radiation accidents and catastrophes as social and legal phenomena, scientific analysis and synthesis, with the help of which the legal structure of environmental harm is developed and its concept is determined. The characteristics of ecogenic harm are revealed on the basis of the structural-functional method. Use of such private scientific study methods as specific sociological and comparative legal methods allowed the author to reveal the legal means of compensation for the harm caused to a person, his/her life and health, and to determine the effectiveness of their legal regulation. Wherein, all the selected methods for obtaining specific results complement and develop fundamental knowledge on the problem under consideration.

3 Results and Discussion

The modern world has become an atmosphere where people cannot feel reliably protected. Inability to live peacefully, work productively, have a family and children, not worry about their health and the health of their loved ones and many other factors create a stressful situation that significantly affects a person's life. The growing number of attacks on human life and health is due to various reasons. One of them was the use of such an energy source as radiation. Despite the long period of its presence in the life of society, the consequences of its influence on a person have not been fully studied. Unfortunately, we are forced to admit that when using it, the state cannot guarantee absolute security. Therefore, the main task is to develop effective ways to minimize the possible harmful effects caused by the adverse effects of radiation. In this case, a special role belongs to legal regulation, since through creation of legal norms, real and proper fulfillment of the requirements prescribed by the legislator is achieved. One of them is compensation for harm caused.

It shall be noted that there is extensive legislation governing the rules and procedures for causing harm by negative changes in the natural environment caused by human activities. Formulation of the issue of compensation for harm in the rule of law shall begin with an analysis of the Constitution of the Russian Federation. The Basic Law securing the rights of citizens proclaims the right to a favorable environment, reliable information about its condition and to compensation for damage caused to their health or property by an environmental offense (Article 42).

The Federal Law "On Environmental Protection" in the provisions of Article 11 reproduces the constitutional rule supplementing it with Article 79 that establishes the rules for compensation for harm caused to health and property of citizens as a result of violation of legislation in the field of environmental protection. So, according to part 1 of this Article, harm caused to the health and property of citizens by the negative impact of the environment as a result of economic and other activities of legal entities and individuals is subject to compensation in full.

In the scientific literature, the harm caused to the environment is commonly referred to as environmental harm. Wherein, from the point of view of adverse consequences for humans, it is defined as ecogenic. However, at present there is no consensus in the legal doctrine on the definition of this type of harm. From the standpoint of differentiation by the object of impact, some researchers distinguish humanitarian environmental harm, suggesting that it is inflicted on a person for whom environmental protection activities are carried out [3]. A number of authors, for example, do not consider ecogenic harm as an independent type and include it in the composition of environmental harm, considering them synonymous with each other [4]. This point of view seems to us not objective, since it is obvious that initially harm is caused directly to the environment, and the negative changes already caused by this, which have occurred as a result of unfavorable anthropogenic impact (and radiation exposure is no exception here), significantly affect the life, health of people living in the affected area, their property. This, in turn, allows us to consider environmental damage to be derived from environmental damage. Diseases, any other health disorders or a threat to life, including potential ones, reduced working capacity, forced consumption of contaminated products and natural resources, the need to change the place of residence and settle in a new place, etc., are characteristic of ecogenic harm. In this aspect, we share the point of view about the ecological nature of ecogenic harm [5]. Based on such a statement, we believe that among the specific features inherent in ecogenic harm is the fact that it is caused, first of all, to a person, i.e. human life and health. In addition, it is expressed in losses and damages (including non-material ones) that arise from the victim. It is not unreasonable to consider that moral harm can also be included in the composition of environmental damage. In fact, moral harm shall be considered as a form (or element) of environmental harm, since as a result of exposure to radiation, the victims, of course, suffer moral suffering associated with moving to another area, the risk of having children with congenital anomalies or the inability to have children at all, or physical suffering caused by diseases, both their own and their children, or, even worse, their death. There are enough examples of such court cases over the long history of combating radiation consequences. Although among them there are both positive, which allows us to come to the above conclusions (the Nazhmutdinovs' case on the recovery of moral damage from the Mayak PA in the amount of 50 million rubles, the decision of the Ozersk City Court of the Chelyabinsk Region on April 7, 1997), and negative (refusal to recover moral damage in the amount of 5 million rubles in the Khasanovs' case, 2013).

Considering the above circumstances, we believe that harm caused to a person as a result of negative changes in the environment, resulting in damage to health, its loss or loss of life, the costs of restoring health and compensation to the victims of the losses incurred, as well as their moral experiences and physical suffering, shall be recognized as ecogenic.

The problem of compensation for environmental damage caused by radiation exposure is legally complex. On the one hand, it is characterized, as already mentioned, by ecological nature; on the other hand, it is anthropogenic, since caused to a person, his/her health, the state of future generations.

Focusing on the general issues of compensation for environmental damage, we note that it is carried out according to the rules of several branches of Russian law, in connection with which it is considered one of the most important intersectoral institutions, due to the fact that it includes the norms of civil and environmental legislation [6]. From the standpoint of civil law regulation, Articles 1064-1098 of the Civil Code of the Russian Federation give us an idea of the harm caused to the person and property. However, considering the specifics of the harm caused to the environment, one cannot but emphasize use of the norms of environmental legislation, which directly indicate its form, inflicting subjects, sizes, methods of calculating compensation, and etc. Therefore, it is possible to clearly trace the ecological specifics of the harm that is compensated in the manner of civil law relations. Environmental legal regulation enshrines the principle of compensation for damage caused in full, which indicates that the victim shall receive full compensation, regardless of the limits of civil liability.

The general provisions of civil law also establish the rule that damage is compensated by the person who caused it. Wherein, being formed as a result of radiation exposure as a special factor, further individualizes it. Therefore, according to Articles 56 and 57 of the Federal Law "On the Use of Atomic Energy", a specific feature of the damage caused by radiation exposure is its provision not only by the inflicting enterprise, but also by the state. Wherein, the operating organization itself shall be obliged to have financial security for the limit of civil liability and compensates for the damage caused within this scope, and compensation for damage to the full amount (i.e., over the volume established for the economic entity) is provided by the state.

When solving the issues of compensation for harm associated with the adverse impact of the environment, the methodology traditionally used is the evaluation of harm caused to human health. When assessing the physical impact, the damage to the population is monetized, i.e. the cost equivalent is determined. For example, in the EU countries the Impact parthway appoarch or IPA method is used as an official method for determining the cost of the equivalent of damage to public health from adverse environmental factors [7]. An analysis of the current legislation shows that the Russian Federation has also followed the path of determining the damage caused in monetary terms. Although, at first glance, this looks somewhat illogical - while proclaiming that life and health are priceless, the price, nevertheless, is determined by it. However, this is precisely the specificity of the ecogenic damage caused by radiation, when the state assumes the responsibility to provide assistance to the victims.

According to the Resolution of the Plenum of the Supreme Court of the Russian Federation of November 30, 2017 No. 49 "On Some Issues of the Application of Legislation on Compensation for Harm Caused to the Environment", cl. 17 defines two forms of compensation - in kind, if it is objectively possible, and if irreplaceable and (or) hard-to-recover losses – in monetary form. Unfortunately, we have to state the fact that the monetary form of compensation for environmental damage has long been a source of criticism. First of all, because the funds received as compensation for harm from violators of environmental legislation are not isolated and often, having become part of the federal budget, can be spent on purposes other than environmental restoration. Its opinion on this issue was expressed by the Constitutional Court of the Russian Federation, which considers that the harm caused to the environment is often irreparable, and the state has the right to send the received sums of money to the budget as compensation, and not the restoration of the environment (Resolution of the Constitutional Court of the Russian Federation of June 2, 2015 No. 12-P). Wherein, the monetary form is additional and compensates only that part that cannot be reimbursed in kind.

The specificity of the harm caused by radiation exposure is its classification as irreparable and incalculable. This circumstance determines the need for a special approach to ensure the victims of the restoration of violated rights. In view of this, based on the principle of the maximum possible use by the state of the funds at its disposal, it can be said that the main way to compensate for environmental damage is monetary and other material benefits and compensation for damage caused. The lack of legal means possessed by civil, administrative, criminal and other sectoral legislation allowed the highest judicial body to offer a constitutional justification for the legal mechanism of social protection of citizens affected by radiation as aimed at fulfilling the constitutional and legal obligation arising from the right to a favorable environment, reliable information on its condition and on compensation for damage caused to health or property by an environmental offense [8]. Due to the fait accompli of radiation exposure, we are talking purely about compensation, because it is almost impossible to prevent social risk. Raising the question of prevention is permissible only when its impact (including on humans) is minimized. That is why the choice of such a legal mechanism is not accidental and is due to the fact that state guarantees are possible only within the framework of public law models obliging the state to provide a citizen with social benefits and services in the manner and under the conditions established by law [9]. The targeted orientation of these legal means most of all ensures compensation for the unfavorable consequences that have arisen among citizens, rather than other methods. Protection in this case acts as a measure of legal compensation for lost resources (material, spiritual and others). To ensure the effectiveness of such measures, they are enshrined in the rule of law, normative legal acts are adopted that provide for provision of compensation and benefits to victims. Along with environmental legislation, social legislation has been formed, the main purpose of which is to compensate for the damage to the health of victims of radiation and provide them with additional benefits (medical, housing, labor, etc.). The choice of such legal means is not accidental. So, for example, according to the All-Russian classifier of information on social protection of the population, benefits are ranked among the most effective measures for its implementation [10]. Support through provision of benefits aims to provide protective measures in order to preserve the life and health of the victims. Moreover, according to the Constitutional Court of the Russian Federation, the state, based on the principle of the maximum possible use of funds for compensation, can choose any methods (benefits, monetary or material compensation). The main thing is that it is unacceptable to cancel or reduce the previously recognized scope (Determination of the Constitutional Court of the Russian Federation of June 14, 2006, No. 273-O). In this understanding, the achieved level of protection of the rights and freedoms of victims, guarantees of their social protection will be ensured.

The main regulatory legal act which embodied all known types of social protection of affected citizens was the Law of the Russian Federation of March 15, 1991, No. 1244-I "On Social Protection of Citizens Exposed to Radiation as a result of the Chernobyl Disaster" (hereinafter referred to as the "Law on the Chernobyl NPP"). Initially, its norms were calculated for all known cases of negative radiation impact that took place in the history of our state. In fact, it can be called a universal act, the effect of which extends both to citizens affected by the results of the activities of enterprises producing radiation substances (for example, the Mayak Chemical Plant), and to participants in nuclear weapons tests or radiation accidents (for example, tests at the Totsk test site in the Orenburg region, the

Novaya Zemlya archipelago, and etc.). Only in 2002, in connection with adoption of Federal Law of January 10, 2002 No. 2-FZ "On Social Guarantees to Citizens Exposed to Radiation as a Result of Nuclear Tests at the Semipalatinsk Test Site", the relevant category of victims was removed from the scope of the Law on the Chernobyl Nuclear Power Plant. Despite more than 30 years of practice in application of the relevant legal norms, their effectiveness is one of the problems of ensuring full protection of victims. Its application revealed insufficient and irregular funding of social protection measures, obvious failures in the programs of medical assistance and treatment of victims, providing them with housing, and etc. Therefore, a huge number of citizens' appeals to the courts for protection of their rights has become an important step towards improving the legal means of compensating for environmental damage caused by radiation exposure.

The specificity of the type of harm under consideration shall also be recognized that Clause 22 of the above-mentioned Resolution of the Plenary Session establishes an exception for this type of harm in terms of the limitation period. If, according to the general rule, the limitation period for compensation for environmental damage is 20 years (Clause 22), then the damage caused by radiation exposure is compensated for within three years from the day when the person learned or should have learned about the violation of his/her right (Article 58 of Federal Law of November 21, 1995, No. 170-FZ "On the Use of Atomic Energy").

4 Conclusion

The state of the environment is a matter of increasing concern every year. Its pollution as a result of anthropogenic impact poses a real threat to the population. Among the most dangerous factors causing negative changes in the environment and adversely affecting human health and life, radiation exposure should be called. The radiation accidents and catastrophes that occurred in different years of the development of our state became the starting point in understanding the danger of radiation for humans. Society has come to understand that modern technologies and production do not forgive the lack of control over risk. The radiation threat as a powerful negative factor undermines the vital resources of the territories and the people living in them. Moreover, in its overwhelming majority, the effect of radiation is already a fait accompli, which was either impossible to prevent from the very beginning (as, for example, in the case of the disaster at the Chernobyl nuclear power plant), or the very fact of radioactive training received by people was not given importance and no measures were taken to its prevention (the activities of Mayak Production Association, testing of atomic weapons at the Semipalatinsk or Totsk test sites, etc.).

Features of compensation for harm to a person caused by adverse changes in the environment, first of all, lies in the specifics of harm. Such harm is ecogenic. It is derived from environmental harm and is characterized by the fact that it is caused directly to a person, leads to damage to health, its loss or loss of life, causes the cost of restoring health and compensation for victims of losses incurred, and is also expressed in moral experiences and physical suffering.

The main task facing the legislator in the event of ecogenic damage is development of effective legal means to compensate as much as possible for the losses incurred. According to general rules, damage caused to the environment by radiation exposure is compensated for under the rules of civil, environmental, administrative and other legislation. However, due to the special danger of such an impact, which is recognized as a social risk for a person, his/her life and health, for the most complete compensation, public law is used.

Recognizing responsibility to citizens for the consequences of radioactive contamination of the environment, the state adopts normative legal acts aimed at protecting them, in the provisions of which legal means are enshrined that make it possible to compensate for environmental damage. However, not all of them can be considered effective and efficient. There are many reasons for this. In this connection, the legal problems of compensation for environmental damage caused as a result of exposure to radiation continue to remain complex, both in theory and practice. Most significantly, because they affect the most valuable thing for a person – his/her life and health.

References

1. A. V. Korsakov, A. S. Domahina, V. P. Troshin, E. V. Geger, Human Ecology, **7**, 4-14 (2020)
2. S. V. Dudarenko, S. N. Lopatin, O. V. Leonteva, Human Ecology, **1**, 11-16 (2020)
3. E. A. Zhulay, E. G. Cherkashina, Bulletin of the AmSU, **42**, 32-36 (2008)
4. M. I. Vasilyeva, State and Law, **10**, 26-36 (2008)
5. M. M. Brinchuk, *Ecological Law*, 472 (2005)
6. A. Ya. Ryzhenkov, Bulletin of the University named after O. E. Kutafina, **3**, 77-85 (77) (2020)
7. V. V. Dyadik, N. V. Dyadik, E. M. Klyuchnikova, Human Ecology, **2**, 57-64 (2021)
8. Yu. S. Cherepantseva, Legal positions of the Constitutional Court of the Russian Federation as the basis for development of a system of social protection of citizens affected by the effects of radiation, For Workers' Rights! Protection of the social and labor rights of workers in a changing world: opportunities and limitations: materials of the Sixth International Scientific and Practical Conference, Yekaterinburg, December 9-10, 169-173 (2020)
9. E. A. Istomina, M. Yu. Fedorova, *Legal mechanism of social risk management*, 240 (RANEPA, Yekaterinburg, 2018)
10. A. G. Repyev, Bulletin of St. Petersburg University. Right, **12(1)** (2021)

Ecology mediatization: strategy of biological resource management (on the example of the RNO-Alania)

Sergey Golovko[1,*], *Liudmila* Kokhanova[2], *Tatiana* Alekseeva[2], *Yulia* Chereshneva[2], and *Maria* Gubanova[3]

[1]Russian State University for Humanities, 125047, Miusskaya Sq., 6, Moscow, Russia
[2]Lomonosov Moscow State University, 125009, Mokhovaya St., 9, Moscow, Russia
[3]Russian Presidential Academy of National Economy and Public Administration, 119571, Vernadsky Prospect, 82, Moscow, Russia

Abstract. The article shows how the process of the formation of environmental mediatization concept contributes to the implementation of the bioresource management strategy. The research has been conducted in the subject of the Russian Federation, the Republic of North Ossetia-Alania. This region has been selected as rather complicated and environmentally vulnerable due to the fact that natural resource and ethno-cultural potentials pose severe challenges there. The final sample consists of 150 publications of an informational and analytical character published in the periodicals of the Republic of North Ossetia-Alania from 2017 to September 2021. These days, network information resources are considered the most effective tool of mediatizing ecology and developing a strategy for managing biological resources. However, as a result of the spontaneous approach to the formation of the RNO-Alania as an environmentally pristine region, there is an imbalance of both information coverage and coverage of environmental issues in comparison to other topics. The conducted research has made it possible to outline the parameters of the spontaneously developing network media system and highlight their main characteristics. Thus, the media image of the RNO-Alania as an environmentally friendly region nowadays basically does not meet the requirements and challenges of modern times.

1 Introduction

Environmental component is an integral part of mediatization processes which, in their turn, characterize the life of contemporary society and all its social institutions. Academician N. N. Moiseyev claims that "mankind is challenged with the necessity of reconsidering the most profound bases of their world outlook, identifying their niche in nature, their interaction with the environment, reconsidering their civilization paradigm [9].

In order to as soon as possible realize the significance of the place of a human being in nature and the results of their activities, it is relevant to create a highly-developed and

* Corresponding author: Sgolovko@inbox.ru

easily accessible strategy of biological resource management. This is what the majority of environmental programs emerging in the region are aimed at. They set the rational usage of natural resources as their paramount goal. One of the examples of these is the governmental project "Environmental Protection, Ecological Security, and the Welfare of the Republic of North Ossetia-Alania" in 2020-2025 (as amended on November, 26, 2019). The program was approved and adopted in the Republic of North Ossetia-Alania [5].

Practice shows that the prospect of strategy realization stated in the program largely hinges on how thoroughly the information on the republic's environmental state will be presented in the public sphere, and to a huge extent, in network media space. These days, the Internet in various formats plays the key role in forming the information-communicational system of the country. Hence, the impact of the Internet on every citizen and various forms of their activities and society in general is undeniably on the increase.

The goal of this particular research is to show how the emergence of the concept of environment mediatization by network information resources promotes the realization and implementation of biological resource management strategy. The RNO-Alania has been selected as an example because of its complexity and environmental vulnerability. It is the region where "the natural resource and ethno-cultural potentials obviously tend to decrease and decline. An enormous number of conflicts only deteriorate the current situation." [8].

Nevertheless, the RNO-Alania participates in 50 out of 67 federal projects that comprise national projects including the national project "Ecology". The republic came third on the list of the amount of financing in 2019 in the Russian Federation. This fact indicates that the region is in the process of developing the strategy of its biological resource management, which contributes to environment mediatization as an indispensable part of social life.

The set goal consists in resolving a few tasks:

- to outline the parameters of spontaneously forming system of network media covering environmental issues;
- to determine the key topic areas;
- to characterize the theme of ecology as one of the basic components of forming media image of the RNO-Alania as the region which develops and implements the biological resource management strategy.

2 Materials and research methods

The study of environment mediatization in the process of implementing the bioresource management strategy in media space of the RNO-Alania is based on the principles of comparative analysis of complex scientific methodology, historical-cultural analysis, and interdisciplinary approach. The basic ideas of the principle of mediatization were laid in the works of M. McLuhan [7]. In recent years, these ideas have been developed in the research papers of the representatives of various scientific schools. The analysis of special literature on the selected topic has revealed that researchers are mostly interested in the mediatization of such social institutions as the government, law, culture, and religion. Environment mediatization is paid less attention to. Moreover, environmental and ecological information mostly bears a certain amount of negative connotation, which causes tension in society.

Both, journalistic and natural-scientific papers have proved really helpful in the process of developing the methodology of the research below. Above all, these are the papers studying the media facilities and opportunities in creating a media image [6]. Besides the abovementioned works, these are the ones that define the specificity of the mountainous territories of the RNO-Alania, which enables structuring information themes as the method of rational nature management and nature protection activities. These papers include the works of D. V. Abbasov [1], D. V. Bubnov [3], Kh. Kh. Makoyev [8], I. K. Gerter and V.

S. Misakov [4], L. Sh. Akhmedova and G. I. Idziyev [2], N. F. Reimers [10], I. Khuzmiev [13] and others.

The selection of this region by the authors of this paper has not been made randomly. First of all, it is due to the fact that we live in the country where, according to the statistics presented by G. S. Samoylova, "out of 89 subjects of the Federation, 43 have mountains. The area of these regions amounts to 85,5 % of the total area of the Russian Federation, in which mountains form 63% of the area, i.e. mountains make 53,6% of the territory of Russia." [11]. Secondly, Kh. Kh. Makoyev claims that "The Northern Caucasus is the only region in the European part of Russia which maximally meets all the classic criteria and standards of mountainous countries: in altitude and disjunction of landform, landscape-climatic and ethno-cultural diversity, and household and settlement peculiarities." [8].

At present, there take place such processes as rapid development of information-communication technologies, transformation of conventional mass media and creation of their hybrid forms, such as media with online versions as a must-have and evolution of social networks, which consequently leads to a spontaneously forming system of network media or information resources. Ever-increasing amounts of information and audience segmentation are among the characteristic features of this system [12].

The following empirical methods have been applied in the research paper: media space monitoring, online observation, structural and visual analysis of network media and their content, expert survey. Methods of random sample have been employed in selecting the empirical material. The base of media texts has been formed to include the name of the RNO-Alania as the subject of the RF mentioned together with the keyword "ecology" in headlines and abstracts. From this media text base there have been selected 15 network information resources later united into 5 structural media groups:

- online versions of conventional periodicals;
- network periodicals;
- specialized scientific resources;
- information resources including official websites of events;
- websites of organizations and state institutions.

The first group includes both federal periodicals ("Rossiiskaya Gazeta", "Kommersant", etc.) and local mass media (the republican newspaper "Severnaya Ossetia", television and radio of the RNO-Alania, and others). The final sample is presented by 150 information and analytical articles published in these periodicals from 2017 to September 2021.

3 Research results

The article below partly presents the results of the research based on the analysis of the information-communication space of the RNO-Alania and the trends stipulating its evolution. A number of tasks have been formulated and accomplished during the study. First of all, the primary condition for environment information to be considered as the method of rational nature management is its presence in the region's information-communication space. In this particular case, it has been determined that there is such information in the RNO-Alania's informational sphere.

The second, no less important, is the condition of structuring information-communication space in which environmental information functions. In the republic it includes such traditional segments as print media, radio, and television. During the past few decades this condition has changed dramatically due to the emergence of online versions of conventional mass media. Thus, the republic's daily newspaper "Severnaya Ossetia" comes out not only in the paper version, but also has its website (http://sevosetia.ru/Article/Index/229994). On this platform it is possible to see the paper's previous issues and how often it covers environmental topics. Among such articles the following can be mentioned: “Awards to Environmental Activists” (07.12.2018), “How Do We Respond to Environmental Challenges?” (07.02.2019) by S. Suanov, and others. It is noteworthy that the correspondent of “Severnaya Ossetia” has been awarded the first prize for the best article on this topic in print media. This environmental contest has been held by the North Ossetia Republican Department of the All-Russian Environmental Protection Society.

Let us give a little example of the article published in “Severnaya Ossetia” on the environmental issues (section “Society”. The question of the day. 08.09.2021).

How do you estimate the state of environment in the region?

Murat KUCHIYEV, student of GSAU (Gorsky State Agrarian University), Kosta Village, Ardon Disrtict:

– I believe that in our republic it is enough to spend some time on the bank of any river to understand that nature needs protection: protection from people who by using all the beauty of their local nature shamelessly destroy it. You may ask, “How?” – by piles of rubbish that they do not only leave after an outing near the river, but also throw into the water. Lots of times I’ve seen bottles, watermelon rind, and plastic bags full of dirty disposable dishes. I feel ashamed of people like these, and the state of our rivers makes me feel depressed. Not only rivers, by the way. The same scenes can be witnessed in the mountains, and in this case I feel ashamed of the tourists who see our beautiful nature and our attitude to it.

Yulia AKSYONOVA, the resident of Mozdok who cares, environmental activist:

– Unfortunately, we are insufficiently informed by competent authorities about the environmental situation in the republic to be able to evaluate it. High sickness rate is one of the attributes of poor state of the environment.

Alexandra RADCHINSKAYA, student of North Ossetia State Institute of Pedagogy:

– Of course, in comparison to major regions of Russia the state of environment in North Ossetia is more or less favorable. But this fact cannot serve as a consolation. We lack in modern education in the sphere of ecology, our mountain rivers – the pride of the republic – will soon become plastic, a dustbin or rubbish container is still scarce in many of our streets. And this list of problems is far from being complete. I think we ought to be ashamed of violating the simplest environmental rules.

Valentin KISIYEV, a veterinary:

– I consider the state of environment in our republic satisfactory. At present there are no big factories which could contribute to air and atmosphere pollution. Moreover, factories producing alcoholic beverages are practically non-functional. So, there is no one to dump distiller’s grains into the rivers.

At the same time, the issue of recultivating the territory near the former plant “Electro-Zinc” still remains really acute, and it will probably take long to solve it. Other areas of concern include automobiles which emit harmful poisonous substances into the atmosphere and some people who litter where they like.

Yu. SLANOVA, journalist:

– I’ve recently noticed that dustbins have been removed from many public transport stops. As far as I understand, they are supposed to be replaced by new ones, but their absence makes people drop litter on the ground. I would also like to touch upon the issue of big rubbish containers installed throughout the city. They are frequently overfilled, rubbish falls out of them, and the smell spreads all over the street. The issue of smart rubbish recycling and utilization deserves immediate attention. Despite the fact that special containers are installed throughout the city and in NOSU (North Ossetia State University) as well, not everyone takes the trouble of sorting rubbish out.

Network space is also used by Ossetia’s radio and television that have their electronic addresses (http://alaniatv.ru), which enables interpreting their category characteristic of connection with the time and place in a somewhat different way. Now watching a particular program on TV or listening to it on the radio is possible on the Internet depending on an individual's wishes and preferences. Thus, for instance, the news about the "children of North Ossetia who contributed to environmental protection by their drawings" posted on the Internet at 17.38. on November, 23, 2018, can be accessed at any time convenient for every person.

It goes without saying that the functioning of such resources as "Vladikavkaz Region Online" (news-r.ru), North Ossetia-Alania News (https://news-r.ru/news/north_ossetia_alania/), Region 15 (https://region15.ru/); electronic network periodical "Vercherniy Kurier" (https://vk-smi.ru/); "Kavkazsky Uzel" (https://www.kavkaz-uzel.eu/), and others make information on environmental issues more open and accessible to residents of the region. To some extent, they work to foresee events by informing about topical ecological issues before this news is published in a newspaper or broadcast on TV.

The research paper below has once again confirmed the fact that the Internet is a living creature. As it can be inferred from the network space analysis of the RNO-Alania, currently there are about one hundred information resources in the region which, to a certain extent, deal with the issues of nature management in the republic. Some of these resources are locally based, some operate outside the region. Environmental news of the republic is covered far less frequently by federal information resources.

At the same time, not only press but regional local authorities realize the crucial role of new channels of disseminating information and are gradually starting to use them in their work. Moreover, sometimes periodicals lose to them in terms of the speed of posting on the Internet. Among them it is necessary to mention both servers of state institutions and websites of public organizations. The information-communication space is represented by the official site of the RNO-Alania (http://alania.gov.ru), the Parliament of the RNO-Alania (http://parliament-osetia.ru), and other governmental bodies and institutions. Practically all the ministries and departments of the republic are represented on the Internet by their press offices. Websites of the Ministry of Interior of the RNO-Alania (https://15.мвд.рф) and press office of the Ministry of Environment of the RNO-Alania (http://mpr.alania.gov.ru), Federal State Department "Central Water Resources" of the RNO-Alania (http://fgu-alania.ru), and others give a wide coverage of topics related to the environment.

The fact of using the information of press offices by other information resources proves the statement that managing by means of information as a method is quite efficient. For example, the website "Ossetia-Iryston" publishes the article headlined "Commission on Investigating Cause of Fire at JSC "Electro-Zink" Plans to Complete Work on December, 20". Further on it continues: "If only the results of expert examination of the police and the Ministry of Emergency Situations are ready by this time, according to the press office of the government of the RNO-Alania." (07.12.18 at 17:54).

"Region 15 (North Ossetia)" provides information with reference to information sources – press offices of various departments. For instance, it reports: "Power engineers of North Ossetia have given an eco-class within the all-Russian project "Nature Concerns Children". Press office of the subsidiary of JSC IDNC NC (Interregional Distribution Network Company of the North Caucasus) – "Sevkavkazenergo" (21.04.2017 at 14:51).

One more example of articles published by "Region 15 (North Ossetia)": "At "The Forum of Action" the activists of PNF (the Public National Front) of North Ossetia have mentioned the problem of worn-out water pipes in the region. Press office of PNF of the RNO-Alania." (19.12.2017 at 10:29).

It is worth mentioning that due to the Internet, the information-communication space of the RNO-Alania is dynamically developing, thus expanding its opportunities of information management in the sphere of environmental usage and protection. Today, there are permanently renewed sites, portals, blogger communities on social networks and other new platforms such as Telegram messenger or YouTube. So, "Region Vladikavkaz Online" has its own telegram channel where it posts "topical news interesting for everyone" (https://tele.click/news_r_vladikavkaz). On YouTube the following videos can be found: "Discussing Environmental Protection Issues" (https://youtube.com, 25.10.2016), "Open Dialogue: "Electro-Zinc" Plant, the RNO-Alania (https://youtube.com, 25.10.2018), "In the Ministry of Natural Resources and Ecology of the RNO-Alania" (https://youtube.com, 25.10. 2019).

The official page of the Ministry of Interior of the RNO-Alania on the "Instagram" is two years old. The republic has created its own network "My Caucasus" as the social net of the Caucasus (MyKavkaz) where photos, videos, and blogs are posted and regional forums are stated.

At the same time, the spontaneously forming system of network media possesses, in the authors' opinion, the following peculiarities:

- uneven coverage saturation;
- wave-like frequency;
- prevalence of information genres;
- low emotiveness of media texts;
- communication strategy deficiency;
- explicit nature of publications.

Let us take a more thorough look at all of the above mentioned characteristics.

Uneven coverage saturation. System saturation is determined by the permanence of parameters conditioned by the equality of speed of opposite processes taking place simultaneously.

The survey results have revealed utter irregularity of information product coverage in different regions of the RNO-Alania (Table 1).

Table 1. Saturation with information product.

No.	Region	Number of publications
1	Vladikavkaz	119
2	Alagir District	3
3	Ardon District	7
4	Digor District	1
5	Iraf District	3
6	Mozdok District	7
7	Kirov District	1
8	Pravoberezhny District	6
9	Prigorodny District	3
Total:		150

Wave-like frequency. The number of publications increases in a wave-like manner to react to the new decrees of the President and the government and gradually decreases shortly after another political event is over.

Low emotiveness of media texts. Ecological issues coverage in contemporary mass media is distinguished by an emotionally neutral character. Transmitting aggressively colored news makes a huge impact on the audience, and thus enables interpreting events by manipulating public perception and imposing positive or negative attitudes to the audience.

However, these days, evaluative vocabulary and stylistic devices of all kinds have given way to dry balanced coverage of events and facts. Thus, in most articles on environmental issues studied with the application of content-analysis methods emotional expressive vocabulary (either positive or negative) was not found.

Prevalence of information genres. Genre diversity of contemporary environmental discourse is extremely scarce. The overwhelming majority of publications are presented in the genre of informational notes (no more than 500 words) characterized by dry unemotional narration. Analytical interview has given way to an information one. Lack of analytical materials may account for the authors' unwillingness to seriously tackle the problem of resolving the region's challenging environmental issues.

Communication strategy deficiency. If the discourse of federal periodicals is basically formed by the current governmental policy of the region and aims at the maximal audience scope, the topics of local media are, as a rule, narrow and are not intended for large groups of readers.

Staff reporters "on the spot" do not usually associate their activities with the news about this particular community and local journalists deeply absorbed in reality are unable to "raise and develop the issue" due to the narrowness of editorial policy.

Internet periodicals, especially federal news agencies (TASS, RIA "Novosti", Russia Today, etc.), can be referred to as a powerful information resource with a vast network of news bureaus, technologies, and sources of information. But loose, scattered structure of thematic presentations hinders the formation of any real viable structures.

The same refers to the specialized scientific resources which, unlike news agencies, possess a deeper approach to understanding themes. In this particular case, the depth of analyzing together with thematic narrowness does not create the synergetic effect of communicative influence and, consequently, the effect of a complete media image.

Explicit nature of publications. Absence of the rigid "agenda" cannot contribute to the creation of a full information picture of the region's environmental problems: that is why the majority of materials which form the information discourse core are explicit and non-conceptual in character.

4 Conclusions

Nowadays, network information resources are considered as the most efficient tool of ecology mediatization and biological resource management strategy development. However, as a result of a spontaneous approach of network information resources to forming the image of the RNO-Alania as a pristine area, there arises imbalance of both information coverage and environmental issues coverage in comparison to other topics.

The conducted research has enabled outlining the parameters of the spontaneously forming system of network media and defining their basic characteristics: uneven coverage saturation; wave-like frequency; prevalence of information genres; low emotiveness of media texts; communication strategy deficiency; and the explicit nature of publications.

Media activities are not coordinated, information strategies are more declared rather than implemented in practice. The process of shaping the media image of a unique republic by means of network resources is rather spontaneous guided mostly by momentary interests of a particular editorial management.

According to statistics, environmental topics are termed as of paramount significance in the State Program "Environmental Protection, Ecological Security, and the Welfare of the Republic of North Ossetia-Alania in 2020-2025 (as amended on November, 26, 2019)" which is adopted in the Republic of North Ossetia-Alania [5]. However, this program has so far not been fully understood by those who are supposed to implement it.

The conducted analysis of articles devoted to environmental issues has revealed a rather one-sided prejudiced approach to forming the thematic core of media texts. So, most of the publications are created in the genre of information note, there is lack of analytical materials, and not all the trends of social policy are given prominence to. The scarcity of analytical articles may account for the authors' unwillingness to seriously tackle the problem of resolving complicated environmental issues.

Thus, the media image of the RNO-Alania as an ecologically pristine area created by the network information resources does not fully meet modern challenges and requirements of the present day.

References

1. D. V. Abbasov, Terra Economicus, **12**, 202 (2014)
2. L. Sh. Akhmedova, G. I. Idziyev, Regional Economy: Theory and Practice, **18**,12 (2013)
3. D. V. Bubnov, Thes. for Ph.D. in Economics, Volgograd SU (2014)
4. I. K. Gerter, V. S. Misakov, Terra Economicus, **10**, 123 (2012)
5. State Program of the Republic of North Ossetia-Alania "Environmental Protection, Environmental Security, and the Welfare of the Republic of North Ossetia-Alania" in 2020-2025 (as amended on 19.12.2017), https://docs.cntd.ru/.
6. M. V. Gribok, INTERCARTO. INTERGIS. **26**, 94 (2020)
7. M. McLuhan, Hyperborea. Kuchkovo Field (2007)
8. Kh. Kh. Makoyev, Izvestiya TulSU: Theor.Sciences, 2, 234 (2009)
9. N. N. Moiseyev, MGVP COX (1995)
10. N. F. Reimers, Mysl (1990)
11. G. S. Samoylova, Izv. RAN. Series: Geography, **2**, 110 (1999)
12. A.V. Chizhik, Communicology, **5**, 55 (2017)
13. I. Khuzmiev, https://helpiks.org/

Cluster approach to production and consumption waste management in regional socio-economic systems

Olga Timofeeva[1], *Irina* Minakova[1], *Tatyana* Bukreeva[1,*], *Svetlana* Starykh[1], and *Lucreţiu* Dancea[2]

[1]Southwest State University, Faculty of Public Administration and International Relations, 305040 Kursk, Russia
[2]Banat's University of Agricultural Sciences and Veterinary Medicine 'King Michael I of Romania' (USAMVBT), 300645 Timişoara, Romania

Abstract. Nowadays, the 'greening' of the economy based on implementation of modern development priorities such as increasing the value of nature, natural resources, as well as a human being, human life and health has the prime importance for achieving sustainable development. Solving the problem of production and consumption waste is becoming one of the priority areas, since it simultaneously contains the following components: economic (disposal cost, saving natural resources), environmental (reducing releases of harmful substances into the environment) and social (new jobs, reducing the burden of diseases, fostering an attitude of care towards nature among the population). The creation and development of regional clusters of waste processing can be a key element in solving the problem.

1 Introduction

Modern world development is fraught with negative impact on the environment, depletion of natural resources, imbalance in the biosphere, and high rate of waste generation of various origins. It exacerbates social and environmental problems, and limits opportunities for further development. According to official data, 65 million tons of waste were produced in Russia in 2019 [1]. The volume of waste discarded annually increases by 1-2%, and by 2050 it will reach 100 million tons annually [2].

The current situation both in Russia and in the world with generation of production and consumption waste determines the priority of the problem for ensuring sustainable development.

Sustainable development is based on a harmonious combination of the triple priorities in sustainability, i.e. economy, nature, and society. However, throughout the 20th century, the focus was exclusively on economy, it led to a significant increase in anthropogenic impact on the environment. Although the term 'sustainable

* Corresponding author: tnbinchina@yandex.ru

development' has been actively used since the 1990s, it most often means long-term stable socio-economic development, relegating the problem of preserving natural ecosystems to secondary importance.

As a result, it is necessary to deepen the 'greening' of the economy based on implementation of modern development priorities such as increasing the value of nature and its resources, as well as a human being, human life and health. The implementation of sustainable development goals based on a market economy includes ensuring the harmonization of market requirements for economic benefits while maintaining a favourable environment.

It is essential to support production modernization in the so-called 'double win' way, ensuring both economic efficiency and reducing the impact on the environment. The world experience of such production modernization shows that this approach is a way to ensure not only the growth of economic indicators, but also to improve the quality of life.

2 Methods and Materials

The study is based on general scientific research methods (analysis, synthesis, induction and deduction) and built upon the works of Russian and foreign scholars: B. R. Alzamora and R. T. de V. Barros [3] studied municipal waste management charging methods in different countries, D. E. T. Cervantes, A. L. Martinez, M. C. Hernandez, and A. L. G. de Cortazar [4] analysed indicators for evaluating municipal solid waste management, J. E. Santibanez-Aguilar, J. M. Ponce-Ortega, J. B. Gonzalez-Campos, M. Serna-Gonzalez, and M. M. El-Halwag [5] researched optimal planning for the sustainable utilization of municipal solid waste, C. Cheng, R. Zhu, R. G. Thompson, and L.-H. Zhang [6] examined multiple-stage solid waste management systems, B. D. Azevedo, L. F. Scavarda, R. G. G. Caiado, and M. Fuss [7] studied the German experience in improving urban household solid waste management. L.A. Mochalova and M.V. Polezhalova [8] substantiated the basic principles of solid municipal waste management; finally, H. Wiesmeth, and N.V. Starodubets [9] considered the solid municipal waste management in compliance with circular economy criteria.

3 Results and Discussion

Considering waste as 'rubbish' that should be disposed of, waste disposal costs have a major importance. However, this approach leads to national economic losses of potential secondary raw materials, natural resource depletion for future generations and rise in environmental impacts.

At the same time, the occurrence of advanced waste management system, where waste can be a resource for further production processes, raw materials for energy, etc., can serve as a platform for achieving the initial sustainable development goals. Moreover, the existing waste management systems have demonstrated their low economic and social efficiency, and led to the aggravation of a number of problems.

Waste generation can cause the following main irreversible damage:

- moreover, the use of land for siting landfills, that could be developed for construction and agricultural purposes, establishment operation, reclamation of landfills, and its maintenance after closure all these require significant capital expenditures;

- waste is a valuable energy raw material, reuse of waste saves natural resources, as well as the energy required for their extraction. In this regard, it is necessary to form the attitude towards waste as a potential resource in the society;

- costs for mental and physical health care;
- resource use fees, a pollution fee, environmental monitoring charges, etc.

Secondary flow of waste will increase the sustainability of regional socio-economic systems due to:

- economic effect (lower cost of secondary raw materials in comparison with primary ones, reduction of territories allocated for landfills and taken out of circulation);
- social effect (new jobs, fostering an attitude of care towards nature among the population, reducing the burden of diseases);
- ecological effect (saving non-renewable natural resources, reducing landfilled waste and releases of harmful substances into the environment).

The sustainable development of regional systems is ensured by a systematic approach to the development of their potential. Scientific literature identifies the following types of potentials for regional development:

- material and technical segment (ecological, natural resource, economic-geographical, demographic), i.e. block of basic resource potentials of the territory;
- financial and economic segment (labor, production, budgetary, socio-infrastructural, export-import) combines local potentials, designed to contribute to the basic resource potentials;
- innovation and institutional segment (scientific and innovative, investment, regulatory support) determines the region's readiness for socio-economic transformations.

At the same time, due to the fact that waste as secondary raw material expands regional resource potential, the authors suggest adding 'resource potential of production and consumption waste' into the material and technical segment. It should be highlighted that considering municipal solid waste (MSW) as 'natural resource potential' is not correct, since it is not a natural resource, but anthropogenic, generated as a result of human activity. In addition, some aspects of waste flow as a part of economic turnover do not allow including WSB to any other type of regional potential, but it is quite possible to determine the relationship between the segments.

The use of secondary raw materials i.e., the resource approach to solving the waste management problem will expand the regional resource potential and increase its ecological capacity and social capital.

The cluster approach, *widely implemented* by the economies of many countries, provides the business entity with a number of advantages: sustainable development and advanced economic competitiveness; solving the problems of modernization and technological development; dealing with the insufficiency of investment, personnel, energy, raw materials and other resources; acceleration of innovative development processes; stimulation of active interaction between industrial enterprises, state and end-users [10].

Thus, according to E. S. Zakharchenko, 'a cluster of waste processing enterprises is a territorial system that provides coordinated interaction of the subjects of production and consumption waste management for the implementation of secondary material resources flows in order to ensure an eco-friendly and safe living environment' [11].

According to the authors, this definition should be expanded by taking into account not only the flows of secondary material resources, but also hazardous waste, otherwise, the authorities' interest in this cluster decreases. In general, a cluster of waste processing enterprises is a complex of enterprises of different types of business industries involved into life cycle of waste. 'Clustering partnership occurs initially on an intersectoral basis. Waste generated in one industry can be used as a secondary resource in the same industry and in other industries. The circulation of secondary resources between business industries contributes to the intersectoral integration of knowledge and the occurrence of fundamentally new innovative solutions' [11].

Federal Law No. 488-FZ dated December 31, 2014 'On industrial policy in the Russian Federation' marks the industrial cluster as 'a set of industrial entities, tied together with relations in the certain industry due to territorial proximity and functional dependence, and located on the territory of one or more subjects of the Russian Federation'[12].This confirms the need and prospects of such a tool at present, and the current regulatory framework for the development of industrial clusters points to the readiness of the state to implement it.

A potential regional waste treatment cluster can become an effective tool for implementing sustainable development policy of a state. Among the advantages of establishing regional waste recycling clusters are:

1. At the macro level: 'transparency' of waste management; 'convenience' of developing tools to encourage resource conservation; creating a single space for waste generation and its treatment; active development of resource conservation; expanding awareness of resource conservation as a natural stage in the progress of modern society within the framework of sustainable development; development of territories with special legal status and the PPP (Public Private Partnership) mechanism (establishing special economic zones).

2. At the meso-level: increasing production efficiency through resource conservation and the use of secondary raw materials; optimization of 'search and sale' of secondary raw materials; development of industrial symbiosis (waste from one industry is used as raw materials in another industry); development of a new processing industry or a complex.

3. At the micro level: reorientation of production processes to resource-saving technologies; adaptation of a waste collection system to new technologies; transformation of organizational culture under the influence of resource conservation; sanitation and cleaning of territories; development of innovations in the waste and secondary resources sector.

The following principles are fundamental for establishing a waste cluster [13]:

1. Resource principle of cluster formation. Waste is considered as raw material elements of the cluster.

2. The principle of regional zoning, taking into account the natural resource potential (natural and climatic conditions, regional primary resource base, and economic infrastructure).

3. The principle of the primary role of regional authorities (the authorities of the subjects of the Russian Federation and the local authorities) in developing environmental policy on waste management.

4. Improving extended producer responsibility in waste management (for controlling the raw material flows in a waste cluster).

In addition, I. R. Makarova and V. M. Tarbaeva in their work have distinguished the fifth principle: the allocation of two subclusters, production waste and consumption waste.

The authors do not share this opinion, since the main task of the cluster is to ensure cost-effective waste management and environmental protection through the collection, sorting and processing of highly profitable secondary raw materials (regardless of waste generators). Thus, it is more expedient to single out a separate subcluster of hazardous waste (medical waste, batteries, accumulators, mercury lamps, etc.) and a subcluster of organic waste (the technology of their processing is significantly different from other types of waste).

It is necessary to maximize waste collection during the waste treatment process. Each regional cluster should install sorting stations for disinfection, sorting, packaging and delivery of waste to processing plants. Sorting stations accumulate all waste generated in the region: SMW, production and consumption waste, industrial waste, construction waste, and medical waste.

The authors do not include waste from the extractive industries in the proposed clusters. According to GOST 30775-2001 '*Resource saving. Waste management. Waste classification, identification and coding. Basic Provisions*' [14], which identifies types of wastes (Appendix A1), first of all, the cluster is supposed to include enterprises belonging to the following groups: 2, 3, 4, 11, 12, 15, 17, 18, 20.

Part of industrial process waste generated by enterprise can be used as an additional material to the main raw materials (a certain percentage of the main raw materials), and waste that cannot be used as an additive in the main technological process should be sent to waste sorting stations or enterprises for recycling (these enterprises operate only on waste from other industries or receive waste from waste sorting stations).

The authors deem to follow the principle of 'cluster specialization' expedient; it allows circulation of resources / waste between regional clusters.

The main participants of the waste processing cluster (structural elements of the cluster system) include government agencies, universities and research institutes, private investors, enterprises engaged in waste accumulation, sorting, collection and transportation, processing into intermediate raw materials, production from secondary raw materials, and disposal.

Universities and research institutes take part in the process of cluster concept formation, the development of new technologies and equipment, and training of specialists to work within the cluster. The organizations responsible for waste collection and transportation can be different: municipal services collecting household waste, enterprises generated waste, business entities interested in collecting secondary raw materials; burial of waste can be done at landfill sites or special enterprises, depending on the waste collection system. Enterprises that provide waste sorting and waste primary processing into secondary raw materials also differ._Thus, the selective waste collection provides the supply of high-quality secondary raw materials (at the first stages of introduction of the selective collection system, this seems impossible); it allows managing without sorting lines. If the overall waste collection system is in place or separate collection is poorly organized, then sorting lines for the incoming waste should be used; it is also necessary to add to the cluster enterprises that use secondary raw materials in the production process, and operate at the expense of non-state funding sources or on the basis of PPP (public-private partnership) [15].

In addition, in order to create conditions for the competitive regional development and strengthen the competitiveness of the national economy as a whole, government authorities should be involved in organizing effective interaction within the cluster.

In our opinion, clustering should have a targeted character, when the initiative for its formation and development belongs to the state. Government authorities will perform the following functions: create legal and regulatory framework for the activities of subjects-participants in the cluster, focused on the regulation of technical and technological processes and organizational and economic relations avoiding planned and administrative characteristics; form a stable structure of the cluster linked on a long-term basis with the federal center; monitoring, analysis and forecasting of macro- and meso-environment; targeted state investment in financing cluster activities, and direct financing of innovations; create environment conducive to attracting private investment (for example, issuing so-called 'green' government securities). Federal bonds, as one of the lowest-risk investment options, can attract significant financial capital through public-private partnerships. The role of the state in attracting investment for waste management is extremely important, since this industry is capital-intensive and low-margin, which creates barriers to the independent participation of private investors. Thus, state 'green' bonds, like 'eco-deposits' in state banks, should be targeted to ensure the transparent use of attracted funds. Since the guarantor of the fulfillment of obligations under federal loan bonds is the state, the bankruptcy risk of these securities is low.

The creation and functioning of regional clusters will contribute to the development of digitalization, infrastructure support, legislative regulation, human resources, technological research, and economic, environmental and social effects.

The authors suggest carrying out a set of activities for successful implementation of the proposed tool.

1. Legislative regulation: development and implementation of economic-stimulus instruments of clustering at the legislative level, coordination of the cluster development strategy with strategic directions.

2. Infrastructure creation: informational (create databases of enterprises using secondary raw materials and being currently ready to process it; configure electronic document flow to track the movement of waste from a producer (collector) to a processor (including blockchain technology)) and production (create and develop waste processing enterprises, collection points for secondary raw materials and hazardous waste, establish communication lines).

3. Staffing: advanced resource conservation and resource efficiency training for management personnel of enterprises, courses for other specialists from enterprises involved in the waste recycling cluster, revision of educational programs.

4. Technological research: participate in R&D in the field of improving resource-saving technologies, commercialization of production methods based on secondary raw materials.

4 Conclusion

The creation and development of waste management and recycling clusters, where waste is considered as a resource, is an important direction of strengthening sustainable development of territorial socio-economic systems. At the same time, it is necessary to realize the high capital intensity of this project with obvious results, but of a mixed nature, i.e. at insignificant economic effect, a high level of social and environmental benefits will be achieved. In this case, it is advisable to develop mechanisms for public-private partnership.

Acknowledgments

The publication was carried out within the framework of the State Task of the Ministry of Science and Higher Education of the Russian Federation (topic No. 1.13.20 F 'Conceptual foundations for ensuring economic security of the Russian Federation in the conditions of digitalization: contours of spatial transformations').

References

1. Open-loop economy, https://greenpeace.ru/
2. Garbage reform, Bulletin of the Accounts Chamber of the Russian Federation, 9 (274) (2020)
3. B. R. Alzamora, R. T. De V. Barros, Waste Manage **115**, 47-55 (2020)
4. D. E. T. Cervantes, A. L. Martinez, M. C. Hernandez, A. L. G. de Cortazar, Waste Manage **80**, 51-63 (2018)
5. J. E. Santibanez-Aguilar, J. M. Ponce-Ortega, J. B. Gonzalez-Campos, M. Serna-Gonzalez, M. M. El-Halwagi, Waste Manage **33 (12)**, 2607-2622 (2013)

6. C. Cheng, R. Zhu, R. G. Thompson, L.-H. Zhang, Waste Manage **120**, 650-658 (2021)
7. B. D. Azevedo, L. F. Scavarda, R. G. G. Caiado, M. Fuss, Waste Manage **120**, 772-783 (2021)
8. L.A. Mochalova, M.V. Polezhaeva, ETAP: Econ Theory, Analysis, and Practice **5**, 108-123 (2020)
9. H. Wiesmeth, V. Starodubets, Econ of Region **16**, 725-738 (2020)
10. T. S. Narolina, S. A. Akulinin, *Creation of regional clusters as a means of increasing the competitiveness of regional enterprises* (Progressive technologies and equipment in electronics and mechanical engineering: interuniversity collection of scientific papers, Voronezh, 137-142, 2005)
11. E. S. Zakharchenko, Principles for the formation of an innovative cluster of waste disposal enterprises, https://studylib.ru/
12. Federal Law No. 488-FZ dated December 31, 2014 'On industrial policy in the Russian Federation', http://www.consultant.ru/
13. I. R. Makarova, V. M. Tarbaeva, Petroleum Geology – Theor. Appl. Studies, 4 (2009), http://www.ngtp.ru/
14. GOST 30775-2001 Resource saving. Waste management. Waste classification, identification and coding. Basic provisions
15. I. Minakova, T. Bukreeva, O. Timofeeva, J. Eng. Appl. Sci. **16(1)**, 99-103 (2018)

E3S Web of Conferences **311**, 04001 (2021) https://doi.org/10.1051/e3sconf/202131104001
EPSD 2021

Ecological and radiation safety of Central Asia

Victor Yurov[1,*], *Evgeniy* Eremin[2], and *Sergey* Guchenko[1]

[1]Karaganda University E.A. Buketova, st. Universitetskaya, 28, Karaganda 100028, Kazakhstan
[2]Department of Mechanical Engineering and Material Science, Machine-Building Institute, Omsk State Technical University, Mira Ave. 11, Omsk 644050, Russia

Abstract. Previous calculations showed that in the Kyrgyz Republic 7.2% of the total population are exposed to the "radiation risk", in the Republic of Kazakhstan - about 3%, in the Republic of Tajikistan - 1.7% and in the Republic of Uzbekistan - 0.3%. The number of radiation-contaminated territories in the Republics of Central Asia is equal: the Kyrgyz Republic contains about 132 million m^3 of waste, the Republic of Tajikistan contains about 180 million m^3 of waste, the Republic of Uzbekistan contains 640 million m^3 of waste, and the Republic of Kazakhstan contains approximately 10,000 million m^3 of waste. Total: in the four Republics of Central Asia, the total amount of radioactive waste is 10952 million m^3. Taking as a basis the limited liability company "Vismut GmbH" from Germany, which carried out the reclamation of uranium tailings with a volume of 160 million m^3 of tailings at a cost of about 8 billion euros for about 37 years. Using a thermodynamic approach, it is shown that the costs of reclamation of uranium tailings for the four Central Asian Republics will be 548 billion euros, and the duration will be about 40 years.

1 Introduction

Uranium mining operations have been carried out in Central Asia, including the Republics of Kyrgyzstan, Kazakhstan, Tajikistan and Uzbekistan, since the 1940s. After the collapse of the USSR (in 1991), these works were stopped. Tailings ponds, dumps, quarries, mines, wells of uranium deposits pose a serious environmental hazard. The main contribution to the effective dose of human exposure at uranium mining enterprises, as well as in the radiation-contaminated territory of spent uranium deposits, quarries and dumps, is made by three radiation-hazardous factors: volumetric activity of short-lived daughter products of radon in the environment; specific activity of long-lived radionuclides of the uranium-radium series in the environment; dose rate of external gamma radiation. If the areas of radiation-contaminated territories are left without reclamation for ten years, then they will be sources of radioactive effects for a long time on the health of animals and people.

In mid-2017, the reclamation plan was presented at the UN General Assembly as its basis and was approved at the 61st session of the IAEA already at the end of 2017.

In [1], we proposed a model for assessing the "radiation risk" of residents of territories where dumps and tailings of uranium mines are located. For the number of inhabitants Nm who have a probability of radioactive contamination, we obtained the formula:

*Corresponding author: exciton@list.ru

$$N_m = \frac{mnS_I^2}{S} \cdot \left(1 - \frac{E_f}{E_I} \right). \quad (1)$$

Here S is the total area of the region with population density n, m is the number of sources of ionizing radiation, S_I is the area of radioactive contamination from all sources of ionizing radiation, E_f is the natural background radiation, and E_I is the background radiation generated by one source of ionizing radiation.

Calculations have been carried out for four Republics of Central Asia. In the Kyrgyz Republic, 7.2% of the total population is exposed to "radiation risk", in the Republic of Kazakhstan - about 3%, in the Republic of Tajikistan - 1.7% and in the Republic of Uzbekistan - 0.3%.

In [2, 3], we propose a thermodynamic approach to radiation-stimulated processes in solids. This is the approach we use in this article.

2 The area of radioactive contamination of Central Asia

The total number of tailings in the Kyrgyz Republic [4] is $m_1 = 35$, which contain 48.31 million m^3 of waste. The total number of mining dumps is $m_2 = 37$, which contain 83.58 million m^3 of waste. Thus, the total number of radiation sources is 72, which contain about 132 million m^3 of waste. The average height of tailings and dumps is $h_I = 20$ m, so the occupied area by ionizing radiation sources (IRS) is estimated as $S_I = 6.6$ million m^2. The background radiation from the Earth and cosmic radiation ranges from 5 to 25 μR/hour, i.e. the average value of $E_I = 15$ μR/hour. The radiation background from the dumps of uranium mines ranges from 150 to 250 μR/hour, i.e. the average value of $E_f = 200$ μR/hour.

Radioactive waste in the Republic of Tajikistan [4] is concentrated in 10 tailings (m = 10) with a total area of 180 hectares ($S_I = 1.8$ million m^2). Area S = 142 million m^2. The average values of the radiation parameters E_f and E_I are similar to those given above.

In the Republic of Uzbekistan [4] there are m = 58 tailing dumps with a total area $S_I = 6.4$ million m^2. The average value of the radiation background of IRS is equal to: $E_I = 1100$ μR/hour. The territory of Uzbekistan is 447,400 million m^2, and the flat part is S = 89,480 million m^2.

In the Republic of Kazakhstan [4], the conducted studies revealed m = 127 sites of contaminated sites from uranium processing and mining in three regions: North Kazakhstan, South and Central Kazakhstan, West Kazakhstan. The total area exposed to radioactive waste from the uranium industry is currently estimated at SI = 100 million m^2. The average value of the radiation background of IRS is equal to: $E_I = 250$ μR/hour. The area of the three regions of Kazakhstan noted above is S = 1,846,328 million m^2.

3 Thermodynamics and Economics

To carry out the reclamation of the radioactively contaminated territory of Central Asia on such a scale, it is necessary to calculate the economics of all these processes using the example of one enterprise, which we will do in this section. Let's start with an approach to thermodynamics in economics. In [5], on the basis of the Carnot cycle, the concepts of thermodynamics and economics were compared and they were expressed between macroeconomics and thermodynamics. The work [6] uses a comparison between microeconomic and thermodynamic systems. We used the analogy method in work [7]. Many phenomena in technology and physics have analogies between thermal and electric fields, between electricity and magnetisms, and much more. This method was used by J.

Maxwell, R. Feynman and many others. Often these analogies lead to the same mathematical equations. A special case of the analogy method is the theory of similarity.

The work [8] demonstrated thermodynamics to different systems that are in a nonequilibrium state. This applies to solids and liquids, information systems and any other complex system. The authors considered the particle system as a system of non-interacting particles immersed in a thermostat. Quantum transitions caused by the interaction of particles with a thermostat will be dissipative (with probability P), in contrast to interaction with the external environment (with probability F). As a result, the following expression is obtained for the probability of dissipative processes:

$$P = \frac{2\Delta S}{k\tau} exp\left\{ -\frac{E_m - G^0/N}{kT} \right\}, \quad (2)$$

where ΔS is the change in entropy in the dissipative process; Em is the average energy of particles; τ is the relaxation time in the dissipative process; G^0 - thermodynamic potential of the thermostat; N is the number of particles. Using the method of analogies and work in [7], for the efficiency of the economic activity of the enterprise, we find:

$$\eta = \frac{\tau}{2\sigma U} \cdot \left(\frac{M}{N} - E \right), \quad (3)$$

where σ is the dissipation of capital (production costs), τ is the production cycle time, E is the profitability of the enterprise, M is the basic resource, N is the resource stock, U is the total capital (U = M + F).

From the obtained expression (3) it can be seen that the efficiency of the economic activity of the enterprise depends on a small number of basic parameters. This makes it possible to analyze the financial and economic activities of an enterprise in a short and long time interval of the production process.

4 Radiation safety of Central Asia

Thus, the number of radiation-contaminated territories in the four Central Asian Republics is: the Kyrgyz Republic contains about 132 million m^3 of waste, the Republic of Tajikistan contains about 180 million m^3 of waste, the Republic of Uzbekistan contains 640 million m^3 of waste, and the Republic of Kazakhstan there is approximately 10,000 million m^3 of waste. Total: in the four Republics of Central Asia, the total amount of radioactive waste is 10952 million m^3.

Comparison of the state of the environment and applied technologies in the places of spent uranium facilities on the territory of Germany (mines of SGAO "Vismut") will be carried out according to [11]. During the cessation of production, Vismut State Joint-Stock Company had 2 enrichment enterprises, 10 tailings dumps with a content of more than 160 million m^3 of tailings. The Government of the Federal Republic of Germany has assumed responsibility for overcoming the huge legacy of Vismut State Joint-Stock Company. For this purpose, a limited liability company "Vismut GmbH" was established in 1991. At the initial assessment of the cost of reclamation of the object in the budget of the Federal Republic of Germany, it was planned to allocate about 13 billion German marks for this work. By the end of 2015, more than 85% of the remediation of the legacy of the uranium mining industry in East Germany had been successfully completed. In total, about 8 billion euros will be spent until the end of the reclamation. The remediation objectives will be fully completed in 2028. Thus, the production cycle time τ for the remediation of uranium tailings in Germany will be about 37 years at a cost of about 8 billion euros for 160 million m^3 of tailings. In other words, the cost of 1 ton of radioactive waste for its reclamation will amount to approximately 50 euros.

For the four Central Asian Republics, the costs of reclamation of uranium tailings will amount to approximately 548 billion euros, i.e. almost 70 times more than Germany's costs for almost 40 years. For each Republic: in the Kyrgyz Republic, the costs of reclamation of uranium tailings will amount to approximately 6.6 billion euros, in the Republic of Tajikistan - 4.0 billion euros, in the Republic of Uzbekistan - 32 billion euros, in the Republic of Kazakhstan - 500 billion euros.

Let us turn to equation (3) and find out the production costs (σ) for the reclamation of uranium tailings for the four Central Asian Republics under the following approximations: the efficiency of the economic activity of the enterprise is set equal to $\eta = 1$, the profitability of the enterprise is $E = 0$, the resources are $M = N$, the total capital $U = 548$ billion euros, production cycle time $\tau = 40$ years $= 480$ months. Will have:

$$\sigma = \tau/2U. \tag{4}$$

The monthly costs for the four Central Asian Republics will be $\sigma \approx 0.438$ (billion euros)$^{-1}$ = 438 (million euros)$^{-1}$.

7 Conclusion

The analogy method can be used to analyze and diagnose the financial and economic activities of an enterprise. In particular, the basic resource is proportional to the square of the price of the product, and not just the price, as is usually taken into account when calculating the available resources of the enterprise. The price changes periodically, but of course according to a more complex law. With the growth of the basic resource, the associated capital also grows. It decreases when prices rise in the market. The more tied capital, the greater the economic losses (costs) of the enterprise. They decrease with the increase in the operating time of the enterprise.

From the obtained expression (3), it can be seen that the efficiency of the economic activity of the enterprise depends on a small number of basic parameters. This makes it possible to analyze the financial and economic activities of an enterprise in a short and long time interval of the production process. The monthly costs for the four Central Asian Republics will be $\sigma \approx 0.438$ (billion euros)$^{-1}$ = 438 (million euros)$^{-1}$.

Acknowledgements

The work was carried out under the program of the Ministry of Education and Science of the Republic of Kazakhstan Grant No. 0118RK000063.

References

1. V. S. Portnov, V. M. Yurov, A. D. Maussymbayeva, S. S. Kassymov and N. R. Zholmagambetov, International Journal of Mining, Reclamation and Environment, **12**, 205 (2017)
2. V. M. Yurov, E. N. Eremin, S. S. Kasymov, V. CH. Laurinas and A. V. Chernyavskii, IOP Conf. Series: Materials Science and Engineering, **168,** 012024 (2017)

3. V.M. Yurov, LXXVII international scientific readings (in memory of N. A. Dollezhal): collection of articles of the International scientific and practical conference (Moscow: EFIR, 2020)
4. *Uranium tailings in Central Asia: national problems, regional implications, global solutions* (Bishkek, 2009)
5. A. Krause, R. Raikhlin, *Economy as an irreversible (irreversible) thermodynamic system*, http://finansbibl.ru
6. A. M. Tsirlin, *Mathematical models and optimal processes in macrosystems* (Moscow, Nauka, 2006)
7. V. M. Yurov, V. S. Portnov, A. K Tursunbaeva, International journal of applied and fundamental research, **4**, 18 (2012)
8. V. V. Yavorskiy, V. M. Yurov, *Applied problems of thermodynamic analysis of nonequilibrium systems* (Moscow, Energoatomizdat, 2008)
9. E. N. Kamnev, V. P. Karamushka, A. V. Seleznev, V. N. Morozov, A. Hiller, MIAB. Mining Informational and Analytical Bulletin, **5**, 26 (2020)

Agroecological risks of the digital economy

Michael Chernyakov[1,*], Olesya Usacheva[1], Maria Chernyakova[2], Kamal Akberov[3], and Ivan Shuraev[2]

[1]Novosibirsk State Technical University, 20, Karl Marx Ave., 630078 Novosibirsk, Russia
[2]Siberian Institute of management-branch of the Russian presidential Academy of national economy and public administration, 6, Nizhnegorskaya, 630102 Novosibirsk, Russia
[3]Novosibirsk State University of Economics and Management, 56 Kamenskaya str., 630099 Novosibirsk, Russia

Abstract. The article is dedicated to the study of risks and building a digital profile, both at the level of economic entities and at the regional level. The factors that determine eight types of risks inherent in agricultural organizations in the context of digital transformation are analyzed. A hypothesis is put forward that the digital profile makes it possible to assess the readiness of companies for further digitalization and to highlight areas of increased risk that require attention from the state. Our calculations can be used in the future to assess the sustainability of agricultural organizations, both in Russia and abroad.

1 Introduction

The last decades have been characterized by the active use of digital technologies in all areas of human life, including the agro-industrial complex. The growing value of information, on the one hand, and the risks associated with consumption of natural resources, determine the need to study the possible mutual influence of agroecological risks on the digital economy. The quality and level of digitalization entering into the agro-industrial complex are studied by scientists from different countries [1-6]. African scientists have studied the impact of the climate change risk on the condition and further development of rural households. They identified six archetypal climate risk profiles, considering their various components. Archetypes show the variability of climatic risks depending on the strategy of obtaining financial resources and the availability of capital [7]. The study of the sustainability of the New Zealand dairy complex in order to minimize exposure to risks depending on agroecological, social and economic factors is provided in the paper of N.A. Cradock-Henry [8].

Italian scientists have investigated the digital, socio-economic and environmental impacts of the agriculture digitalization. Their offered taxonomy will help predict negative outcomes in order to reduce the unintended impact of digitalization. They refer the potential dependence of farms on digital service providers, and the possible increase in electricity consumption by digital tools to the potential digitalization risks. Also, the risks of reducing digitalization arise in the case of uneven ownership of electronic skills among rural workers

* Corresponding author: mkacadem@mail.ru

and the lack of IT infrastructure in rural areas. The possibility of serious problems with confidentiality, data ownership and use exacerbates social asymmetries [9].

Article is dedicated to study of research trends and opportunities related to digitalization of agriculture [10]. According to the authors, it is necessary to develop IoT devices that consider the advanced encryption algorithms in order to increase the security of the collected data. Big data technologies can be viewed as a solution for various applications and used in decision-making and extraction of new ideas and knowledge for agriculture.

Spanish scientists B. Garske, A. Bau and F. Ekardt also speak about development of IT infrastructure, a legal framework that clarifies the issues of data protection and security and the availability of technologies for small and medium-sized farms as a prerequisite for development of agriculture digitalization. They also note that development of digitalization itself generates risks of growth in electricity consumption, environmental risks due to increased pressure on renewable energy sources and on measures to compensate for the remaining inevitable emissions [3].

Domestic scientist E.A. Bubenok notes the insufficient level of digitalization of agriculture, linking this with the lack of scientific and practical knowledge about innovative modern technologies and methodology, a high level of production costs. As a result, only a small number of agricultural producers have the ability to use IT equipment and platforms. At the same time, she notes a change in the very paradigm of creating and developing a successful agribusiness through use of digital platforms as the cores of global digital information ecosystems [9].

The vision of extremely positive effects of digitalization is noted in the work of T.M. Yarkova. So she points out that the expected positive effect of digitalization for the Russian agricultural economy is enormous and is not associated with any risks and threats. It is possible to minimize possible risks and threats of various origins by revising the existing specialization in all territories of the country. As noted, the medium term is aimed at a significant effect from introduction of digitalization elements in the economy of the agro-industrial complex, which consists in increasing labor productivity more than three times, reducing the cost of production and sales by 20-25 %, expanding the boundaries of agricultural business and increasing profits from agricultural production [4].

Paper [11] highlights the risks associated with the digital transformation of the agri-food sector in the countries of the Middle East and North Africa. The most significant potential risks are: displacement of the workforce (digital technologies may displace agricultural labor with labor-saving technologies), different potential for digitalization depending on the size of the company, rejection or low level of adoption of digital technologies due to fear of losing competitiveness and growing lagging behind other countries, reducing privacy and security, generating new flows of resource use and waste, including greenhouse gases associated with increased energy use.

Despite the lot of studies conducted to examine the impact and risks of digitalization on agriculture, there is insufficient empirical evidence to assess the risks posed by this digitalization.

This article develops the concept of studying the risks of digitalization in the agro-industrial complex [12] and reveals the methodology for assessing their impact on the sustainability of companies, forming a digital profile of both individual districts and the region as a whole.

2 Materials and Methods

In the process of study, general and particular methods of scientific study were used [13], such as observation, analysis and synthesis, deduction and induction, which made it possible to identify the risks affecting the economic activity of economic subjects.

 https://doi.org/10.1051/e3sconf/202131104002

To implement the offered algorithm for assessing the impact of agrarian risks of the digital economy (ARTsE) on economic activity, an assessment method (MOARTsE) was offered, and it justifying the need to collect initial information for analysis using questionnaires. The best option for collecting is filling out by experts (managers and specialists of economic entities, as well as third-party specialists) the questionnaires for all 8 units (K_{Te} — process [14] , K_{Pe} — social [15] , K_{El} - environmental (Fig. 1), K_{Sa} — sanitary, K_{En} — economic, K_{Po} — political, K_{In} — infrastructural, K_{Ma} — psychological and management). The number of questionnaires is limited only by the number of possible experts who have expressed a desire to participate in the survey. The criteria for scoring internal risk assessment of the MOARTsE analysis for the "Ecology" unit are shown in Fig. 1.

Since there is information about typical representatives of the general population (managers, specialists and outside experts), the quota method of spontaneous sampling based on the "snowball" principle was preferred. To assess the districts of the Novosibirsk region (30), these are the heads and specialists of the administrations of the districts and the region, and for the assessment of organizations (selectively from more than 390 thousand), these are the heads and specialists of enterprises and experts from the district administration, who agreed to take the survey, and also recommended new respondents matching the sample parameters.

Depending on the situation, the survey can be conducted both in the form of a questionnaire and an interview, both in a contact and remote format. Given the unfavorable epidemiological situation, the survey was carried out mainly remotely by telephone, while it is necessary to use all the advantages and reduce the disadvantages of this method.

E1-1. Presence of factors for introduction of innovations over the past 5 years	0-0.4 – introduction of new methods of caring for crops and/or livestock ("precision farming", "precision animal husbandry"); 0.4-0.7 – partial mastering of the above technologies; 0.7-1 – work in this direction is not being carried out
E1-2. Innovations related to soil improvement	0-0.4 – introduction of proprietary technologies and developments in soil cultivation, use of biotechnological methods to improve the properties of plants; 0.4-0.7 – application of advanced experience; 0.7-1 – no innovations related to soil improvement
E1-3. Using intelligent technology to manage land and biological resources	0-0.4 – the farm regularly uses technical means, navigation, geoinformation and telecommunication technologies, which allow registering, processing and applying information linked to coordinates in order to optimize agro-technological solutions in agriculture; 0.4-0.7 – the technology is under development, but not yet applied; 0.7-1 – work in this direction is not being carried out
E1-4. Application of specialized software products	0-0.4 – software product used to control technological processes (land and process resources); 0.4-0.7 – software product used in the farm management system; 0.7-1 – specialized products have not been purchased for 5 years;
E1-5. Use of manned (unmanned) aircraft and space vehicles	0-0.4 – fully mastered and applied technology for remote monitoring of the agro-ecological state of soil and crops, observation of herds of animals; 0.4-0.7 – partially mastered, but not yet applied technology; 0.7-1 – work in this direction is not being carried out.

Fig. 1. Criteria for scoring internal risk assessment of MOARTsE analysis for the Ecology unit

3 Results and Discussion

According to the offered algorithm, possible ARTsE were estimated for the districts of the Novosibirsk region (NSO) using the MOARTsE analysis. Based on the results of the scoring MOARTsE analysis, graphic models of the digital profile of the districts and the Novosibirsk region as a whole were built (Fig. 2).

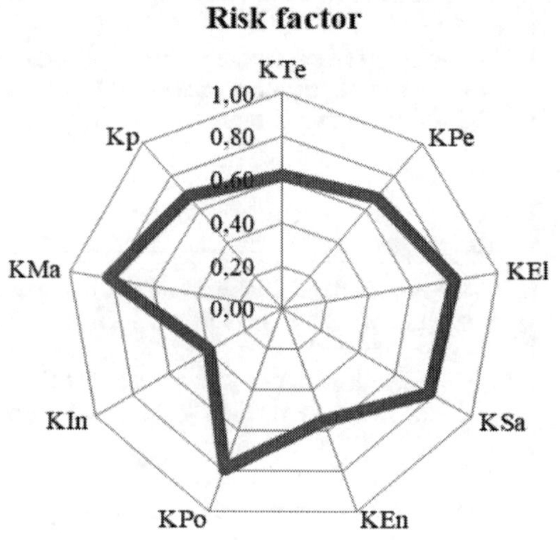

Fig. 2. ARTsE digital profile for the Novosibirsk region.

Digital profiles of districts and the region (Fig. 2) allow to identify the relative impact of the consequences of the potential impacts of the risks of each analytical unit. It was found that infrastructure risk is of the least importance, which indicates the significant potential of economic entities in all districts of the Novosibirsk region to transform into a digital economy and improve financial stability.

Errors made in making managerial decisions (on the inclusion of the agricultural insurance system in the "single subsidy") led to an increase in psychological and managerial risk and a massive refusal of agricultural organizations from agricultural insurance, which affected the growth of environmental and sanitary risks, especially in the field of crop production and a little less in animal husbandry. This can explain the high level of these three types of digital risks.

Regarding the analyzed risks, special attention shall be paid to agro-ecological and sanitary risks, which are specific to the agricultural sector and require special measures to compensate them.

Agriculture refers to a high-risk sector of the national economy, since it largely depends on the conditions and final results of agricultural activities, which are characterized by random factors, most often environmental and sanitary ones. Therefore, use of agricultural insurance programs takes a decisive place in the management of agricultural risks and minimization of unforeseen losses. The stability and efficiency of agricultural insurance systems has been proven by its many years of experience in the global insurance market. The very idea of agricultural insurance is not new for the Russian Federation, and the state has adopted a number of legislative and organizational and economic measures, but today the system of domestic agricultural insurance with state support, due to errors in making managerial decisions, does not correspond to the tasks set and requires improvement.

4 Conclusion

Digitalization that has affected all sectors of economic activity, including agriculture, carries not only benefits, but also risks. The study made it possible to identify and differentiate the risks affecting the sustainability of agricultural enterprises. We have established a high level of influence of such types of risks as: political, sanitary, environmental and psychological and managerial (more than 0.8). The low value of infrastructural risk indicates a high degree of readiness of enterprises for digitalization. Use of MOARTsE analysis makes it possible to form a digital profile of both individual companies and regions as a whole. On its basis, it is possible to develop measures to reduce risks, including programs of state support for agricultural producers. The versatility of the analysis methodology reveals a wide range of areas of its application. The performed analysis gave us grounds to single out the risks that reveal industry specificity — these are sanitary and agroecological, and also to assume that the leveling of the influence of these risks lies in the field of agricultural insurance. We believe that our study shall be further developed. Namely, the risks arising from the digitalization of agriculture itself require a quantitative and qualitative assessment.

Acknowledgements

This paper was supported by the Novosibirsk State Technical University (project C21-11).

References

1. R. Bezner Kerr, H. Nyantakyi-Frimpong, L. Dakishoni, Lupafya, L. Shumba, I. Luginaah, S. S. Snapp, J. Renewable Agriculture and Food Systems, 238–251 (2018)
2. S. Rolandi, G.Brunori, M. Bacco, I. Scotti, J.Sustainability, **13(9)**, 5172 (2021)
3. B. Garske, A. Bau, F.Ekardt, J. Sustainability, **13(9)**, 4652 (2021)
4. M. T. Yarkova, J. HELIX, **10(3)**, 60–63 (2020)
5. O. Tarariko, T. Ilienko, T. Kuchma, I. Novakovska. J. Agricultural Science and Practice, **6(1)**, 18–27 (2019)
6. R. B. Pickson, E. Boateng, J. Environment, Development and Sustainability (2021).
7. F. Paumgarten, B. Locatelli, E. T. F. Witkowski. J. Weather, Climate, and Society, **12(3)**, 545–560 (2020)
8. N. A. Cradock-Henry, J. Ecology and Society, **26(1)**, 3 (2021)
9. E. A. Bubenok, Vestnik MIRBIS, **1(17)**, 90–95 (2019)
10. J. Iaksch, E. Fernandes, M. Borsato, J. Bibliometric and Systemic Analysis (2020)
11. R. A. Bahn, A. A. K. Yehya, R. Zurayk, Sustainability, **13(6)**, 3223 (2021)
12. M. Chernyakov, O. Usacheva, S. Gromov, E3S Web of Conferences, 273, 08081 (2021)
13. E. Dadayan, A. Storozheva, E. Letyagina, E3S Web of Conferences, 161, 01065 (2020)
14. M. K. Chernyakov, M. M. Chernyakova, Journal of Corporate Finance Research, **12**, 99–109 (2018)
15. M. K. Chernyakov, M. M. Chernyakova, K. C. Akberov, Advances in Economics, Business and Management Research, **81**, 373-378 (MTDE 2019)

Biofouling as one of the Factors of Sustainable Development of Shipbuilding

Andrey Mosunov*, and *Vladislav* Evstigneev

Sevastopol State University, 33, Universitetskaya, Sevastopol, Russia

Abstract. Sea transportation is one of the cheapest and most widespread types of cargo transportation, and the process of fouling of the vessel's hull has an extremely negative impact on its maneuverability and running characteristics. This leads to excessive consumption of fuel, an increase in the cost of delivery of goods and environmental risks. Namely, when additional fuel is burned, the emission of the greenhouse gas CO_2 increases. Considering the accompanying environmental risks in the long-term planning of the shipbuilding industry will ensure the sustainability of its development. In the study we suggest a simple way to assess quantitatively the development strategy of shipbuilding from the point of view of the accompanying economic and environmental consequences by constructing a matrix model for the industry development. The matrix model was applied to the Strategy for Development of the Shipbuilding Industry in Russia until 2035. In the future, on the basis of the offered model, it is possible to develop optimal environmentally neutral scenarios for shipbuilding development.

1 Introduction

Strategies for development of sectors of the economy are developed based on the expected needs of the state or business for the forecasted period of years. Often such strategies are aimed only at meeting increasing needs without detailed analysis of related environmental risks. Underestimation of such risks in the future may lead to an additional set of measures to counter the adverse consequences provoked by a given vector of the industry development. Against the background of the trends observed in the global climate system, the problem of mitigating the effects of climate change by reducing industrial greenhouse gas emissions (GHG) has gained additional focus [1]. Considering the commitments made by many countries to reduce GHG, strategies for development of economic sectors and industrial sectors shall consider the possible environmental and climatic risks within their implementation. In Russia, at the state level, a number of national initiatives have been adopted, enshrined in the relevant Strategies, Presidential Decrees and Government Orders.

Introduction of an approach to developing strategies not only for development, but also for managing the risks generated by this, will make it possible to get closer to implementation of "sustainable development" concept [2]. Sustainable development is about meeting the needs of the present generation without threatening future generations to

*Corresponding author: mosunovandrey@mail.ru

meet their own needs. A measure of this development is preservation of the stability of physical and ecological systems achieved through preservation and rational use of natural resources.

One example of an industry in which a sustainable development strategy is necessary is shipbuilding. According to the International Maritime Organization (IMO), the merchant fleet provides up to 90 % of the global transcontinental transportation of goods [3]. Shipbuilding has a multiple effect on development of related industries such as metallurgy, machine tool building, engine building, and etc. Russia has developed a Strategy for Development of the Shipbuilding Industry until 2035 that envisages the annual production of civil vessels and marine equipment with displacement of more than 80 tons [4]. However, strategies of this kind lack a prospective assessment of possible environmental risks associated with this development. Moreover, some of these risks are not obvious and are not considered as the main ones that shall be considered. The factor of marine fouling of ship hulls is among those risks.

Marine biofouling of ships and hydraulic structures is a global problem that has not yet been effectively resolved. The urgency of solving this problem is due, first of all, to the economic feasibility of operating vessels and offshore hydraulic structures. Namely, biofouling of the hulls of vessels leads to increase in the surface friction of the hull up to 70 % [5] and decrease in its running performance up to 50 % of its nominal value. A direct consequence is increase in fuel consumption that leads not only to increase in the cost of cargo delivery, but also to environmental consequences in the form of additional fuel combustion. This, in its turn, leads to increased GHG CO_2 emissions. Total GHG emissions associated with vessel transportation are considered to be about 930 million tons [6].

Controlling marine fouling has traditionally been associated with development of biocide-releasing coatings that is still the most reliable way to prevent biological fouling. However, after prohibition of environmentally toxic organometallic (lead-, mercury- and tin-containing) antifouling coatings, research and the search for new effective biocidal materials and principles for suppressing the vital activity of biofouling have become a significant scientific problem that still awaits its solution. Use of various nanocomposite materials is currently considered one of the most promising components of anti-fouling coatings [7].

In this paper, the goal was to develop a simple way to quantify the strategy for development of shipbuilding from the point of view of the accompanying economic and environmental consequences by constructing a matrix model of strategic planning for the industry.

2 Materials and Methods

The data for analysis were taken from the Strategy for Development of the Shipbuilding Industry until 2035, Appendix 1 "Target Indicators for Implementation of the Strategy", the item "The Volume of Production of Civil Vessels and Marine Equipment with Displacement of More than 80 tons for the Domestic Market and for Export on Accrual Basis from 2019" [4].

The data for the period of 2019-2035 were used to estimate the planned rates of vessel release, and to build a matrix model that includes both the planned release rates and considers the rate of fouling on the surface of vessels and excessive fuel consumption in relation to the nominal value.

3 Results and Discussion

In our paper, we set the task of creating a matrix model of a strategic plan for development of the shipbuilding industry. The main prerequisites for building a matrix model are as follows:

1) discrete moments of time (years) are considered, during which one of the following events occurs:

- a certain number of new vessels are being built and put into operation in accordance with the Strategy;

- early commissioned vessels are moving from one type of surface condition to another, more "heavy" type in terms of the composition and severity of fouling and changes in running characteristics;

- in a certain year, the vessel is docked to carry out maintenance and repair work and restore running performance;

2) the standard duration of an interdock period of 5 years is considered, at the end of which works to clean the (usually mechanically) the hull from fouling are planned;

3) deterioration of the propulsive qualities of vessels with increase in their age is not considered;

4) the discrete (annual) inter-dock fouling cycle of the vessel is considered. According to the data of [8, 9], such a cycle and increase in power with increase in the roughness of the hull to maintain the speed of the vessel is presented as follows:

Stage 1 – primary fouling, microalgae; additional increase in power to overcome friction up to 10 %;

Stage 2 – formation of the dense mucous fouling film; increase in power up to 20 %;

Stage 3 – lime fouling leads to increase in power up to 35-80 %, depending on the degree of fouling "severity".

4) in the first rough approximation, we assume that the increment in power to maintain the speed of the vessel is proportional to the increment in fuel consumption. In its turn, the amount of additional CO_2 emissions into the atmosphere is proportional to the amount of excessive fuel consumption.

5) as a target characteristic in modeling, the total increment of engine power, expressed in %, can be used for all operating vessels. Such total increment will be proportional to the excess fuel consumption and carbon dioxide emissions into the atmosphere.

Therefore, it is assumed that the grouping of new vessels has some internal structure depending on the degree of fouling of the underwater part of the hull. Let us define an interdock period of 5 years, then the structure of the grouping of vessels contains 5 possible states. This detailing of the structure leads to the class of matrix models. Let at the initial moment of time the grouping be described by the column vector$X(t_0) = \left(1, x_1(t_0), \dots x_N(t_0)\right)^T$, where x_i is the number of vessels in the ith state of fouling, a single first element of the vector is necessary to introduce an external factor into the model – commissioning the new vessels (see below). Vector $X(t_1)$ characterizing the state of the grouping of vessels at the next moment in time, for example, in a year, is associated with the vector $X(t_0)$ through the transition matrix L:

$$\boldsymbol{X}(t_1) = \boldsymbol{L}\boldsymbol{X}(t_0).$$

Let us establish the form of this matrix. The first type of vessels is obtained considering two processes. The first is the commissioning new vessels in accordance with the Strategy, the second is the return to service of vessels after docking:

$$x_1(t_i) = F\{i\} + (1 - b_1)x_1(t_0) + b_5 x_5(t_0).$$

$F\{i\}$ plays the role of a certain strategy operator associated with the current year i; it can be either a tabulated function with specific planned indicators, or an analytical function (for example, a linear one) that determines the trend in development of shipbuilding volumes. The second term reflects the fact that due to the high-quality coating on the hull and the anti-fouling compound used, some of the vessels ($1-b_1$) may remain in the first most effective state, while the other fraction of b_1 vessels will go into a worse second state. The equation also assumes that all vessels in a state of 5 years of operation and corresponding fouling are sent for docking, their return to service is equivalent to contribution to the first group of vessels.

The second group of vessels at time i is determined by transition to this state of vessels from the first group and the probability that some of the vessels of the second group will retain their state in a given year:

$$x_2(t_i) = b_1 x_1(t_0) + (1 - b_2) x_2(t_0).$$

Further, all subsequent groups of states are described similarly. It shall be noted that the factors in the equations in the general case $b_i \in [0,1]$. Based on the equations written out, we construct the transition matrix $\boldsymbol{L}$ for 5 states of the grouping of vessels:

$$\boldsymbol{L} = \begin{pmatrix} F\{i\} & 0 & 0 & 0 & 0 & 0 \\ F\{i\} & 1-b_1 & 0 & 0 & 0 & b_5 \\ 0 & b_1 & 1-b_2 & 0 & 0 & 0 \\ 0 & 0 & b_2 & 1-b_3 & 0 & 0 \\ 0 & 0 & 0 & b_3 & 1-b_4 & 0 \\ 0 & 0 & 0 & 0 & b_4 & 1-b_5 \end{pmatrix}.$$

Therefore, knowing the structure of the matrix $\boldsymbol{L}$ and the initial state of the grouping — the column vector $\boldsymbol{X}(t_0)$ — you can predict the state of the grouping at any given moment in time: $\boldsymbol{X}(t_i) = \boldsymbol{LX}(t_{i-1}) = \boldsymbol{LLX}(t_{i-2}) \ldots = \boldsymbol{L}^i \boldsymbol{X}(t_0)$. Knowing the structure of the grouping in a specific forecast year, it is possible to calculate the total excess of fuel consumption due to fouling, for example, using information from the 4th prerequisite (see above). The excess consumption is expressed as a percentage of the nominal fuel consumption, multiplying by which, the excess consumption (or additional CO_2 emissions) can be obtained in absolute terms.

Let us show the result of applying this model using the example of the Strategy for Development of the Shipbuilding Industry in Russia until 2035. According to the Strategy [4], it is planned to build new vessels in Russia with a cumulative total of 358 units by 2035. However, the rate of new vessels construction decreases annually (Fig. 1). According to the estimate of the linear trend in Fig. 1 such a decrease in the rate of construction by one unit per year. Let's introduce this regularity inherent in the Strategy into the model. Let's assume that the initial conditions for 2019 are commissioning 30 vessels. Further, it is planned to annually reduce the number of new vessels by 1 unit.

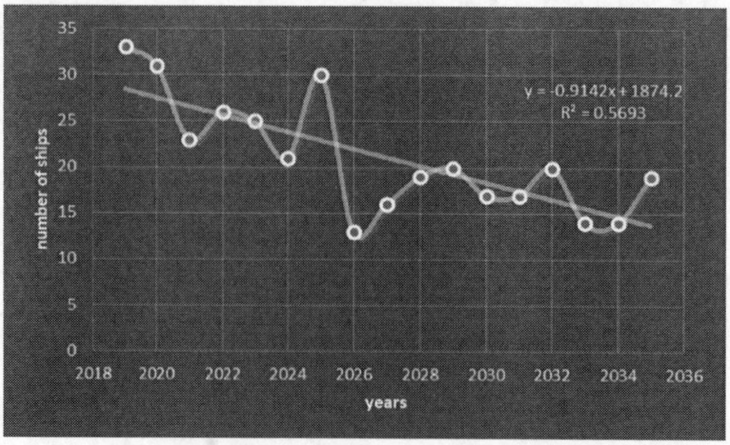

Fig. 1. Planned annual volume of vessel production until 2035 according to the Strategy [4].

To introduce this regularity into the model, we transform the transition matrix $\boldsymbol{L}$ and the column vector $\boldsymbol{X}$ so as to include the linear decrease in the number of vessels introduced into the matrix model:

$$\boldsymbol{X}(\boldsymbol{t_0}) = \begin{pmatrix} 30 \\ -1 \\ 0 \\ 0 \\ 0 \\ 0 \\ 0 \end{pmatrix}; \quad \boldsymbol{L} = \begin{pmatrix} 1 & 1 & 0 & 0 & 0 & 0 & 0 \\ 0 & 1 & 0 & 0 & 0 & 0 & 0 \\ 1 & 0 & 1-b_1 & 0 & 0 & 0 & b_5 \\ 0 & 0 & b_1 & 1-b_2 & 0 & 0 & 0 \\ 0 & 0 & 0 & b_2 & 1-b_3 & 0 & 0 \\ 0 & 0 & 0 & 0 & b_3 & 1-b_4 & 0 \\ 0 & 0 & 0 & 0 & 0 & b_4 & 1-b_5 \end{pmatrix}$$

The first two elements of the initial vector $\boldsymbol{X}(t_0)$ contain information about the strategy for introducing new ships, the transition matrix $\boldsymbol{L}$ *is* also adapted to this pattern. To calculate the excessive fuel consumption (and proportionally to GHG emissions), the following scenario was adopted: 1 year – 0 %, 2 years – 10 %, 3 years – 20 %, 4 years – 35 %, 5 years – 50 %.

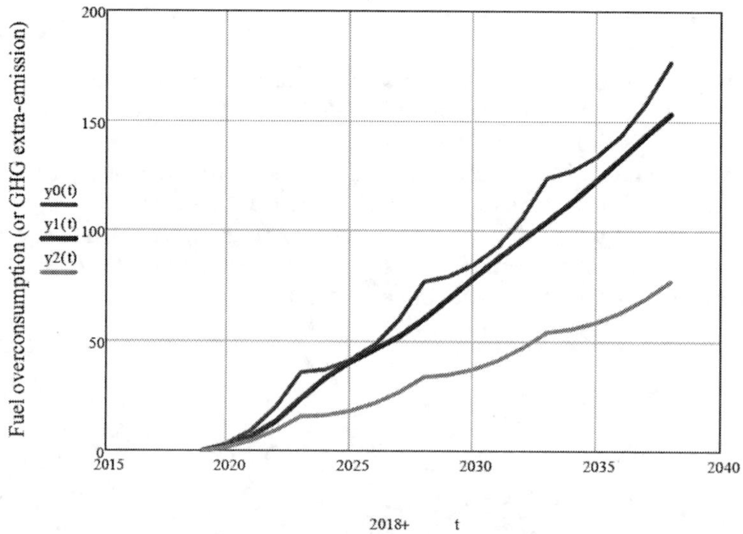

Fig. 2. Annual excess fuel consumption or additional GHG emissions due to fouling of the vessel's hulls according to the Strategy [4]. The values are provided in conventional units. The red line – under the condition $b_i = 1$, blue – under the condition $b_i < 1$, green – with $b_i = 1$ and when using anti-fouling coatings.

The results of calculations using the matrix model are shown in Fig. 2 in 3 versions. In the first option (red line), the probability of transition of all ships from one state to another is equal to one. In this case, the graph shows periodic peaks in the excessive consumption of fuel and GHG emissions associated with the simultaneous deterioration of the fouling condition on all vessels, but after docking the values slightly decrease. In the second option (blue curve), it is assumed that a certain proportion of vessels may remain in the same state (b_i <1). It can be seen that in this case the curve becomes substantially smoothed. The third option shows the effect of use of innovative anti-fouling coatings that reduce the loss of the vessel's course and, as a result, excessive fuel consumption. The scenario of excessive fuel consumption depending on the year was adjusted as follows: 1 year – 0 %, 2 years – 5 %, 3 years – 10 %, 4 years – 15 %, 5 years – 20 %. The general conclusion follows from the results that the increment in fuel consumption will grow steadily even with a decrease in the dynamics of launching new vessels.

4 Conclusions

The paper offers the matrix model for strategic planning of the shipbuilding industry that allows a rough estimate of the accompanying economic and environmental risks caused by fouling of the vessel's hull within its operation. On the basis of the offered model, it is possible to develop optimal environmentally neutral scenarios for construction of new ships, the choice of docking periods in combination with scenarios for introduction of innovative antifouling coatings. The indicator of optimality is the minimum of the total excessive consumption of fuel and GHG emissions. The matrix model is simplified and requires further development to increase its applicability in real practice.

Acknowledgments

The study was funded by Sevastopol State University (project № 47/06-31 "Biologically active bicomponent metal oxide nanoparticles as the basis for alternative compositions of antifouling coatings for ships and hydraulic structures").

References

1. R. K. Pachauri, L. A. Meyer, *IPCC. Climate Change 2014: Synthesis Report. Contribution of Working Groups I, II and III to the Fifth Assessment Report of the Intergovernmental Panel on Climate Change* (2014)
2. Take Action for the Sustainable Development Goals, https://www.un.org/
3. C. Hellio, D. Yebra, *Advances in Marine Antifouling Coatings and Technologies* (2009)
4. Strategy for the development of the shipbuilding industry until 2035, http://government.ru
5. B. S. Galil, C. McKenzie, S. Bailey, M. Campbell, I. Davidson, L. Drake, C. Hewitt, A. Occhipinti-Ambrogi, R. Piola, *ICES Viewpoint background document: Evaluating and mitigating introduction of marine non-native species via vessel bio-fouling* (2019)
6. N. Olmer, B. Comer, B. Roy, X. Mao, D. Rutherford, *Greenhouse gas emissions from global shipping, 2013-2015* (2017)
7. D. Inbakandan, et al., Colloids Surf., B. **111**, 636 (2013)
8. M. P. Schultz, Biofouling, **23**, 331 (2007)

9. M. P. Schultz, J. M.Walker, C. N. Steppe, K. A. Flack, Biofouling, **31**, 759 (2015)

On the comprehensive indicator of the effectiveness of programs in the field of environmental safety and environmental development

Anatoly Rybakov[1,*], *Evgeniy* Ivanov[1], *Alexander* Kuzmin[2], and *Larisa* Gorina[3]

[1]Scientific Research Center, Civil Defense Academy, Khimki, Russia
[2]Kazan National Research Technical University named after A. N. Tupolev - KAI, Kazan, Russia
[3]Togliatti State University, Zhigulevsk, Russia

Abstract. The paper offers a comprehensive indicator for evaluating the effectiveness of the implementation of state policy in the field of environmental safety and environmental development. The indicator under consideration can be a flexible tool for justifying the effectiveness of strategic planning measures in the field of environmental development, taking into account current development trends. The idea of solving the optimization problem (bringing the values of the complex indicator to the maximum) of forming a rational action plan for the implementation of state policy in the field of environmental safety and environmental development is proposed.

1 Introduction

The events of the beginning of the XXI century forced the world community to come to an awareness of the problems associated with ensuring sustainable development not only of countries, regions, but also of the entire planet as a whole. It is worth noting that the solution of sustainable development issues is on the global agenda. It is worth noting that the documents of international law [1-3] in this area are being developed quite intensively.

The fact that the most attention is paid to environmental issues in the documents under consideration indicates the awareness at the intergovernmental level of the fact that our planet is a very vulnerable to changes (both internal and external) closed system, the preservation of the balance of which is possible only through the systematic implementation of a number of interrelated programs [4, 5].

At the same time, mechanisms that allow for a reliable assessment of the effectiveness of the implemented set of measures in the field of environmental safety are not fully developed in detail [6, 7].

* Corresponding author: anatoll_rubakov@mail.ru

2 Materials and methods

Our country is no exception, from the point of view of awareness of the issues of ensuring environmental safety and environmental development (including for the long term until 2030 [8]), in particular, it is currently envisaged to solve 13 main tasks in this area. It should be noted that the proposed structure of the Strategy is systematic, since for each task there are from 3 to 12 mechanisms for implementing the provisions of the Strategy. As a disadvantage, it should be noted that there are no indicators for evaluating the effectiveness of the implementation of activities carried out within the framework of the Strategy.

The key idea of the work is to present an approach to determining a comprehensive indicator for evaluating the effectiveness of the implementation of state policy in the field of environmental safety and environmental development.

At the same time, the idea of considering a number of private criteria that characterize private indicators for each of the tasks of environmental development is proposed (1):

$$P_i = 1/k\sum_{k=1}^{n} c_i^k, \tag{1}$$

where P_i – is a complex efficiency indicator for the i-th environmental development task; i is the number of the environmental development task (13 tasks in total); c_i^k – is a particular efficiency indicator for the k-th event of the i-th environmental development task; k is the number of the implemented event (from 3 to 12, depending on the task).

At the same time, in order to move to dimensionless values of partial indicators for each task, the normalization operation is performed [9] by the following transformation:

$$c^k_i = \frac{c_i - c_{crit}}{c_{targ} - c_{crit}} \tag{2}$$

$$c^k_i = \frac{c_{crit} - c_i}{c_{crit} - c_{targ}} \tag{3}$$

where c_i – absolute values of the measured indicator; c_{targ} – target values of the measured indicator; c_{crit} – critical values of the measured indicator.

As follows from the meaning of the normalization operation, expression (2) is used for cases when the purpose of the implemented measures is to increase the absolute values of the indicator, otherwise expression (3) is used.

For the transition to a single complex indicator for evaluating the effectiveness of the implemented environmental development policy, it is possible to combine the i-th number of complex indicators of the effectiveness of environmental development tasks in the form of a multiplicative convolution [10].

$$P_\Sigma = \sum_{i=1}^{13} P_i a_i, \tag{4}$$

where P_Σ – a comprehensive indicator for evaluating the effectiveness of the implemented environmental development policy; a_i – is a weighting factor for the i-th task of environmental development.

In turn, the weighting coefficients can be found by interviewing the appropriate group of competent experts and processing the results of the survey using the hierarchy analysis method [11].

3 Results and discussion

Thus, the proposed indicator for evaluating the effectiveness of the implemented environmental development policy will allow for a comprehensive assessment of the current state of the environmental situation, and when tracking the dynamics of changes in the indicator, evaluate the effectiveness of the implemented measures.

At the same time, efficiency will be taken as the value of the increment of the values of the complex indicator, relative to the financial resources allocated within the framework of environmental programs (5).

$$E = \Delta P_{\Sigma} / M_{\Sigma}, \tag{5}$$

where E – is an indicator of the effectiveness of the implemented environmental development policy; ΔP_{Σ} – is the difference between the values of the integrated indicator for evaluating the effectiveness of the implemented environmental development policy for the reporting period (1 year); M_{Σ} – financial resources allocated within the framework of environmental programs.

The presented indicator is a tool for carrying out an ongoing assessment of the state of affairs, which allows assessing the degree of implementation of the provisions of strategic planning documents in the field of environmental safety. However, the assessment should not be an end in itself, for the decision-making manager, forecasting the situation and planning measures to correct it will be of greater interest. In particular, the definition of such measures can be carried out by establishing relationships between the volume of their implementation and the change in the indicator (5) [12-14].

The obtained complex indicator of the effectiveness of the implemented environmental development policy (4) allows us to form a tool for justifying the state policy in the field of rational planning of measures. At the same time, the optimization problem itself looks like this (6):

$$P_{\Sigma} = f(c_i, M_j, a_i) \to max, \tag{6}$$

where M_j – financial resources allocated within the framework of the implementation of the j-th event of the i-th environmental development task.

In other words, it is necessary to define such an action plan in the field of environmental safety and environmental development, which will allow achieving the maximum values of the integrated indicator for evaluating the effectiveness of the implemented environmental safety and environmental development policy.

At the same time, the allocated resources should not exceed the total amount of allocated allocations for the implementation of state policy measures in the area under consideration (the maximum and minimum volumes of measures should also be fixed so that the optimization task does not degenerate).

4 Conclusion

The paper proposes an approach to the formation of a rational action plan for the implementation of state policy in the field of environmental safety and environmental development, based on the solution of the optimization problem.

This task has some difficulty in establishing a "feedback" between the activities performed and the values of the efficiency indicator, for which a regression analysis method can be recommended, due to the complex structure of the system of environmental safety and environmental development measures, or big data technology.

 https://doi.org/10.1051/e3sconf/202131104004

Bringing to the maximum of the target function (a complex indicator of the effectiveness of the implementation of state policy in the field of environmental safety and environmental development) can be carried out through the use of gradient methods (5), which consist in a step-by-step search for the largest values of the target function, subject to restrictions on the maximum of financial resources and the permissible limits of variable private indicators.

References

1. Website of the United Nations, https://www.unisdr.org/
2. Protocol on Strategic Environmental Assessment to the Convention on Environmental Impact Assessment in a Transboundary Context, Website the United Nations, https://www.un.org/
3. Framework Convention on Climate Change. The Paris Convention, Website of the United Nations, https://unfccc.int/

A. V. Kukushkina, *Int. legal aspects of environmental safety at the present stage*, 218 (2020)

4. D. D. Melnikov, S. D. Zaburdin and E. G. Rakovskaya, The Newman in Foreign policy, **52(96)**, **1**, 49 (2020)
5. Utepbergenov, M. Daumenov, B. Nizamatdinov and et al., Paideuma, **7**, 45 (2021)
6. B. Khrustalev and et al., IOP Conf. Ser.: Mater. Sci. Eng. (2019), 471 092005
7. Fundamentals of the state policy in the field of environmental development of Russia for the period up to 2030, https://base.garant.ru/
8. E. V. Klyushnikova and E. M. Shitova, Innocenter, **1**, 4 (2016)
9. V. I. Borisov, "Ways to improve the effectiveness of modern scientific research," in *the Int. Scientific and Practical Conf.* (Ufa, 2019), 78
10. V. A. Titov and I. G. Khairulin. Basic research, **10**, 2020 (2013)

A. V. Yandritskaya, The paradigm, **1**, 17 (2019)

11. N. A. Istomina, National interests: priorities and security, **37(178)**, 26 (2012)

A. N. Biryukova and O. I. Glushchenko, Bulletin of the Perm University, **2**, 24 (2011)

E3S Web of Conferences **311**, 04005 (2021) https://doi.org/10.1051/e3sconf/202131104005
EPSD 2021

Conflict mobilization communities in social media as a soft power tool: problematization environmental risks

Larisa Rebrina*

Volgograd State University, pr. Universitetskij 100, 400062 Volgograd, Russia

Abstract. The object of the study is the protest communication practices of conflict mobilization in 2012-2021 by environmental communities in Russian-language social networks. Civil society is a significant actor in environmental policy. Social media as a soft power tool plays an important role today in informing, educating society about the environment and shaping a protesting regional environmental agenda. The system-communicative approach explores the patterns of discursive problematisation of environmental risks within the framework of civic political participation. Authors describe the types of publications constituting the content of environmental mobilization communities; analyse thematic dominants to identify frequent environmental threats constructed by the addressants of the publications; identify and systematise the strategies, tactics and relevant language tools used to problematise fragments of environmental reality; examine regular means of creating and maintaining online solidarity in environmental communities of conflict mobilization.

1 Introduction

Along with the state and business, civil society is a significant actor in all politics. Today, civic participation in the environmental sphere is developing actively, and environmental movements are building their political capacity. However, in our country, the very system of civic participation in this sphere, designed to complement and develop government regulation and to help protect the right to a healthy environment, is still under construction, nevertheless reflecting the greening of public consciousness. We should note the important role of civic participation in informing, educating society about the environment and shaping the environmental agenda, especially the protest agenda and the regional agenda. The object of the study is protest communication practices of conflict mobilization (consolidation mobilization against) 2012-2021 in Russian-language social networks, bringing participants together around an environmental problem that demands a solution to an environmental situation, against environmentally damaging decisions and plans. Social networks are nowadays a soft power and political influence tool, a space for effective implementation of political technologies for the generation and transmission of given

* Corresponding author: LNRebrina@volsu.ru

meanings within the framework of retinal communication and the formation of public consciousness. The research from the perspective of the systems-communicative approach aims to describe the patterns of the discursive construction of the environmental agenda and environmental problems in protest communication practices in social networks. Linguistics seems capable of contributing to the study of these issues, including the impact of new media and communication technologies on the political participation of citizens.

2 Materials and Methods

We consider protest communication as a communicative system, implementing the conjunction of the media and political spheres, characterized by specific genre features, collective-personal (composition of participants, ways of construction of the addressee, addressant, solidarity, relevant strategies and tactics of participants) and subject (proactive-reactive nature, the set of themed objects, ways of agenda formation, problematization of reality) parameters. [1; 2; 3; 4]. Contemporary researchers note the systemic-communicative autonomy, self-validation of protest (substituting problematisation of fragments of objective reality from the socio-cultural positions of protest actors observing reality, and the autorepresentation of protest as eliminating a constructed threat in various forms of competition for resources with diverse institutions), which involves a combination of traditional structural-critical and resource-mobilization approaches [1]. Protest can have a constructive/destructive, autonomous/mobilising, active/passive, conventional/unconventional character, reflecting different levels of interaction between citizens and authorities. Protest practices are institutionalised/non-institutionalised practices of conflict resolution as an acute divergence of the parties' interests; conflict, if properly communicated, can be a constructive force that explicates contradictions and optimises the search for solutions [5]. Contemporary protest reflects the link between socio-cultural transformations and technological changes, primarily information and communication technologies [3; 6; 7; 8].

Conflict mobilization builds on social conflict and negative identity, aims to solidarize and induce the addressees of influence to take joint desirable social action against the constructed and presented danger; it features a number of developmental stages (an information wave generating protest sentiment; formation of emotional internet solidarity; articulation of a decision on joint online/offline protest activity and its form; organisational action; post communication) [see 9; 10].

The formation of mobilization communities in social networks stems from the need to target the consolidating effect [11; 12; 13] against the background of the existing selective, self-imposed filters (limitation by the referential field of the Internet user) informing network users [9] and the advantages of communities in social networks (wide coverage and rapid circulation of information due to voluntary association and horizontal self-organisation, effective interactive tools offered, multiplication of addressants producing content, cross-border) [7; 14]. As for the system of subscriptions, self-imposed filters, we should mention that they contribute to the cultivated avoidance of contradictions by erecting an individual comfort zone, adjusting content, and thereby constructing a social reality to suit oneself. Mobilization communities on social media facilitate segmentation and aggregation of participants according to similar positions, interests, aggregation of news and content sharing of certain topics, individualisation of addressee and information. Such communities are heterogeneous in composition, bringing together consumers and active content producers, participants who differ in their perceptions of the Internet and civic political participation, the regularity of participation, and the prevalence of weak and strong ties [3; 15].

 https://doi.org/10.1051/e3sconf/202131104005

Table 1 presents the environmental mobilization protest communities on social media, whose content served as a source of practical material.

Table 1. Researched environmental mobilization protest communities on social media

social network, community name, purpose of the protest	date of organisation, status, reference	number of subscribers /participants
VKontatke, We are against stink and incinerators in the north of St. Petersburg (against air pollution in north-west St. Petersburg; Novoselki landfill, Severny sewage sludge landfill, asphalt plants in the Kolomyagi zone, sludge incineration plant in Lakhta, waste processing plant placement in the Kamenka zone)	2016, open, https://vk.com/spb_stop_reek	9353
Vkontakte, Sychovo is against (against the construction and commissioning of a new landfill in the vicinity of Sychovo and Shchelkanovo)	2018, open, https://vk.com/sch_poligon	2961
Against Titanium Polymer (against the construction of a polyethylene terephthalate plant in Moglino)	2019, public, https://www.facebook.com/groups/pskovprotiv/about	280
VKontakte, Titan-Polymer - PSKOVICIANS ARE*AGAINST*! (against the construction of a chemical plant)	2019, open, https://vk.com/protivtitana	2359
Facebook, Brest against the construction of a leaded battery plant (against the construction of a battery plant near Brest)	2018, public, https://www.facebook.com/groups/168788820292870/about	5068
Facebook, For a Healthy Vladikavkaz - Against Electrozinc! (against Electrozinc's lead-zinc plant)	2016, public, https://www.facebook.com/groups/protivec/about	804
VKontakte, Vorotynsk Kaluga AGAINST THE DEATH PLANT (against the construction of a recycling plant)	2019, open, https://vk.com/vorotynsk_kaluga_prozavod	1678
Vkontakte, Ghost of the Future against the incinerators. Kolpinsky district (against emissions from the Sveza Ust-Izhora incinerators)	2020, open, https://vk.com/no_trash1	3756
Vkontakte, Against the landfill in Pervomayskoye (against the organisation of the landfill north of Ogonyki village)	2016, open, https://vk.com/protiv_svalki_v_pervomaiskom	2862
Vkontakte, We are against the landfill on Shies (against the landfill in Leninsky District of Arkhangelsk Oblast, construction of a solid waste processing plant)	2019, open, https://vk.com/great__food	13575
Vkontakte, Against the construction of a port in Primorsk (against the construction of a universal transhipment complex in Primorsk and Ermilovo village)	2019, open, https://vk.com/protivporta	4435
Vkontakte, Against transit through Nizhnyaya Yeltsovka! (against the construction of a motorway bypass in the Sovetsky District of Novosibirsk)	2019, open, https://vk.com/net_tranzitu_ne	860

Table 1. Continued

Facebook, Against incinerators, FOR separate waste collection and recycling! (against the construction of incinerators)	2012, public, https://www.facebook.com/groups/stop.burning.trash/	158

The sample size was 1,020 publications.

We consider the practices of conflictual environmental mobilization in social media as an integrative hyper-genre: polyfunctional (informing, evaluating, inciting, agitating and propaganda), combining features of different discourses (media, political, everyday, internet discourse), stylistically and genre-syncretic (bookish and colloquial speech, signs of open writing and internet forum).

This study aims to characterise the subject systemic-communicative dimension of environmental protest, describing patterns of discursive problematisation in the example of environmental communities of conflict mobilization in social media. It addresses the following tasks: describing the types of publications, constituting content communities; correlating solidarity and non-solidarity reactions of community members to publications for assessing the implementation of discursive environmental problematisation in the communities under study; analysing thematic dominants to identify frequent constructs of environmental threats from the observer perspective (ordinary citizens – group participants); describing strategies and tactics for problematising fragments of environmental reality; identifying regular means of creating/maintaining online solidarity in communities.

3 Results and Discussion

The content of the environmental conflict mobilization communities studied is constituted by different types of publications: non-community publications that are shared within the community; publications produced by community members and aimed at consolidating information against the opponent, building emotional solidarity; and direct mobilization publications on behalf of the community, encouraging participation in protest action and organising participation. The interaction relies on informational (administered information of selective nature regarding the objects of thematization), communicative (making public the object of thematisation and the product of solidarity), psychological (influencing motivation, emotions), educational (influencing values, behaviour, awareness and political thinking), organisational (providing necessary conditions and tools for consolidation) modes [see 16].

The analysis, systematization of thematic dominants and quantitative calculations of the publications of the studied environmental mobilization communities in social networks indicate that from the position of the most active citizens (community initiators, content producers) there are the following areas-sources of problematic objects: construction of harmful production (incinerators, asphalt and concrete, waste processing plants, solid waste processing, lead-zinc enterprises, plants for production of lead-coated batteries, etc.) –68.42%; organization of landfills/waste dumps –26.32%; construction of highways –5.26%. At the same time, an account of the number of participants united in a community by a particular problem/threat, demonstrating a similar negative identity, and an analysis of the interactive reactions of group members in support of publications systematised according to named thematic dominants indicates approximately the same interest in two main areas: from the perspective of the mass addressee–consumer of content, the following situations/circumstances of construction/exploitation are the most environmentally problematic, causing concern and protest sentiments of citizens are the following situations/circumstances– construction/operation of harmful production (support 58.34% of the total number of participants in the communities studied); organisation of landfills, rubbish/waste dumps (40.45%). The targeted and successful discursive construction of an environmental situation with a threat index in communities of conflictual environmental mobilization is evident from the unambiguous dominance of solidarity reactions by addressees, indicating a planned perception, sharing of knowledge and attitude among community participants.

Establishing the environmental agenda (justifying and presenting a circumstance, phenomenon or situation as dangerous and requiring a solution) involves a strategy of supplying the object of thematization with a threat index.

One of the most frequent tactics of the described strategy is a direct negative assessment of the current state of affairs (23.61%), characterised by the use of evaluation units with negative semantics, verbs in the present tense, lexemes with the temporal semantics of "present": *Landfills* are ***monstrously maintained*** *by municipalities* [17, 02.09.2021]; Dmitry Baranov: *They will put some kind of class 4 hazard instead of class 1.* ***Now they do not hesitate*** *and* ***are making a mess of things****. All these Vedernikovs, Moscow* ***bandits*** *...* [18, 23.07.2021].

The next tactic actively used is to indirectly indicate dissatisfaction with the state of affairs by directly addressing/calling on members of the mobilization community to change the situation, to counteract (19.33%). The relevant language means – appeals, incentive sentences, performatives, personal and possessive pronouns in the 2nd person plural and 1st person plural, some determinative pronouns (for example: *all, every*), words with the semantics of jointness: ***We call upon all*** *of you to write a statement to the Prosecutor's Office of the Komi Republic concerning the legality of the OMON and police actions to escort fuel tankers to the #Shies* [19, 13.11.2019]; *Dear* ***supporters****! We* ***ask*** *for help* ***from all those involved*** *in defending Shies* [19, 13.11.2019].

The next tactic used – the explication of the dangerous consequences of the phenomenon/situation (14.86%). Its realisation presupposes the use of lexical units (LUs) expressing negative evaluation, future tense verbs, adverbs and prepositions with the temporal semantics of consequence, and LUs belonging to the semantic field of "harm". For example: *With such cargo turnover, the ecology of our city and neighbouring territories* ***will be irreparably damaged****, and there* ***will be a serious danger*** *to people's health...* [20]; ***if there is a problem*** *at this plant or a minimal leak of sulfur vapour or lead dust -* ***it will cover*** *the whole city, the critical area is up to 15 km... and* ***in*** *10 years near this plant,* ***we will*** *build a hospital for* ***oncology patients*** [21].

To justify a negative assessment of the state of affairs, they also use as an indirect argument the tactic of a direct negative assessment of their opponent's qualities (17.16%). The relevant linguistic means are LUs with negative semantics, anthroponyms, and the designation of professions and positions; the references to the opponent are also frequent: *Once again, we observe* ***the ignorance of the staff of the oversight body*** [19, 14.08.2019]; *Volunteers ... negotiate with processors, but officials are clearly* ***not capable of this****. But they* ***are capable of polluting*** *rivers, soil and air* ***with insane decisions****. ...* ***Gentlemen officials****, are you really so* ***dull*** or do you have only *mercantile* ***priorities****?* [22, 07.06.2021].

Related to the previous tactic is the tactic of blaming the opponent, initiating the addressee's assessment of the situation as threatening, dangerous (14.21%). This tactic involves the use of anthroponyms, the designation of a subject of power, LUs with negative semantics, and LUs from the semantic fields of "guilt" and "condemnation". For example: *Neither* ***Parfenov*** *nor* ***Shapsha has done anything*** *to make rubbish segregation mandatory ...* ***Neither of them did anything*** *to conclude contracts with recyclers...* [22, 07.06.2021].

2021]; *Everything happened precisely because the* ***swindlers*** *from "MPBO-2", under the watchful eye of the profile* ***vice-governor Nikolay Bondarenko*** *and* ***Sergey Malinin, head of the Committee for Public Works,*** *continued to quietly transport* ***duds*** in *the form of combustible waste to the landfill - and as a result* they ***themselves created ideal conditions for a fire*** [23, 17.08.2021].

A regular tactic to implement this strategy is a declaration on behalf of the community against the threat as a community goal (7.38%), which contributes to the mass recipient sharing a negative assessment of the presented situation (usually offered in the "information" tab or a fixed entry at the top of the page in the "discussion" tab). The relevant means within this tactic are lexical units (LUs) of the semantic fields "achievement" and "opposition", the inclusive 1st person plural pronoun, LU with the semantics of oughtness, synonyms of the word *problem*, community/group designations. For example: ***Community*** *of residents of Primorsky ... and other districts of St. Petersburg and the Leningrad region, who* ***are tired of enduring*** *the regular stench of rotting waste ..., the fumes from asphalt plants and sludge incineration.* ***We want to get*** *the authorities* ***to solve*** *these* ***problems*** *and* ***prevent*** *the emergence of ... even more dangerous sources of pollution ...* [23]; ***GROUP'S GOAL****: ELECTROCINIC* ***MUST*** *KEEP THE REPUBLIC!* [24].

An indirect argument in presenting some circumstance to the addressee as a threat also uses the tactic of stating the opposing interests and needs of citizens and the opponent (3.45%); the relevant linguistic means – antithetic conjunctions, syntactic parallelism, antonyms, LUs with opposite evaluative semantics and other means that implement the contrast: ***for someone*** *this pipe* ***is a symbol of future cash flow****, and* ***for someone*** it is ***a symbol*** of a ***vanishing hope*** *for good ecology and strong health* [22, 26. 07.07.2021]; *Vladimir Kozlov: Kremlin* ***guys*** *are doing a good job -* ***money*** *from natural resources trade* ***to Moscow, while dumpsters*** *are taken* ***away****... It's time to end it))) POMORIE IS NOT A* ***DUMP***! [19, 06.11.2019].

Alongside the informative, propagandistic, directive and regulative functions, the evaluative function is the most important obligatory feature of the communicative practices described, realised through the strategy of presenting evaluations. We should note that assessment can be offered to community members as subjective, individual (unsupported) and as relying on other resources, such as norms, traditions, expert opinion, etc. (supported).

One of the most frequent tactics of this strategy is the tactic of broadcasting evaluation as shared by multiple subjects (34.67%), which implies extensive identification of the addressant with the addressee, for example, through the use of inclusive pronouns, the first person plural possessive pronoun, a number of definitional pronouns (*all, whole, every, any, whatever*), collective signs, collective subject names and the use of LUs with an evaluative semantics. For example: ***We all*** *know about* ***the lawless*** *acts committed there on October 23* [19, 13.11.2019]; ***Residents*** *are convinced that the connivance* ***of the landfill reclamation*** *organisation is to blame for the fire* [23, 21.08.2021]; ***"Time Bomb"****, or* ***"Monster"****. This* is how Vladikavkaz ***residents*** call *the Elektrotsink plant, which stands right in the city centre* [24, 31.08.2021].

A common tactic is to present the addressee with an individual evaluation (26.81%); frequent are personal and possessive pronouns in the singular, LUs with evaluative semantics, verbs of speech activity and exclamatory sentences: Karina Korzhova: ***I am sick*** *of this* ***lie and disorder****! I won't vote now on principle!* ***No adequate*** *people in power! Just* ***to steal*** *and* ***line their own pockets****!* [20, 04.09.2021]; Natalia Ivanova: ***I*** *don't know who the guilty person is,* ***but I know*** *that it's* ***wrong****, it would be nice to find out who is responsible* [23, 17.08.2021]

The next way of presenting evaluation – the tactic of broadcasting evaluation as a moral axiom (23.13%), which involves appealing to common moral experiences, norms and widespread perceptions. Relevant linguistic means – LUs with evaluative semantics, abstract nouns, cliches, proverbs, idioms, stable expressions. For example: Tatiana Chulkova: *they* ***do not know what they are doing****. They will* ***be damned to the seventh tribe****. They cut everything down. Petersburg* ***will suffocate****. These forests and lakes* ***are protected Do not touch them****. Where is the prosecution to put them all in prison?* ***Devils what will they leave behind****...* [20, 05.09.2021] (appeal to moral norms: one should not touch the untouchable, one should leave behind the good, one should think about consequences); Dmitry Poplavsky: *a country* ***of barbarians who value*** *nothing... neither* ***freedom****, nor* ***nature****, nor* ***the homeland****...* ***only dough on their mind****...* [20, 05.09.2021].

The last tactic revealed is the reinforcement of the evaluation presented to the addressee with a reference to authority (15.39%), which implies the use of LUs with evaluative semantics, anthroponyms, designations of professions, positions, organizations, verbs of speech-thought activity: *Conclusion* ***of public ecological expertise*** *of the project documentation "Primorsky universal transhipment complex". ... brief paraphrase: ... the project* ***will be neither safe*** *ecologically, nor technically* [20, 01.07.2021]; Marina Viktorovna: *the presenter had better remember what* ***academician Yablokov*** said - *the anthropogenic load on the Gulf of Finland has long* ***been exceeded****!* [20, 01.07.2021].

The strategies and tactics mentioned in the description of the strategies and tactics, identification with community members, LUs with the semantics of joint action and unification, appeals, messages with broad addressing, as well as compliments from the community and its members to participants and active post communication are regular means of creating/maintaining online solidarity in the studied environmental communities of conflict mobilization.

4 Conclusion

Environmental communities of conflict mobilization in Russian-language social networks, whose content is constituted by external and internal publications of indirect and direct mobilization, use informational, communicative, psychological, educational and organisational modes of mobilization, correlating with the main stages of the latter; the impact is framed by the concept of "I-we-they", involving the construction of negative and positive group identities, through solidarising group self-representations and consolidating oppositions. From the perspective of the most active citizens producing their own content and content consumers, the most problematic situations correlating with the frequency objects of thematization are the construction/operation of harmful production and the organisation of landfills or dumping sites for rubbish, hazardous waste. Frequent ways of forming virtual solidarity are wide identification of the addressant with community members, building semantic fields of jointness, unity; wide addressing of appeals and messages; and active post-communication. The discursive construction of the ecological situation with a threat index in the described communicative practices is confirmed by the unambiguous dominance of solidarity reactions of community participants, who share the transmitted knowledge, attitudes and emotional experiences. The discursive problematisation of environmental risks in conflict mobilization in social media is predominantly realised through supported or unsupported direct negative assessment of the current situation, with a predominant appeal to grassroots positions, moral norms and values, individual worldviews or, less frequently, –to the opinion of authority; through broad appeals to community members "to stand against"; negative evaluation and blaming the opponent; pointing out the dangerous consequences of the current state of affairs; declaring the goal of environmental community conflict mobilization.

Acknowledgements

The reported study was funded by the Russian Foundation for Basic Research, project number 20-012-00193.

References

1. R. E. Barash , A. Yu. Antonovsky, Philosophical Journal, **11(2)**, 91-105 (2018)
2. S. V. Paulov, Bulletin of the Chita State University, **1(4)**, 26-30 (2008)
3. L. N. Rebrina, N. L. Shamne, Nauchnyi dialog, **3**, 151-167 (2020)
4. Z. Oniszczuk, Studiów Medioznawczych, **4(47)**, 1-15 (2011)
5. R. Fisher, W. Ury, B. Patton, *Getting to Yes: Negotiating Agreement Without Giving*, New York, Penguin Books (2011)
6. L. V. Kazimirchik, Theory and practice of social development, **1**, 99-103 (2014)
7. V. I. Kurbatov, Humanitarians of the South of Russia, **6(6)**, 173-181 (2017)
8. G. Small, G. Vorgan, I-*Brain: Surviving the Technological Alteration of the Modern Mind* (New York : Harper-Collins, 2009)
9. A. Vanka, I. Ksenofontova, I. Tartakovskaya, Inter, **7**, 44-73 (2014)
10. L. N. Rebrina, Nauchnyi dialog, **5**, 115-139 (2021)
11. I. A. Gladchenko, Scientific and Technical Bulletin of SPbSPU. Humanities and social sciences, **10(2)**, 30-43 (2019)
12. A. S. Sherstobitov, A. Bryansk, Questions of theory and practice, **10(36)**/I, 196-202 (2013)
13. O. N. Yanickij, Sot͡siologicheskie Issledovanii͡a, **6**, 3-12 (2012)
14. A. V. Kirka, Vestnik Gosudarstvennogo universiteta upravleniya, **4**, 47-51 (2012)
15. S. V. Dokuka, *Communication in social online networks as a factor of protest mobilization in Russia* (2014)
16. Yu. B. Kinyasheva, Humanities, **3**, 3-10 (2018)
17. PTch – VKontakte, Against transit through Nizhnyaya Yeltsovka, https://vk.com/net_transitu_ne
18. PTP – Facebook, Against "titanium-polymer", https://www.facebook.com/groups/pskovprotiv
19. MPS – VKontakte, We are against the landfill on Shies, https://vk.com/great__food
20. PSP – VKontakte, Against the construction of the port in Primorsk, https://vk.com/protivporta
21. BPS – Facebook, Brest against the construction of a lead-acid battery plant, https://www.facebook.com/groups/168788820292870/about
22. PMZ – Facebook, Protiv musoroszhigatel'nyh zavodov, Against incinerators, FOR separate collection and recycling of waste, https://www.facebook.com/groups/stop.burning.trash
23. MPV - VKontakte, We are against the stench and incineration plants in the north of St. Petersburg, https://vk.com/spb_stop_reek
24. ZZV – Facebook, For healthy Vladikavkaz - against "Electrozinc"!, https://www.facebook.com/groups/protivec/about

Ecological dimension of conflict in Nagorno-Karabakh

G. R. Basharov*, *F. S.* Shedko, *D. P.* Okolota, *A. L.* Kotova, and *A. A.* Volkova

Saint-Petersburg State University (SPBU), Saint-Petersburg, Russia

Abstract. This article examines the conflict in Nagorno-Karabakh not only from the point of the usual historical, ethnic and religious context, but also from the point of the environmental and resource dimensions, which are often ignored by researchers and practitioners of conflict resolution. The so called Six Weeks War significantly changed the current state of the conflict, and through focus group interviews and scientific analysis, the authors try to assess the impact of the environmental aspect on this conflict and substantiate its importance. The authors also hypothesize that the importance of environmental and resource factors is ignored by researchers and local residents, although these factors could become one of the foundations for building relationships, building trust and long-term conflict resolution.

1 Introduction

The conflict in Nagorno-Karabakh is one of the longest lasting and has one of the most ancient historical and cultural roots in the entire post-Soviet space. This leads to the multilevel conflict and the presence of many intertwining dimensions. The origins of the conflict go back centuries, but the beginning of its modern history can be considered the 1980s, the last years of the existence of the Soviet Union.

Although the population of Nagorno-Karabakh is predominantly Armenian, the Nagorno-Karabakh Autonomous Oblast (NKAO) was part of the Azerbaijan Soviet Socialist Republic. In February 1988, an extraordinary session of the regional Council of People's Deputies of the NKAO petitioned for the transfer of the NKAO from the Azerbaijan SSR to the Armenian SSR [1]. The refusal to satisfy this petition provoked demonstrations of Armenians not only in Nagorno-Karabakh, but also in the capital of Armenia - Yerevan. Tensions have sparked a wave of violence and ethnic cleansing in Armenia and Azerbaijan. In a referendum held in Nagorno-Karabakh a few weeks before the official collapse of the Soviet Union, the overwhelming majority (99.98%) voted for the complete independence of the Nagorno-Karabakh Republic from Azerbaijan [2]. The Azerbaijani government declared this act illegal and abolished the autonomy of Nagorno-Karabakh [3].

In the ensuing armed conflict, Azerbaijan tried to regain control of Nagorno-Karabakh, while Armenian troops defended independence with the support of the Armenian government and the Armenian diaspora. The internal separatist conflict gradually became internationalized.

* Corresponding author: basharov.galim@mail.ru

Losses on both sides, according to various estimates, reached up to 25,000 people [4]. Armenia gained control over the territory of the Nagorno-Karabakh Republic and the adjacent seven regions of Azerbaijan (approximately 15% of its current territory) [4]. The hostilities ceased on May 12, 1994 after the signing of a ceasefire agreement [5], but a political solution was not agreed, despite the efforts of the parties of the conflict and numerous international mediators. [6] As a result, Armenia and Azerbaijan remained in an ambiguous state for a long time - without war, but also without peace. Peace talks continued within the framework of the Minsk Group of the Organization for Security and Cooperation in Europe (OSCE), which was established in 1992 by France, Russia and the United States as co-chairs. [7]

In 2016-2017, several outbreaks of armed confrontation between the conflicting parties took place, and the whole year actually passed under the sign of expectation of a possible repetition of those previous events [8, 9]. Then, the parties, with the mediation of Russia, managed to avoid the implementation of the most unfavorable scenarios.

However, the situation repeated itself and sharply worsened in the summer-autumn of 2020. The confrontation has again turned into a hot phase. The military actions were stopped thanks to the intensification of trilateral negotiations and the establishment of a Russian peacekeeping operation on the territory of Nagorno-Karabakh on November 10, 2020 [10]. A peace agreement was signed, providing for «the entry of Russian peacekeepers, the withdrawal of Armenian forces from the areas adjacent to Nagorno-Karabakh and the abandonment of part of the territory of Nagorno-Karabakh, including the city of Shusha, under the control of Azerbaijan» [11].

The peacekeeping contingent in Nagorno-Karabakh is stipulated by the agreement on the cessation of hostilities, which was signed on November 9 by Armenia, Azerbaijan and Russia [12]. It was reported that «a total of 1960 military personnel, 90 armored personnel carriers and 380 units of automobile and special equipment will be sent to the region» [13]. The Russian Defense Ministry's website also publishes daily infographic reports showing that the territory under the control of Russian peacekeepers is divided into two zones – «North» and «South». There are 16 observation posts in these zones, and military police patrols are organized in two areas. [13] According to the decree of the President of the Russian Federation, the peacekeeping contingent will change at least twice a year [14].

This conflict has many dimensions. Against the background of historical, ethnic, religious contradictions, environmental factors often fade into the shadows and are not considered by researchers and local practitioners as a significant factor influencing the entire conflict as a whole. However, it seems that this aspect has a huge impact on the conflict and the course of the settlement. Talking about the relationship between environmental scarcity and conflict, Homer-Dixon et al. (The so-called Toronto Group) argued in the early 1990s that environmental scarcity causes violent conflicts in many parts of the developing world, that these conflicts are usually observed within the country and that they have expansionary characteristics and can spread very quickly. Homer-Dixon's work indicates that environmental deficiency is one of the factors that can lead to armed conflicts within the country [15, 16, 17]. The Nagorno-Karabakh conflict introduces some changes to the above statements. Thus, the conflict in Nagorno-Karabakh was internationalized very quickly, therefore, environmental and resource factors become not only the cause of armed internal, but also internationalized conflicts.

The resource issue is one of the key ones - there are about 155 different deposits of gold, copper, mercury, lead and zinc in Nagorno-Karabakh. According to the estimates of the Johns Hopkins University there are three hydrocarbon deposits with total reserves of at least 150 million tons of oil and up to 250 billion m3 of gas in the Araks Valley of Nagorno-Karabakh, part of which previously fell on the territory of the Nagorno-Karabakh Republic. The water factor is no less important - Azerbaijan has relatively scarce water resources. More than 80%

of river sources are outside the country, which means a certain dependence on the country's water security. NKAO, while still being a part of the Azerbaijan SSR, played a significant role in this dimension, since on its territory there is one of the largest reservoirs in the Sarsang region, commissioned in 1975, the total volume of which is more than 601 million m3 of water, as well as the Tartar and Khachen. At that time, the Sarsang reservoir irrigated 128 thousand hectares of farmland, of which slightly less than 90% (110 thousand hectares) were outside the borders of the NKAO.

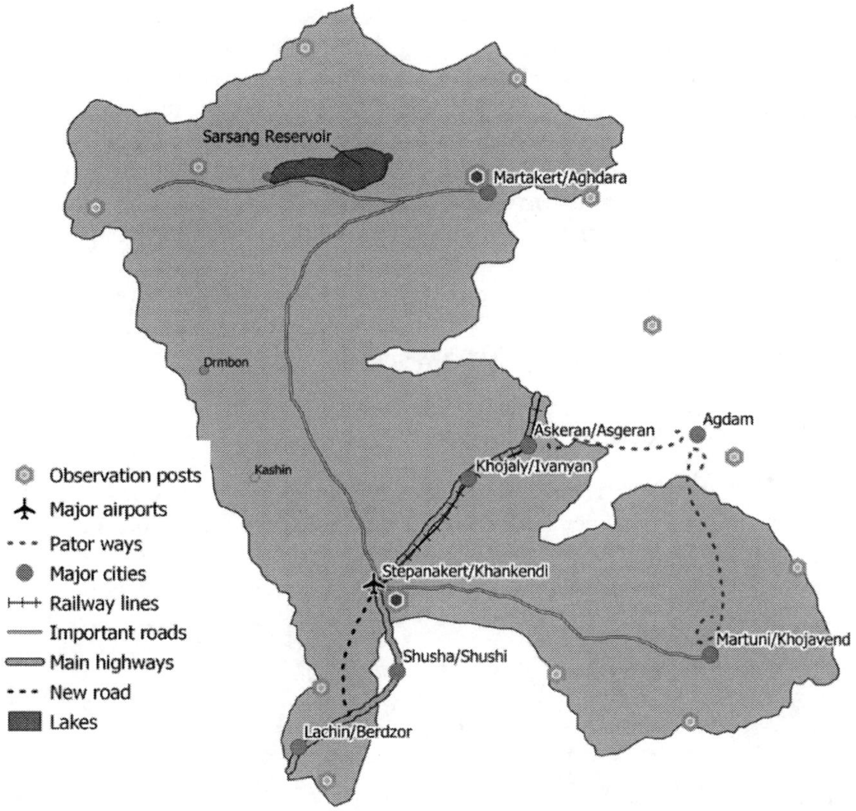

Fig. 1. Infrastructure map of the Nagorno-Karabakh Republic (the map was compiled by the authors independently).

Following the results of the last «six-week war», a significant part of the NKR territory passed to Azerbaijan. Significant water resources came under the control of Azerbaijan. These are, firstly, the sources of the Tartar and Khachen rivers, located in the Kalbajar region. Their average annual flow is about 828 million cubic meters. Secondly, this is a significant part of the territory of the Kashatagh Province and Karvachar region, the average annual flow of which is 2588.5 million cubic meters.

However, NKR is not left without water resources at all. The Sarsang reservoir with a number of key rivers remained under the control of the NKR. At the same time, the population of Nagorno-Karabakh (about 150.000 people) needs about 247 thousand cubic meters per year for food security and about 16 million for drinking needs. This is less than 40% of the volume of the Sarsang reservoir.

2 Materials and Methods

Such general scientific methods as analysis, generalization, comparison, description, content analysis were used in the work. The visualization and mapping method (using the QGIS software) was used to compile the map.

The first step for conducting the research was the analysis of the existing literature on the subject of the Nagorno-Karabakh conflict.

To assess the impact of the environmental factor on the Nagorno-Karabakh conflict, the authors, together with the Eurasian Youth Assembly, conducted interviews with focus groups consisting of residents of Azerbaijan, Armenia and the NKR. In total, the opinions of 3 focus groups were analyzed, each of which consisted of 12 people. In each focus group, residents of Azerbaijan (12 people), Armenia (12 people) and NKR (12 people) were equally represented. The age composition of each of the 3 focus groups was as follows: 4 people aged 18 to 28; 4 people from 29 to 40 years old and 4 people from 40 to 60 years old. Thus, the total number of people who took part in the focus groups is 36.

The interviews were converted into text format with subsequent content analysis. To conduct content analysis, the authors used NVivo and MAXQDA software. The total number of text characters of the focus group interviews was 450,000 characters (excluding spaces). As part of the interview, the authors of the study asked questions about the history of the conflict, its current state and the factors influencing it.

3 Results and Discussion

Based on the results of the content analysis of the focus group interviews, the following conclusions can be drawn. When answering the question: "What are the main factors of the Nagorno-Karabakh conflict?":

Fig. 2. Infographics based on the results of the focus groups (compiled by the authors independently)

- Only 18% of respondents mentioned fossil and rare earth resources in one way or another;
- 10% of respondents noted the importance of environmental and water factors;
- The majority of respondents in the focus group singled out the significant role of the identity and ethno-religious factor (83%) as the main factor in the conflict;
- The historical factor of confrontation also stands out as significant, 67% of the respondents noted its importance;
- Political and economic factors of the conflict were also noted as important, 61% of the respondents highlighted their importance.

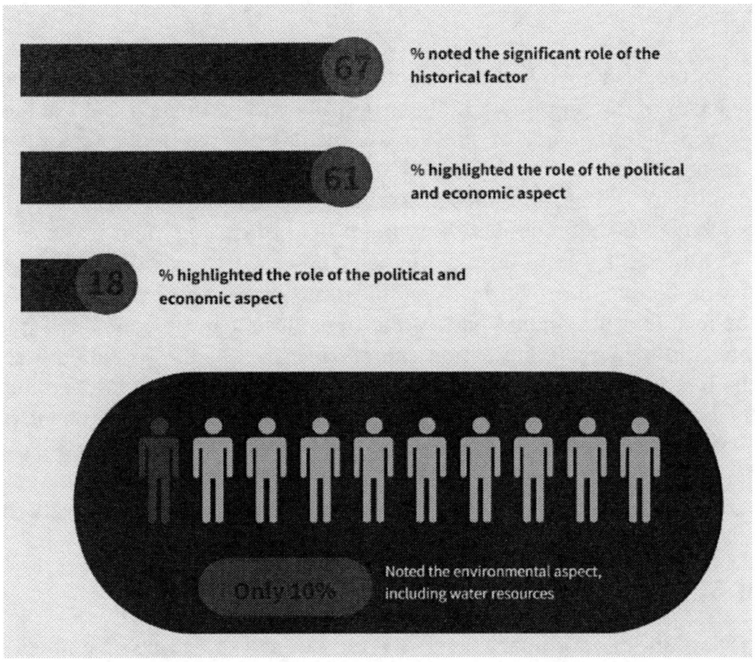

Fig. 3. Infographics based on the results of the focus groups (compiled by the authors independently)

After the initial survey, 91% of respondents believed that the role of the environmental and resource factor was absent or completely insignificant. After a short briefing for the respondents with the provision of information about water and other resources in this conflict, some of the respondents changed their minds. When asked again about the importance of the environmental and resource factor, 43% of the respondents said that cooperation in the field of ecology and water resources could be the first step towards a long-term settlement. It should be emphasized that the authors of the study in no way tried to influence the opinion of the respondents, but only gave them brief factual information, also described in this study.

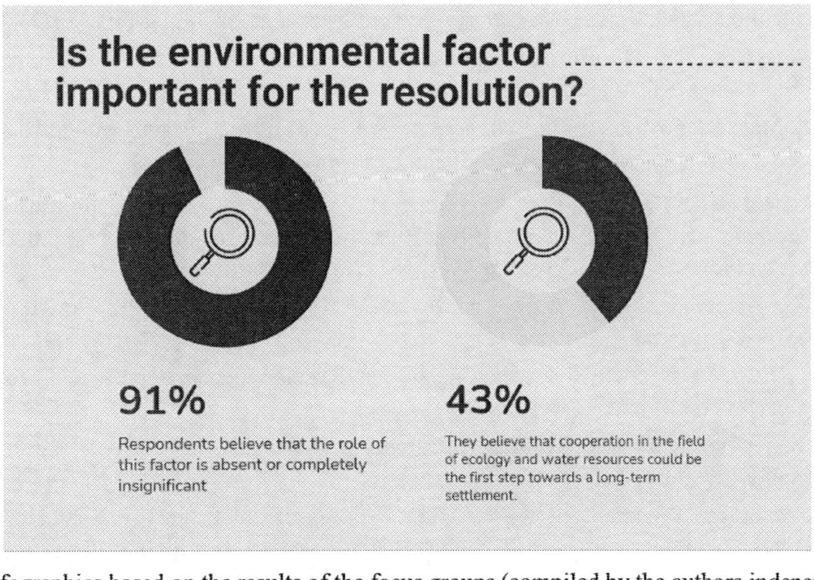

Fig. 4. Infographics based on the results of the focus groups (compiled by the authors independently)

4 Conclusion

The first part of the authors' hypothesis about ignoring the importance of the environmental and resource factor of the Nagorno-Karabakh conflict was confirmed, 91% of the respondents initially believed that the role of the environmental and resource factor was absent or completely insignificant.

The second part of the hypothesis is confirmed by the fact that the percentage of respondents who believe that cooperation in the field of ecology and water resources could be the first step towards a long-term settlement increased from 9% to 43% after providing respondents with factual information about this dimension of the conflict. The authors of the study also believe that the second part of the hypothesis can be confirmed by the fact that environmental and resource cooperation can become part of the so-called «tactics of small steps» method, when the parties gradually take steps towards confidence-building. Small steps usually contribute to the formation of an atmosphere of trust between the parties, which is extremely necessary for the normalization of relations [18]. Environmental and resource cooperation between the three parties to the conflict could, in our opinion, really contribute to confidence-building and the most effective activation of the long-term settlement process.

References

1. Judgement of the extraordinary session of the Council of People's Deputies of the NKAO "On a petition to the Supreme Soviets of the Azerbaijan SSR and the Armenian SSR on the transfer of the NKAO from the Azerbaijan SSR to the Armenian SSR" dated 03/20/1988, ANI Armenian Research Center, https://www.aniarc.am
2. Results of the referendum on the independence of Nagorno-Karabakh from Azerbaijan, dated 10.12.1991, https://www.sudd.ch
3. A. A. Melik-Shakhnazarov, Post-Soviet studies, **2,** 139 (2018)
4. E. Klimenko, *Protracted Armed Conflicts in the Post-Soviet Space and their Impact on Black Sea Security, SIPRI Insights on Peace and Security* (2018) https://www.sipri.org
5. Agreement on a ceasefire in the area of the Nagorno-Karabakh conflict (1994)
6. V. A. Kolosov, M.V. Zotova, Geography, environment, sustainability, **13(1),** 84 (2020)
7. S. Minasyan, The Nagorno-Karabakh conflict in the context of South Caucasus regional security issues: An Armenian perspective, Nationalities Papers, **45(1),** 131 (2017)
8. S. M. Markedonov, *RIAC analytical articles* (2018) https://russiancouncil.ru
9. A. Bayramov, Security and Human Rights, **27(1-2),** 116 (2016)
10. Statement of the President of the Republic of Azerbaijan, Prime Minister of the Republic of Armenia and the President of the Russian Federation dated 11/10/2020, Official network resources of the President of Russia, http://kremlin.ru
11. RBC, Armenia made peace and went to concessions to Azerbaijan, rbc, https://www.rbc.ru
12. RBC, Experts assessed the weapons of the Russian peacekeepers in Karabakh, rbc https://www.rbc.ru
13. Russian peacekeeping contingent in Nagorno-Karabakh, Ministry of Defense of the Russian Federation, http://mil.ru
14. Decree of the President of the Russian Federation dated 11/10/ 2020 No. 695 "On measures to maintain peace in Nagorno-Karabakh", Official Internet portal of legal information, http://publication.pravo.gov.ru

15. T. Homer-Dixon, International Security, **19(1),** 5 (1994)
16. A. Rwabizambuga, Conflict, Security and Development, **7(1),** 203
17. P. Wallensteen, Journal of Peace Research, **13(4),** 277 (1976)
18. N. V. Shtanski, International processes, **36–37,** 33 (2014)

Data centers environmental impact assessment features

Andrey Semenov[1,*], and *Ekaterina* Oganesyan[2,†]

[1]Moscow State University of Civil Engineering (MGSU) National Research University, Moscow, Russia
[2]Mendeleev University of Chemical Technology of Russia, Moscow, Russia

Abstract. Data centers became significant sources of environmental impact: each year global data centers consume TWh of electricity, generate comparable thermal emissions to the atmosphere and/or hydrosphere, create wastes of electronic equipment and life-expired batteries, and create other types of direct and indirect ecological footprint. In conformity with the sustainable development concept data centers environmental impact of all types should be numerically assessed to compare to the environmental capacity and move towards sustainability. It requires ecological footprint (carbon footprint in particular) to be assessed. Existing xUE Effectiveness Metrics used for data centers are all relative, so data centers' environmental impact cannot be calculated directly from it. Methods of payment calculation for negative environmental impact, used in Russia, do not take into account data center features and can hardly be used for the assessment tasks. Data centers need to adapt existing and develop new assessment methods for its environmental impact, considering all the resources consumed and all the emissions generated.

1 Introduction

Data centers consume large amounts of electricity. Sources may differ in their estimates: according to [1] datacenter annual global electricity consumption in 2010-2018 increased from 194 TWh to 205 TWh, while [2] states that it already achieved 286 TWh in 2016. However, sources agree that datacenter consumption is hundreds of TWh per year, which is 1-1.5% of total world electricity production. There is a trend to growth of datacenter energy consumption, and one of the reasons is data centers number increase caused by the ICT development in general and the Internet of Things spreading in particular. Thus, the largest Russian data center (Rostelecom, Udomlya) with its 1st stage of 40 MW power consume 0.65 TWh annually. Total number of data centers increases all over the world including Russia; new facilities appear and some existing facilities expand. Based on existing trends [2] forecasts world datacenter consumption will achieve 321 TWh/year by 2030.

The major part of the electricity consumed by data centers converts to thermal energy and comes to the environment: to the atmosphere when using air cooling, to the atmosphere

* Corresponding author: SemenovAB@mgsu.ru

and/or the hydrosphere when using liquid cooling systems [3]. In addition to thermal environmental pollution directly generated by data centers, these facilities are responsible for emissions occurring at electricity generating plants, energy from which is then consumed by data centers. As carbon quota develops, this aspect should be estimated and taken under control.

In case of centralized energy supply failure data centers use backup power supplies, which also have an environmental impact. Diesel generators able to power datacenter facilities have the output comparable to the same MW data center power, which means dozens tons of diesel fuel consumed per day. Hydrocarbon fuel combustion results in direct CO_2 emissions. Also some less significant emissions of carbon monoxide, nitrogen oxides, unburned hydrocarbons etc. take place.

In case of cooling agent leakage or fire extinguishing system activation liquid, gaseous and solid (powder) emissions/wastes can take place. Some of them are considered pollutants. Depending on systems used these can be: ethylene glycol, "waterless water/dry water" organic compounds (fluorinated ketones), halocarbons (CFCs, HFCs) etc. In case of carbon dioxide fire extinguishing system activation direct emission of CO_2 takes place.

Data centers also create solid wastes: electronic equipment (out of order or obsolete), storage media (including deliberately destroyed to keep confidentiality of information, although their lifetime has not expired yet), and life-expired/lost capacity batteries.

Sustainable development principles applied to data centers functioning mean that all the forms of datacenter environmental impact must be numerically assessed to compare it to environment capacity. Meanwhile numerical criteria data centers currently use do not correspond to this task, despite the fact some of them are called *Sustainability Metrics* [4-5].

2 Materials and methods

PUE (*Power Usage Effectiveness*) metric offered by The Green Grid [6] in 2007 is widely used to assess data center efficiency. Initially it was calculated as follows:

$$PUE = Total\ Facility\ Power\ /\ IT\ Equipment\ Power \quad (1)$$

Later sources [7] use corrected formula:

$$PUE = Total\ Facility\ Energy\ /\ IT\ Equipment\ Energy =$$

$$= 1 + Non\ IT\ Facility\ Energy\ /\ IT\ Equipment\ Energy \quad (2)$$

Created to assess the efficiency of datacenter energy consumption, PUE assumes the closer is the metric to 1 the greater is the share of energy consumed by the target IT equipment (and the less by the auxiliary engineering subsystems). But neither the initial nor the corrected PUE metric report anything on the energy consumption absolute value and the associated environmental impact. All PUE metrics are relative. By the time the PUE was proposed, the ecological footprint and carbon footprint terms had already been developed. But none of them can be calculated or assessed from the PUE metric as it is.

Different data centers calculate PUE in different ways, excluding some engineering subsystems from consideration, which leaves a wide field for manipulation [8]. For example, the numerator may exclude the energy consumed by cooling systems (which is comparable to the IT equipment consumption itself) if the data center "receives cold" from an external supplier. The denominator can be artificially inflated by indicating the consumption declared in the IT equipment specifications instead of the actual consumption. Such manipulations make it difficult to compare different data centers to each other, thus

PUE cannot be used for practical purposes and cannot serve as a basis for sustainable development implementation [9-10].

WUE (*Water Usage Effectiveness*) and CUE (*Carbon Usage Effectiveness*) metrics offered by The Green Grid in 2010 are intended to assess respectively the efficiency of the water use and carbon fuel use in data centers [4-5]. But these as well as other xUE (*X Usage Effectiveness*) metrics are relative, all based on the ratio of the water consumed per year, carbon dioxide emissions per year and other material flows attributable to the entire facility, to the IT equipment energy consumption:

$$WUE = Annual\ Water\ Usage\ /\ IT\ Equipment\ Energy \quad (3)$$

$$CUE = Total\ CO_2\ emissions\ caused\ by\ Total\ Data\ Center\ Energy\ /\ /\ IT\ Equipment\ Energy \quad (4)$$

Despite the units these metrics use – l/kWh for WUE and kg or g of CO_2/kWh for CUE – none of them contain the information on the material and energy resources consumed by the data center. So these metrics cannot be used to assess datacenter environmental impact. At the same time it should be mentioned that WUE and CUE formulas include coefficients that characterize the sources of energy consumed by the data center:

— EWIF (*Energy Water Intensity Factor*), l/kWh, depends on the energy generation facility, which can be own (situated on the datacenter site) or external (owned by third-party suppliers);

— CEF (*Carbon Emission Factor*), kg or g of CO_2/kWh, depends on carbon fuel used for energy generation.

$$WUE_{source} = (Annual\ Source\ Energy\ Water\ Usage + Annual\ Site\ Water\ Usage)\ /\ /\ IT\ Equipment\ Energy = EWIF \times PUE + Annual\ Site\ Water\ Usage\ /\ IT\ Equipment\ Energy \quad (5)$$

$$CUE = CEF \times PUE \quad (6)$$

If data centers will regularly collect and publish data on the energy sources used and the amount of resources consumed, then EWIF and CEF factors will allow calculating the actual emissions and assessing datacenter environmental impact.

Table 1. CO_2 emissions per kWh of energy obtained from carbon fuel combustion.

Fuel	**CEF, kg of CO_2/kWh** (IPCC data)
Coal	0.863 to 0.941
Brown coal	up to 1.175
Petroleum	0.893
Natural gas	0.491 to 0.891 depending on generation technology
	CEF, kg of CO_2/kWh (IPCC and EPA data, Russian sources, and authors' own estimates)
Unknown source, average	0.59 – 0.8

Recent years the estimates or carbon dioxide emissions associated with hydropower have been revised upward, from close to zero to dozens of grams of CO_2 per 1 kWh. The most reasonable estimate for today is made by Chinese authors [11]: 0.092 kg of CO_2/kWh. One can expect that detailed scientific analysis will change values currently assumed by

International Energy Agency for nuclear power as 0.004-0.012 kg of CO_2 per 1 kWh, to higher values.

Table 2. Water consumption per kWh of energy obtained from different sources.

Method used for electricity generation	**EWIF, l/kWh** (The Green Greed recommendations)	**Note**
Hydropower	0	Water is used for electricity generation but not consumed; all the water is available for future consumption
Coal combustion	2.2 to 2.8	
Nuclear power	3.3	
Natural gas combustion	0.8	

When generating electricity locally at the data center site, emissions to the environment can be estimated based on the amount of fuel consumed.

Table 3. CO_2 emissions per carbon fuel unit.

Fuel	**Specific carbon footprint** (EPA and IPCC data, other international sources)	**Specific carbon footprint** (Russian sources and authors' own estimates)
High-octane gasoline	1.93 to 2.35 kg of CO_2/l	2.39 kg of CO_2/l
Diesel fuel	2.64 to 2.69 kg of CO_2/l	2.69 kg of CO_2/l
Natural gas	1.90 to 1.94 kg of CO_2/m^3	1.98 kg of CO_2/m^3

Main resources for typical data center, its emissions and wastes are shown in Figure 1 as input and output flows.

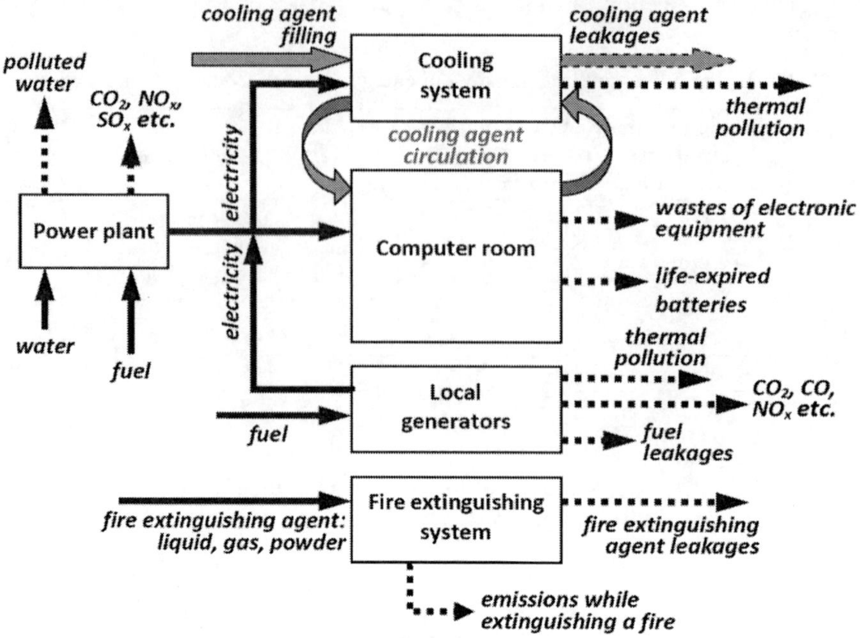

Fig. 1. Data center main resources, emissions and wastes.

Some of material/energy flows and environmental impacts shown in Figure 1 can be estimated by existing methods as shown below.

2.1 Numerical assessment for some types of datacenter environmental impacts

Taking into account different data centers sizes as well as the modular design used for medium and large facilities, it is advisable to perform the assessment per 1 module, which has a certain degree of autonomy: a separate computer room; dedicated cooling and fire extinguishing systems; UPS system based on electrochemical power source etc. The use of local generators can also be assessed per 1 datacenter module. A typical module size of 200 racks with 5 kW power per rack corresponds to total power of 1 MW. The assessment for sites of a different size and power can be performed proportionally.

2.1.1 Environmental impact as a result of electricity consumption obtained from external sources as well from local generation

1 MW datacenter module consumes 8.76 GWh of electricity annually. Current carbon footprint concept assumes that carbon dioxide emissions associated with the electricity consumed are the responsibility of the consumer. If we know the type of power plant delivering the energy to the data center, we can estimate CO_2 emissions in accordance with Table 1. If the source is unknown, the annual emissions can vary from 5,168.4 to 7,008 t CO_2 per 1 MW of datacenter power.

Current payment rates in Russian Federation for pollutants emission into the atmosphere from stationary sources [12] and calculation methods applied do not consider carbon dioxide as an environmental pollutant. A carbon quota and carbon footprint payment system has yet to be developed, both internationally and locally.

As noted before, the major part of the electricity consumed by data centers converts to thermal energy and comes to the atmosphere and/or the hydrosphere. Despite the datacenters heat emission of dozens and hundreds GWh per year, current methods to assess the environmental impact of industrial facilities in Russian Federation do not take heat emissions into account and do not consider necessary any payments for thermal pollution. Payment rates development can base on the ratio of 8.76 GWh/year of thermal emission per each 1 MW of datacenter power.

In addition to CO_2 emission power plants of different types can be responsible for NOx/SOx oxides and other pollutants emissions, as well as for certain water consumption. However, the responsibility for such types of environmental impact, unlike the carbon footprint, is not shifted to the consumer. Energy facilities of the Russian Federation make regular payments to the budget in accordance with the payment rates for negative environmental impact [12].

The calculation methods used for energy facilities can be applied to a certain extent to local datacenter electricity generation processes. Payment rates list for pollutants emission to the atmosphere from stationary sources includes some substances contained in the exhaust gases of diesel generators and other generating units (see Table 4). Calculations are made in rubles per 1 ton of pollutants, and the rates are set each year by corresponding Governmental Regulation. Thus, in 2021 one should use payment rates set for 2018 by a factor of 1.08.

The amount of such pollutants emission per unit of fuel consumed can be estimated based on the specific generating equipment data sheets.

Table 4. Payment rates for pollutants emission to the atmosphere from stationary sources [12].

Pollutant	Payment rates per 1 ton of pollutant (Russian Federation rubles)		
	2016	2017	2018
Nitrogen dioxide	133.1	138.8	138.8
Nitrogen oxide	89.6	93.5	93.5
Carbon monoxide	1.5	1.6	1.6
etc.	…	…	…

Payments taken in past for diesel and other fuels emission to the environment, for emission of unburned hydrocarbons were canceled in Russian Federation due to the use of an excise tax on automobile and diesel fuel, to avoid double payments for negative environmental impact. In fact any data center pays for such an impact at the time of diesel fuel purchase, but it is still necessary to track these emissions in order to plan actions to reduce the environmental impact whatever it is [13, 14].

Based on the average technical specifications for 1000 to 5000 kW diesel generator units, about 280 l/h of diesel fuel is required to support 1 MW target equipment operation, which leads to 0.75 t of CO_2/h emission. Local generation is intended to provide electricity supply in case of emergency and failure of external electricity supplies. Such periods can hardly exceed several hours a year.

Thermal pollution can be estimated through the diesel fuel specific heat of combustion, which is 42.7 MJ/kg: most of the energy will convert directly to thermal form at electricity generation stage; the rest will do the same after the generated electricity is used by data center IT equipment and auxiliary engineering systems. When using diesel generators, 2.8 MWh of low-grade thermal energy will be emitted to the environment every hour per each 1 MW of datacenter power.

2.1.2 Environmental impact as a result of cooling and fire extinguishing agents' leakages

Payment rates [12] for pollutants emission to the atmosphere and discharge to natural water bodies mention only some of the substances used in the data center, which can be subject to leakages during normal operation and emissions in case of failure or fire: ethylene glycol, some freons, the simplest ketones etc. The rates do not list: fluorinated ketones (which include "dry water" Novec 1230), some HFCs/halons and other organic substances used as fire extinguishing agents, powder/aerosol fire extinguishing agents. Payment rates for such substances emission to the environment should be developed in future.

Taking into account the reserve cooling agent containers capacity in datacenter modules, possible leakage/emissions can hardly exceed 30 m^3 (20 – 30 t depending on agent density) per 1 datacenter module for liquids and an equivalent mass for agents converting to gas.

Gas agent emission in case of extinguishing a fire in a datacenter module can be determined through an average normative volume fire extinguishing concentration ~10%. 200 racks datacenter module of 600 m^2 average area and 6 m ceiling height will require more than 360 m^3 of gas fire extinguishing agent.

Noble gases, nitrogen, their mixtures (argonite, etc.) are not considered as environmental pollutants regardless the size of emission. In case carbon dioxide is used as a fire extinguishing agent or forms a part of a mixture (e.g. inergen), this must be taken into account as a direct CO_2 emission to the environment. Emissions can reach 0.72 t of CO_2 per extinguishing a fire in 1 datacenter module. In case of CFCs/HFCs fire extinguishing agents not only payment rates for emission to the environment must be used, but also the probable ozone depleting effect must be taken into account.

2.1.3 Wastes of electronic equipment and life-expired electrochemical power sources (batteries)

Average 5-10 year lifetime of electronic IT equipment, servers, storage systems, battery UPS systems and the batteries themselves means annual depreciation of 20-10% respectively. Thus solid waste flows can be estimates in t/year per 1 datacenter module. Datacenter 45U racks average filling with 15 kg/U IT equipment means 6.8 – 13.5 t of electronic equipment become out of operation each year. The existing payment rates for solid wastes disposal [12] classify electronic boards as Hazard class IV (low hazard) wastes.

Table 5. Payment rates for wastes disposal [12].

Waste types	**Payment rates per 1 ton of wastes (Russian Federation rubles)**		
	2016	**2017**	**2018**
Hazard class I of wastes (extreme hazard)	4452,4	4643,7	4643,7
Hazard class II of wastes (high hazard)	1908,2	1990,2	1990,2
Hazard class III of wastes (moderate hazard)	1272,3	1327	1327
Hazard class IV of wastes (low hazard)	635,9	663,2	663,2

Batteries, depending on their type, are classified as Hazard class II or III wastes. With the average weight of 18 kg/kW for battery equipped UPS systems (based on data sheets), able to maintain the equipment operation for several minutes required for local generators to start and reach a stable operating mode, the laden weight of the UPS system can be determined per 1 MW of datacenter power: 18 t. Thus the weight of out of operation equipment can be 1.8 – 3.6 t/year.

Electronic equipment and batteries recycling is currently at an early development stage. The processing is energy intensive, implies the use of large amounts of acids and other substances. On the one hand, the recycling can reduce datacenter operation ecological footprint and turn the solid wastes to raw materials for processing plants. On the other hand, the economic and resource benefits of precious and rare metals extraction from electronic wastes can be accompanied by a significant negative environmental impact from the processing enterprises themselves, which should be the subject of further study and assessment.

3 Results and discussion

The assessment results are shown in Table 6.

Table 6. Assessment results for datacenter environmental impacts.

Impact originator	**Impact magnitude**				
	Carbon footprint or wastes	**Thermal emission**	**Non CO_2 emission to the atmosphere**	**Discharge to natural water bodies**	**Solid wastes disposal**
Electricity consumption from external sources	5 168.4 – 7 008 t of CO_2/year /MW	8.76 TWh/year/MW	–	–	–
Local electricity generation	0.75 t of CO_2/h /MW	2.8 MWh/h/MW	*Payment included in fuel excise tax*	–	–
One-time cooling agent leakage	–	–	< 30 m³ of liquid/module, max. 20 30 t/module, the substance list in [12] is not comprehensive		–
One-time fire extinguishing agent emission in case of fire	< 0.72 t of CO_2/module	–	> 360 m³, the substance list in [12] is not comprehensive	–	–
Electronic equipment wastes	6.8 – 13.5 t/year/module	–	–	–	Hazard class IV of wastes

Table 6. Continued.

Impact originator	**Impact magnitude**				
	Carbon footprint or wastes	**Thermal emission**	**Non CO_2 emission to the atmosphere**	**Discharge to natural water bodies**	**Solid wastes disposal**
Life-expired batteries	1.8 – 3.6 t/year/module	–	–	–	Hazard class II or III of wastes
	No existing payment rates	No existing payment rates	Payment rates partly developed	Payment rates partly developed	Payment rates do not take into account specific character of datacenter wastes

Values in Table 6 allow assessing the datacenter negative environmental impact based on certain data center power, modules quantity, the number of hours of local generation power supply and one-time cases of leakages/fires. As statistics on datacenter operation accumulate, emissions can be estimated more accurately.

4 Conclusion

Sustainable development requires all the datacenter environmental impacts to be quantified and controlled to create economic and other incentives to reduce datacenter effects on the biosphere. To achieve this goal it is extremely important that data centers regularly publish statistics on resource consumption of all types, indicating their sources, flow sizes in absolute form, as well as technical characteristics of the corresponding equipment.

It is necessary to develop payment rates for the carbon footprint and thermal pollution and complement the existing lists of pollutants, indicating corresponding payment rates for emissions to the atmosphere, discharge to natural water bodies, and solid wastes disposal.

References

1. E. Masanet, A. Shehabi, N. Lei, S. Smith, J. Koomey, Science, **367(6481)**, 984 (2020)
2. M. Koot, F. Wijnhoven, Applied Energy, **291(116798)**, 1 (2021)
3. A. Semenov, E. Oganesyan, Foton-Express, **2(170)**, 8 (2021)
4. M. Patterson (editor), D. Azevedo, C. Belady, J. Pouchet, *Water Usage Effectiveness (WUE^{TM}): a Green Grid Data Center Sustainability Metric* (2011)
5. C. Belady, D. Azevedo, M. Patterson, J. Pouchet, *Carbon Usage Effectiveness (CUE): a Green Grid Data Center Sustainability Metric* (2010)
6. C. Belady, A. Rawson, J. Pfleuger, T. Cader, *Green Grid Data Center Power Efficiency Metrics: PUE and DCiE* (2008)

7. V. Avelar, D. Azevedo, A. French, H. Barrass, C. Belady et al., *PUE^{TM}: a Comprehensive Examination of the Metric* (2012)
8. E. Oganesyan, DCJournal.ru, **9**, 57 (2014)
9. E. Oganesyan, Industrial Ecology, **2(86)**, 77 (2014)
10. E. Oganesyan, Management in Russia and Abroad, **6**, 101 (2020)
11. S. Li, Q. Zhang, Science and Pollution Research, **21**, 13636 (2014)
12. Russian Federation Government Regulation #913, 13.09.2016 "On payment rates for negative environmental impact and additional coefficients" (edition 24.01.2020)
13. A. B. Semenov, V. M. Artushenko, Lecture Notes in Electrical Engineering, **641**, 1167 (2020)
14. M. Bylina, S. Glagolev, A. Semenov, R. Davydov, IEEE Int. Conf. on Electrical Engineering and Photonics (EExPolytech), 5 (2020)

Results on laboratory-instrumental studies of atmospheric air after fire suppression

V. E. Kriyt[1,*], *Yu. N.* Sladkova[1], *M. V.* Sannikov[2], and *A. O.* Pyatibrat[3]

[1]North-West Public Health Research Center of Rospotrebnadzor, 2nd Sovetskaya st., 4, 191036 St. Petersburg, Russia
[2]All-Russian Center for Emergency and Radiation Medicine named after A.M. Nikiforova, Ministry of Emergency Situations of Russia, 4/2 Academician Lebedeva st., 194044 St. Petersburg, Russia
[3]St. Petersburg State Pediatric Medical University, 2 Litovskaya st., 194100 St. Petersburg, Russia

Abstract. The paper presents the results of laboratory-instrumental studies of atmospheric air sampled during the first day after fire suppression in different localization. Carbon oxide, hydrochloride, hydrocyanide, nitrogen oxides, sulfur dioxide, and dioxins were determined in the air samples taken. The comparison was made with the hygienic standards established by SanPiN 1.2.3685-21 "Hygienic standards and requirements to ensure safety and (or) harmlessness for humans of environmental factors". Analysis of the obtained data showed that the concentrations of most of the detected toxic combustion products one day after fire suppression were at a level close to MAC for the working area air, but significantly exceeded MAC established for the atmospheric air of urban and rural settlements. The largest exceedances were obtained at sites such as industrial and residential buildings, which can be explained by the use of a wide range of building materials.

1 Introduction

Atmospheric air is a mixture of gaseous chemical substances of natural and man-made origin. One of the significant sources of atmospheric air pollution, along with the industrial sector of the economy and road transport, are fires [13]. In fires, concentrations of pollutants many times exceed the normative values regulated by sanitary legislation, and their reduction is associated with the reform of a wide range of sectors of the economy and human life. One of the greatest hazards are toxic products of combustion, which should be understood as substances of various aggregate states and compounds formed as a result of oxidation process with the release of significant amounts of thermal energy.

The component composition of combustion products in the air in a fire depends on substances subjected to oxidation process, as well as on conditions of their combustion [1]. The oxygen content plays the main role in combustion process. According to the Methodical Recommendations, in fires in buildings, the reduction of oxygen to 16% occurs within 2-3 minutes, which leads mainly to the formation of incomplete combustion products, among which the leading place is taken by carbon monoxide [14].

* Corresponding author: v.kriit@s-znc.ru

https://doi.org/10.1051/e3sconf/202131104008

In accordance with Art. 9 of the Federal Law dated July 22, 2008, No. 123-FZ, reduced oxygen concentration refers to one of the dangerous factors of fire, the evaluation criteria for which are presented in GOST 12.1.004-91 [15,16]. Thus, the criterion for evaluating the reduced oxygen concentration is the maximum permissible value – 0.226 kg/m^3 [17].

In addition, in fires in confined spaces (especially in residential and public buildings), the process of destruction is mainly exposed to finishing materials and household items, can emit substances such as, for example, acrolein (its source can be polypropylene, cellulose, acrelane (used for carpets)), hydrocyanic acid, which is a strong poison and is formed by combustion of nylon and polyurethane, as well as resins, acrylonitrile, paper, polyamine rubber, etc. [2]. According to the literature, combustion products from fire can contain up to 400 different compounds [3], which places special demands on the development of fire-safe polymeric materials [4], as well as on the control of the use of new and existing finishing materials and decorative items in enclosed (confined) spaces.

Low-temperature fires with temperatures up to 600°C can be identified as a separate group of fires. They are characterized by the release into the air of a complex complex of chemicals, including dioxins and dioxin-like compounds, which are dangerous in any concentration [5, 6].

It should be noted that concentrations of such substances as benzene, vinyl chloride, nitrogen oxide, sulfur dioxide, hydrogen cyanide (hydrocyanide), chloroform, hydrogen chloride (hydrochloride), carbon oxide, formaldehyde in the surface air in fires exceed MAC in 60-100% of cases [7]. Carbon monoxide, formaldehyde, nitrogen dioxide, benzene and toluene are the most dangerous components with the effect of complex or combined action [8].

2 Materials and Methods

In order to assess concentrations of toxic combustion products, single air samples were taken and analyzed during the first day after liquidation of 56 fires of different localization (at industrial, residential, agricultural and forest sites), using both express methods with the use of portable gas analyzers and classical methods with aspirators.

Carbon monoxide, hydrochloride, hydrocyanide, nitrogen oxides, sulfur dioxide, and dioxins were determined in the taken air samples. The comparison was made with the hygienic standards established by SanPiN 1.2.3685-21 [18].

In the testing laboratory, the analysis of selected one-time samples of atmospheric air was carried out using the methods of gas chromatography (gas chromatographic separation), as well as quantitative mass spectrometry (GC-MS) by their sequential use (connecting the separating column of the chromatograph and the ion source of the mass spectrometer) [9, 10].

3 Results and Discussion

Studies were conducted on 6 substances of generally toxic and pulmonotoxic effects, which are most frequently encountered in fires of different localization. Among these substances, it is necessary to highlight carbon monoxide, which is formed in almost all cases of combustion of carbon-containing materials, especially under conditions of oxygen deficiency [3, 11]. The choice of the substances to be determined was also influenced by the possibility of enhancing the toxic effect when combining the typical representatives of destruction products: carbon monoxide and nitrogen dioxide, carbon monoxide and hydrocyanide, as well as the complex effects of exposure to nitrogen dioxide and hydrocyanide [3].

The inclusion of dioxins in the list of analyzed indicators is related to the fact that despite the main source of dioxin intake into the human body with food (up to 95%) and the low

https://doi.org/10.1051/e3sconf/202131104008

level of its intake with air (3.5%), dioxins have the potential to contaminate the entire human food chain [19]. In addition, dioxins are extremely slowly eliminated from the body, their half-life from the body is 10 years or more, have a general toxic effect, have a cytotoxic effect, cause distant effects associated with immunosuppressive, mutagenic, teratogenic, embryotoxic and carcinogenic properties [12].

The results of laboratory-instrumental studies of single atmospheric air samples for carbon monoxide, hydrochloride, hydrocyanide, nitrogen oxides, sulfur dioxide and dioxins during the first day after fire suppression of different localization are presented in Table 1.

Table 1. Results of studies of the air environment during the first day after the fire suppression by individual chemicals, mg/m^3

Objects	Average one-time concentration of chemicals, mg/m^3					
	CO	HCl	HCN	N_xO_x	SO_2	Dioxins
Industrial enterprises	31.2±0.5	10.2±0.6	0.8±0.2	5.4±0.2	24.3±1.4	$(7.6\pm0.8)\times10^{-9}$
Residential buildings and household buildings and	24.5±0.3	8.5±0.3	0.6±0.1	6.2±0.3	17.5±0.5	$(3.4\pm0.7)\times10^{-9}$
Agricultural objects	16.7±0.4	7.2±0.4	0.4±0.1	2.2±0.2	14.5±0.6	$(9.4\pm1.2)\times10^{-10}$
Forest objects	14.4±0.8	6.3±0.5	0.3±0.1	3.1±0.1	9.5±0.4	$(7.2\pm0.7)\times10^{-10}$
Maximum one-time MAC in air of working area, mg/m^3	20$^{\#}$	5	0.3	5*/ 2**	10	-
Maximum one-time MAC in atmospheric air of urban and rural settlements, mg/m^3	5.0	0.2	-	0.2**/ 0.4***	0.5	-
Average daily MAC in atmospheric air of urban and rural settlements, mg/m^3	3.0	0.1	0.01	0.1**/ -***	0.05	0.5 pg/m^3
Average annual MAC in atmospheric air of urban and rural settlements, mg/m ^{3}	3.0	0.02	-	0.04**/ 0.06***	-	-

$^{\#}$ MAC for carbon monoxide can be increased with a duration of operation in an atmosphere containing carbon monoxide, no more than 1 hour – up to 50 mg/m ^{3}, no more than 30 min. – up to 100 mg/m^3, no more than 15 min. – 200 mg/m^3 Repeated work in conditions of high levels of carbon monoxide in the air of the working area can be carried out with a break of at least 2 hours; * Of nitrogen oxides in terms of NO_2; ** Nitrogen dioxide; *** Nitrogen (II) oxide.

It should be noted that in the Russian Federation, there is no standardization of dioxins in the air of working area, the maximum occupational exposure to TCDD, equal to 200 pg/m^3 $^{of\ the}$ air in working area with a temporary load of 8 hours, is proposed only in the guidelines of MR 2.2.9.0056-11 [20]. Roughly for evaluating the results obtained, the rate of 0.1 ng TEQ/Nm^3 at 11% O_2 can be applied [21].

The analysis of the data obtained, presented in Table 1, indicates that during the first days after the elimination of the fire, the concentrations of most of the detected toxic combustion products were at a level close to MAC for the air of the working area, but significantly exceeding MAC established for the atmospheric air of urban and rural settlements. The largest exceedances were obtained at such sites as industrial and residential buildings, which can be explained by the use of a large number of polymer construction materials.

4 Conclusions

1. The obtained data confirm the need to develop methodological approaches to laboratory and instrumental control of building materials at all stages of construction and operation of facilities.
2. Although the gastrointestinal tract is the main route of entry of dioxins into the human body, when their high concentrations in the atmospheric air at all stages of fire suppression, it is necessary to consider their entry through the respiratory organs.

References

1. E. V. Chernushevich, P. V. Shirinkin, Siberian fire and rescue bulletin, **1(16)**, 22-28 (2020)
2. S. Kh. Sarmanaev, V. A. Basharin, P. G. Tolkach, K. A. Sherbashov, Russian biomedical journal, **16(2)**, 434-442 (2015)
3. V. A. Basharin, A. N. Grebenyuk, N. F. Markizova, T. N. Preobrazhenskaya, S. Kh. Sarmanaev, P. G. Tolkach, Military medical journal, **336(1)**, 22-28 (2015)
4. N. V. Ermakova, Electronic scientific journal of Kursk State University, **3(15)**, 104-107 (2017)
5. V. S. Rukavishnikov, I. V. Kolycheva, V. B. Dorogova, L. A. Budarina, Bulletin VSNTS SB RAMS, **2(40)**, 7-14 (2005)
6. G. A. Sofronov, V. S. Rumak, N. V. Umnova, D. A. Belov, K. A. Turbabina, Medical academic journal, **16(3)**, 7-18 (2016)
7. I. V. Kolycheva, Bulletin VSNTS SB RAMS, **8(46)**, 133-139 (2005)
8. V. S. Rukavishnikov, I. V. Kolycheva, Occupational medicine and industrial ecology, **6**, 1-5 (2007)
9. V. E. Kriyt, Yu. N. Sladkova, M. V. Sannikov, Pyatibrat A.O. Occupational medicine and industrial ecology, **8(60)**, 494-501 (2020)
10. V. E. Kriyt, Yu. N. Sladkova, M. V. Sannikov, A. O. Pyatibrat, Public health and habitat, **10(331)**, 65-74 (2020).
11. N. V. Martinovich, I. N. Tatarkin, A. V. Antonov, Scientific-analytical journal bulletin of the St. Petersburg University of the State Fire Service of the Ministry of Emergency Situations of Russia, **2**, 1-6 (2014)
12. G. A. Sofronov, V. R. Rembovsky, A. S. Radilov, L. A. Mogilenkova, Medical academic journal, **19(1)**, 17-28 (2019).
13. Air quality guidelines for Europe (Second Edition), Published on behalf of the World Health Organization Regional Office for Europe, 312 (Publishing house "Ves Mir", 2004).
14. A. N. Grebenyuk, V. A. Basharin, N. F. Markizova, T. N. Preobrazhenskaya, *Methodical recommendations for the provision of medical assistance to personnel in case of damage by combustion products*, 32 (GVMU MO RF, 2011)
15. Federal Law dated July 22, 2008 No. 123-FZ "Technical Regulations on Fire Safety Requirements"
16. GOST 12.1.004-91 "Fire safety. General requirements"

17. Order of the Ministry of Emergency Situations of the Russian Federation dated June 30, 2009 No. 382 "On approval of the methodology for determining the calculated values of fire risk in buildings, structures and structures of various classes of functional fire hazard"
18. SanPiN 1.2.3685-21 "Hygienic standards and requirements for ensuring the safety and (or) harmlessness of environmental factors for humans", approved by the Decree of the Chief State Sanitary Doctor of the Russian Federation No. 2 dated January 28, 2021.
19. Air quality guidelines for Europe (Second Edition). Published on behalf of the World Health Organization Regional Office for Europe. – M.: Ves Mir Publishing House, 2004 – 312 p.
20. MR 2.2.9.0056-11 "Preventive measures at dioxin-hazardous industries"
21. Guidelines for the identification and quantification of emissions of dioxins, furans and other unintentionally produced POPs under Article 5 of the Stockholm Convention, January 2013

Risks of social relations ecological assimilation process

Aleksandr Strebkov*, *Gazimagomed* Gazimagomedov, *Abdurashid* Musaev, *Andrei* Aleinikov, and *Artem* Sunami

St. Petersburg State University, St. Petersburg, Russia

Abstract. Nature is the essential life-giving condition for the humans and their activity. And this fact as no other implies sustainable management of the natural potential in order to produce and meet human needs. Ever more increasing importance of the ecological relations and their transformation into the autonomous relations and, therefore the establishment of their impact on the other social relations and subsequent submission of the social relations to the ecological ones, turns into the irreversible process which means that the ecological relations demand their inclusion in the industrial and property relations. Nonetheless, this process of the ecological assimilation proceeds controversially. The battle over the environmental agenda does not subside and this is the evidence of the interdisciplinary conflict, subject matter of which is the state of the environment that is predetermined by the competitive modes of cooperation which are natural for the free-market economy. The containment of the industrial relations' ecological assimilation process from the market and state accelerates consolidation of the general public in their resistance to this tendency which results in the reduction of the environmental risks and disasters as their logical manifestation.

1 Introduction

Environmental agenda did arise not today or yesterday but well before our time, namely when a shovel was replaced by highly productive mechanized means of labor, when humanity turned science into such a productive force which allowed it not only to take nature-made ready to be used products from the nature but also synthesize new materials which were not created by it. This revolution in human environmental interactions has constituted various relations, attributable primarily to property relations, in different eras throughout its history. Industrial production, and with it industrial relations shift focus to the crafted means of production and after becoming a basic element of property predetermined all the battles of the capitalist dominance. Nature has become the pantry of industry while in its anarchy of human exposure it begins to defer to consciously designed environment that is already affected by vagaries of nature. Large-scale impacts on the nature led to negative repercussions for the very nature and for a human alongside with it.

* Corresponding author: strebkov.com@mail.ru

For example, the new Intergovernmental Panel on Climate Change (IPCC) report indicated that atmospheric CO2 concentrations in 2019 were higher than at any time in 2 million years, temperatures during the most recent decade exceeded those of the most recent multi-century warm period, around 6,500 years ago; global mean sea level has risen faster since 1900, than over any preceding century in the last 3,000 years. Over the next 20 years, global temperature is expected to reach or exceed 1.5°C of heating [1] which will cause an increasing occurrence of extreme sea level events that can occur at least annually at more than half of tidal gauge locations by 2100 while the Arctic is likely to be practically ice-free in September at least once before 2050 [2].

Ecology, value of which is socially recognized whereas respect for the environment is established, is becoming crystallization of the public's moral requirements both for the economy and politics. Ecological assimilation of the social relations is the prospective trend which prejudges the "social construct" of the socialized humanity [3].

2 Materials and methods

The process of the ecological relations' incorporation into the industrial relations has not been finished neither in the particular communities nor in the world at large as have not been finished natural calamities which are the result of the human active interference in the life of nature. Environmental disasters that are dependent to a greater or lesser extent on human, are repeating themselves on an ongoing basis. These aspects of the relationships of humanity and nature are highlighted in the research papers of contemporary foreign authors, such as S.H. Nelson [4], A. Huber, S. Gorostiza, P. Kotsila, M.J. Beltrán and M. Armiero [5], A.M. Stoner [6], C.E. Sica [7], C. Morinville and N. Van Lier [8], M.J. DeBoom [9], D. Faber [10], S. Schindler and F. Demaria [11], J. Szabo [12] and also domestic researchers such as M.N. Vlasenko and Yu.N. Shedieko [13] et al.

Yet incompleteness of the ecological assimilation process is coupled with the conflictual interaction risks of the ecological relations' entities i.e. State, business and citizens and as A. Huber and other coauthors of the aforementioned scientific article claim that "...power relations, economic pressures and profit influence "risky" dam management decisions, often disregarding the vernacular knowledge of concerned communities and silencing critical voices..." [5]. Conflict between the entities of the environmental relations inevitably poses casualty risks to these entities' economic, social and political positions and the severity and imminence of the risks do hinge on the extent of this confrontation. In view of this, conflict approach in the analysis of the industrial relations' ecological assimilation is considered to be most fruitful.

3 Study outcomes and discussion

Humanity does not expect bounties from the nature but instead humanity takes them from it although on a scale far greater that it is physiologically and socially essential if we make calculations based on the consumption which is unburdened by the capitalist exploitation of the natural resources and labour force. Scientists estimate that today foodstuff is produced twice as much as is necessary [14]. Almost 2 billion of grown and produced food on the earth per year, which is up to 30-50%, is not being consumed. At the same time every eighth on the planet (12%) malnourished during 2018. In Russia 8% of the population has admitted to the scarcity of food [15]. For example, United States discards nearly 40 million tons of food every year, representing up to 30-40 percent of the entire US food supply and costing the country around 218 billion dollars per year. Meanwhile 35 million people across USA (12%), including 10 million children, are suffering from the food insecurity. Food wasting has irreversible environmental effects as it wastes the water and energy resources it took to produce and generates greenhouse gases up to 11 percent of the world's emissions [16].

This greatest catastrophe in which hundreds of millions of people die all over the world is the result of the free market reign, the greatest competition shaped on the food market. And the State is to blame for the fact that the agricultural business of the developed countries destroys agricultural production of the developing countries where famine is people's constant life companion. "Observers take note that subsidies to European and American farmers is a veritable scourge for developing countries, especially for Africa" [15]. States with the developed economies maintain domestic business by setting all the political stage for its dominance in the global food markets. And in this case we can hardly agree with the foreign authors who seek on the basis on the analysis of the particular dam disasters – Vajont (Italy) and Ribadelago (Spain), induce public consciousness to debunking "the apolitical notion of "socially constructed disasters" rather than speculating about "capital-driven destructions" [5]. Beyond "socially constructed disasters" lies not re-politicizing of the debates around the man-made or indeed not the re-politicizing of anything but the abolishment of that base on which capitalism rests as ever that is the competition which gives rise to a false and not actual rationalization. "…The truth is that an adequate food supply is produced in the world, as it was noted at the XXIV International Union of Food and Supporting Industries congress in 2002, to satisfy 6 billion people and at the same time thousands of kids are starving to death. This is not the technical issue but rather political. It is the issue in the disproportional food distribution and not the issue of the insufficient food production" [15]. The two entities of the ecological relations – business and State, particularly in the developed countries, are in close alliance and make every endeavor not to politicize ecological relations but to provide them with separate existence and by doing so to rid themselves of the responsibility towards citizens of both their own country and the rest of the world. Thereby the blame for the various natural or man-made hazards is shifted on to either single representatives of a particular business or State officials that do not cope with the risks and cannot prevent possible disasters. While it is the States, especially in the developed countries, who are the spokesmen for the total capital interests and this fact makes them already fully responsible for the actions of business and its catastrophic consequences.

Market relations cannot incorporate in themselves environmental relations. Market rationality is beyond ecology and is even alien to it. Hence Sara H. Nelson does not talk about the market rationality towards the nature and assumes that "rationalizing environmental values was not simply about applying market rationality to the natural world, but entailed reexamining the nature of that rationality itself and its relevance to social behavior" [4]. The search of the new meanings in actual relations is the lot of consciousness which cannot accept the existing reality but also expects to find a meaning which far from reality in order to view reality in even more illusory way giving environmental relations more autonomy which is independent from market and modern political constructions created by that same market. Market rationality takes precedence over the environmental values' rationality by turning emissions and environmental pollution in the area of "the financial gain only" [17]. Market is not in rush to incorporate environmental relations or give them the leading role since capital is getting natural resources at no cost. In social terms it is not the nature that determines any profit of any business but it is the business profit which determines the attitude towards the nature and those effects of the reverse impact of the nature on a human. Therefore disasters, and that are the environmental risks in its real manifestation, are the result of the so-called "social constructing" and it is the unfortunate but inevitable and typical result for the industrial relations which are not striving to incorporate environmental relations in themselves.

State-business alliance confronts third entity – the public, that suffers from the environmental calamities and is unable to cope without the State-business support. However, at times this assistance from the state appears as mockery. Russian State comes to the rescue with 10,000 rubles of compensation for those citizens who were affected by fire and had property destroyed. Notably, Russian State acts even more cynically towards the Russians who died from vaccination during Covid-19 pandemic by paying 30,000 rubles of compensation to the relatives of the deceased [18]. But the more solid state-business alliance and more distinctively shaped their common interest in giving an autonomous character to environmental relations is, which means that the state of environment is recognized as not a political issue, the more active the public movement for preservation of the environment in its original form and its politicization is. Thus, establishment of the construction debris landfill in Shies of the Arkhangelsk region and the authorities tenacity in the implementation of their decision led to "...the rise of the interregional environmental coalition "Stop Shies" which was joined by 36 movements from Arkhangelsk, Vologda regions and Komi Republic [19]. This kind of civil society organization against State-business consolidated interests is common not only for Russia, as a result of which it can be observed slow but permanent strengthening of the intersocietal environmental movement which is blending into general social movement for the ecological civilization [20].

4 Conclusions

Capitalist exploitation of the natural world, in which competition is the primary interaction form of the economic agents, strives to maintain environmental relations over the side of the industrial relations as long as the risks arising from the competition had not become this obvious fact although they are increasingly manifesting themselves in the natural and man-made hazards. As long as competition or conflict determine social visage of the nations and peoples and their social soul filled with the desire for enrichment, environmental relations will be self-contained despite how voluntarily business attempts to include them within the rage of its liability and as much as the state seeks to legitimize them and environmental movements exercise public scrutiny. Environmental relations' self-reliance will not benefit them but only to the detriment of them. Excluded from the industrial relations, they become sporadically updated and develop only from one environmental event to another receiving bigger and smaller support from business, State and public. It can be stated that if the formal recognition of the need for the strengthening of the environmental relations at the State level and level of inter-State agreements is fulfilled, then it is done with difficulty and there is no such power that would coerce the business, especially large business, which could even be part of a corporate social responsibility pool, into meeting all the environmental conservation's requirements.

Acknowledgements

The study was carried out by a grant from the Russian Science Foundation (project No. 19-18-00115) and with the financial support of the Russian Foundation for Basic Research (project no. 20-011-00393.

References

1. United Nations UN News Global perspective Human stories, https://news.un.org/

2. BBC NEWS, https://www.bbc.com/

3. Z. Yunfei, Capitalism Nature Socialism, **30(1)**, 11 (2019)

4. S. H. Nelson, Capitalism Nature Socialism, **28(1)**, 118 (2017)

5. A. Huber, S. Gorostiza, P. Kotsila, M. J. Beltrán, M. Armiero, Capitalism Nature Socialism, **28(3)**, 48 (2017)

6. A. M. Stoner, Capitalism Nature Socialism, **32(2)**, 77 (2020)

7. C. E. Sica, Capitalism Nature Socialism, **31(4)**, 34 (2020)

8. C. Morinville, N. Van Lier, Capitalism Nature Socialism (2021)

9. M. J. DeBoom, Annals of the American Association of Geographers, **111(3)**, 900 (2021)

10. D. Faber, Capitalism Nature Socialism, **31(4)**, 1 (2020)

11. S. Schindler, F. Demaria, Capitalism Nature Socialism, **31(4)**, 52 (2020)

12. J. Szabo, Capitalism Nature Socialism (2020)

13.M. N. Vlasenko, Yu. N. Shedko, Business Strategies, **2**, 21 (2017)

14. Twice as much food as it needed is produced in the world, https://animalworld.com.ua/

15.How many people are suffering from hunger in the world. Mortality from famine, https://fkrchr.ru/

16. Food Waste in America in 2021: Statistics + Facts, https://www.rts.com/

17. A. Vlachou, G. Pantelias, Capitalism Nature Socialism, **28(2)**, 84 (2017)

18. RBC, https://www.rbc.ru/

19.What does the example of Shies teach, https://www.vedomosti.ru/

20. A. Gare, Capitalism Nature Socialism, **32(1)**, 37 (2021)

Conflict potential of risk reflexivity: information management under the conditions of pandemic

Andrei Aleinikov, *Daria* Maltseva*, *Vladimir* Miletskiy, *Olga* Safonova, and *Aleksandr* Strebkov

St. Petersburg State University, St. Petersburg, Russia

Abstract. The article problematizes a range of issues related to the emergence of new mechanisms of risk-reflective legitimization / delegitimization of information control over the spread of a pandemic in the context of Covid-19, taking into account the antagonistic definitions of its threats and dangers. The main substantive task of the study is to analyze the conceptual model of classification of various types of risk reflections - communicative, autonomous, critical, and disturbed, which adjust the hierarchies of the significance and social acceptability of risks. The authors focus on the features of the behavior of a modern social subject in the conditions of uncoupling of risky communications depending on the interpretation of information about risks. The conflict potential of information flows that impose forms of risk perception and risk behavior, the mechanisms of formation of information inequality of reflection of risks and social information exclusion in the policy of distribution and compensation of risk are revealed. The article argues that the inability to appropriate the constructive elements of risk reflections in the strategy of information management of threats destroys the security. The article proves that in the information field, the risk subspace is an asymmetric structure of production, reproduction, ranking and distribution of threats and risks, functioning simultaneously as a tool and a goal of struggle. The authors argue that incorrect attributive judgments and interpretations of the threats and dangers of the pandemic can escalate social and political conflicts.

1 Introduction

The COVID-19 pandemic has redefined the institutional parameters of informational and communicational constructs. The title of the article *This time is really different* by Carmen Reinhart, Chief Economist of the World Bank, has become a symbol of social and political reflexivity upon the crises and risks of the pandemic, perceived in theoretical and political-instrumental discourse [1]. According to V. Mau [2], the poor predictability and uncertainty of the virus' effects as well as the potential of a general existential threat make the current situation all the more similar to a war, necessitating the search for risk management and

* Corresponding author: buenafiesta@mail.ru

decision-making informational solutions amidst the newfound uncertainty and changes in the nature of communication.

The emergence of new mechanisms of communication under the conditions of a structural crisis, that is, new ways and forms of informational exchange, has revitalized the conflict between scientific and social knowledge within the risk society, as identified by W. Beck, involving the fierce debate of whether something is risky or not. The inability to apply constructive elements of risk reflexivity as part of an informational threat management strategy, when managerial decisions are made in uncertainty, disrupts the state of security. Sociologist Victor Vakhshtayn has pointed out the following: "Information about threats influences people's everyday routine ... What matters is the metaphor, the logic, the frame in which information about the epidemic is presented ... The virus is not just a coronavirus, but a language that's being actively spoken ... The choice of format entails far more serious consequences than public opinion influencers would think." [3].

Thus, the interrelation between the formation of trust in political-administrative institutions and information management represents an essential theme of theoretical and expert discussions in the context of the COVID-19 pandemic [4, 5, 6].

The search for informational parameters of pandemic risks revolves around debates about the problems of responsibility and accountability for risks, defining the characteristics of socially approved behavior in the context of pandemic, as well as adequate crisis communication strategies. This new situation is most comprehensively reflected in political-sociological theoretical programs that address ways of conceptualizing risk communication in the context of combating the epidemic. From this perspective, it can be noted that descriptions of communicative responses to the unique combination of exogenous shocks and political-managerial and economic crisis transformations are quite contradictory, as described in scientific literature. One hand, at the forefront is the dialectic of legitimization or delegitimization of COVID-19 spread control that is, gathering scientific and expert information on the tools of proactive response, taking into account the antagonistic definitions of the dangers of COVID-19 in both scientific and social discourse. The most important factors for making sense of the COVID-19 pandemic experience are the weak informational and political responses to social anxieties in risk production, distribution, and management. The factors of social change are related to the crisis of informational credibility.

2 Materials and methods

The Nobel Prize winner Douglass North has methodologically considered fundamentally different types of uncertainties, emphasizing the following: "The world we live in is non-ergodic, that is, with new changes constantly emerging.... new, fundamentally different types of uncertainties." He distinguished five levels of uncertainty depending on society's response:

1. Building up information while maintaining the existing body of knowledge leads to an increase in predictability and a transformation of uncertainty into risk.
2. Building knowledge within the existing institutional framework.
3. Changing the institutional framework within the existing structure of beliefs.
4. Changing the structure of beliefs.
5. Residual uncertainty, which acts as the basis for "irrational" beliefs - in the case of a truly new phenomenon, we are faced with uncertainty, the consequences of which are simply unknown to us. And in this case, the likelihood of successfully reducing uncertainty depends only on luck, and players will act on irrational beliefs." [7].

Importantly, risk in this research tradition is understood as the danger of loss when the degree of uncertainty or the probability of damage can be measured. In this situation, risk communication can determine the emergence of forms of information product alienation, or "communicative exploitation."

The basic provisions of J. Dean's concept of "communicative capitalism" [8], which outlines a number of the following theoretical and methodological assumptions, seem significant for the present study. In our opinion, the toolkit for the analysis of information flows during the pandemic is determined by:

— competition in informational volumes, shaping inequalities and hierarchies of communicative capital;

— reflexivity, that is, the ability to generate many different and equally legitimate opinions, or "unlimited uncertainty." [9];

— the "fantasy of abundance of information," characterized by an inability to distinguish what is important and necessary in a vast flow of information;

— the "fantasy of inclusion," when information technology becomes a fetish that replaces real political activity with simple communicative actions (polemics, commentary) using politicized information technology platforms;

— the "fantasy of unity and integrity," which refers to the transformation of information platforms into "global zero institutions," where one of them politically presents itself as unified and all-knowing, excluding everything foreign [8].

In conceptualizing the phenomenon of risk reflection, we rely on Margaret Archer's sociological model of different types of reflexivity, which define different positions towards risks and threats. The theoretical premise seems to be fundamentally important, according to which risk reflexion is when a social subject poses the question "What next?" when under threat, while risk reflexivity is "What will I do next with this?" At the same time, the boundaries between them are not clear, and reflexion easily turns into reflexivity.

Therefore, it seems promising to use the distinction of four types in the study of risk reflexivity [10, 11]:

1. Communicative reflexivity is a type of reflexivity whereby before taking a risky step, a social subject needs to obtain approval for their actions from others.

2. Autonomous reflexivity is when a subject makes independent decisions about actions under conditions of uncertainty.

3. Meta-reflexivity is a pronounced "inner conversation," a critical dialogue where one critically evaluates the effectiveness of their own actions in risky circumstances, during positive experiences, as well as their ability to choose forms of risky behavior.

4. Broken or impaired reflexivity is a deficient inability to use power over oneself, a lack of meaningfulness in action under conditions of risk, disorientation, and situational expressiveness.

It may be possible to explain how interpretations of information about hazards and the recognition of certain risks as socially acceptable are related, and how this relationship affects the effectiveness of policy decisions to manage threats, based on the concept of lifestyles by H.-P. Müller [12]. It involves making sense of structured patterns of behavior under uncertainty, depending on the information available to producers and consumers of risk and their principles. Following Müller, four basic dimensions of behavior can be distinguished, depending on the interpretation of risk information:

— *expressive* (tendency to search for intense experiences, when the possibility of loss is justified by the subjective value of pleasure),

— *interactive* (taking risks while being able to independently choose the course of action),

— *valuing* (a conscious and appreciative attitude to risk, an assessment of the possibility of loss of something and the value of something else),

— *cognitive* (self-identification in the perceived information-based risk society when making a decision in a dangerous situation).

Based on this, explanatory models of risk perception and acceptability are built upon informational flows, either via assessments of political-emotional emotional perception on a scale of "distribution fairness," or via analysis of decisions built upon certain normative principles and ways of justifying evaluation of threats and dangers.

It is important that "reflexive control" [13] corrects the risk significance hierarchies in the multilevel system of "feedback loops" by "determining the dependence of conditions, principles, norms, etc. (including controlling influences) on previous and current performance and uncertainty factors." [14].

3 Study outcomes and discussion

Ortwin Renn argued that the perceived magnitude of risk depends on a subjective assessment of the likelihood of an event occurring. [15] If information targets quantitative measures of risk (such as vaccinations or severe cases of disease), then avoidance, escape, and risk evasion strategies become evident. Conversely, the more the information flow allows for reasonable risk or considers it the norm, the higher the tendency to take risks.

Controversial results of express tests on the effectiveness of risk and threat management point to the need to update the heuristic analysis toolkit both at the level of theoretical identification strategies and at the level of development of technological mechanisms and methods for solving the problems of political risk management.

The Italian philosopher Giorgio Agamben emphasized the following: "Fear is a poor advisor, but it causes many things to appear that one pretended not to see. The problem is not to give opinions on the gravity of the disease, but to ask about the ethical and political consequences of the epidemic. The first thing that the wave of panic that has paralyzed the country obviously shows is that our society no longer believes in anything but bare life" [16].

Having significantly expanded the conceptual scope of describing societies in a pandemic, Slavoj Žižek stated: "When we react in panic, we do not take the threat seriously—we, on the contrary, trivi-alize it. Just think how ridiculous is the notion that having enough toilet paper would matter in the midst of a deadly epidemic." He justifiably put forward as a subject of analysis in scientific discourse the following questions: "So what would be an appropriate reaction to the coronavirus epidemics? What should we learn and what should we do to confront it seriously?" [17].

In our opinion, all participants of the pandemic risk perception discussion somehow confirm W. Beck's thesis about the modern world society's shift from the "distribution of wealth" logic to the "distribution of risk" logic [18].

Failure to approve the constructive elements of risk reflexivity within the strategy of informational threat management, when management decisions are made in situations of uncertainty, disrupts the state of security. The search for informational parameters of pandemic risks revolves around debates about the problems of responsibility and accountability for risks, defining the characteristics of socially approved behavior in the context of pandemic, as well as adequate crisis communication strategies. This new situation is most comprehensively reflected in political-sociological theoretical programs that address ways of conceptualizing risk communication in the context of combating the epidemic. From this perspective, it can be noted that descriptions of communicative responses to the unique combination of exogenous shocks and political-managerial and economic crisis transformations are quite contradictory, as described in scientific literature. One hand, at the forefront is the dialectic of legitimization or delegitimization of COVID-19 spread control that is, gathering scientific and expert information on the tools of proactive response, taking into account the antagonistic definitions of the dangers of COVID-19 in both scientific and social discourse. The long-term psychological and financial consequences, structural inequalities, deepening social divisions, discrimination under the influence of COVID-19 against socio-economic strata, ethnic minorities, women, the elderly, are no less difficult to address. [19].

A reasonable question would be: how does a modern social subject act under conditions of "risk communication decoupling", the combined "abundant communication" and "communicative poverty", the selectivity and politicization of information flow, social inequality, as well as ways of using digital technologies? [20]

Researchers have put forward reasonable observations about the subject's propensity for personal emotional confidence, rather than rational argumentation, when dealing with the dangers of coronavirus and contemplating ways of prevention and treatment. Facts are displaced by meanings, with the risk consumer in a given informational environment either constructing their own meaning or focusing on the already existing and dominant meaning [21].

We have already noted that the intensification of the search for new approaches to the interpretation of the influence of informational conflict on risk reflexivity is largely an informational culturing of pathological affects that determine the orientation of the communication process of constant risk perception exchange.

The destructive potential of disruptive information, according to some researchers, is not inferior to the viral threat and even surpasses it, being reflected in the society's weariness of the epidemic ("bad infinity") and being an additional source of tension, leading individuals to loss of the sense of belonging to society and the society's loss of control by blurring social norms and "turning norm into absurdity" (W. Beck) as well as by ignoring behavioral rules. This condition is defined as social anomie, "normlessness," where "some structures of society exert a certain pressure on its individual members, pushing them to the path of disobedience rather than to the path of behavior consistent with generally accepted rules." [22]

The coronavirus pandemic has exacerbated the problem of the hierarchy of risks - in today's world; the logic of the "distribution of wealth in society" has been replaced by the "logic of risk distribution" [18].

The situation is complicated by the fact that in modern society, there is always a conflict between different risk interpretations, while "the cognitive equipment of thinking with images of Other-ourself and Other-another allows us to identify a situation as conflictual, to problematize it, and to make decisions." [23].

4 Conclusion

The pandemic, as we can see, has introduced many changes to the problematic of reflexive information management in high-risk situations.

Information and disinformation in risk reflexivity are elements of reflecting the dominant power discourse that determines what is risky and what is not, as well as whether a person belongs to groups that are in a special risk zone, that is, imposing forms of risk perception and risky behavior, when seemingly reasoned risk descriptions come into conflict with managerial practices that provide reinforcement of risk prevention and threat distribution functionality, which are often interpreted as something irrational and fictitious. In the system of risk and threat stratification of the bipolar "informational capitalism" society, social polarization of producers and consumers of threat information, the question of distributive justice of risks, of who should bear the burden, is particularly acute. An analysis of Manuel Castells' theory shows that these institutional practices are "fundamental social fault lines in the age of information" [24].

In fact, we are concerned about the processes and mechanisms of the formation of information inequality of risk reflection, the growth of social information exclusion in the policy of risk distribution and compensation, or "second-order exploitation" (Andre Gorz) [25], which entails the preferences provided in a situation of threat by a social group to the detriment of others, because risks are hierarchical and are most exposed to the social strata least capable of minimizing its consequences [26].

At the same time, in the "subspace of risk", which we consider as an asymmetrical structure of production, reproduction, ranking and distribution of threats and risks, functioning simultaneously as a tool and goal of struggle in the field of information [27], the blocking and replacement of informational feedback is ever increasing; the same goes for the increasing prominence of information dumps and provocations, as well as threats and manipulation of dangerous attitudes. We can agree with Evstafiev D. that "the networked information society has demonstrated the absence of built-in mechanisms to protect society from informational manipulation. The public involved in informational processes has largely come to terms with the manipulative nature of today's information society, which is confirmed, among other things, by the absence of efforts to neutralize the processes of fake information creation, as well as by the signs of the political legalization of such fake information" [28].

Hence, there is reason to believe that "human communication activity is to a significant extent characterized by the intentional production of informational noise and even of misleading or false information. This generates waves of unreliable and low-quality information with potentially very dangerous impacts on the social capacity to respond adaptively at all scales by rapidly adopting those norms and behaviors that may effectively contain the propagation of the epidemics. Spreading false information or even conspiracy theories that support implausible explanations of the causal forces at work behind the crisis may create serious confusion and even discourage people from taking the crisis seriously or responsibly, all the more so, the more such signals receive social validation and spread across social groups and communities." [29]

The key idea of our theoretical construct is that "in a subspace of risk," incorrect attributional judgments and conflicting interpretations of threats and dangers can lead to further outbursts. Such attributions can serve as the basis for, as we have already noted, "riskophobia" or "riskophilia." Russian researcher S.A. Kravchenko notes that "with the help of changing interpretations of risks, produced by scientific and everyday knowledge, the flowing fear is reinforced. As a result, both objective riskogenic realities and subjectively constructed risks and risk perceptions have actually become normal, contributing to permanent self-reproduction of riskophobia... Along with the trend of riskophobia, there exists and develops riskophilia, or craving for riskogenic activity, everything that causes positive emotions from activities associated with an increased degree of risk."[30]

Given the "reproducible algorithm of successful extreme behavior without rules" [31] characteristic of the COVID-19 pandemic, posttraumatic syndrome is intensified by risk-reflexivity determining patterns and strategies of conflict behavior, as well as informational "riskification" of social problems. According to Tedros Ghebreyesus, director general of the World Health Organization, there is a struggle not just with an epidemic of a disease, but also with an informational epidemic. "Fabricated news spread faster and easier than the virus, being no less dangerous." [32, 33]

A number of researchers have introduced the all-important concept of "infodemia" in this regard, or an overabundance of information of varying degrees of accuracy about the virus, coupled with political discourse, including fake, partially true, false and intentionally misleading information, which provokes risky behavior and distrust of the authorities and the healthcare system [29, 34,35]. Some also note the parallel phenomenon of "datademia" (excessive data on epidemics, disease statistics, etc.) [36]

Reflecting the current state of the problem in the context of the global pandemic, it should be emphasized that the conflict of interpretations arises when the two main information models of pandemic control clash - "solidarist" and "absentist" [37]. The "solidarist" model entails purposeful formation of public support in response to the coronavirus-induced crisis, based on trust in information about the scale and consequences of the epidemic;

The "absentist" model entails vague or fatalistic threat reflexivity, inconsistent information flows on the part of the authorities about the risks and dangers of the epidemic, manipulation of statistical data and restrictive measures.

Thus, there are reasons to believe that informational risk management should be studied from the position of analyzing the peculiarities of perception and the risk reflexivity of informational flows about the coronavirus crisis.

Acknowledgments

The study was carried out by a grant from the Russian Science Foundation (project No. 19-18-00115) and with the financial support of the Russian Foundation for Basic Research (project no. 20-011-00393.

References

1. C. Reinhart, Project Syndicate (2020)
2. V. A. Mau, Economic issues, **3**, 5 (2021)
3. Will the Wuhan threat change our lives and ourselves? A sociologist on the impact of the epidemic on society, https://naukatv.ru/
4. L. Bennett, S. Livinston, European J. of Communication, **33(2)**, 122 (2018)
5. S. Guriev, D. Treisman, J. of Economic Perspectives, **33(4)**, 10 (2019)
6. A. V. Aleinikov, A. V. Kurochkin, D. A. Maltseva, Scientific and Technical Information Processing, **1**, 38 (2021)
7. D. North, *Understanding the process of economic change* (NRU HSE, Moscow, 2010)
8. J. Dean, Cultural Politics, **1**, 51 (2005)
9. D. Gavra, V. Dekalov, The J. of Sociology and Social Anthropology, **1**, 27 (2018)
10. M. S. Archer *Structure, agency and the internal conversation* (Cambridge univ. press, Cambridge, 2003)
11. A. M. Kuchinov Method: Moscow Yearbook of Social Studie, **7**, 365 (2017)
12. H.-P Müller. Russian sociological review, **3**, 111 (2017)
13. V. Lefebvre, *Conflicting structures* (Publisher: Soviet Radio, Moscow, 1973)
14. M. V. Belov, D. A. Novikov, Control Sciences, **1**, 43 (2021)
15. O. Renn, Applied Systems Analysis, **2**, 50 (1983)
16. *Coronavirus, Psychoanalysis, and Philosophy Conversations on Pandemics, Politics and Society* (Routledge, London, 2021)
17. S. Žižek *Pandemic! COVID-19. Shakes the World New Yorkand* (OR Books New, London, 2020)
18. U. Bek, *Risk society. On the way to another Art Nouveau* (Progress-Tradiciya, Moscow; 2000)
19. R. Rollston, S. Galea, *Harvard medical school primary care review* (2020)

20. J. Keane, *Democracy and Media Decadence* (NRU HSE, Moscow, 2015)
21. A. N. Chumikov, Russian School of Public Relations, **6**, 32 (2015)
22. R. Merton, Sociology of Power, **4**, 212 (2010)
23. A. I. Makarov, *The image of the enemy in territorial conflicts: methodological aspects* in *Social conflict in different normative-semiotic systems* (Publisher: Kazan University, Kazan, 2012), 77
24. M. Castells *The information age: Economy, Society and Culture* (NRU HSE, Moscow, 2000)
25. A. Gorz *Intangible. Knowledge, Value and Capital* (NRU HSE, Moscow, 2010)
26. K.J. Tierney. Sociological Forum, **2**, 215 (1999)
27. A. V. Aleinikov, G. P. Artemov, A. G. Pinkevich, RUDN J. of Sociology, **4**, 847 (2020)
28. D. Evstafiev, Free Thinking, **1**, 163 (2021)
29. R. Gallotti, F. Valle, N. Castaldo, P. Sacco, M. De Domenico, Nat Hum Behav, **12**, 1285 (2020)
30. S. A. Kravchenko, Sociological Studies, **2**, 3 (2017)
31. G. Pavllovsky, *Ironic Empire. The Risk, the Chance, and the Dogmas of the Russian Federation System* (Europe, Moscow, 2019)
32. T. Gebreyesus, Speech by the WHO Chief Executive Officer at the Munich Security Conf., World Health Organization (2020)
33. A. M., Klimova, K. S. Chmel, N. Y. Savin, Monitoring of Public Opinion: Economic and Social Changes, **6**, 266 (2020)
34. N. Mheidly, J. Fares, Public Health Policy, **41(4)**, 410 (2020)
35. J. Zarocostas.Lancet, **395**, 676 (2020)
36. *COVID-19 in International Media. Global Pandemic Perspectives* (Routledge, London, 2021)
37. *The COVID year: preliminary results and challenges of the decade* (The Liberal Mission, Moscow, 2021)

Evaluation of the effectiveness of the use of heat pumps of the "water-water" type in the wastewater disposal system of the building

Sergey Kuzmin and *Leonid* Kudelkin

Federal State Budgetary Educational Institution of Higher Education "Angarsk State Technical University", 665835 Angarsk, Russia

Abstract. The paper presents the results of the analysis of the effectiveness of the use of heat pumps of the "water-water" type for the extraction of thermal energy from the wastewater of wastewater disposal systems of residential buildings. A system of equations characterizing the operation of heat pumps in the system - "drains-heat consumption" - is obtained. The thermal potential of the wastewater and the cost of the extracted energy were estimated.

1 Introduction

The consumption of thermal energy by the engineering systems of the building (heating, ventilation, hot water supply) is transit in nature - the working environment (indoor air, supply ventilation, hot water) is heated to the design temperature and then removed from the building, irretrievably giving excess heat to the environment. Thus, the energy leaving the building is a potential source of heat for secondary use.

Currently, the most widespread systems for extracting heat from the outgoing ventilation air by means of various heat exchangers [1] or air-to-water heat pumps [2-4]. However, as the experience of operating such devices and the analysis of work shows. [5-7], the efficiency of these systems depends on the changing state of the heat-containing environment - outdoor air.

Water-to-water heat pumps have higher thermodynamic efficiency and operational stability [4]. But their practical application is limited by the presence of a source of natural water near the object and the difficulty with the laying and operation of heat exchangers in water bodies. At the same time, almost every person in operation has a source of relatively warm water - a drain liquid. With a relatively stable flow rate and temperature throughout the year, wastewater can prove to be an effective source of heat energy for heat pumps and a useful component in the overall balance of heat generators for a building.

2 Materials and methods

In this paper, the thermal potential of wastewater in buildings is estimated. The possible effect of the extraction of thermal energy from the waste fluid of residential buildings by heat pumps was determined and the cost of this method of heat production was estimated. The technique is based on the use of mathematical models of the characteristics of technical devices of engineering equipment, statistical models of economic indicators of equipment.

3 Results and Discussion

The reserve amount of heat that can be extracted from the wastewater depends on its amount for the calculation period and temperature potential. The composition of effluents in residential and public buildings is determined by economic and fecal fluids. It is obvious that the utilization of thermal energy is rational to produce from economic (or production) effluents. Given that the proportion of fecal fluid does not exceed 10% of the total flow, the remaining amount of wastewater can be considered as a potentially heat-containing medium.

It is known that the mode of water consumption in the building, although subject to daily regularity, but the hourly (and instantaneous - second) is significantly uneven. But bearing in mind the availability of technical possibilities for the use of regulating tanks in the wastewater disposal system, we use hourly averages to determine the amount of extracted thermal energy of wastewater:

$$W_w = Q_w \cdot \tau \tag{1}$$

where is Q_w - average hourly power of heat extracted from wastewater, kW;

τ - period of operation of the heat pump, hour/year.

Thermal power of wastewater Q_w $[W]$ is determined by its number and potential differenties of the medium up to t_{w1} and after $t_{w.3}$ heat pump:

$$Q_w = 0{,}28 \cdot G_w \cdot \rho_\varpi \cdot c_\varpi \cdot (t_{w1} - t_{w3}), \tag{2}$$

where is G_w – estimated amount of wastewater removed by sewerage systems, $m^3 / hour$

c_w and ρ_w - accordingly heat capacity ($kJ / кg°C$) and water density ($кg / м^3$) at temperature $t_{w.1}$.

Estimated average hourly amount of wastewater G_w will be determined from the daily rate of water consumption of cold g_c and hot water g_h and the number of consumers in the building U:

$$G_к = \frac{U}{24}(0{,}9 \cdot g_c + g_h) \cdot \tag{3}$$

The number of consumers is represented through the rate of usable area per inhabitant f_n and geometric characteristics of the building:

$$U = \frac{A_П}{f_П} = \frac{S * k_П * n_{эт}}{f_П} = \frac{V * k_П * n_{эт}}{H * f_П} = \frac{V * k_П * n_{эт}}{h_{эт} * n_{эт} * f_П} = \frac{V * k_П}{h_{эт} * f_П}, \tag{4}$$

where is $A_{П}$ - total usable area in the building, m^2; S - horizontal projection area of the building, m^2; $k_{П}$ - a factor that determines the share of usable space in a building in its total area; $n_{эт}$ - number of storeys of the building; V - useful volume of the building, m^3; H - height of the useful volume of the building, m; $h_{эт}$ - floor height, m.

The weighted average temperature of the effluents will be determined from the calculated temperatures of the hot t_h and cold t_c water and the ratio of water consumption rates:

$$t_{w1} = \frac{g_h t_h + 0{,}9 \cdot g_c t_c}{g_h + g_c}. \quad (5)$$

Taking into account (10-13) the average estimated hourly amount of heat to be recycled (W) can be determined by the expression:

$$Q_w = \frac{0{,}28}{24}(g_h + 0{,}9 \cdot g_c)\frac{V * k_{П}}{h_{эт} * f_{П}} \cdot \rho_{\varpi} \cdot c_{\varpi}\left(\frac{0{,}9 \cdot g_c t_c + g_h t_h}{0{,}9 \cdot g_c + g_h} - t_{w3}\right) \cdot \quad (6)$$

The graphs (Figure.1) show the calculated values of the heat and energy reserve in the waste fluid at different rates of water consumption and temperature conditions in relation to the corresponding values of the heating system of buildings, located in a "moderately cold" climatic area with a design temperature for heating $t_{co} = -35°C$ and "moderate" with design temperature $t_{co} = -25°C$.

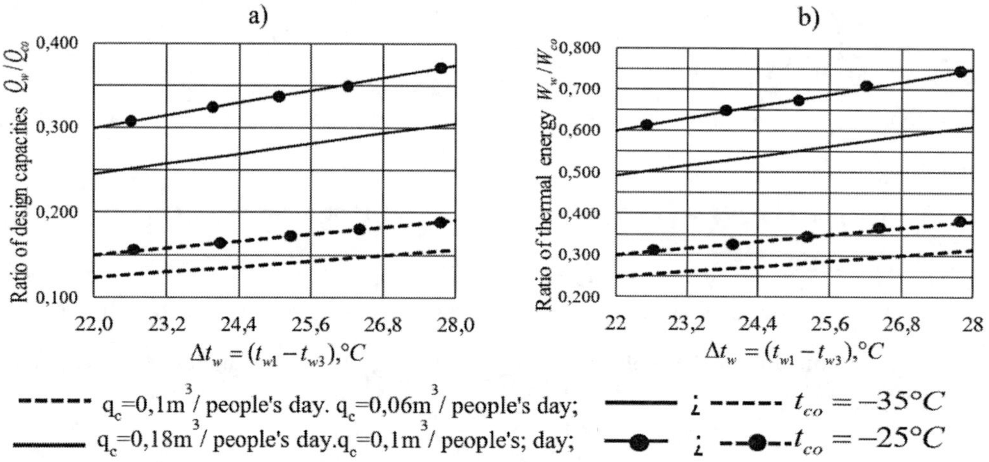

Fig. 1. Ratios between the capacity of the potential of the effluents and the design capacity of the heating system Q_w / Q_{co} - a) and the amount of heat contained in the runoff and the heat requirement for heating the building W_w / W_{co} for the billing period - b).

Calculations show that the reserve of thermal power of the wastewater reaches from 12 W to 25 W per 1 m² of the heated room, which is a significant share of the design capacity for heating the building: from 0.15 to 0.3 for "cold" areas and even up to 40% for "moderate". An even greater proportion of wastewater heat in the necessary energy for heating buildings, which can reach from 25% to 60% for "cold" areas and even 80% for temperate climates. This effect is achieved due to the relatively constant reserve of the energy capacity of the effluents during the year with a changing heating demand during the heating period.

Thus, the analysis of the thermal energy reserve of the waste fluid provides the basis for considering this low-potential environment as a heat source for the engineering systems of the building. The problem lies not so much in the technical implementation of the process of heat extraction with an increase in temperature potential, but in the possible costs and, accordingly, the profitability of this method of energy supply. The efficiency of using wastewater as an energy source will be estimated by the cost of the heat extracted C_W (c.u./MW*h):

$$C_W = \frac{C_Q}{W_w} = \frac{C_{TH} \cdot k_{TH} \cdot K_M + C_N}{W_w}, \quad (7)$$

where is C_{TH} - cost of heat pump, c.u.;

k_{TH} - coefficient of self-sufficiency of heat pump, year^{-1};

C_N - heat pump operating costs per year, c.u./year;

$K_{M.TH}$ - correction factor, taking into account the cost of installation work;

W_w - the amount of heat extracted from the effluents during the billing period, kW*h/year.

The cost of a heat pump is represented by dependence on its calculated (nominal) thermal performance $Q_{TH.o}$:

$$C_{TH} = c_q * Q_{TH.o}, \quad (8)$$

where is c_q - cost indicator (unit cost) attributed to the unit of heat power of the pump, c.u./kW.

As the analysis of prices for water-to-water heat pumps on the market shows, their cost varies widely from different manufacturers. [8,9]. For the universality of the analysis, the cost of equipment is expressed in conventional units (c.u.) and is presented in figure 2.

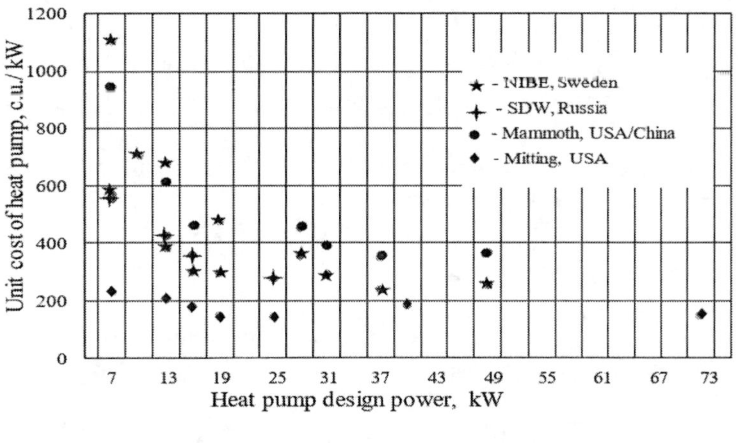

Fig. 2. Distribution of unit cost of heat pumps Cq, c.u./kW by design power $Q_{TH.o}$, kW.

This distribution defines expressions for the maximum $c_q^{\wedge}$ and minimum $c_q^{\vee}$ unit cost values

$$c_q^{\wedge} = 2100 \cdot (Q_{TH}^{m} - 4)^{-0,531}; \qquad c_q^{\vee} = 248 \cdot (Q_{TH}^{m} - 4)^{-0,147}. \tag{9}$$

Installation costs are usually correlated with the cost of equipment with some reducing adjustment for more expensive systems and range from 50% to 80% of C_{TH}. Therefore, the coefficient $K_{M.TH}$ in (7) can be represented as:

$$K_{M.TH} = 1{,}825 - 0{,}004 \cdot Q_{TH.o} \tag{10}$$

The operating costs of the heat pump will be determined through the current electrical capacity $N_{TH.t}$ and opening hours τ_{TH}:

$$C_N = c_N \cdot W_K = c_{эл} \cdot N_{TH.t} \cdot \tau_{TH}, \tag{11}$$

where is c_N - electricity tariff, c.u./kW*h.

Current electrical capacity of the heat pump compressor $N_{TH.t}$ depends on the rated power $N_{TH.o}$, temperature of heat-containing medium t_{w1} and heat conversion efficiency:

$$N_{TH.t} = N_{TH.o} \cdot [1 - 0{,}054 \cdot \ln(0{,}2 \cdot t_{w1} + 1)]. \tag{12}$$

The ratio of the rated heat power of the pump to the (installation) power of the compressor shows the design efficiency of heat energy conversion COP_o. For water-to-water heat pumps, the design conditions usually correspond to the temperatures of the heat-containing medium $t_{w1} = 20°C$ and secondary coolant $t_{w2} = 60°C$ and on average $COP_o = 4{,}2$. Current value COP_{tt} decreases compared to nominal with decrease t_{w1} and increases as the temperature decreases at the outlet of the system t_{w2}. This ratio can be estimated by the expression (8):

$$\begin{aligned} COP_{tt} = {} & (0{,}0021 \cdot t_{w2} - 0{,}2 \cdot t_{w2}^2 - 5{,}38) \cdot \ln(0{,}2 \cdot t_{w1} + 1) - \\ & - 0{,}0028 \cdot t_{w2}^2 + 0{,}232 \cdot t_{w2} - 1{,}048 \end{aligned} \tag{13}$$

Accordingly, the current electrical capacity of the compressor will be:

$$N_{TH.t} = \frac{Q_{TH.o}}{COP_{tt}} \cdot [1 - 0{,}054 \cdot \ln(0{,}2 \cdot t_{w1} + 1)] \tag{14}$$

Amount of heat W_w, extracted from effluents during the billing period τ_{TH} under stable conditions, will be determined from (15):

$$W_w = Q_w \cdot \tau_{TH}. \tag{15}$$

Thus, equations (6-15) make up a system that allows solving the problem.

As an illustration of the obtained model of profitability of the method of extracting heat from household wastewater, consider the effectiveness of the method on the example of a residential apartment building. For a residential building with a high degree of improvement, equipped with all the basic sanitary appliances (bath with shower, sink, sink in the kitchen, toilet), the rate of water consumption per inhabitant is hot water $(g_h = 120l/day)$, cold $(g_c = 180l/day)$ [10]. Maximum cooling temperature of wastewater t_{w3} accept not lower than $+5°C$. Then at the height of the floor of the building $h_{эт} = 3м$, utility factor $k_П = 0{,}8$ and usable area $f_П = 18m^2 / peopl$ formula (6) will take the form:

$$Q_w = 0{,}117 \cdot V \cdot (t_c + 0{,}74 \cdot t_h - 8{,}7)\,. \tag{16}$$

Taking the value Q_w as the rated power of the heat pump $Q_{TH.o}$ let's analyze the possible cost of energy extracted by a heat pump from wastewater under various conditions.

On figure 3 shows the dependence of the cost of heat obtained from wastewater on the temperature of cold water t_c at hot water temperatures $t_h = 60°C$ a) and $t_h = 40°C$ в) for a building with a volume of 1000 m3 and the use of high cost heat pumps (9). A feature of the behavior of the objective function is a relatively weak dependence not only on the temperatures of the initial media, but also on the level of potential of the energy produced. With an increase in the temperature of cold water on $10°C$ (three times), and the temperature of hot water on heat cost is reduced by less than 1%. Also below expectations, the effect of increasing the efficiency of the heat pump is manifested while reducing the level of potential of the heat produced.c $t_{w2} = 60°C$ before $t_{w2} = 40°C$ the cost of energy is reduced by (1-2)%.

The obtained values show that with a relatively short period of self-sufficiency and low cost of electricity, the cost of the extracted heat is obtained about 4 times higher than the selling price, for example, the energy system of JSC «Irkutskenergo» under any initial conditions.

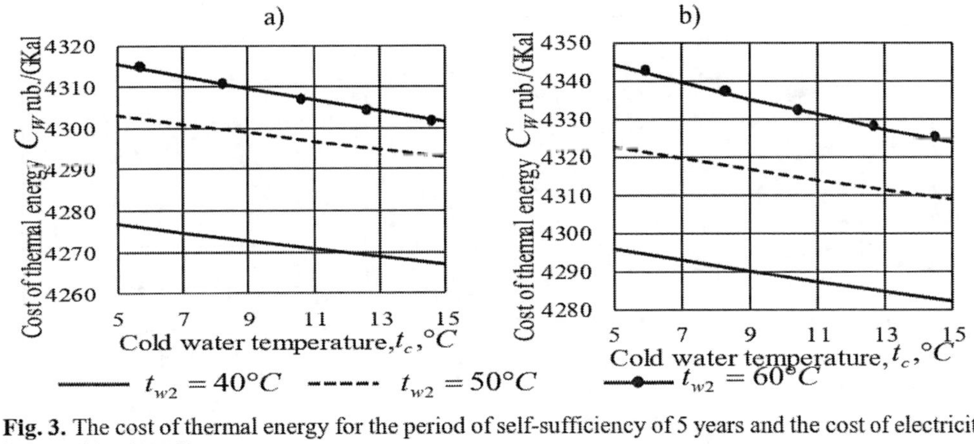

Fig. 3. The cost of thermal energy for the period of self-sufficiency of 5 years and the cost of electricity is 1.5 rubles / kW*h.

For more "expensive" energy systems, the situation is somewhat improving. With an electricity tariff of 5 rubles / kWh and 10 rubles / kWh, the cost of heat, although it increases on average to 5050 rubles / Gcal and 6200 rubles / Gcal, respectively, which is only 2 -2.5 times higher than the selling price («Mosenergo» JSC). The high cost of thermal energy is determined primarily by the large capital costs of the equipment and the short payback period adopted. If it is permissible to increase this figure to 10 years, then the efficiency is significantly improved (Fig. 4) and the cost of heat produced by the heat pump, although it remains higher than in the district heating system, but this excess is not more than (40-50)%. Although under this condition, the mode of operation of the pump has little effect on the overall economic efficiency of the system.

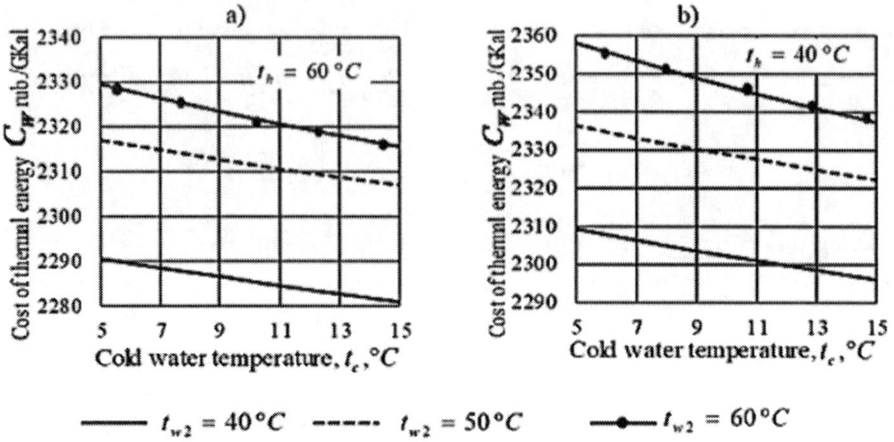

Fig. 4. The cost of thermal energy for the period of self-sufficiency of 10 years and the cost of electricity is 1.5 rubles / kW*h.

An even more significant reduction in cost is achieved for a set of cheaper equipment. The distribution of prices for heat pumps (figure 1) shows a significant price difference, especially in the range of nominal capacity from 4 to 30 kW. Without considering the consumer qualities of the equipment and focusing on the lower price level, it can be shown that the extraction of heat from the waste fluid is more profitable than traditional methods. On figure 5 shows the cost of heat production when using cheap equipment under the conditions already considered (figure 2,3). For buildings of small volume (up to 2000 m3 it is possible to reduce the cost of the heat pump by about 5 (!) times. And this immediately increases the efficiency of heat pumps, even with a short payback period.

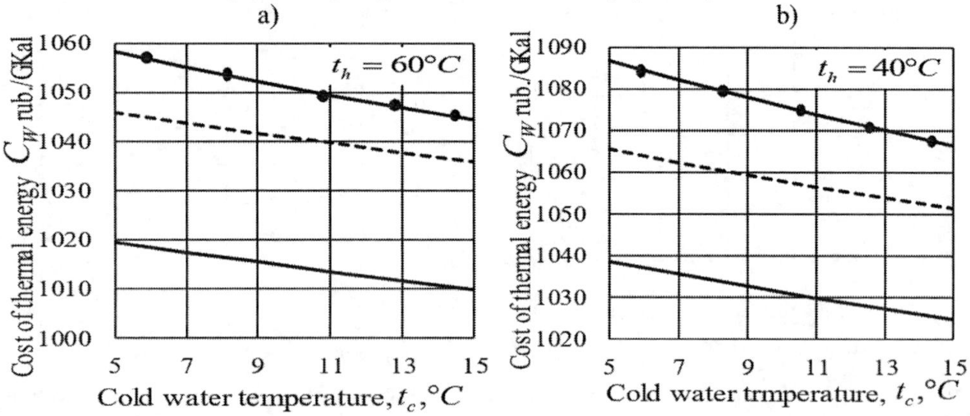

Fig. 5. The cost of thermal energy for the period of self-sufficiency of 5 years and the cost of electricity is 1.5 rubles / kW*h at the lower level of prices for equipment.

The consequence of a decrease in the unit cost of heat pumps with an increase in nominal capacity is a noticeable decrease in the cost of energy produced with an increase in the volume of the building. On Fig. Fig. 6 shows the change in the cost of heat for buildings of different volumes relative to the cost of heat for a building of 1000 m3 under the same conditions.

A 7-factor increase in volume leads to a decrease C_W almost 65% almost regardless of the temperature of cold water, $t_c = +15°C$ with "cheap" electricity. The influence of cold water temperature is noticeably is noticeably manifested at high electricity tariffs (figure 6 b).

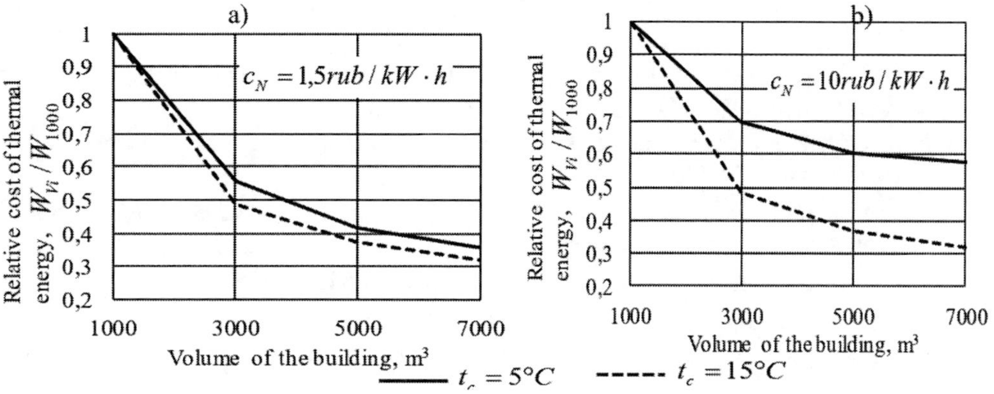

Fig. 6. The relative cost of thermal energy for the period of self-sufficiency of 5 years and the cost of electricity 1.5 rubles / kW*h - a) and 10 rubles / kW*h - b).

Although with the inflated cost of electricity for large buildings, the use of a heat pump can be economically beneficial, especially if you focus on a long payback period. At the same time, the cost of thermal energy may be significantly lower (by 25-35%) of the tariff of the energy system.

4 Conclusion

The analysis shows that the use of water-to-water heat pumps using wastewater as a heat source can be quite a competitive option for a heat generator in building heating systems. Under certain conditions, the cost of energy extracted from wastewater is either comparable or even lower than the tariffs set by regional heat supply organizations.

Of course, the implementation of such a scheme will require a change in the design of the wastewater disposal system associated with the separation of fecal and household wastewater, the accumulation of the latter in an intermediate tank, etc. But these changes do not seem fundamentally feasible even in existing buildings, and when designing new facilities are quite feasible, especially since most buildings provide for two or more sewer risers within one apartment.

The increased cost of thermal energy in small buildings or with the assigned low payback periods of the equipment also does not seem critical. Not to mention the environmental friendliness of this method of heat production, it should be noted that even under adverse conditions, the cost of recoverable energy (4300-6200) rubles / Gcal is not higher than the cost of energy generated by other renewable sources - wind generators, solar stations.

The identified dependencies that establish the efficiency of heat pumps in the system under consideration showed a weak effect of the variability of the state of the environment and the temperature potential of the generated energy on its cost. This suggests a wide possibility of applying this scheme in various objects.

Thus, the above calculations allow us to conclude about a certain prospects of this method of using heat pumps of the "water-water" type. Of course, this requires the behavior of work to change the design of wastewater disposal systems of buildings and even new planning solutions for objects, but given the results obtained and the relevance of the energy saving aspect, such efforts seem quite justified.

References

1. *Designer's Handbook. Internal sanitary devices part 3. Ventilation and air conditioning,* 416 (1992)
2. S. K. Luneva, A.S. Chistovich and I. H. Emirov, Tech and Technology Service Problems Magazine, **2**, 15 (2013)
3. Y. V. Petin *New generation of heat pumps for heating purposes and efficiency of their use in Russia*, 175 (2004)
4. *Heat pumps in modern industry and municipal infrastructure. Information and methodical edition,* 204 (2016)
5. S. I. Kuzmin and N. A. Ivshina, IOP CS: MSaE, **880** (2020)
6. S. I. Kuzmin and A.A. Zelenina, IOP CS: MSaE, **880** (2020)
7. S. I. Kuzmin S I and A. V. Zateeva, IOP CS: MSaE, **880** (2020)
8. Price list for heat pumps Meeting, https//solar-dom.com
9. Price list for heat pumps Mammoth, https:// mammoth-russion.ru
10. SP 30.13330. 2016 *Internal water supply and sewerage* (2016)

https://doi.org/10.1051/e3sconf/202131105002

Some aspects of the problem of energy efficiency and energy saving in Russia

Aleksandr Aleksandrovich Nikiforov[1], *Vera Dmitrievna* Nikiforova[2], and *Mark Mikhailovich* Khaikin[1]

[1]St. Petersburg Mining University, St. Petersburg, Russia
[2]St. Petersburg State University of Industrial Technologies and Design, St. Petersburg, Russia

Abstract. The article is devoted to the analysis of the current state of the Russian energy sector in the context of the gradual reform of the industry. The authors focused their attention on topical issues of energy efficiency and energy saving, using the methodology of space-time and comparative approaches to study. The following are highlighted as topical study tasks: 1) identification of trends and factors of changes in energy consumption of traditional and renewable energy sources, achievement of energy efficiency in Russia and abroad; 2) determination of the potential for investment attractiveness of traditional and renewable Russian energy; 3) substantiation of directions for increasing the level of energy efficiency and energy saving by reducing investment risks in the industry, improving the policy of tariff regulation, gradually eliminating cross-subsidizing, completing the consolidation processes in the power grid market, creating conditions for development of small-scale energy. In accordance with the tasks set, domestic and foreign scientific study in the field of energy conservation and energy efficiency, data from industry statistics, Rosstat are summarized. The results obtained are an intermediate stage in the study of the raised problem and may require clarification.

1 Introduction

Energy forms the basis for the world civilizational development characterized by cyclicality throughout the history of mankind. The regularity of cyclicality is the change of civilizations, the emergence of new goals, meanings and potentials of socio-economic development. [1] At the same time, spatio-temporal fractality can be traced, which manifests itself, first of all, in the trends and commonality of the evolution of civilizations. As criteria for periodization of the stages of development of the world fuel and energy complex, the following can be specified: (1) general and per capita growth / decrease in energy consumption; (2) an increase in the energy intensity of the gross domestic product; (3) the curtailment of energy-intensive industries; (4) gradual change of the dominant energy resource, etc.

While possessing the property of technological unity in the world space, today the fuel and energy complex makes the subjects of the energy market interdependent. The post-industrial economy of developed countries demonstrates a gradual transition from non-renewable to renewable energy, an orientation towards energy-saving and energy-efficient technologies. In developing countries that are still at the stage of industrial development, there is an increase in energy consumption. As world statistics show, in general, world energy consumption is growing for all types of raw materials, mainly due to countries such as the United States, China, India. Wherein, energy consumption per capita is also increased. According to some calculations, in recent years, for every 1 % of population growth, on average, there has been an increase in energy consumption by 1.5 %. Per capita energy consumption will decrease in almost all developed countries. [2]

2 Materials and Methods

It shall be noted that energy consumption indicators are largely determined by demographic factors: the dynamics of population growth/decrease, urbanization processes (mainly in developing countries), an increase in the proportion of the population of retirement age, and etc. However, it would be a mistake to assume that only demographic factors affect energy consumption. Such economic factors as the price of energy resources, GDP growth rates, energy conservation and energy efficiency policies, economic crisis, and etc., can have a certain influence on it. Analysis of the structure of energy consumption of non-renewable energy sources indicates that about 40 % is accounted for by oil, 31.6 % - by natural gas, 24.6 % - by coal and about 4 % - by nuclear energy. The threat of depletion of the reserves of these sources increases the intensity of the struggle and conflicts over energy resources, especially from the United States and the Western countries, highlighting the country's energy safety policy [3].

Among renewable sources, the main share falls on hydropower (43 %), and wind energy (26 %). Europe, North America, as well as Russia and Turkey are actively working in the direction of increasing renewable energy capacities. Above average growth rates of renewable sources in the production of electricity are noted in countries with significant reserves of water resources (China, Russia, Sweden, Norway, Brazil, Japan). It can be assumed that further strengthening of the geographic factor of renewable energy shall be expected. In the global energy balance, the share of renewable energy sources at the moment is more than 28 %, and for individual countries of the world this indicator has the following values: 39 % in the energy balance of the European Union countries, 28 % - in China, 23 % - in India, 20 % - USA, Russia, Japan. [4]

As you know, Russia has a very high natural potential for development of renewable energy. However, the country is also interested in maintaining the existing rates of growth in the production and supply of oil, gas, coal, nuclear energy, since its economy is heavily dependent on the export of hydrocarbon raw materials. The high level of diversification of modern world energy allows maintaining high rates of traditional energy and not allocating the basic energy carrier in the energy balance. In developing countries (Russia, China, India), renewable energy is an element of the diversification of the energy balance, while they maintain high rates of industrial development of the energy sector [5].

According to the World Bank, Russia has been holding a leading position in the production of electric power for a number of years (fourth in the world); at the same time, in terms of GDP per capita (in terms of purchasing power parity), it ranks 50th, in terms of electricity consumption per capita – 28th place (2020). A distinctive feature of 2020 was a significant decrease in the consumption of non-renewable energy sources (by 4.4 %) in the global economy as a whole against the background of an increase in the consumption of renewable energy sources. Russia was among the leaders of the countries where the largest drop in energy consumption of resources was noted. As a regularity of the last decade, we can note the gap between GDP growth and the dynamics of energy consumption mainly in developed countries (Germany, Great Britain, USA, Japan) and China, where the volume of primary energy consumption has decreased or stabilized. For the Russian Federation, the analysis revealed the outstripping growth of energy consumption in comparison with GDP, which demonstrates the extensive direction of development of its economy. Up to 26 % of the total energy consumption falls on the electric power sector, 22 % - on the manufacturing industry, 17 % - on the population, 16 % - on transport; the smallest consumption among these groups of consumers is observed in budgetary organizations.

3 Results and Discussion

The synchronous growth of these indicators, characteristic of the period before the 1970s, today cannot be evidence of sufficient energy efficiency in use of resources. Forecasts of a decrease in the energy intensity of Russia's GDP for 2009-2020 for 40 % were fulfilled by less than 33 % (the reduction in energy intensity was only 13 %). Of course, the cold climate in a large part of its territory, long distances and the associated high transport and infrastructure costs have a significant impact on energy consumption and energy intensity in the country. However, if a comparison is made with a country similar in climatic conditions, for example, with Canada, then the excess of the energy intensity indicator is 17 %, and in terms of the "specific energy consumption in the residential sector" indicator for individual regions may differ by a factor of 3. [6]

It can be assumed that in Russia there is still sufficient potential for increasing the efficiency of energy use not only of primary, but also of secondary forms. According to World Bank estimates, our country can achieve savings of up to 45 % of total primary energy consumption. To move in the direction of global trends in energy efficiency, the following is imperative: (1) a more significant increase in investment in energy conservation; (2) an increase in the number of high-tech industries and introduction of advanced technologies (steam and gas installations, combined generation of electricity and heat, etc.); (3) reduction in the share of exports of first-stage products, especially energy-intensive ones; (4) widespread implementation of energy management and lean manufacturing standards; (5) introduction of energy saving standards in the field of housing and communal services; (6) to abandon all-inclusive tariff policy in the field of electricity production and distribution; (7) to expand tax incentives and measures to stimulate energy efficiency and energy conservation in energy-intensive sectors of the economy; (8) broader information support for state energy efficiency programs; (9) increasing the literacy of the population in the field of energy conservation; (10) a decrease in the level of market concentration, especially in the field of electric power transmission, and other measures. Although much has already been done in this direction, more remains to be done.

Within the framework of this article, the authors understand energy efficiency as the ratio between the beneficial effect (result) obtained from the use of fuel and energy resources and the cost of this resource for its production. Energy saving is understood as a set of measures (organizational and legal, technical and economic, etc.) aimed at the economical and rational use of non-renewable fuel and energy resources and at involvement of renewable energy sources in the economy while achieving the target level of beneficial effect from their use.

According to Rosstat, in 2018, investments in energy saving increased by 26 % compared to 2017 and 2016, but at the same time they did not exceed 0.2 % of the Russian Federation's GDP and were insufficient for structural changes. If we consider this indicator in the context of the regions of the country, then for the constituent entities of the Russian Federation, the spread is about 300 times. Moreover, the coronavirus pandemic had a strong impact on the pace and scale of investments in the Russian fuel and energy complex: design work was cut, investments were postponed, which led to a disruption in supply chains in a number of industries. In its turn, the fall in oil prices intensified the negative impact on the investment activity of the energy sector. Wherein, it shall be noted that despite the general decrease in investments in the electric power industry in some areas, their growth is noted due to measures of state support for renewable energy sources, modernization of thermal generation, small-scale energy projects, construction of new nuclear power plants and hydroelectric power plants (2020). The regional aspect of investment, which is characterized by the modernization of distributed power generation in remote and isolated regions of the Arctic and the Far East, the development of small generation in the Urals is of particular note [7].

In Russia, about 50 % of fuel and energy resources are consumed in industrial units for the direct use of these resources, the rest – in the electric power industry for the production of electricity and heat. The high level of depreciation of fixed assets in the fuel and energy complex largely determines the high costs of energy production (see Table 1) which affect the cost of industrial products. Structural analysis of the depreciation of fixed assets by type of economic activity "provision of electricity, gas and steam, air conditioning" showed that the highest level of depreciation (more than 64 %) falls on vehicles, about 50 % - on machinery and equipment, the share of which in the total volume of fixed assets of organizations of this type of activity reaches approx. 50 %.

Table 1. Depreciation rate of fixed assets of the Russian Federation at the end of the year by type of economic activity, % (according to Rosstat data)

Economic activities	**2017**	**2018**	**2019**	**2020**
For all types of economic activities	47.3	46.6	37.8	38.0
Including:				
- mining	57.7	55.6	55.9	55.9
- provision of electricity, gas and steam, air conditioning	45.2	45.6	45.7	45.7

Use of outdated machinery and equipment, technologies leads to significant technological losses, excessive consumption of equivalent fuel, and an increase in the cost of electricity production. According to some estimates, the efficiency of all fuel and energy resources in the country's economy can reach 40 %.

The current situation with energy efficiency and energy saving is quite alarming and requires an improved approach to the system of regulation of the electric power industry and the entire fuel and energy complex. Wherein, it is important to carry out an objective expert examination of organizations, which shall be greatly facilitated by an energy audit. However, as practice shows, the energy audit has not yet coped with the task set before it. Often it is carried out formally, auditors give various assessments regarding the state of energy efficiency and energy saving at enterprises. The lack of a complete and real picture of the state of the energy industry, unified methods for evaluating innovative projects creates additional risks for private investors, whose share has been decreasing in recent years. [8]

High risks for investors also arise in connection with the policy of tariff regulation. The geography of the electricity market from the point of view of this issue makes it possible to single out several zones, among which there are non-price zones (northern regions, the Kaliningrad region, the Far East). The low level of infrastructure and its isolation do not allow the transition to market mechanisms in non-price zones. For investments, a freer field is formed in the sales sector, however, systematic non-payments by consumers for the supplied electricity and losses make this area unattractive for private investors, and also create problems for development of distribution networks closely related to sales in the supply chain links. Private investors' investments in grid and distribution assets are limited by the territorial monopoly position of operating companies. The government's consistent policy to curb tariffs on the basis of the inflation minus principle does not stimulate the attraction of private investors. The current situation favors the use of bank loans to a greater extent, rather than private investment, under guarantees laid down in tariff regulation. Therefore, the state and consumers become the main investors.

It shall be noted that the policy of low prices for energy resources can give competitive advantages to the domestic industry, but only in the short term, and not in the long term, since prices will not be able to provide a return on investment in new technologies of energy organizations. Moreover, as the analysis has shown, the containment of energy prices, if it had, an insignificant effect on the cost of expenses in most energy-intensive industries. Wherein, it is important to consider that the fuel and energy complex is closely related to other sectors of the economy as a customer and underinvestment in it can negatively affect the economic development of such industries as construction, mechanical engineering, and etc. With regard to the population, it is also necessary to change the policy of general containment of tariffs to expand the practice of targeted support based on the level of income. At the moment, only 15 % of the subjects of the federation use a differentiated approach to the pricing of energy resources for the population; differentiation is carried out not according to income, but on the basis of "rural population", "urban population", equipment of the house with electric stoves. In 7 % of regions, a social norm of energy consumption has been established.

No less problem for the fuel and energy complex today is cross-subsidization, which arose in Russia in the early 90s of the last century as a measure of social protection of the population, financed by industrial enterprises. Since 2013, cross-subsidization has been used to support and develop the field of renewable energy sources, and since 2017 - to compensate for the costs of utilization programs for solid utility costs (intersectoral cross-subsidization). According to some estimates, the volume of cross-subsidization in distribution networks alone reaches 240 billion rubles per year, and in general in the power industry - about 368 billion rubles. [9] Given the growth in electricity consumption, we can expect a natural increase in the volume of cross-subsidies. Wherein, the problem lies not only in the fact that enterprises have to pay for low tariffs for the population and curtail their investment programs, but also that the practice of cross-subsidization makes it difficult to obtain complete and transparent information about the structure of the established tariffs for different groups of consumers.

The measures taken by the state to reduce the volume of cross-subsidization are designed for a long period. [10] These measures are ultimately aimed at bringing tariffs for the population to an economically viable level, which can be partly facilitated by the "inflation plus" policy. Introduction of Rab-regulation aimed at stimulating investment activity in the power grid sector through the application of the cost-plus method for a long period, in conditions of economic instability (due to crises, pandemic) does not allow many companies to receive guaranteed income. Those companies that switch to long-term indexing (instead of the Rab method) have restrictions in the investment tariff in the amount of 12 % of the required gross revenue. According to the calculations of the Ministry of Economic Development, the elimination of cross-subsidization will require approximately 402 billion rubles budgetary funds annually, however, there are no sources for financing this activity in the current situation.

The reform of the fuel and energy complex also includes implementation of the policy of consolidation of the power grid complex, which is designed to reduce the financial burden on the consumer by reducing scattered small grid enterprises, eliminating ineffective costs, as well as duplicating functions, combining sources of financing for all types of activities (operating, financial and investment). As a result of the first stage of consolidation measures, the number of territorial grid organizations decreased by 44.2 % compared to 2014 and amounted to 1,623, which increased the degree of market concentration. However, despite a significant reduction in the number of local grid operators, consumer spending has decreased insignificantly. Wherein, negative consequences arose due to emergence of "ownerless grids" due to the lack of procedures for transfer of electric grids by organizations deprived of the local grid operator status to management of qualified organizations.

Completion of the consolidation process is constrained by gaps in the legal framework. It is important at the legislative level to determine the legal status of the owners of power grids that have been deprived of the local grid operator status, as well as to determine the procedure for reimbursing the costs of servicing power grid facilities that do not have the local grid operator status. It is clear that the local grid operator incurs significant costs for consolidation of power grid assets, which it would like to compensate in the tariff. From the point of view of the Federal Antimonopoly Service of the Russian Federation, inclusion of these costs in the tariff is inexpedient, since it will lead to a significant and unjustified increase in tariffs; therefore, it is offered to use, first of all, owners' funds and other non-tariff sources. A number of experts believe that the cost of consolidation can be offset by the tariff, if the reliability and quality of the power supply is improved. [11] Ultimately, effectiveness of consolidation from the standpoint of tariff consequences will largely depend on the forms of its implementation: purchase, lease, leasing, gratuitous receipt of power grid facilities, concession, establishment of a joint venture, and etc.

4 Conclusion

To a certain extent, small-scale energy is capable of solving the problems of reforming the fuel and energy complex, its adaptation to the market with a lack of large investors, as well as the problems of energy efficiency and energy conservation. Its production capacities are closer to the consumer, losses during power transmission are minimal, higher efficiency; it does not require significant investments compared to large energy systems and does not have long payback periods, which reduces investment risk. Moreover, small-scale energy entities can offer a relatively low price for energy resources due to exclusion of network and sales surcharges from the tariff. However, small-scale energy cannot guarantee the reliability and consistency of energy consumption by analogy with centralized energy systems. It is important to understand that small-scale energy is not a big competitor; it is mainly represented in the regions of the Far North, in the Far East, in Siberia, where it makes up for the lack of electricity and heat through the Unified Energy System of Russia.

In the EU countries, small-scale energy (decentralized generation) accounts for up to 50 % of all electricity generated in the country. In our country, this figure does not exceed 10 %; [12] at the same time, the reasons for its development in Russia are different than in European countries. If, for example, in Germany the main reasons are transition from non-renewable to renewable energy sources and the desire for independence from energy imports, then in our country the development of small-scale energy is mainly associated with the high price of energy resources, as well as the impossibility of obtaining a network resource due to remoteness of the source. In the regulatory documents of the Russian Federation, small energy is usually understood as power plants with a capacity of up to 25 MW; limiting the production capacity limit allows small generation facilities to carry out sales in the retail market and not enter the wholesale market, where activities are associated with certain difficulties for them. Wherein, it shall be noted that at the legislative level, the small power industry does not receive proper support. Its subjects face a number of legislative obstacles when they want to sell electricity to the wholesale network and are often forced to transfer surplus free of charge. Wherein, some experts express concern that development of small generation can undermine the financial base of the country's single power system, if there is a massive consumer shift to decentralized energy.

References

1. V. L. Ulanov, E. Yu. Ulanova, Notes of the Mining Institute, **238**, 474 (2019)
2. Energy Bulletin, **85** (2020)
3. N. Kirsanova, O. Lenkovets, M. Hafeez, J. of Marine Science and Engineering, **8(12)**, 1 (2020)
4. Key trends in energy consumption in the XXI century, https://yearbook.enerdata.ru
5. Modern technologies and economics in energy (MTEE-2020), Materials of the international scientific and practical conference (2020)
6. V. S. Litvinenko, P. S. Tsvetkov, M. V. Dvoinikov, G. V. Buslaev, Notes of the Mining Institute, **244**, 428 (2020)
7. L. Makhova, Academy of Accounting and Financial Studies J., **22(3)**, 1 (2018)
8. P.A. Kruglikov, Yu.V. Smolkin, Innovatics and expertise, **1(16)**, 167 (2016)
9. Recommendations of the round table on the topic "Issues of cross-subsidization in the electric power industry, its impact on the development of the industry and the country's economy, measures to minimize it", http://komitet2-13.km.duma.gov.ru
10. S.N. Osipov, V.M. Pilipenko, Proceedings of higher educational institutions and energy associations of the CIS", **2** (2015)
11. Expert and analytical note on the topic: "Consolidation of the power grid complex of the Russian Federation: legal framework and basic mechanisms", https://em.ranepa.ru
12. Small-scale energy in Russia: advantages and practical problems, https://advocates.su

Relationship between the energy and the environment at the current stage of energy market development

Yulia Anatolyevna Antokhina, *Galina Yuryevna* Peshkova, and *Elena Grigoryevna* Bondar

Saint Petersburg State University of Aerospace Instrumentation, Saint Petersburg, Russia

Abstract. The authors analyzed the current situation related to the global consumption of fossil fuels, as well as environmental problems caused by their extraction, production and consumption. The official statistical data were used, reflecting the volumes of export of fossil raw materials from the Russian Federation, the directions of raw materials flows, characterizing the demand for oil, gas, peat coal in international trade. The authors note that negative factors caused by currency fluctuations, pandemic and other factors did not lead to global changes in the fossil fuel market. The article presents the provisions of international documents adopted in order to prevent climate warming by reducing greenhouse gas emissions. In addition, attention is focused on distribution of responsibility between the developed and developing countries of the world. The authors analyzed the measures implemented in the territory of the Russian Federation, and identified positive trends to reduce emissions of pollutants into the atmosphere. Wherein, the article reflects the environmental risks associated with use of alternative energy sources and nuclear facilities, the assessment of which must be carried out during commissioning the corresponding facilities.

1 Introduction

Currently, the search for optimal solutions regarding use of energy sources that satisfy both environmental and economic interests is relevant for all countries of the world [1, 2]. Importance of such solutions is due to the problem of global warming, associated, among other things, with the active use of fossil fuels in the energy and other industries. Slowing down global warming requires stabilizing the concentration of carbon dioxide (CO2), as well as other greenhouse gases in the atmosphere. Wherein, the fact of stabilization of the concentration of these gases in the atmosphere does not entail a simultaneous cessation of the temperature rise, since there is a certain "lag". Therefore, the temperature will continue to rise for a certain number of years, including provided that atmospheric emissions are stabilized [3]. Wherein, the difficulty of working out the only and correct decision regarding the use (non-use) of fossil fuels is objectively difficult, but it is necessary due to a combination of various reasons, including: insufficient development of alternative energy sources, interruptions in energy production using renewable sources, dependence of national economies on fossil fuels, exhaustion of fossil fuels, presence of emissions into the environment as a result of the use of fossil fuels, ecosystem disturbances accompanying the process of mining, the need for land reclamation, and etc. It is obvious that the search for a solution to the corresponding problems is interesting for the Russian Federation - a country where there are significant deposits of oil, gas and other types of organic fuel (for example, coal, peat), which make it possible to meet both internal needs for heat and energy production, and to supply these products to the world market, receiving the corresponding income from their sale to the budget. Accordingly, certain changes in the spheres, one way or another affecting the extraction, use, sale of organic fuels, are tangible for the Russian economy.

The increased attention to energy issues in the modern period of time is additionally associated with such categories as "green energy", "decarbonization", "carbon tax", "cross-border carbon regulation (CarbonBorderAdjustmentMechanism, CBAM)", which affect the economic interests of a wide range of entities: companies , whose activities are related to geological exploration, mining, export of various types of fuel, production of electricity (heat energy), as well as for public authorities monitoring the situation and developing response measures adequate to real conditions.

2 Materials and Methods

This paper analyzes the current state of affairs in the Russian Federation in areas related to traditional and alternative energy, based on the results of which conclusions are formulated about their further development prospects, considering the measures taken to minimize the negative impact on the environment. Within the framework of the work carried out, a systematic approach was applied, as well as methods such as: normative and legal analysis of international treaties, federal laws and other legal acts, statistical analysis, comparative analysis.

3 Results and Discussion

An analysis of the official statistical information shows that the commodity structure of the Russian Federation's exports, despite a certain instability of oil prices, the epidemiological situation (COVID 19), as well as the European Union's policy on introduction of a mechanism for cross-border carbon regulation, continues to be represented by goods from the fuel and energy complex (FEC), which are exported to various countries of the world. Therefore, in January-June 2021, the volume of value exports of this category of goods (group 27 of the EAEU TN VED) to non-CIS countries amounted to 105,028,667.9 thousand US dollars (the growth rate to the same the period of the last year is estimated at 124.6 %), which is 58 % of the total value of exports of the Russian Federation to the far abroad in the period under review; to neighboring countries, the value of exports of the fuel and energy complex is estimated at 7,191,675.2 thousand US dollars (the growth rate compared to the same period last year is estimated at 134.8 %) and is 26 % of the total value of Russian exports in January-June of 2021 in neighboring countries [4].

Detailing the value of the fuel and energy complex goods exported to the far and near abroad, there is an obvious advantage of oil and petroleum products. Therefore, the value of crude oil exports (commodity item 2709) in January-June of 2021 amounted to 47,632.7 million US dollar (growth rate over the same period last year is 121.5 %), or 22.8 % in total value exports; 31,960.7 million US dollar or 15.3 % of the value of export falls on petroleum products (commodity item 2710), the growth rate compared to the previous period is 122.4 %. In the structure of exports there is also a large share (9.6 %, or 20,010.3 million US dollar) of natural gas (commodity item 2711), the growth rate of which in relation to the comparable period was 172.9 %. The value of the export of other goods of the fuel and energy complex in the analyzed period is significantly lower and in aggregate is no more than 10 %, while it shall be noted that the value of the growth rate of their physical volume in relation to the base period has a positive trend, for example: 109% - coal (commodity item 2701, physical volume of export - 101,886.4 thousand tons), 148 % - coke and semi-coke from coal, lignite or peat, retort coal (commodity item 2704, physical volume of export - 1658.7 thousand tons), 102 % - liquid fuels, not containing biodiesel (commodity item 2710, the physical volume of export - 29,553.1 thousand tons) [5].

The countries of destination of the exported fuel and energy complex products are different, for example [6]:

- crude oil (commodity item 2709) was exported to 32 countries of the world, the largest volumes of deliveries were made to the Netherlands (18,144.1 tons), Germany (8,806.9 tons), the Republic of Korea (7,637.6 tons), Belarus (6,949.6 tons), Poland (5,313.9 tons), Italy (4 543.2 tons), USA (3,444.5 tons), Finland (2,653.6 tons), Japan (2,356.4 tons), United Kingdom (1,616.3 tons), Lithuania (1,575.9 tons), etc.;
- petroleum products (commodity item 2710) were exported to more than 140 countries of the world, for example, to the Netherlands (16,154,985.0 tons), the USA (6,243,319.9 tons), Turkey (3,783,421.2 tons), China (3,388,516.1 tons), France (2,694,233.6 tons), the Republic of Korea (2,671,459.4 tons), Belgium (2,433,973.5 tons), Greece (2,285,024.3 tons), Germany (2,079,151.5 tons), the United Arab Emirates (1,684,932.3 tons);
- other gas and hydrocarbons (commodity item 2711) were exported to 52 countries of the world, including Japan (3,325,677.8 tons), China (2,748,304.9 tons), France (1,887,557.16 tons), the United Kingdom (1,620,596 tons), Spain (1,437,976 tons), the Republic of Korea (1,181,701.5 tons), the Netherlands (885,329 tons), Belgium (881,464 tons), Poland (625,749.8 tons), Portugal (441,643.0 tons);
- hard coal (commodity item 2701) was exported to 74 countries of the world, including China (20,145,241.5 tons), Turkey (7,558,176.2 tons), Ukraine (6,802,512.9 tons), Germany (3,185,327.5 tons), Brazil (2,090,151.7 tons), France (1,377,862.3 tons), Italy (1,004,799.8 tons), the United Kingdom (788,559.8 tons), Belgium (697,935.6 tons), Indonesia (692,754.0 tons);
- peat (commodity item 2703) was supplied to 44 countries of the world, for example, to Germany (41,328.0 tons), Poland (33,016.0 tons), Slovakia (12,588.0 tons), China (11,166.5 tons), Finland (5,661.6 tons), Belgium (5,428.0 tons), Belarus (4,153.0 tons), the Netherlands (3,875.7 tons), Ukraine (2,411.8 tons), Latvia (2,342.2 tons).

In general, in the past period of 2021, the Russian Federation recorded higher export rates of fuel and energy complex products compared to 2020. It shall be noted that the events of 2020 significantly influenced not only companies exporting oil and gas products, but also companies engaged in other types of activities associated with production and sale of oil, gas and other types of fossil fuels. For example, fluctuations in world oil prices in 2020 had a certain impact on Rosgeologia's orders for geological exploration of hydrocarbons; however, they led to a slight decrease in the company's revenue (by 7 %) from the planned one. A fairly optimistic result was achieved thanks to a shift in focus towards solid minerals, as well as thanks to a "multimodal strategy", which, among other things, involves the company's participation in the global market. Namely, in 2020, an international contract was signed with India for implementation of seismic exploration. Wherein, Rosgeologia plans to further expand international cooperation in carrying out this kind of work and to ensure that the company will receive in the near future up to 30 % of the proceeds from the implementation of international contracts [7]. The forecasted plans for cooperation in geological exploration with foreign countries indicate the continued demand on the world market for the extraction and use of fossil raw materials, including for fuel purposes. Representatives of the state authorities in the Russian Federation also give appropriate assessments about the continuing leading positions of oil and gas, despite the beginning of the energy transition, for example, according to available forecasts, by 2050 the share of gas in the world's energy balance is expected to increase to 27 %, and gas is considered as "an integral part of the energy transformation" [8].

Modern realities are such that even in the face of aggravating environmental problems, society is not ready to abandon the use of hydrocarbons, including due to the fact that alternative sources do not currently allow satisfying the existing demand. In this regard, the search for ways to reduce the harmful impact on the environment is acquiring additional relevance.

Climate warming is a global problem, therefore, its solution is possible only with participation of all countries of the world. Such work is currently being carried out under the United Nations Framework Convention on Climate Change (UNFCCC) 1992 (the text of the convention is in Arabic, Chinese, English, French, Russian and Spanish, which are authentic). The main goal of the convention is to ensure a safe level of concentration of greenhouse gases in the atmosphere. In the provisions of the UNFCCC, the significant role of developed countries in the volume of greenhouse gas emissions into the atmosphere is noted. As a result, differentiated responsibility of states is established to participate in the work on adoption of response measures in connection with global climate change. Developed countries are given a special place in taking appropriate measures, this approach seems to be justified in terms of their use of organic fuels in the past and current trends, as evidenced by the statistics on export volumes by country of destination given earlier. The role of developing countries in the negative impact on the environment is significantly lower, however, according to experts' forecasts, it will increase over time. The provisions of the UNFCCC explicitly state that when solving environmental problems, the needs of developing countries shall be taken into account, including the eradication of poverty as a priority. In addition, it is noted that the policies and measures being implemented to prevent climate change shall consider the various conditions of the socio-economic development of states. Also, the convention prescribes the need for adoption at the national level of legal acts in the field of the environment with the obligatory consideration of the interests of other states. For development of the UNFCCC provisions, the Kyoto Protocol was signed in 1998, and in 2015 the Paris Agreement on Climate was signed that was ratified by the Russian Federation in 2019. One of the priority tasks of this agreement is to contain the temperature rise to 1.5 °C [9, 10, 11]. The countries participating in the UNFCCC submit reports on the activities carried out and their results in accordance with the established procedure; the Russian Federation also provides relevant information. From the official reporting of the Russian Federation it follows that: the main share of greenhouse gas emissions in the country is associated with extraction, transportation, processing and use of fossil fuels (mainly oil, natural and oil (associated) gas, coal, peat and oil shale in an insignificant volume); to ensure objective monitoring of the volumes of relevant emissions in the territory of the country, interdepartmental interaction of state authorities is being implemented, which provides a comprehensive assessment of emissions and absorption of greenhouse gases; the country's leadership approves legal acts (federal laws, presidential decrees, government orders, etc.) defining specific measures aimed at reducing greenhouse gas emissions, with establishment of certain deadlines for achieving the planned results. For example, on December 30, 2021, the Federal Law "On Limiting Greenhouse Gas Emissions" will enter into force, the provisions of which, among other things, regulate the issues of setting targets for reducing greenhouse gas emissions, provide for state accounting of relevant emissions, and regulate implementation of audits of reports reflecting emissions volumes, provide for creation of a register of carbon units and its maintenance. In April 2021, the Ministry of Economic Development of the Russian Federation prepared a draft legal act, according to which the Sakhalin Oblast provides for approbation of measures to achieve carbon neutrality, including the provisions of the draft document establish special requirements for companies which production results in greenhouse gas emissions are 50 thousand tons CO2-eq. and above, and it is also supposed to monitor the introduced new technologies to reduce emissions of pollutants, absorb greenhouse gases, and etc. In addition, among the activities carried out in Russia the following of them are important: projects for utilization of associated petroleum gas, improving the functioning of thermal power plants, improving the main infrastructure, developing renewable energy sources, increasing the sustainability of the forest sector [12,13,14,15,16, 17]. It shall be noted that according to the information of the officials of the Russian Federation, the country managed to ensure a reduction in greenhouse gas emissions by 40 billion tons, this volume is comparable to the annual emission of pollutants from all countries of the world, based on this, the global warming of the climate directly by Russia has been restrained for one year [18]. Therefore, within the framework of implementation of obligations in accordance with the UNFCCC, certain positive shifts on the part of the Russian Federation in terms of reducing emissions into the environment shall be noted.

Renewable energy sources (RES) have been given significant attention in maintaining the existing result. In 2020, despite the fact that this year was not easy for all countries of the world, new renewable energy facilities were launched in Russia: generating electricity using the sun (SPP), wind (WPP), water (HPP), the capacity of which exceeds the value of 2019 year by 82 % and is estimated at 1 GW, including the Gukovskaya Wind Farm – the capacity of 198 MW, obtained by means of 52 turbines with the capacity of 3.8 MW each; Sulinskaya Wind Farm - the capacity of 100 MW, obtained by means of 26 units with the capacity of 3.8 MW each; Staromaryevskaya SPP – the power of 100 MW, obtained by means of 83,334 photovoltaic modules; Perovo SPP – capacity of 105 MW, obtained by means of 440,000 crystalline solar photovoltaic modules, the use of which prevents 105 thousand tons of carbon dioxide emissions. In addition, there is a significant number of renewable energy facilities under construction or design, the feasibility of commissioning of which is justified by the subsequent reduction of greenhouse gas emissions into the atmosphere. For example, in the Stavropol Territory, it is planned to build a Windpark Rodnikovsky Wind Farm, the generated electricity of which shall ensure a reduction in release of 180 thousand tons of carbon dioxide into the atmosphere. In the territory of Siberia, in conditions of unfavorable climatic conditions (high seismicity, permafrost), it is planned to build and commission the Telmamskaya HPP, on the Iturup Island (Sakhalin Region) the Okeanskaya-2 Geothermal Power Plant will be restored, the operation of which will be based on the energy of the Baransky volcano [19, 20]. Therefore, despite the wide distribution of fossil raw materials in the territory of Russia, the renewable energy sources are functioning and their number tends to increase, which is favorable in the context of the tasks being solved to reduce pollutant emissions. Wherein, despite the obvious importance of measures to introduce renewable energy sources into large energy in order to reduce emissions into the atmosphere, leading to a warming of the climate, it is necessary to take into account the existing environmental problems associated with the construction and operation of facilities based on renewable energy sources, for example, energy production through hydroelectric power plants is associated with flooding and salinization of lands, disappearance of fish, climate change due to evaporation of water from reservoirs; energy production by means of SPP is associated with the death of birds due to exposure to sunlight, the use of components for SPP made with use of chemicals that are dangerous to living organisms; energy production through wind farms is associated with climate change due to the slowing down of air masses, and etc.) [21,22]. Therefore, construction of facilities for alternative energy sources is also associated with environmental risks, the assessment of which is necessary to make reasonable forecasts regarding their negative impact on the climate.

Nuclear power plants play an important role in energy production in relation to reducing CO2 emissions. In Russia, they account for about 19 % of generated electricity. Similarly, in some countries of the world, for example, Japan, nuclear power occupies one of the leading positions and does not tend to decline [23]. Considering the specifics of the raw materials used at these facilities (nuclear fuel), the considered power plants require increased attention in terms of the safety of their operation, since accidents at such facilities bring irreparable harm to all living things both in the adjacent territories and far and beyond (as an example, Fukushima-1 NPP, Chernobyl NPP). In this regard, in the territory of the Russian Federation, construction and operation of nuclear power plants is carried out considering the negative experience of disasters and relying on new modern technologies.

4 Conclusions

Providing the population with energy and heat is the basic task of any state, the solution of which is individual for each country and is determined by the physical and geographical location, the presence of minerals in the territory, climate characteristics and other conditions for construction of facilities based on renewable energy sources or fossil raw materials. However, when solving this problem, it is necessary to consider possible environmental consequences and take measures to minimize the predicted risks. Currently, according to international treaties, the priority is to contain all countries of the world of climate warming. Among the proposed instruments, the European Union is working on introduction of a "carbon tax" and use of "green energy". Wherein, when using alternative energy sources, there are negative environmental consequences, which indicates an insufficient justification for abandoning some energy sources in favor of others. Without use of new technologies in the operation of facilities based on energy sources of all types, comprehensive monitoring of negative consequences on the environment, it will be difficult to achieve positive changes and keep them.

References

1. Dr Angela Wilkinson, Addresses world energy week 2020 on next world energy congress theme (2020) https://www.worldenergy.org
2. Declaration on Green Growth Adopted at the Meeting of the Council at Ministerial Level (2009) https://www.oecd.org
3. Hannah Ritchie and Max Roser.CO_2 and Greenhouse Gas Emissions, https://ourworldindata.org
4. Commodity structure of exports for January-June of 2021, https://customs.gov.ru
5. Russian exports of the most important goods for January-June 2021, https://customs.gov.ru
6. Customs statistics of foreign trade of the Russian Federation, http://stat.customs.gov.ru
7. Additional capitalization is important for us. Head of Rosgeology Sergey Gorkov on new orders and old problems, https://www.kommersant.ru
8. A. A. Novak, International cooperation is the path to the energy security of the planet, https://energypolicy.ru
9. United Nations Framework Convention on Climate Change, UNFCCC, https://unfccc.int
10. Kyoto Protocol to the United Nations Framework Convention on Climate Change, https://unfccc.int

11. Paris Agreement on Climate, https://unfccc.int
12. First biennial report of the Russian Federation submitted in accordance with Decision 1, CP.16 of the Conference of the Parties to the United Nations Framework Convention on Climate Change, https://unfccc.int
13. Second biennial report of the Russian Federation submitted in accordance with Decision 1, CP.16 of the Conference of the Parties to the United Nations Framework Convention on Climate Change, https://unfccc.int
14. Third biennial report of the Russian Federation submitted in accordance with decision 1, CP.16 of the Conference of the Parties to the United Nations Framework Convention on Climate Change, https://unfccc.int
15. Fourth biennial report of the Russian Federation submitted in accordance with decision 1, CP.16 of the Conference of the Parties to the United Nations Framework Convention on Climate Change, , https://unfccc.int
16. Federal Law, 296-FZ "On Limiting Greenhouse Gas Emissions" (2021)
17. Draft Federal Law "On Conducting an Experiment to Establish Special Regulation of Emissions and Absorption of Greenhouse Gases in the Sakhalin Region"
18. M. Reshetnikov, Russia has slowed down global warming by a year, https://www.vedomosti.ru
19. Power plants, https://energoseti.ru
20. The leader in the commissioning of solar power plants in Russia is the Southern Federal District, https://www.eprussia.ru
21. L. V. Nefedova, A. A. Soloviev, Energy Policy, **4,** 99 (2018)
22. Alternative energy: prospects for development of RES market in Russia, https://www.eprussia.ru
23. Z. S. Podoba, Japanese Studies, **1** (2021) http://www.japanjournal.ru

E3S Web of Conferences **311**, 06001 (2021) https://doi.org/10.1051/e3sconf/202131106001
EPSD 2021

Logistics of urban transport infrastructure development as a mechanism for the formation of a comfortable urban environment (in the example of Yaroslavl)

Alexander Kiselev[1,2,*], *Anna* Savicheva[1], *Valery* Firago[1], and *Sergey* Shkiotov[1]

[1]Yaroslavl State Technical University, Yaroslavl, Russia
[2]Financial University under the Government of the Russian Federation, Yaroslavl branch, Yaroslavl, Russia

Abstract. Currently in Russia the most serious attention is paid to the development of a comfortable urban environment. And there are already specific tasks of the President of the Russian Federation V.V. Putin on increasing the urban environment quality index by 2024 and reducing the number of cities with an unfavorable environment. However, the main problem in the development of a comfortable urban environment today is the state of the urban transport infrastructure and the complexity of its development. This is due to the constant growth in the number of urban residents, with an increase in the number of personal, public transport and freight vehicles necessary to ensure the life of cities, as well as the often poor condition of city roads and dense development of old city blocks, etc. As a result, today we need scientific approaches to solving the problems outlined in the article on the development of urban transport infrastructure, using for this logistics as a tool for effectively solving problems for the development of an urban comfortable environment, including using the experience already available in this area of cities such as Moscow, Kazan, etc. The article outlines topical problems associated with the implementation of the scientific concept of logistics for the development of urban transport infrastructure as a mechanism for solving the problems of forming an urban comfortable environment on the example of Yaroslavl, which are typical for most Russian cities, determined by the weak use of modern technologies in regional management planning.

1 Introduction

At present, the idea of forming a modern and comfortable urban environment is being actively pursued in Russia. So, for example, the President of the Russian Federation V.V. Putin instructed to improve the quality of the urban environment in Russian cities by 2024. The Decree (National goals and strategic objectives for the development of the Russian Federation for the period up to 2024) provides for an increase in the urban environment quality index by 30%, and a halving in compliance with this index of the number of cities with an unfavorable environment [1]. The most striking example of the implementation of this work is such large cities as Moscow, Kazan, etc. But concerning statistical data, there are 1117 cities in Russia today [2]. And the main problem in the development of a comfortable urban environment today is precisely the difficulties in the development of urban transport infrastructure. At the same time, the number of the population in cities is constantly growing. At the same time, the number of personal vehicles in cities is increasing. At the same time, the existing road infrastructure and the state of urban roads negatively affect the operation of urban transport. So, for example, "in 2020, almost 1.6 times more funds (694 million rubles) were spent on the maintenance of the streets of Yaroslavl than allocated in 2021 (431 million rubles). At the same time, both city budget funds and subsidies from the region decreased" [3]. In general, all this makes it difficult for people to move around the city, especially during rush hour in the morning and in the evening, creates inconveniences for people in using transport and makes cities problematic from the standpoint of the comfort of the urban environment. There is no doubt that many programs are being implemented in Russia to develop the transport infrastructure of cities, but, unfortunately, they often do not solve existing problems. Today there are clearly problems associated with the architectural design of urban development, the lack of professionals in construction organizations for the construction of high-quality roads and other modern transport infrastructure of cities, as well as the desire of a number of officials to "save" on the construction of various objects of transport infrastructure in cities and tenders (competitions) for work on its development, which can be won by organizations that are not able to ensure the required quality of performance of the planned work. As a result, today we need scientific approaches to solving the problems outlined in the article on the development of urban transport infrastructure, using logistics for this as a tool for effectively solving problems for the development of an urban comfortable environment. However, developers, road builders and transport infrastructure development officials, municipal government and financial authorities have their own logistics in this process, that is, lobbying, first of all, their own interests, which indicates that the problem requires a scientifically grounded solution.

The relevance of the topic of the article under consideration is determined by the fact that today, in conditions of the constantly increasing traffic load on cities, new approaches to the development of their transport infrastructure are required from the standpoint of the concept of forming a modern comfortable urban environment and the use of modern technologies in regional management planning. At the same time, this approach should be based on the principles of logistics, and logistics itself should be the main mechanism in the development of transport infrastructure as a key factor in the formation of a modern comfortable urban environment.

2 Materials and methods

To study the problems of urban transport infrastructure development in order to form a modern urban comfortable environment, a review of scientific publications on the topic under consideration was carried out, analytical and statistical information on the research topic was studied, interviews were conducted with specialists of various levels related to the development of transport infrastructure in the city of Yaroslavl, as one of the developing regional cities of Russia. Research materials on this topic were repeatedly submitted for discussion at various scientific and practical conferences and "round tables" held jointly by researchers, logisticians and practitioners in the field of transport infrastructure development from various organizations.

3 Results and discussion

Currently, many Russian cities are not ready to rebuild their urban transport infrastructure to meet the modern requirements of forming an urban comfortable environment.

Firstly, this is due to the fact that initially the construction of cities was not designed for such a large number of cars that the population and services for the life support of cities have today. For example, statistics show that already 7 years ago, "137 thousand vehicles, including 80 thousand cars, drove around the city of Yaroslavl, one of the regional centers of the central part of Russia. And "there were 226 cars per 1000 inhabitants of the city." According to the traffic police, their number is increasing by 12 percent annually [4]. Consequently, we can assume that today there can be from 480 to 500 cars per 1000 inhabitants of Yaroslavl.

It should be borne in mind that buildings, especially in the historical part of the city, do not allow expanding the carriageway of roads. At the same time, urban development in Soviet times did not provide for such a large capacity of the city's roadways, which is required today, which negatively affects the development of a comfortable urban environment in modern conditions. In addition, the existing approaches to the development of the city's transport infrastructure exacerbate the problems. For example, during these years, the development of courtyards in the city center with new residential and office buildings was actively carried out. As a result, territories where it was possible to make parking for cars disappeared. The Resolution of the Yaroslavl Mayor's Office No. 1777 of December 29, 2017 "On the municipal program" Formation of a modern urban environment for 2018-2024 "states that one of the main problems in the city is the lack of specially equipped parking spaces for vehicles in most courtyards" [5]. At the same time, many residential and administrative buildings were fenced off or the entrance to their courtyards was limited by barriers. As noted by a member of the Yaroslavl branch of the Union of Architects O.A. Mazanov, "they build just any way, violating the urban planning structure of quarters, the scale and nature of the historical environment, violating the historical building line", etc. [6]. We need a comprehensive program for the development of the city's transport infrastructure based on logistics approaches, taking into account the prospect of creating a comfortable urban environment. In other words, all administrative structures of the city must solve the problems of transport infrastructure development in a coordinated and comprehensive manner in the interests of creating a comfortable urban environment. So, for example, the city's development plans today include issues on the organization of road traffic. In particular, we are talking about the creation of dedicated lanes for public transport, the allocation of additional lanes in front of intersections, the introduction of an intelligent traffic management system and the development of parking space, including through the introduction of paid parking in the UNESCO zone [7]. But given the condition of the roads and the poor quality of their repair and construction, these measures will not ensure the expansion of the traffic flow. At the same time, various administrative structures and organizations most often work based on corporate interests. For example, in Yaroslavl, in the city center, there were two parking lots for personal vehicles of residents, which the city administration decided to make paid. As a result, they now stand empty, and all the streets of the city in the city center are filled with cars on both sides of their carriageway. This reduces their throughput and increases the number of congestion and traffic jams for vehicles. In winter, due to problems with snow removal, which is hindered by cars parked to the side of the road, the roadway in some places is reduced so that buses and trolleybuses of the opposite direction cannot pass. Ultimately, traffic jams, especially during rush hour, are becoming common in cities. These problems are typical for most other Russian cities. This problem needs to be solved through the construction of new roads. However, in Yaroslavl, for more than 10 years, they have been talking about the construction of a new modern road, following the example of such a road, which is being done in Moscow, allowing to reduce the traffic load on the city center. "The authorities of Yaroslavl have scheduled for years how they will build the largest transport interchange in the city. In early documents it was called Karabulinskaya, in the last published strategy of socio-economic development of Yaroslavl from 2021 to 2030 it is called simply a transport interchange" [8]. But "On May 11, 2021, the mayor of Yaroslavl V. Volkov signed a decree according to which it is possible to build again on the lands reserved for the Karabulinskaya interchange" [9]. At the same time, a second road bridge across the Volga River was additionally built in Yaroslavl. For some time the severity of the problem was reduced, but now there is a need to build a third bridge, given the amount of transport in Yaroslavl. Indeed, back in 2016, at that time, the first deputy mayor of Yaroslavl, S. Yastrebov, argued that with 250 cars per 1000 inhabitants, "a bridge must be built, otherwise traffic in Yaroslavl will stop" [10]. However, there is still no final decision on the place of construction of the bridge in the next 10 years [6] . In other words, the development of transport infrastructure today is still being addressed not in anticipation, but as the problems worsen. And the main reason for this situation, referred to by the city authorities, is the lack of funds for construction. Today this is the most "convenient" excuse for those problems of transport infrastructure development that are not being resolved. And in such conditions, there is no need to talk about the formation of a modern comfortable environment in the city in the near future, although the municipality of Yaroslavl has approved a strategy for the socio-economic development of the city for the period 2021-2030.

Secondly, considering the strategy of socio-economic development of the city for the period 2021-2030 approved by the municipality of Yaroslavl, it can be considered that it often talks about the resolution of existing problems in the long term, but does not take into account future problematic issues in the growth dynamics. As the experts noted, the city's socio-economic development strategy did not include some projects important for the city, including those on which public discussions were held and even their projects were ordered. O.A. Mazanova notes that the program does not say anything about limiting the numerous construction in the historical center, about excluding the vicious practice of small-scale construction of the city. As a result, it is not possible to create equipped parking spaces for personal vehicles of citizens and their cars occupy a significant part of the carriageway of city roads, hindering the movement of others, including public transport. At the same time, in 2013, studies were carried out in Yaroslavl to optimize public transport routes. The money for the research was spent, but the result of the work turned out to be formal and did not lead to real practical results. And now the issues of logistics of public transport are relevant. At the same time, the chief architect of Yaroslavl A. Tsymbalov believes that the organization of regional centers in the largest districts of Yaroslavl will help to solve this problem. The abandonment of the microdistrict type of development and the return to quarter development, with the transfer of the first floors of residential buildings to public institutions and organizations would help, in his opinion, to overcome the functional imbalance in the development of urban areas, with the transfer of the first floors of residential buildings to public institutions and organizations, since the division of the city into sleeping areas and the business part forces many citizens make daily trips to the place of work, which leads to the formation of congestion and traffic jams during peak hours [6]. But how to implement this idea in existing cities, there are still no specific practice-oriented proposals aimed not only at preventing future problems, but even many existing problems in the development of the city's transport infrastructure.

Thirdly, there is a large flow of freight transport through Yaroslavl, including transit, but there is no full-fledged road bypass, and in the near future it will not be. At the same time, given how "upset" Yaroslavl is, the old bypass road actually passes through the city's territory, which creates an additional load on the city's transport infrastructure. And the road itself almost every year in the spring requires its repair, which leads to congestion and traffic jams on the road of freight, urban and individual vehicles. The issue of organizing a deep road bypass of the city today is the most acute for the region and requires an urgent solution. Of course, according to the plans such a road will be built, but anyway these are only intentions [11]. At the same time, even the repair of the existing bypass road is going on with difficulty. For example, the YAROSLAVIA State Television and Radio Broadcasting Company, in its program in March 2021, noted that "today the pits are covered with dubious crumbs, and tomorrow the process will be repeated. The southwestern district, practically the face of the city for motorists from all over Russia, is falling apart before our eyes. The speed of movement here is human, 5 km per hour. And not only because of one thousand and one pits" [12]. At the same time, the repaired bridge had to be repaired anew due to the cracks formed, since "after the construction and installation works, tests and opening of traffic for road transport, structural defects were revealed on its left side. It was necessary to close the traffic again, carry out examinations, take measurements. As a result, the experts came to the conclusion that there was a mistake in the design" [13]. The work was carried out by the AO EKS GC contractor under a contract worth over RUB 490 million. In June 2020, an additional 292 million rubles were allocated to complete the work. In general, about 1 billion rubles were spent on the work on the bridge. It was planned that the bridge will be commissioned in October 2020 [14]. This indicates, in our opinion, that the organization that won the tender for its renovation did not have the necessary professional specialists, hiring cheaper labor from among the migrants. This also applies to the repair of other urban roads. Thus, a situation is created when a lot of money is spent annually on poor-quality repair of transport infrastructure, which will ultimately be more costly than high-quality construction of new transport infrastructure, which ultimately determines the comfort of the urban environment.

Fourth, the development of cities, associated with an increase in their territory and population, requires the improvement of public transport logistics. So, for example, according to the Yaroslavl mayor's office, in 2019 more than 800 units of passenger transport of all types went to the roads of the capital of the Golden Ring [15]. Unfortunately, today we can assert that constant reforms in organizing the work of public transport do not give the result people need. Reforms and optimization of the transport infrastructure of public transport in Yaroslavl are ongoing. But we must remember that it is very important not to overdo it with the level of provision of logistics services, since the rule always works: "Leave well enough alone" In 2013, in Yaroslavl, comprehensive studies were carried out to optimize public transport routes with the involvement of students from Yaroslavl universities. The money was spent, but there was no practical result from the study. After that, a reform was carried out in 2019, when the authorities intend to optimize the route network, transfer all routes without exception to a regulated tariff and revise the type of rolling stock "in terms of availability and capacity" [15]. From July 14, 2021, changes are coming again related to the reform of public transport, which raise many questions among the population. So, for example, residents of the city do not understand why they remove through routes, and introduce new ones that require people to change during their trips, assign new numbers to existing routes, remove minibuses, etc. [16]. And all this is linked to the formation of a modern comfortable environment.

Fifth, it is the scientifically grounded principles of logistics that should lie in the reform of the transport system of cities. But when planning the reform of the transport infrastructure of the city of Yaroslavl, all the organizations involved solve only their own tasks, which increases the financial costs of various activities. So, for example, the fact that the regional government intervened in the implementation of the developed reform and the scheme for its implementation was altered, raised questions. And they did it, according to experts, badly, since the transport reform turned out to be unreasonably costly for the budget [16]. The head of the city passenger transport department of the Yaroslavl mayor's office S. Volkanevsky believes that "further formation of the route network should be carried out taking into account the transition to long-term municipal contracts based on payment for the transport work of carriers" [16]. At the same time, the solution of the problems of development of transport infrastructure by "separate", one-off measures with the appearance of work does not provide the required result. So, for example, the expansion of individual sections of city roads does not solve, but only exacerbates the problems of road infrastructure, since the traffic capacity of roads does not increase, but creates a bottleneck effect, when a lot of traffic jams are formed on the roads in which it stands. including public transport, disrupting the traffic schedule, and, accordingly, difficulties are created for people, increasing the time of their movement around the city. At the same time, even new sections of roads today are often made in such a way that after a while they become unusable or require constant repair, especially in the spring after the snow melts. People often joke bitterly that in spring, along with snow, asphalt leaves the roads. So, for example, in April 2021, the media noted that "in Yaroslavl, the asphalt on Uritskogo Street collapsed. Repair of the roadway worth 53 million rubles. spent here four years ago. The work was carried out by the contracting organization OOO Trassa. The company had to replace the 2.1 km asphalt pavement, repair sidewalks, replace curbs, install new road signs and apply markings. As a result, the terms of the street repairs were disrupted, and the asphalt pavement began to deteriorate. The local residents and the authorities of Yaroslavl were outraged by the quality of the repairs" [17]. And this is not an isolated case. Unfortunately, such news is given systematically in the media. So, in the fall of 2020, information about the repair of a section of the road near the village Volochkovo near Yaroslavl was reported on the social networks of the federal national project "Safe High-Quality Roads" . The cost of repairs cost 368 million rubles. However, after winter in the spring of 2021, local residents were outraged by the condition of the repaired road. They wrote indignantly on social networks: "A layer of asphalt is like butter on bread. So much for safe and high-quality roads!" [18]. And similar problems today are relevant for most Russian cities. We must not forget that the increase in the number of cars in the city is due to the fact that the amount of payment of the transport tax is increasing, and, consequently, the funds that should be used to improve the transport infrastructure of the city. This suggests that it is impossible to explain the problems of poor development of the transport infrastructure of cities only by the lack of funding for the planned work. And now there are many complaints from the population of Yaroslavl about the inconvenience of traffic. As a result, in order to solve these problems, the governor of the Yaroslavl region was forced to intervene so that the problems of carrying out the transport reform were resolved.

Thus, we can assume that today we need not just separate solutions in resolving the identified problems of individual leaders of regional and local authorities. Their solution should be based on the scientific foundations of logistics, first of all, on the development of urban transport infrastructure, since it characterizes the urban environment to a greater extent as comfortable or not comfortable.And it is precisely the logistics of urban transport infrastructure development that should act as a mechanism for the formation of a comfortable urban environment in Russian cities in modern conditions.

4 Conclusions

Today, in many Russian cities, the task of creating a comfortable urban environment is being solved, taking into account the development of society and the needs of people. The most problematic in solving this problem is the creation of an appropriate transport infrastructure in cities, which should be developed on scientific principles of logistics, comprehensively and in close interaction of all structures and organizations of the city involved in its development. At the same time, all possible risks associated with the development of the city's transport infrastructure should be taken into account and thought out, including both objective ones, due to the need for effective use of the allocated financial resources, and subjective ones, determined by the human factor [21].

In the example of Yaroslavl, the article highlights the main problems of forming a comfortable urban environment based on the development of the city's transport infrastructure. At the same time, it is substantiated that it is the development of transport infrastructure that is a key factor in the formation of a comfortable environment in Russian cities. And individual one-time, albeit large projects for the expansion of individual road sections, road construction without scientific approaches from the standpoint of logistics, and not a formal possibility or impossibility. There is not doubt that the basis for decisions on the development of urban transport infrastructure should be predicted indicators for the development of the urban environment as a comfortable urban environment from the standpoint of strategic management and strategic planning of this work [22]. Accordingly, the measures for the development of the transport infrastructure of cities today should be ahead of the possibility of the onset of problem and risk situations.

But, today problems in this area often begin to be resolved only after a critical situation has been created, when further ignoring it leads to social tension among city residents. At the same time, they are still trying to implement their solution by holding separate measures, justifying this by a lack of funding.

At the same time, statistics show that the use of allocated funds for the development of transport infrastructure is very often ineffective and does not bring the desired result. At the same time, the measures taken do not ensure proactive development of the transport structure.

Thus, the time has come to address the issues of forming a comfortable urban infrastructure from a scientific standpoint, taking into account the trends of its development, and using modern technologies in regional management planning. At the same time, in modern conditions, the main problems of creating a comfortable environment are directly related to the development of urban transport infrastructure. According to that, the logistics of urban transport infrastructure development as a scientific concept can be considered a mechanism for its formation, taking into account the social development of Russian cities, which requires the use of modern technologies in regional management planning.

References

1. Putin instructed to radically improve the comfort of the urban environment by 2024, https://tass.ru
2. List of cities in Russia by population, https://chislennost.com
3. In Yaroslavl in 2021, the cost of road maintenance has been reduced, https://regnum.ru
4. In Yaroslavl and the region, there are already 226 cars per 1000 inhabitants, https://www.yarcom.ru
5. Resolution of the Mayor's Office of Yaroslavl No. 1777 of December 29, 2017 (The municipal program: Formation of a modern urban environment for 2018-2024), https://docs.cntd.ru
6. A third bridge across the Volga and a new ring road will be built in Yaroslavl, https://www.kommersant.ru

7. A third bridge across the Volga and a new ring road will be built in Yaroslavl, https://76.ru
8. The Mayor's Office of Yaroslavl drew up plans for the construction of the Karabulinskaya junction, https://76.ru
9. The refusal of the Karabulinskaya junction in Yaroslavl will be checked by the Prosecutor General's Office, https://r76.su
10. How many cars are there in Yaroslavl? https://yarportal.ru
11. Around Yaroslavl they plan to build a new bypass road, https://r76.su
12. In Yaroslavl, the South-Western ring road is falling apart, https://vesti-yaroslavl.ru
13. Repair of the Dobryninsky bridge in Yaroslavl will be completed in 2021, https://www.kommersant.ru
14. Half-bridge with a crack, https://www.kommersant.ru
15. Everyone is under one rate, https://rg.ru
16. There will be no transitional period: how will transport in Yaroslavl change after the reform, https://76.ru
17. In Yaroslavl, the road to Uritsky collapsed. The repair cost 53 million rubles, https://ufocar.ru
18. "A layer of asphalt is like butter on bread": a road for 368 million is falling apart near Yaroslavl, https://progorod76.ru
19. A. A. Kiselev, *Fundamentals of strategic management and the essence of strategic planning in organizations* (Direct-Media, Berlin, 2020), 343
20. A. Kiselev, M. Ugryumova, A. Strausova and M. Majorova, The Int. Scientific And Practical Conf. "Ensuring The Stability And Security Of Socio-Economic Systems: Overcoming The Threats Of The Crisis Space"

Applying a consulting system in the management of construction companies' interactions

Larisa Selyutina[1], *Elena* Pesotskaya[2,*], *Maria* Ivanova[3], and *Tatiana* Maleeva [4]

[1]Emperor Alexander I St. Petersburg State Transport University, 190031, Moskovsky pr., 9, St. Petersburg, Russia
[2]Saint-Petersburg State University of Economics, 191023, Sadovaya str., 21, St. Petersburg, Russia
[3]Saint Petersburg State University of Civil Aviation, 196210, street of Pilots, 38, St.Petersburg, Russia
[4]Pushkin Leningrad State University, 196605, St. Petersburg sh., Pushkin, 10, St. Petersburg, Russia

Abstract. The article reveals the interaction system features of construction companies, substantiates the need to introduce special management means into the organizational and management practice of construction companies, which include management consulting. The features of management consulting, its methodology and methods are considered. The author's approach to the formation of managerial decisions in the system of interactions between construction companies, focused on the implementation of the principle of joining internal and external consultants' efforts within the framework of combined consulting, is stated. It is shown that combined consulting acts as universal mean that provides a full suite of services in management, creating preconditions for the successful implementation of the principles of the concept of interaction in the management of construction companies, and is the most important factor that activates the construction companies' interaction with the subjects of the regional investment and construction market.

1 Introduction

Under modern conditions of the ongoing reform process across all sectors and realms of the Russian economy, the improvement of the construction business management methods is important, aimed at increasing the level of adaptability of construction companies to the dynamically developing competitive environment [1, 2]. It must be emphasized that the adoption and implementation of management decisions in the field of rationalization of interactions (for example, the transition to new forms of cooperation between actors operating in the investment and construction market) can be legitimately identified as a complex process that requires significant innovative experience and the volume of innovative potential of the enterprise, and that most of these decisions are quite appropriate

*Corresponding author: epes2@mail.ru

to develop through the involvement of professional knowledge and experience of consultants [3].

Taking the above mentioned into consideration, it is advisable to take the following hypothesis as a basis: a systematic approach should underpin the process of strategic decision making on managing the interactions between construction companies, firstly, it assumes the full involvement of all the available advantages of interaction with entrepreneurial business, and secondly, aimed at intensive development of interaction in order to identify new, additional opportunities within the issues under investigation of partnership and interaction. In this connexion, *the goal* of the research carried out in this article is the justification of necessity and possibility of attracting means of consulting to fulfill the tasks set before construction companies in the process of their interaction under the conditions of complicating competitive environment.

2 Materials and Methods

In this research it is highlighted the thesis of the concepts of the interaction theory which are the conceptual and categorical framework that guarantees the reality of an adequate and enough impartial, objectively suitable description of the features, type and motives of interactions between business entities within a competitive market environment. The category of interactions between business partners in the construction market should be recognized as predominant in the interpretation of the economic nature and essence of competition under current conditions. The positive role of economic rivalry between participants in the market economy is manifested in the quality of the formation and improvement of competitive interactions and the building up of competitive advantages at all levels and in the spheres of economic activities.

Interaction as a process of direct and indirect affections of business entities to each other establishes a concrete basis for the development and preservation of the competitive advantages achieved by the company, which can be manifested in a peculiar way as a general category, on the one hand, and as the local characteristic of a specific feature that favorably distinguishes a particular enterprise from other similar entities in the market, on the other hand. During the last decades, the construction sector as a complex economic system has undergone a number of substantial alterations associated with the acceleration of scientific and technological progress, the need to introduce innovative management methods.

The use of the principle of market adaptation in the process of interactions between entities orientates construction companies to improve the organization of production and management within the framework of strategic and tactical actions. Moreover, the competitive struggle as a distinctive feature of the market economy is not weakening under current conditions [4]. The measures aimed at attracting special management facilities to the managerial and workplace problems solving, comprising management consulting, helps enterprises to function and develop under such conditions.

3 Results and Discussion

The problems of management consulting are covered in many scientific works of domestic and foreign experts: E. Pesotskaya, L. Selyutina, L. Bazarova, V. Biswas, D. Twitchell and others. Their interpretations of management consulting are somewhat different from each other, but there are no fundamental differences between them. Based on the results of the analysis of these interpretations, it seems appropriate to take as a basis the definition formulated in the work [5].

A characteristic feature of modern research on the methodology of consulting and methods of its implementation is the predominant focus on management consulting [6], which is probably due to its widespread use in practice. To confirm this thesis, one can cite the classification of consulting services proposed by D. Litvin [7]. It should be noted that, in fact, all of the types of consulting presented in this classification cover many aspects of the management process, which, by the way, made it possible to identify consulting as an activity with a multidisciplinary specialization. Of interest is the classification of consulting proposed by L. Bazarova, also focusing on management consulting [8]. In the system of interaction between construction companies, of course, all types of management consulting can be used. Expert consulting is widely applicable, it has versatility and is suitable for solving any problems. The typology proposed in [9] generalizes a number of typological criteria and highlights the most significant of them (Figure). This method of structuring represents systematic assessments that are useful for various fields of activity, including the construction sector.

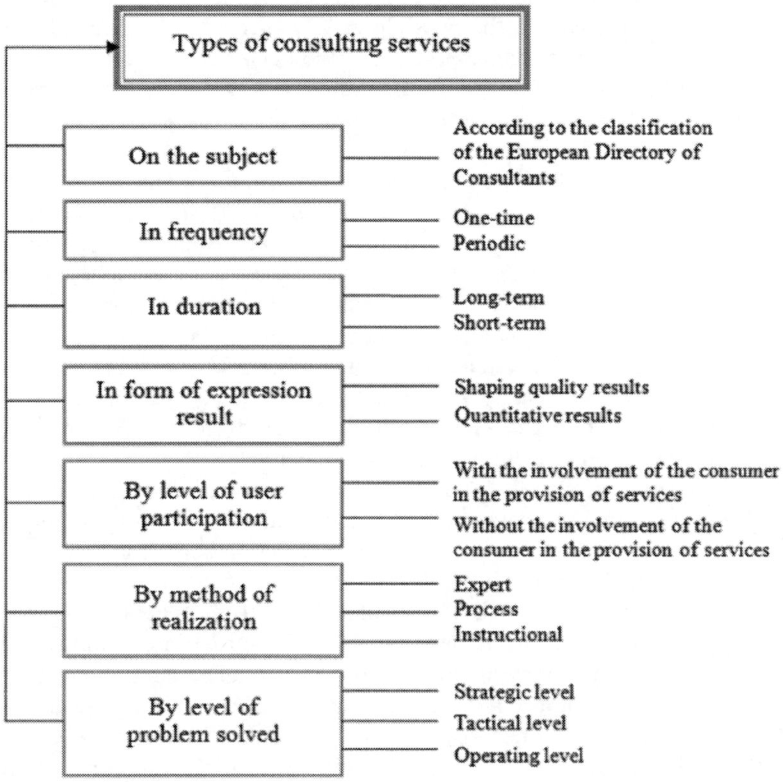

Fig. 1. Types of management consulting services.

In attempts to adapt these features to the specifics of the construction industry, it is important to note that articulation in one system (a subsystem of the control system) will probably require the use of special organizational techniques. The basis for the formation of such organizational techniques should be the separation of the management consulting function and the creation of appropriate organizational structures.

Organizational support of management consulting (in its theoretical models) includes two types of it:

- External consulting;
- Internal consulting.

External consultants include invited experts, who are always assigned specific, well-defined tasks. Such consultants are invited to design and implement specific projects or to the rationalization of processes. The project-based approach is predominantly used. The consultants can be invited on any of the project stages and for projects of any scale. When developing large projects, the involvement of consultants (working individually or as part of a group) is a common practice [10]. Moreover, its success directly depends on the correct choice of consultants as representatives of the consulting services market that has developed in Russia and its regions. Internal consultants are staff members of an organization whose experience and knowledge determine their authority and leadership (often informal) positions. Specialists with the largest volume of necessary information should also be involved as internal consultants.

Practice shows that in most cases the leadership of organizations seeks to attract external consultants from among recognized authorities, that is fully justified. The management personnel of organizations, as a rule, cannot abstract the mind from internal organizational problems and contradictions, which means that they can objectively assess neither the existing problems, nor the possible ways to solve them. External specialists are devoid of this disadvantage, and their independence becomes the key factor in achieving the justification of the decisions made. In the system of organizing the interaction process (in the economic and managerial interpretation), the situation is somewhat different. Connections within organization in this system are weaker, and the information flow is sufficiently complete. Essentially, each of the organizations included in the system can provide professional consultants for other such organizations, who will be external to them. At the same time, the interest of the consultants in the results of the activities of the consulted organization will be ensured (to a large extent) and their independence will be preserved, although not completely. Intra-organizational ties in this system are weaker, and the information flow is quite complete. Essentially, each of the organizations included in the system can provide specialists-consultants for other such organizations, who will be external to them. At the same time, the interest of the consultants in the results of the activities of the consulted organization will be ensured (to a large extent) and their independence will be preserved, although not completely.

The use of external consultants, for all their usefulness, is not free from shortcomings.

First, it is obvious that they have no direct interest in the success of the organization of interactions, and in the absence of a clearly fixed connection between goals and interests, it is practically impossible to ensure the achievement of the latter [11].

Second, the involvement of external consultants in the management of the interaction system is always associated with the dissemination of internal information to the external environment. In the organization's internal processes of a construction company, not only positive, but also negative information is created that one would not want to broadcast. Negative information, for all its necessary for the analysis and selection of certain response measures, should be accumulated in the internal environment, and its spread to the external environment must be prevented [12].

Thirdly, external consultants, despite their efforts, cannot characterize in a fairly complete manner all the elements of the internal environment of the interaction system organization of the construction company. Only formalized information is available to them. But organization's internal processes are never fully formalized. Non-formalized, accurately hidden processes are also emerging, interpersonal communications (with all their uncertainty) are actively manifesting themselves, influence groups are formed that do not have fixed boundaries etc. [13]. Non-formalized processes, as a rule, are represented at insufficiently understandable multitude in construction companies, but their impact is always felt. The task of the management system in such a case is reduced to the study and compensation of non-formalized processes that give rise to the organization's internal problems, and sometimes leading to the destruction of the organization and, at the same time, to support processes that are capable in implicit form to stimulate development.

The foregoing forms the basis for the formation of the following proposal: when developing and implementing a system of interaction, it is advisable to follow the principle of combining the efforts of internal and external consultants in the framework of combined consulting [14]. Combined consulting, as a phenomenon, has clearly manifested itself in Russian practice. In the economic literature, it has not yet received sufficient reflection. However, its main features can be characterized quite clearly if we rely on the typology of consulting as a producer of consulting services.

4 Conclusions

The aforesaid allows us to characterize management consulting in construction business organizations as the symbiosis of positive and negative qualities. However the influence of negative qualities can be reduced through the use of special type of consulting, so called a combined consulting. Combined consulting as an organizational and a managerial method in the fulfillment of consultation exercise has the following feature - it is intended to combine the efforts of internal and external consultants. Combined consulting is among the phenomena are initiated by practical experience. It is still insufficiently provided with theoretical and methodological substantiation. But some theoretical and methodological research studies have been carried out and, based on their results, it is possible to identify combined consulting as a producer of services special type.

References

1. B. Kuznetsov et al., E3S Web of Conferences **263**, 05017 (2021)
2. E. Trushkovskaya et al., IOP Conf Ser: Mater. Sci. Eng. **753**, 032044 (2020)
3. T. V. Maleeva et al., IOP Conf. Ser.: Mater. Sci. Eng. **687**, 044002 (2019)
4. L.G. Selyutina, *Methodological problems of optimization of the structure of the housing stock and housing in a major city in modem conditions*: *Diss. Doc. Econ. of Sciences* (2002)
5. O. Egorova et al., IOP Conf. Ser.: Mater. Sci. Eng. **698**, 077029 (2019)
6. S. Biswas, D. Twitchell, *Management Consulting: A Complete Guide to the Industry*. (2004)
7. D. V. Litvin, *Russian banks in the consulting services market: Dissertation* (2000)
8. L. A. Bazarova, *Socio-technological and situational variables of management consulting. Dissertation, Belgorod State University* (2002)
9. E. Pesotskaya et al., IOP Conf Ser: Mater. Sci. Eng. **962**, 022082 (2020)
10. A. H. Hakimov et al., IOP Conf Ser: Earth Environ. Sci. **751**, 012173 (2021)
11. M. Ivanova et al., E3S Web of Conf **281**, 08001 (2021)
12. E. Rybnov et al., E3S Web of Conferences **217**, 11010 (2020)
13. E. Pesotskaya et al., IOP Conf Ser: Mater. Sci. Eng. **698**, 077030 (2019)
14. M. Ivanova et al., E3S Web of Conferences **291**, 05010 (2021)

Ecosystem methodology for the development of the scientific potential of the network interaction of economic agents in the digital environment

Nataliia Simchenko[1], *Svetlana* Tsohla[1], and *Viktor* Fursov[2]

[1]V.I. Vernadsky Crimean Federal University, 295097 Simferopol, Russia
[2]North-Caucasus Federal University, 355017 Stavropol, Russia

Abstract. The paper considers the study of the problem of the development of the scientific potential of the network interaction of economic agents in the digital environment based on the ecosystem methodology. The study of this problem was based on the use of analytical methods for reviewing the scientific interdisciplinary literature of international databases, methods for analyzing the digital transformation of management relations "ecosystem is a management system of an organization", as well as methods for integrating heterogeneous economic data on the development of economic relations in the development of platform ecosystems. The most important result of the research is the conceptual substantiation of the formation of the scientific potential of the network interaction of economic agents based on the ecosystem development of forms of technological entrepreneurship and lifelong learning. The ecosystem methodology proposed by the authors is based on the identification of three types of relations "ecosystem is a management system": intersystematic relations; relations of intra-organizational regulation; platform management relationship. The significance of the results obtained is due to the importance of developing the economy of ecosystems on the basis of interdisciplinary network integrativity of the interaction of economic agents in the digital environment.

1 Introduction

The modern development of digital technologies radically changes the concept of sustainable development of society, makes serious changes in the structure of economic systems, which leads to the need to rethink the possibilities of science and technology based on the ecosystem development of the scientific potential of interaction between economic agents. The growing contradictions between the increase in the level of development of digital technologies and the loss of the sustainability of the development of economic systems requires the platform introduction of new business models, stimulating the development of new forms of technological entrepreneurship in industry based on breakthrough information technologies, which will create new jobs. The relevance of solving the scientific problem of developing the scientific potential of the interaction of economic agents in the digital environment is due to the need to address the escalating social and economic threats of the loss of stability of systems, the growth of intellectualization and dematerialization of technologies in the digital environment, the applied nature and needs of modern society in ensuring sustainable economic development on the principles of the new consistency and transformation of models for the development of the scientific potential of the interaction of economic agents in the digital environment.

2 Materials and Methods

The study of the prerequisites for the formation of an ecosystem methodology for the development of the scientific potential of the network interaction of economic agents in the digital environment was based on the use of analytical methods for reviewing the scientific interdisciplinary literature of the Scopus, Web of Science databases for the development of ecosystems and their impact on the economic development of organizations. Methods of institutional analysis of digital transformation of management relations "ecosystem is a management system of an organization", as well as methods of integration of heterogeneous economic data on the development of economic relations in the development of platform ecosystems are used. The application of the AI-methodology to the formation of models of technological entrepreneurship as an institutional-environmental toolkit for the development of the economy of ecosystems has been substantiated.

3 Results and Discussion

The scientific significance of this problem is due to the development of the fundamental principles of the new systemicity and conceptual provisions of the economy of ecosystems in the digital environment based on the integration of scientific achievements in the field of general systems theory, management theory, institutional theory, evolutionary economic theory, the paradigm of sustainable development, the paradigm of the digital economy, theory entrepreneurship, the theory of platform economics and the formation on this basis of a new pioneering theory of ecosystems, which provides, among other things, a new systemicity as the ability of the system to long-term sustainable independent functioning due to the network circulation of resources.

In economic science, the concept of an ecosystem arose as an intermediate between the concepts of "the economic agent" and "the market" [1]. In general, agent models are widely used in complex environments. According to the Oxford American Dictionary an agent is a person or someone who plays an active role or produces a certain effect. Also, an agent is understood as a firm, organization or socio-technical environment that ensures the receipt and increment of the effect [2].

The socio-economic ecosystem is a localized complex of organizations, business processes, innovative projects and infrastructural formations, capable of long-term independent functioning due to the circulation of resources, products and systems [3].

In this paper, we consider the ecosystem as a platform model for the interaction of economic agents in ensuring integrative sustainability based on breakthrough digital technologies.

The problematic of studying the principles of new consistency in the development of economic systems is due to the prevailing importance of the platformization of economic relations. New systemicity is characterized by permanent systemic changes in the structure of the system, which allows us to assert, based on the essential characteristics of the methodologies of hierarchical and network topology of systems, about deep changes in the classical postulates of systems theory regarding the primary functions and secondary structure. Consequently, the traditional approach to generating the economic value of goods in the economy based on the development of network effects is radically changing. It is important to take into account the principles of the new systemicity in the process of regulating the activities of ecosystems in ensuring their sustainable independent functioning over a long period.

The concept of ecosystems has received a lot of attention around the world by leading scientists, international organizations and national governments. On the one hand, in many foreign publications the ecosystem is considered from the standpoint of sustainable development of a circular economy, or circular economy. According to experts, due to population growth, the number of consumers is increasing and more and more resources are required for life support. Opportunities for economic growth should be separated from the consumption of raw materials and energy resources. New technologies and the transition to a circular economy provide the opportunity and benefits of environmentally responsible growth with limited resources [4, 5, 6].

The studies carried out by BEROC scientists in the development of the circular economy made it possible to note the emergence of various risks in the implementation of R-imperatives, which are inevitable and due to the different level of development of the countries of the world. An obvious way to resolve possible conflict situations is the differentiation of responsibility of countries - the most developed are assigned a dominant role in the design and development of circular systems, and the developing ones need to revise the strategies and ways of solving their urgent challenges from the perspective of including circular models [7, 8].

On the other hand, in modern scientific publications, the problems of the formation and development of entrepreneurial ecosystems that create new consumer value and represent intermediate structures between the firm and the market are considered [9, 10, 11, 12, 13, 14, 15].

In our opinion, the approach of A.A. Morozova, O.V. Staroverova regarding the consideration of the digital ecosystem not only from the point of view of information technology, but also as a targeted transformation of the economy [16]. In this regard, scientists distinguish three layers of the digital ecosystem: technological; managerial; consumer [16].

Let us consider the features of the application of the ecosystem methodology to the development of the scientific potential of the interaction of organizations through the prism of the formation of models of technological entrepreneurship in the field of the introduction of digital twins as an institutional and environmental toolkit for the development of the economy of ecosystems. Figure 1 shows the structural and functional relationships in the interaction of economic agents in the digital environment.

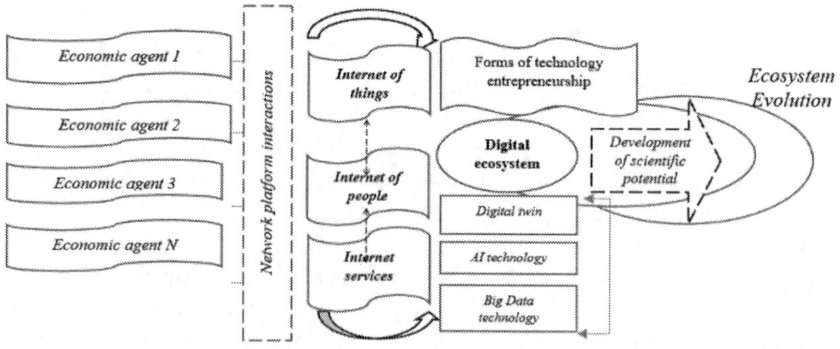

Fig. 1. Structural and functional ties in the interaction of economic agents in the digital environment
Source: compiled by the authors.

The interaction of economic agents in the digital environment is carried out on the basis of breakthrough technologies using the platforms of the Internet of Things, the Internet of People, and the Internet of services. It is advisable to ensure the development of the scientific potential of the interaction of economic agents on the basis of ecosystem methodology using forms of technological entrepreneurship. The characteristic features of technological entrepreneurship are the following: 1) supply creates demand; 2) innovations predetermine the development of society, technologies and services; 3) transformation of the criteria for the effectiveness of innovation in the direction of generating new services and products, and not exclusively in reducing the manufacturer's costs.

In the context of digital transformations in the management of organizations, data, not real assets, become a strategic asset for a company. The development of digital twin platforms in a company predetermines the formation of an ecosystem as a natural system environment of a company, which allows economists to study both the market as a whole and the intermediate structure between the market and the company [2, 26]. The digital twin, as a virtualized model of a physical object, plays a critical role in the digital transformation of an organization. Based on the need to process a significant amount of big data, digital twin technology has given a kind of impetus to the development of artificial intelligence (AI) technologies and big data technologies.

Digital transformations of management relations determine the feasibility of conducting further research in the field of applying the AI methodology to the formation of models of technological entrepreneurship in the field of introducing digital twins as an institutional and environmental toolkit for the development of the ecosystem economy.

The author's interpretation of the ecosystem methodology for the development of the scientific potential of the network interaction of economic agents in the digital environment is based on the principles of a new systemicity, reflecting the variability of the manifestation of the postulates of the general theory of systems regarding the primary functions and secondary structure. Of course, this provision does not imply the exclusive dominance of the "structure" category over the "functionality" category in the systemic methodology for managing organizations, but it shows the importance of the dominant trends in digital transformations in the management of facilities.

4 Conclusion

The problematic of studying the principles of new consistency in the development of economic systems is due to the prevailing importance of the platformization of economic relations. New systemicity is characterized by permanent systemic changes in the structure of the system, which allows us to assert, based on the essential characteristics of the methodologies of hierarchical and network topology of systems, about deep changes in the classical postulates of systems theory regarding the primary functions and secondary structure. Consequently, the traditional approach to generating the economic value of goods in the economy based on the development of network effects is radically changing. It is important to take into account the principles of the new systemicity in the process of regulating the activities of ecosystems in ensuring their sustainable independent functioning over a long period.

Acknowledgements

The reported study was funded by RFBR according to the research project №19-010-00346.

References

1. J. F. Moore, Predators and Prey: A New Ecology of Competition. Harvard Business Review, 75-86 (1993)
2. C. Stary, Sensors, **21(4)**, 1-24 (2021)
3. G. B. Kleiner, Economic revival of Russia, **1(59)**, 40-45 (2019)
4. G. Bressanelli, F. Adrodegari, M. Perona, N. Saccani, Sustainability (Switzerland), **10(3)**, 639 (2018)
5. A. Best, L. Duin, M. Chelminske, *Macroeconomic and societal impacts of mainstreaming the circular economy* (2018)
6. J. Brinkman, Waste to Wealth. Creating Advantage in a Circular Economy, 24, http://ec.europa.eu/
7. M. A. Gur'eva, V. V. Butko, Journal of International Economic Affairs, **9(4)**, 2367 (2019).
8. N. Tura, Jy. Hanski, T. Ahola, Journal of Cleaner Production, 90-98 (2019)
9. V. A. Karpinskaja, Materials of the Second conference of the Department of Modeling of Production Facilities and Complexes of the Central Economic and Mathematical Institute of the Russian Academy of Sciences, 2, 125–141 (2018)
10. G. B. Klejner, System analysis in economics - 2018: collection of works of the V International scientific and practical conference - biennale (November 21-23, 2018) / under total. ed. G. B. Kleiner, S. E. Shchepetova, 5-14 (2018)
11. Ju. N. Androsik, Economics and Management: Proceedings of BSTU, **7**, 38–44 (2016)
12. S. V. Doroshenko, Journal of Economic Theory, **4**, 212–221 (2017)
13. S. K. Mazhitova, B. K. Dzhazykbaeva, I. V. Denisov, M. A. Polozhishnikova, E. S. Petrenko, Economics, Entrepreneurship and Law, **10(3)**, 601-614 (2020)
14. A. Balli, İşletme Araştirmalari Dergisi Journal of Business Research-Turk, **12(2),** 1058-1071 (2020)
15. L Griessbach, K. Ettl, Journal of East European management studies, 171-150 (2020)
16. L. Satalkina, G. Steiner, Sustainability, **12**, 4018 (2020)

Assessment of risks and threats to the development of artificial intelligence technologies

Elena Shevchenko*, *Angelika* Efremova, *Dmitriy* Titarenko, *Aleksey* Voloshin, and *Inga* Artyukhova

V.I. Vernadsky Crimean Federal University, Simferopol, Russian Federation

Abstract. The paper analyses the risks and threats caused by the development and implementation of artificial intelligence technologies. We believe that these risks and threats need research in the long term. Importantly, the use of intelligent information systems has a twofold effect: it can lead to both positive and negative results. The paper also considers the influence of artificial intelligence technologies on the various activities. It proposes a classification of risks and threats caused by the development and implementation of artificial intelligence technologies by the main spheres of human activity.

1 Introduction

Artificial intelligence as a research area emerged in the 1950s thanks to the development of information and communication technologies and computer machines. This made scientists confident that it was possible to create artificial intelligence. Now the prospects for creating artificial intelligence seem to be achievable for both scientists and practitioners. Computer science and information technology are developing rapidly, with discoveries every day in different parts of the world.

As an example, Alan Turing was the first to publish a research paper on artificial intelligence in 1950, entitled "Computing Machinery and Intelligence" and looking into the problem of the machine's ability to think. Later, in 1997, IBM's Deep Blue chess supercomputer won a six-game match against the world chess champion Garry Kasparov. Then, in 2016, Google's AlphaGo beat Go's absolute champion Lee Sedol. As you know, this game belongs to the category of very complex, highly intelligent games. The technology of deep learning allowed achieving this victory [1].

Currently, in artificial intelligence, several scientific areas are forming and developing, such as knowledge representation, automation of reasoning, knowledge acquisition, machine learning and automatic hypothesis generation, data mining and processing of image information, multi-agent systems, dynamic intelligent systems and planning, natural language processing, fuzzy models and soft computing, and development of tools [2].

* Corresponding author: shevchenko_lena_@list.ru

Artificial intelligence technologies have found their application in various information systems, known as intelligent information systems. The use of intelligent information systems has a twofold effect: on the one hand, it gives significant economic benefits to those who use it, and on the other hand, it can bring significant negative aspects for society as a whole, various risks and threats. Moreover, in the future, these negative aspects may exceed the total positive effect of artificial intelligence technologies.

We need to think about the risks and threats of artificial intelligence right now and design measures to overcome and prevent them so that in the future we do not face a crisis without a possible change for the better. This problem is complex and needs a system analysis methodology.

Like any discovery, artificial intelligence can be used for different purposes. For example, dynamite was created for mining, and no one thought that it could find itself inhumane use.

The purpose of the article is to analyze and classify possible risks and threats of the development of artificial intelligence technologies in various spheres of human activity.

2 Materials and methods

The study is based on the specific scientific literature on artificial intelligence. Using the methods of analysis and synthesis, we have designed a classification of risks and threats of the development and implementation of artificial intelligence technologies.

We have grouped the risks and threats of the development and implementation of artificial intelligence technologies according to the main spheres of human activity: economic, social, political, cultural, spiritual, and military.

3 Results

Currently, a national economy's competitiveness is impossible without modern information technologies. In its turn, artificial intelligence is the crown of information technology development. Economies will have to add capacities in this direction as well as in the military sector since lag in this area threatens with economic lag in the future.

The use of artificial intelligence technologies leads to a significant positive economic effect: it reduces by 30% the time spent on the formation of warehouse stocks, leads to a 20% increase in electricity production, speeds up the delivery of materials in industrial production by 30%, increases the productivity of the healthcare sector by 30-50% [3]. We can provide many more such facts.

Besides, large corporations (in particular, transnational companies) are currently interested in artificial intelligence technologies, which, as you know, can significantly influence many areas of human activity, including the economy and politics.

There are the following groups of subjects in the global market for artificial intelligence [3]:

— companies that develop artificial intelligence technologies and have their own data (for example, Google, Microsoft);

— companies whose activities are related to the processing of large amounts of data (for example, IBM, Oracle);

— innovative companies that solve specific problems but do not have their own data (for example, Two Sigma Investments).

E. N. Smirnov and S. A. Lukyanov point out that "artificial intelligence development can not only modify business but also modify the competition in the world market, contribute to further differentiation of the countries in the world by levels of economic and technological development ..." [3]. Moreover, in the current hyperglobalization, the competitive advantages of individual countries, markets, and companies can be lost as quickly as acquired [3].

The development and implementation of artificial intelligence technologies can give rise to significant risks and threats in various spheres of human activity. As K. M. Manheim and L. Kaplan [4] note, artificial intelligence is just a technology with many possible applications. "Rather it is how the tool is used, by whom, and for what purpose that generate concern" [4].

Table 1 shows the classification of risks and threats of the development and implementation of artificial intelligence technologies. We have grouped these risks and threats by the main spheres of human activity: economic, social, political, cultural, spiritual, military, and informational.

We have identified several risks and threats that can belong to all basic areas of human activity:

— the emergence of complete dependence on computers. Any serious failure in such a system can lead to its full stop, and in some cases to destruction;

— the unpredictability of self-conscious systems (intelligent robots), and the influence of various accidents on them. A system endowed with artificial intelligence can face failures and unforeseen factors, which can change the behavior of such a system and lead to the loss of control;

— errors and failures in the work of intelligent information systems. Any information system is created and programmed by people, who can make various kinds of errors in the process of its creation, therefore failures are possible during the operation of these systems;

Table 1. Classification of risks and threats of the development and implementation of artificial intelligence technologies.

<table>
<tr><th>Sphere of human activity</th><th colspan="2">Risks and threats of the development and implementation of artificial intelligence technologies</th></tr>
<tr><td>Economic</td><td>Malicious and criminal use of artificial intelligence to obtain various material benefits. Financial risks arising from the use of artificial intelligence technologies. Differentiation of countries by development level.</td><td rowspan="4">The emergence of complete dependence on computers. The unpredictability of self-conscious systems (intelligent robots), and the influence of various accidents on them. Errors and failures in the work of intelligent information systems. Unreliable, unsafe, or unsatisfactory results. Threats to information confidentiality. The impossibility of bringing to justice the parties responsible for negative consequences.</td></tr>
<tr><td>Social</td><td>Unemployment. Growing poverty and the emergence of a strong economic stratification of society (increased marginalization and inequality). Increased discrimination. Physical security threat.</td></tr>
<tr><td>Political</td><td>Loss of control over the national economy. Loss of sovereignty. A decline in democracy. Manipulation of political systems.</td></tr>
<tr><td>Cultural</td><td>The disunity of people (disintegration of social ties). Loss of natural human skills.</td></tr>
</table>

Table 1. Continued.

Spiritual	Loss of spirituality. Moral degradation.	The creation of an artificial superintelligence that will surpass humans and destroy the human civilization.
Military	The threat to national security. The use of combat robots can lead to additional human losses, among civilians as well.	
Informational	Fake news reports with realistic fabricated video and audio. Automated, hyper-personalized disinformation campaigns. Automating influence campaigns. Denial-of-information attacks. Manipulation of information availability.	

[a] Formed by the authors based on [1; 3–11].

— unreliable, unsafe, or unsatisfactory results. "Irresponsible data management, negligent design and production processes, and questionable deployment practices can, each in their own ways, lead to the implementation and distribution of AI systems that produce unreliable, unsafe, or poor-quality outcomes" [9];

— threats to information confidentiality. Global digitalization of all spheres of human activity can lead to the loss of confidentiality and information leaks, as well as to complete control over a human personality;

— the impossibility of bringing to justice the parties responsible for negative consequences;

— the creation of an artificial superintelligence that will surpass humans and destroy the human civilization (the darkest scenario). However, the current state of science and technology does not yet allow this scenario to come true. Still, we should keep this option in mind due to the rapid development of technology. See, for example, [11].

Below we shall consider the risks and threats of the development and implementation of artificial intelligence technologies that belong to specific areas of human life.

Economy. Artificial intelligence technologies may have criminal uses: cybercrime, particularly in finance; terrorism, including an international one. Here we should take into account not only the current level of technology development but also the future one. These technologies may fall into the hands of dishonest people. As D. Yu. Bazarkina and Y. N. Pashentsev note, "readiness to meet unexpected hidden risks is crucial" [5]. There are also numerous risks of using artificial intelligence technologies in the financial sector. Tom C.W. Lin underlines that "… the proliferation of artificial intelligence in finance increases the dangers of systemic risks and major financial accidents" [10]. Moreover, the development of artificial intelligence technologies may result in further differentiation of countries by development level.

Artificial intelligence technologies can pose significant risks and threats in the social sphere. Among them, we would like to highlight an increase in unemployment and poverty, a strong economic stratification of the society (worsened marginalization and inequality), growing discrimination, and a threat to physical security.

In politics, we can list the following risks and threats: loss of control over the national economy, loss of sovereignty, a decline in democracy, and manipulation of political systems.

As K. M. Manheim and L. Kaplan note, "Artificial intelligence could erase many practical advantages of democracy, and erode the ideals of liberty and equality. It will further concentrate power among a small elite if we don't take steps to stop it" [4].

In culture, artificial intelligence technologies can contribute to the disunity of people (disintegration of social ties), and the loss of natural human skills. "Excessive automation, for example, might reduce the need for human-to-human interaction, while algorithmically enabled hyper-personalization, by limiting our exposure to worldviews different from ours, might polarize social relationships" [9]. "Humans may gradually become passive elements of an emerging global socio-technical system – a system composed of machines, algorithms, sensors and actuators, AI programs, and humans interacting in the globally present Internet, and the Internet that is ever present due to mobile technologies and ambient intelligence" [11].

In the spiritual realm, there can be a loss of spirituality and moral degradation.

The development of artificial intelligence can lead to significant and, in some cases, even catastrophic risks in the military sphere: a threat to national security (artificial intelligence used against any sovereign state); constantly improving unmanned weapon systems based on artificial intelligence can have not only defense but also offense applications; the use of combat robots can lead to additional human losses, including among civilians.

In the information sphere, we can distinguish the following risks and threats: fake news reports with realistic fabricated video and audio; automated, hyper-personalized disinformation campaigns; automating influence campaigns; denial-of-information attacks; manipulation of information availability [6].

Existing and future threats associated with artificial intelligence technologies require measures to overcome and prevent them. The paper [6] proposes the following measures:

— closer cooperation between the government and the designers of artificial intelligence to prevent the possible illegal use of artificial intelligence technologies;

— socially responsible behavior of artificial intelligence designers; they must be constantly aware of possible risks and threats;

— application of the world's best security practices (information, economic, social, etc.);

— involvement of the best scientists and experts.

Dr. David Leslie talks about the need of artificial intelligence ethics, i.e. "a set of values, principles, and techniques that employ widely accepted standards of right and wrong to guide moral conduct in the development and use of AI technologies" [9].

B. Cheatham, K. Javanmardian and H. Samandari also advocate the need to manage the risks of artificial intelligence technologies at the company (firm) level. Besides, they suggest the following principles [7]:

— using a comprehensive approach to risk identification;

— introducing reliable risk management tools;

— strengthening control over risks.

In investment decision-making in the field of artificial intelligence, it is necessary to adhere to the ethical principle. As shown in [12; 13], companies' compliance with ethical standards in business operations helps to increase their competitiveness and has a positive effect on national competitiveness.

4 Discussion

A very important question is: can this problem be solved without government intervention? In our opinion, the regulatory role of the state in the field of artificial intelligence is very important since, in the pursuit of profit, companies can ignore aspects of security, including the future one. This, in particular, is stated in [14].

5 Conclusion

The risks and threats caused by the development and implementation of artificial intelligence technologies can group by the main areas of human activity: economic, social, political, cultural, spiritual, military, and informational. It is also possible to single out several risks and threats related to all of the above areas: the emergence of complete dependence on computers; the unpredictability of self-conscious systems; errors and failures in the work of intelligent information systems; unreliable, unsafe, or unsatisfactory results; threats to information confidentiality, etc.

Preventing risks and threats caused by the development of artificial intelligence technologies will require efforts not only by the state but also by businesses, representatives of science, and society as a whole. The state should adopt special laws to ensure human safety and rights, and companies should conduct socially responsible policies and comply with the ethical principle in their activities.

References

1. E. M. Proydakov, Digital Economy, **3(3)**, 50 (2018)
2. G. S. Osipov, *Artificial Intelligence Lectures* (LENAND, Moscow, 2018)
3. E. N. Smirnov, S. A. Lukyanov, Region Economy, **15(1)**, 57 (2019)
4. K. M. Manheim, L. Kaplan, Yale J. of Law and Technology, **21**, 106 (2019)
5. D. Yu. Bazarkina, Y. N. Pashentsev, Russia in Global Affairs, **17(1)**, 147 (2019)
6. M. Brundage, et al., *The malicious use of artificial intelligence: forecasting, prevention, and mitigation* (2018)
7. B. Cheatham, K. Javanmardian, H. Samandari, *Confronting the risks of artificial intelligence* (McKinsey Quarterly, 2019)
8. J. Johnson, *Artificial intelligence & future warfare: implications for international security. Defense & Security Analysis* (2019)
9. D. Leslie, *Understanding Artificial Intelligence Ethics and Safety: a Guide for the Responsible Design and Implementation of AI Systems in the Public Sector. The Alan Turing Institute* (2019)
10. D. Lin, Fordham Law Review, **88** (2019)
11. A. Nowak, P. Lukowicz, P. Horodecki, "Assessing artificial intelligence for humanity. will AI be our biggest ever advance — or the biggest threat?" in *IEEE Technology and Society Magazine* (2018)
12. M. Bakumenko and A. Sigal, Int. J. of Risk Assessment and Management, **23(1)**, 80 (2020)
13. M. Bakumenko, A. Sigal, Int. J. of Risk Assessment and Management, **21(1/2)**, 111 (2018)
14. C. Wheelus, X. Zhu, IoT, **1**, 259 (2020)

 https://doi.org/10.1051/e3sconf/202131106005

Role and significance of global bilingualism in the convergence of real and virtual spaces (within their communicative segment)

Emil Sarkisov, and *Sergey* Nikolaev

Southern Federal University, Bolshaya Sadovaya Str., 105/42, 344006 Rostov-on-Don, Russia

Abstract. This paper investigates communication in its bilingual expression as a factor enhancing subjective convergence of real and virtual spaces by the sign system. The aim of the investigation is to analyze language and speech which both, under global bilingual conditions, are influenced by cross-linguistic interference and facilitate in their interaction the mechanism of coding those new concepts of reality that arise and consolidate due to the mergence of real and virtual spaces. The basic research methods used are contrastive analysis and conceptual analysis. From the practical standpoint the obtained results can be used as complementary in, say, sociological and/or psychological studies of the impact produced by the real space upon social and individual con-science. The authors come to a conclusion that global bilingualism has created a special pattern in the interaction between language and speech where the coded information concerning real and virtual spaces is conceptually metaphorized. This type of metaphorization is nothing but simulacrization leading to faulty judgments and attitudes towards reality.

1 Introduction

Global bilingualism of the 21st century opens up new opportunities in contact linguistics [1, 2, 3]. This is stipulated by the fact that the key components of the theory of languages in contact, bilingualism included here first and foremost, may be viewed now from a variety of perspectives: linguistic only, sociolinguistic, psycholinguistic and neurolinguistic, pragmatic and cognitive, the latter being of paramount importance within the context of this paper. At the same time cross-linguistic interference is in our case rather an indicative phenomenon. Those cognitive processes occurring within the language partly reflect the other processes occurring within the extralinguistic reality. Thus, psychologists use a cognate notion of associative interference defined by Matsumoto as "the interference of either learning or recalling an association due to another association which is incompatible in some way with the first association" [4].

Linguistic contextualization of the above process under today's conditions of global bilingualism places even greater emphasis on the problem of studying the theory of associative interference as the one that extends to quite a number of spheres of human life and activities, particularly the sphere of communicative. Indeed, while staying unaltered in its essence, the phenomenon of cross-linguistic interference has changed in its formal embodiment, which is evidenced by the emergence of such idioms as Spanglish and other alike lingua francas [5]. It should be borne in mind that the above parallel between the two types of interference is to indicate here elements of a joint cognitive matrix of an alike functioning mechanism. These elements move even closer together in case this is in reference to such base concepts as space and time.

The concepts fixed within the language and actualized in speech not only exist in the dimensions of space and time, but also in the social [6] and technological [7] dimensions. Sure enough man originally existed in a rather limited, yet quite essential for the development of the rest, language and speech too, space-time continuum. Throughout their history, secondary dimensions were transformed because their existence has always hinged and still hinges on the fundamental dimensions of space and time which are changeable by their very nature. Hence language and speech, both emerging with the appearance of secondary dimensions, in their development actually depend the same way (or perhaps even more) on the space-time continuum like the social and technological dimensions. Thus, language contacts would hardly be possible if people did not move in space and their interaction were not characterized by its extension in time. Rise of bilingualism was thus stipulated by subjective space contraction due to the technological progress, the latter promoting and intensifying language contacts. W. Labov, in the second book of his *Principles of Linguistic Change*, describes those factors influencing the language and belonging to the social dimension [8]. Thus, the capitalist social system left such an imprint on language and speech under which the base concept of time is perceived as a result of metaphorization of the formula "commodity – money – commodity". However, such perception was to a great extent supported by long lasting language contacts.

Table 1. Metaphorized perception of time as a commodity

English example	Spanish example	Russian example
You're wasting my time.	Me estás haciendo perder el tiempo.	Ты попусту тратишь мое время.
This gadget will save you hours.	Este artilugio te ahorrará horas.	Это устройство сэкономит тебе много времени.
That flat tire cost me an hour.	Esa rueda deshinchada me ha costado una hora.	Эта спустившаяся шина стоила мне часа.
I've invested a lot of time in her.	He invertido mucho tiempo en ella.	Я вложил в нее много времени.
You need to budget your time.	Tienes que calcular el tiempo.	Тебе необходимо рассчитывать свое время.

George Lakoff calls such metaphors cognitive since they serve as perception prisms for the society, hence change attitudes to various phenomena [9]. To illustrate the latter statement, we intend to analyze here the impact of global bilingualism upon the convergence of real and virtual spaces within the context of a changing attitude to the phenomenon of death.

Subjective changes in the space-time continuum and the associated global bilingualism promoted this metaphor as cognitive in the languages united by the capitalist system, which can be seen in Table 1. This hints upon interferential influence of one dimension on the other, which is reflected in different languages as metaphorization of reality. Another testament to the fact that the above metaphorization is strongly stipulated by language contacts, may be the following: this cognitive metaphor could be traced neither in language nor in speech in Russia before the dissolution of the Soviet Union.

These days, interferential influence of technological dimension on the space-time continuum is so strong that reality is more and more becoming virtual. This happens due to the appearance and development of the new virtual space. In fact, back in 1970s George Lakoff confirmed dynamic existence in mass consciousness of such a cognitive metaphor as *the mind is a machine*. What is viewed today in both language and speech is continued interference between the technological and пspace-time dimensions. Convergence of the real space-time continuum and virtual space dimensions is also coded in individual and mass consciousness and metaphorically decoded. The reason is that interference in language and speech as metaphorization of reality through sign systems – a verbal sign system or a mathematical structure – is the only possible mind-implanted method of reality perception. The basic change characterizing the convergence of real and virtual spaces, is a still stronger "shrinkage" of the real space-time continuum and, therefore, intensification of language contacts and total bilingualization of the society in the globalization era.

In its turn, the change in the formal embodiment of lingua franca idioms testifies evident manifestations of the regularities characteristic of interference in general. It shows that nonassociative cognitive processes do not exist, hence interinfluence of associations is inevitable. If in communication this principle is most evident due to convergence of languages and intensification of contacts among the same, yet prior to the codification stage of metaphorization it might be stipulated by information surfeit and involvement of an individual into a greater number of activities, as well as conditions of their performance. Virtual space is a brand new variant of such conditions here, and it is extraordinary in its characteristics too. Communication in virtual space can be defined as perceptual-linguistic interaction associated with acts of cognition and the creation of meaning-forming systems manifesting itself in a whole set of principles, namely:

1) dialogueness and communicativeness;
2) visual-perceptual experience;
3) interpersonal communication;
4) a tool for cross-cultural dialogue;
5) linguistic component [10].

Analyzing the above-listed features and principles of virtual space communication, also taking into account the all-inclusive expansion of virtual space into various social and communicative spheres that earlier only existed in real space, it is logical to assume that the communicative specificity formed by it determines speech-associated, hence language-associated too, cognitive processes and changes attitudes towards various phenomena. Language and speech, being means of communication, are also means of reflection of some new reality which in its turn defines the development of these two phenomena. Speaking of language and speech as subjects within this scheme, one should again note the global character of bilingualism, which promotes unified metaphorization of reality. The analysis of language and speech as the object of metaphorization under global bilingualism enables to illustrate convergence of real and virtual spaces. This analysis is making the aim of the present investigation within the context of changing attitude towards death in the latter.

Ony special function in the interaction between language and speech as subjects within the bilingual system of communication in contacting real and virtual spaces is performed by cross-linguistic interference in speech. Being affected by the situation of bilingualism, it leads to the appearance and development of various lingua francas depending on the dominant norm-exporting language (a term introduced by W. Labov). Among these of special importance for the present investigation are conceptual norms first and foremost formed at the lexical-semantic level. The interferemes that influence a second lingua franca, turn into a norm due to a stable interferential picture.

In fact, any language under specific extralinguistic conditions may perform the functions of a full-fledged lingua franca, and today we have a lot of examples confirming this statement: such are English, Arabic, Russian, French and other languages [11, 12]. In spite of this, a "classical" lingua franca is characterized by a somewhat different specificity: it implies vaster, deeper and wider penetration in the communicative system of the society. Under existing conditions, from out of all the idioms that can be rightfully qualified as lingua francas, English as lingua franca fully conforms to the said specificity.

In virtue of a special interferential picture existing in a "classical" lingua franca, there is no language (it being opposed to speech in Saussure's model) preserving and consolidating norm in speech. As a matter of fact, norm is transformed, and it transforms the life of the society and its ideas of reality through speech as a thinking tool.

Topicality of this investigation is in the need of a deeper study of the theoretical component of the phenomenon of convergence of real and virtual spaces under global bilingualism, as well as the influence produced on the linguistic consciousness, both individual and social, of cross-linguistic interference, and search for the regularities in this process.

2 Materials and methods

To carry out the practical part of the study, we have analyzed utterances selected in the communicative segment of virtual space, the central feature of such utterances being metaphorization of individual's inner world in software terms. The object of research in this material are lexical-semantic interferemes, i.e. minimum interference units, belonging to the lexical-semantic level of the language system. Interferemes, while they pass to the second lingua franca (Spanglish in our case), get fixed in this idiom as norms under a reversal impact of contacting dimensions and then lead to integration between real and virtual spaces by way of unification of metaphoric perception of the space-time continuum as an idea taken as a whole.

Perception of space seems to be the most significant one because it is the one that makes the basis of perception of reality: it is naturally linked with time and, above all, in many languages it is semantically linked with time [13, 14].

A combination of two types of analysis, contrastive [15] and conceptual [9], makes the basic research methods. This enables to reveal traces of interference at the conceptual level through the lexical-semantic level. Availability of such traces, interferemes, is linked with the bilingual medium, the latter stipulating convergence of real and virtual spaces.

Contrastive analysis of interferemes in the virtual space communicative segment enables to reveal similarities and differences in the norms of the lexical-semantic levels in the involved languages as compared to the real space. This is done with the aim of detecting availability of elements of metaphorization and other conceptualization of reality since the source of cross-linguistic interference is specifically discrepancy of norms belonging to different languages. Conceptual analysis affords ample opportunities for the interpretation of the data obtained during the analysis of the units referred to the lexical-semantic level.

3 Results

Below can be seen a comparative table with utterances in two languages, English and Spanish, which clearly shows metaphorization of language consciousness in the terms applied in the field of *software*. In some cases, the fact that the result of such metaphorization is global bilingualism, cannot cause any doubt: to convey conceptual senses speakers probably use different varieties of calquing, the latter being the major formal type of cross-linguistic interference. Hence the human body under such metaphorization must be perceived not as a "mortal" biological organism, but rather as a *hardware* with its own peculiarities and features.

Table 2. Metaphorized perception of reality in the terms of software

English example	Spanish example
Semantic calque: borrowing the transferred meaning of a lexical-semantic unit of the source language of interference	
Moving your body can disconnect yourself from the dream you were probably having just minutes ago.	Cómo "desconectarse" de WhatsApp sin desconectar tu celular de internet (article title)
Hack Your Mind: Unleash the Hidden Power of Your Subconscious Mind, Harness Brain's Neuroplasticity, Learn How to Bend Reality and Become Limitless (book title)	Aprende a hackear tu mente para mejorar tu estilo de vida (YouTube video title)
Word-formation calque: morpheme-by-morpheme translation of a lexical-semantic unit of the source language of interference	
Are you going from one thing to another without a pause to decompress or reboot your brain?	El zumbido de la cafeína revitaliza instantáneamente tus reservas de energía agotadas y reinicia tu cerebro.
Update yourself with global activities (article title)	Actualiza tu software mental (article title)

4 Discussion

Thus, global bilingualism and a generated "classical" lingua franca with its specific interferential picture promote the following tendency. Signs develop metaphorization of their originally (real) representing character not only at the cross-linguistic, but also, quite consequentially, conceptual level. This happens because both language and speech in their bilingual expression are a coding tool distributing, diversifying in intra-linguistic and unifying in cross-linguistic relations and dynamicizing those conceptual norms resulting from convergence of real and virtual spaces. Since the results of the present study are obtained by the example of concepts coding the perception of individual's inner world in software terms, hence forming the dynamic conceptual norm of perception of death as some reversible process, they confirm the data obtained by the scholars of the University of Barcelona. Using the technology of virtual reality to study mental implications of out-of-body experience (OBE), their influence on death anxiety among other things, they put forward the following hypothesis: "If we could place people in a situation illustrating a possibility of their mind to stay out of their bodies, then this could offer them implicit evidence (need not evident conviction) of the fact that out-of-body life is possible, which would therefore lessen their death anxiety" [16].

In most cases man is unable to overcome the information stream, hence man has no other choice but to try and search for a changeable conceptual norm.

According to M. Poster, expansion and world-wide penetration of (bilingual today) communication in virtual space has crucially changed social ties. This scholar introduces the notion of "information modes" by which he understands various types of symbol exchange creating a definite social order pattern. M. Poster proposes the following periodization in the development of information modes: 1) the stage of communication by way of oral face-to-face language expression where signs correspond to the conventional disposition of social relations and steady lifestyle at which a personality is linked with a definite group; 2) the stage of communication by way of written language expression where signs represent definite concepts and the individual acts reasonably and responsibly; and 3) the stage of communication with electronic innovations where signs obfuscate the reality losing their denotative essence, and the individual feels difficulty in telling reality from metaphors [17]. As is shown by the results of this study, this process acquires special importance under global bilingualism where the sign, instead of denoting definite people, articles, actions, features, etc., is combined in individual and mass consciousness with some flow of the signified.

In fact, M. Poster's concept testifies that interference penetrates all cognitive processes, language and speech included, starting with their meanings and up to the signs shaping them. Thus, under the conditions of global bilingualism it would be worth speaking of the intensification of a bilingual informational flow as one of the major reasons of such changes. A consequence here is decentralization and dispersion of personality that constantly changes at a new communicative stage and is involved in the process formation and, synchronically, establishment of a variety of identities.

Convergence of real and virtual spaces leads to such a reflection of the reality in the language consciousness of a bilingual at which, in the terms of Jean Baudrillard's model, there occurs endless circulation of images and signs that does not permit truthful information. The French scholar introduces the notion of simulacre (image, analogy) denoting "a sign that has turned into a virtual object" [18]. Baudrillard assumes that the meaning of signs is so much "mixed up" that many such signs fail to correspond to the reality. This again refers to metaphorization of the reality that distorts the worldview that had existed before virtual space communication. The notion of simulacre is used by Baudrillard to characterize quite a wide range of phenomena from metaphysical problems to social life.

In our opinion, it is metaphorization of the worldview supported by multifaceted and integral effect of global bilingualism, that leads to what is named by M. Poster "the stage of electronic communication" and by Baudrillard "a simulacre". Metaphorization means simulation resulting in the replacement of the real world by a system of symbols. Modernity for Baudrillard is connected to total simulation of social and cultural phenomena: power, information, communication. Some conceptual metaphors have so deeply penetrated human life that their perception is now weakly recognized or totally unrecognized.

4 Conclusions

As one can see from the above, language and speech under the present conditions of global bilingualism function as a converter of integrated information from those being in the state of convergence, hence contacting dimensions, the most significant of which from the perspective of general essence are real and virtual spaces. Using elements of two languages in coding information under the conditions of convergence of real and virtual spaces, man exposes those conceptual norms that have hardened at the previous stages of information mode development to dispersion and simulacrization.

Acknowledgments

The reported study was funded by RFBR, project number 20-312-90057.

References

1. Y. Matras, Revue française de linguistique appliquée, **18**, 7 (2013)
2. B.B. Kachru, World Englishes, **13**, 135 (1994)
3. T. Ishikawa, Englishes in Practice **4**, 31 (2017)
4. D. Matsumoto, *The Cambridge dictionary of psychology*, 56 (2009)
5. E.E. Sarkisov, Bulletin of Pyatigorsk State University, **1**, 109 (2021)
6. N. Dempsey, G. Bramley, S. Power, C. Brown, Sustainable Development, **19**, 289 (2011)
7. Y. Dimitriadis, Educational Technology, **52** , 50 (2012)
8. W. Labov, *Principles of linguistic change. Volume 2: Social Factors (Language in Society)*, 1 (2001)
9. G. Lakoff, M. Johnson, *Metaphors we live by*, 4 (1980)
10. S.G. Nikolaev, E.E. Sarkisov, E3S Web of Conferences, 273 (2021)
11. A. Mustajoki, Pragmatics & Beyond New Series, **274**, 55 (2017)
12. S.G. Nikolaev, *Slavic philology and culture in the intellectual context of time*, 72 (2019)
13. K.M. Jaszczolt, L. Filipović, Cognitive Foundations of Language Structure and Use, **15**, 13 (2012)
14. C. Wegener, Language, **81**, 1012 (2005)
15. H. Bussmann, *Routledge Dictionary of language and linguistics*, 102 (1996)
16. P. Bourdin, I. Barberia, R. Oliva, M. Slater, PLoS ONE, 12, 1 (2017)
17. M. Poster, *The mode of information: Poststructuralism and Social Contexts*, 1 (2013)
18. J. Baudrillard, *Symbolic exchange and death (Published in association with Theory, Culture & Society)*, 1 (2017)

 https://doi.org/10.1051/e3sconf/202131106006

Institutional support of promising sectors of the Chechen Republic's economy

Tamila Tasueva, and *Vera* Borisova

Federal State Budgetary Institution of Science Complex Research Institute of the Russian Academy of Sciences named after Kh.I. Ibragimov
Federal State-Owned Publicly-Funded Institution of Higher Education Grozny State Oil Technical University named after acad. M.D. Millionshchikov, 364064, Grozny, Isaeva Ave., 100, Russia
Federal State-Owned Publicly-Funded Institution of Higher Education St. Petersburg State University of Economics, 191023, St. Petersburg, 21 Sadovaya Street, Russia

Abstract. In the context of the digital transformation of the region's economy, the focus of attention of scientists and practitioners in terms of institutional changes is on rationalizing the economic relations of business entities. Institutions form the organizational-economic and regulatory-legal backbone of support for promising sectors of the economy, uniting not only social institutions, but also the structures that implement them. If laws can be changed quickly enough, informal regulations change gradually. Wherein, it is informal institutions that create the basis for evolutionary changes in legislative and regulatory rules. Nowadays, the set of regulations and means that are designed to provide institutional support to economic entities in the region is supplemented by the theme of sustainable development and the principles of environmental, resource-saving, social and corporate cooperation (ESG, Environmental, Social, Governance). From the point of view of these principles implementation, it is assumed to create and implement the modern sustainable ecologically-oriented ecosystem objects [1]. In our study, we proceed from the fact that the basis of the digitalization of the regional economy and its sustainable development is the institutional support of the economic system. It is the institutions that are provided with the most important role in ensuring the digital transformation of the basic sectors of the region's economy. In its turn, the digital transformation of economic sectors depends on the market potential of the infrastructure, development of financial, legal and other aspects that implement the integration relations of economic entities. Taken together, these terms determine the competitiveness of the region's economy. There is an objective need for development of theoretical and methodological provisions for formation of institutional support for the region's economy in new post-COVID normality. Special attention of the authors was focused on the study of the institutional support of the digital sector of the economy of the Chechen Republic, which is a driving force and a tool for solving a whole range of economic, social, environmental and cultural problems that are urgent in the region. This circumstance determines the relevance of this study, the significance of the tasks set and the recommendations provided.

1 Introduction

In the context of digital transformation and the search for ways out of the global crisis caused by the COVID-19 coronavirus pandemic, it becomes vital to combine economic thought, scientific and practical conclusions with those real processes that are taking place in the economy of the country and its regions.

The problem of the conditionality of institutional changes and economic growth is a complex and controversial socio-economic phenomenon. Laws, rules, codes and types of moral and aesthetic behavior constitute the structure of institutions. Global transformations of institutions are the result of historical transformations in the economic system that are taking place slowly. However, the greater the uncertainty and the slower institutional transformations are implemented, the higher the risk of stagnation in the economy.

Under these conditions, it is necessary to master actively the digital innovations, while avoiding mistakes in terms of institution building. The goal of institutional support for the region's economy is to ensure the sustainability and consistent reproduction of the socio-economic integrity of the economic system, to achieve the maximum economic effect within the framework of limited resources, diverse transformations, and instability of the market environment.

The designated goal is to some extent universal for all basic institutions. Wherein, each institution performs its specific functions, in accordance with the assigned role and position on the socio-economic landscape of the region. The economic institutions under consideration form the material basis of the economic system, and the viability of the society development.

The situation is complicated by the fact that market efficiency turns out to be inadequate to social, economic and environmental efficiency [2]. A high price has been paid for "pure" market efficiency in the form of deep and often irreparable environmental destruction. Therefore, Rosprirodnadzor estimated the environmental damage from the accident in Norilsk at 148.2 billion rubles [3]. The issues of ecology, health care, science and introduction of digital innovations into economic practice have become the main agenda for development of multilateral institutions.

The coronavirus pandemic has dramatically changed the world. Many events that were expected to occur decades later took place in less than one year. The coronavirus crisis tuned the region's economy to accelerated transformation of industries and areas of activity, predetermined the need for structural changes. A situation has arisen when the established institutions must adapt as quickly as possible to the new conditions, and in a number of cases the creation of new institutions is required.

Adaptation of the institutional support of the region's economy to the new normalcy, on the one hand, opens up additional opportunities for the growth of the region's economy, but at the same time, presents new challenges to business entities. Adjustment of market behavior must penetrate all the components of new business models. [4] First of all, this is associated with construction and recovery of business after the crisis on the principles of cooperation between partners in the coordinates of digital, environmental, social and corporate interaction.

It is obvious that dynamic development of the region, an increase in production capacities, modernization of production processes, increase in export-import opportunities largely depends on the driving industries. This article presents the results of the study of the institutional support of promising sectors of the economy of the Chechen Republic. The authors' attention is focused on the digital sector of the republic's economy and the assessment of the prospects for its development.

2 Research Methodology

The methodological basis for the disclosure of the problem under study is consideration of the economy of the Chechen Republic in the context of the institutional support of the digital sector as a driver of innovative transformations in the economic system of the region. Identification of the features of the emergence and functioning the digital innovations in the economy of the Chechen Republic was carried out on the basis of the scientific groundwork formed to date [5-7], use of methods of institutional design, stratification and focus group analysis [8-10].

The study was based on the Legislative acts and policy documents on development of the digital sector of the economy in Russia and its regions, the provisions set out in the Strategy for the socio-economic development of the Chechen Republic until 2035, research materials on institutional development, the concept of sustainable development and environmental conservation [11- 16]. Study conclusions regarding the institutional support of promising sectors of the regional economy are formed on the basis of generalizing the existing practices and theoretical views on formation of a new institutional framework of infrastructure in the digital economy of the region [17-20]. The set of analytical conclusions based on the above scientific basis determines the author's position and the theoretical and applied significance of the problem under study.

3 Research Results

The socio-economic development of the modern region is determined by the level of digitalization of the areas of its economy and the society life-sustaining activity. The mission of the digital economy formation in the Chechen Republic is defined by the Strategy of its socio-economic development until 2035. Due to digitalization, it is planned to ensure the quality of life of citizens, increase the competitiveness and security of the region's economy, and in the future – to form an independent competitive digital industry in the republic.

Problems of sustainable development have become the agenda of multilateral institutions in the world and in Russia. In our study, we relied on the basic provisions of the concept of sustainable development, the recommendations provided in the "Strategy for the socio-economic development of the Chechen Republic until 2035".

The study found that the priorities, goals and objectives of the state administration of the Chechen Republic are consistent with the basic provisions of the concept of sustainable development and the principles of ecological, environmental, socially-oriented cooperation of partners (ESG, Environmental, Social, Governance). The state governing bodies of the republic have to deal with a group of various parties, including investors, business entities, public organizations and other structures, the balance of interaction between which is quite difficult to find. The best way to measure progress that the authorities can use is summarized in the fourth edition of the Strategy. Formation of institutional and infrastructural conditions for development of the digital sector in the region's economy are among the targeted results outlined by the leadership of the republic.

Development of recommendations regarding implementation of institutional adjustments in the regional economy was preceded by a comprehensive analysis of the challenges and opportunities for digitalization. Table 1 shows the results of this paper on the example of the Chechen Republic [11].

Table 1. Results of the SWOT matrix (strengths and weaknesses of the economy of the Chechen Republic).

Strengths	**Weaknesses**
A good managerial groundwork in the regional leadership of the republic, the ability of personnel to carry out large-scale transformations in strategically important sectors; favorable natural and climatic resources for development of the agro-industrial complex, tourism; availability of mineral raw materials for production of building materials and development of the construction industry; good provision of hydro and geothermal resources for energy development.	Insufficient development of institutions that facilitate interaction between the state, private business and science. High subsidization of the region with a shortage of financial resources. Lack of subjects that are drivers of socio-economic development and insufficient information integration of business entities. Weak level of competition and entrepreneurial initiatives, low labor productivity with a significant share of the shadow economy and the informal sector.
Possibilities	**Threats**
Development of interregional network communities and forms of communication with them. Interregional exchange and export of agricultural products. Development and implementation of new instruments of public-private partnership, entrepreneurial potential and interregional innovation projects. Improving the quality of banking and insurance products as part of creating a digital system in the fintech finance sector and realizing the potential of the capital of the Republic, Grozny - the driver of the region's economy.	Presence of foreign innovations on the market and leakage of domestic software products to foreign companies. Reducing financial opportunities for maintaining research centers and reducing the motivation of qualified personnel; migration of the economically active population. The danger of increase in unemployment, both as a result of demographic growth and as a result of the impact of possible man-made, natural-climatic and sanitary-epidemiological factors; the likelihood of interethnic conflicts and hotbeds of international terrorism.

Experts note that further development of the Chechen Republic is impeded by a set of interrelated problems, including: lack of factors contributing to economic development; low level of development of sectors that are drivers of the economy; predominance of the public sector in the overall structure of the economy (the share of the public sector is 16.5 %); low level of labor productivity (35 % lower than the average for the North Caucasian Federal District); weak level of competition and entrepreneurial initiatives [21].

The analysis showed that institutional and infrastructural problems affect the creation of quality jobs and the level of their accessibility for young people. The consequence of this is structural imbalances in the economy of the republic and low competitiveness.

The analytical review prepared for the fourth edition of the program of strategic socio-economic development of the Chechen Republic until 2035 noted that the structure of the economy of the Chechen Republic remains uncompetitive. In 6 of the 14 key sectors of the economy, the Chechen Republic lags far behind in productivity. Therefore, in the "Mining" sector, productivity in the Chechen Republic in 2015 amounted to 739 thousand rubles/person compared to 3,764 thousand rubles/person on average for Russia, in the energy sector, respectively, – 497 and 993 thousand rubles/person, in the "Transport and Communications" sector – 494 and 929 thousand rubles/person, in the "Manufacturing" sector – 171 and 905 thousand rubles/person, in the "Agro-Industrial Complex" sector – 138 and 565 thousand rubles/person, in the "Finance" sector – 54 and 128 thousand rubles/person" [21].

Among the reasons for this state of affairs, experts name the following: depreciation of fixed assets and depletion of wells in resource extraction and in the energy sector; limited logistics schemes for goods movement in view of a small internal market and underdevelopment of interregional trade; weak integration of the industrial and agricultural sectors of the economy; lack of modern banking and insurance products. This situation negatively affects the investment attractiveness of the republican economy. Investors interested in access to natural resources need institutional support, provision of modern infrastructure, digital products and services.

We have highlighted the weaknesses and threats to development of the regional economy allow us to identify the following urgent problems:

- insufficient financing of development of subjects that are drivers of socio-economic development and weak information integration of business entities;
- weak level of technical and process development of the digital sector of the republic's economy;
- insufficient introduction of the mechanism of public-private partnership and cross-functional management methods into economic practice;
- underdeveloped transport and logistics infrastructure;
- lack of competition among business structures and a high share of the informal sector of the economy.

Solution of these problems predetermines the need for further analysis of the institutional support for the digital economy of the region as a backbone factor.

4 Discussion of Study Results

Development of promising sectors of the Chechen Republic's economy (agro-industrial complex, construction, tourism, digital economy, transport, logistics, healthcare, education) has always been in the center of attention of the leadership and the scientific community of the region. Let's pay attention to the digital sector of the Chechen Republic's economy.

The strategic task has been set to significantly increase the contribution of the digital industry to the gross regional product of the republic and increase the number of participants in this sector, as well as to develop the export of information and communication technologies, products and services. In the future, on the basis of the Chechen Republic, it is planned to develop and introduce a cross-channel information system for the credit and financial sector of the North Caucasus Federal District.

The subject-object composition of the digital sector of the Chechen Republic's economy is characterized by: suppliers of digital products and services; infrastructure; human resources; integration links of business partners and the overall development of the digital environment.

Development and production of digital products and services is aimed at servicing information and telecommunications infrastructure in the state, industrial and consumer segments. Key participants: external IT companies, Technopark of the Chechen State University, State Unitary Enterprise Center for Information Technologies.

Among the barriers to achieving these goals, experts note: "the lack of its own IT companies and effective demand for digital products and services in the republic, as well as low digital literacy of the population" [11].

Another problem of the progressive development of the digital sector of the Republic is insufficient level of its infrastructural support. Participants of the infrastructure digital subsector: MTS, Megafon, Beeline, Vainakhtelecom, note that infrastructure development is hampered by: "low 4G coverage (only the central part of the Republic is covered); insufficient quality and coverage of cellular GSM communication; lack of effective monitoring of network performance by telecom operators and modern infrastructure for storing and processing data; insufficient level of security of data transmission and storage in the public sector" [11].

An important factor in development of the digital sector of the economy is its staffing. Training of highly qualified specialists for the digital economy in the Republic is carried out not only by higher educational institutions. The digital environment and business climate of the Chechen Republic includes "specialized platforms for IT companies; institutions for the development and support of startups; regulatory services and institutions for the protection of intellectual property. Key players in the digital environment: the Government of the Chechen Republic, the Ministry of Transport of the Chechen Republic, Technopark of the Chechen State University. Among the barriers to development of the IT sector, there are limitations associated with development of an innovative business climate; weak motivation of investors and large IT companies; insufficient number of business incubators, accelerators and venture funds; lack of support for commercialization of IT products (including registration of patents for IT startups); high monopoly prices for cable broadband Internet" [11].

Wherein, it can be noted that certain prospects for development have emerged in the digital economy of the Chechen Republic. According to experts, they are associated with an increase in digital transformations in the state and social sectors of the region's economy, in healthcare and education institutions, in creation of state registries and use of blockchain technology. "One of the largest fiber-optic networks in Russia in terms of density operates throughout the territory of the Chechen Republic. 4G mobile communications are deployed in the central part of the Republic. New Veduchi ski resort is provided with 4G communication and cable high-speed Internet. There is one local operator Vainakhtelecom and all major federal operators (Megafon, MTS, VimpelCom) in the region."

"Another opportunity for innovative development is formation of digital skills among the population of the Chechen Republic, the use of new online services. The city of Grozny has chosen the digital format of urban management and transport logistics ("smart city") is incorporated in the concept of development of the city of Grozny.

Among the target results of digitalization of the republican economy, the following are indicated: further development of information and communication technologies in the public sector, health care and education, including the potential of the Chechen State University, Grozny State Oil Technical University named after acad. M.D. Millionshchikov; creation of key growth points based on big data, introduction of blockchain technology in the agro-industrial complex, construction, finance and credit and logistics.

The leadership of the republic has set ambitious tasks - achieving leadership in the field of digitalization in the regional market; "export of digital products and services, development of the regional center of digital competencies, IT infrastructure operating throughout the North Caucasus Federal District. Joint efforts of the state and business in this direction are concentrated on interaction with international partners; support of companies exporting IT products and services; creation of mechanisms for preservation of successful IT startups in the territory of the republic; building a flagship IT educational institution; modernization of the previously created IT infrastructure" [11].

In the Chechen Republic digital economy sector, it is planned to create levels of interaction between business entities and the Ministry of Digital Development and Public Administration. According to the results of the focus group survey conducted within the framework of this study, a "digital valley" in the republic is singled out among the visible effects of the institutional provision of infrastructure.

According to forecasted estimates of experts, "the share of products of IT companies in the export structure in the next five years will reach 1.5 %, more than 1000 IT startups will be launched; coverage of the region with broadband internet will be 100 %.

Investments in these digital initiatives are provided at the expense of federal and regional budget funds, as well as at the expense of Russian and foreign partners of digital transformation (Sberbank PJSC, Yandex LLC, Abu Dhabi Development Fund), funds of non-profit organizations. Such cooperation presupposes the development of environmental institutional conditions: of legislative, regulatory, informational, infrastructural, and personnel nature.

On the regional digital institutional landscape, the prerequisites for networked cooperation of partners have emerged. This is the interaction of participants in interregional cooperation within the framework of digital platforms, "joint use of marketing and advertising programs by partners; business inter-network communication; promotion of joint interregional brands" [11]. Higher educational institutions of the Chechen Republic and the campus of the Chechen State University as a communicative environment for interaction of doctoral students, teachers and researchers can act as the initiator and coordinator of the activities of the participants of the digital platform." The problem of creating infrastructure to ensure universal access to digital services is related to the issue of institutional support for infrastructure development. Primary one is the infrastructure development itself, as a condition for creating the possibility of digital transformation of the region's economy. From the point of view of creation and development of the modern infrastructure of the region, the latest technologies and tools are necessary to improve the level of its quality.

5 Conclusion

Therefore, we approached the analysis and assessment of the institutional support of the region's economy critically, considering both the strengths and weaknesses of the ongoing transformations. We concluded that the institutional support of the digital economy in the Chechen Republic has growth potential, primarily in the state and social sectors of the economy. This determines the feasibility of formation and development of the ecosystem of the Republic's digital economy aimed at achieving synergy of cross-channel cooperation of its participants. Institutional support for promising sectors of the region's economy plays a key role among the factors contributing to formation of the digital economy ecosystem.

Acknowledgment

The article was prepared with the financial support of the Russian Foundation for Basic Research (project No. 20-010-00141/20 "Formation of the institutional framework of the regional infrastructure in the digital economy". "The reported study was funded by RFBR, project number 20-010-00141\20").

References

1. V. Borisova, T Tasueva, European Proceedings of Social and Behavioural Sciences EpSBS, 230 (2021)
2. O. V. Trofimov, V. Ya. Zakharov, V. G. Frolov, Bulletin of the Nizhny Novgorod University named after N.I. Lobachevsky. Series. Social sciencies., **4(56),** 43 (2019)
3. Interfax 2020, https://www.rbc.ru
4. V. Borisova, M. Bilczak, Olsztyn Economic Journal (uwm.edu.pl) **15(3)** (2020)
5. A. A. Auzan. World Economy and International Relations, **61(10),** 96 (2017)
6. A. E. Shastitko, Problems of Economics, **8** (2016)
7. D. North, Institutions, I*nstitutional Changes and Economic Performance*, 17 (1997)

8. V. V. Borisova. Bulletin of the St. Petersburg State University of Economics, **2(120),** 86 (2020)
9. V. V. Borisova, T. S. Tasueva, B. K. Rakhimova, Lecture Notes in Networks and Systems, **91,** 631 (2020)
10. V. V. Borisova, K. G. Gordei, Science of the XXI century: problems and prospects of researches, **2,** 3 (2017)
11. Strategy for the socio-economic development of the Chechen Republic until 2035, https://economy-chr.ru
12. L. M. Grigoriev, I. A. Makarov, A. K. Sokolova, V. A. Pavlyushina, I. A. Stepanov, Bulletin of international organizations, **15(1)** (2020)
13. I. A. Makarov, I. A. Stepanov, Bulletin of Moscow University, Series 6: Economics, **6** (2017)
14. V. I. Danilov-Danilyan, Bulletin of the Russian Academy of Sciences, **4** (2013)
15. *Higher School of Economics Report "Turn to Nature: Russia's New Environmental Policy in the Context of Green Transformation of the World Economy and Politics: A Report on the Results of a Series of Situational Analysis,* 97 (2021)
16. On the state and protection of the environment of the Russian Federation in 2019. State report. M.: Ministry of Natural Resources of Russia; Moscow State University named after M.V. Lomonosov (2020)
17. N. A. Osintsev, E. V. Kazarmshchikova. Factors of sustainable development of transport and logistics systems. SPTKR, **1** (2017)
18. I. G. Rzun, N. O. Starkova, Bulletin of NGIEI, **11**, 89 (2016)
19. A. D. Moiseev, I. F. Narizhny. Central Scientific Bulletin Magazine, **19(36)-2** (2017)
20. V. V. Borisova, Bulletin of the Russian State Economic University (RINKh), **2**, 45 (2020)
21. T. S. Tasueva, V. V. Borisova, Bulletin of GGNTU. Humanities and socio-economic sciences, **16-4(22),** 15 (2020)
22. Indicators of social and economic development of the Chechen Republic, https://economy-chr.ru

Digital innovations as a development driver for regional management and consumer interaction in the post-pandemic period

Diana Yampolskaya*, and *Sergey U.* Chernikov

Peoples' Friendship University of Russia, Moscow, Russia

Abstract. The development of enterprises and organizations in the new conditions created by the COVID-19 pandemic raises many problems and sets new challenges. This article examines issues related to the changing operating conditions of regional companies, their use of digital technologies and the development of e-commerce. The relevance of the work is dictated by extensive discussion of these issues in the media and field research to determine the readiness of companies to transition to digital technologies. The need for this transition is apparent. Both positive and negative aspects of the transition to new working conditions have been identified based on the results of a content analysis of open sources of information. Digital technologies enable more flexible and dynamic development of companies, comprehensive services to their customers through the creation of service ecosystems. Electronic commerce is developing rapidly. At the same time, digitalization can divide the team, increase control (and therefore pressure) on personnel, and cause emotional burnout of employees. Consideration of these areas and a set of digitalization tasks is the main topic of this publication.

1 Introduction

The COVID-19 pandemic affected not only the structure of production and consumption, but also the management of organizations. It required organizational changes and a re-examination of interests and values within the organization. Management models are emerging associated with the new position of the employee in the organization and new principles of communication with clients. There are groups of issues that are important in the formation of new relationships both within the organization and outside it: rethinking values, the emergence of new organizational structures, the development of innovative approaches to collecting and analyzing data [1].

Any organizational changes require significant financial resources, especially if they are associated with the introduction of new technologies (since it becomes simply impossible to work on the old ones). Sarah Nagem in her article "5 business strategies for achieving success during a pandemic" [2] cites the opinions of various experts, which boil down to the following: you cannot reduce costs without taking into account strategic changes, you

* Corresponding author: yampolskaya_do@rudn.ru

must continue to engage in innovative development, cooperate with the company's customers and keep keep them up to date, look for optimal ways to cut costs and ... stay calm.

The COVID-19 pandemic has changed the direction of social and economic life. In the social sphere, this is the strengthening of individualism and self-sufficiency, the manifestation of “discrete” thinking, when, for example, instead of dialogue, you can simply “press the button” and not bother with arguments if you don’t like what your opponent says. In the economic sphere, the slogan "increasing efficiency and reducing costs" prevails. Many traditional industries are facing declining consumption and are forced to look for ways to maintain financial sustainability.

The new conditions in which many territorial and industry associations find themselves stimulate the search and development of new technologies. The concept of the 4th technological revolution has appeared, which presupposes (like any revolution) a radical change in all processes and technologies. Almost all of these technologies are in the field of digitalization. The pandemic has accelerated digitalization. If we consider the development of companies as participants in the technological revolution, then there are two main directions that need to be radically changed.

The first would be personnel management problems, especially outside of state capital and major cities. The impact of the pandemic on the internal environment of companies seems to have changed the scope of management. Traditional management functions (planning, organization, motivation and control) had to be adjusted severely to be carried out in a “distant mode”. For example, imagine a boss who brings up a negligent subordinate on Zoom using foul or negative language. The subordinate can simply turn off the sound and the whole "educational" process will have no effect. On the other hand, teleworking leads to a "burnout" effect, since you cannot get up and go home – one is at work non-stop. And there are a lot of such examples appearing on a daily basis in the management knowledge base today.

Digitalization is in many ways a transformation of governance. Management moves to a new level, with a great decrease of the emotional component. It becomes more formal and deterministic. The main challenge facing many companies during a pandemic is resilience. According to many management specialists, the main element of sustainability is the organization's personnel and relationships in the new conditions of remote and in-person work. Think about how to maintain and develop the company's most important asset: employees and customer relationships. [3].

The management features that appear during a pandemic include:

1. the speed of decision-making, which technically depends on the automation of workflow and the communicating procedures with employees;
2. organization of online procedures for interviewing employees in various areas, which are more formal in nature - the procedure for relationships becomes more formal, while interpersonal relationships are weakening;
3. tracking contacts of work with clients and highlighting the so-called “points of contact”;
4. stimulating employees to work more creatively (real-time employee experience or employee engagement);
5. the control functions change, the emotionality and subjectivity of the assessment disappears;
6. redevelopment of assessing criteria the work done by personnel.

Another problem is the introduction of new technologies that are associated with the processes of active digitalization of business processes. The list of these technologies is well known: robot process automatization, Big Data, chat bots, virtual and augmented reality (VR / AR), optical recognition, artificial intelligence (AI), Internet of things (IoT), blockchain. Companies actively implement digitalization of various processes once they understand what benefits it will bring in terms of cost reduction and in terms of creating added value for the client (for which the client will agree to pay more). The process of creating and implementing these technologies is always associated with providing added value and ensuring financial stability. The development of e-commerce as a comprehensive tool for sustainability and business development in the new conditions is possible only with the use of digital technologies, the development of which was accelerated by the COVID-19 pandemic.

2 Research results

2.1 Digitalization in Russian regional companies: current situation and prospects

Digitalization of technological and business processes involves work in the following areas. First of all, it is the operational efficiency associated with production technology. As mentioned above, the introduction of digital technologies should be justified by obtaining added value and increasing the efficiency of the company. Russian regional companies began to implement digital technologies even before the COVID-19 pandemic. The digitalization leaders are telecommunications companies, IT, banks and financial institutions, retail and oil and gas industry.

In 2019, a Russian partner of the consulting company KPMG conducted a study aimed at studying the use of digital technologies in Russian companies. According to the study, the majority of companies (77%) expect digitalization to increase the productivity and efficiency of management processes.

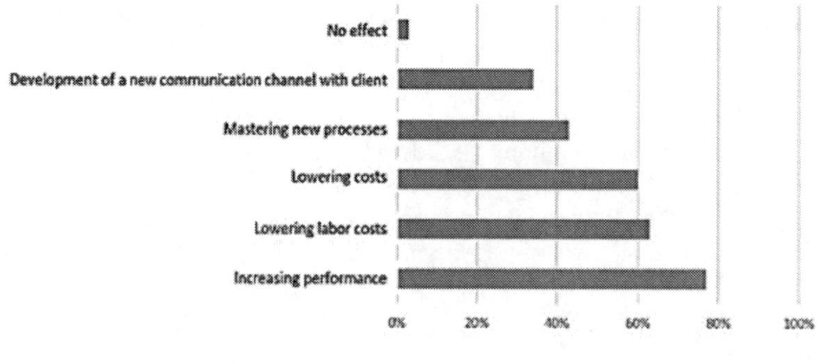

Fig. 1. The effect of using digitalization in companies.

Source: Digital technologies in Russian companies. Study results. KPMG, 2019

Figure 1 shows the benefits that companies have already seen from digital adoption. The first three positions are related to the economic performance of companies: increased productivity, reduced labor costs, and reduced costs. All this leads to a reduction in the cost of production and an increase in price stability. The effect of mastering new processes is associated with an increase in the efficiency of personnel and the speed of adoption of a new technological process. All this is also reflected in the cost price. The fifth point was the improvement of work with clients. This position characterizes the company's interest in developing demand for its products. Attracting and retaining customers, increasing the speed of their service, etc. create additional competitive advantages for companies. Although the absolute cost will increase in this case. And only about 4% of respondents answered that they did not notice any effect.

The popularity of technology shows the directions of organizational changes that are most in demand: cost reduction and optimization of work with clients. The first change in terms of cost savings is staff reduction and / or more efficient work. The introduction of robots allows, on average, to replace 4.5 employees and simplifies the personnel management process. Another popular area is analyzing the state of the company and planning work with clients. To optimize these processes, predictive analytics tools are used, which allow not only highlighting target groups of consumers, but also making the most balanced commercial offer in terms of price and quality. In this regard, big data tools and chat bots have become the technology leaders. Figure 2 shows a rating of technologies that are already being used by Russian companies. In terms of technology, the most popular are big data tools (68%), robots (50%), and chatbots (51%). [4].

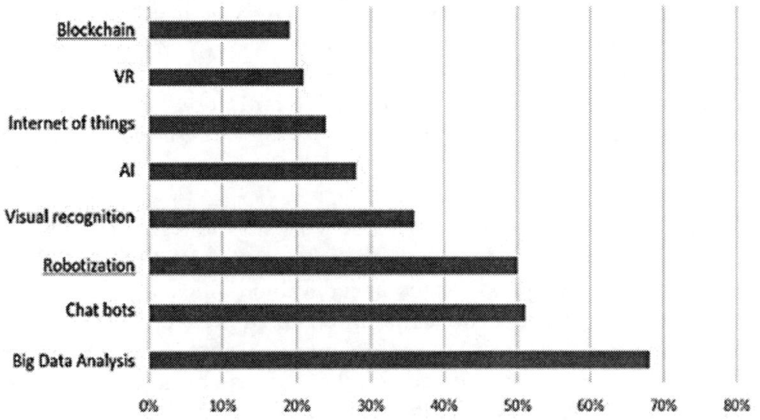

Fig. 2. Technologies used by Russian companies

Information source: Digital technologies in Russian companies. Study results. KPMG,2019.

What trends are observed in Russian business during the pandemic? Let us consider the main directions of the introduction of digital technologies in the Russian Federation during the pandemic (2019 - 20 years) [5]. The priorities remain the same as before the pandemic: developing technologies that reduce labor costs and open up new channels of interaction with the customer / supplier, i.e. the processes of cost reduction and demand generation are relevant [6]. Optimization of business processes can be considered both in the economic context and in the aspect of organizational changes.

The introduction of major innovations requires significant funds and retraining of personnel. The theory and practice of organizational change in companies tells us that the transition to new technologies is very often impeded by technical personnel (middle managers - key performers), who must change their skills and improve their knowledge. It is difficult to motivate experienced workers. "Old" and "new" experience collide. For example, a salesperson who has good connections and contacts with consumers will not want to master predictive analytics techniques. Even without her, he communicates well with customers and sales are going well. Analytical conclusions, which will be sent to him by experts in the field of working with big data, he will consider as superfluous information. Therefore, the introduction of new technologies occurs mainly in stages. The process begins with the improvement of individual operations and, as you master them, moves on. In the conditions of most enterprises, the digitalization process can be called evolutionary. Revolutionary reorganization actions carry great risks, both in the field of finance and in personnel policy. Difficulties can arise from a legal point of view, when innovative processes are introduced in which regulations have not been developed and the regulatory authorities simply cannot check it due to the lack of certification. From the point of view of organizational processes, in general, innovation processes are not separated into a separate function or a separate subdivision, which confirms the model of implementation of innovations "upwards".

The development of digitalization has its positive aspects and problem areas. Digital technologies can be applied in any field of activity. Digitalization leads to business diversification and the entry of non-core companies into the market, which intensifies competition and the struggle for a buyer. For example, transportation and delivery, whose role has grown significantly during the pandemic, can now be handled by completely non-core companies. An example is the diversification of Yandex, which delivers everything from food to passengers to the airport. The largest financial organization in Russia, Sberbank, has organized its own service for the delivery of goods from retail and parcels to the population. Russian Post has opened its own bank. There are trends in business merger - processing centers and their management. The regional companies inevitably would follow these trends.

Despite the fact that digitalization is announced as a strategic direction for business development, the importance of applied technologies and real assets still remains higher. The use of new digital technologies brings added value, which is significantly higher in service organizations than in real manufacturing companies. The tasks of digitalization in production are the financial sustainability of the enterprise and the development of skills for mastering new technologies, i.e. management tasks. The use of new technologies develops the creativity and communication skills of employees during the period of work remotely. The processes of knowledge exchange, learning and joint activities are also being developed to achieve the common goals of the organization [7].

3 E-commerce as the main engine of digital technologies

One of the main tasks of any commercial organization is to maintain and develop demand for its products. This requires offering consumers “convenient” products and building appropriate communication channels. E-commerce is now the main engine driving the growth of innovation and digital technology. Digital technologies are divided into two areas: digital products and digital channels used by consumers.

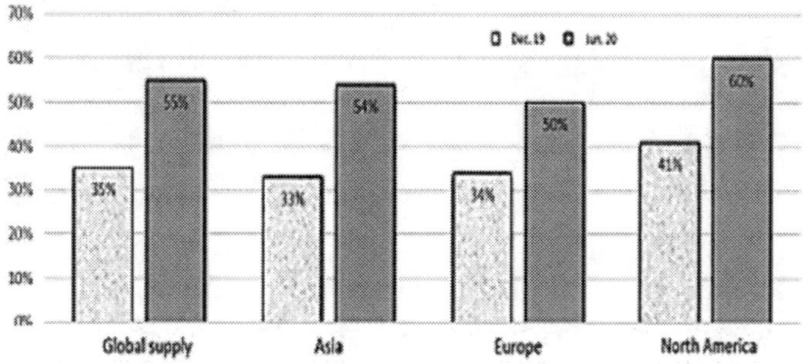

Fig. 3. Relative share of products / services that are partially or fully digital.

Source: How COVID-19 has pushed companies over the technology tipping point – and transformed business forever. Survey. McKinsey & Company. https://www.mckinsey.com/b.

Figure 3 shows data from McKinsey's research on how digital product offerings have changed over the six months of the pandemic, from December 2019 to June 2020. In almost all regions, this supply has increased in leaps and bounds. According to the same study, purchasing behavior has also changed in leaps and bounds. Shoppers all over the world are embracing digital communications - consumers have abruptly shifted to online channels. The study also showed that the degree of digitalization of the supply in different industries is different. Manufacturing businesses have a relatively low percentage of digital products, while healthcare and pharmaceuticals, financial services, and the like, companies report nearly double the growth of companies like CPG (consumer packaging goods).

The SARS-CoV-2 pandemic and the measures taken by governments in many countries to contain the infection have accelerated the growth trend of e-commerce and prompted many consumers to shop online for the first time. For example, based on an analysis of 1.4 billion banking transactions in Spain since 2019, online purchases are up 22%, up 7% from before lockdown. Opinions vary as to the sustainability of this growth following the lifting of the stringent restrictive measures, however, e-commerce can be expected to maintain its momentum, given the trend observed before the pandemic. [8].

Recently, the term "mobile commerce" has appeared. It can be defined as digital trading using applications on mobile devices (smartphones and tablets). The availability of mobile devices is a key driving force in the development of the digital economy. In developed countries in Europe and the United States, about 70% of B2C e-commerce transactions take place through mobile smartphones. China and India are the world leaders in the size of the smartphone market in the world, surpassing the United States in 2019. Smartphones and social media are indispensable elements of e-commerce development, and the availability of 4G / 5G services and low-cost tariff plans accelerates the growth of e-commerce.

The fastest growing segment of the Internet is the number of users of mobile social networks such as WhatsApp, Telegram, Twitter, Facebook, WeChat. The integration of social media and e-commerce forms social commerce. The growth rate of social commerce is expected to accelerate over the next 5 years, and it will occupy over a quarter of the e-commerce market. For example, in India alone there are about 310 million active users of social networks, including 59.6 million unique Facebook users and 200 million. active WhatsApp users per month.

The number of Internet users in Russia totals almost 97 million. people with a penetration index of 78%. The audience of the mobile Internet in the Russian Federation totals 85.2 mln. users, the number of smartphone holders in 2020 reached 106.23 million. people [9]. The share of e-commerce in GDP is 1.3%, and the market volume reached 30.6 billion. dollars in 2019. Online sales to the end consumer (B2C) amounted to 24.9 billion. dollars, which corresponds to an increase of 39% in orders and 24% in money turnover [10]. 30% of digital purchases are made using mobile devices ($ 9.2 billion) [11]. The online market in Russia last year amounted to 23 billion rubles, while the entire food market amounted to 15 trillion. rubles. Expected in 2021 about 40 trillion rubles.

Technologies for digital marketing, electronic data exchange and automated collection of such data, processing of online transactions, logistics management systems are essential conditions for the growth of e-commerce. The focus should be on the aspect of effective innovation management. Automation of the processes of selling goods and services is becoming a tool for attracting customers. Ecosystems are being actively developed to provide basic needs in one place. Also, services are linked in time and, using artificial intelligence systems, the problem of reducing the choice is solved.

Ecosystems can be classified as global trends in digital commerce. Sberbank, with its recent rebranding into Sberbank, VK and Yandex, is a good example of the formation of a successful brand ecosystem. Sberbank and Yandex combine such components as virtual voice assistants Salyut and Alisa, respectively, streaming services Okko and Kinopoisk, music and podcast service SberSvuk and Yandex Music and complex B2B solutions. Both brands offer their users a comprehensive subscription to all digital services, including music and video services, delivery of food and goods from the store, taxi and delivery services, car sharing and the Dark Store format (only from Yandex). The latter service belongs to the food delivery market, which will drive e-commerce in 2020 and 2021, with projected growth rates of 80% and 50%, respectively [12]. VK is the most popular social network in Russia (market share - 30.3%). Created within the brand, the VK ecosystem is also integrated with many companies providing various services and services. It is worth noting that, by analogy with WeChat in China, the company began transforming its mobile application into Super-App, which already contains many mini-versions of its services (Yula, Citymobil) and third-party applications (AliExpress Russia).

The Yula C2C regional platform is also part of the VK ecosystem. The total number of sellers on C2C channels is 13.9 million. people, the number of sellers - 11.8 million. The total market volume in 2019 reached 568 billion rubles from 117 million rubles. sales. The leader of the private trading market is the Avito platform, where 2/3 of all transactions are made. In the first half of 2019 alone, logistics companies delivered 218 million orders made in online stores by customers. The most common way to pay for an online order is with a debit card (53.3%); 65% upon advance payment, 47% upon receipt. The most common types of delivery are self-pickup (69%), courier delivery (40%) and parcel terminals (20%). The volume of the market for cross-border online purchases in 2019 amounted to 5 billion dollars (1% of retail sales in the country), which is the 6th indicator in the world [13].

As you can see from these examples, digital technologies collected in customer service ecosystems are the main trend in the development of communication between companies and customers. Leading companies are developing business management models that include elements of internal organization, production, service and communication services to customers.

4 Conclusion

The Covid-19 pandemic has shifted the focus to finding the best ways to manage an organization and communicate with customers using digital technologies. E-commerce is, in fact, becoming a new form of communication between companies and their customers.

Global brands are going through the process of digital transformation in a standard mode, using technologies and innovations of the digital economy for this. The development of the digital economy is expressed in the increased use of such innovations as Data Analytics, Artificial Intelligence, which have become widespread in voice assistants. They, in turn, are often used in conjunction with the Internet of Things and the Robotization process, as well as the practical use of drones and virtual reality (VR). Digital platforms are developing, such as "Facebook", "Amazon", "AliBaba", operating in the framework of e-commerce.

From the point of view of optimizing the management processes of companies, digital technologies make it possible to reduce operating and overhead costs, reduce costs, increase manageability, speed up the solution of R&D tasks, and simplify business diversification. The complexity of the implementation of digital technologies in the management process is also associated with the increased importance of control over the activities of personnel and their communication with customers. Formalization and individualization of decision making is increasing. The disunity of the team, which works remotely, leads to emotional burnout of employees. And this becomes a problem for the management of the company.

New technologies make it possible to provide the client with new opportunities in solving his problems. Here we can speak about solving problems, not about a "simple" purchase. From the point of view of customer service, digital technologies increase the security and agility of processes, inspire confidence in automated algorithms (decreasing so-called "human factor"), individualization (segmentation) of offers increases, new business models are developed (example of CRM). In conclusion, I would like to emphasize that with all the positive aspects of the introduction of digital technologies, it is not yet entirely clear how this will affect the social life of a person and society.

References

1. J. M. Amis, R. Greenwood, J. of Management Study
2. N. Sarah, *5 business strategies for success during the pandemic*
3. *Leading and Managing a Company During the COVID-19 Pandemic* (Walden University, 2021)
4. *Digital technologies in Russian companies. Results of the study. Survey* (2019)
5. Research on digitalization and innovative openness of Russian enterprises. Analytical report. KPMG 2020 (2020)
6. KPMG – media portal, https://mustread.kpmg.ru/
7. *The Fourth Industrial Revolution Targets the development of industrial technologies and innovations* (2019)
8. *Effects of e-commerce on prices and business competition* (Bancode Espana, 2020)
9. *Ranking of the number of smartphone users in Europe by country 2020* (2020)
10. *E-commerce in Russia Data Insight* (2019)
11. *The Growth of Russia's E-commerce Market. Data Insight* (2020)
12. *Results of the year: RAEC data* (2019)
13. *Russian e-commerce market: results of 2019, trends of 2020* (HSE, 2020)
14. *How COVID-19 has pushed companies over the technology tipping point—and transformed business forever. Survey* (McKinsey & Company, 2020)
15. *KPMG media portal. Technologies, Interviews*

Sentiment analysis as a tool for assessing the negative impact of tourism on a destination

Aleksandr Vetitnev1, *Dmitriy* Chigarev[1,*], and *Vladlena* Enushevskaya[2]

[1]Faculty of Economics and Management Processes, Sochi State University, 354000 Sochi, Russian Federation
[2]Environmental Engineering Faculty, Sochi State University, 354000 Sochi, Russian Federation

Abstract. This article presents the deliverables of the analysis of the experiences of the visitors to the Sochi resort, which is one of the criteria for assessment of the risk of problems caused by excessive tourism pressure. The authors highlight the influence of perception and experiences of the visitors and their informal web communication on the creation of the area's image. The visitors' reviews of the resort on TripAdvisor were studied using sentiment analysis. The article outlines the essence of the sentiment analysis method, the reasons for its choice to study the visitors' attitude towards the Sochi resort, and the relevance of determining the experiences of its visitors. The authors describe the process of studying 33,475 city visitors reviews in 2011-2021 of 58 Sochi tourist attractions and makes conclusions about the visitors' attitude towards the city, the assessment of its attractions, areas, the level of satisfaction of the tourists from various regions with their vacation. The authors underline predominance of positive reviews and confirms broader possibilities of the sentiment analysis method in comparison with traditional methods of studying public opinion and the possibility of obtaining an objective view of the area image with this method.

1 Introduction

The effectiveness of the activities of regional governments accurately determines the satisfaction of the population with the quality of life.

In the last decade, the experts' attention has been attracted by the problem of the deterioration of the quality of life of the local population of the most demanded destinations, caused by the excessive tourism pressure, which in 2017-2018 even spawned a wave of protests against tourists in Europe [3, 11]. For destination management organizations, especially those that "suffer from their attractiveness", planning issues - scientifically based foresight and forecasting of tourism development are of paramount importance.

Many academic publications have been devoted to the study of overtourism; in the topic of congestion of the areas demanded by visitors, which has been discussed in the tourism discourse for more than 50 years, the emphasis is shifting from a quantitative approach to a

*Corresponding author: dvchigarev@gmail.com

psychological component, perceived pressure, moreover, perceived not only by the locals but as well by the tourists [6].

Perception as a subjective psychological aspect of the experience is the commencement of the creation of the area's image in the eyes of visitors, its assessment, which can be used, inter alia, when forecasting the development of this specific destination, taking into account its tourist overload. Informal travelers communication, their comments, reviews and messages on the Internet play an important role in shaping the destination image.

Originated in the early 2000s, the analysis of the text sentiment (sentiment analysis), which belongs to the group of content analysis methods in computational linguistics, is one of the major methods of identifying the emotional attitude of the text author towards the object under study. It is a method for quantification of qualitative data; its purpose is the automated extraction of emotional content from the text and determination of the polarity of emotional assessments in the text: positive, negative, and neutral [2]. And, since the experience is a sensual emotional reflection of reality, it is practicable to study the tourists reviews using sentiment analysis of these reviews, which allows to better understand the opinions of visitors about leisure facilities.

In Russia, domestic tourism is becoming increasingly sought after. The leading positions of the Sochi resort, the increasing number of tourist arrivals at this destination, their sharp growth after the lifting of quarantine restrictions in 2020 which is observed already now, in 2021 may cause a number of problems in the city that would adversely affect the quality of life of the locals and the quality of the experience of visitors [21, 22]. Therefore, we believe that the issue of studying the Sochi guests experience through sentiment analysis of their reviews on the Internet, which allows to understand the emotional attitude of the authors of messages towards the resort city, is currently quite relevant.

The purpose of this work is to assess the experiences of the visitors to the Sochi resort destination; its task is to test the sentiment analysis method in determining their attitude towards the city.

2 A Literature Review

In the model proposed by Therkelsen & Halkier [20], the perception of people and areas is based, firstly, on the dichotomy of "we" and "they, others" and, secondly, on the fact that the attitude towards the area can be both positive and negative. Based on this, four types of an area's image are possible (which in one of his definitions Philip L. Pearce [12] considers the perception and experience of a site), namely – "friendly", "exotic", "boring", and "hostile". The first two types shape a positive attitude towards the area, while the second create a negative one.

We believe that the Sochi resort belongs to the areas with an exotic image, full of allure and appeal; the attitude towards such sites is most likely positive. In order to confirm or deny the positive attitude of tourists towards Sochi, we have studied their experiences, which they share in their web reviews of the resort.

In the academic community in recent years, the analysis of reviews on the Internet is becoming increasingly more popular. It helps determine travelers satisfaction and a destination image [5, 9, 10, 19], segmentation of the tourist market [1, 8, 13], the assessment of the quality of tourist services [4, 18, 24], the evaluation of the attitude of the locals and guests of host destinations towards mega-events [7, 16].

TripAdvisor, the world's largest travel website, available in 49 countries in 28 languages, was launched in 2000 and immediately became the most popular social network. This platform is visited monthly by 463 million current and potential travelers. The service contains over 884 million reviews and opinions about tourist attractions ("Things to Do"), which number has the largest share in comparison with other websites. Therefore, this platform was selected to study the tourists reviews of Sochi.

As already noted, the method of sentiment analysis of the text allows us to reveal the emotional attitude of the author of the text to the object in question. The Sochi tourist attractions have become such an object in our study. According to Sazykin & Glushko [15], a tourist attraction is "a tourist and recreational resource localized in space, having a structure, attractiveness, hierarchy, informational support, involved in tourist production activities", the main quality of which is attractiveness – the ability to attract the attention of tourists.

3 Materials and Methods

The sentiment analysis of the reviews of visitors to the Sochi resort destination was carried out using the Python programming language. The data was saved in the .csv format. 58 CSV files were prepared for analysis; then, they were encoded for subsequent statistical processing in SPSS (IBM SPSS Statistics v.21). After all the reviews had been collected, a sentiment score was calculated for each review.

Reviews can have negative, neutral and positive score. The summary index ("compound") is an integrated assessment of the first three scores, which varies from -1 to 1. In order to determine the sentiment of each review, we set the threshold of the compound score as ± 0.2. If the compound score of the review is more than 0.2, then the review is positive; if less than 0.2 - negative; if from -0.2 to 0.2 - neutral [23].

4 Results

This study covered 33,475 reviews of visitors to the Sochi resort destination in 2011–2021 published on TripAdvisor. The perception of the whole always depends on the perception of the parts; therefore, the empirical basis of this study was the tourists' reviews of 58 Sochi sights (attractions). Of them, two sights received from 90 to 100% of the total number of positive reviews, 35 – from 80 to 90%, 15 – from 70 to 80%, 5 – from 60 to 70%, and only one – 42%. In total, there were 80% positive reviews, 10% negative, 10% neutral, and 0.576 compound ones.

The attractions under consideration were grouped by category and city districts; the reviewers were segmented by their place of residence. The number of reviews in each group and their share of the total number of reviews are presented in Table 1:

Table 1. Segmentation of 58 Sochi attractions by category and district. Segmentation of reviewers by their place of residence

Segmentation of attractions by category and district	**No. of reviews**	**%**
By category		
Alpine ski resorts	5,990	17.9
Nature and man-made attractions	6,405	19.1
Theme amusement parks	7,340	21.9
Culture venues, embankments	4,215	12.6
Sports facilities, infrastructure	5,310	15.9
Retail facilities, shopping malls, casinos	1,395	4.2
Water parks, dolphinariums, aquariums	2,560	7.6
Other	260	0.8

Table 1. Continued

By district		
Krasnaya Polyana	10,445	31.2
Adler	13,285	39.7
Khosta	5,260	15.7
Central	3,480	10.4
Lazarevsky	1,005	3.0
Segmentation of reviewers by their place of residence	**No. of reviews**	**%**
Central Federal District	3,135	11.8
Northwestern Federal District	2,562	9.7
Southern Federal District	4,711	17.8
North Caucasian Federal District	362	1.4
Volga Federal District	2,190	8.3
Ural federal district	1,363	5.2
Siberian Federal District	1,140	4.3
Far Eastern Federal District	649	2.5
Foreign state	1283	4.8
Moscow	9068	34.3

Based on the number of the tourists' reviews, grouped by thematic criterion, the leading attractions are theme amusement parks – 21.9%, followed (in popularity) by natural and man-made attractions – 19.1% and ski resorts – 17.9%. Visitors to the city contributed the largest number of reviews of the attractions of the Adler district – 39.7%, Krasnaya Polyana – 31.2%, and Khosta district – 15.7%. One third of all comments was made by Moscow residents (34.3%), one fifth – by the residents of the Southern Federal District (17.8%), and one eighth – by the residents of the Central Federal District (11.8%).

Based on this segmentation of reviews, we've analyzed the experiences of the visitors to the Sochi resort destination by two indicators: the overall index (compound) and the negative connotation (Tables 2, 3, and 4).

Table 2. Deliverables of the sentiment analysis of reviews of 58 Sochi attractions, grouped by category

Category	**Compound score**			**Negative connotation**		
	No.	**Average**	**Standard deviation**	**No.**	**Average**	**Standard deviation**
Alpine ski resorts	5,824	0.637*	0.402	2,211	0.075	0.050
Nature and man-made attractions	6,233	0.548	0.465	3,169	0.082	0.055
Theme amusement parks	7,148	0.593	0.429	3,250	0.075	0.048
Culture venues, embankments	4,063	0.602	0.407	1,571	0.077	0.049
Sports facilities, infrastructure	5,104	0.599	0.415	1,915	0.081	0.054
Retail facilities, shopping malls, casinos	1,353	0.504	0.483	632	0.084	0.060
Water parks, dolphinariums, aquariums	2,482	0.406*	0.566	1,329	0.093	0.061
Other	248	0.531	0.467	118	0.076	0.047
TOTAL	32,455	0.576	0.445	14,195	0.080	0.053

Notes: One-way ANOVA test: $F = 83.184$, $p = 0.000$; *- $p < 0.05$ using Games-Howell test

The data in the table show that the highest summary index (compound score – 0.637) and the lowest negative connotation (0.075) were received by alpine ski resorts. The group, which included water parks, dolphinariums and aquariums, received the lowest compound score, 0.406, and the largest negative connotation, 0.093.

The deliverables of the sentiment analysis of the reviews by the city districts showed (Table 3) that the reviewers were most satisfied with Krasnaya Polyana (compound score – 0.608) and Adler district (compound score – 0.575); least of all were they satisfied with the Lazarevsky district (compound score – 0.372), which also demonstrated stronger negative connotation compared to Krasnaya Polyana and Adler district (0.106 and 0.078, respectively).

Table 3. Deliverables of sentiment analysis of reviews of 58 Sochi attractions, grouped by the city districts

Group	Compound score			Negative connotation		
	No.	Average	Standard deviation	No.	Average	Standard deviation
Krasnaya Polyana	10,154	0.608*	0.427	4,027	0.078	0.051
Adler	12,834	0.575	0.437	5,637	0.078	0.051
Khosta	5,139	0.562	0.453	2,532	0.078	0.052
Central	3,353	0.563	0.447	1,464	0.086	0.055
Lazarevsky	975	0.372*	0.601	535	0.106	0.067
TOTAL	32,455	0.576	0.445	14,195	0.080	0.053

Notes: One-way ANOVA test: F = 66,952, p = 0.000; *- p < 0.05 using Games-Howell test

The greatest satisfaction with the Sochi resort (Table 4) was noted by residents of the Southern Federal District (compound score – 0.612), the least – by residents of the North Caucasian Federal District (compound score – 0.548); the latter showed the highest negative connotation (0.082). The reviews of tourists from the Ural Federal District contain the least negative assessments (0.073).

Table 4. Deliverables of the analysis of the satisfaction of visitors to the Sochi resort, residing in different regions of Russia and the world

Reviewer's place of residence	Compound score			Negative connotation		
	No.	Averag e	Standard deviation	No.	Avera ge	Standard deviation
Central Federal District	3,035	0.591	0.430	1,312	0.076	0.049
Northwestern Federal District	2,474	0.556	0.453	1,118	0.080	0.052
Southern Federal District	4,582	0.612*	0.426	1,871	0.079	0.052
North Caucasian Federal District	353	0.548*	0.460	158	0.082	0.058
Volga Federal District	2,116	0.582	0.427	919	0.078	0.050
Ural federal district	1,325	0.575	0.444	590	0.073	0.046
Siberian Federal District	1,093	0.590	0.412	449	0.079	0.052
Far Eastern Federal District	623	0.596	0.423	272	0.076	0.044
Foreign state	1,246	0.568	0.444	554	0.081	0.055
Moscow	8,787	0.580	0.438	3,835	0.079	0.052
TOTAL	25,63 4	0.584	0.435	11,07 8	0.078	0.051

Notes: One-way ANOVA test: F = 66,952, p = 0.000; *- p < 0.05 using Games-Howell test

After the 2014 Winter Olympic Games, Sochi has become very popular among Russian travelers. The number of TripAdvisor reviews of the resort has skyrocketed with the creation of the pages dedicated to the Olympic venues. This is also confirmed in Figure 1 (the Y-axis indicates the number of the tourists' reviews). The figure shows that the number of reviews in 2015–2017 has increased significantly compared to 2012–2014, which positively correlates with the growth rates of tourist arrivals during this period: in 2014, the city was visited by 4.6 million people, in 2015 – 5 million, in 2016 – 5,2 million, and in 2017 – 6 million [17].

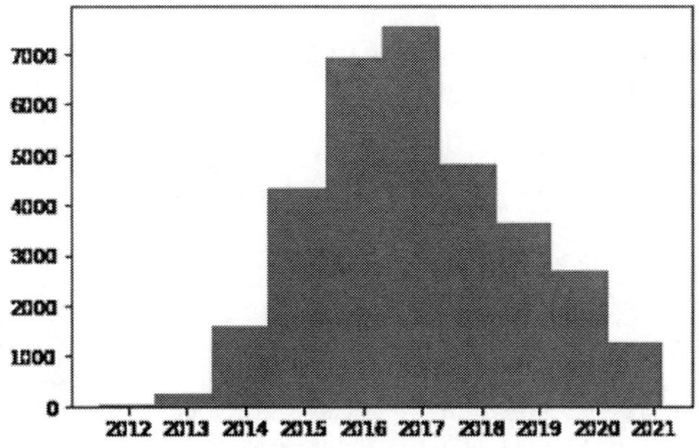

Fig. 1. Distribution of the reviews from 2012 to 2021

In order to visualize the most common words in 33,475 tourist reviews, we've created a cloud using the wordcloud package (Fig. 2). The almost complete absence of negative words confirms the positive attitude of the reviews authors.

Fig. 2. Word Cloud

5 Discussion

Sentiment analysis of the reviews of visitors to Sochi showed their positive attitude towards the city. The tourists rated alpine ski resorts, as well as cultural venues and embankments better than all other attractions (compound score 0.637 and 0.602, respectively). Resort guests were less satisfied with water parks, dolphinariums, and aquariums (compound score – 0.406).

Although the Lazarevsky district is the largest in the city (79.59 km^2) and, for example, it exceeds the area of the Adler district together with Krasnaya Polyana (35.25 km^2) more than twice, visitors contributed the least number of reviews of this district (1,005 and 23,730, respectively) which proves lesser popularity of the district, which is confirmed by its lower compound score (0.372) compared to the Adler district (0.575) and Krasnaya Polyana (0.608) and higher negative connotation in the reviews of the attractions of the Lazarevsky region (0.106) vs. 0.078 for the Adler district and Krasnaya Polyana.

The XXII Olympic Winter Games boosted interest in the Sochi resort, which caused an increase in the number of tourist arrivals (Vetitnev et. al., 2021). Figure 1 shows a skyrocketed number of tourists' reviews of Sochi after 2014 compared to, for example, pre-Olympic 2012. A joint study by the social media reputation agency Digital Guru and Brand Analytics [14] also showed that the number of positive messages about winter vacations in Krasnaya Polyana increased significantly.

In the 33,475 reviews of Sochi sights studied, words with positive connotation prevailed (Figure 2). The tourists' reviews of the resort city, which have positive connotation, encourage potential travelers to visit the Sochi resort destination.

6 Conclusion

For tourism researchers, the Internet, as a global network that allows its users to exchange information, serves as the most important and objective source of information.

Our study confirmed the possibility to get a general idea of the Sochi resort image using sentiment analysis of the visitors' reviews: the positive image of the city is characterized by the prevalence of positive reviews and comments of the tourists over negative ones. The attractiveness of Sochi as an area with a positive image encourages excessive tourism, which may cause certain negative consequences for the Sochi resort destination. This conclusion, drawn, in turn, on the base of sentiment analysis of the visitors reviews can be considered the practical relevance of this study, and the method of sentiment analysis is a tool for assessing the negative impact of tourism on the destination.

The practical significance of this work also lies in the fact that it has shown the efficiency of sentiment analysis of the texts in comparison with sociological surveys and other traditional methods of studying public opinion. Besides considerable saving in time, money, and labor resources, the advantage of this method was manifested in the speed of the study, the possibility of processing a large number of reviews, and, most importantly, in determining their emotional substance.

However, the following limitations should be noted: the general sentiment of the review could be influenced by certain factors that distort the meaning of the text, such as, for example, incorrect translation of some words, the impossibility of recognizing ironic and sarcastic statements, the presence of many negatives in the text, etc. All this could change the statement's sentiment to the opposite. The use of just one website, TripAdvisor, has significantly reduced the scope of the study which could have been expanded if using other travel websites and social networks.

In the future, in order to determine the main reasons for the attractiveness of tourist areas, it is recommended to apply other methods, for example, topic modeling. It is also possible to expand the empirical base of the study with such TripAdvisor items as hotels, restaurants, and entertainment venues.

Therefore, sentiment analysis of tourist reviews could become both an alternative and a supplement to the study of the experiences of tourist destinations visitors.

References

1. A. Ahani, M. Nilashi, O. Ibrahim, L. Sanzogni, S. Weaven, International Journal of Hospitality Management, **80**, 52-77 (2019).
2. C. Bhadane, H. Dalal, H. Doshi, Procedia Computer Science, **45**, 808-814 (2015).
3. I. Egresi, Proceedings of the TAW2018 International Scientific Conference, Tiranë, Albania, from 20th to 23rd September 2018 (Albania: POLIS University, 2018), 701-714.

4. M. Geetha, P. Singha, S. Sinha, Tourism Management, **61**, 43-54 (2017).
5. M. González-Rodríguez, R. Martinez-Torres, S.L. Toral, International Journal of Contemporary Hospitality Management, **28**, 2609-2627 (2016).
6. V. Gonzalez, L. Coromina, N. Galí, Tourism Review, **73**, 277-296 (2018).
7. A. P. Kirilenko, S. O. Stepchenkova, Tourism Management, **63**, 54-65 (2017).
8. A. P. Kirilenko, S. O. Stepchenkova, J. M. Hernandez, Tourism Management, **72**, 400-410 (2019).
9. S. Kladou, E. Mavragani, Journal of Destination Marketing & Management (2015).
10. E. Marine-Roig, Sustainability, **9**, 1425 (2017).
11. M. Martins, Journal of Tourism, Heritage & Services Marketing, **4(1)**, 3-7 (2018).
12. P. L. Pearce, Bristol, Blue Ridge Summit: Channel View Publications (2005).
13. J. A. Pesonen, A. Tuohino, Journal of Vacation Marketing, **23(2)**, 145–158 (2017).
14. RBC, Nice, but ... What people like and dislike about Krasnaya Polyana (2018) https://kuban.rbc.ru/
15. A. M. Sazykin, A. A. Glushko, Service and Tourism: Current Challenges, **11(2)**, 7–16 (2017).
16. A. Seilsepour, R. Ravanmehr, H. Sima, Journal of Advances in Computer Engineering and Technology, **5(3)**, 143-160 (2019).
17. Sochi Municipal Administration, Socio-economic development of the city of Sochi (2021) https://sochi.ru/
18. Y. Su, W. Teng, Tourism Management, **69**, 214-222 (2018).
19. L. G. Tamajón, G. C. Valiente, Current Issues in Tourism (2015).
20. A. Therkelsen, H. Halkier, *Umbrella Place Branding. A Study of Friendly Exoticism and Exotic Friendliness in Coordinated National Tourism and Investment Promotion*, SPIRIT (2004).
21. A. M. Vetitnev, D. V. Chigarev, Journal of Environmental Management and Tourism, **12(1)**, 218-232 (2021).
22. A. M. Vetitnev, D. V. Chigarev, O. E. Afanasiev, European Journal of Tourism Research, **28**, 2812 (2021).
23. J. Wu, Sentiment Analysis for Hotel Reviews (2019) https://www.octoparse.com/
24. Y. Yang, S. Park, X. Hu, Tourism Management, **67**, 248-260 (2018).

Phenomena of digital divide and ways of its bridging

K. I. Tkachenko[1], *A. L.* Zolkin[2,*], *D. E.* Lomakin[5], *A. I.* Pakhomova[4], and *V. V.* Pomazanov[3]

[1]St. Petersburg State University, St. Petersburg, Russia
[2]Povolzhskiy State University of Telecommunications and Informatics, Samara, Russia
[3]Kuban State Agrarian University named after I.T. Trubilin, Krasnodar, Russia
[4]Admiral Ushakov Maritime State University", Novorossiysk, Russia
[5]Orel State University named after I.S. Turgenev, Orel, Russia

Abstract. This article is concerned with the research of digital divide challenge and its social and economic impacts. Authors review the formation of definitions and their development. Because of research, the views of famous scientists on divide, its main aspects and role in social and economic relation system are mentioned; definition of “digital divide” is reviewed; author typologies of digital divide are reviewed; five main areas of digital divide concept are highlighted; its reasons and impacts are defined. Phenomena of digital divide started to be researched extensively at the end of XX century; the following concepts can be highlighted from the main ones: total access availability, capacity development, economical change, digital public goods development, global digital cooperation. Therefore, digital divide is the newest type of divide, which is still in theoretical justification infancy; there is still no universal definition of this phenomena.

1 Introduction

Onrush of information and communication technologies (hereafter ICT) and improvement of information role in people’s life result in significant changes in social and economic public life becoming the source not only for new possibilities but also for new challenges. Implementation process of ICT in public life is accelerator of new forms of divide, which separate people on those who have skills and possibilities to use technical tools and on those who do not. Digital divide is increased not only between the states but also between cities and countries, the haves and have-nots, the youth and the aged, disabled and healthy people. Lack of access to ICT result in malfunction of educational, healthcare and other cultural and social institutions. Modern states review the building of digital society as the base of its social and economic, political and cultural development, providing the well-directed national policy in this area.

* Corresponding author: alzolkin@list.ru

2 Problem statement

In 2020, COVID-19 identifies the enormous divide in our society, which is in forced transformation. Over the past twelve months the significant restrictions in such areas like our communication, work, travelling, education and value creation had a major impact on our life. At the same time the technologies often came to help to mitigate or get around the restrictions. Remote work, teleconferences and increasing reliance on telehealth, remote education and e-commerce became the part of "new normal «which is still changed every day. From the beginning of coronavirus attack the digital divide start to improve from one side but it negatively effects on social life from other one. Lack of possibility to perform responsibilities remotely and on quality level and loss of income, decreasing of educational quality (and access difficulty to education for part of population) and other results including development of digital divide phenomena are the negative results of digital divide. State, society and business structures consolidated their powers for providing the possibility to use the digital devices and Internet technologies by people. The number of cloud services, services for remote work, teleconference, on-line meetings and webinar was significantly increased with moving to remote education and work was significantly increased but not all challenges were solved. Experts of Pro education Institute during express monitoring of remote education during pandemic identified that main claims are directed to technical failures of network, technical non-availability of Internet resources and lack of digital skills. So more than quarter of percent of claims were directed to the lack of digital behaviours of teachers/ lecturers. More than thirteen percent of claims were directed to availability of part of lessons on educational portals at a fee. Moreover, every ten claim was directed to lack of necessary equipment for on-line education. As a result, analysis of divide fact, its results and ways of its negotiation is the part of the actual challenges today.

3 Research questions

Urgency of scientific challenge associates with lack of knowledge of digital divide study in the context of economical processes as early the digital divide was studied only as sociological and partially psychological challenge. The challenge of digital divide has special interest for Russia: the average value of population digital literacy is 7.25 points from total possible, gap in educational levels is specified in generation measurement therewith the low levels of digital literacy are specified in 60 age and up audience - 6.80 points.

The following scientists put special attention on study of digital divide challenge: B. Acharya, P. Gilster, P. Himanen, M. Castells, D. Tapscott, D. Lining, G. Y. Soldatova, E. L. Vartanova, E.V. Morozova, N. V. Plotichkina, I.V. Miroshnichenko, N. A. Ryabchenko and others.

P. Gilster was among the first to develop the digital divide challenge. He for the first time presented the "Digital divide" term ans his work "Digital literacy" became the first world monograph dedicated to digital literacy and digital divide. He gave attention to skill to understand and use the information represented in various formats and wide range of sources with computers and Internet.

P. Himanen and M. Castell referred to study of "digital divide" phenomena in work "Information society and state of prosperity. Finnish model". Authors developed the theory of state information gap, which based on information economic development. Information economic connects to its network those who is of importance but disconnects those who is not of importance (therewith even more reducing their possibility to get it).

In fundamental work "Media age: economic, society and culture" M. Castell defined the unequal access to Internet as "digital gap" from sociological point of view. He emphasised that technological revolution covered the whole planet and united the world through information technologies.

4 Materials and methods

The following research tasks should be covered:
1. To study the digital divide phenomena and define its social and economic results;
2. To develop and justify the set of associated measures of government and business on reducing the digital divide in Russia.

Mutually conditioned challenges related to restriction of categorising and timing frames of research are occurred in this work with detailed study of digital divide. During study of digital divide it is required to address to the beginnings of divide origin in society, when the development of definitions was found in early works of Ancient philosopher-thinkers who endeavoured to sensemaking the dividing in society. Therewith the rich research history effects on discipline differences in study of divide as well as on terminological variety of analysed object. That is exactly why the challenges of divide should be reviewed in the light of history.

5 Results

There were endeavours to dispute this challenge because of its fundamental nature and ambiguity. Therefore, Ancient Greek philosopher Plato associated the problem of divide with values of freedom and justice. He identified the natural causes of people divide associated with difference in soul types established by nature and developed in process of ability training [1]. Partly, Aristotle, follower of Plato, takes the same position. He reviewed the divide as the natural conditions of people depends on their ability to act rationally, manage their fates and be responsible for their lives therefore "one men are naturally free, the others are slaves, and for last ones to be slaves is usefully and justly" [2].

New surge of interest to challenge of divide was in Modern Age when the classic political ideologies were arisen: liberalism, conservatism and socialism.

The liberals review the challenge of divide through light of freedom. They think that total elimination of divide in society is impossible; the endeavours of elimination of divide are disastrous. As English philosopher, John Locke thought people subjection to any state power is possible only on the base of mutual agreement of free, equal and independent personalities [3], i.e. on the base of social contract. Due to this, any political system assumed the violation of equality of people rights and freedoms, discrimination of one group at expense of others must be rejected.

One of the famous concepts of social divide affected on practice of digital divide diagnostic was the Marxism. Therefore, in Karl Marx's social stratification theory the main attention is placed on economic aspects, and it is clear that this opportunity (and its implications) is more brightly performed in digital divide research area. Marx described in classic way the property as the base of class differences but social stratification is unavoidable associated with economy class. Digital divide can be reviewed from this point of view as the way by which the social divide is used in digital era with application of new communication technologies.

Social stratification under Max Weber's approach includes three independent factors each of which has its own structure and therefore plays role in social ranging formation: economy class, social status and political power. The interaction between three stratification aspects corresponds to way of social structure development. Authors partly follow the Weber's views in modern representation: each of these elements might be a digital divide because access to new communication technologies, digital skills, computer literacy and ability to receive the profit in network facilitate the growth of political power, social status and economy impact. Due to this approach the digital divide creates the social divide in network society because it effects in social status, increasing this for those who able to use the new communication technologies, master new specialities or skills and increase the intelligent abilities Internet is powerful tool of social status support and such conclusions can be immediately identified in Weber's tradition. For example, direct connection is occurred between education (status marker and economy impact) and ability to transform the knowledge (by digital fluency) to social, economic or political impact.

The UN notes that need of reducing the digital divide is "life-or-death issue" for achievement of stable development. According to the UN's general-secretary, we still cannot totally understand what will be the social results of pandemic for world "after COVID-19". "One thing is clear: in process of recovery the digital technologies will be important more than ever before" [4], - Guterres has declared.

In the early 1990s, the "digital divide" term started to review in scientific paradigm based on knowledge gap theory and describing the digital divide through political, economic and social restrictions of access to Internet.

For the first time digital divide as a challenge was declared in report of Administration on national telecommunications and information of the US government in 1995 [5].

In Russia "digital divide" definition arisen in the early of XX century, the international seminar "Challenges of "digital divide" negotiation in Russia and CIS" was in House of Government in Novemeber, 2000. Still there is no unified definitions of abroad "digital divide" in Russian practice.

"Digital divide" term was the reason of World Summit of the UN on informational society (WSIS) in Geneva in 2003 and in Tunis in 2005. This term (and "digital gap") was widely accepted by representatives of civil society and in official documents.

Not only differences between current network access and deleted from digital-environment but also through practising the on-line-cooperation are meant under digital gap. For more complete characterization of reviewed challenge the types, mechanisms and theoretical models of digital divide were studied. Currently the generally accepted structure and model of digital divide are not defined. In addition, the classification of digital divide is based on traditional criteria as well as on conceptual ideas followed by ones or others authors. E. Hargittai proposed the first of such scientific typology.

In 2001, E. Hargittai highlighted two types of digital divide:

—First-order digital divide, i.e. The divide of access/ usage;

—Second-order digital divide, i.e. Differences in specific of Internet using which are became actual on the latest stages of internetization.

It was assumed that the second type of divide is possible on more wildly entry of ICT in people daily life. For our opinion, exactly this gives the base of digital divide study from different knowledge areas - sociology, economic, political science, media environment and marketing [6].

M. Hilbert developed the typology proposed by E. Hargittai; the forms defined by M. Hilbert are not completely described the digital divide conditions in XXI century.

He defines three forms of digital divide:

—Divide in access;

—Divide in ICT usage;

—Divide in Internet impact on behaviour and beliefs [7].

The same typology is actual for Vern Harper; he defines two types of digital divide in his works:

—Digital divide of access;

—Social digital divide which includes motivation barrier, barrier of skills and knowledge and social networks barrier [8].

The problem of scientific researches obtains the highly specialized nature. P. DiMaggio and E. Hargittai highlighted five basic characteristics for population dividing in updated definition of digital divide:

—Availability of technical equipment and Internet connection;

—Access autonomy;

—Skills on ICT's usage;

—Availability of social support in informational technologies capturing and improving the digital literacy level;

—Tasks for informational technologies using.

Native tradition of digital divide conceptualization is mostly affected by discussions in foreign literature.

T.P. Shariphyanov and D.A. Gaynanov propose summary three-level model of digital divide which includes two sublevels of skills associated with mastery of environment and four sublevels of work with content [9].

6 Findings

Aspect of ICT development is reviewed as development driver in report on global development "Digital dividends", 2016 [10]. In response to new technologies, it is easier and more comfortable to communicate and receive the information; free of charge digital products and new forms of leisure time were arisen. Therefore the life of people who do not use the ICT due to different reasons (do not have access or specific skills) become more complicated. We cannot but agree as people excluded from digital environment more often lose a job, ability to participate in process of society management and become politically and economy weak. Two models of informational and digital divide can be highlighted: in interpersonal social area and global economy area (countries, regions and territories).

In report of human development of the UN, 2019, where the divide in human development in XXI century is studied, the experts highlighted two types of divide associated with extension of new technologies: divide in basic abilities and divide of access to advanced ones. Currently it is not enough to have the basic set of abilities of access to ICT because advanced ones become key determinant for self-selection of life path and provide with free actions in life. The number of concluded contracts on mobile services is part of basic abilities, and the number of concluded contracts on fixed wide access to Internet is a part of advanced ones. As far as we come in 2020s the wide set of abilities become the basic for life in XXI century. Divide in these advanced abilities demonstrates starkly other dynamic in compare with divide in basic ones. Their lay in base of new generation of divide aspects [11]. World situation with digital divide is sufficient tragic because if it is reduced in basic ability area but it is increased in advanced ones including Russia.

In May 2020 the General Assembly of the UN developed the roadmap on digital cooperation where the recommendations of High level group on digital cooperation were described [12]. This document is basic world mark, recommendations developed direct to approach the tasks of stable development: the gender gap is reviewed, human rights through modern challenges of digital world are analysed.

There is still no unified definition of "digital divide" term but it is in common view:

"Digital divide" or "digital gap" [13] means the divide of abilities in physical, technical and social access to modern communication techniques, especially Internet.

In the author opinion, the digital divide is cross-cutting issue which effects on all areas of social life.

In modern conditions business-structures are the main factors in digital divide negotiation but it is not possible without government support; therefore, the need of motivation of business-structures from government is existed. Authors propose the development of social business as subject of digital divide. The reason of social business underdevelopment in Russia is still the meaning of people that social problem solving is the responsibility of government. Business in all sections can improve the capacity, income; extend the markets; decrease the prices on operations and more effective control the reserves, specially the small and medium businesses, based on its large scale. Regards to consumer the informational and communication technologies can make benefits in labour policy, improvement of people prosperity and life quality. Business plays huge role in production of digital content and providing the services. Commercial players, from their side, perform the experience of network infrastructure for public programs.

For business-structures interest in minimization of digital divide the development and agitation of social business in Russia are required. If the companies will be motivated with social bonuses from government, they will have more abilities to work in this direction. The social business began to develop only recently in Russia. Nevertheless that this fact has novel character, it already takes place in one line with commercial initiatives, venture philanthropy, corporate social responsibility and welfare work. "Social business activity stands at the nexus of social and commercial sections; it is a kind of hybrid form. Now the social business is developed under uncontrolled conditions. Positions of various agencies and society regards to "social businessmen" term are different. Therefore the conditions of draft legislation is reviewed regularly."

For our opinion, the possible ways for solving the social business problem in Russia can be the following aspects as additional to common increase of "social threshold" and principles of its equal distributive between "centre" and deprived "periphery" what can become the factor of improvement of list and rate of bonuses and preferences for subjects of social business:

1. Development of social business theory;
2. Filling with real content of constitutional position about Russia as social government;
3. Legalization of social business on federal level and support to its development in critical and social important areas of activity. It is assumed to give the status of non-commercial organisation to production cooperatives as well as state enterprises, which allow to increase the prices and rates for population and economic units throughout all technological chain under condition of "open management" and establishment of control from civil society institutions for expenditure of performance results. Economic effect of such transformations is the increase of their competitiveness on internal and external markets;
4. Socialization of important enterprises in critical and social important areas of activity, first of all, in environmental management, housing and utilities infrastructure, MIC by transformation of its organizational legal forms to social business and socialy-owned enterprises.
5. Extension of direct and indirect tools of government, region and territory on politics of social business's subjects including the various bonuses and preferences in frame of indicative and contract planning with directive elements.

7 Conclusion

Implementation process of ICT develops unevenly that creates digital divide. Firstly, digital gap is considered as social and economic challenge. It is expressed in that the ones have technical abilities, materiel resources and necessary educational level for using the ICT but the others do not have access to global network due to the absence of resources on gadget purchasing, low qualification in informational technologies area. This issue is not enough studied, there is no necessary scientific base for this challenge justification. Knowledge deficit blocks the extension of individual's access to ICT.

References

1. Brockhaus and Efron Encyclopedic Dictionary, **31(61)**, 305 (1900)
2. Politics, Aristotle. Writings, **4**, 384 (1983)
3. J. Locke, Locke J. Writings, **3**, 317 (1988)
4. World can not reach the Task of stable development without reducing the digital divide: The UN's News, https://news.un.org/
5. Connecting the Nation: Libraries and Health Care Organizations in the Information Age: National telecommunications and information administration, http:// www.ntia.doc.gov/
6. T. P. Shariphyanov, D. A. Gaynanov, Regional Economics: Theory and Practice, **22**, 2 (2014)
7. T. S. Martynenko, D. E. Dobrinskaya, Monitoring of public opinion: economical and social changes, **1**, 171 (2021)
8. V. Yumashev, T. N. Gorobets, O. I. Admakin, G. G. Kuzminov, I. V. Nefedova, J. of Science and Technology, **9(19)** (2016)
9. Abdryashitova, E. V. Gracheva, M. Yu. Kazakov, *Management tools for the implementation of municipal functions in electronic form* (2018), 251
10. V. Zhukovskyy, N. Zhukovska, A. Vlasyuk, A. Safonyk, 2019 IEEE 2nd Ukraine Conference on Electrical and Computer Engineering, UKRCON 2019 - Proceedings, 2019, 1179 (2019)
11. E. A. Lavrov, P. I. Paderno, A. A. Volosiuk, N. B. Pasko, V. I. Kyzenko, Proceedings of 2019 3rd Int. Conf. on Control in Technical Systems, CTS 2019, 144 (2019)
12. Zolkin, E. A. Domracheva, A. N. Losev, Yu. M. Avdeev, IOP Conference Series: Materials Science and Engineering. Krasnoyarsk Science and Technology City Hall. (2021), 12094
13. V. S. Tormozov, A. L. Zolkin, K. A. Vasilenko, 2020 Int. Multi-Conf. on Industrial Engineering and Modern Technologies (2020), 9271536

Inter-settlement territories, their realities and prospects

Valeria Vylegzhanina[1,*], *Ilgiz* Giniyatov[2], *Nadezhda* Dobrotvorskaya[2,3], *Anastasia* Ilinykh[2], and *Victor* Timonov[4]

[1]Department of the Federal Service for State Registration, Cadastral Records and Cartography for Novosibirsk Region, Novosibirsk, Russia
[2]Siberian State University of Geosystems and Technologies, Novosibirsk, Russia
[3]Siberian Federal Scientific Center of Agrobiotechnology RAS, Novosibirsk region, Russia
[4]Kryachkov Novosibirsk State University of Architecture, Design and Arts (NSUADA), Novosibirsk, Russia

Abstract. The paper presents the results of research and analysis of the state of agricultural land in the Novosibirsk region, located in inter-settlement areas, considered the main reasons contributing to the degradation of agricultural land. The need for the formation of scientific approaches in solving the problems identified by the Government of the Russian Federation within the framework of the Doctrine of Food Security of the Russian Federation, which provides for the development of the agro-industrial complex, which allows providing the population of the country with high-quality, affordable and safe agricultural products, has been substantiated. It is proposed to form "exchange zones" in each region of the country, which should include the scientific community represented by representatives of specialized higher educational institutions with the involvement of production structures, Rosreestr (Russian State Register), regional ministries of agriculture, ensuring the implementation of the state agro-industrial policy. The concept of a system of step-by-step work is proposed, aimed at the rational use of agricultural land in the framework of managing the development of inter-settlement territories.

1 Introduction

The agro-industrial complex (AIC) of the Russian Federation requires close attention not only from the Government of the Russian Federation, legislative and executive authorities, and the economic entities themselves. It also requires a scientifically grounded approach to solving problems and problems that arise and are formed as a result of the ongoing socio-economic processes and phenomena in the agricultural sector [1]. In recent years, there has been a tendency to enlargement of agricultural enterprises, the formation of latifundia, as well as the creation of agro-industrial holdings producing various agricultural products. Large peasant farm enterprises occupy a certain place in the competition [2]. At the same time, the regulatory authorities often face unscrupulous owners of agricultural land,

* Corresponding author: valeria741974@mail.ru

ineffective management, and often their inaction leads to land degradation due to downtime and overgrowing of cultivated areas with bushes and weeds [3].

The purpose of this article is a general analysis of the state of agricultural land on the territory of the Novosibirsk region and, in particular, the study of the impact of the activities of agricultural enterprises on the development of inter-settlement territories. As a result of the research, proposals were prepared for the formation of scientific approaches within the framework of managing the development of inter-settlement territories.

2 Materials and research methods

The object of research is agricultural land located on the territory of the Novosibirsk region, including lands within the Novosibirsk agglomeration, located in the Kochenevsky district of the Novosibirsk region. As a result of the analysis of the state of agricultural lands and the presentation of the research results, the materials of remote sensing of the Earth were used; data of the public cadastral map posted on the Rosreestr website; official information published on the websites of Novosibirskstat (Federal State Statistics Service for the Novosibirsk region), the Ministry of Agriculture of the Novosibirsk Region, the Government of the Novosibirsk Region; information posted on the official website of the administration of the Kochenevsky district in the Novosibirsk region; official information posted on the Rosselkhoznadzor(Federal Service for Veterinary and Phytosanitary Surveillance) website for the Novosibirsk region.

3 Results and discussion

The Doctrine of Food Security of the Russian Federation (Doctrine) [4], which was approved by the President of the Russian Federation in 2020, is aimed at developing and implementing comprehensive measures to provide the country's population with high-quality, affordable and safe agricultural products, which also implies the country's independence in this area. All this once again confirms the need for the rational use of agricultural land, laying this aspect in the tasks of rule-making and the principles of implementing state policy in this area.

The tasks of the Doctrine are, among other things, to provide conceptual scientific approaches that provide for the stage-by-stage development and implementation of complex measures using new modern technologies. At the same time, the optimal performance of work implies, in our opinion, the study and analysis of the state of agricultural land, the quantitative and qualitative characteristics of economic entities, the collection and study of information on the natural, landscape and climatic characteristics of the territories, ground and underground facilities of the existing infrastructure that ensures the economic activity of agro-industrial enterprises. In addition, it is necessary to pay attention to those territories where there was a decrease or termination of agricultural activity in order to assess the economic efficiency of the creation of new enterprises and associations or the revival of lost agricultural activities. These works will allow you to prepare and update all the necessary information for planning and designing the development of territories in general.

Considering the general picture in the agricultural sector, the following can be noted.

The difficult socio-economic situation that has developed in recent years has significantly contributed to the ruin of small peasant farm enterprises (PFEs), the creation of large agricultural holdings and the emergence of latifundia [2]. More or less stable and strong PFEs in the process of carrying out agricultural activities monitor the condition of the lands they cultivate. But the ruined PFEs, as well as the latifundia formed in their place, cause the degradation of lands in this category. As an example we can cite the official information posted on the website of the Rosselkhoznadzor Departments for the Tver and Yaroslavl Regions, according to which the latifundists, who bought up agricultural land located in these regions for a pittance, use no more than 5% of the total land area for their intended purpose. The rest of the cultivated areas are overgrown with shrubs and weeds [3]. On the territory of the Novosibirsk region there are also lands of this category, which are in an unsatisfactory condition. According to the Rosselkhoznadzor Directorate for the Novosibirsk Region, as a result of scheduled inspections of several separate municipal units in the region's districts in 2019, about 29,520 hectares of uncultivated agricultural land were identified, where the overgrowth process is in full swing, and about 152 hectares of land are occupied by unauthorized dumps of solid household waste [5]. According to the information on the results of work of the agro-industrial complex of the Novosibirsk region in 2020, the number of sown areas increased by 46.3 thousand hectares compared to 2019. The increase in acreage is a consequence of the policy pursued in the framework of the Doctrine [4] by the Government of the Novosibirsk Region to support and develop PFEs, including small forms of farming [6]. At the same time, the decline in agricultural activities of personal subsidiary plots continues [6]. This circumstance, in our opinion, is due to the outflow of the local population to larger settlements for the purpose of employment and more comfortable living and, accordingly, the impoverishment of rural areas by the able-bodied population. A detailed analysis of some individual agricultural enterprises and личных PFEs in the territory of the Novosibirsk region shows the following picture. According to a press release from the Rosselkhoznadzor Directorate for the Novosibirsk Region, in September 2019, inspectors of the State Land Supervision Department of the Rosselkhoznadzor Directorate for the Novosibirsk Region carried out a scheduled field inspection in relation to the administration of the Kremlin Village Council of the Kochenevsky District. As a result of the control and supervisory measures, no violations of the land legislation of the Russian Federation were revealed [5]. However, let us turn to public data. According to the public cadastral map of Rosreestr, within the boundaries of the specified village council, there is a land plot with cadastral number 54: 11: 017016: 593, which belongs to agricultural land and has a permitted use - for agricultural production, the area of which is 16845.493 hectares [7]. Until 2017, the farmland was leased by the federal state unitary enterprise FSUE "Kremlevskoye" of the Russian Agricultural Academy, which specialized in seed production and cattle breeding. In 2019, it was liquidated as a result of bankruptcy proceedings [8, 9]. These lands are state property, in 2017 they were leased to a legal entity - "Agrosib" Limited Liability Company, whose main activity is the cultivation of cereals (except for rice), legumes and oilseeds. General analysis of the data indicating the economic activity of the lessee confirmed the stability of the entity in this sector of the economy [8].

For many years of active farming, the considered agricultural lands were systematically plowed without taking into account their agroecological properties [10]. However, the farming system in the Federal State Unitary Enterprise "Kremlin" in the Kochenevsky District of the Novosibirsk Region has always been focused on the preservation and reproduction of soil fertility by sowing perennial grasses in large areas. The relatively high grain yield was ensured by the use, albeit limited, of fertilizers and plant protection products. At the same time, according to remote sensing data, part of the land has been removed from cultivation (Figures 1, 2).

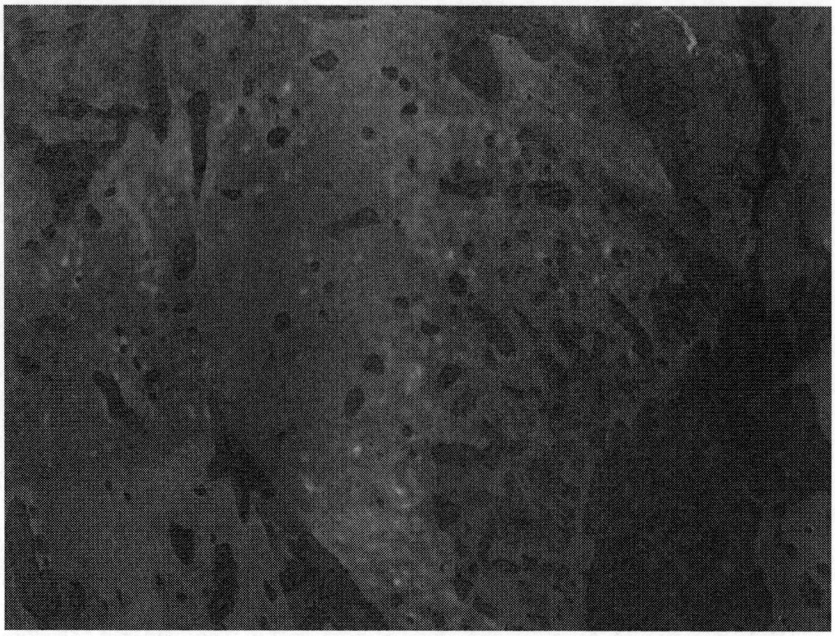

Fig. 1. Untreated arrays of fields on the territory of the Federal State Unitary Enterprise "Kremlin" of the Kochenevsky district of the Novosibirsk region (snapshot of 2021, Google maps program, scale 1:40 000).

Fig. 2. Overgrowth of agricultural land on the territory Pervomaisky settlement of the Kremlin Village Council of the Kochenevsky District Novosibirsk region (snapshot of 2021, Google maps program, scale 1: 5000).

Thus, the problem of agricultural land overgrowth is urgent and there are discrepancies between official statistics and reality. In order to eliminate such inconsistencies, it is proposed to carry out an agricultural microcensus in all regions of the Russian Federation [11]. This work has already begun in August 2021. Its goal is to cover all agricultural organizations and all farms. In addition, a sample survey of personal subsidiary plots is supposed [11]. As the head of Rosstat said, "a census taker will not come to a private household if it is located in the city or in an urban-type settlement, in small villages (where there are less than 10 households), as well as in private households located in remote areas of the country" [12]. At the same time, a survey of hard-to-reach territories can be carried out using satellite monitoring data obtained as a result of overflights of unmanned aerial vehicles [11]. Thus, for the first time, a survey of agricultural land will be carried out using this technology. According to the results of the agricultural microcensus, there will be up-to-date information on the number of operating agricultural entities. The collection and accumulation of the rest of the above information is possible within the framework of the "exchange zone" [13], which may include both specialized higher educational institutions and research institutes. At the same time, one cannot discount the need for regional divisions of Rosreestr to participate in this process, since this department is working to update information in the Unified State Register of Real Estate, including in relation to agricultural land and land users of this category. These measures also need to be coordinated with the regional Ministry of Agriculture, since this executive authority carries out state management and legal regulation in the field of the agro-industrial complex.

4 Conclusions

The results obtained in this work allow us to draw some conclusions. The negative impact of socio-economic processes and phenomena occurring in the agricultural sector of the Russian Federation caused a decrease in the number of employable population in rural areas and its outflow to large cities, the destruction of medium and small peasant farms, which, in turn, contributed to an active increase in the degradation of arable lands as a result of the lack of necessary agricultural measures on them. The formation of latifundia also negatively affects the state of agricultural land, since in this case the land becomes a means of capitalization, and not a means of production in the implementation of agricultural activities, which directly contradicts the basic principles of land legislation.

The activities of the Government of the Russian Federation are aimed at developing and implementing a policy for the development of the agro-industrial complex and sustainable development of territories. However, for obvious reasons, it cannot cover all the necessary areas of activity and, in particular, the regional and local level. For the successful implementation of these activities, in our opinion, it is necessary to carry out the following works:

— creation of "exchange zones", in terms of which complex measures will be carried out to study the territory of municipalities and inter-settlement territories, each separate district of the region in terms of the availability of agricultural land, the location of the existing and lost engineering infrastructure, and the collection of information on landscape-adaptive farming;

— within the framework of land management design, it is necessary to prepare design solutions for the development of inter-settlement territories from the point of view of the rational use of agricultural land and recommendations on the quantitative and qualitative optimization of agricultural enterprises and organizations of each type of agricultural activity;

— preparation of design solutions for on-farm land management of the above agricultural enterprises;

— preparation of design solutions for the development of agritourism and attraction of the employable population, including pensioners, in order to strengthen the labor force in the development of agricultural enterprises;

— preparation of design solutions for the logistics of processing, storage and marketing of agricultural products produced by enterprises within the region and beyond;

— preparation of an economic feasibility study to optimize the quantitative and qualitative state of agricultural enterprises with an address reference to the territory of each district of the region;

— preparation of research projects for their participation in competitive selections for the right to receive grants and subsidies in the framework of government measures to support the agro-industrial complex.

Thus, the implementation of an integrated scientific approach to solving problems and problems that have emerged in the agro-industrial sector will not only meet the needs of society within the framework of the Food Security Doctrine [4], but also manage the development of inter-settlement territories in terms of conservation and rational use of agricultural land, referred to the category of lands specially protected by the state.

The above activities can be considered as a kind of concept of a system of step-by-step work aimed at the rational use of agricultural land within the framework of managing the development of inter-settlement territories.

References

1. O. I. Rusakova, Izvestia BSU, 1 (2006)
2. D. M. Khomyakov, Agriculture failed to remain the main driver of the Russian economy (2018)
3. Rosselkhoznadzor in Tver and Yaroslavl regions latifundists, http://rshn-tver.ru/
4. Decree of the President of the Russian Federation as of January 21, 2020 N 20 "On Approval of the Food Security Doctrine of the Russian Federation", http://www.consultant.ru
5. Website of the Rosselkhoznadzor of the Novosibirsk region, http://rshn-nso.ru/
6. Official website of the Ministry of Agriculture of the Novosibirsk Region, http://mcx.nso.ru/
7. Official website of Rosreestr. Public cadastral map, https://pkk.rosreestr.ru/
8. Electronic service "SBIS", https://sbis.ru/
9. Electronic version of the newspaper "Kommersant", https://www.kommersant.ru/
10. N. Vlasenko, N. I. Dobrotvorskaya, L. N. Iodko, I. N. Sharkov and et al., A model of adaptive landscape agriculture and agricultural technologies (on the example of the Federal State Unitary Enterprise "Kremlin" of the Kochenevsky district of the Novosibirsk region), 200 (2012)
11. Official portal of the Government of the Rostov region, https://www.donland.ru/
12. Electronic version of the business newspaper, https://www.donland.ru/
13. E. I. Avrunev, V. V. Vylegzhanina, I. A. Giniyatov, V. A. Timonov, International scientific and methodological conference “Topical issues of education. A model of problem-oriented project-based education in a modern university " (2021)

E3S Web of Conferences **311**, 06011 (2021) https://doi.org/10.1051/e3sconf/202131106011
EPSD 2021

Improvement and implementation of modern strategic management tools on the example of a commercial enterprise of the Kyrgyz Republic

Elena Zenina, *Anna* Kramarenko, and *Maria* Stoyka

Non-commercial educational institution Academic Consortium "International University of Kyrgyzstan", Bishkek, Kyrgyz Republic

Abstract. This study examines the objective foundations of strategic management as a necessary tool for studying the existing experience of the company, its indicators, as well as subsequent conclusions and justifications for the future chosen development strategy. The relevance of the research topic determines the modern challenges studied and considered in the work, and the external threats that companies face during the global economic crisis, taking into account macro and microeconomics, financial accounting and company marketing. The analysis of the existing main indicators of the activity of a commercial enterprise of the Kyrgyz Republic (on the example of "Energy Save" LLC) allows drawing conclusions and substantiating the most effective and available tools for drawing up a new development strategy for the company. The purpose of this study is to consider the foundations of strategic management, its role in the development of an enterprise, as well as to develop recommendations for improving the strategic management of "Energy Save" LLC.

1 Introduction

In a market economy, high competition and saturation of markets with goods and services, commercial enterprises need to go through a constant path of development. Today it is necessary to declare yourself, work on your promotion, constantly update and improve your products, expand markets, look for new opportunities for implementation. Domestic companies are no exception and go through their own "natural selection".

In 2020, the corona virus pandemic has tightened the conditions for the "survival" of domestic companies. Closing borders, restrictions on movement within the country, control by government agencies on compliance with sanitary standards, as well as uncertainty in the context of political changes are just the most obvious tasks that managers and owners of companies need to solve every day.

The management of companies strives to anticipate changes in the political and economic environment and are in constant search of effective tools to create favorable conditions for development. One of the ways to effectively plan, systematize processes and manage the company as a whole (especially in crisis conditions) is strategic management.

The relevance for strategic management has already been proven. I would like to emphasize that in the new conditions of existence, it is important for companies to use classical management tools, a small part of which will be considered in this study.

In the conditions of instability in which the whole world is located for the second year already, the owners of the companies have more and more questions for strategic managers. Thanks to a correctly developed strategy of the organization, its development takes place, the planned results are achieved.

However, in conditions of turbulence, all existing principles of strategic management are not fully applied as an effective tool for managing an enterprise and its long-term development.

Strategic management allows you to assess risks, prospects and create a development strategy from different angles, as well as, owing to a correctly chosen strategy, opportunities open up to overcome possible crises. Modern strategic management should take into account such fields of companies as marketing, financial management and human resource - the company's team.

2 Materials and methods

To date, about 70 advertising agencies of various forms and structures are registered on the territory of Kyrgyzstan. In 2019, the advertising market was estimated at 19 million US dollars and a significant part of the market - 54.1% - of the country's advertising market is occupied by television, in second place is outdoor advertising (15.5%), in third place is the press (13.9%), then radio (12.9%) and the Internet (2.7%)[1] ... It should be noted that these figures are given without discounts, the study was guided by the amount of time sold and the official price lists. However, due to the very large number of market players, about 60 TV channels, 20 radio stations and the rapid development of the Internet, the competitive environment among media carriers is becoming very tough. A limited number of advertisers are forcing media and advertising agencies to operate under dumping conditions.

The stages of implementing an effective strategy include the following algorithms (PDCA - Plan, Do, Check, Inform) laid down in the ISO standards:

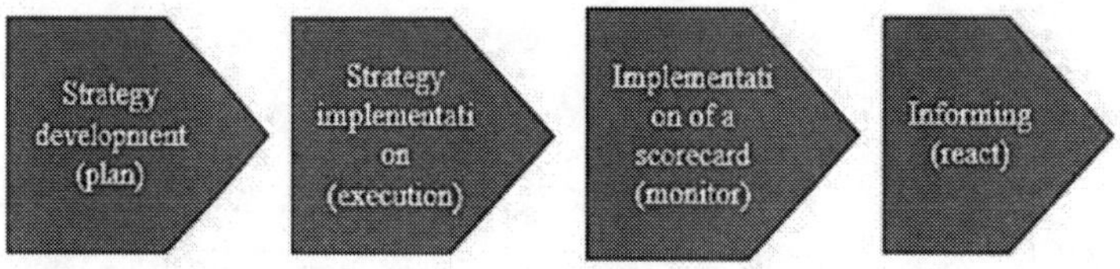

Fig. 1. Strategy implementation algorithms.

The Charter of "Energy Save" LLC reflects the goals of creation and the subject of the company's activities, it is said that the purpose of the creation of the Company is to make a profit in the Kyrgyz Republic through the implementation of commercial activities by the Company, which is not prohibited by the legislation of the Kyrgyz Republic. Initially, the main activity of the Company was the construction of other civilian facilities. However, since 2018 "Energy Save" LLC, owint to the skills and experience of the founder, reoriented itself to an advertising agency.

When the company entered the market, the company used the classic SWOT analysis model presented in Table 1 and described at the following link.

"Modern processes represent the stages of strategic analysis, forecasting, planning, implementation of strategies, as well as other strategic decisions, as well as strategic control. Each of the stages includes many sub-stages and specific implementation methods used within their framework. For example, at the stage of strategic analysis, methods of benchmarking, competitive analysis, SWOT, EFAS analysis, in-depth interview methods, focus groups, positioning methods, etc. can be used. It is advisable to take into account that the consistent implementation of the processes allows the formation and improvement of the very system of strategic management as a whole. "[2] ...

Table 1. SWOT analysis of "Energy Save" LLC in 2017.

Strengths of the company: Mobility, Versatility, Flexibility, Knowledge and understanding of the country's advertising and media market, Individual approach to each client	Weaknesses: Small personnel, Dependence on the client's budget, Freelance work system of employees - waiting for their free time
Possibilities Loyal customer base, New and small companies with small needs for advertising services, The need for companies to produce quality promotional materials for a small budget	Threats: Highly competitive environment, Dependence of advertising budgets on political and economic situations in the country

As such, the company did not have a strategy, the team of "Energy Save" LLC used the current client base, which had been transferred to them owing to the experience and competence of two managers. The company's managers were recruited to the company on terms of remuneration above market.

The strategic goal of the company was to attract clients, at the time of the opening of the company there were 3 clients. One of the regular clients - a construction company is placed out of season, but with a small budget, another client - a new travel operator, is accommodated seasonally 1-2 times a year, depending on their strategic directions.

When developing its strategy for entering the advertising services market, "Energy Save" LLC took into account aspects of the external environment, including political and legal factors, economic, socio-cultural, technological factors. The regulatory framework of the Kyrgyz Republic promotes the development of small and medium-sized businesses - the Law on Advertising of the Kyrgyz Republic, which regulates the activities of the company, the Law on Competition of the Kyrgyz Republic, in which the government supports competition as a free competition of economic entities in the market, stimulating the production of goods required by the consumer[3] The foreign economic policy of the government in the National Development Strategy of the Kyrgyz Republic for 2018-2040 is aimed at promoting the export of domestic products to foreign countries, attracting foreign investment, advanced technologies and innovations into the national economy of the country.[4] , which also corresponds to favorable conditions for the emergence and development of international goods and services in the country, as well as investment and social projects.

Unfortunately, our country is highly dependent on the world political and economic situation, as well as crises. One of the most important factors for business development is tracking the economic factors of the country. So, according to the National Statistical Committee of the Kyrgyz Republic, in January 2021 compared to January 2020, the highest inflation rate (10.1 percent) in the consumer sector of the economy was recorded in the Kyrgyz Republic, in January this year, compared to January 2020, prices increased the most and tariffs for food products (including alcoholic beverages and tobacco products) - by 16.2 percent [8]. In general, inflation in the Kyrgyz Republic in 2020 amounted to 9.7%, this is the highest indicator among the countries of the Eurasian Economic Union, where the inflation target at the end of 2020 amounted to 8.7%. The main influence on the acceleration of inflation in the Kyrgyz Republic, as noted on the website of the Eurasian Economic Commission, is the growth rate of food prices - almost 5.7%. Food inflation in Kyrgyzstan increased largely due to a sharp rise in prices for meat and meat products in the fourth quarter of 2020 (1.8 p.p.) due to problems with the import of livestock. According to forecasts of central (national) banks, inflation in the EAEU countries will peak in the first quarter of this year, by the end of the year will gradually decrease and will remain close to national targets.[5] ...

In addition to analyzing external political and economic factors, an analysis of the competitive environment is required, which is the main factor for a company to enter the market. In Kyrgyzstan, there are two types of profitability for advertising agencies. The first type is when a client pays an advertising agency an agreed percentage for the service. For example, a client does not have a local representative office, marketing department, advertising department, then an advertising agency performs this function. Such clients are usually large with global brands such as P&G, Mars, Henkel, Nivea, Loreal and large pharmaceutical companies. These clients are served by network advertising agencies. For example, in Kyrgyzstan there are offices of Quattro Media Bishkek - an international full-cycle advertising agency that is part of the Dentsu Aegis Network[6] , DAS media Kyrgyzstan - part of the global network WPP and Group M, the largest group of advertising companies in Kazakhstan, Mongolia, Ukraine and the Caucasus. At the moment, there are not many network advertising agencies, only 4.

Among with network advertising agencies, there are local large advertising agencies serving cellular operators and other large advertisers. Such agencies exist at the expense of marginal income. The pricing policy of such companies is to provide lower offers for the placement of advertising information materials of the customer in the media.

Acknowledging the small market size and high internal competitiveness, "Energy Save" LLC decided to develop its vision of the company at the stage of formation: We are Energy Media! - a team of like-minded people with more than 15 years of experience in the media market of Kyrgyzstan, we have extensive experience in media planning, production of films and TV projects, in the construction and implementation of new strategies and media communications, in the development of fresh ideas to promote your products and services[7] Focusing on your mobility, and the ability to work with a client on a trusting, friendly note.

According to the developed SWOT analysis in Table 2. the company developed at that time the slogans: 1) We are able to create advertising concepts, graphic design and video products; 2) We work to achieve an ideal result 7 days a week, 24 hours a day and value every minute of your and our time.

Table 2. SWOT analysis of "Energy Save" LLC in 2017.

Stage No.	**Stage characteristics**	**Methods used**
1.	Analysis of the external environment	Development of SWOT analysis, analysis of the competitive environment, diagnostics of the company's direction with key employees
2.	Assessment of the needs of regular and potential customers based on the results of the survey	Questioning of existing clients and partners of the company
3.	Defining strategic goals and objectives	Brainstorming with key employees
4.	Defining the company's mission	Development of a mission based on customer expectations, diagnostics are carried out during an interview with a client

It should be noted that today, in conditions of high rates of change in the external environment and the associated uncertainty, the urgent problem is the constant improvement of strategic management. "Energy Save" LLC, to define a strategic position in 2017, constantly updated it, focusing on the strengths of the company - mobility and flexibility. The company has not expanded its staff, since 2017 the company has been cooperating with two advertising managers, with two designers, one audio engineer, and one moviemaker. A small personnel allows the director of the company to instantly react to risks and threats, as well as to take part in all processes himself. This strategy allowed the company to quickly respond to the global changes that occurred in 2020 and ensured the company's resilience in a changing environment.

3 Results and discussion

The main task of any strategy is to improve the current results of the company and achieve the desired results in the future. The director of "Energy Save" LLC chose high motivation of employees as the basis for the company's internal strategy. The company's employees are a team of like-minded people, individuals, and professionals. Everyone treats their work responsibly and works both independently for themselves and in a team, if necessary. The high motivation of the team is due to the fact that the company immediately recruited professionals who value their name and time. Payment terms for work are transparent and depend on the value of the contract. The customer's budget is not confidential information for employees, so the wages are piecework. Under such working conditions, employees have the possibility of additional income, the deadlines for the delivery of work for the company are negotiated with each just as individually.

Modern business with its fierce competition and periodic shocks requires managers of all levels to search for such a strategy that would guarantee the enterprise, on the one hand, the most efficient use of resources, on the other hand, the maximum stability of the enterprise in conditions of instability of environmental factors. The choice of enterprise strategy is determined by a combination of external and internal environmental factors.

Stages of the formation of the basic elements of the strategy of economic development of the enterprise in the usual predicted conditions[8] are shown in Figure 2.

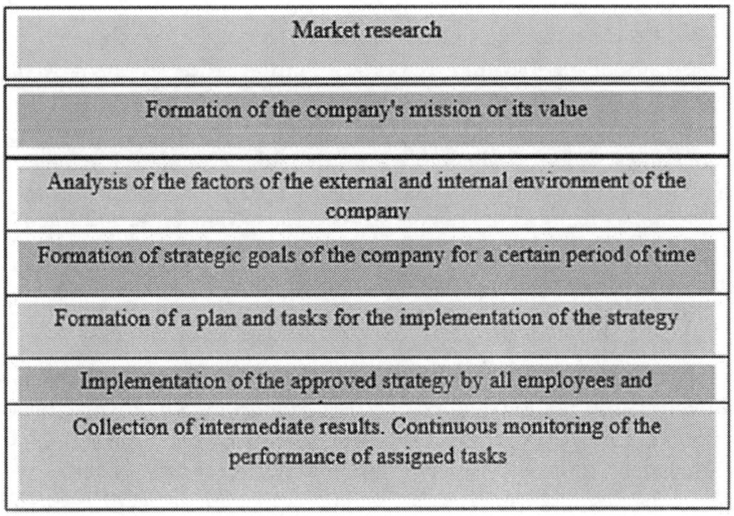

Fig. 2. Basic elements of strategy formation.

However, these steps may be unacceptable in the turbulence environment faced by Kyrgyz companies. Rapid events 2019-2020 prove that in a world of unexpected threats and opportunities, strategic decisions must be made quickly, regardless of approved planning cycles. Methods are needed for a more sensitive and receptive analysis of the external environment, constant monitoring of the information field, an increase in strategic flexibility, the development of an entrepreneurial spirit and innovation will have a positive effect on the company's results. Strategies should not obey the environment, but they should not take it for granted either. They are obliged to anticipate possible changes, to influence possible changes inside and outside the company.[9]

The attention of "Energy Save" LLC is directed to the constant study of the external environment, because the threat to the development of the company is the political and economic crises in the country. Therefore, the company decided in 2020 to use the decision-making method based on weak signals. Early detection of threats represents the time it takes to react to it. Risks are also possible here, since with the increase in the frequency of possible sudden changes, the degree of their predictability decreases. This means that by the time sufficient information is received for a thoughtful response, there is a lack of time to implement them, which can harm the company or lead to the loss of a profitable economic opportunity. This fact obliges companies to use several tools when developing a strategy.

The qualities of a successful company in the modern world are flexibility and reaction to the external environment, the ability to focus on the strengths of the company and constantly be in development. So that the reaction is not belated to the external environment, "Energy Save" LLC builds trusting, partnership relations with the team, when each employee is a manager - the manager of his work, department, and aims to take responsibility for the quality and timing of the services provided. In such a team it is easier to keep a finger on the pulse, information is not delayed, not hidden, but is openly discussed within the team.

The political crisis in the Kyrgyz Republic in October 2020, as well as news about the Covid-19 coronavirus pandemic, increased the influence of the media as a source of receiving the first signals about possible threats.

So it was decided that at the company's weekly planning meetings, account managers would share latest news. The practice of identifying new opportunities or threats helps managers make appropriate strategic decisions.

"Energy Save" LLC, recognizing external threats by weak signals, developed an action plan during the period of political unrest in the fall of 2020.

In Figure 3., it can be seen that online news outlets, telegram channels of news outlets, where news spreads faster, become a source of information about new opportunities or threats. The signal strength is determined when analyzing publications in the media, if the same question or problem is published several times in different sources of information, then the problem becomes serious and obvious. Further, the company's management begins to take measures to reduce strategic vulnerability. If the news is broadcast in all media, including in the news blocks of the evening news, the company moves to an increased level of flexibility and begins to develop preparatory protective programs. With clear signals of possible threats, the company has already developed plans of practical actions and measures. There is a redistribution of resources, savings in current assets, and the expectation of an improvement in the external environment.

The attention of company management in conditions of weak signals requires, first of all, that when collecting information about the external environment, details are not omitted, and the information is correctly classified. Further, based on the information received, it is necessary to make correct calculations of the possible impact of the hazard and the time of its occurrence. The less information is available, the greater will be the scatter of the calculated data on the impact of the hazard and the timing of its approach.

<table>
<tr><td rowspan="2">Signals</td><td colspan="6">The nature of measures to increase reality</td></tr>
<tr><td>Permanent observation</td><td>Signal Strength Determination</td><td rowspan="2">Reducing strategic vulnerability</td><td rowspan="2">Greater responsiveness</td><td rowspan="3">Development of preparatory plans and programs</td><td rowspan="4">Action plans and implementation</td></tr>
<tr><td colspan="3">Political aggravation in the country, October 2020 News is first published on Telegram channels and sent out via WhatsApp groups</td></tr>
<tr><td colspan="5">Political instability began to be described in online news outlets</td></tr>
<tr><td colspan="6">The scale of the danger or new opportunity takes on concrete outlines - the problem is talked about in the traditional media, news in the evening on TV</td></tr>
<tr><td colspan="7">Ways to solve the problem have been established, the company has a plan B. Employees go to remote work. Signals come from all media.</td></tr>
</table>

Fig. 3. Assessment of the first signals about possible threats to "Energy Save" LLC in 2020.

The general economic downfal of the country, a systematic decline in GDP and the dependence of the Kyrgyz economy on the economies of neighboring countries obliges Kyrgyz companies to be sensitive to external changes. Dynamics of GDP decline in 2020 became a record for the country's economy and among the countries of the Eurasian Economic Union (EAEU). The gross domestic product in January-February 2020, according to preliminary estimates, amounted to 69 billion soms and compared to January-February 2019. increased by 4.3%. However, already in May 2020. According to the National Statistical Committee, the Gross Domestic Product in January-April 2020, according to preliminary estimates, amounted to about 149 billion soms and compared to January-April 2019 decreased by 3.8%. The decline continued and already in July the Gross Domestic Product in January-July 2020, according to preliminary estimates, amounted to 282 billion soms and compared to January-July 2019 decreased by 6.1%[10] ...

Against the background of the economic downfal of the country in 2020, the competition is intensifying, prompting Energy Save LLC to create a new development strategy in the long term. There is a need to strengthen the company's position in the market, preserve current circulating assets, retain existing customers and attract new customers.

4 Conclusions

In this study, the issues of improving the strategic management system were considered on the example of "Energy Save" LLC, taking into account international experience and, as a result, a project was developed to introduce effective goals and objectives.

The Materials and Methods section examines the importance of strategic management in the current conditions of globalization and a rapidly changing external environment. The main principles and tasks of strategic management are considered. The first chapter also explained that the strategy determines the results of the organization, and with the help of which goals and action plans these results will be achieved , what the goals and mission of the company should be. The main approaches to the development of strategies proposed to the world by the graduates of the Harvard Business School were considered. The emphasis was placed on the study of SWOT analysis, analysis of Porter's 5 competitive forces and growth strategies using the Ansoff matrix. Based on the results of the studied material, at the end of this section, it was concluded that the external environment is a factor in the formation of a new strategy, as well as that a strategy can never be thought out and calculated completely, and its constant adjustment as external and internal changes conditions - the procedure is necessary.

In the Result and Discussion section, it was determined that at the time the agency was launched, "Energy Save" LLC had defined its strategic position in 2017 as a mobile and flexible company. The strategic goal of the company was to attract customers, in 2018 the company had only 3 clients.

After analyzing the sales data in 2019, it was decided to apply the market development strategy / extensive growth strategy proposed by Ansoff. The company's existing services were offered to a new market - international projects. "Energy Save" LLC is reorienting itself and launching a new strategic direction of its activity - services for the creation of media products for an international project. The analysis carried out in the second chapter showed the effectiveness and timeliness of the new direction for the development of the company, owing to the contract with an international project, the company retained its financial performance. Energy Save LLC uses a strategy for recognizing external threats based on weak signals and has developed an action plan during the period of political unrest in the fall of 2020.

In the course of the study, the set tasks were solved, namely, a strategic analysis of the external environment of the advertising agency "Energy Media" was carried out, an analysis of the current state of the company - the theory, international and domestic experience of improving the strategic management system in the company were analyzed. The priority goals and objectives for improving the strategic management of "Energy Save" LLC were considered, the sales plan project was planned to achieve the set results, and the economic efficiency of the proposed strategy was assessed.

References

1. Financial portal Akchabar, https://www.akchabar.kg/
2. V. M. Zhigalov N. V. P., Bulletin of St. Petersburg State University, **3**, 74 (2016)
3. Law of the Kyrgyz Republic on Competition (2021)
4. National Development Strategy of the Kyrgyz Republic for 2018-2040, https://mfa.gov.kg
5. Eurasian Economic Commission Inflation in the Kyrgyz Republic in 2020, http://www.eurasiancommission.org/
6. Leading mass media about advertising, marketing and PR in the Russian Federation, https://www.sostav.ru/
7. Energy Save LLC website, https://energymedia.kg/
8. Digital around the world. Data reportal
9. M. Porter, *Competitive strategy: Methodology for analyzing competitors' industries*, 232 (2016)
10. National Statistical Committee of the Kyrgyz Republic, http:/www.stat.kg/

Do circular economy and green investments create value at the regional level? Evidence from Russian regions

Ekaterina Popova* and *Nikolay* Strikh

Surgut State University, Surgut, Russia

Abstract. Green investments and principles of circular development have become one of the prominent dimensions of regional development policies. The purpose of this article is to study the impact of green investments on circular economy indicators and contribution of these indicators to economic growth at the regional level. The study examines two key indicators of the circular economy: the share of captured and neutralized emissions into the atmosphere and the share of recycled water in the regional manufacturing systems. To test the proposed hypotheses, methods of regression analysis are used. Gross regional product is used as a predictor of value creation. The empirical base is Rosstat data for the period from 2015 to 2019. The results show that the circular economy indicators do not have a visible impact on economic growth at the regional level, but green investments increase the volume of resources used in circular manufacturing systems. It is probable that in Russian industrial economy, implementation of the circular economy principles is associated only with an increase in costs, which do not provide linear returns at the macroeconomic level.

1 Introduction

Economic development that relies on a linear logic of resource utilization and allocation inevitably faces environmental challenges. Various activities of households and industrial organizations lead to the generation of a significant amount of waste, which, in the context of a rapidly growing population, may leave future generations worse off. Circular economy concept is the product of a synthesis of interdisciplinary scientific thought in economics, industrial design, environmental chemistry and biology. Ideas of a circular economy involve restoration and regeneration of resources within closed flows, where materials are used repeatedly with minimal losses [1]. As a result, the impact of human activities on the environment is significantly reduced, in addition, less valuable non-renewable resources are involved in the manufacturing processes [2]. Despite this, there are a number of barriers to the practical implementation of a circular economy principles and methods. Therefore, a clear demonstration of the value that circular use of resources can offer to regional and

* Corresponding author : k121time@yandex.ru

national economies is an important tool for dealing with objections and justifying the effectiveness of green investments.

The process of creating economic value refers to all positive outcomes created in service and production systems that relate to saving resources, expanding market opportunities and increasing the competitiveness of organizations. Thus, the concept of a circular economy is attractive for both industrial organizations and service sector of the economy, since it makes a significant contribution to maintaining a long-term business development strategy [3]. By introducing the principles of rational recovery and reuse of resources, companies demonstrate socially responsible behaviour that enhances their image and investment attractiveness. Due to the circular nature of resource flows, payments for environmental impact are also reduced. A number of researchers also note that the creation of circular regional clusters allows the development of local manufacturing and processing companies that provide additional jobs [4].

Responsible allocation of capital in the form of green investments is of a strategic importance for Russia, since the country is one of the economies that actively exploit its own resource base [5]. Green investment primarily implies allocating capital for environmental protection, developing assets and business models that support a responsible attitude of consumers and producers towards the environment [6]. In a circular economy, companies are investing in green technologies that are fundamentally changing key business processes, such as renewable energy sources, advanced waste recycling, product sharing, and development of environmentally friendly transport infrastructure. Consumers also participate in green investment initiatives by choosing products that reduce the impact on local ecosystems.

In existing literature there is insufficient evidence of the regional contribution of green investment to perpetuating the principles of recycling and reuse of resources in developing regional economies. In addition, more attention should be paid to the impact of environmental concepts on macroeconomic growth performance. The purpose of this article is to study the impact of green investments on the circular economy indicators and the contribution of these indicators to the gross regional product formation. To identify the degree of circular economy development, the following indicators are considered: the level of captured and neutralized atmospheric emissions and the share of recycled water consumption by households and organizations. Taken together, these indicators may indicate the development of rational use of resources and a decrease in environmental impact.

2 The role of the circular economy in sustaining value creation: literature review and hypotheses development

The concept of circular economy originates in the works of the 1960s, when interest in harmonious development of production systems, considering principles of environmental responsibility, increased significantly [3]. Economists have shown that closed models can explain production processes and reflect the movement of resources in an interfirm environment to achieve savings [7]. The concept focuses not only on the reuse and recycling of materials, but also on product and process design, supply chains configurations and consumer behaviour patterns [8]. The process of creating value in a circular economy therefore transcends the internal production system of the company and focuses attention of the responsible parties on the network interaction for non-renewable resources distribution and consumption.

The value proposition is created by applying the circular economy principles. They refer to all positive effects and results that are created in resource supply chains, production systems and consumer behaviour. The increasing complexity of supply chains creates both threats and opportunities for the regeneration and reuse of resources. At the initial level of environmental maturity, opportunities of a circular economy are closely related to raising awareness among stakeholders, strengthening marketing communication, developing innovative activities of companies, and reducing environmental pressure from the public [3]. The value-creating strategies for the circular economy are, first, based on industrial symbiosis. Its primary aim is to create partnerships in value chains for knowledge sharing. Second, these principles introduce new design principles that encourage engineers to avoid waste and material leakage wherever and whenever possible. Third, the circular economy is characterized by re-use of products, as well as deep processing of waste, which makes it possible to maximize extraction of useful components in accordance with available advanced technologies. Fourth, the production should use environmentally friendly materials that are subject to rapid decomposition or are easily recycled [2].

Among the significant natural resources that are intensively used by industrial and service enterprises, one can single out water resources and the atmosphere. It is predicted that water consumption in the first half of the 21st century will increase by 50% [9]. Therefore, the consistent use of recycled water is an important strategy to support the implementation of a circular economy. Air pollution also has a significant impact on the environment and biodiversity. More than 80% of air emissions are associated with industrial companies, in addition, solid household waste also has a negative impact on atmosphere, since only about 4% of the total volume of waste is sent for processing in Russian regions [10]. For the broad practical implementation of the circular economy principles in Russian conditions, social capital is a significant resource. Raising public awareness of resource reuse, deep recycling and sharing of things is supposed to encourage the value creation process [11].

At present, the widespread implementation of the circular manufacturing principles in the industrially developed regions of the world is at the initial stage of development, especially in the primary sector, which is focused on the extraction of natural resources [12]. Traditionally, it is the primary sector that is associated with a significant impact on the environment and imbalance in ecosystems. Since the Russian economy gains a significant competitive advantage in international markets from the extraction, processing of metal ores, and the sale of oil and gas, the task of developing a circular economy becomes even more complicated. Circular economic models suggest that resources should be used by local producers with maximum efficiency to create added value in a circular chain of production and distribution [8]. Therefore, the expansion of technologies that make it possible to reuse water resources and neutralize emissions into the atmosphere, leads to the emergence of additional income streams in the economy, which in turn provides creation of consumer value. Therefore, the following hypotheses are formulated:

Hypothesis 1.1. An increase in the share of captured and neutralized air emissions in the regions has a significant positive impact on the gross regional product.

Hypothesis 1.2. An increase in the share of water used in a closed cycle has a significant positive impact on the gross regional product.

Green investment projects are designed to support all types of technological and organizational innovations and improvements that provide a responsible attitude towards the environment [13]. Modern investors demand from companies not only financial performance, but also the development of projects to introduce environmentally friendly technologies [6]. The principles of a circular economy create a stricter framework for environmental activities in companies, they stimulate them to revise the technological underpinnings of business processes, choose strategies for closing resource flows and using renewable energy sources. The main financial results of green investments are lower waste charges and an additional stream of income from the supply of green products to the market. Therefore, the following hypothesis is formulated:

Hypothesis 2. Green investments have a significant positive impact on increasing the share of resources used in a closed cycle of production and consumption.

3 Methods and data

Multiple linear regression method is used to test the hypotheses put forward. Regression coefficients show the contribution of each indicator to change in dependent variable; the least squares method is used for the assessment. The dependent variable for testing the first hypothesis is the natural logarithm of the gross regional product (GRP) per capita in roubles (LN_GRP). It is assumed that this variable reflects the process of creating value through the additional flows of investment in various types of activities specific for a circular economy. The independent variables include control variables such as the share of innovatively active companies in the region (INN_ACT) and the volume of innovative products shipped (INN_Prod). Target variables, such as the share of captured and neutralized air pollutant emissions (CE_WCap) and the share of recycled sequentially used water in total consumption (CE_Wat_Coef) are used.

To test the second hypothesis, the authors calculate and use the circular economy rating (CE_RATE) as a dependent variable. The proposed indicator reflects the ratio of the amount of materials and water used in closed cycles to the maximum observable value of this indicator for all regions in the current year. The indicator is measured in the range from 0 to 100 units, it reflects the average share of resources that are used in a circular production flow. The target independent variable is the natural logarithm of investments in environmental protection in roubles (LN_CE_Invest). This indicator demonstrates the placement of green investment in the economy and its impact on the implementation of circular production and consumption systems.

4 Results and discussion

Descriptive statistics for the variables used in the study are presented in Table 1. From the data presented, it can be concluded that the circular economy rating has not changed significantly over the five years. This is explained by the fact that the share of consistently used water and neutralized emissions into the atmosphere also changed insignificantly over the reviewed period. Despite this, the volume of green investments in the Russian economy has grown significantly.

Table 1. Descriptive statistics. Obtained by the authors.

Variables	2015		2017		2019	
	Mean	S.D.	Mean	S.D.	Mean	S.D.
LN_GRP – Logarithm of GRP per capita	12,8	0,7	12,9	0,7	13,2	0,7
CE_RATE – Circular economy rating	32,3	18,7	33,0	18,6	33,2	17,9
INN_Prod – Volume of shipped innovative goods and services as a percentage of the total volume	5,9	6,0	5,7	5,9	4,6	4,6
INN_ACT – Share of innovative organizations	8,6	4,5	13,1	6,3	8,4	4,3
LN_CE_Invest – Logarithm of environmental protection investments	21,8	1,4	21,9	1,3	22,2	1,3
CE_Wat_Coef – Share of recycled and sequentially used water	54,7	29,2	55,7	28,6	56,6	28,8
CE_WCap – Share of captured and neutralized air pollutants in the total amount of waste from stationary sources	55,7	27,7	54,7	27,6	53,4	27,8

The results of regression analysis for testing the first hypotheses are presented in Table 2. The results show that green investments make a significant contribution to the formation of GRP, because they are a component of capital investments of industrial and service companies. The variables reflecting the proportion of recycled water used consistently and the proportion of captured and neutralized air emissions show statistical significance. However, although the regression coefficients have a relatively high statistical significance, their actual contribution to the formation of the final result is relatively insignificant. In addition, the proposed models explain no more than half of the observed variance (the values of the determination coefficients do not exceed 40%). An increase in the share of resources used in a closed production cycle by 1% leads to an increase in GRP by less than 1%. Thus, hypotheses 1.1 and 1.2 are rejected, since the indicators of the circular economy do not make a significant direct contribution to the formation of the financial performance of the regions.

Table 2. Non-standardized regression coefficients and t-statistics. Dependent variable is the natural logarithm of GRP per capita. Obtained by the authors.

Dependent variables	**2015**		**2017**		**2019**	
	Coeff.	t-stat.	Coeff.	t-stat.	Coeff.	t-stat.
Constant	6,987***	6,622	6,984***	5,968	7,675***	5,797
INN_Prod	-0,019*	-1,763	-0,017	-1,485	-0,024	-1,474
LN_CE_Invest	0,273***	5,285	0,283***	4,962	0,264***	4,177
CE_Wat_Coef	0,006***	2,420	0,004*	1,673	0,004	1,458
CE_WCap	-0,007***	-3,082	-0,008***	-3,000	-0,009***	-3,466
R^2	0,411		0,360		0,286	
Adjusted R^2	0,381		0,328		0,251	
F-statistic	13,9***		11,3***		8,0***	
Number of observations	85		85		85	

Note. *** – significant at 1% level, ** – significant at 5% level, * – significant at 10% level.

The results of regression analysis for testing the second hypothesis are presented in Table 3. The rating of the circular economy proposed by the authors reflects the share of consistently used water resources and neutralized atmospheric emissions. The high value of the index is typical for the central regions of Russia, while the southern regions show no inclination to introduce closed production flows (Figure 1). The leader in the proposed rating is the Sverdlovsk region, located in the Urals: more than 90% of all water resources in the regional industry are in a closed cycle of consumption, the share of captured and neutralized emissions into the atmosphere is also high.

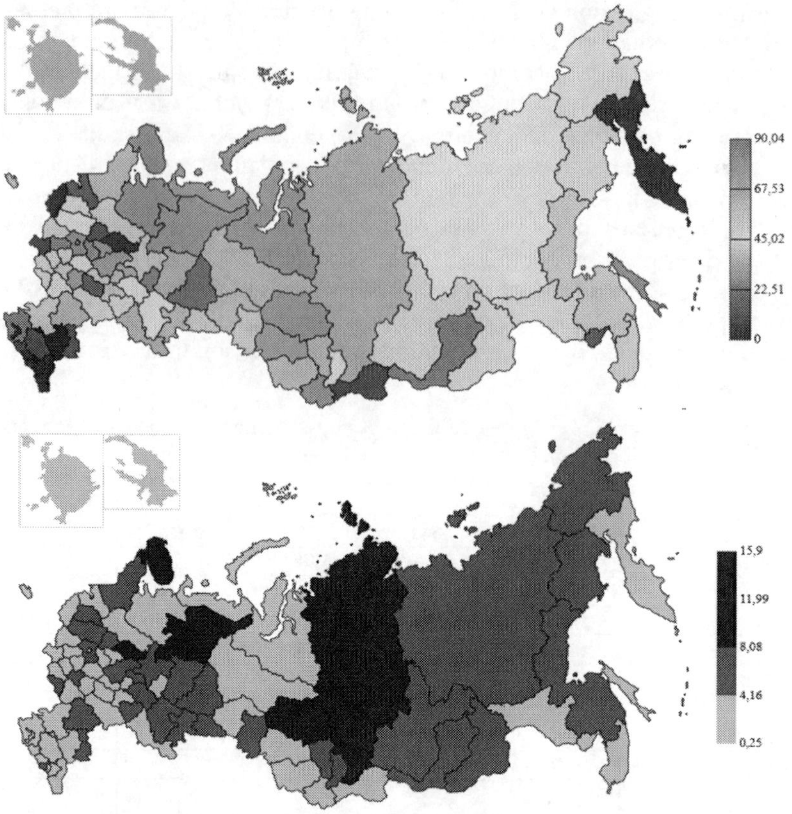

Fig. 1. Territorial distribution of the proposed circular economy index in 2019 (top). The share of green investments in the regions of Russia in the total volume of fixed asset investments in 2019 (bottom). Obtained by the authors.

According to the results of regression analysis authors conclude that green investments have a significant positive impact on the use of technologies inherent in the circular economy in the Russian regions. Therefore, the second hypothesis is supported. The innovative activity of companies does not demonstrate a stable relationship with the consistent use of water and the neutralization of atmospheric emissions; a significant positive relationship is observed only in 2019.

Table 3. Non-standardized regression coefficients and t-statistics. Dependent variable is the circular economy rating. Obtained by the authors.

Dependent variables	**2015**		**2017**		**2019**	
	Coeff.	t-stat.	Coeff.	t-stat.	Coeff.	t-stat.
Constant	-148,4***	-4,5	-144,0***	-4,3	-135,0***	-4,1
LN_GRP_P	2,547***	0,884	-0,577	-0,211	0,916	0,399
INN_ACT	0,200	0,471	-0,053	-0,178	1,389***	2,764
INN_Prod_D	0,300	0,925	0,104	0,329	-0,305	-0,651
LN_CE_Invest	6,636***	4,349	8,415***	5,418	6,586***	4,715
R^2	0,357		0,353		0,388	
Adjusted R^2	0,325		0,321		0,357	
F-statistic	11,1***		10,9***		12,7***	
Number of observations	85		85		85	

5 Conclusions

The results show that the principles of a circular economy play a very moderate role in creating direct value at the regional level. The contribution of technologies for the consistent use of recycled water and neutralization of atmospheric emissions does not significantly affect the formation of the gross regional product. Meanwhile, green investments that are attracted in the regions to update and improve environmental technologies have a positive effect on the consistent use of resources in circular production systems. Probably the circular principles do not create value for the industrial economy, because capital investment is seen as an expense rather than an investment for the process of creating value at the regional level. Such costs limit the activities of manufacturing companies under the influence of the authorities and local communities, which express concerns about the emerging situation with the extensive use of resources and the increase in amount of irrecoverable waste.

The results obtained have practical implications, since they can be used to analyse the level of circular economy development in the regions of Russia and to identify the value added. The limitations of the study are related to the application of a broad regional approach that summarizes the contribution of environmental indicators to development.

References

1. X. Vence, Á. Pereira, Contaduria y Administracion, **64** (2019)
2. R. Salvador, M. V. Barros, F. Freire, A. Halog, C.M. Piekarski, A.C. De Francisco, Journal of Cleaner Production, **299** (2021)
3. O. Ungerman, J. Dědková, Environment, Development and Sustainability, **22** (2020)
4. Z. Wen, X. Meng, Journal of Cleaner Production, **90** (2015)
5. V.A. Borodin, K.A. Melekhova, S.I. Kazachenko, *Development of Economy of a Multicultural Region by the Circular Principles: Directions and Tools of State Management* (Springer Nature Switzerland AG, 2020)
6. M. Indriastuti, A. Chariri, Cogent Business & Management, **8** (2021)
7. W. Leontief, Structural Change and Economic Dynamics, **2** (1991)
8. S. Geisendorf, F. Pietrulla, Thunderbird International Business Review, **60** (2018)
9. O. Saritas, L.N. Proskuryakova, Foresight, **19** (2017)
10. O. Fedotkina, E. Gorbashko, N. Vatolkina, Sustainability, **11** (2019)
11. B. Nikitina, *Waste Management and Circular Economy in the Public Discourse in Russia* (Springer Nature Switzerland, 2021)
12. A. Upadhyay, T. Laing, V. Kumar, M. Dora, Resources Policy, **72** (2021)
13. G. Fortune, C.C. Ngwakwe, C.M. Ambe, International Journal of Sustainable Economy, **9** (2017)

Cluster analysis of variability of key parameters of the economic environment in the framework of diffusion of blockchain technologies

Marat Safiullin[1,2], *Leonid* Elshin[1,2,3], and *Aliya* Abdukaeva [1,2,3,*]

[1]Kazan Federal University, Kazan, Russia
[2]Center of Advanced Economic Research in the Academy of Sciences of the Republic of Tatarstan, Kazan, Russia
[3]Kazan National Research Technological University, Kazan, Russia

Abstract. Prospects for the development of the national economy in the context of integration of blockchain technologies into the system of economic processes indicate the need to develop methods and algorithms for a formalized assessment of their impact on the key parameters of socio-economic dynamics. Meanwhile, if in the scientific and expert space, although it is extremely rare, but still one can find works devoted to certain aspects of this scientific and methodological problem, then, unfortunately, it should be stated that the study of the impact of the technologies under study on certain sectors of the economy does not occur in information-analytical and scientific spaces at all. Unfortunately, the methods of empirical study of the influence of blockchain technologies on the parameters of economic development at the current time not only have not found their consolidated solution, but are completely characterized by fragmentation and patchiness. This study is an attempt to strengthen the positions of formalized approaches to the study of the scientific and practical problem. The paper proposes an algorithm for studying the dynamics of the gross value added of the sectors of the Russian economy through the prism of a possible transformation of the key parameters of the functioning of the financial and real sectors of the economy as a result of the diffusion of blockchain technologies. Relying on the hypotheses put forward, cointegration models were built for the studied types of economic activity, which make it possible to identify the contribution of exogenous factors, corrected under the pressure of penetration of distributed data storage technologies into the economic environment, on the degree of added value growth. This, in turn, made it possible to carry out a cluster analysis of the studied sectors of the economy in terms of their level of sensitivity to institutional transformations caused by the integration of blockchain technologies into the economic environment.

[*] Corresponding author: Aliya.Abdukaeva@tatar.ru

1 Introduction

It should be noted that in the conditions of the gaining momentum of the fourth industrial revolution, it is increasingly possible to observe the active integration into the system of economic relations of new, progressive forms of organization of models of socio-economic development. Among them, undoubtedly, it is necessary to include blockchain technologies that have been intensively penetrating the socioeconomic environment in recent years and falling into the "orbit" of attention of many experts and scientists. Meanwhile, with regret, we have to state that, despite the progressive interest on the part of the expert and scientific community in distributed data storage technologies and the problems of studying their impact on the development of the national economy and its individual sectors, there is both disunity of views on the solution of the question posed, and lack of unified approaches to a formalized assessment of possible generated opportunities and risks. As a rule, existing works devoted to the problems of researching the impact of blockchain technologies on economic dynamics are limited either by qualitative characteristics, or implemented through the prism of expert assessments, as well as reasoning of a general logical order. At the same time, in the overwhelming majority of cases, the positions of authors, both Russian and foreign, are inclined to believe that studies of this kind of issues are extremely relevant, significant from a practical and scientific point of view, and require the development of an appropriate methodological apparatus [1-8].

In this regard, taking into account the very high interest in the problems posed, there is a need to develop methodological solutions that provide the process of substantiating the influence of distributed data storage technologies on the possible growth dynamics of individual sectors of the national economy. The solution to this problem will allow not only to move from qualitative assessments to formalized ones based on the use of special methods of economic and mathematical modeling, but also to form a foundation for the creation and development of the theory of blockchain economics.

2 Research methodology

It is important to emphasize that the development of economic sectors is characterized by heterogeneity and differentiation, both in terms of dynamics and in terms of their sensitivity to transformations in the external and internal environment. The integration of blockchain technologies into the economic environment of the national economic system will undoubtedly affect the key parameters of the development of economic activities in the same uneven and disproportionate way.

Developing this research paradigm, the authors make an attempt to develop an appropriate methodological toolkit, the testing of which is implemented on the example of certain types of economic activity corresponding to the OKVED codes (Russian National Classifier of Economic Activities).

The work used quarterly data for the period from 2008 to 2019 from official sources [9]. The calculations were carried out using the Eviews statistical package. Table 1 shows the variables of the developed model, their symbols and data sources. The choice of these exogenous factors used in the construction of the corresponding models is due to their priority in terms of sensitivity to the penetration of blockchain technologies into the economic environment [10, 11].

Table 1. Description of the variables of the developed model.

Variable	**Designation**	**Data source**
Dependent		
Gross value added (in the context of each analyzed sector of the Russian economy), billion rubles	GVA	Rosstat
Undependent		
Stock market trading volume, billion rubles	V_{trades}	Moscow Exchange
Money transfers through the payment system of the Bank of Russia using translation services / settlement systems, billion rubles	$V_{transactions}$	Central Bank of Russia
Total profit / loss of operating credit institutions, million rubles	$V_{finresults}$	Central Bank of Russia

An important methodological aspect that predetermined the order of constructing a model is that in the case of studying stochastic time series, the use of traditional methods of correlation-regression analysis can lead to problems expressed in bias, inconsistency and inefficiency of the estimates obtained. This means that such a model may be unsuitable for further analysis and forecasting.

The study of dependencies between financial (stochastic) time series can be carried out using the method of cointegration analysis, where the cointegration equation has the form:

$$\Delta Y_t = \rho Y_{t-1} + \sum_{i=1}^{m-1} T_i \Delta Y_{t-i} \, \varphi X_t + \varepsilon_t \quad (1)$$

$$\rho = \sum_{i=1}^{n} A_i - I \text{ and } T_i = -\sum_{j=i+1}^{n} A_j . \quad (2)$$

In a concentrated form, Figure 1 shows a graphical interpretation of the algorithm for assessing the impact of diffusion of blockchain technologies on the development of sectors of the national economy.

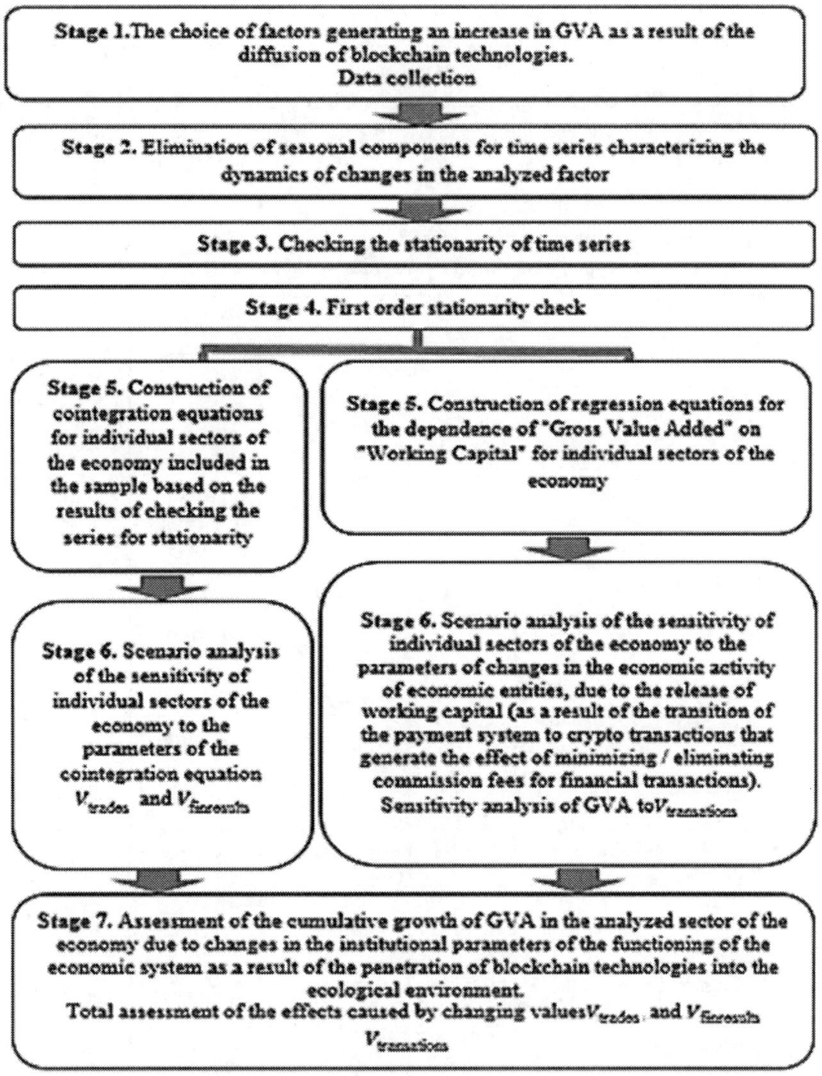

Figure 1. Graphical interpretation of the algorithm for assessing the impact of diffusion of blockchain technologies on the development of sectors of the national economy.

3 Research results

According to the results of testing the analyzed time series for their stationarity, the integrated series of the first order are the series characterizing the development parameters of the following sectors of the Russian economy:

— agriculture,
— mining,
— manufacturing industries,
— trade,
— financial and insurance activities,
— transportation and storage,
— public administration and military security; social security,

— activities in the field of health care and social services.

The identified sectors of the economy are suitable for applying the method of cointegration research and formed the basis for further iterations.

In accordance with the proposed research algorithm (Figure 2), for each type of economic activity that fell into the investigated, according to the results of evaluating the stationarity of the series, the sample, the corresponding cointegration equations are constructed. The results are shown in Table 2.

Table 2. Results and data of the statistical significance of the cointegration equations in the context of the analyzed types of economic activities included in the studied sample.

	Cointegration equation parameters			R^2
	Stock market trading volume	**The number of money transfers made through the payment system of the Bank of Russia, million rubles**	**Balanced financial result of organizations' activities, billion rubles**	
agriculture	0.000852	0.652552	0.04299	0.79
mining	0.140868	1.844097	0.300956	0.7
manufacturing industries	0.061163	7.250821	0.155	0.80
trade	-0.003355	5.800934	0.070829	0.87
financial and insurance activities	0.034893	0.472347	0.047427	0.76
transportation and storage	0.000803	0.635689	0.005392	0.62
public administration and military security; social security	0.014	0.07	0.007	0.75
activities in the field of health and social services	0.004177	0.637559	0.009911	0.84

All the constructed equations included in the sample have a high coefficient of determination (R-squared), which indicates their statistical significance.

In accordance with the data published earlier by the authors [10, 11], the possible effects caused by the adjustment of the analyzed factors (V_{trades} and $V_{finresults}$, $V_{transactions}$) as a result of the penetration of blockchain technologies into the system of economic relations are shown in Table 3.

Table 3. Possible effects for the national economy of the Russian Federation caused by the adjustment of the studied factors ($V_{finresults}$, V_{trades}, $V_{transactions}$) as a result of the penetration of blockchain technologies into the system of economic relations.

No	The exogenous factor of the cointegration model	Expected, in accordance with the scenario analysis, increase in the factor value, in billion rubles.
1	$V_{finresults}$ - the total amount of profit / loss received by operating credit institutions	+ 88.5 per year; + 22.125 average quarter
2	V_{trades} - the total amount of profit / loss received by operating credit institutions	+ 462.61 on average per quarter
3	Increase in working capital, intensification of business activity (effect of 1 factor $V_{transactions}$)	+128.0 per year (baseline scenario 1, Table 9)

The following is a sequence of calculations assessing the level of growth in gross value added as a result of the penetration of blockchain technologies into the national economy, as applied to a specific sector of the economy, was carried out on the basis of the following sequence of iterations (for example, the type of economic activity "Agriculture"):

1. Based on the data of the scenario analysis of the exogenous factors of the cointegration model $V_{finresults}$ and V_{trades} in accordance with the previously calculated effects (Table 3), the sensitivity of the gross value added of the studied sector to changes in the analyzed parameters is determined.

2. To determine the parameter assessing the impact of the transformation of the payment system as a result of its transition to crypto transactions, a separate regression equation is constructed, where the GVA of the studied sector of the economy acts as an endogenous variable, and its working capital (in billion rubles) as an exogenous variable.

Table 4. Assessment of the impact of blockchain technologies on the parameters of gross value added (GVA) growth in the context of the studied set of sectors of the Russian economy, %.

Economic activity	GVA growth per year, %	GVA growth per year, billion rubles
agriculture	0.24	3.1
mining	2.16	73.8
manufacturing industries	0.87	32.9
trade	0.0023	0.1
financial and insurance activities	1.7	18.1
transportation and storage	0.17	2.4
public administration and military security; social security	0.58	9.3
activities in the field of health and social services	0.43	2.7

4 Discussion of the results

The obtained results, revealing the peculiarities of the influence of the penetration of blockchain technologies into the system of economic and operational processes of the national economy, form the basic foundations for conducting derivative forms of analysis, including clustering the studied types of economic activities by the type of their response to the penetration of distributed data storage technologies into the economic environment.

In accordance with the rules of statistical clustering, below are the main results of this type of work. The research was carried out on the basis of the use of hierarchical cluster analysis. When carrying out calculations and assessments, the information-automated package SPSS was used.

The preliminary results of the calculations made it possible to determine the number of enlarged groups (clusters) with similar characteristics, revealing the peculiarities of the influence of the analyzed three effects (growth of business activity as a result of the release of working capital; socialization of access channels to stock markets and growth of financial results of credit institutions) caused by "blockchainization" of national economic system (Figure 3).

The initial data of the parameters (factors) used for clustering the analyzed set of sectors of the national economy are presented in Table 5.

Table 5. Values of the factors used in the framework of cluster analysis aimed at determining homogeneous sectors of the economy by the degree of response to GVA growth as a result of diffusion of blockchain technologies.

Economic activity	GVA growth per year, billion rubles			
	$V_{finresults}$ - total profit / loss received by operating credit institutions	V_{trades} - stock market volume	Increase in working capital, revitalization of business activity	TOTAL
agriculture	0.95	0.39	1.79	3.1
mining	6.65	65.16	1.98	73.8
manufacturing industries	3.43	28.29	1.15	32.9
trade	1.567	-1.55	0.065	0.1
financial and insurance activities	1.05	16.14	0.86	18.1
transportation and storage	1.95	0.37	0.119298	2.4
public administration and military security; social security	0.16	6.8	2.36	9.3
activities in the field of health and social services	0.21	1.93	0.54	2.7

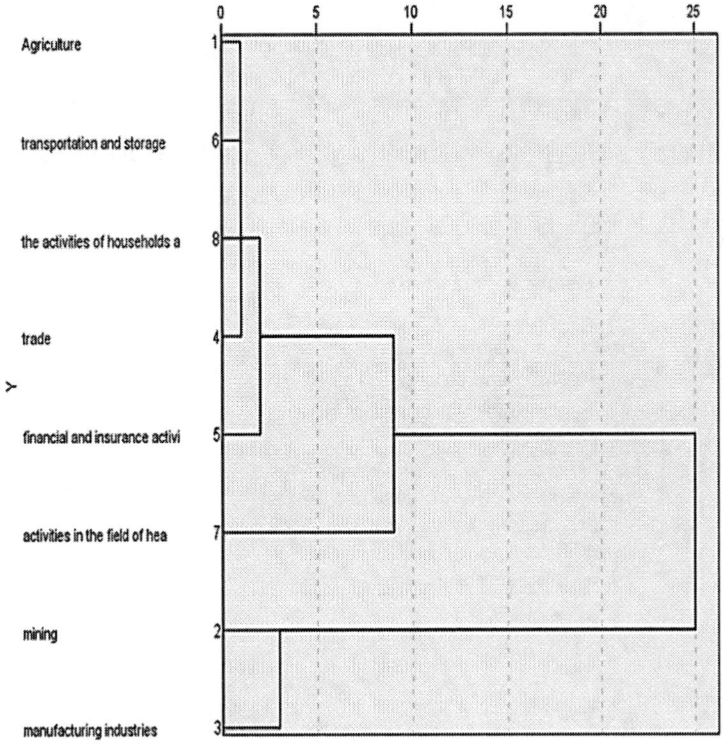

Figure 2. Dendrogram using the intragroup data method.

The data presented on the dendrogram demonstrate the main regularities in the formation of clusters located at a relatively large distance from each other. Interpretation of the results of hierarchical analysis demonstrates the need to create two clusters.

Having carried out the final stage of cluster analysis by the k-means method, the values of the final centers of the clusters were obtained (Table 6), as well as the belonging of the analyzed types of economic activities to clusters characterizing the influence of the studied externalities caused by the "blockchainization" of the economic environment on the gross value added growth. As a result, Table 7 shows the grouping of the studied types of economic activity, in accordance with the generalizing signs of the GVA growth as a result of sensitivity to the analyzed factors V_{trades}, $V_{finresults}$, $V_{transactions}$.

Table 6. Final centers of clusters.

Factors	**Cluster**	
	1	**2**
V_{trades}	0.0288	0.289
$V_{finresults}$	0.7069	6.5259
$V_{transactions}$	0.0641	0.1129

Table 7. Belonging of the analyzed types of economic activities to clusters characterizing the influence of the studied externalities caused by the "blockchainizatoin" "of the economic environment on the GVA growth in rubles.

Economic activity	Cluster belonging
agriculture	1
mining	2
manufacturing industries	2
trade	1
financial and insurance activities	2
transportation and storage	1
activities in the field of health and social services	1
the activities of households as employers; undifferentiated activities of private households for the production of goods and services for their own consumption	1

5 Summary

The implemented assessments demonstrate the formation of two key clusters that unite the sectors of the Russian economy in terms of the degree of influence on them by externalities caused by the integration of blockchain technologies into the socioeconomic environment. The first, the broadest, cluster includes types of economic activities characterized by a moderate response to the processes of "blockchainization", the second includes sectors of the economy that demonstrate a very noticeable sensitivity to the integration of the studied technologies into the ecological environment (manufacturing, mining, agriculture). The main distinguishing characteristic of the sectors of the economy belonging to the second cluster group is a high level of response to changes in the financial results of credit institutions caused by the processes of "blockchainization". In addition, these sectors of the economy demonstrate significant sensitivity to changes in the processes of socialization of stock markets under the pressure of blockchain technologies. Apparently, this is due to the fact that economic entities of these types of economic activities are very strongly integrated into the system of stock markets, and therefore react extremely positively to the growth of exchange trading activity. At the same time, these sectors show a less noticeable reaction to the transformation of the institutional environment caused by the transition of the payment system to the crypto environment. This is most likely due to their lower capital turnover ratios.

6 Conclusion

In general, despite the heterogeneous degree of response to penetration into the system of economic processes, the analyzed sectors of the economy demonstrate an overwhelmingly positive reaction to digitalization in the form of diffusion of blockchain technologies (in the context of the studied areas).

Acknowledgements

The study was supported by a grant from the Russian Science Foundation (project No. 19-18-00202).

References

1. E. A. Pekhtereva, Economic and social problems of Russia, **1(37)**, 71 (2018)
2. R. K. Nurmukhametov, Financial analytics: problems and solutions **11**, **2(344)**, 179 (2018)
3. Yu. A. Konopleva, Economics and Management: Problems, Solutions, **5(4)**, 136 (2018)
4. H. Vranken, Current Opinion in Environmental Sustainability, **28**, 1 (2017)
5. K. J. Kim, S. P. Hong, Int. J. of Security and its Application, **10(11)**, 201 (2016)
6. A. F. Bariviera, M. J. Basgall, W. Hasperué, M. Naiouf, Physica *A 484*, 82 (2017)
7. L. Cocco, G. Concas, M. Marchesi, J. of Economic Interaction and Coordination, **12(2)**, 345 (2017)
8. G. Pieters, S. Vivanco, Information Economics and Policy, **39**, 1 (2017)
9. Federal State Statistics Service of the Russian Federation, https://rosstat.gov.ru/
10. M. R. Safiullin, A. A. Abdukaeva, L. A. Elshin, M.V. Savushkin, University Bulletin, **7**, 154 (2020)
11. M. R. Safiullin, R. T. Burganov, L. A. Elshin, A. A. Abdukaeva, Theoretical and Applied Economics, **3**, 105 (2020)
12. N. G. Borisova, M. F. Grigoriev, L. Yu. Dragileva, L. A. Elshin, L. A. Zbinyakova, A.V. Ivanov, N. R. Izhguzina and et al., *Factors of sustainable development of Russian regions* (Novosibirsk, 2015)

Problems of residential migration in the framework of the formation of "smart agglomeration"

Aidar Asylbayev[1], *Kunduzkul* Niiazalieva[2], *Natalia* Brovko[1], and *Malik* Borbugulov[3,*]

[1]SEIHPE Kyrgyz-Russian Slavic University named after the first President of the Russian Federation B. N. Yeltsin, Bishkek, Kyrgyz Republic
[2]Educational institution autonomous non-profit organization of higher education "Kyrgyz-Russian Academy of Education", Bishkek, Kyrgyz Republic
[3]Ala-Too International University, Bishkek, Kyrgyz Republic

Abstract. The relevance of the work is in solving problems arising from the disproportionate development of migration flows and natural and technical resources. The paper examines the development of migratory residential areas in the context of the need to form an integral system of smart satellite cities around the capital – Bishkek city. There is also a study of the environmental and social problems that arise due to the lack of a systematic approach to the management of the city and its satellites. The existing today, numerous individual housing buildings around the capital cause a serious problem for the rational use of environmental and engineering resources and balanced demographic development of the whole country. To solve the impending problems, it is recommended to build a fundamentally new model of an urban and environmentally friendly agglomeration. Thus, the purpose of the article is to identify problems among new immigration residential buildings and rationale for the need to build city-forming "smart zones of urban development" with a control center in Bishkek. Conclusions and recommendations are given for the government and administrative services of the Kyrgyz Republic.

1 Introduction

Nowadays, the world is facing not only with the problem of the quantity of housing, but also with the necessity of building environmentally friendly, safe and affordable housing. There is no secret that the global trend is the concept of "environmentally friendly, safe and smart city", which is aimed at increasing the level of environmental safety and comfort of living for all segments of the population in cities of different scale and purpose. For example, the presence of green areas, undeveloped spatial areas, smart housing and transport systems help to improve the health and safety of the population and protect people's interests from various of risks and threats.

There is a big problem for the Kyrgyz Republic, which is a high and active internal migration to an economically developed region, Chui oblast, where the capital of the country,

[*]Corresponding author: malik.borbugulov@alatoo.edu.kg

Bishkek, is located. It is obvious that such migration will predict the formation of demand for real estate, and first of all for land plots for individual housing construction. It is necessary to take into account that these plots are belonged to green zones or agricultural zones with high quality land. It is worth noting that at current conditions the internal migration is represented by residents moving to Bishkek from the rural areas of the southern region. Internal migration is present in many countries, where compatriots from rural settlements move to urbanized cities with a high demand for labor. Such picture of the internal movement of the population is a natural internal migration process. However, it should be taken into account that the excessive level of internal migration shows the disproportionate economic development of the regions of the country. The world practice shows that it often leads to an excessive overpopulation of capital and business cities.

In our case, overpopulation should be understood not as an excessively large accumulation of newcomers, but as the failure of the city to provide a decent environmental condition, engineering and social services and other infrastructure services to new individual developers. At the same time, migration residential areas quite successfully solve the demographic problem of big cities, a small birth rate among the local urban population. As a result, the absence of an influx of population, causes a shortage of labor resources for the functioning of the city itself.

The relevance of the article at the time of the active influx of the population is beyond doubt. The unorganized appearance of housing estates and, at the same time, the unorganized registration of the population in them leads to the adoption of inadequate management decisions on the provision of engineering and other infrastructure communications. Without knowledge about the exact size of the population and the density of residence in new individual buildings the state is unable to provide this population with social facilities in a decent amount. Also, the lack of legal status, due to the absence of registration, creates additional barriers to the successful development of a smart city. Not managed migration residential areas around the capital can cause a disproportionate demographic growth of the population. In particular, such migration leads to the appearance of regions with demographic failure and abandoned settlements, which, in turn, create additional economic and environmental problems.

In that regard, the authors' assertion that the problems of active growth of individual new buildings have led to the fact that the capital cannot adequately adapt to the large influx of internal migrants is true. Therefore, an undeveloped infrastructure and the absence of basic conditions for life, new immigration buildings eventually become the center of environmental pollution and social injustice with a high percentage of dissatisfaction with the country. Thus, the social problems, accumulated over the years and unresolved in time, of internally displaced people contribute to the intensification of the already exacerbated social tension in the society [1].

The purpose of the article is the identification of problems among new residential buildings of internal new immigrants and justification of the need to build city-forming "smart belts of urban development" with a control center in Bishkek.

Analysis of the latest international and domestic publications. Foreign studies concerning migration and the housing issue of migrants include the works of the following authors: C. H. Mulder (2006, 2009) [2; 3], T. De Vroome (2010) [4], F. Van Tubergen (2010) [4], Thomas J . Cooke (2009) [3].

In the process of preparing the article, the following researches of Russian and domestic scientists related to the problems of housing and demographic development of the country were studied: A. N. Asaul. [5; 6], K. Isaeva [7], S. Borsokbaeva [7], G.V. Kumskova [8], K. N. Niyazalieva [8], R. B. Salmorbekova [9], N. A. Israilova [9], U. Zh. Ergeshbaeva [10], etc.

2 Materials and methods

It should be noted that the fact of identifying the connection between housing problems and urbanization of the population will make it possible to form an effective unified demographic and housing strategy within the framework of the development of "smart" cities and environmentally friendly satellite cities. At the same time, the improvement of the housing situation among the citizens of Kyrgyzstan creates the foundation for the development of positive dynamics of demographic processes by eliminating or minimizing factors, including environmental ones, that have a negative impact on the dynamics of reproduction and migration of the population.

Based on the foregoing, the research methodology of the topic was based on general methods of systematizing information, analysis and synthesis, methods of classifying and isolating information, as well as methods of survey data analysis on the problems of an agglomeration planning.

3 Results and Discussion

If the internal migration is viewed through the lens of housing aspects, then it is a change in the ways, models of citizens' life with the obligatory change of housing located in more than one city or town [11]. The availability of affordable, decent and environmentally friendly housing in a prosperous region attracts migrants and at the same time serves as a reason for the permanent abode of local citizens. And as the reverse process, a limited, for any reason, access to a decent housing in the proposed area leads to a sharp decrease in migration to this area and, moreover, increases the outflow of the population from the area [12]. Thus, there is a relationship between the dynamics of the migration flow and disproportionate housing conditions at the point of residence and arrival. More appropriate hypotheses to explain this situation is the following: the smaller the number of the population needs internal migration, the more closely it is connected with living conditions. A decent housing and living conditions are an essential factor for returning to the home area. This means that people satisfied with their living and environmentally friendly living conditions on the spot do not seek to change their place of residence. Thus, the demographic effect of "ageing-in-place" is manifested in a favorable natural environment.

However, in the context of independent Kyrgyzstan, the internal migration, which varies around 25%, has caused an unregulated urban growth, especially in the capital Bishkek. Thereby putting at serious risk a systematic and predictable development of the capital, a demographic development of waste regions, an agrarian potential of villages, villages and other settlements [13].

The state policy of Kazakhstan towards the capital Nur-Sultan can serve as an example of an organized migration flow. The dynamics of the numerical and the urban population there has increased dramatically, but at the same time the volume of housing construction and infrastructure support to the social objects is closely linked to the dynamics of growth of internal migrants. In general, it is an effective program for the movement of citizens from a region with a surplus of labor force to the rapidly developing urbanized center of the country. Such a centralized migration policy ensured social stability and economic growth through the proportional distribution of labor and production resources [14].

In Kyrgyzstan due to the absence of systemic state regulation of housing construction and demographic policy, the capital's authorities did not have the opportunity to fully respond to the modern challenges of migration flows. The authorities did not fully appreciate the scale of the migration mobility of their compatriots.

In this way, new housing estates have become hotbeds of social instability and environmental danger. The source of instability was represented by the issues of inadequate provision and maintenance of migratory housing estates with infrastructural and socially significant facilities. Household waste and the exploitation of natural resources are not regulated by a centralized system, which leads to an unfavorable environmental situation in the new housing estate. It turns out, as if they live in the capital, but at the same time do not have basic living conditions, since there are no schools, hospitals, clinics, kindergartens, shopping centers and there are electricity, water and roads in some housing estates. City service organizations refuse to provide them with services due to a systemic collapse. Infrastructure facilities of some new buildings do not meet the needs of this residential area. Meanwhile, many residents already have all permits for residence and construction in the metropolitan area.

The current situation in Bishkek reflects that the existing housing problems in the context of the demographic growth of the population have begun to negatively affect the quality of life of all citizens. It became obvious that the entire engineering and social infrastructure of the city is under excessive pressure, because it wasn't designed for such intensive work, and requires an early remodeling.

Recently, the capital has been struggling to cope with car traffic, the work of urban utilities is wearing out quickly, social facilities cannot cope with the increasing flow of people, and the migration wave not only does not subside, but is also difficult to control. The pace of construction of socially significant and housing facilities doesn't meet the consumer needs of the population.

But the danger of unregulated development of suburban migrant settlements cannot be underestimated. Such concepts as "uncontrolled areas", "black belt", "shahid's belt", "disadvantaged belt" and similar names appeared as a result of active internal migration not controlled by the state. Some migratory settlements appeared with the tacit consent of the state, others out of objective social necessity in order to meet their needs with minimal financial resources. As a result, households are organized around the capital that cannot afford expensive housing within the city.

Today, therefore, due to the active internal migration it is an important task not only to develop migratory residential areas and small towns, but also to organize a modern residential area - the formation of a modern Bishkek agglomeration.

It is not a secret that the core problem of Bishkek is the territorial restriction in the expansion of the city due to the natural landscape. There are many well-grounded opinions about not expanding the construction zone, but finding ways to build higher-rise buildings, including residential complexes. However, such a decision only postpones the decision to expand the city's territory, but in general doesn't solve the issue of providing citizens, especially migrants, with housing space. With the expansion of the boundaries of the urban area, it is necessary to modernize management activities with a certain delegation of authority.

It is important to note that the internal migration situation continues to this day, this trend began in 1989. The problem of internal migration lies in the disproportionate development of the country's regions [15]. So, since then, more than 50 new housing estates have been organized by migrants around Bishkek. In most of them social facilities and engineering communications haven't yet been built. This is indeed a violation of citizens' rights to own decent and affordable housing.

At the same time, the statement that such characteristics are inherent in all new buildings is absolutely erroneous. Among the new housing estates, there is a clear distribution of them according to the profitability of the homeowners themselves. Basically, the differentiation by value is based on the geographical and climatic location of the settlement [13]. The owners of housing located in new buildings in the southern part of the capital, which is considered more expensive and prestigious, are citizens with high incomes. And, conversely, individual new buildings in the northern part of the capital, mainly the middle and lower middle stratum of the population live, with the exception of some households. The overwhelming majority of internal migrants settle down precisely among the poor housing estates.

If at the beginning of the active period of self-development in residential new buildings housing was mainly built by city dwellers, then in subsequent years these lands for individual housing became attractive for rural migrants. At the same time, it is difficult to track the number of internal migrants, due to the lack of a systematic registration of the living in migration areas.

According to R. Salmorbekova (2010) and N.A. Israilov (2010) the residential areas are home to 250.0 - 300.0 thousand people [9]. The situation is aggravated by the fact that some dwellings haven't been commissioned, so residents cannot register here and obtain a legal status.

Thus, the urban area has increased by 2,500 hectares due to the last 24 migratory areas. The supply of water there is 37.0%, electricity - 56.0%, roads - 57.0%. Let us especially note the problems of the "Ala-Archa" housing estate, housing was built near the "sarcophagus", where the carcasses of animals infected with dangerous diseases, such as foot-and-mouth disease, anthrax and so on, were disposed of. Accommodation is not permitted within a 500-meter radius.

The residents of migration settlements don't have an access to important socially significant services. For example, due to the lack of numbering or names of streets and house numbers, it is not possible to call an emergency medical assistance. Only in 18 large new buildings there are full-fledged schools, and in new housing estates there are no secondary schools at all, and therefore 7.0% of children living in such new buildings don't attend school on a regular basis. Meanwhile, the existing schools in that area are a three-fold overcrowding, the school is designed for 380 students, but over 1000 is studying and children sit down for 3 children at one desk [16].

According to the statistics of the Kyrgyz Republic, 74.4% of migrants are between the ages of 25-60. At the same time, 16.5% came from Naryn region, 5.4% from Issyk-Kul region, 5.5% from Talas region, 9.4% from Jalal-Abad region and 11.5% from Osh region. The reasons for migration were the unemployment - 41.2%, future prospects - 23.1%, standard of living - 12.1%, education - 11.6% and family circumstances - 9.6% [17].

According to the theory of territorial organization of the migrant population by J. Gibbs, A. Lesh, W. Izard [18], today's capital of Kyrgyzstan belongs to the fourth class of urbanization, there is an active movement of the population to large cities with a simultaneous demographic failure in smaller settlements.

However, there is an undesirable feature of urbanization - the excess concentration of migrants, local population and other people in the capital. More than 20.0% of the total population lives in Bishkek. Nowadays the population of Bishkek, according to independent experts, is about 1.4 million people with a total population of more than 6.5 million people. Taking into account the growth of the population by 2-3%, the forecast of the city's population in 25 years will be more than 2.5 million citizens [19].

The concept of the so-called discontinuous/continuous spatial development is proposed with such growth rates of Bishkek. It refers to the discontinuity of the living environment with the continuity of communications and the natural environment. This form of spatial growth of the city is attractive, firstly, by the preservation and implementation of natural resources between residential formations, and secondly, by an increase in the speed of movement depending on the frequency of necessary stops of high-speed public transport [19].

The versatile structure of the capital is attractive for its importance for future expansion, using various technical and planning solutions and with the possibility of correlating with the surrounding natural and ecological conditions [20]. The capital's scheme has a significant land reserve without active development of any real estate objects. The improvement of the structural scheme of the city will occur due to a gradual increase in the functional saturation of residential and recreational spaces. And the roads and railways must ensure the accessibility of the main functional areas of the capital and nearby microdistricts. Such an urbanistic scheme makes it possible to organically combine the natural discontinuity of urban construction and the compactness of individual migratory settlements.

It is important to note that the combined and diverse structural scheme of the future development of the agglomeration affects the planning of the landscape environment, engineering communications, control and planning of the demographic situation and socio-cultural planning.

The spatial expansion of the city and migratory housing estates within the framework of the chosen urban planning model will allow the city's growth to "loop back", redirecting it to large-scale development. At the same time, new inconvenient territories will be developed, and the looking, diagonal-toothed form of the city plan will gradually be been modernized into a multidirectional model [19].

4 Recommendations and Conclusion

It is proposed to create the Administration of the Bishkek agglomeration with the purpose of centralized and systemic management of the construction of all objects. Further, it is recommended to transfer to the administration the authority to regulate and manage migration housing estates. The Bishkek agglomeration should become the main source of housing distribution and the center of the country's economic development. Alternative district administrations will manage migration flows around cities - satellites of the agglomeration.

Due to such model of agglomeration, it is assumed that the growth of the capital's city dwellers will decrease by 25.0% and the corresponding growth in the population of satellite cities and migratory residential areas of the agglomeration will take place. The transit traffic through the city of Bishkek will sharply decrease and the growth rate of private vehicles will decrease by 50%. The ecological situation in the capital will significantly improve due to the transfer of industrial zones outside the city and their placement in close proximity to the international transport and air corridor [19].

By studying the problems of individual migratory buildings and constructions, the following difficulties were identified: problems of access to and/or lack of quality drinking water, problems of asphalt roads and sidewalks, lack of socially significant educational and public facilities, lack of public hospitals and clinics, absence of recreation and sports leisure centers. Such shortcomings in migration areas create unequal social conditions for migrants and city residents.

With regard to the relationship between the rate of migration and the housing conditions, there is a close relationship between private home ownership and migration. In particular, private homeowners are less willing and inclined to leave their homes than tenants. The main reason for not moving is that the cost of housing in the regions is much lower than in the capital. And also, the costs of moving for homeowners are much higher than for tenants. In this regard, the younger generation is prone to migration than the older settled household. Therefore, the projects of attracting migrants to areas with a shortage of labor resources through affordable and high-quality housing are interesting.

The reorientation of a large city and adjacent settlements towards the formation of an agglomeration and the determination of the legal status of satellite cities is a natural result of development, which manifests itself in compliance with the functional structure of an agglomeration with its changing needs. The development of the agglomeration sets the goal of not only increasing the scale of development, but also improving the quality of life of the population as a whole.

The depressed residential areas of the future agglomeration will revive due to the development of a network of railways and highways. Such transport logistics routes will create a single territorial and market system and increase the mobility of the population. The construction and expansion of the agglomeration is a large and attractive investment object. The entire agglomeration project can be recognized as a desirable investment potential for foreign investors [21]. In order to maintain the attractiveness of the agglomeration, it is necessary to build and launch production facilities with appropriate infrastructure facilities.

It should be noted that the proportional regional development of the country reduces the rate of oriented migration and thereby reduces the demographic burden on Bishkek and its suburbs. The creation of an agglomeration will make it possible to control the number of residential new buildings, the number of migrants and the level of social benefits in them.

The listed circumstances determine the need to expand the state policy aimed at solving the problems associated with oriented internal migration. It is necessary to strengthen the implementation of measures taken by the state bodies and agencies aimed at optimizing and regulating migration processes from the regions of the country.

References

1. A. B. Asylbaev, Bull. of the Taj. Nat. Univ, **2/4-2**, 6 (2017)
2. C. H. Mulder, Demographic Research, **15**, 401 (2006)
3. C. H. Mulder, H. Clara H., and J. C. Thomas, Population, Space and Place, **15(4)**, 299 (2009)
4. T. De Vroome, F. Van Tubergen, International Migration Review, **44(2)**, 376-403 (2010)
5. A. N. Asaul, G. V. Kumskov, M. T. Kasymova et al., *Economics and property management: a textbook for universities in Bishkek*, 428 (2012)
6. A. N. Asaul, M. A. Asaul, P. B. Lyulin, N. V. Chepachenko, Problems of forecasting, **3(174)**, 111 (2019)
7. K. Isaev, S. Borsokbaeva, *Population: Quality of life and health,* 94 (2002)
8. G. V. Kumskov, A. B. Asylbaev, K. N. Niyazalieva, Bull. of the Taj. Nat. Univ. **2/3**, 87 (2017)
9. R. Salmorbekova, N. A. Israilova, *Social tension in modern Kyrgyzstan,* 31 (2010)
10. U. Zh. Ergeshbaev, Scientific research in the Kyrgyz Republic, **1**, 22 (2020)
11. A. B. Asylbaev , K. N. Niyazalieva , A. Bakasov, Bull. of the Kyr.-Russ. Slav. Univ., **18(11)**, 8 (2018)
12. A. B. Asylbaev, Bull. of the Taj. Nat. Univ, **2/4**, 6 (2017)
13. A. B. Asylbaev, Competitiveness in the global world: economics, science, technology, **10(57)**, 134 (2017)
14. K. B. Asylbaev, A. B. Asylbaev, K. N. Niyazaliva, Actual problems of economics. **180(6)**, 231 (2016)
15. A. B. Asylbaev, K. N. Niyazalieva, Bull. of the Taj. Nat. Univ, **3**, 15 (2018)
16. A. B. Asylbaev, Bull. of the Kyr.-Russ. Slav. Univ., **15(1)**, 97 (2015)
17. National Statistical Committee of the Kyrgyz Republic, the Ministry of Health [Kyrgyz Republic] and the ICF International, Demographic and Health Survey of the Kyrgyz Republic 2012, 488 (2013)
18. The main patterns of the development of urbanization. The theory of J. Gibbs, A. Lesch, W. Izard.

19. M. Kurmanbekova, B. Asanakunov, S. Dresvyannikov, The concept of the formation and development of the Bishkek agglomeration and adjacent city. http://analitika.akipress.org.
20. T. I. Turdiev. Financial analytics: problems and solutions, **7(289)**, 17 (2016)
21. Yu. A. Levin, A. M. Platonov, Economy of region, **16(3)**, 884 (2020)

«Green infrastructure» of urbanized territories

Ljubov Maksimenko[1], *Olga* Dudinova[1], and *Olga* Korobova[2]

[1]Siberian State University of Geosystems and Technologies, 10 Plakhotny str., 630108 Novosibirsk, Russia
[2]Novosibirsk State University of Architecture and Civil Engineering, 113 Leningradskaya str., 630008 Novosibirsk, Russia

Abstract. Currently, there is a process of spatial concentration of the population and the spread of urban lifestyle, which inevitably leads to the expansion of the boundaries of cities and their sphere of influence, an increase in the diversity of activities. In this regard, the creation and maintenance of a comfortable urban environment of a megalopolis is becoming an increasingly urgent managerial task. The decisive factors in supporting urbanized territories in a comfortable state for living are the objects of" green " infrastructure. One of the important issues, at the same time, is the information support of the management system for the territories of land plots with green spaces.

1 Introduction

Urbanization as a process of spatial concentration of the urban population and the spread of the urban lifestyle inevitably leads to an increase in the diversity and complexity of infrastructures, the concentration and intensification of production, differentiation and an increase in the diversity of urban activities. In this regard, the creation and maintenance of a comfortable urban environment of a megalopolis is becoming an increasingly urgent managerial task [1]. Urbanization as a natural-historical process has led to the emergence of a wide range of problems, the analysis and evaluation of which becomes the main content of designing the infrastructure of a modern city. The decisive factors in supporting urbanized territories in a comfortable state for living are the objects of" green " infrastructure. "Green" infrastructure in general is a strategically planned network of natural territories with developed and managed innovative processes for creating a comfortable urban environment. One of the important issues, at the same time, is the information support of the management system for the territories of land plots with green spaces [2].

2 Materials and Methods

Mechanisms for the development of a comfortable urban environment in urbanized territories are currently supported by national projects and departmental projects of the government of the Russian Federation [3-5].

Let's consider the main terms and definitions from the regulatory and technical documentation on the essence of the issue under study. The formation of land plots is considered in the Land Code of the Russian Federation [6, Article 11.2.]. The concept of public land plots is fixed in the Civil Code of the Russian Federation [7, Article 262]. In the Urban Planning Code, the term public use territories is disclosed [8]. In SP 42.13330.2016, the following concepts are considered: "natural object", "green zone", "green areas" [9]. The Housing Code of the Russian Federation [10, Article 36] establishes the right of ownership of common property in an apartment building (MCD), including the land plot where the MCD object is located, including elements of landscaping and landscaping.

Legislative acts of federal cities (Moscow, St. Petersburg and Sevastopol [11, article 65], [12]) and the accumulated experience of these cities in the field of preservation and accounting of green areas are of no small importance for the development of other regions of the Russian Federation. Table 1 presents a list and names of documents reflecting the policy of municipalities of the largest megacities of the Russian Federation regarding the conservation and development of green areas. As can be seen from the table, the law of St. Petersburg [11] introduces the concept of "territory of public green spaces", which does not contradict and is consistent with the above definitions of "public land" and "green areas".

Table 1.

Разработанные документы	Установленные приоритеты
"About public green spaces". St. Petersburghttp://docs.cntd.ru/document/8458668	The list and borders of territories of green zones of general use are established [11].
"On the protection of green spaces" the law of the city of Moscow. http://docs.cntd.ru/document/901734936	The registration of green spaces is carried out in accordance with the procedure established by the Government of Moscow. For the purposes of accounting, a register of green spaces is created [12].
The concept of the development of green public spaces of citywide significance "Green Novosibirsk".	"Formation of a comfortable urban environment" [13] in Novosibirsk for 2018-2022

Thus, the object under consideration "territory of public green spaces" has all the characteristics of a real estate object, is subject to cadastral registration and registration of ownership rights in the EGRN [14].

As a result of the conducted research, the authors propose to make changes for the classification of types of permitted use of land plots in the section public land plots as an addition with the code 12.0.3. "Public green spaces". This type of permitted use uniquely and holistically characterizes the territories of green spaces of general use, which are territories (land plots) of general use. This classification is also necessary for the preservation and development of valuable natural territories, for the placement of recreation and recreation facilities, entertainment within urban areas.

The development of urbanized territories in our country is due to the current national project "Housing and the urban Environment" and activities within the framework of the federal project "Formation of a comfortable urban environment" [2]. In 2019, the methodology for forming the urban environment quality index was developed and approved. The list of indicators includes indicators: "The level of landscaping" and "The state of green spaces". A unified methodology has not yet been developed for calculating these indicators, so the analysis and generalization of experience, the choice of software is an urgent and timely task. According to the results of decryption and interpretation of satellite images of open access, such indicators as "The level of landscaping" and "The State of landscaping"are determined. The quantitative assessment of these indicators is made by calculating the vegetation index NDVI (Normalized Difference Vegetation Index), as the share of the territory with planted areas of increased biomass density in the total area of landscaping of the city. The method of calculating the NDVI indicator is one of the research goals of this work. The features of the implementation and calculation of the NDVI indicator value in individual GIS (ArcGIS, SAGA, GRASS, ENVI, ERDAS IMAGINE) were analyzed. To analyze the changes in the NDVI index, Sentinel-2 L1C, Vegetation Indices satellite data were taken for two seasons, from 2018 to 2019. It was found that for the studied territory, the trend of changing the index values is the same from 0.45-0.55 in June to 0.58-0.6 in July. The obtained data can be used to calculate the threshold value of the calculated score of the indicator "state of green spaces». Assessment of the current level of comfort of the urban environment using the indicators "level of landscaping" and "state of green spaces", etc. it allows you to form and predict the development of a megapolis as a "single organism" capable of maintaining optimal parameters that ensure the normal functioning of each of its inhabitants in harmony with the urban environment. The orderly formation of a system of evaluation indicators of a comfortable urban environment, of course, increases the efficiency and attractiveness of investing investment resources in the development of "green" infrastructure facilities.

3 Results and Discussion

The development of urbanized territories in our country is due to the current national project "Housing and the urban Environment" and activities within the framework of the federal project "Formation of a comfortable urban environment" [2]. In 2019, the methodology for forming the urban environment quality index was developed and approved. The list of indicators includes indicators: "The level of landscaping" and "The state of green spaces". A unified methodology has not yet been developed for calculating these indicators, so the analysis and generalization of experience, the choice of software is an urgent and timely task. According to the results of decryption and interpretation of satellite images of open access, such indicators as "The level of landscaping" and "The State of landscaping «are determined. The quantitative assessment of these indicators is made by calculating the vegetation index NDVI (Normalized Difference Vegetation Index), as the share of the territory with planted areas of increased biomass density in the total area of landscaping of the city [17]. The method of calculating the NDVI indicator is one of the research goals of this work. The features of the implementation and calculation of the NDVI indicator value in individual GIS (ArcGIS, SAGA, GRASS, ENVI, ERDAS IMAGINE) were analyzed. To analyze the changes in the NDVI index, Sentinel-2 L1C, Vegetation Indices satellite data were taken for two seasons, from 2018 to 2019. It was found that for the studied territory, the trend of changing the index values is the same from 0.45-0.55 in June to 0.58-0.6 in July. The obtained data can be used to calculate the threshold value of the calculated score of the indicator "state of green spaces". Thus, the most important factors in supporting urbanized territories in a comfortable state for living are the objects of "green" infrastructure. For objective technical and cadastral accounting of "green infrastructure" objects, there is a need for a more in-depth study of the classifier of types of permitted use of public territories. To date, there are no questions about the need for surveying for cadastral registration of land plots of public green spaces and entering information into the unified state register, but the indication of the type of permitted use should be carried out according to the results of the actual purpose of the object, to exclude errors in state cadastral accounting [18,19]. The established status, in relation to land plots of green spaces of general use, excludes the construction of any real estate objects not related to its maintenance within its borders.

4 Conclusions

Objects of "green infrastructure" of various functional purposes should be included in the urban land use system as its main elements, in order to develop new approaches to regulating and planning the development of urban landscaping. During the operation of "green infrastructure" facilities, their effective accounting and inventory are of great importance, which, along with assessment and control, constitute a system of management measures, the result of which is the sustainable development of cities and improving the quality of life of people, a new value of real estate arises, the efficiency of using municipal resources increases, the investment attractiveness of territories increases. A technically competent and methodically prepared approach to the technical accounting of non-residential facilities, including infrastructure facilities, is digitalization, carried out on the basis of a technical inventory.

References

1. Urbanization and Development: Emerging Futures. World Cities Report 2016. Nairobi: United Nations Human Settlements Program (UN-Habitat) (2016)
2. Order of the Government of the Russian Federation of 23.03.2019 N 510-r (ed. From 05.11.2019) «On approval of the Methodology for forming the urban environment quality index»
3. "Protocol of the federal project" Formation of a comfortable urban environment" (approved by Minutes of the meeting of the project committee on the national project "Housing and urban environment" dated 21.12.2018 N 3)
4. Activities of the Ministry of Natural Resources and Ecology of the Russian Federation, http://www.mnr.gov.ru/activity/
5. Basic and additional requirements for smart cities (the "Smart City" standard). Start date of the action: March 04 (2019)
6. Land Code of the Russian Federation No. 136–FZ of 25.10.2001 (as amended on 27.12.2019), http://base.garant.ru/12124624
7. The Civil Code of the Russian Federation (part one) of 30.11.1994 N 51-FZ (as amended on 03.07.2016) (with amendments and additions, intro. effective from 01.09.2016)
8. The Urban Planning Code of the Russian Federation " from 29.12.2004 N 190-FZ (ed. from 03.07.2016) (with amendments and additions, intro. effective from 01.09.2016)
9. SP 42.13330.2016 Urban planning. Planning and construction of urban and rural settlements. Updated version of SNiP 2.07.01-89
10. Housing Code of the Russian Federation No. 188-FZ of 29.12.2004 (ed. From 03.08.2018)
11. " On public green spaces «(version effective from January 1, 2020) », http://docs.cntd.ru/document/8458668
12. " On the protection of green spaces " Law of the City of Moscow No. 17 of May 5, 1999, as amended on May 7, 2014), http://docs.cntd.ru/document/901734936
13. About the measures for the implementation of the priority project "Formation of a comfortable urban environment" in the city of Novosibirsk for 2018-2022, the Mayor's Office of Novosibirsk. Resolution No. 3411 of 19.07.2017, https://novo-sibirsk.ru/
14. Federal Law" On State Registration of real estate " dated 13.07.2015 N 218-FZ (latest edition)
15. Public cadastral map, https://pkk5.rosreestr.ru/

16. Classifier of types of permitted use of land plots Appendix to the order of the Ministry of Economic Development of the Russian Federation on September 1, 2014 540 as amended on April 8, 2019-Order of the Ministry of Economic Development of the Russian Federation of February 4, 2019 N 44
17. EO Browser, https://apps.sentinel-hub.com/eo-browser/
18. L. A. Maksimenko, O. S. Dudinova, Geo-Siberia **3(2)**, 251-258 (2019)
19. L. A. Maksimenko, O. Korobova, O. Dudinova, X. Soskova, 13 International Scientific Conference Architecture and Construction, Novosibirsk, 953, 012043 (2020)

 https://doi.org/10.1051/e3sconf/202131107004

Smart city technologies in the context of ensuring environmental safety and political stability

Oleg Evgenyevich Grishin[1,2], *Olga Alekseevna* Nesterchuk[1,3], *Sergey Ivanovich* Popov[1], *Alexander Vladimirovich* Tolochko[4], and *Olga Viktorovna* Levashova[4]

[1]Peoples' Friendship University of Russia (RUDN University), Moscow, Russia
[2]Russian State Social University, Moscow, Russia
[3]Russian Academy of National Economy and Public Administration under the President of the Russian Federation, Moscow, Russia
[4]Bunin Yelets State University, Yelets, Russia

Abstract. The article analyzes the problems of formation and implementation of technologies of "smart cities" through the prism of ensuring environmental safety and maintaining the political stability of a large city. It has been established that there are various aspects of the development of smart cities. An increased interest on the part of scientists, engineers, ecologists, politicians, government and international structures and institutions in the technologies of "smart cities" is noted. The focus of the study is on the fact that the range of real problems of a modern city includes the solution of problems of safety, ecology, harmonious formation of infrastructure, logistics and other aspects of its effective development. The effect of use of such technologies in the environmental sphere is considered in the context of maintaining political stability. The problem of radiophobia in Moscow is illustrated. The costs of the media functioning when covering environmental issues are indicated. The role of information and communication technologies as a toolkit contributing to development of smart cities is outlined. An example of Moscow metropolis radiation-ecological monitoring system is given. It is offered to replicate technologies of an integrated automated system for environmental monitoring of urban space. The authors complement the existing approaches to the study of smart city technologies in key aspects of ensuring environmental safety and maintaining political stability.

1 Introduction

Relevance of study of implementation and application of smart city technologies is due to the progressive development of scientific and technological progress, including digitalization, robotization, technologization, algorithmization, and automation of the life of the planet's population. Wherein, scientists are concerned with finding solutions to environmental and political problems that can potentially arise as a result of these processes. Currently, the topical objects of study of scientists are becoming – human ecology [1], environmental risks [2], environmental comfort [3, 4], environmental safety [5, 6]. The concept of a livable smart city emphasizes the imperative of functionality and integrated resilience in sustainable cities of the future. It is based on the needs of people and their natural environment, leading to the broader definition of urban sustainability. This concept offers many opportunities for residents of smart cities and is an attractive future market for innovative products and services [7]. Innovative offers in the context of implementation of smart city technologies are developing all over the world. The experience of using such technologies is just emerging. Currently, as a rule, there are no universal standards for classifying the maturity levels of initiatives and transformation models for implementation of a particular smart city. We focus on environmental safety in the context of the Smart City concept through the prism of ensuring political stability.

2 Materials and Methods

When studying implementation of technologies used in the innovative idea of smart cities in the context of ensuring environmental safety and political stability, the authors used a systemic, structural-functional and interdisciplinary approach to the problem under study. The study basis was formed by publications of domestic and foreign authors – specialists in the field of smart city technologies, environmental safety and political stability. The study methods were content analysis of mass media, participant observation, event analysis, case studies.

3 Results and Discussion

Modern society is riddled with many forms of environmental degradation that pose innumerable risks to human life. Hazardous waste, industrial disasters and environmental health risks are the original synthesized complex of this rapidly developing area of interdisciplinary study. An interdisciplinary approach is in demand for the critical study of environmental problems. It is essential to consider the key themes of power, justice and environmental sustainability and safety in the context of maintaining political stability.

It is stated that there are various aspects of development of smart cities, theoretical concepts and empirical studies, as well as examples of implementation of smart city programs and, on their basis, formation of the ability to create environmental values for citizens. There is a growing interest in this topic, supported by both national governments and international organizations [8]. Use of information and communication technologies in urban areas for the sustainable development of social and environmental living space involves implementation of projects to increase mobility, use smart water and energy systems, develop social networks, increase entrepreneurship, protect the environment and improve environmental safety and quality of life, as well as increased participation in political life [9].

The state of protection of the vital needs and interests of the individual and society as a whole determines a complex system of multi-vector tools for providing sanitary-epidemiological, educational, hygienic, technical and technological and others. This state is interpreted as absence of environmental hazard [10]. Environmental safety is becoming decisive conditions for survival of modern civilization. To reduce the environmental load on the Earth, it is necessary to change the strategy of technogenesis and solve many other pressing issues.

We focus on the technological aspect. To improve the efficiency of urban management, many countries use smart city technologies. Introduction and active use of digital technologies in cities leads to an increase, as a rule, of positive effects from network interaction, and minimization of various kinds of costs and obtaining peripheral socio-political, economic, logistic and other effects [11]. Some researchers believe that smart city technologies are rapidly infiltrating the everyday life of city dwellers, allowing them to recreate better performance in urban settlements. Analysis showed that their implementation in some European cities contributes to increase in the overall degree of safety, including environmental safety. Wherein, the amount of harmful emissions into the atmosphere decreases, and the consumption of resources decreases [12]. It is noted that the range of real tasks of a modern city includes issues of safety, ecology, well-coordinated formation of infrastructure, logistics and many other aspects of its effective development. For each of the listed issues, the potential solutions on the information technology platform are possible. Engineering systems responsible for tracking potential threats to environmental safety, as a rule, are formed on the basis of sensors that measure the background radiation, the level of air, soil, water pollution, and etc. The analysis showed that smart city technologies, as a rule, are oriented towards the future provision of sustainable ecological development of cities. The authors believe that an integral part of political stability is the environmental safety of the population and the territories of its residence. Stability is a stable state that is not subject to significant changes. Stable development, therefore – the ability to constancy of any parameters of the dynamic state, preservation of basic elements within transition from one qualitative status to another, from the old to the new. As a rule, political stability is considered as a qualitative condition (property) of the system stability, that is, its state, when the potential is maintained to autonomously realize relative constancy within the system and to remain in its previous specified parameters. "Political stability is an opportunity that remains with the constancy of integratively significant parts (even under conditions of destabilization of some parts of the system), the likelihood of developing and adapting to transformations" [13]. The authors agree with this statement. Political stability, in general terms, implies a staged, progressive, step-by-step growth of certain indicators of the vital activity of society and the state characterizing them as an integral well and jointly functioning organism. Political stability can be viewed as a main factor for socio-ecological stability and safety.

There is a problem of radiophobia in society that is often politicized. So, for example, in February of 2020, Moscow residents, about 400 people, came to a gathering against the construction of the Southeast Expressway. The reason was that its construction took place in potential dumping sites for radioactive waste of the second half of the 20th century. Deputy of the State Duma of the Russian Federation V. Rashkin, as well as supporters of other left-wing forces came to the gathering. The main demand of the inhabitants of the megalopolis was to cancel construction of the Expressway according to the approved project that involved felling of trees, demolition of garages ... [14].

The analysis showed that the quality of the information provided by the electronic media, sometimes due to the free interpretation of facts and figures by journalists, namely, regarding implementation of measures to prevent potentially dangerous situations related to the radiation safety of the population and the city's territory, can create panic, for example, focusing on places of mass recreation of Moscow residents. It can be stated that the electronic media sometimes do not provide solid and understandable explanations for Moscow residents, using vaguely evasive formulations and "everyday" terminological apparatus (for example, "burial ground", "get infected with radiation" instead of being polluted, and etc.).

Achieving environmental sustainability through rapid industrialization is currently a major global challenge. Industrial plants are key factors in the economy, but also major pollutants. There is now an urgent need to find environmentally friendly, socially and politically sustainable solutions for management of industrial waste and to develop sustainable methods for treating waste before it enters the environment.

In modern conditions of progressive development of the information society in Russia, the information policy of enterprises whose operation is associated with ensuring a safe environmental situation and its monitoring is focused on constant and timely information to the public through the media, conferences, inserts, and etc. Consider, in this regard, the activities of Radon Federal State Unitary Enterprise.

Fig. 1. Radon Federal State Unitary Enterprise stand at the exhibition-conference "Clean City", Moscow, June of 2011.

A comprehensive automated environmental monitoring system has been deployed in the capital. So, for example, Radon Federal State Unitary Enterprise also conducts a systematic inspection of construction sites, potentially radiation-hazardous facilities and the state of the urban space of the environment, conducts environmental education work with the population. The radiation safety system created in Moscow on the basis of Radon Federal State Unitary Enterprise is unique and can be replicated in other regions.

In the territory of the city, there are several dozens of stationary radiation monitoring stations, information from which is accumulated in a single information center. Residents of the city can familiarize themselves with the indicators of some sensors on their own, since they are located on the streets, and also online [15] (Fig. 2). This system can become the basis for formation of the state system of radiation safety at the regional and federal levels.

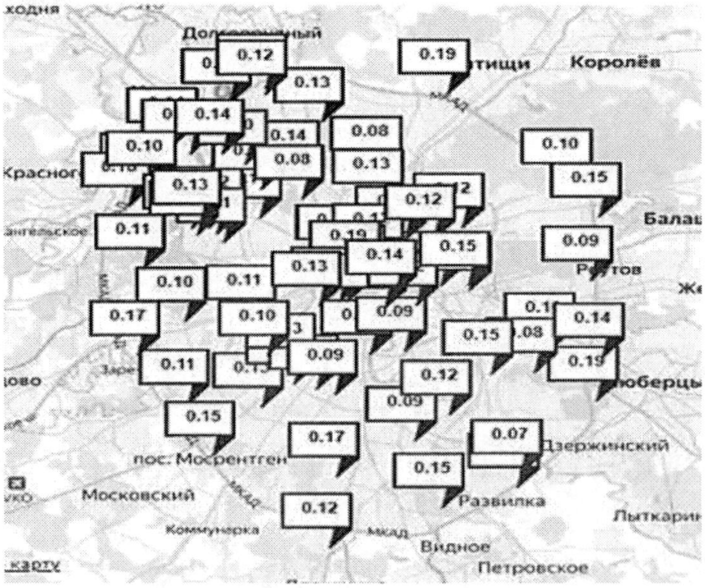

Fig. 2. Online monitoring of the radiation background in Moscow.

4 Conclusion

It has been established that modern researchers analyze the concept of the smart city phenomenon from a global and regional point of view, with an emphasis on its implementation throughout the world. Consider the role of a complex of information and communication technologies as a toolkit for promoting development of smart cities, and the role of such technologies for ensuring environmental safety in the context of ensuring political stability. The concept of an urban smart platform that allows stakeholders to measure human ecology as a value is studied separately.

It was revealed that insufficient research and, as a result, the lack of harmonious formation and implementation in the environmental sphere can lead to impoverishment of the population (poverty) and environmental degradation. A clear conceptual framework is necessary to understand the underlying challenges of sustainable urban growth in developing and developed countries. It is necessary to use interdisciplinary and holistic approaches to assessing the interdependencies between environmental threats, industrial activities, environmental risk, poverty and social vulnerability to facilitate development and adoption of effective socio-political decisions in the context of ensuring the political stability of a particular urban community.

It is shown that environmental risks, as a rule, accompany development of urban space. Modern economic development has accelerated environmental pollution, has resulted in the loss of natural habitats and altered landscapes. These ecological changes in natural systems degrade ecosystem services and, as a result, increase environmental risks to humans. Environmental risks are not only risks to human health from pollution, climate anomalies and natural disasters, but also the degradation of ecosystem services for people. It is impossible to completely eliminate risks, because you cannot achieve zero environmental impact, but it is important to find tools that minimize environmental risks to humans in a sustainable manner.

The analysis showed that environmental safety management requires decision-makers to weigh additional risks with the potential consequences of implementation of environmental policy considering both positive and negative effects that can arise as a result of various socio-political actions of both authorities and government, and the population of cities. Innovative developments in use of economically effective methods for assessing environmental risks, especially in the context of a lack of budgetary resources, are necessary. This is especially important for assessing environmental risks in developing countries, given the underdeveloped regulatory institutions and limited resources. Overcoming environmental risks contributes to preservation and maintenance of the political stability of communities.

It was demonstrated that Moscow uses components of the "smart city" technology (radiation and environmental monitoring system) in the context of ensuring environmental safety. Such technologies are being modernized based on achievements of scientific and technological progress, forming the ecological comfort of the city residents and reducing the degree of socio-ecological conflicts.

Acknowledgements

The study was carried out with the support of the Peoples' Friendship University of Russia. The code of the research and development topic "Technologies for Ensuring Political Stability in Modern Russia" is 100701-0-000.

References

1. A. M. Prentiss, *Human Ecology,* 217 (2019)
2. T. Takatorige, Y. Hirose, S. Nagamatsu, Science of Societal Safety. Trust (Interdisciplinary Perspectives), **2,** 87 (2019)
3. S. Sheina, K. Chubarova, EMMFT 2018. Advances in Intelligent Systems and Computing, **983,** 216 (2019)
4. P. Antoniadou, E. Giama, S. Nižetić, A. M. Papadopoulos, The Role of Exergy in Energy and the Environment. Green Energy and Technology, 833 (2018)
5. M. Yu. Zenkevich, V.E. Prokofiev, K.V. Yanovich, Actual problems of military scientific studies, **1(13),** 186 (2021)
6. R. N. Bharagava, G. Saxena, Springer Nature Singapore Pte Ltd. **XXVIII,** 538
7. Ch. Etezadzadeh, Smart City 2.0 as a Livable City and Future Market, 66 (2016)
8. R. P. Dameri, C. Rosenthal-Sabroux, Springer International Publishing **VIII,** 238
9. A. Meier, E. Portmann, Springer Vieweg, **XXXII,** 346
10. D. S. Velieva, Citizen and Law, **3,** 46 (2012)
11. E. V. Popov, K. A. Semyachkov, Problems of the development of the territory, **5(103),** 53 (2019)
12. E. A. Tsvetkova, A. V. Onishchenko, D. S. Bykov, A. Yu. Luzhnev, V. L. Stepkina. Economy and Entrepreneurship, **12-4(89),** 788 (2017)
13. R. I. Balbek. *Technologies for ensuring political stability in conditions of external destabilizing influences (on the example of the Republic of Crimea)*, 51 (2019)
14. Moscow residents demand to stop the construction of the South-East Expressway through a radioactive burial ground, https://www.leftfront.org
15. Radon Federal State Unitary Enterprise. Radiation and ecological monitoring of environmental objects, http://www.radon.ru

Concepts and principles of the sustainable digital technologies use: process science and activity theory view

Alexander S. Geyda[1,2,*], *Lyudmila N.* Fedorchenko[1], *Igor V.* Lysenko[1], *Aleksandra S.* Svistunova[1], and *Dmitry S.* Khasanov[1]

[1]St Petersburg Federal Research Center of the Russian Academy of Sciences, 198188 14 Line, 39, St. Petersburg, Russia
[2]The North-West Institute of management – branch of the Russian Presidential Academy of National Economy and Public Administration, 197376 Pesochnaya Embankment, 4, St. Petersburg, Russia

Abstract. Using information technologies and implemented with their help information operations to implement sustainable development systems in the economy and society is considered. Based on the previously performed systematic literature review, the conclusions made about future research directions on using information technologies for sustainable development. A set of corresponding theoretical tools is proposed. This set should allow solving the problem based on mathematical models and methods. The concepts and principles of using information operations to study sustainably developed systems are proposed to solve the specified problem.

1 Introduction

The systematic literature review [1] on the problems of digitalization of the economy and society allowed us to conclude: one of the most important and promising areas of using modern (i.e., digital) information technologies is ensuring the sustainable development of the economy, society, and the state, strategic planning in the economy and society, creating a closed-cycle economy. This perspective is because information technologies allow us to justify, predict, and then organize and implement various types of changes in economic, social, and societal systems in general. As a result, by digital technologies, it becomes possible to implement a study of possible future effects of activities where it was previously impossible and, as a result, to make progress in solving the problems of sustainable development. It is possible to implement new, better forecasts with better accuracy for longer planning horizons with digital technologies. Digital technologies allow us to organize new ways of reusing and disposing of resources of different types. As a result, this may enable us to solve the problems of sustainable development of societal systems that were impossible to solve earlier. Digital technologies allow us to organize new ways of reusing and disposing of different types of resources. As a result, this may enable us to solve the problems of sustainable development of societal systems that were impossible to solve earlier. However – on the other hand, based on the conducted review, it was concluded that the existing

* Corresponding author: geida@iias.spb.su

theoretical means of studying the use of digital technologies, especially mathematical models, and methods for solving such problems, are developed insufficiently. In particular, mathematical models and methods for solving sustainable development problems of various systems, as mathematical problems, are developed insufficiently well.

In this regard, within the developed theory of the use of information in systems functioning, it is possible to create theoretical means of studying digital technologies of sustainable development. According to the review results, these tools, and the means of the theory of the use of information in the functioning of systems should be based on the process science and the theory of activity. In addition, it is promising to use extensions of the graph theory, the alternative stochastic networks, the algebra of cyclograms, and the partially ordered sets. The article analyzes the main results of these theories in the application to the formal means of information technologies for sustainable development. It suggests concepts, principles, and basic schemas for using the information in the functioning of systems. Their further formalization should allow us to move on to the mathematical modeling of information technologies for the sustainable development of systems.

2 The concept of the use of information operations for sustainable development

The concept includes a set of interrelated concepts and principles of research of sustainable systems functioning regarding digital technology use. They are generated based on a systematic review [1] of conceptual and mathematical models, methods, and technologies for studying the digital transformation of economic and social systems.

The systematic review of the publications allowed us to draw the following conclusions, which would enable us to justify the direction of research on the use of information operations in the study of the functioning of sustainable developed systems:

First, it is necessary to eliminate the discrepancy between the need to consider the aspects of using information technologies for sustainable development and the limited available theoretical means.

Second, it is necessary to eliminate the discrepancy between the need to use predictive mathematical means for studying the results of activities in different types of systems - from the one hand. And from another hand, the lack of such mathematical means allowing studying activities, considering the predictive mathematical dependencies between the characteristics of the information technologies used, the results of activities obtained, and their compliance with the requirements.

Third, the closest to the required theoretical means for solving the problems of sustainable development regarding digital technology use indicated. According to the review results obtained, these are means, one way or another, based on the study of the role of information and information technologies in activities following the "activity paradigm" of the research. The keywords related to such a paradigm include Process Science, Complexity Science, Applied Activity Theories; Systematic-Structural Activity Theory; Activity Theory; Action Theory; Action Research, Digital Philosophy.

This conclusion is related to the need to link different types and aspects of information and activities based on the "process" and "activity" paradigms. First of all, information activities related to the acquisition and transformation of knowledge, training, social activities, and the psychology of information action — on the one hand, and — on the other hand, subject-transforming, energy-material, and instrumental actions. Theoretical means to link informational and other activities are necessary, in our opinion, to solve the problems noted earlier as unsolved. Examples of using the "activity approach" for different types of activities using information and information technologies are presented in[7, 8].

The "activity approach" (or "activity paradigm") has a much greater generality because it is based on the description of the activity, including information aspects. However, at the time of writing this article, it does not contain theoretical means of predictive analytical evaluation of activities, their results, their compliance with requirements-depending on the characteristics of the information technologies used. When writing, its mathematical tools are mainly limited to project and program modeling tools and partly to project management tools. However, these tools do not describe the use of information technologies and the relationship between IT characteristics and the characteristics of the implemented activity. Thus, it should be noted that the theoretical tools of the "activity approach," on the one hand, are characterized by interdisciplinarity and breadth. Still, largely due to this, they are deprived of well—established mathematical means for solving problems. Thus, it is necessary to eliminate two different deficiencies caused by the current theoretical means for solving problems using information technologies for sustainable system actions.

2.1 Concepts of information operations use for sustainable development systems research

The concepts of using information operations for sustainable systems are proposed to form based on the process science and activity sciences (activity theory, action theory, action research). The concepts are proposed to explain and expand the suggested results with information, information operations, their relations with human activity, the environment, and sustainable development.

Following [2] we define process science as "…the interdisciplinary study of continuous change. By process, we mean a coherent series of changes that unfold over time and occur at multiple levels". Thus, by process science use regarding information actions, performed according to certain information technologies (IT) it is possible to explain value of IT, including but not limited – on the design stage.

Process science emphasizes the following key characteristics: (1) processes are in the focus, (2) we scientifically investigate processes (3) we do that through an interdisciplinary lens, and (4) we intend to influence and change processes to create needed impact.

In accordance with our earlier works [3–8], *it is targeted changes that require information operations,* possibly of different types. We define *information* [9] as a form — for example, a record on some "information carrier" - of the existence of ideas or thoughts (representations, concepts, judgments) in the material world (that is, outside of consciousness). *Information* is necessary for use to obtain the required results of *human activity*. Thus, we distinguish *reflection,* which is observed in the material world and without people, from information reflecting people's ideas. *Informational operation (Information operation, operation with information)* – an operation (an element of activity with specified goals, methods, and predictable implementation results) intended to obtain the required information (because of operating with information). The description of interrelated information operations is contained in *information technology* – a set of descriptions of possible information operations in different conditions and possible relations.

Sustainable development [10–12] (concerning digital technologies, process science, and activity sciences) is defined as follows. It is the ability to create and use information about the possible future results of the implementation of various types of activity (i.e., development, provision, and management) to meet the needs of the near future as well as the possible needs of future generations. The role of IT in such processes of activity is associated with the prediction, organization, regular monitoring, and change of activities in order for this activity to cause as a little negative and as many positive consequences as possible at the planning horizons that are marginal for research in the existing conditions. In this sense, this interpretation is close to the subject of strategic planning. However, it is aimed at greater detail of possible threats to the environment. This role of IT differs significantly from the traditional (and, nevertheless, still insufficiently studied) role. This traditional role consists in providing targeted changes in the processes of activity on limited planning horizons with the specified final goals of a limited number of stakeholders, i.e., as a rule, without considering the consequences for the environment and the development of the system on long–term planning horizons.

Along with the process science various action and activity theory approaches [13–23] to information operations research must be taken into consideration. This is necessary because of process science has not for now genuine concepts and methodologies of action research. Such concepts and methodologies, being adapted by using the information in system action theory, may allow to describe organizational [24, 25], social [26, 27], economic [28], psychological [29], cultural [30, 31], cognitive, behavioral [32], environmental and sustainability [11, 33, 34], educational [30, 35–37] aspects of using information [38] for system action. Principles of the research of using information operations in s [39] system action suggested based on concepts suggested, as well as on process science and action theory results.

2.2 Principles of the research of using information operations in system action

The first principle is: effects of the use of information are changes inside or outside the mind. We should measure the results of such use according to characteristics of the change, their relationships with change and the extent to which changes comply with human demands. This principle corresponds to process science in that respect, measuring information operations results we shall measure changes of different kind and such measurement should be in the core of process science, too. However, when consider information operations levels of change shall be accounted for. Information operations must be used to design, to prepare any process. But change of the processes and higher level of changes made with information operations use too and thus – higher levels of changes shall be accounted for when quality of the information operation is measured.

The second principle is – we can measure quality of information use in a given system according to the characteristics of changes caused by information Im, which changes the relationship between action elements and action effects and their compliance with human demands.

The third principle is: the contour of possible changes due to information actions ends when changes in the material world occur. At that point, we can measure the results of using information.

Thus, after a series of changes in the environment and subsequent reactions of a sustainably developing system using information technologies, it is necessary to consider subsequent possible changes depending on certain alternative events in the environment and system. As a result, sequences of alternative changes or scenarios of possible changes are formed. According to the characteristics of possible sets of such scenarios and their compliance with the requirements, the functioning of steadily developing systems should be evaluated-depending on the conditions and characteristics of information operations (and the corresponding information technologies).

As a result, it becomes possible to assess the quality of a sustainably developing system analytically. That is - assessing using mathematical models of possible scenarios of environmental changes and subsequent reactions of a sustainably developing system based on evaluating the compliance measures of the functioning results of steadily developing systems with (changing) requirements.

3 Conclusion

Based on the review of the current study results of the use of information operations and related technologies in the functioning of sustainably developed systems, the concept of using information operations in the study of sustainably developed systems is proposed based on the science of processes and theories of activity. The basic concepts and principles of using information operations to study the functioning of sustainably developed systems are substantiated. The obtained results allow us to construct mathematical models of sustainably developed systems' functioning regarding information technologies used in changing conditions.

Acknowledgements

The reported study was funded by RFBR, project number 20-08-00649, 19-08-00989 and by budget research work number 0073-2019-0003.

References

1. A.S. Geyda, T.N. Gurieva, V.N. Naumov. Conceptual and mathematical models, methods, and technologies for the study of the digital transformation of economic and social systems: a literature review and research agenda. *Management consulting (2021), In press*
2. J. vom Brocke, W. V. der Aalst, T. Grisold, W. Kremser, J. Mendling, B. Pentland., J. Recker. Process Science: The Interdisciplinary Study of Continuous Change (2021)
3. A. Geyda, Conference of Open Innovation FRUCT 28: FRUCT Oy, 9347589 (2021)
4. A. S. Geyda, *Workshop on computer science and information technologies* (2019)
5. A. S. Geyda, Proceedings of the 25TH conference of FRUCT association, 95 (2019)
6. A. Geyda, *Recent Research in Control Engineering and Decision Making*, 379 (Cham: Springer International Publishing, 2019)
7. A. Geyda, Future Internet, **11**, 62 (2019)
8. A. Geyda, Information, **11(3)**, 164 (2020)
9. A. S. Geyda, *Fundamentals of the theory of the potential of complex technical systems* (Publishing House of the Russian Academy of Sciences, 2021)
1. D. J. Lang, A. Wiek, M. Bergmann, Sustainability Science, **7(1),** 25 (2012)
10. I.-A. Hengst, P. Jarzabkowski, M. Hoegl, M. Muethel, Academy of Management Journal, **63(1),** 246 (2020)
11. F. Popa, Futures, **65(4)**, 45 (2015)
12. J. Aranzadi, *Toward a General Theory of Action*, 9 (2018)
13. G. Goos, J. Hartmanis, J. van Leeuwen, M. Thielscher, *Challenges for Action Theories* (2000)
14. M. Thielscher, *Action programming languages* (2008)
15. M. Staron, *Action Research in Software Engineering* (2020)
16. N. V. Patel, *Deferred Action: Theoretical Model of Process Architecture Design for Emergent Business Processes,* 29 (2011)
17. R. P. Cooper, Topics in cognitive science, **13(1),** 63 (2021)
18. P. J. Ågerfalk, European Journal of Information Systems, **13(1),** 80 (2004)

19. X. Ma, *Data-Information-Knowledge-Action Model,* 1 (2018)
20. V.E. van Reijswoud, H.B.F. Mulder, J.L.G. Dietz, Information Systems Journal, **9(2),** 117 (1999)
21. F. Piovesan, Action Research, **3(2),** 147675032091916 (2020)
22. B. Dick, Action Research, **7(1),** 5 (2009)
23. E. Turban, Information technology for management, 460 (2015)
24. S. Karanasios, Information Technology & People, **31(1),** 134 (2018)
25. K. Umapathy, Information Systems Theory, 59 (2012)
26. R. Miettinen, Journal for the Theory of Social Behaviour, **42(3),** 345 (2012)
27. L. V. Mises, *Human activity: a treatise on economic theory,* 878 (2019)
28. G. Z. Bedny, *Work activity studies within the framework of ergonomics, psychology, and economics* (2019)
29. A. Sannino, Cultural-Historical Psychology, **14(3),** 43 (2018)
30. M. Cole, B. Ferholt, A. Jornet, Mind, Culture, and Activity, **26(4),** 283 (2019)
31. I. Ajzen, Psychology & health, **26(9),** 1113 (2011)
32. W. C. Clark, *Linking knowledge with action for sustainable development*, 122 (2006)
33. J. W. Cook, *Sustainability, Human Well-Being, and the Future of Education,* 425 (2019)
34. T. Aagaard, *Digital agency in higher education* (2020)
35. A. Emejulu, Critical Studies in Education, **60(1),** 131 (2019)
36. S. Pal, Q. C. Ton, R. S. S. Nehru, *Digital education pedagogy* (2021)
37. S. Karanasios, Information Systems Journal, **28(3),** 439 (2018)
38. C. J. Martin, Ecological Economics, **121**, 149 (2016)

Environmental aspects of designing urban infrastructure for smart cities

Alexey Dubrovsky, *Anastasia* Ilinykh*, *Galina* Yurina, and *Tatyana* Mezhueva

Siberian State University of Geosystems and Technologies, Novosibirsk, Russia

Abstract. The article discusses the need to develop elements of smart rational design of urban space with recreational areas and focus on creating an environmentally friendly environment. The authors proposed a sequence of measures to control the recreational provision of the territory of the settlement. The authors highlight the factors influencing the development of underground urban space and the principles of the formation of an ecocity as promising directions for organizing urban space. The article presents the principles of rational organization of urban space.

1 Introduction

Recreation at present is an important element of the modern territorial arrangement, it helps and contributes to the health of a person, as well as the restoration of strength and health [1]. This term is over 30 years old. Developers of plans for the development of the territory began to include green recreational areas in their projects. The population needed places within settlements with a clean, unpolluted environment, landscaping, and other constructive elements conducive to recreation. Garden plots, which are for the personal use of citizens play the role of recreational areas for the population in many settlements of Russia and large metropolitan areas [2]. This approach does not solve the problem of deteriorating well-being of the population of large cities. Often people have the "urbanist" syndrome - a youthful lifestyle, fatigue.

Smart design technologies apply to various complex projects for the development and arrangement of the territory [3]. Geodesign technology allows us to minimize the cost of building a real estate object because of the optimal placement of this object in the natural landscape [4, 5]. "Smart" design saves money at the stage of the primary development of the territory, for example, it is minimizing the volume of earthworks and reclamation works, minimizing landscaping work, since most often natural vegetation is preserved. As a result, the costs of operating the territory are reduced, caused, for example, by the need to carry out additional landscaping or landscape design of the territory.

* Corresponding author: ilinykh_al@mail.ru

2 Materials and methods

In the practice of urban planning, the methods of geoinformation analysis and geomodeling are used in the organization of a rational land use system and in measures to maintain optimal living conditions for the population. Recreational land use is a complex geographically distributed multicomponent system. The urban environment is a complex of diverse functional areas. The primary task of the recreational zones of the city is to ensure the restoration of strength and health of the population using the natural-climatic, landscape-composition and engineering-technical features of the territory.

3 Results and discussion

The concept of smart building design and construction has developed over the years in modern urban planning. Smart home technologies are multifaceted and cover the entire range of all kinds of improvements in the quality properties of a real estate object.

A smart home is a system that ensures safety, resource saving and comfort for all users [6].

A smart city is an urban planning concept for integrating a variety of information and communication technologies (ICT), including the Internet of Things systems. The purpose of Smart cities is to manage urban infrastructure, healthcare, education, transport, security, housing, and communal services.

The British Standard Institution (BSI) views a smart city as "the effective integration of physical, digital and human systems in an artificially created environment to ensure a sustainable, prosperous and inclusive future for citizens" [7].

New York is one of the safest megacities in America for recreation [8]. According to experts, 97% of city residents need only 10 minutes to walk from their house to the nearest park or garden. This figure for a city with 8 million inhabitants is one of the best in the world.

Each “smart city” has standards and properties that should correspond to the direction of development of such cities. The properties of a “smart city” include [9] urbanization, smart technologies, quality of life, mobility, virtualization, personification, and infrastructure modernization.

Earlier studies of the recreational provision of the territory of Novosibirsk (Russia) showed that in some estimated blocks the deficit is 85% or more of the standard of provision for 1 person. The results of the study are presented in detail in [10].

Even on the territory of Novosibirsk, there are many examples when developers build up recreational zones to the detriment of the aesthetic. Development of forest park zones in cities exacerbates the problem of recreational provision of the territory of settlements.

The authors present a sequence of measures to control the recreational provision of the territory of a settlement (Figure 1).

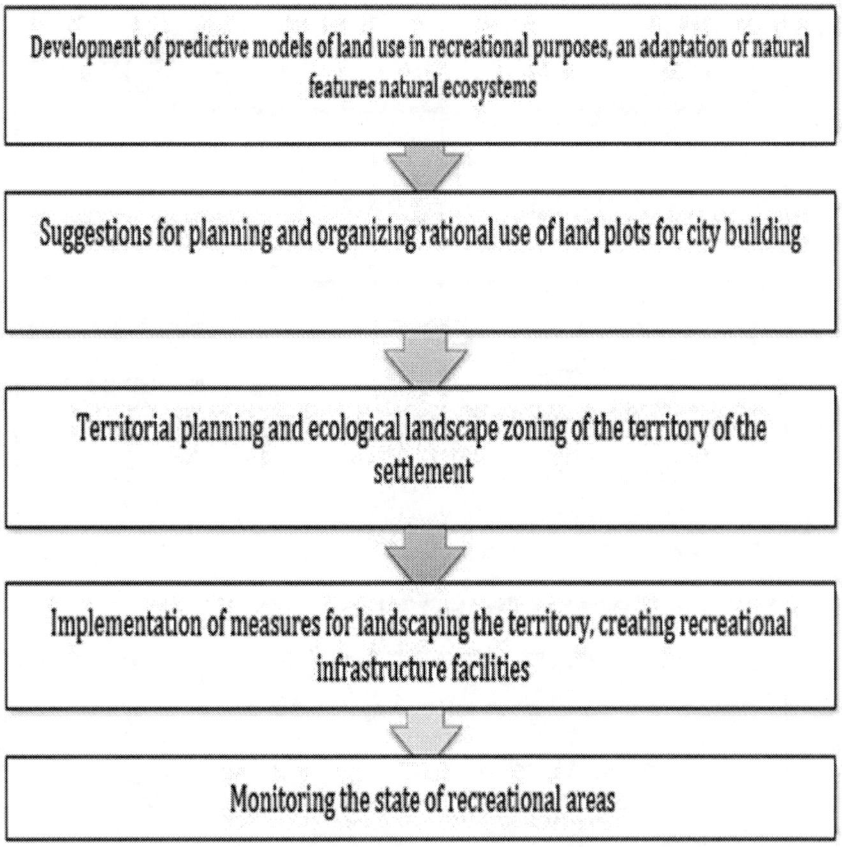

Fig. 1. The sequence of measures to control the recreational security of the territory of the settlement.

One of the promising areas of rational organization of urban space is the use of urban underground space. Analysis of the goals and directions of the development of urban underground space allowed us to highlight the promising factors shown in Figure 2.

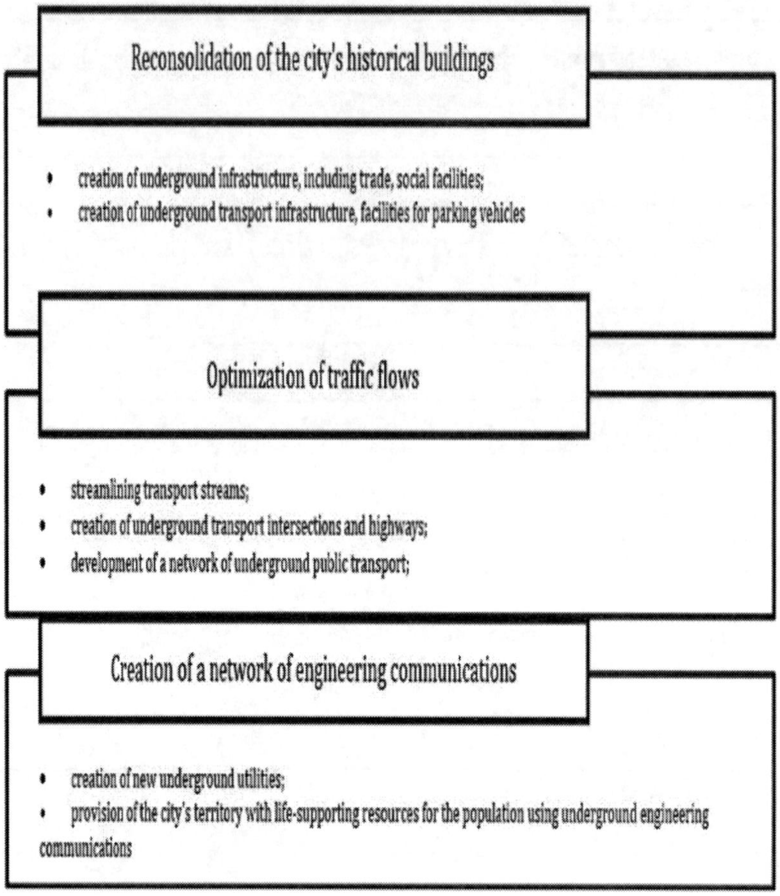

Fig. 2. Factors influencing the development of underground urban space.

An updated concept of ecocity appeared in the 70s of the last century. 3 main provisions for the formation of ecocity were presented (Figure 3).

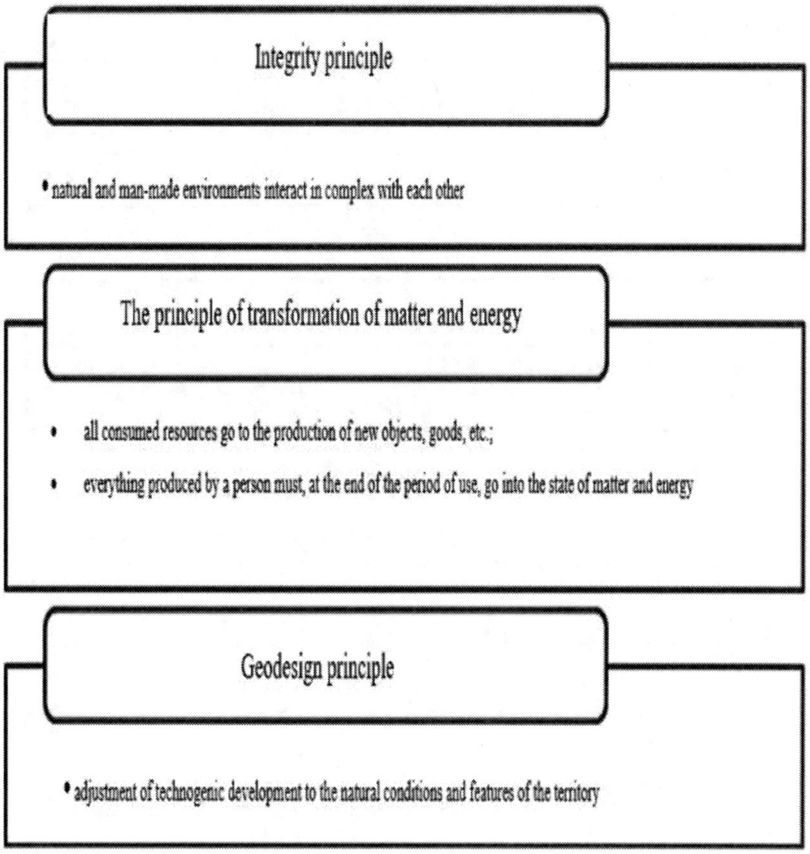

Fig. 3. Ecocity formation principles.

The concept of a «smart city» contains three basic parameters: adaptability, intellectualization, focus on lifestyle.

The areas that primarily need intellectual modernization are the economy, infrastructure, and urban management (Table 1) [11].

Table 1. The most important directions for the development of a «smart city».

Innovative economy	**Urban infrastructure**	**Public administration**
Innovations in industry, clusters, city districts	Transport	Administrative services to citizens
Smart Workforce: Education and Employment	Energy / Utilities	Representative and direct democracy
Creation of knowledge-intensive companies	Environment / safety	Services for citizens: quality of life

We found advanced developments in the planning and organization of the territory of settlements in the concept of urbanism. The concept of urbanism is based on sustainable and environmentally-friendly planning of urban space. The key principles are the following provisions (Figure 4).

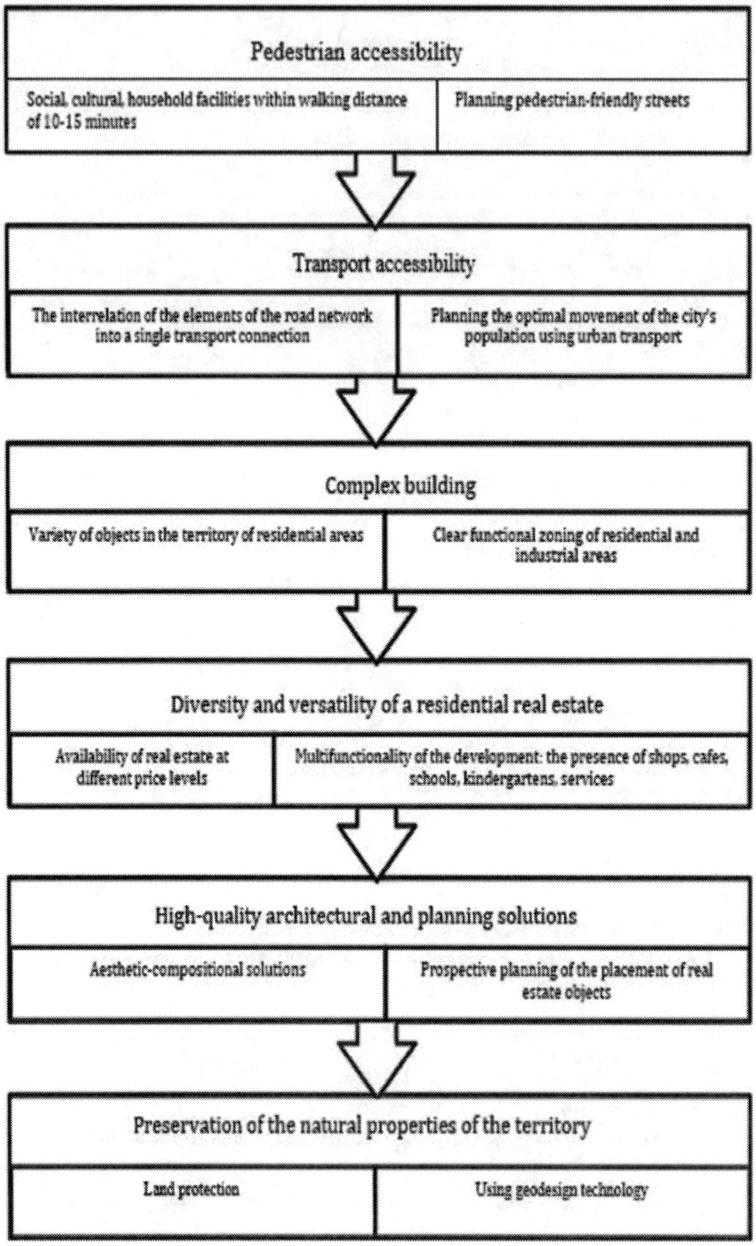

Fig. 4. Principles of rational organization of urban space.

The key principle of a "smart city" is the reuse of resources, reuse of data in various departmental information systems and infrastructures (server capacities, sensors, video cameras, etc.).

As a result of their implementation, end-to-end technologies will give an incentive to improve the quality of life and management of various sectors of the urban economy while reducing resource consumption and the comfort of the urban environment.

4 Conclusion

The population of modern megalopolises, medium and small settlements needs green forest parks, places for recreation in nature and in open space.

Such a not comforting tendency occurs in the territory of many modern developing, intensively built-up settlements. The lack of recreational areas is not compensated, and the existing recreational facilities are being reduced because of intensive development.

The goal of creating a "smart city" is to ensure the most comfortable life, environmental friendliness, energy intensity, safety, and a wide range of opportunities.

References

1. J. D. Blower, A. L. Gemmell, G. H. Griffiths, K. Haines, A. Santokhee, X. Yang, Environmental Modelling & Software, **47**, 218 (2013)
2. A. V. Dubrovsky, I. T. Antipov, A. I. Kalenitsky, Int. J. of Advanced Biotechnology and Research, **9(1)**, 1185 (2018)
3. Five real solutions for a smart city: an example from Russia, Tass, https://tass.ru/
4. G. G. Kurierova, Ecology of the object world as a design strategy in the post-industrial period, 132 (2008)
5. V. I. Tatarenko, E. A. Popp, Izvestiya Vuzov «Geodesy and aerophotosurveying», **4**, 165 (2014)
6. G. V. Esaulov, Architecture and construction, **5**, 56 (2017)
7. M. A. Shneps-Shneppe, Int. J. of Open Information Technologies, **2**, 11 (2016)
8. V. A. Gorokhov, Greening of foreign cities Landscape architecture and green building, http://landscape.totalarch.com/
9. Designing smart homes, Habr, https://habr.com/
10. A. V. Dubrovsky, Int. scientific. conf. "Remote sensing methods of the Earth and photogrammetry, environmental monitoring, geoecology" (SGUGiT, Novosibirsk, 2019), 135
11. Smart city. Concept, standardization and implementation of smart city, Portal about modern technologies of mobile and wireless communications, http://1234g.ru/

Improving energy efficiency through the use of waste heat products in the production of building materials

Galina Medvedeva*, *Kamilla* Giniyatova, and *Kamilia* Minikaeva

FSBEI HE "Kazan State University of Architecture and Engineering", Kazan, Russia

Abstract. Concern for the sustainable management and recycling of solid waste is becoming more visible in all sectors of the economy. This study explores the possibility of using coal ash residue (waste from Kazan CHPP-2) as a substitute for fine-grained aggregate in sulfur concrete. The trend towards an increase in the level of utilization of waste heat power engineering is an important task. The chemical composition, microstructure and mechanical properties, including density, water absorption, compressive strength and thermal conductivity of sulfur concrete, including coal ash with partial and complete sand replacement, were investigated and the results were compared with those for standard cellular concrete. The authors studied modern heat-insulating materials and materials from industrial waste products. The article analyzes the estimated thickness of the insulating material depending on the type of structure. Outside walls made of sulfur concrete, in addition to high strength properties, have high thermal and economic performance.

1 Introduction

Worldwide, waste production continues to grow due to the constant demand for resources used by people, and this is considered one of the most serious environmental problems we face. Therefore, the need arose for the management of industrial, household and other waste and their processing, as well as for the reasonable and efficient use of non-renewable resources [1]. Certain types of waste can be incorporated into concrete either as part of the cement mix or as aggregate. This helps to improve the sustainability of building materials. Some industrial wastes or by-products have been successfully disposed of in this way, including the bottom ash of coal used as a component in the production of Portland cement. In addition, waste or by-products larger than cement particles can be used as aggregates in mortar or concrete [2]. After coal is burned in the furnace of a coal-fired power plant, smaller particles of non-combustible ash are recovered in electrostatic precipitators. Part of the molten ash accumulates on the walls of the boiler and solidifies, forming mass particles, which fall to the bottom of the furnace and are cooled in water. Coal ash collected from electrostatic precipitators and from the bottom of the furnace is called fly ash and coal bottom ash, respectively [3].

* Corresponding author: medvedevaga79@mail.ru

Efficient utilization of ash and slag waste from coal-fired power plants can significantly reduce the negative impact on the environment and improve their economic performance [4]. The mineralogical composition of ash-and-slag wastes obtained during the combustion of coal-water fuel based on fine coal washing wastes in a specially designed boiler with a vortex combustion system has been investigated [5].

Possibility of effective use of ash and slag waste for the production of building materials, primarily building mixtures, widely used in mining operations in mine workings, backfilling of mined-out space, and others [3]. (High content of silicon oxide and aluminum oxide). combined with low carbon content, in other words, negligible loss of unburned carbon) [6]. The optimal percentage of the initial components of the casting mixture based on ash and slag waste and crushed stone (granulated slag) has been established. The results of experimental tests of quenching inserts for strength under uniaxial compression are presented [1]. It was found that a sample containing 18% ash and slag waste, 33% granular slag and 19% cement corresponds to the required technological parameters in terms of strength and cement content [7].

In recent years, the use of ash in concrete has received considerable attention due to its potential use in civil engineering for many reasons, including its reduced heating during hydration, low permeability and resistance to sulfate attack. However, only a limited number of studies have been carried out on the use of bottom ash from coal as a partial replacement for Portland cement or as a partial replacement for a small unit [8].

1.1 Comparison of materials of the building envelope

Concrete, an artificial stone-like mass, is a composite material that is created by mixing a binder (cement or lime) together with aggregate (sand, gravel, stone, brick chips, etc.), water, impurities, etc. In certain proportions. Strength and quality depend on the mixing ratio.

The formula for obtaining concrete from its components can be represented in the following equation:

Concrete = binder + fine and coarse aggregate + water + admixture (optional).

Concrete is a very necessary and useful material for construction work. Once all of the ingredients - cement, aggregate, and unit of measure of water - are mixed in the required proportions, the cement and water begin to react with each other to bind themselves into a hardened mass. This hardening rock-like mass is concrete.

Aerated concrete is a special engineered concrete made by mixing Portland cement, sand, fly ash, water and preformed foam in various proportions to form a hardened material. Lightweight aerated concrete is replacing traditional materials in the construction industry due to its lightness, high quality and affordability.

It is highly versatile as it can be tailored for optimal performance and minimum cost.

Sulfur concrete is a composite building material composed primarily of sulfur and aggregate (usually a coarse aggregate of gravel or crushed stone and a fine aggregate such as sand). Cement and water, important compounds in normal concrete, are not part of sulfur concrete. After cooling, the concrete achieves high strength, without the need for long-term curing like conventional concrete. Sulfur concrete is resistant to some compounds, such as acids, which attack ordinary concrete, however, unlike ordinary concrete, it cannot withstand high temperatures, so it is not fire resistant. This concrete was developed and marketed as a building material to get rid of the large amount of accumulated sulfur produced during the hydrodesulfurization of gas and oil. Table 1 shows the main physical properties of cellular and sulfur concrete.

Table 1. Physical properties of concrete.

	Thermal conductivity λ, W / m * 0C	Density, ρ, kg / m3	Cost, rub	Water absorption, %	Strength σ, MPa	Water permeability	Fire resistance
Foam concrete	0.14-0.38	400-1200	3200	10-16	0.4-1.2	-	+
Aerated concrete	0.18-0.28	600-800		25		-	+
Sulfur concrete on porous aggregates	0.05-0.11	800-1200	1500 - 2000	1	1.3-2.2	+	-

1. 2 Ash and slag waste, its' properties, composition and formation

In Russia today there are approximately 350 coal-fired CHP and power plants. 172 of them produce more than 100 thousand tons of ash and slag in 1 year, and about 100 are potential sources of ash.

More than 22 mln. tons of ash and slag waste. Of these, 4 million tons are allocated to the consumer market in the form of ash and slag materials.

Approximately 1.5-1.8 billion tons of ash and slag are stored in ash dumps, which indicates serious environmental disasters in Russia and the need to dispose of these by-products soon.

In November 2013, a decision was made based on a four-year program of annual 10% reduction in the intrinsic value of energy companies partially owned by the state. These factors affect the relevance of solving the problem of ash and slag waste at Russian coal-fired power plants and coal-fired thermal power plants.

Ash and slag are coarse, granular, non-combustible by-products that are collected from the bottom of furnaces for power generation or steam generation. Like most ash and slag, a significant portion of coal cork products is produced in power plants. The type of furnace in a power plant has a large influence on the type of plug product formed. Table 2 provides a comparison of the compositions of Portland cement, sand and bottom ash.

Table 2. Chemical composition of Portland cement, sand and ash of coal residue.

Oxide	Portland cement (%)	Sand (%)	Ash residue of coal (%)
SiO_2	20,18	95,19	36,84
Al_2O_3	4,98	3,05	18,28
CaO	63,74	-	18,43
Fe_2O_3	3,44	0,57	15,46
MgO	2,66	-	2,62
K_2O	0,64	-	2,29
N_2O_5	-	-	0,17
Na_2O	0,45	-	0,75
TiO_2	-	-	0,90
MnO_2	-	-	0,20
Loss on fire	1,45	1,19	2,02

The waste enters the incinerator and is transported through the incineration zone to discharge using moving grates. In the combustion zone, metal parts that were present in the composite with plastic or other organic material are released as the organic matter burns.

 https://doi.org/10.1051/e3sconf/202131107007

The mineralized non-combustible residue from incineration is called ash and is essentially a mixture of minerals and metals.

A significant contribution to the instability of the material composition of ash dumps is made by the presence of unburned coal, the so-called unburned carbon. The amount of unburned carbon in ash varies widely and averages 10–15% of the ash dump mass. This is completely or partially unburned coal of various sizes. The amount of carbon loss depends on the stability of the furnace, temperature, gas composition and dust removal. It is well known that the combustion process is an intense decomposition of organic matter, accompanied by the disappearance and formation of new phases. Phase formation processes occur in solid, gaseous and liquid phases, i.e. they are heteropolar, and the interaction of these processes is so great, and the number of influencing factors is also significant, that it is practically impossible to effectively predict the composition of the final combustion products. It is known that up to 97% of ash materials are oxides of silicon, aluminum, calcium, magnesium, iron and other impurities 3%. This data is correct for ash obtained from complete combustion of coal. In practice, this ratio is most often violated. In addition, the composition of the ash varies depending on the method of precipitation and ash removal from the reaction zone and transportation to the disposal site. At present, ash removal is carried out mainly by the wet method, which, on the one hand, promotes the solidification of the main phases and, on the other hand, the development of leaching of certain phases formed during combustion.

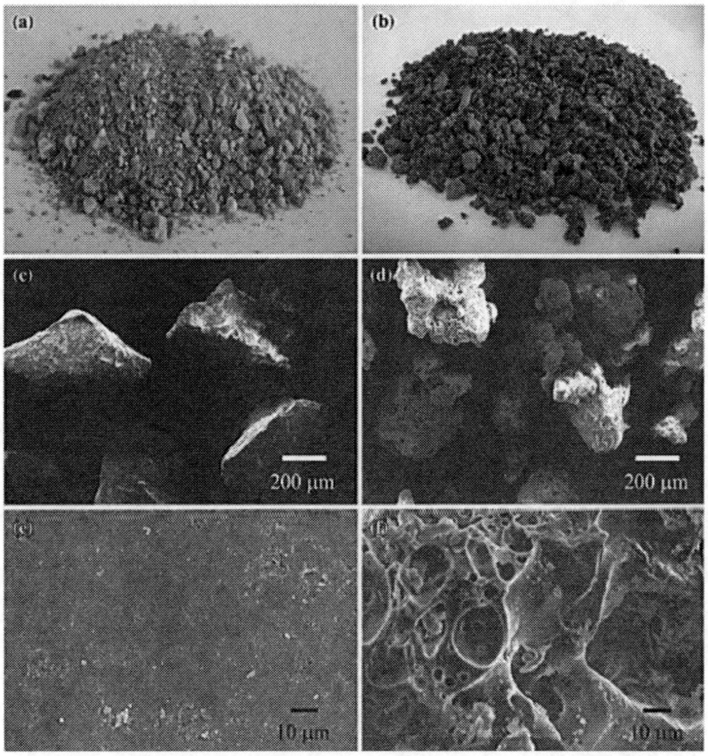

Fig. 1. a - sand, b - coal ash, c - SEM illustration of sand particles, d - SEM illustration of coal ash particles, e - SEM image of natural sand surface, f - SEM image of coal ash surface.

The ash is discharged from the grate and passes through the exhaust duct into the hood, which is a siphon filled with water. The siphon prevents air from entering the combustion chamber, which could otherwise disrupt the combustion process. Ash is removed from the water bath using a reciprocating piston. Some incinerators use "dry extraction", that is, an extractor that is not filled with water and therefore prevents wetting of the bottom ash.

The air required for combustion is blown from below through the passages in the grate elements and carries away very fine ash particles ("fly ash"). After heat recovery in the boiler, fly ash and other contaminants are removed from the flue gases using a series of filters and / or washers. The cleaned flue gas is then discharged through the chimney.

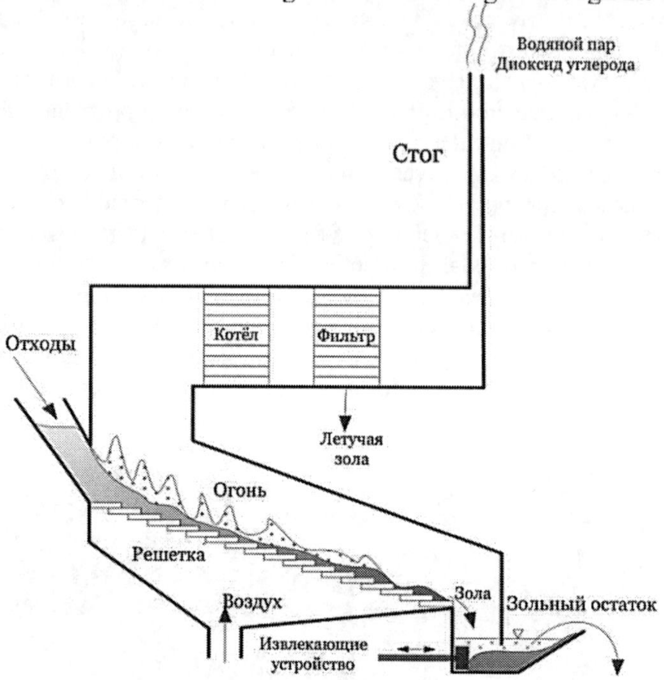

Fig. 2. Simplified scheme of the stoker type, where: водянойпар - water vapor, диоксидуглерода - carbon dioxide, стог–stack, отходы–stack, котёл–boiler, фильтр–filter, летучаязола - fly ash, огонь– fire, решетка–lattice, воздух–air, извлекающееустройство–extractor, зола–ash, зольныйостаток - ash residue.

Sulfur concrete has been analyzed in this article. To begin with, in order to impregnate the concrete elements with sulfur, they were immersed in a sulfurous melt and kept for some time. Then the impregnated elements were cooled to room temperature. The pores of the cement stone are impregnated with sulfur, which ensures that the performance requirements are met through complete water permeability.

In this article, the object of research is the comparison and analysis of the properties of heat-insulating materials.

2 The main part. Calculation of the thickness of thermal insulation

The article calculates the thickness of thermal insulation for a facade system with a layer of decorative plaster according to the method described in 1. Two options were considered: the first - the walls were made of brickwork (solid brick), the second option - from utopian masonry (gray concrete). The following brickwork options have been calculated: 120 mm, 250 mm, 510 mm, 640 mm. The building's thermal protection requirements have also been verified.

Wall compositions are shown in Figures 3 and 4.

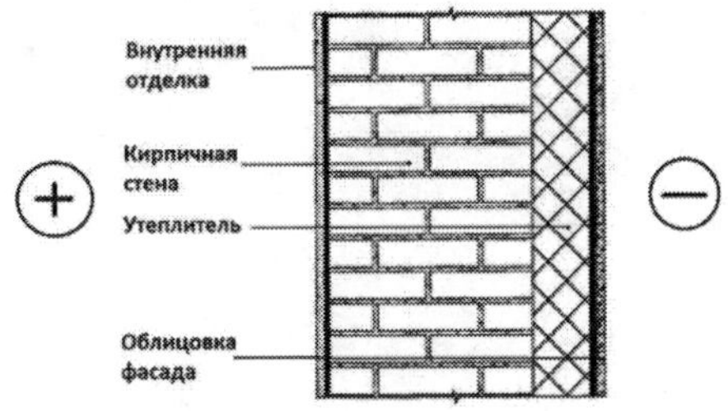

Fig. 3. The scheme of the enclosing structure, where: внутренняя отделка - interior decoration, кирпичная стена - Brick wall, утеплитель – insulation, облицовка фасада - facade cladding.

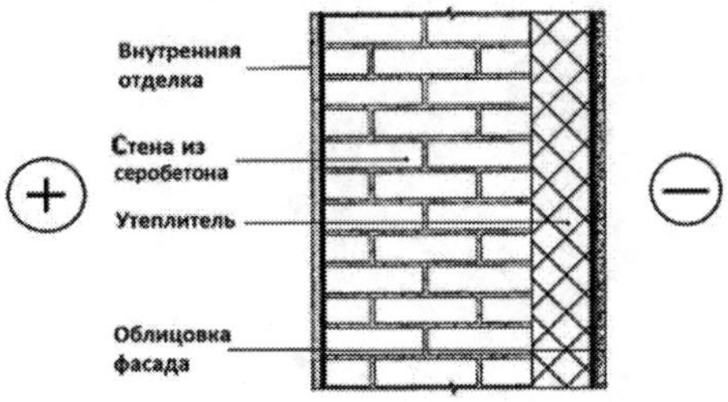

Fig. 4. The scheme of the enclosing structure, where: внутренняя отделка - interior decoration, стена из серобетона - gray concrete wall, утеплитель – insulation, облицовка фасада - facade cladding.

2.1 Description of the solution. Formulas

Degree-day of the heating period (GSOP):

$$GSOP = (t_{in} - t_{from})\, z_{from} \quad (1)$$

where: t_{in} - calculated internal air temperature, ° C(GOST 30494-2011 "Residential and public buildings. Indoor microclimate parameters (corrected)"); t_{from} - average temperature of the period with an average daily air temperature below or equal to 8 ° C, ° C(SP 131.13330.2020 "Construction climatology. Updated edition of SNiP 23-01-99 * (with Amendment No. 2)").

z_{from} - the duration of the period with the average daily air temperature below or equal to 8 ° C, days.

Heat transfer resistance of the enclosing structure required R_{0tr}', (m ^{2} · ° C) / W.

Taking into account the coefficient of heat engineering homogeneity of the enclosing structure r, we finally obtain the required resistance to heat transfer of the enclosing structure R_{0tr} :

$$R_{0tr} = R_{0tr}' / r, \text{ m}^2 \cdot {}^\circ\text{C / W} \quad (2)$$

where: r is the coefficient of heat engineering uniformity in the enclosing structure, which takes into account the influence of joints, slopes of openings, framing ribs, flexible ties and other heat-conducting inclusions(GOST R 54851-2011 "Non-uniform enclosing structures".).

Thermal resistance of the wall layer without taking into account thermal insulation:

$$R_{st} = \delta_{st} / \lambda_{st}, \text{ (m}^2 \cdot {}^\circ\text{C) / W} \quad (3)$$

where: δ_{st} is the thickness of the wall layer without taking into account thermal insulation, m;
λ_{st} is the coefficient of thermal conductivity of the material from which the masonry is made, W / (m · ° C).

Thermal resistance of the internal plaster layer:

$$R_{pcs} = \delta_{pcs} / \lambda_{pcs}\ ,\ (\mathrm{m}^2 \cdot {}^\circ \mathrm{C}) / \mathrm{W} \tag{4}$$

where: δ_{pcs} - thickness of the layer of internal plaster, m; λ_{pc} - coefficient of thermal conductivity of the layer of internal plaster, W / (m · ° C).

Thermal resistance of the thermal insulation layer:

$$R_{ut} = R_{0tr} - R_{st} - \mathrm{R}_{pcs},\ (\mathrm{m2} \cdot {}^\circ \mathrm{C}) / \mathrm{W} \tag{5}$$

Thermal insulation layer thickness:

$$\delta_{yt} = R_{ut} \cdot \lambda_{ut}\ ,\ \mathrm{m} \tag{6}$$

where: λ_{ut} is the coefficient of thermal conductivity of the thermal insulation material, W / m · ° C.

2. 1.1 Initial data

Indoor temperature and outdoor temperature for Kemerovo: t_{in} = 20 ° C, t_n = -39 ° C;
Duration of the heating period: z_{from} = 228 days;
Average temperature of the period: t_{from} = -7.9 ° C;
Thermal conductivity coefficient of solid brick masonry under operating conditions B: λ = 0.59 W / (m · ° C);
Thermal conductivity coefficient of sulfur concrete:
λ = 0.07 W / (m ° C);
Thermal conductivity coefficient of thermal insulation:
For foam concrete λ = 0.14 W / (m ° C);
For aerated concrete λ = 0.18 W / (m ° C).
Internal plaster layer thickness: δ = 0.015 m.
Coefficient of thermal conductivity of the layer of internal plaster:
λ = 0.21 W / (m ° C).

3 Solution

Let's calculate the thickness of the thermal insulation layer made of aerated concrete with a brickwork thickness of 120 mm:
$GSOP$ = (20 + 7.9) 228 = 6361.2;
R_{0tr}' = 3.34 (m ^{2} · ° C) / W;
R_{0tr} = 3.34 / 0.85 = 3.93 (m ^{2} · ° C) / W;
R_{st} = 0.12 / 0.59 = 0.20 (m ^{2} · ° C) / W;
R_{pcs} = 0.015 / 0.21 = 0.07 (m ^{2} · ° C) / W;
R_{ut} = 3.93 - 0.20 - 0.07 = 3.66 (m ^{2} · ° C) / W;
δ_{ut} = 3.66 · 0.14 = 0.510 m.

The following calculations are performed in a similar way. Calculations of the thickness of thermal insulation made of foam concrete and aerated concrete for walls made of brick and sulfur concrete are given in tables 1-3. Graphs of the dependence of the thickness of thermal insulation on the thickness of the walls - in Figures 5-7.

Table 3. Dependence of the thickness of the required layer of thermal insulation from foam concrete on the thickness of brickwork and sulfur concrete masonry.

Foam concrete		
Masonry thickness, mm	**Thermal insulation thickness for wall material, m**	
	Brick	**Sulfur concrete**
120	0.510	0.300
250	0.481	0.040
380	0.450	-0.220
510	0.419	-0.480
640	0.389	-0.740

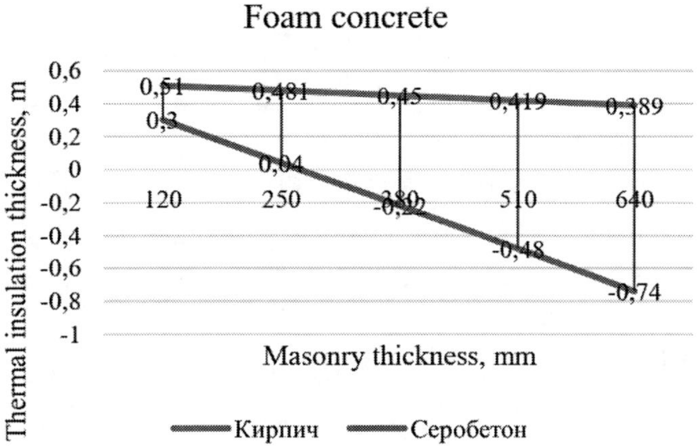

Fig. 5. Dependence of the thickness of thermal insulation on the thickness of the wall, where: кирпич - brick,серобетон - gray concrete.

Table 4. Dependence of the thickness of the required layer of thermal insulation from aerated concrete on the thickness of brickwork and sulfur concrete masonry.

Aerated concrete		
Masonry thickness, mm	**Thermal insulation thickness for wall material, m**	
	Brick	**Sulfur concrete**
120	0.658	0.386
250	0.619	0.052
380	0.579	-0.282
510	0.539	-0.617
640	0.500	-0.951

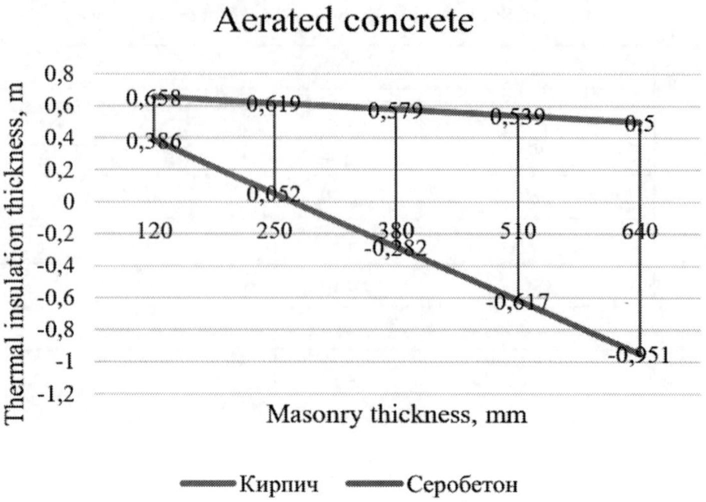

Figure 6. Dependence of the thickness of thermal insulation on the thickness of the wall, where: кирпич - brick, серобетон - gray concrete.

Table 5. Dependence of the thickness of the required layer of thermal insulation from sulfur concrete on the thickness of brickwork and sulfur concrete masonry.

Sulfur concrete		
Masonry thickness, mm	**Thermal insulation thickness for wall material, m**	
	Brick	**Sulfur concrete**
120	0.256	0.150
250	0.241	0.020
380	0.225	-0.110
510	0.210	-0.240
640	0.194	-0.370

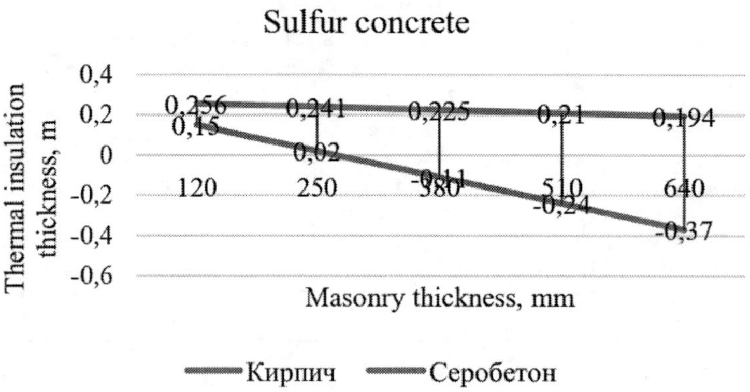

Fig. 7. Dependence of the thickness of thermal insulation on the thickness of the wall, where: кирпич - brick, серобетон - gray concrete.

Analyzing the graphs shown in Figures 5-7, we can conclude that for brick wall and sulfur concrete masonry, the thickness of the thermal insulation layer is different for different materials.

Consider a masonry with a thickness of 380 mm (1.5 bricks). When insulating a brick wall with foam concrete, an insulation layer of 442 mm is required, and when insulating a wall made of sulfur concrete, an insulation layer is not needed. Similarly, when insulating with foam concrete 568 mm, an insulation layer is not needed, with sulfur concrete 221 mm and an insulation layer is not needed, respectively. Thus, the thickness of the insulation layer depends both on the masonry material and its thickness, and on the characteristics of the thermal insulation itself - density and thermal conductivity.

Sulfur concrete is a good insulating material, as we understand from the graphs (the line has a large slope). If the masonry of sulfur concrete is more than 380 mm, then an insulating material is not required. For bricks, the slope of the line is less. So, if we take into account the above properties of sulfur concrete, we can come to the conclusion that walls made of sulfur concrete have high thermophysical properties. You can significantly save on wall laying using sulfur concrete.

4 Conclusions

1. The density of concrete decreased markedly with an increase in the ash residue of coal. This is due to a lower specific gravity of coal slag and an increase in pore volume.

2. The use of coal bottom ash as a replacement for fine aggregate improved thermal insulation as thermal conductivity decreased significantly and linearly with increasing coal bottom ash. This is probably due to the porous structure of the coal slag and the large volume of permeable pore space in the test specimens. Therefore, in the future, they can be used as energy-efficient building materials such as thermal insulation blocks or precast concrete walls.

3. Replacement of fine aggregate with coal ash did not adversely affect the compressive strength of concrete.

References

1. 1.G.A. Medvedeva, O.S. Sekarina, Science of Science,**9**(2017)
2. 2.V. I. Vereshchagin, Building materials(2000)
3. O.V. Kazmina, N.A. Kuznetsov, V.I. Vereshchagin, V.P. Kazmin, Bulletin of the Tomsk Polytechnic University, **319** (2011)
4. A.S. Kosarev, V.A. Smoliy, A.V. Skorikov, Izvestiya vuzov. North Caucasian region. Technical sciences (2018)
5. L.I. Khudyakova, A.V. Zalutsky, P.L. Paleev, Technosphere safety(2019)
6. E.R. Barieva, E.A. Korolev, N.Kh. Galimullina, M.A. Feshchenko, Energy Problems(2008)
7. S.L.Buyantuev, V.D. Sultimova, Building materials(2004)
8. A.P. Semke., O. V. Kazmina, National Research Tomsk Polytechnic University(2018)
9. V. A. Smoliy, A.S. Kosarev, E.A. Yatsenko, B.M. Goltsman, Izvestiya VUZov. North Caucasian region. Technical science, **4** (2015)
10. F.L. Kapustin, V.M. Ufimtsev, B.L. Cherry, V.Ya. Putilov, Ural State Technical University(2017)
11. V.V. Remnev, Concrete technology(2016)

12. E.A. Yatsenko, V.A. Rytchenkova, O.S. Krasnikova, A.V. Ryabova, N.N. Efimov, A.S. Kosarev, Izvestiya vuzov. North caucasian region. Engineering science(2010)
13. L. N. T. Zang, N.A. Epishkin, V.B. Balabanov, V.P. Baryshok, Izvestiya vuzov. Investments. Construction. Real Estate, **7** (2017)
14. J. Moon, P. D. Kalb, L. Milian, P. A. Northrup, Cement and Concrete Composites (2016)
15. P. Torkittikul1, T. Nochaiya, W. Wongkeo, A. Chaipanich, J. Mater Cycles Waste Manag(2015)

Development of human potential as an important condition of the economic security of the enterprise

Ekaterina Viktorovna Gomeleva[1], *Lyudmila Gennad'evna* Ripoll-Saragosi[1], *Viktoriia Sergeevna* Kokhanova[2], and *Natalia Stanislavovna* Grigorieva[2]

[1]Rostov State Transport University, square of the Rostov, Rostov-on-Don, Russia
[2]Private Educational Institution of Higher Education"SOUTHERN UNIVERSITY (IMBL)", Rostov-on-Don, Russia

Abstract. The main concept of the study is based on the study of human potential as one of the conditions for the economic security of an enterprise. The article considers the theoretical aspects of human potential, describes the subject of human potential, reveals assessment tools and factors influencing its human development. The most significant potential threats that directly affect the development of human potential are indicated, including: the risk of burnout of employees working remotely, the prevalence of relatively directive management methods at the level of line managers of the enterprise. The ways of overcoming the existing problems are proposed, the solution of which will influence the development of human potential.

1 Introduction

Human potential and its development in the modern world is increasing its role every year. The statement "Human resources is key" shows that both the development of an enterprise and the possibility of its bankruptcy depend on the personnel. According to many experts, a person is the most difficult link in the security system, since it is the human factor that can have a critical impact on the success of the activity and the existence of the company itself. Employees are one of the most promising and inexhaustible resources of an enterprise that determine its efficiency and success. After all, it is people who play the first fiddle in business processes - they process information, make decisions, and are directly involved in the production process. Their competence and correctness of actions directly depends on how efficiently the means of production and other resources are used at the enterprise, and how successfully the enterprise as a whole works. Competent personnel management is a contribution to the future income of an organization by providing the enterprise with highly qualified and motivated employees, effectively using their potential, skill and creative capabilities, and meeting their social needs. One of the goals of economic security in terms of personnel management is the selection of qualified personnel with high potential, its effective use and development, as well as minimization of risks and threats from employees. To ensure the economic security of an economic entity, its personnel policy should be structured in such a way that the development of human potential is one of the priority tasks.

Theoretical and conceptual aspects of human potential and its development were the subject of scientific research by Russian scientists: Avdeeva E., Babina S.; B urkalcevoy D. Davydova T., A. Kazarcevoy Makeeva T., Mishchenko V., Nesterova A., Plaksa S., Reznikova O., Sadovnikova I., Sirotkina N. and a number of others who proposed different approaches to the study of economic phenomena and human potential that affects economic security, and also foreign scientists Amartya Kumar Sen, Tokatly H.,; Dorneich, MC; Moran, S., Krustik F., Kechek D., Polgar I., Gardner H., Dai D. and others.

The topic of human development as an important condition for the economic security of an enterprise has recently become a trend in the modern economy, so the choice of a research topic is relevant.

2 Research methodology

The studies carried out show that Russian economists do not have a unified approach to the definition of “human potential”. In addition, there is no unambiguous understanding of what human potential is. In different sources, there are different interpretations: "human factor", human capital assets, human resources potential assets, labor potential assets, etc. In general, Human potential assets is the ability of an individual (a set of individuals) to show theoretical capabilities in practice [1, 2, 3, 4].

A number of authors believe that human potential assets should be understood as the qualities of people (a person) that fundamentally affect the results of the activity in which they are involved in a particular sphere (area) of activity [5, 6, 7]. From the standpoint of the authors of this study, it is interesting and quite complete to consider the definition of human potential in the context of the economic security of an enterprise, namely, certain qualitative characteristics and potential capabilities of an employee, which have a fundamental impact on the effectiveness of his production activities in a certain area of the economy.

The research carried out made it possible to single out a tool for assessing human potential - competence [2, 5, 8]. The most important key competencies of human potential assets, which play a decisive role in ensuring the success of an organization, include the following: learning ability - the ability to use, create, save, transfer new knowledge; flexibility, adaptability - the ability to quickly rebuild without losing its integrity and identity; leadership - the ability to inspire and lead; communication skills - the ability to build communications both within the organization (between colleagues, between management and personnel, etc.) and outside the organization; innovativeness - the ability to use creative potential to find new solutions, rationalization proposals, improvements, etc. [3, 4, 9, 10].

So how is it possible to identify high-potential employees? A successful employee showing good work results does not always have high potential, which gives him the opportunity in the future to successfully realize himself in new projects and roles. In modern companies, effective employees are divided into highly productive (High Professionals - HiPro) and high potential (High Potential - HiPo) [11]. The former are highly effective in their current role. The latter also show good results, but this is not their limit, since their internal potential can be much wider than the demonstrated one, and they are able to master new functionality faster than others to be successful when changing their role. As mentioned above, such employees can quickly adapt and be involved in performing tasks from a new direction of professional activity without special difficulties, work effectively in multitasking mode, are distinguished by creativity and the ability to non-standard problem solving. They have pronounced leadership and organizational skills, as well as the desire and readiness for continuous development and self-improvement.

To attract, retain and motivate HiPo employees, it is necessary to create an effective system for the identification, support and development of talents in the organization. High-potential employees should be treated with particular care by the company's management, including investing in their development. The implemented personnel policy should be aimed at forming a special system for working with HiPo in order to develop it, including the following tools:

– the use of modern methods, tools and approaches in the selection and selection of personnel, both in the labor market when filling vacancies, and in the formation of the organization's personnel reserve;

– maintaining the professional level of knowledge and skills of employees for effective work, taking into account the need to solve both current and future business problems;

– informing employees about the development strategy of the enterprise, its structure and technology of activity, involving personnel at the stage of formulating business goals, thereby creating and maintaining a sense of involvement in the activities of the organization;

– ensuring an adequate level of income for workers and strengthening the incentive role of payment for work;

– implementation of diverse programs for the development of professional, personal and managerial competencies and analysis of changes in the competencies of employees after each training;

– development of individual plans for the development of talents, taking into account their potential and needs for self-realization, conducting regular assessment interviews for developmental feedback;

– preparation of employees for potential relocation or promotion, possible replacement of colleagues on vacation, business trip, absence due to illness, etc.;

– implementation of various stimulating and motivational programs aimed at maintaining a positive attitude in employees towards the work performed and increasing interest in the results of their work;

– development of social programs stimulating an increase in personnel loyalty and satisfaction with working conditions.

In the context of ensuring the economic security of enterprises, it is important to establish the most significant potential threats that directly affect the development of human potential and, accordingly, the level of economic security of enterprises. These, in our opinion, include:

– First, the risk of burnout for telecommuters can be identified.

– secondly, the problem may be the prevalence of relatively directive management methods at the level of line managers of the enterprise.

The study of the above threats is extremely important in the modern conditions of society.

3 Research results

Examining the external and internal factors of the emergence of threats to the economic security of the enterprise, the management, the personnel management service in the development of the main directions of personnel policy should be guided by the timely determination of personnel needs, both in quantitative and qualitative aspects, to apply current forms of attracting and further use of human potential, to carry out activities on the timely identification of factors affecting the effectiveness of the use of human potential assets.

Based on the results of the study of factors affecting human development, one of the most significant potential threats was substantiated is the risk of burnout of employees working remotely [1, 12].

In connection with the introduction of a self-isolation regime associated with the coronavirus pandemic, office workers were urgently transferred to work from home in April 2020. Modern technologies allowed the company to quickly and successfully organize work from home and business managed to cope with the forced massive transition to remote work. At the same time, as an advantage for the organization, it can be noted a decrease in the costs of maintaining and servicing the office, delivering employees to the enterprise on corporate buses, etc. Employees also saw a number of advantages in the remote work : saving time on commuting to the office, reducing the number of distractions when working in the office, and more. The productivity of employees in performing current tasks (analyzing data, preparing reports and presentations, etc.) remained at the same level, and sometimes even increased slightly due to the high pace of work in a crisis. However, over time, certain problems associated with telecommuting began to appear: the home became a permanent workplace, and for many it turned out to be more difficult than it seemed earlier [13].

To solve the existing problem, it is proposed to implement an additional set of activities at the enterprise.

Due to the fact that the office for employees has become their home, the boundaries of "work-home" have been erased, and even with fewer tasks, they can appear at any time in an e-mail or communicator. At the same time, no one set the task for employees to stay in touch around the clock, for the most part this is their independent decision, due to the blurring of the boundaries separating work and home, their loyalty and a high sense of responsibility. One of the activities to mitigate this problem is prioritizing and scheduling cases in terms of importance and urgency, establishing a clear daily plan of the day in an online calendar with obligatory breaks for lunch and rest. It allows to visualize tasks and distribute the charge. It can be used the "Tomato Method" by Francesco Cirillo, when 25 minutes are allocated for solving current problems, and then a break is taken for 3-5 minutes; to keep track of time, a regular timer or special applications are used [14]. It is also necessary to take breaks from online meetings to complete work tasks: close email and all communicators at the scheduled time and work on a specific task. Otherwise, the whole day will be spent on small questions and meetings, and you will have to work at night when no one is distracting.

To solve production issues with colleagues, it is proposed to give preference to phone calls or communication in communicators with the inclusion of video communication (Teams, Zoom, etc.), rather than correspondence by e-mail or in a communicator, because emotions are not visible in the chat. Since you have incomplete communication without verbal contact, there is a chance to miss something. Without seeing each other, it is difficult to understand what kind of person is, what his look and intonation is, what gestures the interlocutor has at the moment, or how he changes poses. Moreover, not everyone use the same terms, concepts, etc. Therefore, more efforts are spent to understand what information they want to convey, internal instincts are turned on, which make you think out the situation and not always in the right direction.

The lack of personal contact with the manager worries many employees that their professional efforts will go unnoticed. Timely face-to-face communication is the main way to deal with burnout. It is very important to be able and practice to express your concerns, objections, worries orally and discuss them. Therefore, in addition to general meetings, managers are advised to regularly hold individual online meetings with subordinate employees to discuss their success, difficulties in work and personal communication. It is also necessary to discuss and plan the training and development of employees, their participation in any projects, since a person physically sits in one place at home, and it seems to him that there has been a pause in his development. All this will help employees gain confidence, and managers can track changes in mood in time and jointly develop a solution plan.

For informal communication with colleagues, you can envisage regular (for example, once a week on Fridays) online meetings for half an hour for conversations on everyday topics. A series of online stress management and neurographics courses could be launched to teach employees how to help themselves under stress. Case study: a Chinese Internet technology company has set up a 24-hour emergency counselor and psychologist service for all telecommuting workers (including interns) and their families, as well as a distance learning course on emotion management during quarantine and COVID-19 pandemic [15].

Often , an employee does not have the opportunity to organize an isolated workplace at home, since many live in small apartments, some have small children who find it difficult to explain that mom or dad are working now. In this case, the remote work format turns out to be quite difficult and requires very high self-discipline from the employee. In any case, it is recommended to equip the most equipped workplace so that everything is at hand, and households can go about their business without interfering or distracting with questions. If the house is noisy, you can use headphones or earplugs. Some employees can use the help of the ritual of "imitation of the office", when they put on "formal clothes" in the morning, tea or coffee during the day is brewed only in a separate "office mug", etc. That is, to attach specific activities to objects and places at home: work only at a laptop, dine at the kitchen table, and only relax on the couch, without mixing these chains with each other. This can help emphasize work-home separation at the start and end of the work day.

You can also recommend that employees adhere to the following rules when working remotely:

– monitor your biorhythms and pay attention to which hours the activity is maximum - during these hours you can complete the maximum number of tasks;

– Take time out of work, eat well, and do physical exercises during the breaks (some companies encourage healthy habits, for example, by linking to videos that promote physical and mental well-being, and by encouraging the use of fitness applications and services;

– while working at the computer, remove the smartphone as far as possible so that there is no temptation to look into social networks or be distracted by chats with friends;

– adjust the dim monitor screen, give rest to the eyes and do special exercises for the eyes;

– during a break, if possible, go for a short walk near your home and try not to scroll through tasks in your head or look into your phone and work chats at this time.

When studying the factors affecting the development of human potential, the predominance of relatively directive management methods at the level of line managers of the enterprise was noted. This management style can be very effective when a crisis occurs in the company, as well as for newcomers or employees with a rather low level of professional maturity. Then the manager sets the task with maximum detail, stage by stage describes what exactly needs to be done. His instructions and orders are subject to strict execution by his subordinates [4, 16].

In view of the active phase of growth and development of companies and the sufficient professional maturity of the team of employees, it is proposed to more actively use the coaching approach in personnel management as a method of increasing human potential. Coaching is a well-established business technology. Today this technique is successfully applied in business and management. With its help, modern leaders unleash the potential of both the entire team and each team member, which ultimately helps to awaken the internal motivation of employees and achieve maximum efficiency of each individual, overcoming obstacles and making optimal use of resources. A coach does not solve problems for a person, but only by observing a sequence of certain actions directs a person so that he independently finds the most beneficial solutions for himself and for the organization.

The task of a leader using a coaching approach is to motivate the employee to bring his idea to life, to support him.

At the same time, the use of a leader as a coach will bring success only if a certain number of conditions are met:

- relations of mutual trust, respect and mutual understanding are established between the coach and the ward;
- both parties feel responsible for the result they are striving for;
- both the coach and his student have a great desire to learn and improve;
- the coach will try to be as impartial as possible (although in reality it can be difficult to get rid of the influence of his point of view).

Thus, the use of coaching in work with personnel allows you to reach a qualitatively new level of interaction and make progress in improving the professional skills of each employee. All this will greatly contribute to increasing the motivation of personnel and the disclosure of human potential in order to maximize its efficiency, which in the end should have a positive effect on the effectiveness of the entire business.

In order to develop human potential, it is recommended to implement a comprehensive training and development program for personnel, including both various advanced training programs for production workers, and seminars and trainings for office and "field" employees. For the first category of workers, an assessment of the effectiveness of training is a quite measurable indicator - the result of an exam after mastering a new / related profession, increasing the category or mastering a new type of equipment. At the same time, it turns out to be quite difficult to measure the effectiveness and efficiency of the second type of training. However, evaluation is necessary to ensure the effectiveness of the process, because if the training is not evaluated in any way, then it is difficult to prove that it turned out to be useful, and then also convince others that money should be further invested in this direction.

Learning effectiveness is a measurable business benefit. A large number of factors influence the effectiveness of training: the motivation of employees, the professionalism of the trainers, the correspondence of the training program to the customer's needs, pre-training diagnostics, the number of people in the group, the optimal balance of theory and practice, etc.

Assessment of the effectiveness of training makes it possible to:

- answer the question of whether the choice of the training organization / trainer was right;
- to assess how the advanced training of employees affects the quality and dynamics of the enterprise's development;
- monitor the skills and abilities of employees, growth / decline of the level of their competencies;
- help to achieve a greater economic effect of the organization's activities.

To assess the effectiveness of training, you can use a fairly well-known four-level model proposed in 1959 by Donald Kirkpatrick, which is now classical.

This model is built on 4 levels for assessing the effectiveness of personnel training:

Level 1 - Assessment of the participants' reaction to the training program.

At this level, through a questionnaire, the assessment of the training by its participants is recorded: their emotional state, attention and interest in the topic of training, opinion about the quality of the training (usefulness, the work of the trainer, the level of complexity of the material and exercises, etc.).

The results of the questionnaire are usually used by the training organizers to assess the correct choice of the training topic and the level of the trainer's work.

2nd level - assessment of knowledge and experience gained by participants in the training program.

At this level, it is assessed to what extent the knowledge and skills of the participants have been assimilated, which they can apply during their work. It traditionally uses assessment methods such as professional and behavioral tests, thematic questionnaires, control exercises, observation, etc., according to the results of which the participants are given marks or points.

The results of the assessment at this level are important both for the participants (to quantify the progress in the acquired knowledge) and for the trainer (how effective the chosen teaching strategy and methods were).

3rd level - assessment of the level of behavior in the workplace.

At this level of assessment, we mean the practical results of the training: are the knowledge, skills and attitudes acquired by the participants used / implemented in work situations (as opposed to educational ones) and how effectively they are used in professional activities. This stage is one of the most difficult to assess, since the effect of training is usually not noticeable immediately. Therefore, this type of assessment is carried out only some time after the completion of the training, when the participants will be able to practice applying new knowledge and skills in the workplace. The direct supervisors of the training participants are necessarily involved in this stage of the assessment.

To compare the new work behavior with the previous one (or with the established corporate standards), both formal methods (qualification exam, certification, performance of control tasks in the workplace) and informal methods (structured observation, keeping work diaries, etc.) are used.

4th level - evaluation of results.

At this level, the impact of the learning outcomes on the activities and / or financial results of the entire organization is assessed. Key business metrics are identified, which are indicators of the effectiveness of training (increased productivity, improved quality, reduced staff turnover, etc.). At the same time, it is important to know the exact data of these indicators before training, since only those changes that have occurred after the training of employees are related to the results.

Collecting this kind of information requires serious analytical work, time, financial resources and coordinated efforts of many departments of the company. Therefore, this level of assessment is not used for individual typical trainings. It makes sense to use this tool to assess the business effect of expensive complex training programs [9, 12].

4 Discussion of results

Taking into account the commensurability of labor costs for assessing the effectiveness of training and the corresponding business effect, it can be concluded that according to the 1st level of the Kirkpatrick model, it is advisable to evaluate 80-100% of the programs, according to the 2nd - 40-60%, according to the 3rd - 30%, on the 4th - 10-20%.

If you try to integrate the Kirkpatrick model to assess the effectiveness of training, you can propose the following algorithm of actions:

For the 1st level of assessment, which is usually used immediately upon completion of the training, several methods can be proposed.

DISCUSSION. Group discussion, in which the trainer identifies the feedback structure (for example, what was useful at the event, whether the participants' expectations from the training were met, was there enough theory / practice, etc.) After that, everyone takes turns expressing their point of view, and the trainer captures information for consolidation. This format is quite emotional and makes it possible to quickly get feedback.

QUESTIONNAIRE. Participants fill out a questionnaire with a standardized set of questions, including the following aspects: the usefulness of the acquired knowledge and skills, the quality of the presentation of the material, the level of complexity of the exercises, the informativeness of the handouts, the optimal balance of theory and practice, the level of the coach's work, etc. The data of the questionnaires are usually quite rational and informative.

DARTS. A method of visualizing assessments, in which a classic target is depicted on a flipchart, divided into sectors with a marker, reflecting the aspects described above. Participants mark on the target crosses corresponding to their assessment of each parameter. This methodology allows you to get a visual informative picture of the assessment of the training: in what aspects the training has achieved the goal, and in what aspects missed the target. This method of group rapid assessment is used when there is a shortage of time, when there is very little time left for summing up the results.

For the 2nd level of assessment, you can also offer the TESTING methodology. Tests with questions on the topic of the training are prepared in advance. Participants take a test before and after training, based on a comparison of the results, a conclusion is made about their assimilation of the material.

For the 3rd level of assessment, which implies the degree of application in practice of the knowledge gained and their transformation into skills, the Method 360 can be used - when managers, subordinates and colleagues of an employee fill out a questionnaire about changes in his behavior in the workplace due to the application of new knowledge and skills. This method appeared back in the 1980s. as a complement to the method of personnel assessment through management by objectives. The organization of this method requires some effort, however, specialists in the field of personnel assessment define it as one of the most progressive.

It should be noted that in order for the skills to become skills, they must be repeated at least 21 times. Therefore, it is advisable to use this level of assessment no earlier than 1-2 months after the training.

To consolidate the acquired knowledge and skills, it is possible to instruct employees who have undergone training, conduct mini-presentations using the material covered for colleagues from other departments, or participate in the adaptation and training of new employees. There is no better way to properly learn something than to try to teach it to another.

It is also recommended for employees to suggest improving any business processes using the information obtained during the training. An employee can submit for the management's consideration his proposals for improving the activities of both the employee himself and the organization as a whole (for example, an action plan to improve any direction of the company's activities, creating a new work standard, developing and implementing a project, improving or optimizing a business process etc.).

The 4th level of assessment is the most difficult, since it is necessary to identify, formulate and evaluate key performance indicators (KPIs), which are potentially influenced by learning outcomes.

These business indicators of the company include: improving quality, reducing the number of complaints or customer complaints, increasing labor productivity, increasing production rates, reducing costs, reducing staff turnover, increasing sales, developing new market segments, reducing financial costs, reducing the number of accidents and industrial injuries, reduced customer service time, reduced production costs, fewer conflicts, etc.

The list of KPIs for assessing learning outcomes are selected individually in each specific situation, taking into account the topic of the training, the relevance of the business indicator and the goals of the company.

In conclusion, I would like to emphasize that organizing a training system is not enough. It is also necessary to ensure that employees use the acquired knowledge and skills in practice in their work, only then the results of training will be development and investments will be justified.

5 Conclusion

It has been found that of all organizational resources, it is human potential that hides the greatest reserves for increasing the efficiency of an organization and its economic security, since the company's profitability, its capitalization, an increase in the share of intangible assets and other business results directly depend on the competence and correctness of employees' actions. Therefore, one of the main tasks is the selection of qualified personnel with high potential, its effective use and development, as well as minimization of risks and threats from employees.

The main directions of personnel policy when working with high-potential employees and effective tools for their development and motivation have been determined. The main goals and approaches for the formation of the training system are outlined: providing employees with the necessary knowledge and skills and their constant support at a high level; preparation of employees for promotion or replacement of colleagues during their absence; maintaining employee engagement and a positive attitude towards work.

Instruments are proposed to improve measures for the development of human potential, contributing to an increase in the level of economic security of the enterprise. Also, proposals aimed at reducing the risk of burnout of remotely working employees, the introduction of a coaching approach in personnel management, proposals for improving the system for assessing the effectiveness of employee training.

Thus, the procedures for the development of human potential, in order to ensure the economic security of the enterprise, are primarily aimed at continuous work with the employees of the company as a whole, individual employees to increase the status and loyalty of the company, to form a team spirit based on common values and interests.

References

1. D. Burkaltseva, O.Reznikova, S.Plaksa, Economic security: conflict in the organization **7(2)**(2020)
2. D. Kechek,I. Polgar, *Overview of operational research in Croatia* (2020)
3. H. Gardner, *About Human Potential: A 40-Year Saga*(2020)
4. D. Y. Dai, Magazine for the education of the gifted(2020)
5. T.E.Davydova, E.A. Avdeeva, T.I.Makeeva,2nd international scientific conference of Volgograd state University at a competitive, sustainable and secure development of the regional economy (CSSDRE) (2019)
6. N.V.Latova,Sociological research(2019)
7. S.I.Babina, A.Y. Nesterov, I.Y.Sadovnikova, 2nd International Scientific Conference of Volgograd State University on Competitive, Sustainable and Safe Development of the Regional Economy (CSSDRE) (2019)
8. O.V.Loseva, N.M. Abdikeev, M.V. Melnichuk,Revista Inclusions(2020)

9. E.N.Klochko,A.A.Adamenko, O.V.Brizhak,2nd International Scientific Conference of Volgograd State University on Competitive, Sustainable and Safe Development of the Regional Economy (CSSDRE) (2019)
10. N.Sirotkina, V.Mishchenko, A. Kazartseva,33rd Conference of the International Association of Business Information and Management (IBOMA) (2019)
11. G. Tokatly, M.C. Dorneich, 38th IEEEAIAA Conferenc e on Digital Avionics Systems (DASC) (2019)
12. Ya. Epifanova, I.Dzhedzhula, International scientific electronic Journal «Economics of Agriculture and Resources» (2020)
13. S. Moran, Magazine for the education of the gifted(2020)
14. Ya. Rudskaya, S. Kuznetsov, P.Salnikova,34th Conference of the International Association of Business Information and Management (IBOMA) (2019)
15. L. Boronina, A. Balyasov, I. Visnevsky, The 13th International Days of Statistics and Economics (2019)
16. Z.Zhivko, Ya.Gorban, M. Korzh, Management theory and research for the development of rural business and infrastructure (2020)

https://doi.org/10.1051/e3sconf/202131108002

Personnel training in Russia: response to modern challenges and threats in the context of digitalization of the economy

Anatoly Gretchenko[1, 2], and *Alexander* Gretchenko[1]

[1]Plekhanov Russian University of Economics, 117997, Stremyanny lane 36, Moscow, Russia
[2]Financial University under the Government of the Russian Federation, Russia, Leningradsky Prospekt 49, Moscow, 125993, GSP-3, Russia

Abstract. The article discusses the results of a study conducted at the Research Institute "New Economics and Business" of the Plekhanov Russian University of Economics. University science, the system of university research is a modern form of integration of science and digital technologies. The development of the research methodology and the transformation of personnel training for the Russian economy, taking into account the emerging risks and threats to business, society and the state, is one of the priority areas of university research. The rapid development of digital technologies not only opens up opportunities for achieving the goals of sustainable socio-economic development, but also creates threats and risks, including of an ethical nature. It allows us to link positive trends and the growth of sustainable development from the widespread use of digital resources, tools and technologies, as well as possible risks and threats associated with it. According to the authors, the answer to the risks and threats that have arisen can only be a deep complex transformation of the entire domestic personnel training system. Only a digital educational ecosystem is able to form a modern specialist with high adaptability and self-learning. The circumstances considered in the article determine the relevance of such studies and indicate the need for digital transformation of personnel training for the Russian economy in the current socio-economic conditions.

1 Introduction

The modern realities of the development of international relations give rise to new challenges and threats to the national economy. This is largely due to the increasing independence and activity of the Russian Federation on the world stage every year. Russia has a serious competitive impact on traditional economic leaders. It occupies leading positions not only in the commodity markets, but also in the markets of computer technology, microelectronics and robotics. On the one hand, the response of our partners is natural, but on the other it is paradoxical – the sanctions policy has become an everyday reality. At the same time, there are serious breakthrough achievements of domestic specialists in such end-to-end technologies as artificial intelligence, big data analysis, quantum technologies, unmanned transport, etc. There is a situation when the restrictive policy of foreign countries spurs the process of import substitution, Russian advanced research and production facilities pick it up. At the same time, the qualified personnel often does not have sufficient qualitative and quantitative potential to maintain it.

We are deeply convinced that the answer to these challenges can only be a deep comprehensive transformation of the entire domestic personnel training system, which should be based on the following digital solutions [1, 2]:

1. The educational trajectory for each student should be based on an intellectual analysis of the transformation processes taking place in the digital environment of the Russian economy.
2. The availability of global educational resources using high-speed communication channels.
3. End-to-end identification and authentication of the student when implementing any of his own educational requests.
4. Total virtualization of laboratory, experimental and production educational bases.
5. Creating avatars or digital doubles of teachers, mentors, tutors who, based on artificial intelligence systems, will be able to develop basic soft and hard competencies in the student.
6. Digitization of the student himself with the creation of his avatar or digital double for the implementation of routine non-creative search, calculation, sorting and other procedures.
7. Ensuring the priority development of the student's self-competencies, which open up opportunities for self-development and self-learning, again on the basis of recommendation intelligent systems.

It is such a digital educational ecosystem that is able to form a modern specialist with high adaptability and self-learning, which are basic when building artificial intelligence systems. And here we come to a certain futuristic phenomenon – not digital systems are trying to copy wildlife and humans, as it has always been, but a person is equal to neural networks, quantum technologies, etc. Most likely, this trend will increase.

At the same time, the digital transformation of the personnel training system, simultaneously with the response to existing challenges and threats to the Russian economy, generates its own problems, vulnerabilities and dangers.

These circumstances determine the relevance of such studies and indicate the need for digital transformation of personnel training for the Russian economy in the current socio-economic conditions. This issue was particularly acute during the coronavirus pandemic and self-isolation, starting in 2020. When the learning process switched to distance learning in strict lines, and not all educational institutions were ready for such a format of training both in methodological and technical equipment.

2 Materials and Methods

The presented research results in this article were based not only on their own authors, but also on the use of existing both foreign and Russian scientific materials using modern approaches and research methods.

Thus, the authors of the article "Societal and ethical issues of digitalization" come to the conclusion that "... digital transformation is not only technological in nature, but affects the entire established way of business: the nature of interaction of subjects within the business ecosystem, the business model of enterprises, the architecture of enterprise management systems, the system of business processes, the architecture of information systems" [3]. In the work "Economic modeling of digital transformation processes"," ... an analysis of a new trend that is replacing computerization and informatization of society - the global digitalization of society " was carried out [4].

The authors (Kulaji T. V., Babkin A.V., Murtazaev S. A. Yu., 2019) see the use of digital technologies in solving national projects and programs [5], (Savina T. N. 2018, Erokhina E.

V., Gretchenko A. I. 2019) " ... the uncertainties and risks inextricably linked with the transition to the digital economy are identified, the informatization processes in the Russian Federation are analyzed. It is established that digitalization and informatization are an inseparable unity of processes: intellectualization, computerization, mediatization, associated with a high degree of uncertainty and risks. It is concluded that digital transformation should be considered as a new tool in solving problems that requires enormous efforts at all levels" [6, 7, 8].

In his scientific article Domenico Consol in detail " ... analyzes the introduction and use of ICTs, and their impact on organizations. It is noted that the introduction and use of ICT can bring benefits in terms of efficiency, efficiency, innovation, growth and competitive advantages, especially in small and medium-sized businesses" [9].

Don Tapscott (1996) in the work "Digital Economy" "... focuses on three main areas: "the new economy and the factors that shape it; inter-network interaction and its relation to business and government, and, finally, the need for a strong progressive leadership that will be responsible for transformation or will be a conductor of change in this new era" [10].

The transition to process automation requires certain financial costs and investments. It is necessary to conduct a thorough analysis before the transformation and assess the readiness of the transition to digitalization and automation.

These and other studies prove the ongoing transformation by the impact of digitalization. Digitalization can provide an increase in productivity in the work of enterprises [11] in the field of high-tech industry development and digital technologies in the financial sector require a different level of training of personnel with appropriate digital competencies.

It should be noted that in the course of the study, the authors widely used the methods adopted in modern sociological theory. The focus of this study was an ecosystem approach based on a system paradigm. The universality of the ecosystem approach allows us to apply it to the study of socio-economic processes, in particular, for the system of education and training of personnel. To assess the effectiveness of the professional and educational ecosystem, this article assumes the use of the basics of the theory of stakeholders, a direct subjective methodology, as well as the method of situational analysis. The analysis of the effectiveness of the professional and educational ecosystem is based on the assessment of the importance and satisfaction with the knowledge and skills formed in the learning process in its organizations. An empirical study was also conducted in the form of a survey (written; questionnaire; survey; interview) (faculty; students; administrative staff) for the quality and condition of technological equipment with modern digital technologies.

3 Results and Discussion

Personnel is the most important component of the productive forces of society, accumulating a complex of knowledge and experience, as well as the presence of certain personal abilities and qualities. The revolutionary processes taking place in the scientific and technical sphere are rapidly changing production, which is currently a driver of economic growth. In modern industry, a significant share of knowledge-intensive products is created. Moreover, the digitalization of the Russian economy is one of the strategic directions of development. In this regard, the question arises: what will happen to the labor market? Will there be a million-strong army of unemployed people? The concept of work is changing, the concept of work is changing, employment once appeared, but it will disappear sometime. There are more and more free professions. At present, the main problem, oddly enough, is not how to find a job, but to find your subject of activity, it is desirable that this is a personality-developing activity, and it will provide a source of income. This will not be quite a market and not quite a labor in the current sense. There is a need to develop completely new methodological approaches to the labor market, training fundamentally new highly qualified personnel, improving the system of personnel reserves, which is the main driving force of a modern organization.

Work is changing to such an extent that it will be difficult to understand what a profession is. Some Russian researchers claim that mathematics and IT come out on top, as the subject of work that a particular person will face will increasingly resemble the solution of a certain algorithm. The content of the work of each individual person in 30-40 years will increasingly approach universal labor. And the number one question will be finding an occupation and forming a lifestyle. The very concept of a profession will change in 10-20 years, that is, today young people are already facing this problem.

An important problem in the conditions of the formation of the information society is the training of personnel capable of developing fundamentally new information and communication technologies. Information and communication technologies (ICT) and digitization are widespread in modern society. ICTs are also linked to other technologies, such as nanotechnology, biotechnology and neurotechnology. This so-called NBIC convergence has become increasingly noticeable since the late 1990s. Digitization penetrates into all aspects of public life: technology takes root in workers (for example, with the help of brain implants), between people (through social networks such as Facebook), learns more and more about us (with the help of big data and methods such as emotion recognition), and constantly learns to behave like us (robots and software demonstrate intelligent behavior and can simulate emotions). The digitization of society pushes the boundaries of our capabilities and offers all sorts of opportunities, but also challenges our moral boundaries.

According to the results of the annual study of the Norton Cybercrime Report 2019, the damage from cyber attacks on educational information resources in the world amounted to $336 billion (in 2018 - $110 billion). The situation in our country does not look any better. Studies have shown that in 2019, about 80% of all educational resources were victims of cyber attacks, which led to damage of about 203.3 billion rubles (in 2018 - 102 billion).

This fact, on the one hand, clearly illustrates the value of educational content and infrastructure, and on the other hand, their vulnerability to attacks via communication channels from intruders. The alarm is caused not even by the number of attacks themselves, but by the speed of their increase. And this is despite the fact that the market of means and methods of protecting educational information resources is steadily growing.

It is necessary to observe a kind of "scissors" - an increase in the cost of creating and implementing security tools not only does not stop the growth of damage from attacks, but even creates the impression that it contributes to it. In our opinion, this is determined by the essence of the digital transformation of the educational system itself, which consists in using the following advanced information technologies to provide the digital solutions listed above:

1. Cloud computing.
2. Virtualization of information resources.
3. Internet of Things systems.
4. Mobility of users ' devices
5. Artificial intelligence systems.

Cloud technologies, which are widely used nowadays, of course, significantly increase the efficiency of collecting, storing and processing educational content, but at the same time, they lead to its vulnerability, which is very critical, given the high share of intellectual property in its composition. It is enough for an attacker to gain access to the information storage infrastructure, as he has at his disposal its entire array, concentrated in one data processing center.

Some Russian scientists note that "... a sign of the information society, the knowledge society is the increase in requirements for the level of professional training, the quality of education, and an increase in funding for the scientific sphere" [12 - 16].

4 Conclusion

The author's research and the results obtained allow us to draw the following conclusions. For the purpose of digital transformation of personnel training for the Russian economy, taking into account the responses to modern challenges and threats, it is necessary to consistently solve the following tasks:

1. Substantiation of the main elements of the concept of studying the digital transformation of educational institution management systems based on artificial intelligence technologies:

- development and implementation of integration platforms for management systems of administrative and economic activities of an educational institution: accounting, budgeting, personnel accounting of students, personnel management, management of the results of the activities of teachers and researchers, management of the alumni community, management of the enrollment of applicants;
- transition to electronic document management in full with the implementation of digital services for electronic interaction with government agencies;
- management of research projects;
- management of educational buildings, dormitories, sports complexes based on resource-saving technologies "Smart", centralized notification systems, intellectualization of access control systems.

2. Theoretical substantiation of the genesis of the transformation of the IT infrastructure of an educational institution:

- transformation of the data processing center in the direction of a software-defined infrastructure;
- connection of high-speed telecommunication lines;
- introduction of a seamless wireless access system in academic buildings and dormitories of an educational institution;
- implementation of a unified authentication and identification system;
- creation of an intelligent computer resource management system;
- creation of digital transformable classrooms and coworking spaces.

3. Justification of the main elements of the security management concept:

- implementation of intelligent recognition systems when organizing access to an educational institution and its individual premises;
- implementation of an infrastructure security incident detection system, adaptation of the system in order to automatically respond to emerging incidents.

4. Development of a methodology for digital transformation of the educational environment:

- placement of educational materials on disciplines and modules in digital form in the LMS of an educational institution, automated task verification;
- asynchronous interaction of all participants of the educational process, orientation to problem-oriented learning;
- introduction of courses on open national MOOCs platforms, mixed classical and e-learning Blended learning into the educational process;
- introduction of adaptive learning methods, adaptive courses with an individual trajectory of training, depending on the features of mastering the educational material (using big data processing technologies and artificial intelligence in the construction of adaptive tracks);
- introduction of interactive products and technologies: simulators, simulators, VR/AR reality systems;
- integration of online interaction services into the educational platform with support for audio and video communication, collaboration of all participants in the educational process;
- creation of corporate cloud computing resources services;

- development and implementation of virtual project laboratories.

- development of a system for constructing an individual trajectory for each student, taking into account his digital footprint, the recommendations of the teacher and the recommendations of the system based on artificial intelligence;

5. Development of a methodology for the digital transformation of the research environment:

- creation of a cataloged digital platform for interaction in the field of creating innovative products between an educational institution, industrial and academic partners;

- creation of digital research laboratories within the framework of scientific directions and schools of an educational institution;

- creation of a digital infrastructure for intelligent processing of big data in the framework of R & D;

- creation of a spatial and digital infrastructure for an educational factory for piloting innovative projects.

Acknowledgements

This article was prepared as part of the government contract as requested by the Ministry of Science and Higher Education of the Russian Federation on the subject formulated as «Structural changes in economy and society as a result of achieving the target indicators of National projects, which provide opportunities to organize new areas of social and economic activity, including commercial, both in Russia and abroad» (project No. FSSW-2020-0010).

References

1. A.I. Gretchenko, I.V. Gorokhova, O.G. Demenko, A.A. Gretchenko, Journal of Advanced Research in Law and Economics. Quarterly, **IX(4(34))**, 1243 (2018)
2. G. Beliakov, A.Gretchenko, A.Ryzhaya, A. Shpak, S. Belyakov, International Journal of Supply Chain Management, **8(6),** 1035 (2019)
3. Royakkers, Lambèr, et al., Ethics and Information Technology, **20(2),** 127 (2018)
4. V. P. Bauer, V. V. Eremin, S. N. Silvestrov, V. V. Smirnov, Journal of Economic Theory, **16(3),** 428 (2019)
5. T. V. Kulaji, A.V. Babkin, S. A. Yu. Murtazaev, *Digital Economy and End-to-end Technologies: theory and Practice*, 191 (2019)
6. T. N. Savina, Finance and credit, **24(3),** 771 (2018)
7. E. V. Erokhina, A. I. Gretchenko, Vestnik NSUEM, **4,** 41 (2019)
8. E. V. Erokhina, A. I. Gretchenko, Science and practice, **13(1(41)),** 56 (2021)
9. D. Consoli, Literature analysis on determinant factors and the impact of ICT in SMEs, **62,** 93 (2012)
10. D. Tapscott, The digital economy: Promise and peril in the age of networked intelligence, **1** (1996)
11. A. I. Gretchenko, A. A. Gretchenko, Springer, Cham, **161** (2021)
12. A. I. Gretchenko, O. G. Demenko, A. A. Gretchenko, Journal of Advanced Research in Law and Economics. Quarterly, **IX(4(34)),** 1249 (2018)
13. A. Gretchenko and A. Gretchenko, SHS Web of Conferences, **93,** 03003 (2021)
14. V. M. Maslova, *Regulation of the labor market*, 271 (2019)

15. A. I. Gretchenko, A. A. Gretchenko, *The Fourth International Conference "Business Management in the Digital Economy": a collection of abstracts,* 492 (2021)
16. A. I. Gretchenko, A. A. Gretchenko, Proceedings of the International Conference "Scientific research of the SCO countries: synergy and integration". Part 1- Reports in English, 8 (2021)

E3S Web of Conferences **311**, 08003 (2021) https://doi.org/10.1051/e3sconf/202131108003
EPSD 2021

Labor intensity in context of information and communication technologies sustainable development

Elena Krasova*

Vladivostok State University of Economics and Service, 690014 Vladivostok, Russia

Abstract. The information and communication technologies (ICT) impact on labor intensity is one of the actual, but understudied interdisciplinary issues of modern science and practice. Observations and analysis of labor processes in context of the ICT use point out the increase in labor intensity. The purpose of the research is to identify key sources of workers' labor intensity growth under the ICT development, regardless of the field of professional activity. The article highlights such sources as increase in work strenuousness, growth of number of labor functions and operations, lengthening working day, and also change in work and rest regime. A special focus is on the problem of «leisure time colonization». An ∩-shaped correlation between labor intensity and productivity is presented. The general conclusion that unlimited growth of labor intensity is inconsistent with the sustainable development and effective development of human capital is made.

1 Introduction

Over the past decades, information and communication technology (ICT) has significantly changed people's lifestyle and work format. At home and at work, people use various gadgets and devices that enable searching and processing information, expanding business possibilities and personal communications, making transactions in real time, getting an education, etc. The ICT in industry are manifested in automation and robotization processes, in artificial intelligence systems. All this induces many researchers to talk about sustainable development under «a new reality – globalization 4.0» [1], «revolution that will fundamentally change our life, our work and our communication» [2], «future without job» [3, 4], «era of intellectualization» [5], «technological alternation of the modern mind» [6] and free creative labor predominance [7].

Despite the growth in information production and consumption we observe today, modern literature tends to exaggerate the ICT positive effects and underestimate the negative ones. This is especially true for such field of researches as labor organization, where, in our opinion, some faraway and probabilistic problems attract too great interest and attention. At the same time, many urgent everyday problems concerning the computerization and informatization processes in economic activity are rather poorly

* Corresponding author: elena_krasova@rambler.ru

studied. Many experts, inspired by the ICT potential, actively discuss an upcoming technological unemployment, new «digital» professions and a problem of «free time, which will be too much soon» [3]. But now, millions of workers around the world feel burned out due to ultra-high work intensity, computer syndrome, information pollution and permanent compelled communications. «While the tedium of office work is nothing new, all-day, everyday computer work is a fairly recent and growing phenomenon. This is a horrible existence masquerading as the apex of a civilized world» [8].

World Health Organization identifies labor intensity as one of the most important social components that determines the level of public health and the level of socio-economic development. Work-related health problems lead to economic losses of 4-6% of GDP in most countries [9]. Extension of researches of the labor intensity growth factors in context of the ICT use is relevant, firstly, in terms of increasing labor productivity and ensuring the national economy sustainable development. Secondly, it is very important in terms of workers' safety, their physical, intellectual, psychological and social potentials salvation. On the basis of these potentials, a human capital which is so necessary to work with ICT is steadily developing and growing.

The purpose of the research is to identify key sources of workers' labor intensity growth under the ICT development at the present stage. A specific scientific and practical task, the research can help solve, is optimization of labor organization and human capital management in a steadily developing information economy.

2 Materials and methods

The research is fundamentally based on the labor organization theory, which examines the labor intensity as a complex category that includes organizational, economic and physiological components. This category defines amount of physical, mental and psychical energy expended by a worker per a working time unit. On the one hand, high labor intensity can make labor productivity grow. On the other hand, high energy expenses of workers can negatively affect their health, making their work hard and stressful, regardless of the profession and activity type. Labor intensity depends on many factors, but influence of such important organizational and technological factor as the ICT is underexplored. This is because it is a recent problem in science and this problem lacks much attention now. Under studying the ICT impact on labor intensity we can highlight the following main research areas in modern literature.

1. *The ICT influence on workplaces' functions in various fields of activity*. The subject here is replacement of hard and harmful labor with machines; facilitation of working conditions under changing technologies; perspectives of automation, robotization and digitalization in different industries [10, 11].

2. *The ICT influence on labor intensity by changing content of knowledge and varying requirements for workers*. Here, focus is on skills and abilities required in the era of informatization and digitalization; on «intellectualization», «digitalization» and «creativity» of labor; on hard & soft skills combination and brain efforts intensification that is consistent with the skills [12, 13].

3. *Influence of working conditions with ICT on employees' health and well-being through the work & rest regime*. Scientists confirm the health deterioration under the technological, informational and communication stressors impact [14, 15]. A spatial & temporal deformation of boundaries of the work & rest organizational field, including the problem of «leisure time colonization» and «leisure fetishization» is on a special focus today [16, 17].

On the whole, most economists and engineers believe that ICT makes work easier and significantly improves work efficiency. At the same time, medics, psychologists and educators convincingly demonstrate negative consequences of the ICT use.

The research is methodologically based on general methods of economics and institutional analysis, including systematization, generalization, abstraction, comparisons, and expert assessments.

3 Results and discussion

Here we consider a general mechanism of work format change, taking into account the ICT factor.

Information and computer technologies have made the work with information more organized and efficient by assuming a number of complex auxiliary functions, for example, calculations, data processing, algorithming, unification, copying, etc. However, this does not cancel a fact of people's work with information at all. Both in earlier historical periods, and nowadays, any work is based on the perception, processing and use of professional information. The only difference is that today the active circulation of information leads to its rapid obsolescence and to need to obtain new information so that to maintain and/or increase labor productivity. The processing speed and the processed information amount increase proportionally. This naturally increases informational, sensory and psycho-emotional burden on a worker. At the same time the worker bears full responsibility for decisions he makes on the basis of the information obtained on his workplace.

In today's literature there is a prevailing opinion about inevitable labor intellectualization as a result of the ICT implementation. We can find even some apologetics of mental labor as the highest form of human activity. This opinion, as a rule, does not take into account the high degree of mental work monotony because of continuous and steady attention concentration and brain activity mobilization in conditions of physical activity deficit. Moreover, being along with non-standard and creative operations that are visible for intellectual work, everyday routine operations taking the largest share in working process are out of consideration scope, although such invisible, but no less necessary operations are typical for overwhelming majority of professions. «In any profession we can distinguish its iconic social visibility and invisible routine activities which take most of working time» [18]. Thus, the ICT objectively stress a worker's central nervous system, and it leads to labor strenuousness increase.

ICT has accelerated executing a large number of organizational and communication tasks, such as doing correspondence, notifying & sending messages, writing & registering different applications, ordering & paying for goods and services, and other bureaucratic tasks. «Technology has simplified a number of operations so much that now we do them ourselves, although before it used to require separate workers» [19]. However, to simplify is not to abolish. Now most of these work operations are being done by workers themself. Within a working time unit there are a lot of «simple and effective» deeds that create a backlog and work not for the final goal, but for ensuring the labor process. On the one hand, process actions can be standard (formulaic or similar actions), for example, sending a mail, writing & agreeing an application, entering data into a program, etc. In this case, additional work routinization emerges, even if a creative work is formally being done. On the other hand, an individual acts in a multi-situational environment, where templates sometimes do not work, for example, computer is freezing, software is not responding, document approval procedure has been changed, etc. Here a worker has to spend additional efforts to rectify the non-standard situation. Often, the efforts and time wasted for ensuring current labor process are much more than ones spent on a result. Eventually, the worker does more and catches less. This is a kind of Pareto law strengthening, concerning to modern labor.

A duplication of work with traditional and digital technologies is one of the important factors in increasing labor intensity nowadays. This is especially true for jobs not directly related to the IT sector. The current stage of informatization and digitalization is incomplete, transitional; therefore work with ICT is well combined with traditional work format. We don't observe the purely digital economy. We can see a multi-structured economy in most countries, including developed ones. In particular, business experience of Russian enterprises and organizations shows frequent use of computers, software and paperwork in combination. This is due to high costs of the ICT objects implementation for many business entities, to labor organization imperfection and incomplete information security systems [20]. Combination and duplication of old and new technologies increase a number of labor functions and operations. In addition, transition to ICT requires studying instructions for software, administering accounts, registering in various electronic systems, being a member of different professional online communities, etc. This all takes a lot of time and efforts, although it may seem very simple to a modern person.

Today, advantages of implementation of different ICT objects into production are being actively discussed and substantiated. The objects – computers, automatic machines, robots, etc. – reduce costs, improve product quality, eliminate the human factor, optimize the workspace. However, everyday practice often shows us the opposite situation: imperfections of the ICT and/or the ways they use are clearly manifested. The ICT equipment is periodically broken down; the software efficiency & speed largely depend on developer's workmanship, equipment & software compatibility, server power and other factors. Turning off the electricity can stop a work at all. An ordinary worker is able to solve not all of these troubles and not always. If they are out of his control he has to wait for a technical service or take the work home to continue.

Thereby, under investigating the mechanism of the ICT impact on labor processes, we have noticed the significantly growing number of labor process operations and functions per a working time unit. Scientists have fixed a paradox: «We thought that new technologies would free us from work. But they freed us up for even more work» [17]. At the same time, wages often depend on work results, on goals or plans achievement. The disbalance between spent time and paid time makes workers to strengthen their work density and increase work hours amount at the expense of leisure time of course.

Today «leisure time colonization» is a civilizational problem that goes beyond just «I had no enough time to finish it». ICT enabled to break the connection between workers and their workplaces, but have placed workers in a constant and at once intermittent work cycle [17]. We highlight the following most relevant sources of labor intensity growth through the free time colonization:

- *Irregular working hours, multitasking and soi-disant work flexibility*, when a part of the work is taken home in order to redistribute physical and psychological tension.
- *Necessity of permanent learning and training*, because requirements are imposed not on labor operation any more, but on worker. He must not and does not want to lag behind the technology, but he can train only out of his working hours as a rule.
- *Communication in real time*. ICT is erasing the spatial boundaries of business communication. The «always in touch» strategy is good for business efficiency, but it greatly increases the labor strenuousness.
- *Self-identification in the social environment*. Just doing official duties is not enough for modern individual: his look and life style must be consistent with the standards of his professional and social surroundings in order to win in interpersonal professional competition. Desire for a professional and social status has a dominant influence on all other spheres of his life.

We should note that today the 24/7 work principle is presented by business as a way to the highest level of productivity, creativity and innovation. It reflects a conceptualization of work as a hobby, a desire to be «the best version of myself». However, such total mobilization of an individual's professional and personal resources, soi-disant self-entrepreneurship, leads only to sustained constant stress, which ends with professional and personal burnout.

So, the ICT let increase efficiency of some labor functions and operations. But this efficiency increases along with the labor intensity, which has its own temporal and physiological limits. Beyond the limits all normal conditions for human capital's functioning and reproduction are destroyed. This stipulates an ∩-shaped correlation between labor intensity and productivity: intensity growth beyond the physiologically acceptable level can turn into productivity fall (figure 1).

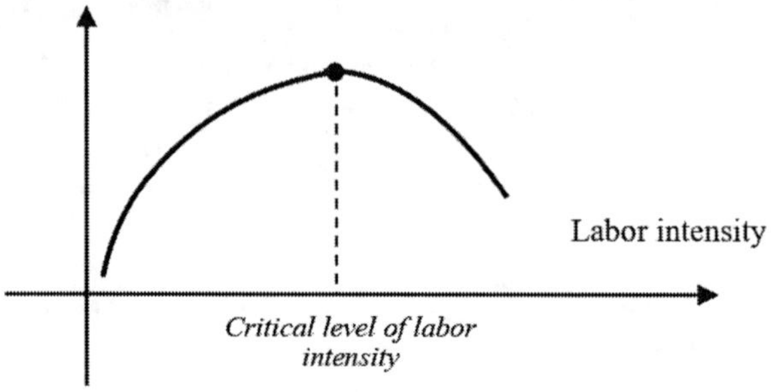

Fig. 1. Correlation between labor intensity and labor productivity.

We can assume that at present humanity is on the ascending segment of this curve: «digital euphoria» is raising force; labor intensity is growing; «work supervalue» principle is encouraged as a factor of human self-realization. However, there is a risk of public health deterioration and even a social tension growth if an acceptable level of labor intensity is not maintained. By the way, according to a Russian research results, only 15% of adults are always ready to learn [21]. And it doesn't reduce business development potential [22]. Today, there is no reason to believe that human labor in any profession will ever become absolutely free and creative, because the most of the labor functions and operations are routine by their nature and are carried out under organizational, technological and communication stressors. Not all jobs can become a hobby: work by vocation is much less common than work for remuneration. Therefore «only a minority works at the craftsmanship level» [18]. As for self-entrepreneurship, this is a kind of social dystopia conceptualized by leading business structures. Its formation will most likely collide with the natural defense mechanism of human survival, since «the self-entrepreneur is in a temporal imbalance that strikes his existence entirely» [16].

4 Conclusions

As a result of the research, we can draw the following conclusions.

1. The ICT use in different fields of professional activity contributes to increase in labor intensity. The ICT influence is manifested through the labor strenuousness growing, adding the number of labor functions and operations, lengthening the working day, as well as changing the work and rest regime.

2. Work with large amounts of information and monotony of mental efforts enhance the labor strenuousness. The working time is lengthened because of growing number of labor

https://doi.org/10.1051/e3sconf/202131108003

functions and operations performed by a worker within his workplace, duplication of work with standard and computer technologies, and the ICT imperfection. The «leisure time colonization» is a special contributor when a worker spends his leisure time on activities directly or indirectly related to work by means of the ICT.

3. Labor intensity and productivity have an ∩-shaped correlation. Unlimited growth in labor intensity is inconsistent with the sustainable development concept and effective development of human capital. In order to prevent the negative consequences of ultra-high labor intensity, it is necessary to develop adequate health-saving technologies for optimal labor organization and human capital management.

References

1. A. A. Porokhovskiy, USA & Canada: Ec., Pol., Cult., **49(8)**, 5-24 (2019)
2. K. Schwab, *The fourth industrial revolution* (WEF Publ., Geneva, Switzerland, 2016)
3. A. Hines, Foresight and STI Governance, **13(1)**, 19-30 (2019)
4. M. Ford, *Rise of the robots: technology and the threat of a jobless future* (Basic Books Publ., NY, US, 2015)
5. O. V. Nedoluzhko, Journal of Social Sciences Research, **3**, 228-234 (2018)
6. G. Small, G. Vorgan, *iBrain: surviving the technological alternation of the modern mind* (William Morrow Paperbacks Publ., NY, US, 2008)
7. W. D. Holford, Futures, **105**, 143-154 (2018)
8. M. Heid (2019) https://elemental.medium.com/
9. Protecting workers' health, WHO (2021) https://www.who.int/
10. A. E. Varshavsky, V. V. Dubinina, MIR, **11(3)**, 294-319 (2020)
11. A. Akaev, T. Devezas, Yu. Ichkitidze, A. Sarygulovd, Technological Forecasting and Social Change, **167**, 120675 (2021)
12. E. V. Levina, Economics and Society: Contemporary Models of Development, **9(2)**, 177-192 (2019)
13. N. A. Tsareva, S. Y. Omelyanenko, Revista De La Universidad Del Zulia, **11(31)**, 131-140 (2020)
14. A. Khedhaouria, A. Cucchi, Journal of Business Research, **101**, 349-361 (2019)
15. N. A. Tsareva, Z. V. Yakimova, A. A. Vlasenko, Dilemas Contemporaneos Educacion Politica Y Valores, **6**, 81 (2019)
16. A. Corsani, Logos, **25(3)**, 51-71 (2015)
17. P. Arora. Logos, **25(3)**, 88-119 (2015)
18. N. Klowait, M. Erofeeva, Logos, **29(1)**, 53-84 (2019)
19. N. B. Afanasov, Galactica Media: Journal of Media Studies, **1**, 43-61 (2019)
20. T. V. Bubnovskaya, M. E. Vasilenko, N. Y. Titova, V. V. Pekhotsky, Opcion, **34(87-2)**, 457-470 (2018)
21. Life Long Learning (2021) https://rg.ru/
22. E. V. Krasova, Azimuth of Scientific Research: Econ. and Adm., **9(3(32))**, 215-218 (2020)

E3S Web of Conferences **311**, 08004 (2021) https://doi.org/10.1051/e3sconf/202131108004
EPSD 2021

Public and corporate human resources management programs as one of the factors of monocities modernization in the framework of sustainable development

L.A. Chikatueva*, and *T.A.* Okhotina

Rostov State University of Economics (RINH), Rostov-on-Don, Russia

Abstract. In this article, the author examines the reasons for social stratification of the country's regions and singles out the difficult socio-economic situation of the majority of single-industry municipalities in Russia as one of the main ones. The aim of the study is to analyze the efficiency of use of human resources in the country as a whole, as well as the current state of single-industry towns and create a step-by-step plan for their modernization and restructuring as a necessary vector for sustainable economic development. The article emphasizes the need to integrate efforts on the part of the state, society and business in solving social problems in certain territories. The study was carried out using the following methods: observation, modeling, method of scientific abstraction, analysis and synthesis. The study result was development of a universal step-by-step plan for development of single-industry municipalities, substantiation of the possibility of its application as one of the factors for the effective use of human resources in a particular territory.

1 Introduction

In September 2015, the UN General Assembly adopted the resolution "Transforming our World: the Agenda for Sustainable Development up to 2030". The document is a definite plan of action for the countries of the UN aimed at eliminating poverty, hunger, social and gender inequality, as well as environmental problems for the period up to 2030. If we consider the sustainable development goals outlined in this resolution, then at least seven out of seventeen are in one way or another related to development and improvement of human resource management [1]. This means that most of the problems that exist today as obstacles to the sustainable development of individual countries and the world economy as a whole can be solved via the more rational use of human potential.

The prerequisites of the study described above determined its goal – to analyze the efficiency of human resources use, as well as the current state of monocities in our country and creation of the staged plan for their modernization and restructuring, as a necessary vector of sustainable economic development. Relevance of the study is due to the need to

* Corresponding author: l_chikatueva@mail.ru

develop new ways to more efficiently human resources use in depressed regions of the country, develop the human potential of these regions, improve the quality of life of the population in the territory of single-industry municipalities. The offered staged plan for modernizing the economy of monocities is of practical importance, since it can be used by any municipality adjusted for the specifics of the particular region. In general, the scheme can be used at a separate enterprise to attract social investment, including in human capital. As a hypothesis, we assume the following: gradual modernization and restructuring of single-industry municipalities in depressed regions will smooth out social stratification between individual territories of the country, as a result, make investments in human resources in all regions more profitable and improve the quality of life of the population.

2 Study theory and methodology

At the moment, our country is significantly inferior to many developed European countries in implementation of social policy, both at the state and corporate levels. So, for example, according to the Human Development Index (HDI) that characterizes the quality and standard of human life in the country, according to the rating compiled annually by the UN, Russia took only the 52nd place in 2019. This index is calculated considering three main directions of the country's development: health, education and GDP per capita. Let us analyze the quantitative HDI values taken from the annual final reports of the United Nations Development Program (UNDP) for the last three years using the example of 12 countries, including Russia (Figure 1).

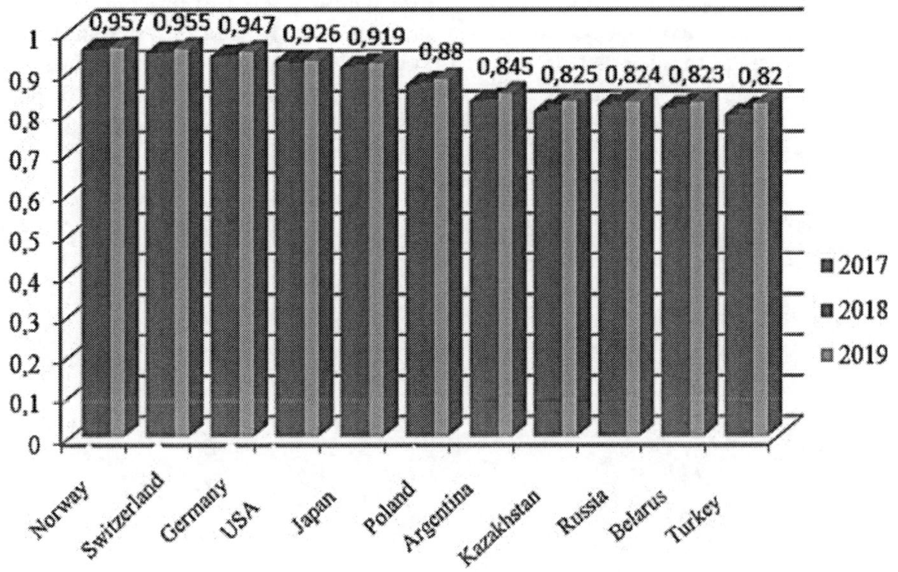

* Drawn up by the author

Fig. 1. HDI values in different countries for 2017-2019.

The figure shows that Norway is the undisputed leader in this indicator, in 2019 the HDI there was 0.957, with the maximum possible value of 1. Russia is ranked between its closest neighbors in the Commonwealth of Independent States Kazakhstan and Belarus, with an indicator of 0.824, yielding not only to developed European countries, the USA, Japan, but also to countries such as Poland and Argentina. The UN divides all countries by HDI into 4

https://doi.org/10.1051/e3sconf/202131108004

groups: countries with a very high level of this indicator, countries with a high level, with an average and low level, while our country belongs to countries with a high HDI. It is worth noting that a year ago Russia was ranked 49th in the ranking, but in 2019 we were outstripped by such countries as Kazakhstan, Palau and Romania. It is quite obvious that despite implementation of national projects as provided for in the May decree of the President of the Russian Federation V.V. Putin "On the National Goals and Strategic Objectives of Development of the Russian Federation for the Period up to 2024" development goals (increasing life expectancy, increasing real incomes, increasing pensions, improving housing conditions, reducing poverty, and etc.), our country needs cardinal changes in the area of using human potential and investing in it. The volumes and quality of investments in human resources directly affect the standard of living of the population, at the same time, according to foreign scientists, an increase in human potential by 1 % leads to an acceleration of the country's GDP growth by 1-2 % [2].

Solution to the problem of effective investment in human capital and creation of prerequisites for the sustainable development of the economy of our country is complicated by a number of economic, social and geopolitical factors, the key ones of which are presence of a large territory and a significant number of entities that differ in the level and quality of life of the population.

In terms of regional inequality, our country is in the third place in the world, behind Venezuela and Thailand only. The per capita income in the richest region of the country - the Yamalo-Nenets Autonomous Okrug, according to Rosstat data for the 3rd quarter of 2020 was 82,749 rubles, and the lowest average per capita income in the same period was observed in the Republic of Tyva - 15,608 rubles [3]. The difference of more than 5 times clearly indicates that serious reforms are necessary on a national scale in order to overcome the existing problems and avoid social "outbursts" and unrest in the future. Wherein, even within the federal districts, there is a colossal gap between the most successful and the poorest regions. For example, in the Urals Federal District, the difference in the level of per capita income between the Yamalo-Nenets Autonomous District and the Kurgan Region amounted to 61,784 rubles in 2019, i.e. almost 4 times (Figure 2) [4].

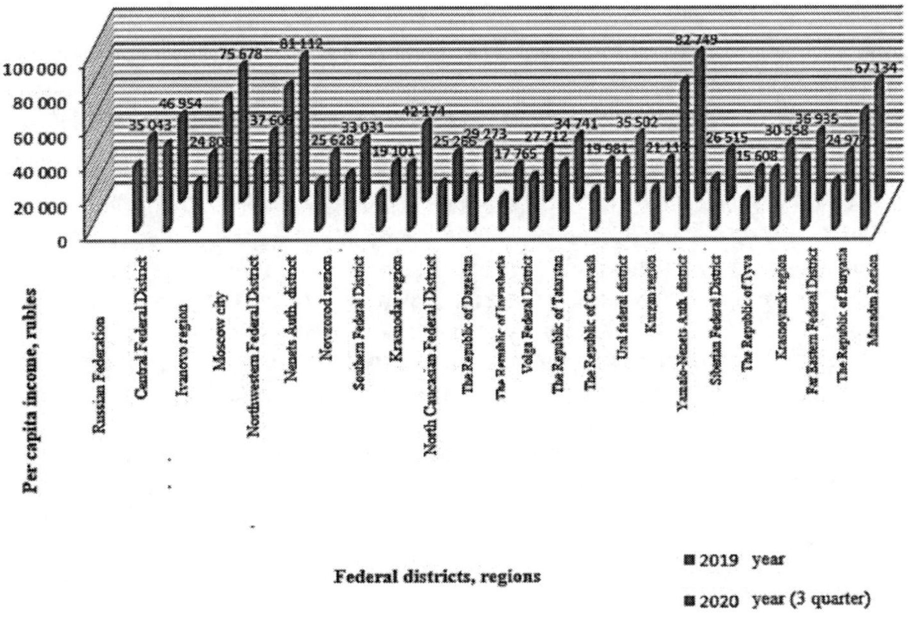

* Drawn up by the author

Fig. 2. Average per capita monetary income of the population by the constituent entities of the Russian Federation (the richest and poorest region of the constituent entity), rubles/month.

The main reason for this inequality is the high concentration of industrial production in the most developed regions. Yamalo-Nenets Autonomous Okrug, Khanty-Mansi Autonomous Okrug and Moscow give the federal budget 45-50 % of all tax revenues.

On the other hand, we see depressed regions, which, even considering the state subsidies to the regional budget, fail to reduce the social gap with the richest entities by more than 3 times. Often the reason for this is the geographic location of the region, the availability of resources and their current demand on the market. Therefore, most regions with low indicators in terms of per capita income, GRP, etc. are located in the Urals and Siberia [5]. Cities here arose around industrial enterprises for the extraction or processing of minerals. Presence of a large number of such monocities, i.e. settlements founded at the city-forming enterprise, "pulls" down the economy of the region as a whole. Therefore, one of the possible options for smoothing social inequality between the regions of the country is diversification of production and the restructuring of the economies of monocities. The need to hold such events arose not so long ago, since the problems of monocities became especially acute within the economic crisis of 2007-2008. In 2010, the Government of the Russian Federation created a working group, the purpose of which was to develop a plan for financing the single-industry municipalities. However, the offers to subsidize these settlements did not help to solve all the problems, a significant part of which lies not only in the financial plane.

Therefore, income inequality across the regions of the country is due to the different level of development of single-industry cities in these regions, while issue of universal strategies for development of these municipalities has received insufficient coverage in the currently existing studies.

3 Study results and their discussion

As of January 21, 2020, there are 334 single-industry municipalities in Russia, with a total population of more than 16 million people that is more than 10 % of the population of the entire country [6]. According to the current decree of the Government of the Russian Federation of July 29, 2014 No. 709, all monocities are divided into 3 categories depending on the risk of deterioration of their socio-economic situation.

The 1st category includes single-industry municipalities with the most difficult economic and social situation (these are 97 monocities, such as, for example, the city of Pervouralsk in the Sverdlovsk region, the city of Karabash in the Chelyabinsk region, and etc.). The second category includes monocities in which there are risks of deterioration of the socio-economic situation (these are 150 monocities, including Yarovoye, Altai Territory, Tynda, Amur Region, Novodvinsk, Arkhangelsk Region, and etc.). The third category includes monocities with the stable socio-economic situation. There are currently about 87 of them (for example, the city of Rossosh, the Voronezh Region, the city of Novokuznetsk, the Kemerovo Region, the city of Pechora, the Pskov Region, and etc.).

This division into categories was not accidental. Each of them requires its own approaches to solving major social and economic problems. Based on world experience, it is necessary to work out certain strategies for development of monocities, based on their economic situation at this stage.

Almost all industrially developed countries faced the problem of single-industry cities and settlements, but in most of them this problem arose much earlier than in our country, approximately in the 60-70s of XX century. In Russia, a rapid outflow of the population and the destruction of the infrastructure of monocities took place in the early 90s, immediately after the collapse of the Soviet Union. By this time, in Europe and America, various strategies have already been offered and are actively used to overcome the economic degradation of single-industry municipalities (for example, the experience of the Ruhr region of Germany) [7]. All these strategies can essentially be reduced to two scenarios: closure and reorganization.

The first option does not solve social problems, but only creates new ones and, moreover, the monocities that remain at the moment are too large to be completely settled apart. Closure of monocities and the resettlement of residents apart can only be used as an extreme measure, for example, in areas with unfavorable environmental conditions or too harsh climatic conditions, since presence of unpopulated areas is fraught with not only social, but also geopolitical threats [8].

The second option is the reorganization of monocities that is the most rational and acceptable for our country. In foreign practice, three strategies have been developed for implementing the rehabilitation scenario: "controlled compression", "stable monocity" and "industrial diversification".

In order to put the first strategy into practice in our country - "controlled compression", which implies a gradual controlled reduction in the population and size of the city, it is necessary to identify a group of monocities that could be "compressed". These are the largest cities from the 1st most problematic category, having numerous economic and social problems, the solution of which on the existing scale of the municipality will be extremely unprofitable. A city-forming enterprise remains in the city, to which, if necessary, the state provides financial support, and the number of residents is brought to the required minimum based on the number of jobs at the enterprise. Therefore, the unemployment rate decreases and the social situation improves.

The "stable monocity" strategy is suitable for the 3rd most economically stable category of monocities. It implies that the city remains in the same size and with the same population, while the city-forming enterprise must operate stably, without reducing production volumes and the number of jobs. The municipal authorities, together with the founders of the enterprise, must provide the city and its inhabitants with the necessary social infrastructure. Enterprises that pursue a competent corporate social policy can be established with reduced rates for regional and local taxes and fees, which will be an additional incentive to participate in the social life of the city.

The third "industrial diversification" strategy is the most complex and costly. It requires an integrated development of the territory, opening the related industries by an already existing city-forming enterprise, as well as the active attraction of investors from other regions and countries. The most appropriate use of this strategy would be in single-industry municipalities belonging to the 2nd category. To implement this strategy, you need to draw up a rough staged plan or roadmap. It can be represented in the form of a table:

Table 1. Plan of the monocities "industrial diversification" strategy implementation.

Stage name	Activities
Stage 1. Determination of the group of monocities in need of comprehensive development and reorganization	• selection of economically stable cities, but with risks of worsening the situation; • selection of cities with the possibility of developing the tourism industry, innovation, science and education;
Stage 2. Drawing up a detailed city development plan	• calculation of all possible options for location of various new industries; • analysis of the possibility of restructuring existing enterprises; • selection of the most rational from an economic and social point of view of options
Stage 3. Determination of the circle of potential investors	• attracting investors with preferential terms for loans and taxes (creation of free economic zones and public-private partnerships); • calculation of the volume of capital investments that can be attracted thanks to each of them; • calculation of the amount of financial resources that can be attracted at the expense of the enterprise itself
Stage 4. Determination of the amount of necessary state support	• calculation of the missing amount of financial resources, considering the cost of the project, possible investments and the company's own resources; • determining the form and type of state support (soft loans, subsidies, creation of free economic zones, public-private partnership)
Stage 5: Project implementation	• staged implementation of the measures outlined in the development project; • control over their timely implementation; • evaluation of the results of the implementation of individual activities; • adjustment of the development and reorganization plan, if necessary; • stimulating the enterprise to develop various areas of corporate social policy and actively participate in social programs of the city

The plan is approximate and can be adjusted considering the certain features of specific territories. Also, in each case, it is imperative to specify the terms for each stage, to determine the responsible executors and control bodies.

Special attention shall be paid to the problem of employment of the population in single-industry municipalities, since it is key and determines the further vector of development.

Municipal authorities shall regularly monitor the main indicators of employment, income levels, and labor movement. Currently, social guidelines in development of the city are increasingly becoming the basis for creating development strategies, goals and missions of urban areas [9]. It is important to note here that the solution to the issue of effective use of human resources shall be carried out by joint efforts of state bodies, local governments and business structures [10]. Therefore, encouraging enterprises by the state to develop corporate social programs aimed at stabilizing the situation on the labor market by means of creating new jobs, attracting highly qualified specialists to the region, additional social packages, training programs and retraining the existing employees will allow to solve the problem of employment of the population in single-industry cities more effectively.

4 Conclusions

As part of implementation of the sustainable development concept, human capital acquires a special role as one of the most important factors of production. The social policy implemented today, both at the state and corporate levels, does not allow our country to even come close to those indicators of the level and quality of life that are currently demonstrated by the developed countries of the EU, the USA and Japan. One of the reasons for the lack of stable development of human resources is the difficult socio-economic situation of single-industry municipalities, which, despite the existing development potential and the Government's attempts of "recover", are indicators of an unfavorable situation in the country's labor market. Application of the industrial diversification strategy and the offered staged plan for its implementation would allow a significant part of single-industry municipalities in our country to change the socio-economic situation for the better. This would help to reduce the unemployment rate, attract qualified personnel from more successful regions and, as a result, would lead to an increase in the level and quality of life of the population, which is a necessary prerequisite for the sustainable development of the economy of the entire country.

References

1. *Transforming Our World: Transforming our World: the Agenda for Sustainable Development up to 2030. Resolution adopted by the UN General Assembly on* (2015)
2. L. A. Shmeleva, K. A. Shtanova, Bulletin of the Altai Academy of Economics and Law, **11(2)**, 369 (2020)
3. *Regions of Russia. Socio-economic indicators*, 1242 (2020)
4. *Regions of Russia. Socio-economic indicators*, 1204 (2019)
5. T. A. Vereshchagina, K. A. Trushkina, Problems and prospects of economics and management: materials of the IV International scientific conference, 172 (2015)
6. On approval of the list of single-industry municipalities of the Russian Federation (monocities): Order of the Government of the Russian Federation of July 29, 2014 No. 1398-r
7. A. O. Sobolev, Management issues, **32**, 75 (2015)
8. V. V. Fauzer, A. V. Smirnov, G. N. Fauzer, Economy of region, **17(2)**, 552 (2021)
9. O.V. Artemova, N. M. Logacheva, and A. N. Savchenko, Economy of region, **17(2)**, 538 (2021)
10. S.V. Kulay, Public Administration. Electronic Bulletin, **73**, 224 (2019)

11. T. A. Vereshchagina, Problems and prospects of economics and management: materials of the IV Mezhdunar scientific conference, 172 (2015)

Alternatives to staff reduction in the context of labour digitalization

Liliya Babynina[1,*], *Larisa* Kartashova[1], *Natalia* Loktyukhina[2], *Ekaterina* Chernykh[3], and *Felix* Akhmedov[1]

[1]Plekhanov Russian University of Economics (PRUE), Moscow, Russia
[2]Financial Research Institute, Moscow, Russia
[3]Federal State Budgetary Institution of Science Institute of Socio-Economic Studies of Population, Russian Academy of Sciences (ISESP RAS), Moscow, Russia

Abstract. The article discusses alternative forms of employment in the process of staff reduction with the development of information and communication technologies in the world of work. Using the results of the research, the authors propose to critically evaluate Russian legislation in terms of the regulation of self-employment, consider new forms of employment and remote work, legislatively enshrined in the Labour Code of the Russian Federation. One of the problems highlighted in the article has related to the lack of necessary digital skills for using the Internet, which limits the possibilities of the population and employers in using alternative forms of staff reduction and interaction with employment services. Changes in the structure of the economy, the transition of personnel to remote work are accompanied by a reduction in the participation in the trade union movement, as a result, of which an employee may be forced to terminate an employment contract without social guarantees and compensation. The authors of the article argue that alternative solutions to regulate employment of the population based on the principles of social partnership and uniting the efforts of all interested parties - the state and society, representatives of employers and workers themselves.

1 Introduction

The digitalization of the labor market and the employment sphere changes the strategy of the behavior of the employee and the employer in connection with potential or ongoing processes of staff layoffs, which have accompanied by the de-standardization of employment. New forms of employment are emerging all over the world: telecommuting, part-time, hybrid, irregular, temporary, freelance, self-employment, work on digital labour platforms, work are using mobile or cloud applications and others. The scaling up of these forms of employment is associated with the use of information and communication technologies, which allows workers to be always in contact with employers and clients, as well as get the opportunity to represent yourself on digital labor platforms.

* Corresponding author: Babynina.LS@rea.ru

2 Materials and methods

The authors used the results of their own research obtained within the framework of the following research projects and government assignments: "Digitalization of the labour market and employment in Russia: trends and development mechanisms", financed from the funds of the Federal State Budgetary Educational Institution of Higher Education Plekhanov Russian University of Economics" (Order No. 867 dated 28/06/2021);

"Components, social standards and indicators of the level and quality of life of the population in modern Russia: qualitative identification and quantitative assessment in conditions of socio-economic inequality" (No. 0137-2019-0032) financed from the funds of the Federal Scientific Research Center of the Russian Academy of Sciences;

"Organizational and financial mechanisms to support the employment of the population in 2021-2023, aimed at reducing the unemployment rate", carried out by the Scientific Research Institute for Basic Research under the state order of the Ministry of Finance of Russia (No. 092-00001-21-00).

Empirical methods base on the analysis of scientific foreign and Russian literature. Field research, a combination of quantitative and qualitative analysis methods, online surveys, personal and telephone interviews used to assess the challenges and opportunities for transforming the employment service of the Russian Federation to support laid-off workers and unemployed citizens.

3 Results and discussion

3. 1 Assessment of legislation on self-employment

Russian studies show that in order to reduce the costs associated with the guarantees to employees provided by the Labour Code of the Russian Federation, employers can use non-standard forms of employment that are outside the scope of labour law, which allows employees not to pay severance pay, compensation for unused vacation, etc. Employers, in violation of labour legislation, massively transfer hired workers to the category of self-employed, concluding civil law contracts with them, or prefer to hire new employees on the terms of similar contracts. Of the 1 million self-employed registered in 2020, 400 thousand people are former employees, from whose salary fund the employer paid insurance premiums and transferred the personal income tax on the amount of 13% from the salary of each employee. The self-employed pay 4% or 6% from their income, depending on who they did the work for - for individuals or legal entities [1].

The transfer of employees to the category of self-employed frees the employer from the obligation to pay insurance premiums for the employee and guarantee the compensation provided for by the labour Code of the Russian Federation. At the same time, an employee in the status of self-employed, and in fact an employee, finds himself in a vulnerable position, deprived of guarantees and social benefits. The signs of labour relations established in Article 15 of the labour Code of the Russian Federation do not allow to unambiguously draw the line between self-employment and labour relations.

The legal status of self-employment is a subject of heated debate [2]. The position of the International labour Organization, based on the fact that "Countries should strengthen, and sometimes adapt, their social protection systems to ensure that all workers benefit from social protection coverage" [3], seems justified. A critical assessment of the Russian legislation on self-employment, which determines the legal status, rights and guarantees of this category of workers, is advisable.

3. 2 Alternative forms of employment in case of staff reduction

The development of non-standard forms of employment allows both the employee and the employer to implement alternative scenarios in case of staff reduction. The employee has more opportunities for part-time jobs, secondary employment, and work via the Internet. The employer, as an alternative to the dismissal of personnel, could be use a remote, part-time or temporary forms of employment [4].

The authors compiled a table of new forms of employment that as alternatives to the staff reduction in Russia (table 1).

Table 1. New forms of employment [5].

Forms of employment	Content
Interim Management	Highly qualified specialists are temporarily engaged by the company to carry out a specific project or solve a specific problem or anti-crisis management.
Job sharing	An employer hires two or more workers to do the work together (with all co-performers taking one full-time job).
Casual work	Casual work refers to a type of work where employment is neither stable nor continuous.
Two main types of casual work	Intermittent work. An employers approach workers on a regular or irregular basis to conduct a specific task, often related to an individual project or seasonally occurring jobs.
	On-call work. An employer does not guarantee regular employment, but calls the employee on demand / demand.
ICT-based mobile work	An employees do their work from any location, at any time, with the support of modern technologies and remote access to the computer systems of the employer.
Voucher-based work	The employer pays the worker by means of vouchers purchased from an authorized organization (generally a governmental authority) which cover wages and contributions to the social insurance system.
Portfolio work	The self-employed and freelancers works for many clients, performing small tasks for each. The portfolio of the candidate with the results of his work plays an important role. It is used in some areas, for example, in the creative field and the scientific and technical field.
Platform work	Employers and workers, service providers and consumers of services communicate with each other through an online platform, with large tasks usually spread across the workers' virtual cloud.
Collaborative self-employment	Freelancers, self-employed, micro-enterprises have cooperate to fulfill the minimum size criteria for enterprises-applicants, or to overcome the professional isolation (co-working, a cooperatives)
Outstaffing	Removal of employees from the staff of the actual employer and their transfer to the staff of the employer-supplier, which are private employment agencies or recruitment agencies.

However, the use of such forms of employment is associated with the risks of not being in demand in the labour market and loss of income for employees, especially among older people. Psychological stress is growing, many workers experience constant stress and social insecurity with the increase of precarious work.

3. 3 Remote work

According to the Eurostat data, the share of teleworkers only for the period from 2008 to 2019 increased from 12.4% to 16.1% [6]. In the long term, remote work will become an increasingly common alternative to staff cuts. ILO experts estimate the potential of remote work in 18% of jobs (from 12-13% in low-income countries to 25% in developed countries), which is about six times more than the proportion of those who worked remotely before the pandemic [7].

In Russia, the conditions to create for the development of this form of employment - the labour Code of the Russian Federation regulates the features of remote work, including a hybrid form, which implies combining work at home and in the office. The transfer of workers to remote work as an alternative to the dismissal is suitable for some companies, since it has a limited character and more focused on office personnel [8].

With all the advantages of remote work as an alternative to the reduction of staff, questions remain about the organization of remote work at the company level. Companies and employees need to responsibly approach the remote work format and conduct trainings and courses on managing and organizing work remotely (time management, communication skills, stress management skills when working with large flows of information) in order to increase work efficiency, exercise control, create a comfortable environment for the interaction of colleagues, promote corporate culture and minimize the level of stress [9].

3. 4 Digital Skills Assessment

The researches by foreign authors have shown that employers searching for workers in more concentrated labour markets, demand higher the cognitive, social, and organizational skills. They write that upskilling is more prevalent among low-skill workers than high-skill ones and these workers are also more vulnerable to employment instability and low wage growth [10].

As shown of our researches, in the context of the digitalization of the economy, those who do not have the necessary digital skills prevail among the dismissed workers. Mastering digital technologies allows you to increase the competitiveness of personnel, maintain social status, but they require personal motivation, as well as accessibility to the digital environment.

According to statistics, in 2019, 77% of the surveyed households had access to the Internet, which is lower than in the member states of the European Union, with the exception of Bulgaria, where this figure is 75%. In 2019, 72.6% of households in the Russian Federation used the Internet on a daily basis. At the same time, statistics indicate that the non-use of the Internet in households (as a percentage of the number of households without access to the Internet) occurs for the following reasons: the lack of need to use the Internet, unwillingness to use it or lack of interest in it (in 2015 - 66.3%; in 2019 - 70.7%); insufficient skills to work on the Internet (16.8% in 2015; up to 32.5% in 2019; lack of sufficient financial resources to connect to the Internet (18.6% in 2015; 20.8% in 2019). Only 7.3% of respondents indicated the lack of technical capabilities to connect to the Internet in 2019 [11]. Therefore, professional retraining programs are important, which will allow to maintain the activity of labor resources in situation of their limited supply and the search for talents [12].

3. 5 Transfer to online services of the employment service

New strategies for employers and employees, and alternatives to dismissed employees are accompanied by an accelerated transfer to online services of the employment service. For the employer and for the employee, this means a more convenient format of online interaction on the Job in Russia portal; applying for vocational training in the framework of programs to reduce tensions in the labour market; and etc.

However, the analysis of the functioning of the Internet portal revealed a number of problems: the results of the level of user satisfaction with its work are not published; the probability of failures in its work remains; employers see a restriction of their right to choose the form of interaction with the employment service; the population lacks Internet skills.

3. 6 The role of the social partnership

According to the research results, we revealed a tendency of the reduction of the role of social partnership in regulating labour and employment issues, which makes it possible for the employer to feel “more comfortable” when releasing personnel, and puts the employee in a vulnerable position. Changes in the structure of the economy, the transfer of personnel to "remote employment" and robotization are accompanied by a reduction in the scale of the trade union movement. In such conditions and in the absence of a trade union, an unscrupulous employer, when dismissing employees, can compel the employee to terminate the employment contract "by agreement of the parties" and thereby avoid all compensations required. Digital technologies, while creating new employment opportunities, simultaneously reinforce the challenges of protecting and representing workers' rights in the digital economy. In order to respond to the challenges of digitalization, the Russian trade union movement needs new organizational forms that go beyond specific organizations; the use of online technologies and online collective forms of action for the mass of workers.

4 Conclusion

In the context of the digitalization of the employment sector, the problem of staff reduction remains inevitable. In this regard, the behaviour strategy of the employee and the employer is changing, since the employer has opportunities for alternative ways of the staff reduction and providing them with a new job with the necessary digital skills. At the same time, the worker becomes more vulnerable in social and labour relations, since the de-standardization of employment is often implemented outside the legal framework. In this regard, through legal regulation, it is necessary to use the positive possibilities of the digital transformation of the world of work and ensure the protection of the rights and interests of the employee, as well as maintaining a high level of his quality of working life. The use of non-standard forms of employment, is associated with overcoming new challenges in terms of finding optimal management solutions, and will require new forms and approaches to balancing between a set of preventive employment policy measures and the choice of means of providing maximum support to those who have already lost their jobs, including financial and other incentives to increase their professional and geographic mobility. It is necessary to master digital skills that will increase the competitiveness of personnel, maintain social status, and be able to access the digital environment, including in interaction with employment services and trade unions.

References

1. Labour relations with workers in vulnerable forms of employment must be recognized, https://www.kiout.ru/
2. W. Däubler, J. of the Higher School of Economics (2016)
3. Non-Standard Employment Around the World: Understanding Challenges, Shaping Prospects (2016)
4. M. Charalampous, C. A. Grant, C. Tramontano, E. Michailidis, European J. of Work and Organizational Psychology, **28(1)** (2018)
5. *Eurofound, Overview of new forms of employment – 2018 update* (Publications Office of the European Union, Luxembourg, 2018)
6. Eurostat, https://ec.europa.eu/
7. Working from home: Estimating the worldwide potential, https://voxeu.org/
8. 08.12.2020 No. 407-FZ
9. N. V. Loktyukhina, E. A. Chernykh, Living standards of the population of Russian regions, **17(1)** (2021)
10. B. Hershbein, C. Macaluso, "Labor Market Concentration and the Demand for Skills", *IDSC of IZA Workshop: Matching Workers and Jobs Online* (Bonn, 2018)
11. *Information Society in the Russian Federation 2020. Statistical Collection* (Nat. issled. University Higher School of Economics, SRU HSE, 2020)
12. L. S. Babynina, L.V., Kartashova, P. P. Pilipenko, Lecture Notes in Networks and Systems (2021)

Homeostatic version of the formation of human capital: ecological aspect

Radiy Ibragimov*

Katanov Khakass State University, Abakan, Russia

Abstract. Now it is found that under the capitalized qualities of a person, not only high-tech competencies can be understood. This capital does not exist outside time and space; Its implementation responds to specific social queries. For example, with the current demographic dynamics, the territory of a huge country through a couple of generations will be unnecessary. The article substantiates the thesis that the post-industrial model of the history is an optional development line. In conditions of social entropy, the return to homeostatic equilibrium with an ecological environment is both the most humanistic, and the most clearly predicted, and the least costly of the future scenarios. The consideration of the inevitability of homeostasis, in turn, leads to the need to interpret the concept of «human capital», which will allow the typical carrier of such capital to be the most successful in the new socio-ecological situation.

1 Introduction

At first glance, there is no direct relationship between the patterns of human capital generation and functioning the natural and even anthropogenic ecosystems. It arises when a high-performance engineering intelligence is set, and does not naturally mature, the task of designing or adjusting processes in geobiocenoses of various scales.

In a sense, two problems, human capital and ecology, in their classical understanding are even alien to each other. In absolute terms, the highly developed human capital of the Kuznets-Becker-Schultz paradigm lives mainly in the ultra-urban environment, is much less common in industrial centers and almost never occurs in rural areas. Many researchers put the country level of human capital development in direct dependence on the proportion of the urban population [1, 104]. In this classical and narrow understanding of human capital, many regions of Russia, especially Siberia and the Far East, are a dull spectacle, to put it more academically, they are depressive. Not only is the share of the rural population almost equal to the share of the urban population in Khakassia, Yakutia, Tyva, Buryatia, in the Chita and Irkutsk regions, in the north and southeast of the Krasnoyarsk Territory, this population is also scattered over vast territories.

However, to date, the discourse "Human Capital" has undergone many substantive corrections, additions and semantic layering. [2] It was found that capitalized qualities can be understood not only as high-tech or "digital" competencies, but also sports or aesthetic

* Corresponding author: dison1@mail.ru

talents. Moreover, this discourse has long been tested in social practice, where the range of interpretations and applications is even wider. For example, it is not a secret for anyone that in Russia the "parent capital" has become a specific concretization of human capital; that is, female fertility was designated the capitalized "talent".

In these conditions, it is quite fair to raise the question of practical concretization and localization of the concept of "human capital". This capital does not exist outside of time and space; it, its production and sales respond to specific social needs. The same project of parental capital did not appear out of nowhere, but as a response (belated!) to specifically Russian problems that are not typical of Western Europe or Africa. Namely: with the current demographic dynamics, the territory of a huge country will turn out to be unpopulated in a couple of generations.

However, even if the program of demographic activation turns out to be successful, it does not automatically solve all the problems associated with use of natural resources and nature conservation. The situation with forest fires in recent years shows that even such a rare population manages to inflict catastrophes on a federal and even planetary scale. In other words, in remote Siberian and Far Eastern provinces, it is necessary not only to preserve the population of Russians; it is also necessary to install certain "human qualities" in it. The ecological culture of the population serves here as a very informative social marker.

2 Materials and methods

Our study was carried out within the framework of the study project "The Problem of Effective Use of Human Capital: Regional Dimension". Its methodological feature is approbation of a predominantly macroeconomic concept of human capital by means of empirical sociology. [3]

This explains the features of the axiomatic basis of the study, namely, the emphasis and concentration of our search activity on the motivational component in the structure of human capital, which allows us to make the subject of study not only traditional, but also alternative scenarios for the embodiment of human capital.

Implementation of this program segment was a series of formalized interviews with respondents who are hypothetically carriers of excess motivational energy, the so-called "passionaries". Founding, identification from the methodological and technical point of view was not particularly difficult, since the passionary always stands out against the average statistical social background. Therefore, the "snowball" method was used for the sample, as a result of which 168 interviews were obtained, in the guide of which such meaningful parameters as attitude to work, everyday life, social microenvironment, and time management were revealed. Attitudes towards environmental problems were not specifically revealed, however, almost all respondents from rural areas and a fairly large percentage of respondents from urban areas (113 texts in total) spoke about certain environmental problems in the region. Considering the fact that these were economically and socially active citizens, we considered it necessary to form a special, "ecological" section in the general program of the study project.

In addition, in this project, a large methodological role is assigned to the analysis of documents of various forms and statuses. This shall include statistical reviews, filing of local newspapers of different years (Khakassia, Abakan), materials from electronic media, local scientific and business forums (for example, Big Business held on August 21, 2019 (https://r-19.ru/news/ekonomika/91455/)). It can be seen from the materials that environmental issues practically did not leave the agenda of public discussion.

The study of models of human capital formation in the conditions of the Khakassia region and the south of the Krasnoyarsk Territory allows, in our opinion, application of the extrapolation method both to other regions of the Russian Federation and to the model of their evolution in time, to the "scheme" of their historical development.

3 Results and discussion

So, location of the empirical study is the Republic of Khakassia and the south of the Krasnoyarsk Territory. This region, despite the current administrative division, represents a geological, ecological, economic and cultural unity. In the language of geology and ecology, this unity is called the Khakass-Minusinsk Basin. It is a kind of geological "bowl" bordered on all sides by the spurs of the Sayan Mountains. The basin area is more than 130 thousand square kilometers. The population is about 660,000 people.

Historians believe that during periods of economic and political heyday (for example, during the times of Dinling-go or the Kyrgyz Kaganate), the population was kept near this mark. [4, pp. 28, 39]. For our study, these figures are one of the key points, since they fix the demographic possibilities of the enclosing landscape. Indeed, the statistics of recent decades also show that the threshold of population capacity in the region has not fundamentally changed.

The demographic dynamics shown above is suggestive. Its mechanisms in relation to the eras of antiquity and the Middle Ages are well studied. The social system of the Dinling-go and the Kyrgyz Kaganate rested on the extensive agricultural and nomadic use of the landscape, and the relatively high birth rate was compensated for by child mortality and the death of young people in the constant internal and external conflicts of those times.

It would seem that the current situation is fundamentally different. Child mortality and war are long excluded as demographic factors. The share of the rural population in the region is close to 50 %; there are no large cities at all. The only city whose population exceeds 100,000 is the capital of Khakassia, Abakan (180,000); the population of 3 more cities (Minusinsk, Chernogorsk, Sayanogorsk) fluctuates around 50 thousand. Therefore, the factor of "urban utilization of the population", which is characteristic of cities with a population of one million, does not work here either.

Meanwhile, as mentioned above, there is still no population growth in the region. [5] The most frequently cited reason - the alcoholization of the population, especially the rural population - certainly takes place. In our interviews, the word "soldering" was used to describe problems 48 times, the names of alcoholic beverages as a source of evil ("vodka", "vodyara", "spirtyaga") – 77. However, this factor is not a comprehensive explanation.

Inclusion of the human capital factor in the framework adds more detail to the outlined diagram. The inflow and outflow of the population in the region in the last 20 years has been about 5 thousand people per year. Mostly school graduates leave, whose USE results allow them to apply for places in universities in Krasnoyarsk, Novosibirsk, Tomsk, St. Petersburg and Moscow. These are promising carriers of human capital, the "production" of which is quite well established by the school education system in the region. As well as in the whole country. [6; 7] Our expert from the Ministry of Education and Science of the RH reported (in fact, boasted) that with an annual release of approx. 1,800 only Moscow universities enroll "at least two dozen" of graduates.

Another reserve of outflow, but much less "full-flowing", is labor migration in order to find more promising jobs.

The inflow into the region is formed mainly from two social groups - 1) residents of the northern territories (Norilsk residents) migrating to settle in a more favorable climate than before, and 2) residents of the Central Asian republics who come in search of work.

It is easy to conclude that the dynamics of human capital in the region is most accurately reflected by the terms "erosion" and "homeostasis". Here we need an argumentative inclusion designed to clarify our position. In the structure of human capital, we include a component that is explicitly contained even in the classical theory - namely, the social and energy potential of the individual and the population, which is often called "passionarity". This means that the knowledge, skills and abilities that bring added value to investors do not mean much in and of themselves. Their owner must show increased motivation, activity, ambition in order to bring their knowledge and talents to the stage of productive implementation.

With this argument in mind, from our provincial region (as well as from many others), a social cohort of bearers of not only competencies, but also energy is gradually being washed out. The same is observed in relation to many other regions [8]. The depletion of passionarity, in turn, gives rise to a state of homeostasis – economic, political, demographic and environmental. [9].

Demographic stagnation has already been discussed above. More details shall be said about similar processes in the economy. Let's start with the next news item. Now, literally before our eyes, a spontaneous political and economic experiment is unfolding. Russian Defense Minister S.K. Shoigu, a native of neighboring Tuva, recently announced the idea of creating "on the border of the Krasnoyarsk Territory and Khakassia" a million-plus city called Electrograd on the basis of a powerful industrial cluster also planned to be created. (https://svpressa.ru/economy/article/308314/, date of last visit 09/09/21.) Where to get an additional million in the demographically depressive outback, not just residents, but city dwellers? Where do several high-tech enterprises of federal and even global scale come from there?

As usual, there are two possible answers to the questions posed - optimistic and pessimistic, "Pro" and "Contra". The optimistic answer, as we see it, appeals to two arguments - 1) money and 2) personality.

Financing is the easiest: a country that can build Mordovia-Arena stadiums for 17 billion rubles and Gazprom-Arena for 50 billion rubles can easily afford the construction of a dozen factories in the Khakass steppes.

Item 2 shall be articulated in more detail. Personally, Sergei Kuzhugetovich is a vivid example of the implementation of human capital. However, considering the fact that at a certain moment in the embodiment of the life scenario, he, as a manager, with his qualities and passion, has outgrown the scale of the region. Being, without any doubt, not indifferent to the fate and condition of his small homeland, he looks at these fate and condition through the eyes of a productive passionate who has migrated to a more promising federal center. A passionary is guided by the principle "perspective is more real than reality".

A passionary equipped with a financial resource is, of course, a great power. There are many examples in history when the most fantastic projects were implemented without such a resource. However, this is the problem: the theory of passionarity speaks the language of synergetics, and synergetics does not have predictive power.

Therefore, a pessimistic scenario, "Contra". The same theory asserts that the entire content of the historical process is utilization of passionarity. Homeostasis is an inevitable outcome of any social evolution. The settings of a social system, in which there is only enough energy to maintain the present state, are just enough to get rid of any "growth points" that require excessive efforts. Demography is the most striking example here. If women do not want to give birth even under threat (☺) of payments of maternity capital, it means that the population is not even in homeostasis, but in the stage of decay.

This is evident even in regions with a predominantly rural population, such as ours. Since 2010, monitoring the attitude of young girls to the prospects of childbearing shows that 2 or more children are going to give birth to no more than 20 %, and vice versa, twice as many respondents do not consider such a prospect at all.

The slide of the local community into homeostasis is no less evident in the sphere of economics. The peak of economic development here was the construction of the Sayano-

Shushenskaya hydroelectric power station and the Sayan Aluminum Plant (SaAZ). In 1970-80s, a project for the industrialization of the region with the same geolocation as in the project of S.K. Shoigu, provided for commissioning of 12 more large advanced productions. However, then perestroika (restructuring) and post-perestroika began, and the frames of the unfinished workshops still adorn the right bank of Yenisei river.

Attempts to intensify the industrial life of the region did not end there. In the 2000s, against the background of the extinction of the city-forming enterprise AbakanVagonMash on the initiative of O.V. Deripaska (also a passionate, originally from neighboring Krasnoyarsk, his business empire was born in Khakassia) began to actively discuss the issue of building a polysilicon enterprise in the city of Abakan. Arguments "Pro" were voiced in his interview by the Director General *of Russian Silicon LLC A. Suponenko:* "Silicon production is a very energy-intensive process, and Khakassia has the necessary energy resources. In addition, in Abakan there is a convenient industrial site that is not used in any way, there are roads, communications, there are people who are able to build such a plant and in the future will be able to work on it". (https://marketelectro.ru/content/abakanskiy-polikremniy-ot-sluhov-i-domyslov-k-istine, date of last visit 09/09/21).

However, it was the "people", and precisely for environmental reasons, who actively opposed this project, and in 2010 the local authorities were forced to put an end to this project.

The irony of the situation is that, having won the ecological war against high-tech industrialization, the population of the RK lost it on the front of the struggle against the economically, energetically and technologically unpromising and no less environmentally hazardous coal industrialization. Until the 2000s coal in Khakassia was mined by the mine method; however, over the past 20 years, as many as 8 new open pit coal mining companies have been formed. It is not even worth commenting on how detrimental coal mines affect the ecological situation, which is aggravated by the fact that coal mining is carried out in the steppe ecosystem, which is extremely fragile. Thickness of humus here is on average 10 cm; its growth rate does not exceed 1 cm per 100 years.

The population and spontaneously formed initiative groups are waging a continuous struggle against the coal industrialization disfiguring the nature of the Khakass-Minusinsk Basin, but in vain. The large coal business turned out to be more influential even than the local authorities, since only 1 out of 8 coal companies (Mayrykhsky Razrez LLC) is a local tax resident. The rest deduct taxes either in Moscow or abroad in general. That is, the republic does not even have economic benefits from the environmentally hazardous activities of economic entities.

Since the shortfall in taxes occurs not only on the part of coal companies, but also on the part of almost all major regional players, the population's discontent also extends to local legislative and executive authorities. This also has a direct bearing on the efficiency of using human capital. Thus, Khakassia became the second region in which the population in 2018 elections refused to trust the acting governor M.V. Razvozhaev, who was appointed from the personnel reserve of the President, proved himself well in a few months in office and is showing himself even better at the moment as governor of Sevastopol. Instead, the population stubbornly voted for the local young communist V. Konovalov, whose team brought the republic to the last places in the rating of economic well-being.

Thus, not only in the economy, but also in politics, in the civil service, the efficiency of using human capital is most accurately characterized by the concept of a "problem". Its manifestation was the population's choice in favor of the Communist Party, the desire for innovation, or, conversely, the manifestation of ideological fatigue, i.e. homeostasis? The answer is obvious, although in the desire to "do everything as it once was" the population of the region turned out to be more original than other regions.

So, signs of homeostasis take place in various spheres of social life. From an ecological point of view, homeostasis is a state where a population is in equilibrium with its environment. That is, it reproduces indefinitely the same load on the landscape, the same livestock, the same management methods. It is not difficult to see in this model the signs of a pre-industrial society.

And vice versa, all revolutions, both political, economic and cultural, were initiated by "consortia" of passionaries who received certain resources of influence. Industrial and, further, post-industrial societies are, among other things, the result of the passionate enthusiasm of communities and social cohorts of promoters.

The environmental consequences of such explosive changes are almost always catastrophic. For the inhabitants of our region, this pattern is illustrated by the native steppe cut by coal quarries.

However, for Khakassia with southern Krasnoyar, and for many similar territories of Russia, the socio-ecological situation is somewhat different than, for example, for Western Europe. There, socio-economic changes have gone so far that the city has become an ersatz substitute for "natural" nature. Adaptation to the environment is concretized in the concept of "urban environment". It has been noticed, for example, that populations of hereditary inhabitants of megalopolises are always less energetic than newcomers ("brand-new arrivals"). That is, even falling into "urban homeostasis" [10] acquires irreversible socio-economic forms, which, in turn, creates the illusion of the irreversibility of the historical process, the vector of movement from a pre-industrial society to a post-industrial one.

In remote provinces of Russia with its large undeveloped and semi-developed territories, the loss of social energy (passionarity) by the population can and often proceeds according to an alternative scenario. More precisely, the main one, since according to most parameters, economic management in regions like ours, for the majority of the population, means a return to the original cultural, historical and economic positions - to "pre-industrialism", and not a breakthrough to something new.

The most important argument (both theoretical and practical) is the "cost of homeostasis". Any revolution, whether industrial or post-industrial, requires increased costs, and at the same time generates "artificial natural selection", rejecting, making more and more people unable to enjoy the benefits of post-industrial civilization. The fourth industrial revolution threatens to make the issue of impending unemployment unsolvable, in principle. [11] The "cheapest" solution may be the "traditional homeostatic" scenario. Recognition of this alternative is already taking place, for example, in relation to "pandemic migration". [12]

However, this automatically raises the question of the main typological features of the participants in this scenario, of a new reformulation of the concept of human capital. The coming urbanist lumpen-proletarian is least of all suited for "humanization" of the expanses of Siberia and the Far East.

4 Conclusion

Consideration of the problems of generation and effective use of human capital in remote Russian provinces through the prism of an ecological approach leads to a number of conclusions that are not so obvious within the framework of economic determinism or various sociological paradigms.

The most ambitious conclusion, or rather a hypothesis awaiting further discussion, is the assumption that the three-part model of history "pre-industrial, industrial, post-industrial society" is not the main, but an optional line of development. In the context of social entropy, signs of which can be found both in depressed provincial regions and in the most ultra-urban clusters, a return to homeostatic equilibrium with the ecological environment is simultaneously the most humanistic, and the most clearly predictable, and the least costly of the future scenarios.

The statement of the inevitability of homeostasis, in turn, leads to the need to interpret the concept of "human capital", which will allow the typical bearer of such capital to be most successful in a new socio-ecological situation. Simply said, one shall seriously think about whether German Sterligov is an eccentric exception to the rule or a social archetype of the personality of the future.

Acknowledgements

The reported study was funded by RFBR according to the research project № 19-011-00365.

References

1. D. Didenko. World of Russia, **30(3)**, 100 (2021)
2. N. Tikhonova, A. Karavay, Sociological research, **5**, 84 (2018)
3. R. Ibrahimov, E3S Web of Conferences, 258, 10003, 1 (2021)
4. *History of Khakassia from ancient times until 1917* (1993)
5. O. Lushnikova. Sociological research, **10**, 63 (2020)
6. E. Denisova-Schmidt, World of Russiaю, **30(3)**, 174 (2021)
7. G. Zborovsky, P. Ambarova, World of Russia, **30(1)**, 88 (2021)
8. J. Toshchenko, P. Velikij, World of Russia, **27(1)**, 7 (2018)
9. A. A. Posanenko, World of Russia, **27(4)**, 31 (2018)
10. O. Borisova, I. Pavlutkin, World of Russia, **28(4)**, 128 (2019)
11. K. Schwab, T. Malleret, *COVID-19: The great reset* (2020)
12. N. Pokrovsky, A. Makshanchikova, E. Nikishin, Sociological research, **12** (2020)

Labour market under economy digitalization

Z. B. Kenzhin[1,*], *A.U.* Tulegenova[2], *A. L.* Zolkin[3], *O. V.* Kosnikova[4], and *I.A.* Shichkin[5]

[1]West Kazakhstan agrarian and technical University named after Zhangir Khan, Uralsk, Kazakhstan
[2]West Kazakhstan Innovative and Technological university, Uralsk, Kazakhstan
[3]Povolzhskiy State University of Telecommunications and Informatics, Samara, Russia
[4]Kuban State Agrarian University named after I.T. Trubilin», Krasnodar, Russia
[5]Plekhanov Russian University of Economics, Moscow, Russia

Abstract. The main problems of current labour market under economy digitalization caused by COVID-19 pandemic are reviewed in this article. Current economy system is based on factors of innovative development that effects on labour market and, therewith, reducing employment of population and increasing of structural unemployment. Therefore, the problems of labour market studies are becoming actual according to structural development of economy system itself in the age of economy globalization. Methodological framework has defined against the background of developed procedures and regulatory settings for regulation of labour market and labour power in Kazakhstan. The assessment, composition and statistical analysis methods were used for studies.

1 Introduction

Economy and political events of modern world have strong effect on labour marker in Kazakhstan. The events happening inside country as well as outside events are important.

Full employment of population facilitates the development of economic as main factor of labour capacity for increasing and growing of mobile labour power. Moreover, emerging the new technologies stimulates the economy development; however, their emerging favours in change of social and economic relations including the economic, labour market, business environment and so on.

Apparently, the technologies directly effect on labour market that was significantly changed over the past two decades: some professions passed away due to the lack of their necessity, other became highly wanted, and the new ones were developed. Intensive development of digital, global enter of such leading global companies as Google or Apple in life of society, noticed upheavals in global economic led to significant changes of labour market [1]. Therefore, each country tries to regulate the process and guarantee the employment of population by increasing of government expenditures (transfer payments). This event is temporary and associated with trends of economy cycles but requires the specific measures on labour market regulation.

* Corresponding author: jaksat_22@mail.ru

https://doi.org/10.1051/e3sconf/202131108007

2 Problem statement

According to Atlas of professions, the processes of production systems are changed; at the same time innovations and informative and communicative technologies strongly effect on labour market. The labour power becomes unclaimed and there is no demand on specific specializations, there is a change of new professions that require taking into account the requirements of employers and guarantee the opportunity to get a job under the National Qualifications Framework, which was developed according to the European Qualifications Framework; this approach made it possible to recognize Kazakhstan's education (diploma) in the European labour market. Therefore, there is a need of labour mobility and its demand on global labour market but this lead to brain drain abroad that makes this study is actual.

3 Research questions

Labour market is economy relation system whereby exchange of labour services on material or other profits takes place. Labour market has a set of features. This is not only purchase and sale area or labour exchange but this is labour potential reproduction and labour service area [2,3]. Supply and demand on labour market depends on various factors. At the moment high innovative and technical infrastructure, digitization and automation are the main factors that lead to reducing the demand on labour. However, it might to be considered that reducing of demand on labour takes place in areas with automation and digitization of labour but the demand on labour is increased in areas where new equipment and its maintenance are produced.

4 Materials and methods

Methodological framework has defined against the background of developed procedures and regulatory settings for regulation of labour market and labour power in Kazakhstan. The assessment, composition and statistical analysis methods were used for this study.

5 Results

According to the UN's forecasts, the global population will be increased from current 7.5 billion of individuals to 8.5 billion by the year 2030. The number of people in Kazakhstan will reach 24 million against 18.6 million in 2019. Projecting the current level of economy activity on the advanced mass of our country population, taking into account the rate of increase of life expectancy and retiring age, the part of our economy active people abroad will exceed at the turn of 2030 - 14.4 millions of working age people against 9.2 million [4].

For modern Kazakhstan is important to study the labour market of foreign countries and specially those the economy of which as well as demographic situation are similar to Kazakhstan ones. Moreover, the experience of countries reel from major reconstruction in economic including innovation, re-thinking of ideological and moral values, changes in social policy, and implementation process of digital technology in business's activities has value.

The results of labour market problem studies show that many home manufacturers are still not available for automation and digitization of business's activities. At the same time regarding the implementation of new innovative technologies on enterprises the labour power faces with multiple issues, particularly, would it be the creation of new working places on market or significant their reducing available in different economy areas and what is important source of structural unemployment. However, in Klaus Martin Schwab's opinion some countries with developing economics are actually eliminated from process of development of new technologies and knowledge effect on their societies, and not effected by new industrial revolution. As countries with developed economics are first movers of scientific and technical progress the balance between technologies, society and economics can be easily shifted to their side. Without any activities, the future for developing countries will be developed in random ways not targeted, and technologies will limit the opportunities but not expand [5,6].

6 Findings

The unemployment is specific feature for market system and inevitable one when some specializations will be unclaimed. Under innovative development numerous small, middle and large-scale enterprises contributes to new technology's implementation resulting many workers are leaving out of enterprises. However, the labour power in construction field is still in-demand regards the development of this field in Kazakhstan. The re-qualification and re-training of labour power taking into account the structural changes in economic will be needed due to the demand on other specializations.

The main challenges of unemployment in Republic are connected not only with availability of free vacancies but, first of all, with skilled labour power shortage. The assessment of industrial enterprise's activity shown that many of them do not have the labourpower with appropriate skill level. At the same time, there are too much individuals with unclaimed professions where the competition for working place is very high. The results of assessment shown that only in 2019 142.4 thousands of specialists were graduated that is higher on 9% against 2018. However, from the side of quality manpower preparation this parameter requires the re-assessment of their qualification and skill level according to employer's requirements as in the employer's opinion the graduates do not have the sufficient practical skills and the employers are not ready to employ and guarantee them the high salary.

The unemployment level for 2020 is 5% against 4.8% in 2019. According to data of Ministry of National economic of Kazakhstan Republic, the level of unemployed men and women was 216.3 thousands and 237.7 thousands people respectively [7]. At the same time, the highest unemployment level in 2020 was 13.7%, and the lowest level was up to 4.8% (see Fig.1). These values provide evidence that social and economy costs of unemployment are growing.

In the second quarter of 2020, there were 454 thousands of unemployed individuals that is on 2.8% higher than in the second quarter of 2019. The number of unemployed men and women was 216.3 thousands and 237.7 thousands individuals respectively. The unemployment level for men and women was 4.6% and 5.4% respectively.

If analyse the unemployment level with regards to town and country it is lower for country (174.7 thousands individuals with unemployment level 4.7%) than for town (279.2 thousands individuals with unemployment level 5.1%).

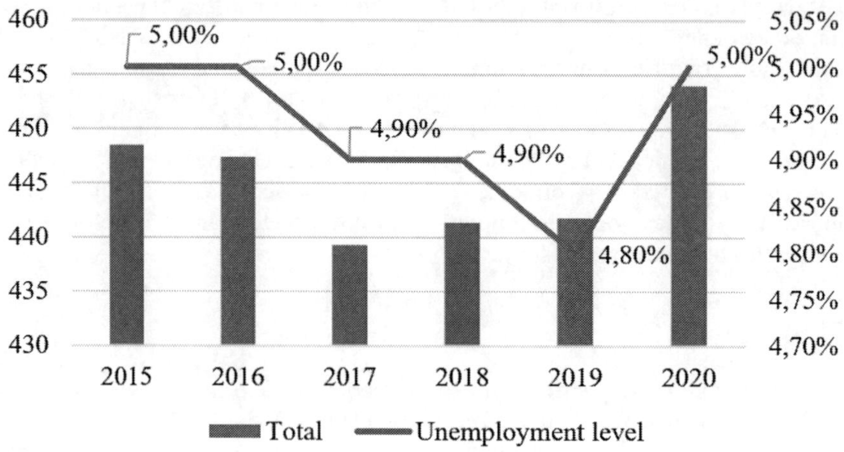

Fig. 1. Unemployment level in Kazakhstan Republic.

Regionally the highest quantity of unemployed people was in south areas of country: Almaty, Almaty and Turkestan regions.

There are 54 thousands of unemployed individuals in Almaty that is 7.3% higher than for last year. 49.4 thousands of unemployed individuals live in Almaty region that is plus 3% over year; 42.3 thousands of unemployed individuals live in Turkestan region that is plus 0.6% over year.

7 Discussion

Due to the COVID-19 pandemic in 2019, the global labour market was hit hard as well as economic totally. Early April 2020 3.3 billion people or 81% from total working power suffered from complete or partial job cut. March-April, 2020 in Kazakhstan, specifically, 4.4 million of people stayed temporary unemployed due to the quarantine measures. Totally, as on January 1st, 2021 the unemployment level in country was 5.1% or 457 600 unemployed people because of negative effect of crisis.

To solve the problem on Kazakhstan labour marker as well as on foreign countries the stimulus measures are intensified for the purpose to assure the complete population employment.

Measures to control the unemployment include the set of social guarantees, social insurance, various forms and methods of social support from unemployment specifically two main social groups - working people who are upon pain of dismission, and unemployed people registered in state employment service [8].

"Action plan for employment" implementation in Kazakhstan favours to solve the problem of social and economic costs of unemployment. Within the action plan, more than 6.5 thousand projects were presented, in which 255 thousand working places were created. A special condition for implementation of "Action plan for employment" infrastructure projects is to ensure the use of at least 90% of goods and services of domestic manufactures; in addition, at least 50% of workers will be hired through employment services. As on July 15, 2019, work on 6295 projects began in the regions, more than 148 thousand working places created, including 77 thousand ones through employment services.

The Employment Services (hereinafter the Services), which function in all regions of the country, remains one of the most important mechanism for labour market regulation.

The Service's activity is addressed at analysis and forecasting the supply and demand for labour power, their employment, re-training and re-qualification, etc.

The services are a mechanism for employment and labour market management, what effects on the indicator of social and economy development of country. It should be noted that the management of processes taking place in the labour market in foreign countries is based on the principles of social partnership. The "social partnership" concept is used in the West mainly by social democrats. Historically, it has opposed to class struggle. Now, the confrontation between labour and capital has not become so acute due to the general rise of the living level of the population of countries with market systems. The middle class in these countries is 70–80% of the population. In such conditions, the slogans of social partnership began to be used to political stability strengthening in the country [9,10].

The analysis of Service's activity shown that according to the results of the first quarter of 2020, more than 3.3 thousand people were employed for permanent work, while the largest number of labour power (651 people) were employed through the employment services of Nur-Sultan city. It should be noted that such difference in the number of employed people is according to a demographic factor, namely the uneven distribution of the country population, what is very actually for Kazakhstan. Therefore, there is a lack of demand in some regions and glut in others. Commonly, salaries in all regions for the same professions are different; the average monthly salaries of Kazakhstan population in the fourth quarter of 2020 were 203.9 thousand tenge, what is 15.8% more than in the same period in 2019. Taking into account the inflation, the real growth of salary purchasing power was a significant 9.9%. However, the lack of demand on labour power in specific regions is a positive side of the labour market, it makes possible of labour power mobility and find the prospects for them in other regions, and is a disadvantage for mature people or citizens of pre-retirement age who have been laid off. Therefore, development of social legislation is needed: the setting of subsistence level, unemployment payments, pension payments, the development of health care, etc.

Modern labour market management has become more multi-faceted and systemic, at the same time trend of educational and innovative development pick up is important, what will help to increase the necessary skills and ensure the functioning of the labour market, taking into account the technological revolution. Such factors effects on the skills of labour power and the ability to achieve them. This requires labour power and companies to adapt to changing conditions and culture, as well as to meet the expectations of a new generation of candidates and employers. At the same time, employment trends are constantly changing, therefore employers and the labour power must adapt to the transformation of the labour market in order to be well informed of changing trends in this area.

8 Conclusion

Hereby, the fourth industrial revolution has a powerful effect on the restructuring of the labour market in Kazakhstan. These processes strongly effect and present in all fields of the economy, since it is a subject to various natural, technological, social and economic and political factors. It is difficult to overestimate the importance of economy fields, but at the same time, their dependence on various external factors is also very high.

Three main factors effect on labour market of the country from social and technological sides: the emergence of a new labour power (economically active), increase in the duration of the active labour power's presence in the market, and technological re-equipment, that frees up the labour force. The last two factors hold down the possibilities of the labour market for stable growth, and the first one, which is key from a social and political point of view, needs it. If ignored, this situation will eventually reach a phase when the labour market will not be able to absorb the ever-increasing flow of new employees.

New technologies lead to transformation of the labour market on an unprecedented scale. Professions disappear and change, new specializations appear. Corporations, governments and economy industries compete for the intellectual capital of the future - people with skills and competencies focused on the economy with innovative development. In Kazakhstan, the implementation of "Action plan for employment" determines the state policy in the field of population employment regulation, that ensures the reducing of social costs of unemployment in the country.

References

1. D.-J. Rahmil, *How is digital technology changing the labour market?* (2014)
2. E. G. Yakovenko, *Labour economic*, 320 (2012)
3. A. A. Kochetkov, *Economy theory: Book for bachelors*, 696 (2013)
4. K. Grass, E. Weber, *The Debateon Digitalisation and the Labor Market in Europe* (2016)
5. F. Delfani, H. Samanipour, H. Beiki, A. Yumashev, E. Akhmetshin, Int. J. of Systems Science: Operations and Logistics, 1 (2020)
6. M. A. Chirkov, T. A. Lachinina, M. S. Chistyakov, Free Thought, 37 (2020)
7. E. Lavrov, Eastern-European J. ofEnterprise Technologies, **2(86)**, 4 (2017)
8. E. A. Lavrov, P. I. Paderno, A. A. Volosiuk, N. B. Pasko, V. I. Kyzenko, Proceedings of 2019 3rd Int. Conf. on Control in Technical Systems, CTS 2019, 144 (2019)
9. I. A. Poskryakov, V. D. Munister, A. L. Zolkin, M. A. Kuzmin, N. F. Zavialova, IOP Conf. Series. Krasnoyarsk Science and Technology City Hall, 32014 (2021)
10. A. L. Zolkin, T. G. Aygumov, A. I. Pakhomova, V. V. Bobkov, IOP Conf. Series: Earth and Environmental Science. Krasnoyarsk Science and Technology City Hall. 22032 (2021)

Supportive mechanisms for redundant employees in the context of digital transformation of the Russian labour market

Liliya Babynina[1,*], *Larisa* Kartashova[1], *Natalia* Loktyukhina[2], *Evgeniya* Shubenkova[1], and *Olga* Nikonova[1]

[1]Plekhanov Russian University of Economics (PRUE), Moscow, Russia
[2]Financial Research Institute, Moscow, Russia

Abstract. The article presents results of the analysis of the current mechanism for supporting redundant employees with the development of digital transformation processes in the Russian labour market. Three directions for improving this mechanism have been identified: the choice of criteria for the release of personnel, the implementation of an active employment policy and a system of social support for the unemployed. The redundant employees are labour resources and the potential for the development of the country, therefore, it is important not only government support for such employees, but also additional measures of socially responsible employers.

1 Introduction

In the context of market-oriented economy, the release of employees (reduction in the number or staff of employees) is due to competition, crises in the development of the company and the search for the most profitable business areas. The existing mechanism for supporting redundant workers in Russia have based on labour legislation, as well as on additional initiative measures of socially responsible employers. To implement the system of national projects in the Russian Federation, adopted to achieve new strategic goals of national development, state programs and subprograms have been develop that provide for measures of an active employment policy. They are aimed at improving the system of social support for the unemployed, retraining, raising the qualifications of enterprise employees modernizing the employment infrastructure in the constituent entities of the Russian Federation, improving the work of the state service to support released workers and unemployed citizens, developing the competencies of employees of the employment service.

In the context of a tense demographic situation in Russia with an aging population and a declining birth rate, redundant employees are labour resources and the potential for the development of the country not only among people of working age, but also among pensioners. In this case, we are talking about a large category of citizens in the labour

* Corresponding author: Babynina.LS@rea.ru

market. In 2020, they left their previous place of work due to layoffs, personnel cuts, liquidation of an enterprises; own business, 856 thousand people, or almost 20% of 4.3 million unemployed (in 2019 - 530 thousand people or 20.2% of 2.6 million unemployed). In the period from January 2020 to June 2021, the highest share of the unemployed was observe in the age groups 20–39 years old [1]. Youth unemployment negatively affects the socio-psychological development of young people, leads to a reorientation or degradation of labour values and to the disappearance of the view of work as a means of personal self-realization, and is also a significant limitation of the implementation of the country's labour and intellectual potential [2]. Thus, the task of improving the mechanism for supporting redundant personnel in the context of the digital transformation of the Russian labour market seems extremely urgent. In this regard, the authors consider it important to present an analysis of the existing challenges of state policy in this direction and formulate proposals for their solution.

2 Materials and methods

The authors have relied for preparing the article on the following materials and research: the results of in-depth interviews with employees of the employment service of the Ryazan region, conducted in 2021 as part of the research work on the topic "Digitalization of the labour market and employment in Russia: trends and development mechanisms", funded from funds of the Federal State Budgetary Educational Institution of Higher Education “Plekhanov Russian University of Economics ”(Order No. 867 dated June 28, 2021); the results of expert discussions and surveys of the heads of employment services of the regions of the central federal district in the framework of the study "Organizational and financial mechanisms for supporting the employment of the population in 2021-2023, aimed at reducing the unemployment rate", carried out by the Research Institute for Basic Research on the state order of the Ministry of Finance of Russia (092-00001-21 -00).

As the initial data, the authors used the official statistics of Rosstat, information on the registered labour market provided by the Ministry of Labour of Russia, reports of the International Labour Organization and the Eurofund, population surveys conducted in connection with the scaling of new forms of employment.

Empirical methods based on the analysis of scientific foreign and Russian literature are used. Field studies, a combination of quantitative and qualitative analysis methods, online surveys, personal and telephone interviews were carry out to assess the challenges and opportunities for transforming the employment service of the Russian Federation to support laid-off workers and unemployed citizens.

3 Results and discussion

3.1 Selection criteria for applicants for redundancy

Studies show that if staff reductions cannot be avoid and all alternatives and opportunities to stimulate the voluntary departure of employees from the enterprise have been exhaust already, the problem arises of the socio - economic justification of measures for forced staff layoffs. At the same time, not enough attention is pay to measures to support employees who remain to work at the enterprise, which reduces the effectiveness of the work carried out to optimize the number of personnel. Not enough attention was pay to the fact that during the period of staff reduction, the support of employees of the human resource management services of the enterprise from the top management is also required to determine the criteria for selecting candidates for the release. The issue of choosing the selection criteria for applicants for release requires careful substantiation from the standpoint of economic feasibility and social justice. In this case, two approaches can be use. The first is to leave those who will find it most difficult to gain a new job. The second approach, which has recently become prevailing, is to keep the most qualified workers capable of quickly restructuring and retraining in accordance with new tasks. In the first case, the main criteria will be work experience, age, marital status, health status, in the second, versatile experience and multifunctional skills, qualifications, work efficiency [3].

3.2 Implementation of an active employment policy

The results of the analysis of the mechanism for the redundancy of personnel in modern conditions showed that it is based on labour legislation and is implemented within the framework of the state program of the Russian Federation "Promotion of employment of the population", in particular, its subprogram "Active employment policy and social support for unemployed citizens." In fact, almost all the activities of the subprogram aimed at unemployed citizens, and the redundant personnel, who are potentially in the labour market from the moment they receive notification of the upcoming reduction, are actually have not embraced by an active employment policy, which provides for the development of labour mobility of the population. The legislative initiation of the expansion of participants in active employment programs was reflected in the draft law prepared by the Ministry of Labour of Russia [5], which proposes to involve in such programs not only registered unemployed. The citizens registers in employment services as job seekers, and ready to relocate temporarily or relocate for employment for permanent residence in another locality. However, in this case, according to the authors, it is required to assess the possibility of employers' abuse of use of labour mobility programs for the movement of personnel.

Additional measures aimed at reducing tensions in the labour market of the constituent entities of the Russian Federation have developed and implemented by the Government of the Russian Federation. Those measures use during economic crises and sanctions, for support the employment on the North Caucasus, due to restrictions to the COVID-19 pandemic and due to consequences of the pandemic. The additional measures that provide support to enterprises and redundant personnel, such as advanced training of those who are under the threat of dismissal, subsidizing wages, internships, equipping jobs for people with disabilities, etc. However, traditional active employment programs, the list of which have established by Article 7.1 of the Employment Law and additional measures, do not form a single system. The main disadvantages of additional measures are associated with the lack of open reviews and feedback, the ability to assess the long-term effect of their implementation. It is difficult to determine whether the training is formal, whether the competitiveness of the personnel is increasing, whether the business is satisfied with the results of the implementation of these programs.

Since 2012, the implementation of active employment policy measures within the framework of the "Employment Promotion" state program have been transferrin to the regional level and are finance from regional budgets. Under these conditions, the federal list of active employment programs should not be mandatory, but recommendatory. Therefore, the region, based on the situation on the labour market, the specifics of job seekers, available jobs, should itself determine which programs it is preferable to implement.

Thus, we see the improvement of the organization of work with redundant personnel within the framework of an active employment policy in additional measures of labour mobility of redundant workers and in providing the region with greater flexibility in the implementation of such programs.

3.3 Problems with the payment of unemployment benefits

The analysis of social support for the unemployed in the payment of unemployment benefits, carried out by the authors, made it possible to note the following. The allowance has become an analogue of social assistance, having lost its connection with the parameters of the previous job and the unemployed earnings. During the COVID-19 pandemic, when unemployment benefits were increase in 2020, it became a key regulator of the registered labour market. As a result, for the first time, the indicators of actual and registered unemployment came close, since the increase on the amount of the benefit stimulated the population, including those working informally or long-term unemployed, to contact the employment services. The number of recipients of unemployment benefits since the beginning of the pandemic has grown fivefold - from 720 thousand to 3.5 million people [1]. According to the Ministry of Labour of Russia: "With the possibility of receiving benefits in an increased amount, including payments for underage children, citizens, in most cases, did not show interest in receiving services to assist in finding a suitable job. The main goal of a fairly significant part of appeals to the employment service was not employment, but receiving unemployment benefits" 4]. Some citizens receive unemployment benefits fraudulently when they work informally and have an unofficial income. The motivation for this behavior is not so much the benefit itself as the benefits. In particular, an unemployed person, due to the small amount of the benefit or after the termination of its payment, can expect the provision of compensation for the costs of housing and utilities services.

Some unemployed are "professional unemployed" who have been registered with the employment service for a long time, receive benefits and other services. Among of them there are both those who receive benefits fraudulently, and the unemployed who are not motivated to find a job, for example, leading a household. The challenges faced by the employment service and which it will have to overcome are related to the fact that at present its role begins to be reduced mainly to the payment of benefits to "indiscriminate". Determining exact amount benefits for the unemployed and informally employed appears to be a difficult task today. Employers are not interested in cooperation with the employment service for fear of hiring unmotivated applicants. The current situation does not allow the employment service to work fully with motivated job seekers, including laid-off workers and citizens who are experiencing difficulties in finding a job.

At the same time, the is task of supporting the form of social payments for the period of unemployment, highlighted for those who are left really without means of subsistence and to aim for finding employment. In this regard, the authors propose to change the concept of social protection against unemployment - to abolish unemployment benefits and introduce a universal poverty benefit, the concept of which was developed by the FRI in 2015 [6]. The concept of a universal poverty benefit based on the idea of reducing small and large risks of poverty, implemented through the mechanisms of providing job search services to the unemployed and social support for low-income citizens who have lost their jobs. The amount of the allowance should ensure the minimum possible standard of living for a low-income unemployed person and his household for the period of job search.

This proposal deserves a special study, since the opinions of scientists and practitioners do not find unanimous support [7]. Thus, a survey of the heads of employment services in the regions of Russia, conducted in order to determine the directions for improving the system of payment of unemployment benefits, showed the following (Fig. 1). The overwhelming majority of heads of employment services (60 people) to propose strictly link the amount and periods of payment of benefits to the activity of citizens for employment. The second most of important proposal is related with implementation of the institution of unemployment insurance system. The experience of the OECD countries shows that in addition to the state, business and citizens participate in financing programs in the labour market through the unemployment insurance system, which fundamentally distinguishes this approach from the state employment policy in the Russian Federation [8].

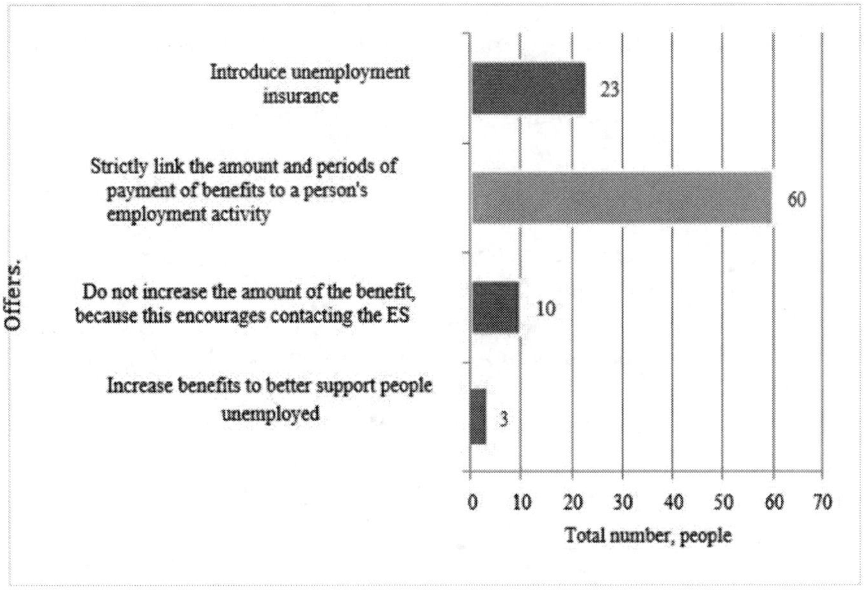

Fig. 1. Results of a survey of heads of employment services of Russian regions

Source: compiled by the author.

4 Conclusion

The results of the analysis showed that the implementation of an active policy of employment of the population requires improving the organization of work with redundant personnel. The authors proposed directions for the development of additional measures of labour mobility of redundant employees and providing the region with greater flexibility in the implementation of such programs. An assessment of the comprehensive effectiveness of an active employment policy should base on the indicators (indicators) of the state program of the Russian Federation "Promotion of employment of the population".

Currently, the socio - economic preconditions are ripe for changing the model of calculating and paying unemployment benefits. At the same time, the incentives to find a job should be significantly higher than the incentives to hide the income that can potentially obtained in the informal sector. A significant drawback of the existing mechanism of social support for laid off employees is that often the efforts of employment services to find jobs for citizens who have difficulties in finding a job are brought to naught in connection with the guaranteed measures of social support. It is necessary to take a comprehensive look at the whole range of support measures and guarantees for citizens who have difficulties in finding a job, from the point of view of their effectiveness.

It seems expedient to continue scientific and practical research of the problem both from the point of view of finding ways to maximize the integration of solutions carried out in this area, and from the point of view of predicting their results and distant socio-economic consequences. In-depth consideration also requires a search for ways to strengthen the social responsibility of managers at all levels of management for the consequences of decisions taken in the field of sustainable development, prevention of unjustified layoffs and support for laid-off workers and unemployed citizens.

References

1. Statistical Bulletin, Results of a sample survey of the labour force. II quarter 2021 (2021)
2. Yu. G. Odegov, L. S. Babynina, Monitoring of Public Opinion:Economic and Social Changes, **4** (2018)
3. L.V. Kartashova, Vestnik of the Plekhanov Russian University of Economics, **4** (2020)
4. Monitoring and assessing the quality and availability of public services in the field of promoting employment of the population, https://mintrud.gov.ru/
5. Draft Federal Law "On Amendments being made to the Law of the Russian Federation" On Employment of the Population in the Russian Federation", https://regulation.gov.ru/
6. E. I. Andreeva, D. G. Bychkov. Scientific Research Financial Institute. Financial j., **6(34)** (2016)
7. V. N. Bobkov, N. V. Loktyuhina, M. V. Simonova, I. A. Shichkin, Advances in Intelligent Systems and Computing, **908** (2020)
8. Unemployment in Russia and foreign countries during the crisis. Social bulletin (2015)

Comparative analysis of demographic potential of rural territories of the central black earth

Evgeny Reutov, *Marina* Reutova, *Svetlana* Vangorodskaya, and *Galina* Gaidukova*

Belgorod State University, 308015 Belgorod, Russia

Abstract. The article presents the substantiation of the methodology for comparative analysis of the demographic potential of rural areas and identifies the main trends in the development of the demographic potential of rural areas of the Central Black Earth Economic Region. Rural areas are considered as an essential factor in the reproduction of the natural, economic, demographic and socio-cultural potential of Russian society. It has been substantiated that the main problems hindering sustainable social development are mainly due to the change in the balance between the urban and rural populations characteristic of the era of accelerated industrialization. The definition of demographic potential as a set of opportunities and conditions for the development of territories lying in the sphere of population reproduction is given, and a great variability of the concept of demographic potential in sociological and demographic discourses is noted. Taking into account two key factors of population dynamics – fertility and migration, it is proposed to measure the demographic potential of Russian municipalities by using indicators such as the proportion of the population of reproductive age, the birth rate and the coefficient of migration inflow/outflow of the population. The expediency of using available statistical data, adapted in relation to the proposed research methodology, is shown. Based on the generalization and analysis of statistical data characterizing the natural and migration movement of the population of rural areas of five regions of the Central Black Earth Economic Region, as well as the results of the sociological study "Mental inequalities as a factor in the social polarization of the Russian province" conducted in the Belgorod and Voronezh regions in 2018-2019, it was concluded that the rural areas of the Central Black Earth Economic Region are characterized by heterogeneity of the demographic potential, which in the Kursk and Tambov regions is acquiring the character of obvious degradation, and the migration attractiveness of the territories (with the exception of suburbs) has rather negative values.

1 Introduction

With a sufficiently high degree of urbanization in the country, rural areas continue to be an essential factor in the reproduction of the natural, economic, demographic and sociocultural potential of Russian society. The share of the rural population of Russia (as of January 1,

* Corresponding author: g_gaidukova@bsu.edu.ru

2021) is 25.34%, and in some regions it is much higher. So, in the Altai Republic the share of the rural population is 70.72%, in the Chechen Republic – 63.12%, in the Karachay-Cherkess Republic – 57.12%, in the Republic of Dagestan – 54.69%. In the regions of the Central Black Earth Economic Region, the share of the rural population is lower than in the regions noted above, but still noticeably higher than the national average: in the Belgorod region – 32.51%, in the Voronezh region – 32.04%, in the Kursk region –31.48%, in Lipetsk – 35.37%, in Tambov – 38.61% [1].

However, due to the process of urbanization, changes in the technological structure in the agricultural sector, degradation of social infrastructure, rural areas are currently losing their population. The only exceptions are mainly suburbs – rural settlements located along the perimeter of large cities and, in fact, are part of urban agglomerations. The change in the existing balance between the urban and rural populations, which is absolutely natural in the era of accelerated industrialization, at the present time, rather gives rise to a set of problems, rather than contributes to the sustainable development of society. The influx of the rural population into the cities leads to an excess supply of labor in the urban labor market and spins up the rise in property prices.

In turn, the decline in the rural population creates problems of a different kind. In the aggregate of resources for the development of rural areas, human and labor resources play a special role. A skilled workforce, along with production capacity and investment, is the backbone of economic growth. High or moderately high population density in rural areas prevents them from desertification and is a powerful driver of public, corporate and private investment in social infrastructure. There are many examples in social history when the disappearance or a sharp decrease in the population of certain territories led to the degradation of natural landscapes and technological infrastructure, not to mention their actual loss for the economies of countries and regions.

2 Materials and methods

The purpose of this article is to substantiate the methodology for comparative analysis of the demographic potential of rural areas and identify trends in the development of the demographic potential of rural areas of the Central Black Earth Economic Region, which includes Belgorod, Voronezh, Kursk, Lipetsk and Tambov regions.

The empirical basis of the article is statistical data characterizing the natural and migration movement of the population in rural areas of five regions of the Central Black Earth Economic Region, as well as the results of the sociological study "Mental inequalities as a factor in the social polarization of the Russian province" conducted in the Belgorod and Voronezh regions (in 2018 there was questionnaire survey, N=1200 respondents, 600 in each region, in 2019 there was a series of semi-formalized interviews (N=50).

In the context of the study of the main factors of simple and extended reproduction of the population of rural areas, the concept of demographic potential plays a special role. Demographic potential must be distinguished from the actual demographic resources. The demographic resources of a territory are, first of all, people living on it, assessed by a set of quantitative and qualitative parameters in order to understand their contribution to the economic, social, cultural and other development of the territory. Unlike resources, demographic potential (from Latin potentia – strength, opportunity) is, rather, a set of opportunities and conditions for the development of territories lying in the sphere of population reproduction. So, for example, when analyzing the demographic resources of a particular territory, it will be necessary to take into account the number and proportion of the population of young people, skilled workers, etc. And when analyzing the demographic potential, along with the fixation and quantitative description of these groups, it will be extremely important to understand their values and attitudes concerning migration outside rural areas or the region as a whole.

At the same time, it should be noted that the concept of demographic potential is still quite varied in sociological and demographic discourses.

For the first time the concept of "demographic potential", according to some sources, was proposed to be introduced by the English demographer R. Fisher in the 1920s. R. Fisher metaphorically considered the birth of a person as receiving life "on credit", and the subsequent birth of his own children as "paying off the debt." Thus, the influence of the demographic potential on the development of society, according to R. Fischer, was traced only in the narrowly demographic sphere and was reflected in the birth rate. Moreover, the concept already introduced by R. Fisher has not received precise indicators in the demographic and statistical sciences of his day. According to other data, the concept of "demographic potential" was introduced by J. Stewart, who studied the factors of migration. He borrowed the concept of "potential" from physics and, using the concept of "gravitational potential, he believed that the interaction between masses of people is determined by connections similar to physical ones" [2].

The study of the demographic situation and the potential of rural areas has become widespread in European countries. Jadwiga Biganska, based on indicators characterizing population density, population structure by sex and age, population migration and vital statistics, assessed the potential of rural areas in Poland in 1999-2011 from a demographic point of view. The k-means method was used to classify and compare rural areas [3]. Trends in the development of economic and demographic processes in the region were outlined. L. Nimets, N. Guseva, K. Segida presented the characteristics of the spatial distribution of the population of the Kharkiv region, as well as the methodology for calculating the demographic potential, taking into account the qualitative and quantitative characteristics of the population and trends in its movement. Based on the results of the study, the authors calculated the demographic potentials of the regions of Ukraine and determined the place of the Kharkiv region among them, including its territorial differentiation in relation to the environment, which includes key components and cause-and-effect relationships of the socio-economic development of cities and districts [4]. Daniele Rama et al., using the example of rural Italy, analyzed the usefulness of the Facebook advertising platform, which offers a digital census of more than two billion of its users, to measure potential inequalities between rural and urban areas [5].

Later, in the study of migration processes in domestic science, the concept of "migration potential" prevailed, under which L.L. Rybakovsky and N.I. Kozhevnikov, for example, understood the "assessment of migration resources located in countries – possible donors for the recipient country" [6].

There have also been repeated attempts to develop a set of indicators for quantitative analysis of demographic potential.

So, outside the migration aspect of demography, it was proposed to evaluate the demographic potential also in person-years according to their number and life expectancy. This proposal is borrowed from the concept of "labor potential", where labor means the person existence in itself [7]. In the generalizing article by O.L. Rybakovsky and O.A. Tayunova notes that "demographic potential can mean both the potential for population growth and the potential for population growth for each, for several or for all components of the demographic movement of the population. Population growth potential shows general changes, possible general results of changes in population parameters. The population growth potential can be considered as the difference between the maximum (conditionally) admissible possibilities and the inertial scenario of demographic development" [2].

With regard to the topic of our research, the concept of the demographic potential of a territory has significant methodological value, which, from the point of view of O.L.

Rybakovsky and O.A. Tayunova is formed, on the one hand, under the influence of fertility and mortality processes, and on the other, under the migration: "the demographic potential of a territory is, firstly, the reproductive capabilities of the resident population of a given territory, including retention or changes in the levels and trends of fertility and mortality, and, second, the possibilities of the population of migration-related regions of the region to change the size and structure of its population by migration" [2]. Also, these authors note the difference in measuring the demographic potential of the territory, depending on the methodology and purpose of its assessment: "depending on the time of assessment, the demographic potential can be either the level of demographic development achieved at the moment, or those opportunities that can be realized in the future" [2].

From the point of view of A.I. Kuzmina, "demographic potential as a synthetic category is not just the size or mass of the population of a country or a macroregion, but a certain system for assessing the potentials of its viability, marriage, fertility, divorce rate, the formation of sex, age and family and marriage structure and territorial mobility of the population (including pendulum migration)" [8].

M.L. Yashin and N.R. Aleksandrova propose to measure the demographic potential of Russian municipalities, using indicators such as population growth/decline rate, population size, share of the working-age population, fertility rate, fertility growth/decline rate, mortality rate, dynamics of mortality rate, natural decline rate of population, the coefficient of migration inflow/outflow of the population [9].

From our point of view, for the calculation and comparative analysis of the demographic potential of territories, it is advisable to use such indicators as the share of the population of reproductive age, the birth rate and the coefficient of migration inflow/outflow of the population. Thus, two key factors of population dynamics are indirectly taken into account – fertility and migration. A certain problem is the lack of a unified methodological and publication model of territorial statistical measurements and the incompleteness of statistical data in relation to the rural areas of the regions of Russia. Proceeding from this, in the future, in the empirical analysis of the demographic potential of rural areas in the regions of the Central Black Earth Economic Region, available statistical data will be used, adapted in relation to the proposed research methodology.

The reproductive age in women, in accordance with the well-established methodology of demographic research, is considered to be the interval of 15-49 years [10]. However, within this interval, the birth rate is extremely heterogeneous and depends on the specific socio-economic and socio-cultural conditions in certain societies. In Russia, the birth rate is sharply decreasing in relation to the group of women over 44 years old (in 2019 it was 3.0 children per 1000 women versus 82.9 in the group of women of reproductive age of 15-49 years old) [11]. Thus, in the calculation and comparative analysis, it is quite permissible to limit ourselves to the age of 15-44 years. For men, it is advisable to use the age interval of 20-49 years. Considered in the context of the demographic potential, the indicator of the proportion of the population of reproductive age allows, with certain reservations (indicators of actual childhood, socio-economic situation, measures of state support for motherhood and childhood, socio-cultural characteristics of the territorial community, etc.) to carried out the fertility forecast.

3 Results and discussion

In the Belgorod region, the aggregate share of men aged 20-49 and women aged 15-44 in the rural population, as of January 1, 2020, is 36.2%, while as of January 1, 2016 it was 32.2% [12]. In the Kursk region, as of January 1, 2020, the share of the population of reproductive age is 32.7%. As of January 1, 2018, it was 33.2% [13, 14] (an earlier value cannot be calculated due to lack of data). In other regions of the Central Black Earth Region, it was not

https://doi.org/10.1051/e3sconf/202131108009

possible to calculate the values of this indicator due to the lack of initial data. Thus, in two of the five regions of the Central Black Earth Region, opposite trends are observed: in the Belgorod Region, the share of the population of rural areas of reproductive age in 2015-2019 is has grown quite noticeably, which is due to the strong regional policy of stimulating individual housing construction; in the Kursk region for 2017-2019 this share, albeit insignificantly, decreased. Thus, in the Belgorod region, rural areas received a rather tangible impetus for the growth of birth rates in rural areas in the short and medium term. However, at the same time, the total fertility rate in rural areas of the Belgorod region has been consistently decreasing since 2015. If in 2015 it was 1.757, then in 2019 it was 1.404 [15]. To a large extent, the decrease in childbirth per woman of reproductive age neutralized the effect of an increase in the proportion of the population in rural areas of reproductive age. Nevertheless, the rate of natural growth of the rural population in the Belgorod region in 2019 was -8.3, which, although it characterizes negatively the dynamics of natural reproduction, is still the highest value since 2010 [15].

Data on the migration movement of the population of the rural areas of the Belgorod Region are not available, it can only be assumed that the migration increase in the population of the region recorded in the last two decades has at least partially affected the rural areas.

In the Voronezh region, the rural population, as well as its share in the region's population from 2009 to 2019. consistently decreased [16]. However, apart from these data, the official statistical publications of the region do not contain other parameters of the natural demographic movement of the population in rural areas.

The migration growth of the population in rural areas of the Voronezh region amounted to 37.2 per 1000 people in 2019, which is the highest rate since 2015. Moreover, from 2015 to 2018 a decrease in migration growth was recorded down to its negative value (-8.3 in 2018) [16]. The lack of actual data allows us to build only hypothetical models, according to which the demographic potential of the rural territories of the Voronezh region is decreasing under the influence of the natural decline of the rural population, but it is partly compensated by an unstable migration increase.

In the Kursk region, in rural areas, the birth rate in 2019 was 7.0. At the same time, it has been consistently decreasing since 2014 (in which it was 11.1) with a simultaneous increase in the rate of natural population decline (from 10.9 to 12.5) [14]. Thus, in the Kursk region, the cumulative negative effect of all components of the demographic potential of rural areas was clearly demonstrated: a decrease in the proportion of the population of reproductive age – a decrease in the birth rate – a natural population decline.

The migration loss in rural areas of the Kursk region in 2019 was 199 people, but in 2018 it was even higher – 1986 people [14]. Nevertheless, migration processes exacerbate the negative dynamics of the demographic potential of the rural areas of the region.

In the Lipetsk region, the birth rate in rural areas in 2019 was 7.7 children per 1000 population. At the same time, since 2015, it has been consistently decreasing. Accordingly, the rate of natural population decline increased from 7.8 to 8.7 [17].

The migration growth of the rural population in the Lipetsk region in 2019 amounted to 689 people, while in 2018 a decline of 811 people was recorded [17].

Thus, we can state a decrease in the demographic potential of the rural areas of the Lipetsk region, albeit at a slower pace than in the Kursk region. The influence of migration processes on the demographic potential of the rural areas of the Lipetsk region is generally ambivalent; nevertheless, in the last year, the migration inflow of the population at least partially compensates for the decline in the demographic potential as a whole.

In the Tambov region, in rural areas, the rate of natural increase/decrease of the population in 2019 was -9.4, in 2014 – -8.5 [18]. Thus, in the mid-term retrospective, the demographic potential of the rural areas of the region has significantly decreased.

At the same time, the migration loss of the rural population of the Tambov region in 2019 amounted to 2558 people [18, 19], significantly increasing the degradation of the demographic potential.

A sociological study we conducted earlier in two regions of the Central Black Earth Region – Belgorod and Voronezh regions – revealed a potential negative effect on the demographic potential of the regions of the existing migration attitudes of the population. Thus, a sociological survey conducted in 2018 (N=1200 respondents, 600 in each region) showed that 64.9% of respondents, in one way or another, have a potential readiness to migrate from the region of residence (the share of respondents who admitted for other reasons, leaving their region permanently or for a long period) and only 24.8% do not consider this option under any circumstances. At the same time, among young people 18-29 years old, the potential for migration readiness is 90.7%, and among respondents 30-39 years old – 71.89% (Fig.1) [20].

Fig. 1. The responses' overview on the question "Under what circumstances are you ready to move from your region permanently, or for a long time (from 4-5 years)?", %

And although the attitudes towards territorial mobility are somewhat more pronounced among residents of large cities, in the assessments of rural residents identified in 2019 during focus groups, the topic of the absence or insufficiency of self-realization mechanisms emerges more clearly: "Of course, the city has more chances to achieve success in life. More opportunities to find yourself and in general... Find a more prestigious job and study, of course. Actually, we have nowhere to get education" (woman, 42 years old, village, Voronezh region); "You can't do it here... I think, in the city" (woman, 69 years old, Voronezh region); "Of course, in the city. You can already come to the village when you are already a professional in some area, for example, in the agricultural one. Then you can come and organize your business or connect to what is already there. But there is nothing to get here yet" (woman, 72 years old, village, Voronezh region); "If my child stays in the village, he will not be able to achieve what I would like, what he would have achieved" (woman, 30 years old, village, Belgorod region); "Well, how can you achieve much in a village with six yards? You need to study, get an education, go to a big city" (man, 50 years old, village, Belgorod region).

4 Conclusion

Thus, the rural areas of the Central Black Earth Economic Region are characterized by a heterogeneous demographic potential. Nevertheless, in most of them (Kursk, Lipetsk and Tambov regions), the processes of a decrease in the demographic potential are recorded, acquiring the character of an obvious degradation in relation to the Kursk and Tambov regions. A certain exception is the Belgorod region, where the proportion of the population of rural areas of reproductive age is growing, but the number of children is decreasing and, although its rates are somewhat slowing down, the natural decline of the rural population remains. At the same time, in the rural territories of the Kursk and Tambov regions, the migration outflow aggravates the process of natural population decline, and in the Voronezh, Lipetsk and, possibly, Belgorod regions, the migration increase only slightly compensates for it. At the same time, the migration attractiveness of rural areas (with the exception of suburbs), according to opinion polls, has rather negative values.

Acknowledgments

The article was carried out within the framework of the state task of the national research Belgorod State University for 2020, project No. 0624-2020-0027 «Socio-demographic determinants of rural development in the Central Chernozem economic region».

References

1. Rosstatб Number and migration of the population of the Russian Federation in 2019б https://gks.ru/
2. O. L. Rybakovsky, O. A. Tayunova, Population, **2**, 19-20 (2019)
3. Ja. Biegańska, Bulletin of Geography. Socio-economic Series, **20**, 7-22 (2013)
4. L. Niemets, N. Guseva, K. Segida, Economic Annals-XXI, **3-4(1)**, 93-96 (2015)
5. D. Rama, Ye. Mejova, M. Tizzoni, K. Kalimeri, I. Webwr, WWW '20: Proceedings of The Web Conference 2020, April, 327–338 (2020)
6. L. L. Rybakovsky, N. I. Kozhevnikova, *Assessment of the possible and necessary scale of attracting immigrants to Russia* (2010)
7. V. M. Medkov, Demography (2004)
8. A. I. Kuzmin, Problems of modeling social processes: Russia and the countries of the Asia-Pacific region: materials of the Second All-Russian scientific-practical conference with international participation, Vladivostok, December 7-8, 183-187 (Vladivostok, 2016)
9. M. L. Yashina, N. R. Aleksandrova, International Research Journal, **5-3**, 86-91 (2021)
10. D. A. Tuzov, Economics, Statistics and Informatics, **1**, 121-126 (2010)
11. Russian statistical yearbook, https://rosstat.gov.ru/ (2020)
12. Statistical Yearbook. Belgorod region (Belgorod, 2016)
13. Statistical Yearbook. Kursk Region (Kursk, 2018)
14. Statistical Yearbook. Kursk Region (Kursk, 2020)
15. Statistical Yearbook. Belgorod region (Belgorod, 2020)

16. Statistical Yearbook. Voronezh region (Voronezh, 2020)
17. Statistical yearbook. Lipetsk region (Lipetsk, 2020)
18. Statistical Yearbook. Tambov region (Tambov, 2020)
19. Statistical Yearbook. Tambov region (Tambov, 2015)
20. E. V. Reutov, M. N. Reutova, I. V. Shavyrina, Sociological Research, **6(46)**, 61-71 (2020)

Sociocultural anthropoecology as an element of human resources management policy in the context of sustainable development.

Larisa Konstantinovna Kruglova*

FSBEI HE "Admiral Makarov State University of Maritime and Inland Shipping", Saint-Petersburg, Russia

Abstract. The article examines the heuristic and methodological potential of socio-cultural anthropoecology as an element of human resources management policy in the context of sustainable development. The author defines socio-cultural anthropology ecology as the science of the laws of structure, functioning and development of the ecosystem "man - nature - society - culture". The article shows that the development of socio-cultural anthropoecology as a scientific discipline makes it possible to get new knowledge in such spheres as general principles of human interaction with the living environment and its individual components; favorable and unfavorable living conditions (natural, social, cultural) typology of personality, the criterion of which construction is the nature of the interaction of the individual with the environment of life. The article concludes that the institutionalization of socio-cultural anthropoecology and its development as a scientific and educational discipline are necessary for strengthen of theoretical base and increasing of effectiveness of human resource management policy in the context of sustainable development

1 Introduction

Solving the problem of sustainable development is of crucial importance for humanity. The legitimacy of this kind of reasoning has its grounds and follows from the understanding of sustainable development as such a development of mankind in which "meeting the needs of the present does not undermine the ability of future generations to meet their own needs" [1]. Accordingly, a negative result in solving the problem of sustainable development, i.e. the exhaustion of the possibilities of satisfying the needs of future generations, in reality, will appear not as some distressing failure, the lessons of which can be taken into account in the future, but as the death of humanity, as a sentence carried out.

The main difficulty and at the same time the main opportunity for a successful solution to the problem of sustainable development is the most powerful and at the same time inexhaustible resource that can be used for these purposes is human resources, i.e. the salvation of man as a species depends entirely on the man himself.

*Corresponding author: lkkruglova@gmail.com

It is gratifying to note that this idea is slowly but steadily acquiring the status of the dominant one in the modern public consciousness. She received a convincing argument in the work of J. Simon "Inexhaustible resource", where the author confidently declares that "the primordial and inexhaustible resources of the earth are the people who are skillful, knowledgeable, full of animation and hope, who strain their strength and imagination to make life better not only for themselves, but also for society as a whole. " [2, p.28]

It is characteristic that the idea of the need for a "human revolution" to solve environmental problems was presented by Aurelio Peccei as the main result of his activities and the activities of the Club of Rome, he was one of the founders whereof [3]. It is quite significant that in the anniversary report of the Club of Rome, timed to coincide with its fiftieth anniversary, the concept of "New Enlightenment" occupies a central place, ie. the problem of a radical change in the spiritual component of what is designated by the concept of "human resource" [4, 5].

The decisive importance of anthropological problems for determining the ways of development of earthly civilization is expressively emphasized in the title of the theme of the XXIV World Philosophical Congress - "Learning to be human" (Beijing, August 13-20, 2018 [5-8].

At the same time, it should be noted that the idea of the inexhaustibility of the human resource is opposed by the idea of an anthropological crisis as one of the global problems of our time [9]. However, it seems that the opposition of these ideas not only does not deny the importance of human resources for determining the fate of mankind, but, on the contrary, tragically emphasizes, sets off, reveals this meaning, since we are talking about the possible exhaustion of the main resource that a person has in resolving the issue " to be or not to be".

But as soon as this is so, and the main life-supporting resource of mankind is contained in itself, then inevitably the problem of managing this resource arises. "The most common management functions are: maintenance and optimization of system characteristics, a conscious impact on internal and external (in relation to the system) processes [10].

The concept of "management" is now accepted to associate with the concept of "management policy". It should be noted that traditionally the concept of politics was used mainly in relation to the sphere of interaction of various social groups and states. At the same time, it was interpreted and is being interpreted in such a different way that, for example, in the publication "New Philosophical Encyclopedia" on the word "politics" there are three articles by different authors (G.Yu. Semygin, A.P. Ogurtsov, T.A. Alekseeva), and in each of them the author's position is supplemented by a broad overview and analysis of other approaches to the definition of policy. Using the presented in these articles and, in particular, in the article by G.Yu. Semigin [11], materials, and taking into account the current trend of an expanded understanding of the phenomenon of politics, it can be stated that the policy of human resource management includes the following points:

- determination of the goals and directions of human resource development in accordance with the interests of subjects of different levels (humanity, country, region, corporation, individual),
- the choice of means to achieve the set goals,
- organization and implementation of practical actions to achieve the set goals.

It is quite obvious that the successful implementation of human resource management policies in the context of sustainable development, which, as shown earlier, is of crucial importance for humanity as a whole, and for an individual human community, and for individual organizations, is possible only on a solid scientific basis.

The result of realizing this need can be considered the formation and development of such a scientific and educational discipline as "Human Resource Management". However, as it turns out, this discipline is currently strictly localized to the sphere of economic sciences, as evidenced, for example, by the composition of the team of authors of a very solid textbook for students of higher educational institutions "Human Resource Management [12]. It was prepared by the staff of the Department of Labor Economics and Labor Resources of the St. Petersburg State University of Economics and members of the editorial board of the journal Creative Economy. The team of authors (20 people) includes only doctors and candidates of economic sciences. To this we can add that the textbook is intended only for students of economic specialties.

Meanwhile, it is quite obvious that the scientific basis of human resource management policy, in view of the complexity that the human phenomenon is, should include knowledge in the field of philosophy, anthropology in all its varieties, sociology, cultural studies, psychology and other humanities. The diversity of sciences and disciplines related to the field of human studies suggests that some kind of integrative discipline is needed that would ensure the meaningful unity of this diversity and would contribute to the translation of the language of the human sciences into the language of practice.

It seems that sociocultural anthropoecology can act as such an integrative, and at the same time innovative discipline. The starting point of its construction is a broader understanding of ecology, proceeding from the initial idea of it as a science of the interaction of organisms with the environment, which is directly related to the etymological meaning of the word - eikos - house. Here it is important to emphasize the meaning of both meaning-forming words - interaction and environment (home, eikos). At the same time, if for an animal it is the environment, the home is nature, then for a person the environment and the home is not only nature, but also the society and culture created by it. Based on this, we can conclude that sociocultural anthropoecology is designed to study the principles and laws of human interaction with the environment of his life, including nature, society and culture.

In accordance with the above-mentioned, the purpose of the presented article is to identify the heuristic, methodological and practical potential of sociocultural anthropoecology as an element of human resource management policy in the context of sustainable development.

The tasks, the solution of which is necessary to achieve this goal, are:

- to define the subject of sociocultural anthropoecology;

- to identify the main points of the content of sociocultural anthropoecology as a scientific and academic discipline and the possibility of its practical use as an element of human resource management policy in the context of sustainable development.

2 Materials and Methods

The content of the presented article is based on the author's works on the innovative discipline "Sociocultural Anthropoecology ", which does not yet have a corresponding place in scientific discourse, educational process and practical activity [13-15].

In addition, the work used the works of domestic and foreign scientists on ecology [16-20], social ecology [21], ecological history [22], human ecology [23,24], ecological aesthetics [29,30], cultural ecology [26-28], human resource management [12], materials of the international commission on the environment, reports of the Club of Rome [1,4], materials of the XXIV World Congress of Philosophy [5-8].

The solution of the problems posed in this work required the use of a number of methods of theoretical research. Among them:

- holistic, the need for which is due to the fact that the subject of sociocultural anthropoecology is the multicomponent integrity "man-nature-society-culture".

- systemic - allowing to consider a person, nature, society and culture as a single ecosystem, to determine the type of this system, its systemic (emergent) properties, the laws of its structure, functioning and development.
- hermeneutic, which is associated with the holistic method by the relationship of complementarity, since it allows to solve the problem of the hermeneutic circle, which is that the whole cannot be cognized without knowing the individual parts, and the individual parts cannot be cognized without knowing the whole; from the standpoint of hermeneutics, this circle can be broken with the help of intuition, imagination, thanks to which it becomes possible to "see" the whole image of the object under study.
- synergistic - allowing to determine the nature and specifics of self-organization processes in the ecosystem "man-nature-society-culture" [30].

In addition to the methods of the theoretical level of research, the work contains references to the results of empirical research, which are found in the above-mentioned works on ecology, social ecology, and human ecology.

3 Results and discussion

Currently, environmental knowledge is represented by a number of sciences and disciplines. This is, first of all, ecology as a biological science about the relationship of living organisms and their habitat. The development of this science, the founder of which is the German zoologist Ernst Haeckel (1834-1919), led to the understanding that man and his environment are also a single ecosystem. Associated with this is the modern interpretation of ecology as a science, the subject of which is the relationship between human society and its environment. In order to distinguish it from ecology as a kind of biological science, some researchers propose to define this branch of ecological science as social ecology [22].

In addition, in the course of the development of environmental science, it became necessary to single out another special environmental discipline - human ecology.

However, it seems that both in modern ecology and in human ecology the content of the concept "environment" is unjustifiably limited.

So, V.N. Movchan sees the task of the discipline "human ecology" in focusing on two interrelated processes: adaptive changes in the human population under the influence of natural and technogenic limiting factors and changes in natural complexes experiencing anthropogenic impact [21, p. 12-13].

Another author, B.B. Prokhorov, expands the subject field of human ecology and includes in it the study of the impact on the population of not only natural, but also "... social, household, production factors, including culture, customs, religion" [24, p. 18]. However, here there is a limitation of a different kind, since we are not talking about interaction, but only about the impact on the population of social and cultural factors, that is, a person appears as a passive element of the ecosystem.

The most widespread understanding of the subject of human ecology is given, for example, by the journal Human Ecology, published in Arkhangelsk. It is intended to publish the results of scientific research in the field of medical, biological and technical sciences and, I must say, basically corresponds to the designated boundaries of the subject field of human ecology. So, in all issues of the magazine for 2021, it was possible to find only one article, the content of which shows that, in addition to natural factors, social factors of the environment also act on a person, in particular, news of terrorist acts. [25]

As for the awareness of the role of culture as an element of the human life environment, it, as expected, took place and is happening among the humanities, who actively supported the idea of "ecology of culture" put forward by Academician D.S. Likhachev. However, it must be admitted that in this area there was and still is a kind of limitation, since the ecology of culture is understood mainly as activities for the preservation of cultural heritage. [25.26]

It must be said that this is also characteristic of ecology as such, with which, in both everyday life and in scientific consciousness, the idea of environmental protection is associated, first of all.

Thus, it can be stated that not only none of the above environmental disciplines (ecology, social ecology, human ecology, cultural ecology), but all of them in total, do not cover the problem field, the boundaries of which are indicated by the posing of the question of human interaction with its environment.

First of all, it is necessary to take into account that a person's living environment, his home - eikos - is not only nature, but also society and culture. Man is incapable of living outside nature, but he is incapable of living outside society and culture to the same extent. As a result, man, nature, society and culture become integral elements of a single ecosystem.

It seems that the laws of the structure, functioning and development of this ecosystem should become the subject of study in a special ecological discipline - sociocultural anthropoecology.

Based on the ideas about the structure of the human living environment, sociocultural anthropoecology can be divided into several sections:

- social ecology - the science of the interaction of society and, therefore, a person and his environment
- social anthropoecology - the science of the interaction of man and society
- cultural anthropoecology - the science of the interaction of man and culture.

Each of the components of the human habitat - nature, society-culture - can be structured at least into three levels: mega-environment, macro-environment, micro-environment. Their specificity also presupposes the specificity of human behavior within the framework of each of them.

If we consider the system "man - nature - society - culture" from the point of view of synergetics, then one should pay attention to the fact that in this system a person, as an element of the system, has the greatest synergistic potential, since he is the bearer of the properties of all other elements of the system - natural , social and cultural.

Pursuant thereto, the primary task of sociocultural anthropoecology, as a unified scientific discipline, is to search for some general principles of human interaction with the environment.

It seems that the most important of these principles is the worldview principle of harmony, which allows a person to evaluate all elements of the environment and build their behavior in terms of the presence of two necessary features: diversity and unity. There is no variety - monotony is obtained, there is no unity - a cacophony is obtained.

The speculative principle of harmony should be supplemented with a principle that would indicate the energy source of human activity. Such, undoubtedly, is the principle of love - to oneself, to another, to one's native land, to nature, to humanity.

And, finally, the principle of harmony, which sets the angle of speculation, and the soul-emotional principle of love, apparently, need to be supplemented with the principle of a practically effective relationship to the environment, that is, the principle of creative action. Its meaning is expressed in the wisdom that Voltaire's Candide came to after his long and painful wanderings: "We will cultivate our garden."

All these principles are deeply rooted in the history of culture and have already found their place not only in philosophical, but also in scientific discourse [14,15].

The basic principles of human interaction with the environment constitute the core of the methodological base of sociocultural anthropoecology. On this basis, using a complex of knowledge about man, nature, society and culture, it becomes possible to solve one of the most important tasks of sociocultural anthropoecology as a scientific and educational discipline - the formation of a body of knowledge about favorable and unfavorable conditions of the natural and sociocultural environment. It should be noted that at present we have sufficiently large data on the favorable and unfavorable conditions of the natural environment for human life, the social and cultural environment in this regard is still poorly studied. Meanwhile, both at the level of scientific consciousness and at the level of common sense, there is a clear understanding that, for example, an aggressive social or cultural environment is just as destructive for a person as an aggressive natural environment. In this regard, one cannot but recall the well-known statement "the environment has stuck". And, on the contrary, a favorable socio-cultural environment is as healing for a person as fresh air, clean water, and the light of the sun.

It is quite obvious that the formation of a block of knowledge about favorable and unfavorable conditions not only of the natural, but also of the socio-cultural environment of a person, considered at the mega - macro - and micro levels, is of particular importance for the development and implementation of human resource management policies in the context of sustainable development, since only on this basis it is possible to determine indicators of the state of the environment and create an appropriate regulatory framework. At present, such indicators have been developed only for the natural and man-made component of the human environment. At the same time, the technogenic component means basically the same natural one, that is, the state of water, air, etc. - only experienced by a person. As for the social and cultural components of the human life environment, here we have to admit that there is no clear system of indicators regarding them, and the existing laws and regulations, as a rule, are not implemented. For example, the Urban Development Complex of Russia in 2004. In the first paragraph of his first article "Principles of sustainable development of the territory", as a regulatory requirement, it is stated that "in new settlements or quarters, a humane number of storeys of residential buildings (no higher than 5 floors) is being created" [32]. Instead of fulfilling this requirement, we see in megalopolises the construction of so-called "human beings", which, both in size and in architecture, have nothing to do with anthropomorphism and humanity.

This problem is of global importance. In particular, it was discussed at the XXIV World Philosophical Congress. So, at the section "Environmental Philosophy" they discussed the destructive influence of high-rise architecture on people, the need to take into account the principles of "cooperation with the sun", "respect for the inhabitant", "respect for the place" in urban planning [6, p.18].

It is quite obvious that these and other principles, united by the concept of "humanism", should form the basis for the development of indicators of the state of the human environment at all levels - from the global one to the level of a separate organization, a separate settlement, and finally, a separate house.

Meanwhile, from the analysis of the content of training courses in the university discipline "Human Resource Management" we can conclude that they do not pose the problem of favorable and unfavorable conditions in the socio-cultural environment and, accordingly, are not solved. One of the reasons for this is the obsession of the authors of these courses on economic issues and underestimation, not to say neglect, of other social and humanitarian sciences. Suffice it to say that, for example, in the "Conclusion" of the already mentioned textbook "Human Resource Management", under the editorship of I.A. Maksimtsev, N.A. Gorelov [12] named "the formation of new economic thinking" as the main goal of human resource management.

The implementation of the methodological and heuristic potential of sociocultural anthropoecology allows us to solve another problem that is directly related to the policy of human resource management. This is the problem of constructing a personality typology [32]. In the context of sociocultural anthroecology, the main criterion for typologization is the nature of the person's interaction with all the components of his living environment - nature, society and culture. This criterion, first of all, makes it possible to single out the main types of personality, the vectors of interaction of which with the environment are directly opposite. They are "creator", "constructor" - on the one hand, and "predator", "destroyer" - on the other. In addition, within the framework of this typology, intermediate personality types can be distinguished: among the positive ones - "doctor", "orderly", among the negative ones - "slob." environment - natural, social and cultural. A “slob” is a type of personality whose activity is not directly aimed at destroying the environment, but leads to its clogging and, accordingly, degradation.

The use of this kind of typology in human resource management policy can be expressed in the form of the development and implementation of programs for encouraging and stimulating individuals of a positive series and neutralizing the undesirable consequences of actions of individuals of a negative series.

The possibilities of realizing the heuristic and methodological potential of sociocultural anthropoecology, as a scientific and academic discipline, are systematically presented in a number of works by the author of this article [13-15]. However, the ideas of these works and numerous articles published in various collections have not yet evoked either positive or critical comments, that is, they have remained unnoticed. In accordance with this, it can be stated that sociocultural anthropoecology has not yet developed as a scientific and academic discipline, but does not even exist at the level of discussions.

However, given the powerful heuristic, methodological and, most importantly, the practical potential of sociocultural anthropoecology as a scientific and academic discipline, there is every reason to believe that its institutionalization, development and use in practice, in particular, in the development and implementation of human resource management policies, this is a matter of the near future.

4 Conclusions

1. Sociocultural anthropoecology as a science about the laws of the structure, functioning and development of the ecosystem "man-nature-society-culture" has an innovative character, since on its basis it is possible to expand the volume of knowledge about man and the ways of his interaction with the environment.

2. As an innovative discipline, sociocultural anthropoecology can perform an integrative function in relation to the entire body of environmental, socio-economic and humanitarian sciences, which are currently part of the theoretical framework of human resource management policy.

3. Institutionalization and development of socio-cultural anthropoecology as an educational and scientific discipline is necessary to strengthen the theoretical base and increase the effectiveness of human resource management policy in the context of sustainable development at all levels of its development and implementation: from the global to the level of an individual organization and settlement.

References

1. S.A. Evteeva and R.A. Pereleta, Report of the International Commission on Environment and Development (ICED), translated from English 376 (1989)
2. J. Simon, *An inexhaustible resource,* 797 (2009) 797 p.
3. A. Peccei, *Human qualities,* 302 (1980)
4. Come On! Capitalism, myopia, population and planet destruction, https://www.researchgate.net

5. D. I. Dubrovsky, Problems of Philosophy, **12,** 77 (2018)
6. A. N. Chumakov, A. D. Korolev, Problems of Philosophy, **3,** 15 (2019)
7. M. T. Stepanyants, Problems of Philosophy, **3,** 22 (2019)
8. B. I.Pruzhinin, T. G. Shchedrina, Problems of Philosophy, **3,** 33 (2019)
9. D. I. Dubrovsky and S. M. Klimova, *Global Future: Anthropological Crisis, Convergent Technologies, Transhumanist Projects* (2014),
10. I. A. Vasilenko Management, New Philosophical Encyclopedia, **4,** 144 (2010)
11. G. Yu. Semigin, Politics, New Philosophical Encyclopedia, **3,** 269 (2010)
12. I. A. Maksimtsev, N. A. Gorelov, *Human resource management,* 467 (2019)
13. L. K. Kruglova, *Sociocultural anthropoecology,* 75 (2000)
14. L. K. Kruglova, Man-Nature-Society-Culture: Sociocultural Anthropoecology, 120 (2013)
15. L. K. Kruglova, SELECTED WORKS. Anthropological principle in cultural studies: theory and practice. Section II, Sociocultural Anthropoecology, 39 (2018)
16. T. A. Akimova, V. V. Khaskin, Ecology, 495 (2007)
17. K. S. Losev, *Myths and misconceptions in economics*, 224 (2011)
18. A. A. Muzalevsky, *Ecology. Tutorial,* 664 (2008)
19. A. A. Popov, *Ecology of the era of globalization of nature management*, 600 (2009)
20. A. F. Bugaev, *Global Ecology: Conceptual Foundations,* 496 (2010)
21. E. V. Girusov, Fundamentals of Social Ecology, 168 (1998)
22. *Man and nature: ecological history,* 349 (2008_
23. V. N. Movchan, *Human ecology,* 292 (2006)
24. B. B. Prokhorov, Human ecology, 320 (2007)
25. O. A. Plyushkevich, R. V. Ivanov, I. A. Zhuravleva, A. V. Zavyalov, Human Ecology, **8,** 42 (2021)
26. D. S. Likhachev, *Selected works on Russian and world culture. 2nd ed,* 485 (2015)
27. D. S. Likhachev, *Selected works on Russian and world culture. 2nd ed,* 21 (2015)
28. E. N. Ustyugova, Bulletin of St. Petersburg University, **3**, 64
29. V. V. Prozersky, Bulletin of St. Petersburg University, **3,** 22
30. T. A. Akindinova, Bulletin of St. Petersburg University, **3,** 58
31. L. K. Kruglova, *Man and culture,* 60 (2017)
32. Urban Planning Code of the Russian Federation of December 29, 2004 N 190-FZ (2021) http://www.consultant.ru
33. L. K. Kruglova, SELECTED WORKS. Anthropological principle in cultural studies: theory and practice. Section II, Sociocultural Anthropoecology, 152 (2018)

Institutional investors in Russia: problems and prospects

Valery Salygin[1], *Irina* Yarygina[1,*], and *Vladimir* Smirnov[2]

[1]Moscow State Institute of International Relations, Moscow, Russia
[2]Financial University under the Government of the Russian federation, Moscow, Russia

Abstract. The objective reasons for assessing the role of institutional investors as valuable participants of the Russian financial market and forecast trends within the framework of various projects is explained by the necessity of optimizing their participation in serving the needs of national and international economy as well as using the tools of environmental policy promotion. The subject of the research is the activity of institutional investors in contemporary environment and possible ways of making them as financial intermediaries more effective in providing the needs of the modern economy, especially in the fuel and energy complex as a key sector of the Russian Economy. The article provides a comparative analysis of the current set of factors and threats that affect the prospects of participation of institutional investors in financial markets. The complex analysis methodology, used in assessing the problems, involves considering the standpoint of the influence of a set of factors, subdivided into internal and external. Internal factors include the national market conditions and environmental issues. External factors reflect the long-term impact of a potentially adverse or favorable impact on the course of projects implementation based on institutional investors' facility in spite of market turbulence, political sanctions or pandemic conditions.

1 Introduction

Modern institutional investors are defined as special financial institutions that manage savings in the interests of private investors, maintain an acceptable level of risk, with the intent of obtaining the maximum investment return, backed by the terms and conditions of an agreement.

Institutional investors comprise collective investment institutions, pension funds, and insurance companies.

Institutional investors participate in the process of direct (on their name) and indirect (as agents) financing and perform the following economic functions:

- they are responsible for the efficient distribution of financial resources among economic entities;
- accumulate savings of the population and reduce transaction costs;
- diversify risks and participate in the environment protection.

* Corresponding author: jiz4@yandex.ru

The term "institutional investors" is interpreted in the Russian economic literature relatively broadly, and includes both non-deposit and deposit intermediaries. Some authors limit themselves in listing most financial intermediaries as institutional investors. At the same time experts of the Bank for International settlements define insurance companies, pension funds and various collective investment schemes as institutional investors. In a number of studies of Russian economists the institutional investors are considered as non-bank financial intermediaries who invest funds, accumulated in course of their activity, in financial market instruments in order to gain profit.

We believe that retail banks cannot be classified as classic institutional investors, since their activity is related to attracting deposits and placing them in loans, while institutional investors are mostly focused on long-term investments.

It is worthwhile mentioning that the term "collective investment" is also widely used in the Russian Federation and related to non-state pension funds and specialized companies. Sometimes, so called collective investors are involved in raising funds from a large number of individuals for their subsequent investment as institutional investors.

Thus, to define prospects and problems of institutional investors, it is important to clear up the issue of identity of the institutional investor and the institution of collective investment

2 Methodology

The research used the methods of comparative analysis, content analysis and complex analysis. The method of comparative analysis is used to determine the influence of institutional investors on contemporary economy and identify the role of institutional investors as participants of financial markets.

The content analysis method made it possible to assess the role of institutional investors as participants of the financial market and forecast trends of developments within the framework of implementing various projects. In general, the collected data allows us to understand the trends of economic and financial situation and form evaluative judgment about the problems and prospects for institutional investors (II) activity in contemporary environment. The complex analysis method is used to systematize a set of factors and threats that affect the prospects of II participation in financial markets. The complex analysis methodology involves considering the object of assessment from the standpoint of the influence on it of a set of factors, subdivided into internal and external. Internal environment factors include the national market conditions. External factors reflect the long-term impact of a potentially adverse or favorable impact on the course of projects implementation on the basis of institutional investors facility in spite of market turbulence, political sanctions or pandemic conditions.

3 Results and discussion

International practice often use the term "asset management industry". It is known that the subject of asset management can be property owners as well as professional managers. Moreover, the management object is private and clients funds. Thus, institutional investors as financial intermediaries, operate in the asset management industry. As a result, we can distinguish the following essential features of institutional investors:

- accumulation of funds of individual and corporate investors;
- accumulated resources management as a single portfolio;
- long-term portfolio investments in securities;

Taking into consideration the mentioned features we come to conclusion, that insurance companies, pension funds (PF) and collective investment institutions (collective investments), endowment funds can be defined as Institutional investors.

Actually, we define the trends of slowing down the process of globalization of financial markets and growing cyber activity of the market.

Globalization was the key factor of the financial markets development for a definite period. Actually, we confess the new stage of its development – economic partnership of globalized environment, manifested in integration of financial instruments, market participants, regulatory authorities, securities trading mechanisms, and unification of rules and regulations.

Having a wide range of investment options, institutional investors, actually, do not limit themselves to financial assets available on domestic markets. The increasing efficiency of information technologies and the easing of regulatory restrictions have contributed to their global activity. The trend towards formation of a global investment environment and integration of national capital markets to the global context has accelerated investments in foreign assets and usually purchase securities, depositary receipts, shares of investment funds, derivative financial instruments.

It is worthwhile mentioning that PF operate in the national and international markets and invest up to 30% of their assets in international instruments and markets. Thus, we come to conclusion, that contemporary stage of market development accelerates the cross-border supply of services by institutional investors and gives way to life insurers and pension funds to attract funds and contribute to the sustainable economic development on the national and supranational level.

Actually, we consider, that the main trend manifest itself in developing the resource base of green projects requests improvement of production and environment protection base. For example, the International Energy Agency (IEA) estimates that prevention of major climate changes by reducing carbon emission, additional $500 billion is required to be invested annually, in addition to the $10.5 trillion of investments in 2010-2030. At the same time BP experts consider, that in the nearest future approximately two thousand hydroelectric power plants will operate worldwide. It is important for strategic development and planning.

Actually, institutional investors have accumulated a significant amount of capital and the variety of instruments, well known to financial professionals that can help to smooth out the consequences of a significant slowdown in global economic growth. Acceleration of II activity will be a positive trend that will contribute to sustainable development of the global economy.

Due to the flexibility and mobility (along with huge assets and other advantages), institutional investors play a significant role in the global financial architecture, that is confirmed by their cooperation with supranational structures, changes in the composition and direction of global capital flows and environmental policy measures which change the role of institutional investors in global economy, turning them into systemic financial institutions.

Analysis has proved that institutional investors control more than 80 percent of the US stock market capitalization. At the same time, private investors, implementing a policy of strategic investment of company assets, control only 10 percent of the UK market. The consequence of such kind of II participation in the market is a trend of their significant control over business and its financing.

We consider it both positive and negative. On the one hand, institutional investors provide companies with broad access to financing, which plays a huge role in modern economy. On the other hand, there is a trend of market monopolization and formation of unofficial cartel agreements due to operational control of different players in the same segment of the market by one group of investors.

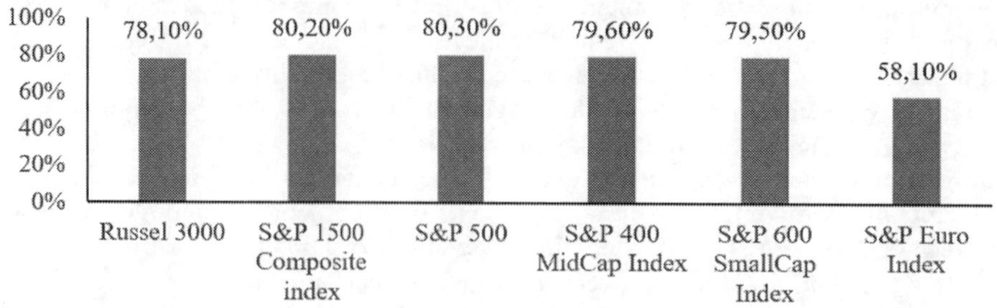

Fig. 1. Institutional investors ownership of the largest US stock indexes, Source: 80% of equity market cap held by institutions / / Pensions and Investment Magazine, 2019

It should also be noted that according to Bloomberg, different kind of institutions own more than 80 percent of the shares of the ten largest US companies.

Actually, investment funds control assets totaling more than $ 18 trillion, which on average is about 24% of the GDP of the organization's member countries.

Pension funds and insurance companies control assets with a total value of 21.4 and 7 trillion US dollars (37% and 15% of GDP, respectively).

Table1. Investment funds assets in relation to GDP.

Volume of assets in relation to GDP	2009	2019
Austria	44%	45%
Belgium	24%	38%
Canada	43%	103%
Chie	0%	0%
Czeck Republic	3%	8%
Danmark	0%	110%
Estoni	3%	4%
Finland	23%	54%
France	69%	73%
USA	69%	116%
Russia	1%	3%

Calculated based on OECD data // https://stats.oecd.org/

Table 1 indicates the very little involvement of institutional investors in the economy of the Russian Federation. It seems appropriate to expand activity of pension funds, endowment funds and other specialized financial institutions - players in the financial market in order to use resources in solving strategic tasks.

As can be seen from the table above, the share of investment funds during 2009 - 2019 significantly increased in many countries. The share of their assets greatly exceeds the GDP of some countries. Thus, we can conclude that the financial sector prevails in the economies of these countries.

The best value of this indicator is observed in the United States, which is due to the level of development of the stock market and the well-established tradition of personal savings for individuals.

The most popular investment tools for institutional investors are shares and units of various investment funds. It is important to note that in the period of 2009 - 2019, the volume of assets placed in the mentioned tools exceeded the share of debt instruments such as corporate bonds and Federal loan bonds. Another important trend is the refusal of investors to increase placement of their funds with banks (deposits), caused by the difference of the interest rates of the banks and global financial market.

Table 2. Value of assets of institutional investors in million US dollars.

Asset value	2009	2019
Deposits	4 181 116	4 450 829
Debt instruments	22 329 853	23 234 58
Loans	3 378 084	3 375 834
Shares and units of investment funds	21 289 707	24 863143

Calculated based on OECD data // https://stats.oecd.org/

The most attractive areas of investments is the infrastructure development. The most popular type of projects are the following.

Roads. The volume of investments amounted to more than 10 billion US dollars. The main forms of participation were concession agreements and investments through market infrastructure funds.

Airports. The volume of investments is about 1 billion US dollars with participation through infrastructure funds.

Gas transportation infrastructure in the United States, Mexico, and Australia. The volume of investments amounted to more than 20 billion US dollars. The main form of participation is investment through market infrastructure funds.

Investment in energy assets. The total volume of this category is about 1.5 billion US dollars. Participation is carried out by using debt financing, direct investments, as well as participation in market infrastructure funds.

Actually, we can distinguish the following classification of mutual funds in infrastructure financial market, depending on the structure (accumulation) of their assets, such as money market funds, bond funds, equity funds, mixed investment funds private equity funds, high-risk (venture) investment funds, fund of funds, rental funds, estate funds, mortgage funds, index funds (with index indication), credit funds, foundations of the commodity markets, hedge funds, foundations of artistic values, environment protection funds.

The UN set benchmarks in 2015 for achieving the 2030 Sustainable Development Goals, stating that they must "meet the requirements of the present without compromising the ability of future generations to meet their needs. For this, it is important to reconcile three pillars: economic growth, social inclusion and environmental protection. These elements are interrelated and critical to the well-being of individuals and society." [9].

Deterioration of the environment directly affects the quality of life of the population. Therefore, consumers began to vote with money in favor of goods that are produced in accordance with the principles of sustainable development. A March 2020 poll by the Capgemini Research Institute found that 79% of consumers change their shopping habits to reflect the impact of their purchases on social or environmental issues [1].

This kind of demand is converted into indicators that are significant for companies. Standard & Poor's notes that from January 2010 to December 2019, the average monthly growth in the shares of companies with the lowest carbon dioxide emissions and accounting for 20% of the total number of companies in the MSCI World Index (developed markets) was 25 basis points higher than the 20% of companies in this Index with the worst such emission [2]. Bank of America Merrill Lynch found that firms with better ESG (environmental, social and governance) performance than their peers were more likely to turn into high-quality stocks, less likely to have significant price declines, and less likely to go bankrupt [3]. In addition, socially responsible companies have the opportunity to raise borrowed funds from banks at lower interest rates compared to companies with similar financial reliability that do not take ESG risks into account in their activities [4]. The cost of capital for companies that adhere to ESG principles is about 10% lower than for other companies, due to the fact that investors believe that they have lower risks in terms of non-compliance with environmental legislation [5].

The potential benefits of investing in projects implementing environmental protection programs are clear. Therefore, the analysis of possible risks associated with environmental pollution is very important, especially for large investment projects that provide for long payback periods and are typical for institutional investors, including portfolio investments of pension funds, national wealth funds, etc.

The topic of climate risks for portfolio and institutional investors, including sovereign funds, has become relevant only in the last 10-15 years. Moreover, in different parts of the world they are perceived differently. If earlier the main topic of the pre-project study was the financial analysis of the investment object, then in 2019 in Europe just over 50% of investments were made taking into account the factors of sustainable development, and in the USA the share of such investments in stocks was about 25% [6].

Currently, the only generally accepted quantitative objectively verifiable indicator of environmental risk is the price of a ton of carbon dioxide emitted by an enterprise. Wherein, investors take into account in the financial analysis the possible costs of greenhouse emissions, regardless of whether such costs are incurred at the national level or not, because the regulatory rules can change quickly. When the project's products are exported, it is very likely that it will be additionally charged with a carbon tax, as already planned in the EU. The market estimates the cost of environmental pollution very high: for example, in the EU, where CO_2 trading has been developing since 2005, the price per metric ton of CO_2 in the European Emissions Trading System was above 60 euros at the beginning of September 2021 [10]. At the same time, hopes for taking into account the high absorptive capacity of the environment at the national level in the exporting country are not entirely justified, since the calculation of the future carbon tax in the EU will be carried out only taking into account the carbon neutrality measures of the enterprise exporting its products.

As you can see, measures to improve the climate significantly affect the efficiency of investments. On the other hand, as the Boston Consulting Group and the Global Financial Markets Association in their fundamental research on climate finance in December 2020 noted that while the need for financing climate projects around the world is 100-150 trillion US dollars, the role of institutional investors is very important: in the equity investments they provide it accounts for 35% of the total funding requirement, and in bonds - 21% [11].

High economic and political risks associated with sanctions pressure, the Russian Federation is on the path of economic development through substitution of European and American sources of financing, as well as optimization of domestic financial resources [7]. An important role in this regard is played by institutional investors who can participate effectively in the implementation of development programs.

According to our calculations, the public and private sectors invested about 5.7 trillion rubles (2019) in infrastructure development, which is 200 billion rubles (3.6%) more than a year earlier. However, insurance companies and pension funds have no experience of participating in big projects, as a result, their share in capital investments is insignificant.

Investments in infrastructure remain attractive for private funds, but their money is getting into the industry too slowly. We consider, that the main reason is the lack of projects structured under bonds.

At the end of 2019, the share of corporate bonds in the investment portfolio of national private funds (NPF) decreased by 6 percentage points, to 45% (1.2 trillion rubles), while the share of Federal loan bonds (OFZ) increased sharply to 28% (0.7 trillion rubles), from 10% in the first half of the year. In monetary terms, NPF investments in Russian OFZs more than tripled, from 235 billion rubles to 739 billion rubles. Thus we can conclude, that the activity of private investment funds should be supported by public authorities, if the invest in social or industrial sector.

VEB (Vnesheconombank of the Russian Federation) since 2003 has been a state management company for the trust management of pension savings and since 2012 VEB performs the functions of a state management company for the funds of the payment reserve.

VEB's activity is carried out in accordance with the legislation of the Russian Federation, regulatory legal acts of the President and Government of the Russian Federation, regulatory documents of the Central Bank of the Russian Federation, the Ministry of Finance of the Russian Federation, and other Federal Executive authorities. Observing the absolute priority of the interests of insured citizens, VEB invests pension savings in order to ensure their growth with a minimum level of risk.

Up to date VEB manages the funds of about 39 million future pensioners. A state-owned management company invests in government securities, international bonds, deposits, etc.

Isn't it limited? We consider infrastructure bonds to be of an interest to public authorities in contemporary economic environment.

Currently, there are about 58 non-state pension funds operating in Russia: 36 funds carry out mandatory pension insurance and 18 funds (2.6 trillion rubles).

The allocation of pension reserves and investment of pension savings are based on the following principles: •

- ensuring the safety of the funds;
- ensuring profitability, diversification and liquidity of investment portfolios;
- defining an investment strategy based on objective criteria that can be quantified;
- accounting for the reliability of securities;
- information openness of the process
- professional management of the investment process and their control.

We consider that pension funds can contribute greatly to the implementation of the environment protection policy by developing financial instruments for sustainable development.

In a contrast to developed countries pension savings legislation in Russia strictly defines the procedure of investment.

Russian clients usually choose conservative strategies, the yield of which is less volatile. More than half of investments are accumulated in Russian bonds [8]. However, over the past two years, shareholders ' appetite for risk has increased – the shares of equity and mixed investment funds assets have increased significantly due to net inflows, as well as outperforming returns from investing in shares.

A significant amount of investment by pension funds is concentrated in the Russian oil and gas sector. For example, last year two large associations of private funds were completed: in the middle of the year, three NPFs that fell within the perimeter of the Bank "FC OTKRITIE "merged on the basis of "LUKOIL-Garant" (now NPF "OTKRITIE"), and at the end of the year "Consent OPS" and "Neftegarant NGO"joined the Fund "Neftegarant". OTKRITIE NPF with assets at the beginning of this year of 565 billion rubles occupied the second place, but in the first quarter it fell to the third place (market assets decreased to 556 billion rubles), losing to Gazfond Pension savings (563.7 billion rubles). NPF Neftegarant took the ninth place (205.6 billion rubles).

The Association of private funds has also continued its activity. In the first quarter, NPF Safmar was joined by the Trust Fund. As a result, the unified Fund (264.7 billion rubles of assets) in the first three months of the year came in sixth place, ahead of NPF "Future" (256.5 billion rubles). as a result the concentration of the top 10 Russian private funds exceeded 90% of the reserves and savings market. The Russian financial market has long been living in a phase of oligopoly. At the beginning of last year, the top 10 players had more than 92.4% of all pension savings. Actually, this figure is 2.6 percentage points higher. The non-state pension provision (NGO) market is more diversified, but it is also approaching an oligopolistic one. At the beginning of the year, the ten largest players concentrated 89.5% of all pension reserves.

These examples vividly depicts volatility of the Russian financial market and the tendency of oligopolistic accumulation economic capital of institutional investors as participants of financial markets in strategic sectors of the national economy. At the same time, there is a variety of types of II and their instruments which can contribute to the sustainable development of the national economy. International experience in this regard is presented above.

4 Conclusions

The results obtained allowed us to draw the following conclusions. There are several trends for institutional investors activity in the financial market. Such as growing competition; passive portfolio management strategies with the predominance of active strategies and demand for new investment instruments to form securities portfolio.

The trend of growing competition is supported by increasing regulation requirements for the volume of funds and structure of assets, as well as the constant growth in the number of mutual funds, offering similar services to the population. At the same time the development of passive strategies is a logical result of continuation.

There are also some specific challenges for each type of institutional investors that are determined by market activity of insurance companies, non-state pension funds, and collective investment institutions. Main challenge for the insurance companies is the reduction of insurance schemes (tax optimization) and non-state pension funds activity. As one of the elements of scientific foresight, forecasting allows you to display possible scenarios for the development of institutional investors in Russia. Current market environment has proved, that the authorities are not interested in the development of domestic non-bank financial intermediaries, that prevents from transforming savings to investments.

Actually, the activity of institutional investors in the Russian financial market is insignificant. The main share of investments is accumulated by the Federal loan bonds and deposits in leading Russian banks. At the same time, based on the analysis of investment strategies and market behavior of key players, we can conclude about passive investment strategies in the Russian financial market. All of the above do not contribute to increasing the role of indirect financing.

It is important to take the following measures, which contribute to sustainable economic growth:

- improve taxation of insurance companies, requirements for the structure of the insurance company's investment portfolio and legislation by creating a unified code of responsible investments;
- promote cooperation between insurance and management companies, as well as increase information transparency of non-state investment funds.
- clarify the rights of depositors and participants in non-state investment funds while using corporate pension programs.

The above mentioned measures contribute to the resources to support national programs for sustainable economic development and environment protection.

References

1. K. Jacobs, J. Robey, K. van Beaumont, C. Lago, M. Rietra, S. Hewett, J. Buvat, N. Manchanda, S. Cherian, Capgemini Research Institute, **6** (2020)
2. R. Falk, K. Wang, T. Morris, E. Chan, S&P Global. Market Intelligence, **5** (2020)
3. S. Subramanian, M. Kabra, S. Chopra, P. Strzelinska, L. Huang, ESG Matters - Global. ESG from A to Z: a global primer. Bank of America Merrill Lynch, **5** (2019)
4. O. Kordsachia, Review of Managerial Science (2020)

5. S. Bernow, R. Nuttall, S. Brown, McKinsey, **3** (2020)
6. R. Eccles, Harvard Business Review IdeaCast, **663,** https://hbr.org
7. R. Brayley, S. Myers, *Principles of corporate Finance,* 1008 (2016)
8. E. I. Shokhin, *Financial management,* 480 (2017)
9. The Sustainable Development Agenda. What is sustainable development? United Nations. https://www.un.org
10. Daily Carbon Prices, https://ember-climate.org/data/carbon-price-viewer/
11. *Climate Finance Markets and the Real Economy. Boston Consulting Group and the Global Financial Market Association,* 178 (2020)

Scientific and educational vector of the social and human capital network component formation

Olga Yarmak[1,*], *Ekaterina* Strashko[1], *Mariya* Bolshakova[1], *Pavel* Deryugin[2], and *Veronika* Yarmak[2]

[1]Sevastopol State University, Sevastopol, Russia
[2]Saint Petersburg State University, Saint Petersburg, Russia

Abstract. The authors raise the problem of the scientific and educational vector of the formation of the network component of social and human capital in Russian regions. The development of the network component is becoming one of the policy directions for the formation of the region's human capital through the creation of certain network centres. The emerging network connections of world-class scientific and educational centres in Russian regions are analyzed, which create conditions for transforming the existing human potential of the territory into the human and social capital of the country. Based on the definition of social capital by P. Bourdieu and M. Paldam as a group resource for obtaining social connections, the authors analyze the process of forming network connections between the educational, scientific environment of the region, its business community and government bodies at the sites of world-class scientific and educational centres operating in Russia, which is the basis for the formation of the social capital of the territory. The conclusions of the study are to determine the structure of ties in the scientific and educational vector of social and human capital through the functioning of network RECs, which become interregional coordination structures for the scientific and educational space of the country.

1 Introduction

Human and social capitals are closely related to each other, since in order to possess social capital, a person must be connected with other people, and communications and social ties are the source of social capital. In this regard, in the modern world, the issue of creating consortia that unites the capabilities of various organizations and groups of people has become topical. It is consortia, both scientific and industrial, that are able to build up social and human capital and form their network component.

The network component is provided by a social mechanism that connects social ties between the internal elements of human capital and carriers of other capitals. It can be defined as the ability of an individual to form and develop social ties between collective

* Corresponding author: olga_yarmak@inbox.ru

and individual bearers of capital. The degree of integration and sustainability of human capital depends on the density of social ties, that is, it is the network component that ensures the social integration of the human capital of the territory.

The network component of human capital functions at four levels: personal, group, regional and country or federal. The integrity of human capital at all these levels is ensured by its network component, which is formed in the process of executing decisions of state policy and implementing national projects of the state.

The development of the network component of capital is becoming one of the directions of the regional human capital development policy through the creation of certain network centres.

The main resource for the formation, development, preservation of the network component of human capital is social capital, which is understood as a set of social ties created and existing in human collectives; norms and rules that govern relationships between people that ensure group cohesion and solidarity. All the listed components of social capital can participate in the formation of its network element of human capital, and, conversely, a developed network component can increase social capital [1-3].

An important priority in the development of the network component of capital is the possibility of forming a system for transforming the existing human potential into human capital. Today in Russia there is a shortage of personnel with such abilities - increasing the "profitability" of the potential by increasing it. A similar situation occurs at the regional level. Consequently, network connections should be formed aimed at creating, preserving, producing the conditions necessary for realizing the potential and converting it into capital in all territories of the country.

The system of education and science is both an environment for the accumulation of social capital and the strengthening of the unity of the nation and civil society. This, in turn, serves as a basis for the formation of public trust [4], which is the result of a person's confidence in the future, guarantees of the availability of public resources and the level of security. Thus, universities are becoming crossroads of social cooperation, forming an atmosphere of social trust around them.

The higher school promotes the development of cooperation both in the learning process and after graduation, forming human consortia and uniting people working in various spheres of the economy, and linking them outside of formal structures. This cooperation ultimately leads to an increase in the effectiveness of state power, promotion of civic engagement, smoothing out inequality, reducing the scale of social exclusion and corruption - for the benefit of society, the state and the market [5].

Such network connections are possessed by world-class scientific and educational centres, which, on the one hand, form network communications between the educational, scientific environment of the region, its business community and authorities, on the other hand, they create conditions for transforming the existing human potential of the region into human and social capital of the country.

2 Materials and Methods

Social capital determines the quality of social ties in a society. Sometimes it is considered as a characteristic of society that determines the quality of life in a particular country.

Social capital is a concept introduced by Pierre Bourdieu to denote social ties that can act as a resource for obtaining benefits. In Bourdieu's concept, social capital is an exclusively group resource [6]. The authors used this approach as the basis for the research methodology.

Indicators for measuring social capital are “proximal” - the results of the influence of social capital on certain aspects of people's lives associated with networks, trust and reciprocity in relationships, and "distal" - the results of the influence of social capital, which are not directly related to its key components. For the measurement, social capital indices are used, which are constructed on the basis of a certain system of indicators. In particular, in quantitative studies, when calculating social capital, the trust index is calculated, as well as the number of groups (associations, public organizations, etc.) and their members, which are united by the civil society development index.

The Danish social economist Martin Paldam proposed three pillars for building social capital: trust, ease of cooperation, and networking. According to game theory, social capital is the propensity for cooperative decisions in games with prisoners' dilemmas. The scientist measures social capital, focusing on social networks as the number of relationships that an individual has, and describes the process of striving for the dream of social capital, which consists in the fact that everyone is trying to capture aspects of the same phenomenon, as a result of which the hidden variables of social connections. He introduces an assessment of ease of cooperation - the number of participants in voluntary organizations (the density of voluntary organizations) and the strength of relationships within the network. This approach to measuring social capital is closely related to measuring trust: how much people trust each other, how much their desire to cooperate and enter into voluntary cooperation and consortia depends [7-9].

Russian economist Alexander Auzan, studying the level of social capital, asks the question why in some countries different groups produce a lot of public goods, while in others – few. He explains this by the costs of social communications, which are an indicator of social capital, indicating the level of mutual trust and honesty in society. The most important way to build up social capital is the growth of social ties and the level of public trust.

Auzan talks about two types of social capital. The first is called bonding, from the English word "bond" - "bond". This is a kind of enclosure within one group of people: I trust, but only my own. Another type of capital is bridging, from the English word "bridge" - "bridge". It is more conducive to economic growth, because it allows you to build connections between different groups of people. But it is very difficult and difficult to accumulate and realize bridging social capital. For the accumulation of bridging social capital, the growth of trust between groups, the harmonization of formal and informal institutions is required, the development of public trust and negotiability as not a manifestation of weakness, but a social formula [10]

The challenges facing the process of "developing", "accumulating" and "investing" human capital in Russian society are "connecting human capital" to all processes of the country's development. To do this, we need to make sure that our human capital becomes the driver of the economy, it is necessary to create its social basis - a network component and communication between students of universities and their graduates, between universities and employers.

The authors identified the main category of analysis as the network social component of the region's human capital or social capital, which depends on a number of factors: available human, natural, material, financial, innovative and other resources, resources to improve the quality of life of the population, socio-demographic indicators, migration situation, the level of health care, qualification characteristics of labour resources, indicators of labor force mobility, employment and unemployment, cultural and historical characteristics of the region, the specifics of the way of life [11].

The purpose of our study was to analyse the process of forming network connections between the educational and scientific environment of the region, its business community and authorities at the sites of world-class scientific and educational centres operating in Russian regions, which serves as the basis for the formation of the social capital of the territory.

To achieve the set goals within the framework of the research project, expert interviews were carried out with representatives of education, science, business and authorities of the Belgorod and Samara regions, where the World-class Scientific and Educational Centre "Innovative Solutions in the Agroindustrial Complex" (Belgorod) and the REC "Engineering of the Future "(Samara). Experts participating in the study are attracted to the sites of these RECs as educational, scientific members, industrial partners, and employees. The expert interview included 4 blocks: an assessment of the national project "Science and Universities", its importance for the functioning of REC, understanding of network interaction within the REC and REC site with the federal centre, REC and universities, analysis of the student's competence profile and his life trajectories, citizenship as a future employee REC.

3 Results and Discussion

The collected expert interviews of participants in world-class scientific and educational centres in Belgorod and Samara reflected the assessments of the significance of the Science and Universities national project for regional development and for the work of REC, the formation of network interaction between participants, industrial partners at the REC and REC with the federal centre.

The significance of the Science and Universities national project is assessed as a new chapter in the history of Russia's scientific breakthrough, which will increase the level of Russian developments in priority areas and increase the population's interest in science, education, and innovative development. As a result of the implementation of the national project, an interuniversity network will be formed and "stronger ties will be established between universities in different regions at all levels, and an open mechanism for regional lobbying of projects will be created."

Speaking about the significance of the national project "Science and Universities" for the Russian regions, the experts talked about "achieving a scientific breakthrough in order to create a socio-cultural environment for high-tech solutions, which will lead to the creation in Russia of not a social, but a high-tech state. a kind of " scientific and technological sovereignty ".

The RECs themselves are assessed by them “not as ready-made ecosystems, but as a kind of manual adjustment of interaction, a system of synergistic interaction between science and education”; “REC is not only about growing human capital, REC is about creating its network component - science, business, education”. At the same time, the format of interaction between REC and the federal centre should not be based on domination, but on reassembling federal processes at the local level: “there should be a reboot, through the REC, tasks that the President entrusts to the government should be broadcast, and REC performs them and provides feedback”. Thus, the REC expert community is perceived as a "mediator" between regional processes and federal requirements, and therefore the format of interaction between REC in the region and the federal centre is most effective precisely in this perspective. REC should work as a "single window" for representatives of education, science, business, which will effectively carry out federal tasks. But at the same time, the interaction between REC and the federal centre should be based on regional processes: all changes should ensure an improvement in the quality of life of most people on a certain territory, and this criterion should be the most important.

The importance of the network component is evidenced by expert assessments about the understanding of the very idea of REC and its further functioning: “it is a system of cooperation between science, education, business and government in the region; one of the REC tasks is to build successful interaction between those who are engaged in the development of innovative products and technologies, those who can introduce these developments / technologies into production and those who regulate the processes of regional development ". Considering the network characteristics of REC, experts refer to it as a "platform for experiments and case creation".

Experts unanimously expressed the opinion that REC is a factor in the formation and development of the region's human capital from the internal potential, that is, a platform for "growing" personnel. Here the youth vector is important: work with schoolchildren, students and young scientists should be carried out in order to actively involve them in science and in the activities of RECs themselves.

Representatives of business and industrial partners consider REC as "a new trend in business and production, as a basis for sustainable economic, resource and social development of the region and a factor in sustainable development of the territory." REC is perceived as "a bridge that can help to obtain new scientific developments, implement their investment and implementation."

In fact, REC is a bridge, bridge build-up of social and human capital of the region and the country as a whole: "REC is not just a concentration of intellectual resources, but an opportunity to establish network interaction"; "REC is not an institution, not an organization, it is rather a platform for the exchange of experience, ideas, contacts, it is a way of interaction, within the framework of which communication takes place on the REC platform itself between representatives of the expert community".

Experts believe that in order to achieve the set goals, REC must "create a new innovation space in regions and cities - interuniversity space, which will be built directly into the city. At the same time, the university / university should become an eco-environment, when not startups, but real companies that set trends work inside the university ". Business participants of scientific and educational centres consider them as a tool for scientific development and investment in a broad sense: "Now REC has become a tool for working with new solutions, a tool to support a caring person, business, student who wants to develop".

For development in the flagship universities of REC, it is necessary:

- design an entrepreneurial environment, which will allow researchers to focus their attention not only on scientific research, but also on the possibility of commercializing finished products;
- carry out technology transfer and successfully apply the results of scientific research in the REC network.

The entrepreneurial environment in a flagship REC institution should be expressed:

- in the number of entrepreneurial programs;
- in the number of companies with which the university has agreements on cooperation in the entrepreneurial sphere (including agreements on the creation of basic departments);
- in the number of participants in university programs aimed at the development of entrepreneurship;
- in the number of university centres in the field of engineering;
- in the development of an entrepreneurial culture and the attraction of successful entrepreneurs to the university.

The implementation of technology transfer in a flagship REC should be expressed:

- in the amount of funds for research and development, attracted from extrabudgetary sources;
- in the amount of funds received through the use of the results of intellectual activity;
- in the amount of investments attracted by operating small innovative enterprises created at the university;
- in the amount of funds received through the provision of services / performance of work for external contractors in the field of engineering.

The pivotal university REC will be able to become a full-fledged factor of the socio-economic environment of the region, provided that its needs are met, "the implementation of the results of activities in the daily life of the inhabitants of the region", an increase in the number of jobs in existing small innovative enterprises created under it.

4 Conclusion

Currently, there are 15 world-class scientific and educational centres in Russia, which include 35 regions, where half of the country's population lives. Scientific and educational centres in the regions have been operating for two years. In the first two years, 10 scientific and educational centres were created, and the first 5 of them are dynamically moving and are already showing the first results, which is expressed in the cooperation of scientific and educational organizations in these regions, the formation of a new scientific agenda, closer ties with business. The expected level of network interaction at REC sites has not yet been achieved, but its different quality is already being recorded and the start of implementation of specific technological projects.

The scientific and educational vector of the formation of the network component of social and human capital through the functioning of a world-class scientific and educational centre in the region should have two results:

1. REC should encourage talented youth to come to the region and gain a foothold there, working in the scientific and educational sector, universities, research institutes, business companies, occupying high-tech jobs; creating conditions for migration flow from region to region, from capitals to regions, from regions to capitals. Therefore, the first task of REC is to create the most comfortable conditions for the development and self-realization of talents, school and university graduates.

2. REC should participate in the socio-economic development of the region through the launch of new technological projects, that is, the specifics of its functioning should consist in creating a different quality of network interaction between the main participants in scientific and technological development - universities, research institutes, business and the region itself.

Thus, the network RECs are becoming interregional coordination structures for the scientific and educational space of the country. But at the same time, it is necessary to take into account the existing mental and scientific and technological differences between regional opportunities and federal tasks.

For the effective functioning of a world-class REC, universities should become centres of regional management: not only representatives of the scientific sphere, but also industrial partners unite around the REC. Networking at the REC site should be filled with specific projects created by consortia and agreements. Networking should not be the development of the human potential of the region, but the augmentation of its social and human capital, while a significant condition is that the region demonstrates a financial model with the participation of a business partner.

At the REC site in the region, strategic plans of universities should be implemented to increase the competitiveness of the region, improve the quality of life of its population, and the university should act as an attraction point for both human resources and technologies and innovations. Scientific organizations, government and business unite around the university, which allows transforming the socio-cultural environment of the region and forming an ecosystem around the university. This is the logic behind the world-class REC model.

Acknowledgements

The study was carried out with the support of the Russian Foundation for Basic Research within the framework of the scientific project No. 19-29-07443/20 "Scientific and educational centres as a factor in the formation of human capital in Russia: the format of creating world-class scientific and educational centres in accordance with the Presidential Decree" On national goals and strategic objectives of the development of the Russian Federation for the period up to 2024 "

References

1. J. S. Coleman, American Journal of Sociology **94**, 95-120 (1999)
2. N. Lin, Social Capital. A Theory of Social Structure and Action (2001)
3. R. Inglehart, M. Basáñez, A. Moreno, *Human Values and Beliefs: A Cross-Cultural Sourcebook: Political, Religious, Sexual, and Economic Norms in 43 Societies* (1998)
4. F. Fukuyama, *Trust. The Social Virtues and the Creation of Prosperity* (1995)
5. Yarmak Olga, Shkaiderova Tatyana, Strashko Ekaterina, Bolshakova Maria, Garas Lyudmila, Institution of higher education transformation and society's response to distance learning during the Covid 19 pandemic, E3S Web of Conferences 244, 11043 (2021)
6. P. Bourdieu, The Forms of Capital. In Richardson J. ed. Handbook of Theory and Research for the Sociology of Education (1986)
7. M. Paldam, G. T. Svendsen, *Trust, Social Capital and Economic Growth: An International Comparison* (2006)
8. M. Paldam, Social capital and social policy. Arusha Conference, "New Frontiers of Social Policy" – December 12-15 (2005)
9. M. Paldam, Journal of Economic Surveys **14(5)**, 629-654 (2008)
10. A. Auzan, *Economy of everything. How institutions define our life*, 160 (2014)
11. P. P. Deryugin, O. V. Yarmak, E. V. Strashko, E. V. Strashko, V. E. Yarmak, V. A. Glukhikh, Revista inclusions **7**, 178-197 (2020)

Humanitarian component of the proactive education model in the age of global transformations

Nadezhda Arkhipova, *Pavel* Shakarenkov, and *Vasily* Makolov*

Russian State University for the Humanities, Miusskaya sq. 6, 125993 Moscow, Russia

Abstract. The article presents and substantiates the humanitarian component of the proactive education model in the context of global transformations given the influence of modern threats and challenges (technological, biogenic, socio-cultural, etc.). In the framework of Russia's contemporary education policy, the authors present key elements of the humanitarian component of the model under survey (professional competences, ideological values, spiritual-patriotic and humanitarian-ecological components as well as digital competences and soft skills).

1 Introduction

Systemic crises and global challenges that the mankind faced at the turn of the 20th and 21st centuries (technological, biogenic and socio-cultural threats, terrorism, ideological extremism, and cyber safety) brought about a new view of the global world and understanding of the sustainable development importance which is ensured solely by the balance of three most important elements: economic growth, social responsibility and ecological balance.

Gro Harlem Brundtland's report "Our Common Future" regards sustainable development as the one that corresponds to the needs of the present and does not threaten the ability of future generations to satisfy their own needs [1].

Global socio-economic changes taking place in the world over the last few decades, including those associated with the spread of the COVID-19 infection, once again prove the importance of education in overcoming crisis developments.

Article 1 of the "World Declaration on Higher Education for the 21st Century: Vision and Action" of October 9, 1998 stresses that it is necessary to preserve, reinforce and further expand the main tasks and values of higher education, specifically – its contribution to the sustainable development and improvement of the entire society - as well as to educate for citizenship and for active participation in society, with a worldwide vision, for endogenous capacity-building and consolidation of the human rights, sustainable development, democracy and peace, in a context of justice [2]. The Declaration also confirms that the role of higher education consists in consolidating human rights, ensuring sustainable development, democracy and peace in a context of justice [3].

* Corresponding author: vmakolov@yandex.ru

The major objective of the educational system is to train personnel possessing knowledge, competencies, values and approaches for the implementation of the sustainable development concept. It is achieved by developing new curricula and modules in the field of sustainable development, introducing ideas and principles of sustainability into the educational process and extra-curricular activities with students.

In E. Korteze's opinion, in order to "achieve a sustainable future, universities should ensure awareness, develop knowledge, skills and values that make it possible to reach life goals maintaining a high level of well-being" [cit.4].

Higher education in Russia has changed significantly in the last few decades. Rapid development of informational and online technologies, expansion of opportunities for students' and teachers' mobility, an increased role of research, structural transformations in the content of curricula, and a larger spectrum and models of their implementation result in the transformation of the entire educational system.

The most important part, in our view, in ensuring sustainable development is played by humanitarian education and knowledge. Humanitarian knowledge in education has very deep roots. Ancient Greek philosopher Plato was the first to put forward an idea that education means teaching communication, understanding senses and ability to think.

The results of the study made within the project of strategic management development in Russian universities have allowed us to make a conclusion about the expressed need of the young generation in socio-humanitarian knowledge and expansion of the humanitarian outlook as the basis for the establishment of the future professional personality. The study also states the important role of socio-humanitarian knowledge as a proactive function of graduates' preparation in the epoch of crises, active changes and new challenges from the outside world [5].

2 Materials and methods

The focus on the role of education in ensuring sustainable development has been increasingly growing in recent decades. Arising issues are discussed by respected experts, politicians, economists, representatives of the world of art, science, business and education. Researchers have developed various models of educational innovations and integration of sustainable development principles into profiles of competences in the field of personnel training [6].

In M. Keiner's opinion, sustainable development is a key principle of human potential development [7]. D. Foster (2002) argues that sustainable development is a process of improving results of human activities through education [6]. H. Singh (2010) considers the main areas of activity in the field of sustainable development to be as follows: promotion and improvement of basic education; refocusing of education at all levels to solve the problems of sustainable development; growth of awareness and understanding of sustainability ideas by society; education and development of appropriate skills for real work [8].

The Association for the Advancement of Sustainability in Higher Education (AASHE) was set up in 2005 to promote education in the interests of sustainable development; at present, it unites over 900 universities in 20 countries of the world. The goal of the Association is to inspire and stimulate higher education to become the basis for a prosperous, fair and ecologically healthy world because it is universities that are conductors of change and moving forces of innovations in the field of sustainable development.

In 2015, the UN adopted the agenda of sustainable development for the period of up to 2030. The program includes 17 goals aimed at reducing poverty, preserving global resources, ensuring quality of education and improving people's well-being. The achievement of these goals requires concerted efforts of power authorities, civil society and business companies. On May 17-19, 2021, Berlin hosted the world UNESCO conference on education in the field of sustainable development, which noted that sustainable development could not be reached solely through technologies, political regulation or financial mechanisms [9].

Considering the importance and urgency of existing problems, many universities in Europe and North America began to implement Bachelor and Master programs developed to create systems of education aimed at sustainable development. Among the most conspicuous programs are the following: Leadership in the Interests of Sustainable Development" (Södertörn University, Stockholm; Queen's University Belfast, UK); "Sustainable Entrepreneurship" (University of Groningen – Campus Fryslan Groningen, the Netherlands); 'Management of Sustainability" (American University Washington DC, the USA); "Sustainable Management and Logical Innovations" (Institut Polytechnique UniLaSalle, France); "Sustainable Development and Environmental Management" (Coventry University, UK); "Policies in the Field of Sustainable Food Supply and Agricultural Development" (Royal Agricultural University, UK); 'Sustainable Development and Innovations" (Utrecht University, the Netherlands); "Leadership and Sustainable Development Guidance in International Measurements" (Academy of Leadership Sciences, Switzerland); "Creative Sustainability" (Aalto University, Finland); "Sustainable Development, Entrepreneurship and Design" (Brunel University London, UK); "Economics for Sustainable Development" (Tilburg University, the Netherlands); "Sustainability and Social Innovations" (HEC Paris France). Russian universities also have similar programs on education for the purposes of sustainable development: "International Programs of Sustainable Development" (Moscow State University of International Relations under the RF Foreign Ministry); "State Management of Sustainable Development" (Russian Academy of National Economy and Public Service); "Sustainable Development of the Hospitality and Tourism Industry" (Finance University under the RF Government); "Economics of Sustainable Development" (Russian University of Peoples' Friendship); "Sustainable Development of Business: Management and Modern Marketing" (A.N. Kosygin Russian State University); "Standardization and Sustainable Development in the Aerospace Industry" (Moscow Aviation University); "Management of Sustainable Development in Modern Corporation" (South Urals State University); "Ecological Management for Sustainable Development" (V. I. Vernadsky Crimean Federal University) and others.

Education is not only a process of acquiring knowledge; it includes all kinds of activities aimed at providing people of all ages with an opportunity to grasp sustainable development problems, professionally participate in solution thereof and undertake responsibility for the sustainable future development. Education for sustainable development relies on several principles, the most important being the principles of social centrism, fundamentality and prognostics that provide for balanced and complex development of social medium, economy, culture and ecology (National Strategy of Education for Sustainable Development in the Russian Federation).

In our opinion, it is essential to understand that education in the interests of sustainable development does not only stipulate the inclusion of new disciplines into curricula – it also implies evolvement of students' responsibility for the sustainable development of society. In reality, it means including into education and training the issues of sustainable development such as: civic-mindedness, peace, ethics, responsibility in the local and global context, safety, human rights, health, gender equality, cultural diversity, production and consumption models, environmental protection, rational use of natural resources and biological diversity (Fig. 1).

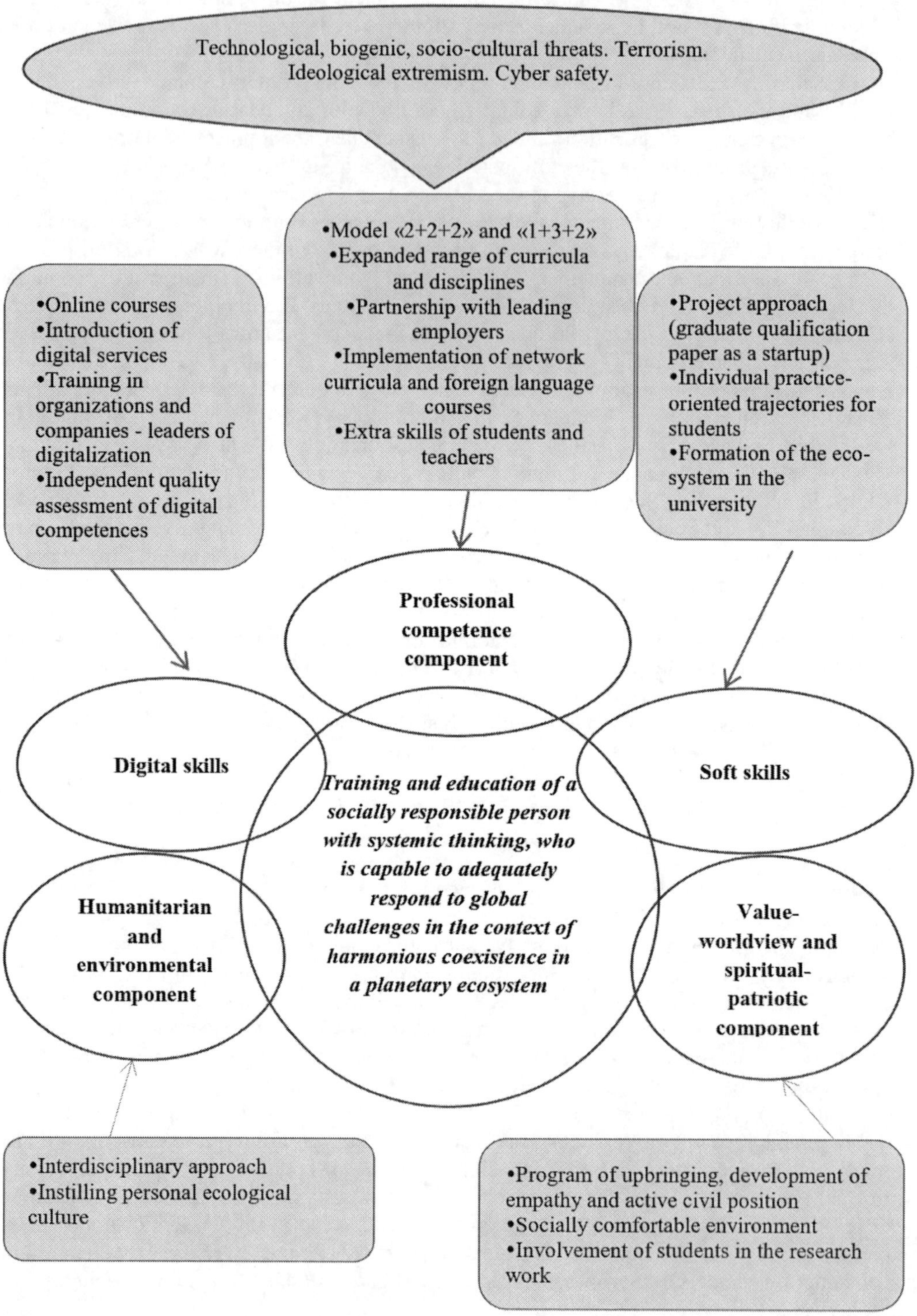

Fig. 1. Humanitarian component of the proactive education model

The successful implementation of the concept of education for sustainable development is impossible without reliance on the humanitarian component as a set of moral, ethical and esthetic norms and values that constitute the foundation for the global cultural development as well as for national and personal self-identification.

The humanitarian component of sustainable development allows for generation of an integral worldview that ensures the integrative perception by a human being of the image of the world he/she exists in. L.S. Vygotsky wrote: 'Education is similar to a tower; it should be built step by step" to help the mankind find constructive and creative solutions for present and future global problems as well as raise sustainability and vitality of society.

3 Results and Discussion

The authors present the humanitarian component of the proactive education model (Fig.1). The key elements of this model are professional competences, ideological values, spiritual-patriotic and humanitarian-ecological components, digital competences and soft skills.

The presented model is aimed at developing and educating a socially responsible personality with systemic mentality, capable of responding to global challenges in the context of harmonious coexistence in the planetary ecosystem. The humanitarian component will become the basis for the model of proactive regenerative education in the future.

Value-oriented and spiritual-patriotic components are to form the contemporary world outlook on the basis of the value-ontological and methodical value of all elements such as cultural diversity, gender equality, social justice, environmental protection, personal enhancement, etc. The educational function includes adherence to general national values: civic-mindedness, readiness for social partnership, negative attitude to any forms of inequality (economic, gender, racial, age, educational) with observance of ecological legitimacy by developing a wide range of competences including interpersonal and communicative ones, team cooperation, coordination, empathy, orientation towards conflict solution and a proactive view. Socialization of students is achieved through molding their active social position by including agreed forms of action on solving complex problems of civilization development into the achievement of sustainable development goals on the basis of civil agreement and democratic forms.

"Soft skills" presuppose developing students' flexibility, creative thinking, business communication and self-organization, emotional intelligence, competences in the field of intercultural communications, teamwork including the process of development and effective implementation of projects, etc. 'Soft skills" have a socio-humanitarian essence and the importance of their development in students of all specialties as strategic competences defines the evolvement of most effective behavioral models and strategies in the student's future professional activities. In the current environment of global threats and risks, these models and strategies are, in our opinion, to account for all aspects of sustainable development. It is essential to maintain continuous curricula of different levels: to start working at these skills at the stage of general education and continue this development in the system of vocational and higher education.

The component of professional competences, being the "gist" of the curricula, helps integrate 'hard skills" in the context of current environment for development of economy, society and harmonious coexistence in the planetary ecosystem.

The development of digital competences is one of the key conditions for competitiveness of graduates at the labor market. Digital competences market possible to develop most effectively the "professional centerpiece" and "soft skills" of any curriculum (specifically, on the basis of Internet-technologies, educational online-platforms and remote courses, etc.).

The most important condition for building the model under discussion is possession of required competences by educators who have a comprehensive vision of sustainable development problems and understand its social, ecological, economic and cultural aspects. These are a real moving force of changes capable of promoting implementation of goals in the field of sustainable development by means of education. The achievement of these goals will require introducing sustainable development principles into the system of training and retraining of teachers.

Curricula at all levels are to be reconsidered and updated based on the interdisciplinary approach with due account for global changes and their local manifestations, which will provide a new impetus to the process of education and promote it by means of the intergenerational dialogue in the ever renovating global environment.

The results of this model's introduction imply readiness and ability for cultural development, based on the contemporary context of historical and social experience of the sustainable way of life and ethical assessment of the consequences of personal action in the socio-natural environment proceeding from the principles of ensuring intergenerational and global justice.

Thus, the implementation of the humanitarian component of the proactive education model provides for changing the content, tasks and methods of education at all its stages and formats for the purposes of acquiring competences in the field of sustainable development. It is essential for designing proactive education for the common good.

The current real educational process is alienated from the formation of the up-to-date scientific worldview, sustainable development ideas, and strategic socio-economic objectives of the government.

With consideration to priorities in the Russian educational policy, we have defined the most important instruments for implementing the components of the proactive education model.

We refer the following to the element of professional competences in the above model:

Compiling new unique curricula of the new generation, including those in foreign languages, corresponding to the Russian and global market requirements and providing for variability of educational treks, multidisciplinary approach and extra competences;

Taking measures to implement curricula compiled within the «2+2+2» model and give students an opportunity of choosing specialization in the course of training at least two times – during the second year of learning and after graduating from the Bachelor Course;

Introducing a system for students, learning by major professional Bachelor, Specialty and Master curricula, to receive additional competences through extra educational programs simultaneously with getting basic education, including skills upraising and professional retraining;

Developing new forms of strategic interaction with leading companies and government corporations – leaders of digitalization in the context of organizing academic internship facilities, practical training and further employment;

Expanding network forms of implementing curricula and applying the best practices of specialists' training including the development of joint programs and modules, practical training, mobility of academic staff and students, and the development of a university's international digital brand.

The component of ideological value and spiritual patriotism may include the following priorities:

Designing programs of personality development, ensuring a sense of involvement and building an active civil position;

Forming a socially comfortable environment;

Enhancing participation of students in scientific research;

Developing a mechanism of coaching, tutorship, mentorship, including "digital" one;

Creating adaptive educational modules, disciplines and courses in the digital format for disabled and persons with health limitations in the curricula to be mastered.

The component of digital competences is focused on the implementation of the following priorities:

Developing online technologies and automated proctoring in the educational process, including introduction of high quality online courses;

Developing students' digital competences through enhancement of IT-training;

Developing and implementing multi-format offline and online programs of skills upgrading for teachers to master new methods and technologies of teaching required in the digital economic environment and associated with the processes of digital transformation, including teaching in foreign languages;

Molding a student's digital profile including a digital portfolio, rating and individual educational trek, and having a through character with an opportunity of receiving additional education;

Introducing digital services for all curricula (an educational space navigator, "digital tutor" for making an individual educational trajectory, individual schedule management, scientific research and design activities, feedback with university teachers and administration, etc.);

Creating new digital models of interaction with graduates to account for their educational needs after graduation and give a flexible response to them, making changes in the existing curricula;

Developing, using comprehensively and storing generally accessible digital collections of educational resources, guidelines, instruments and services to be applied in learning activities on implemented programs;

Forming a teacher's digital profile, upgrading the academic staff's skills in using digital competences in educational activities;

Instilling Data Culture among students as well as Project Culture; teaching them communication skills and work ethics in the digital medium;

Developing learning analytics – a system of monitoring and support of educational activities with the use of big data, students' and teachers' digital footprints. Installing a system of digital valid measuring instruments and services for an independent assessment of education quality. Quality assurance of education with the help of systematic monitoring and using digital footprints;

Introducing up-to-date methods of design and promotion for unique Bachelor, Speciality, and Master Curricula with the use of up-to-date informational resources on leading digital platforms with English language (foreign language) versions, in CIS countries included.

The humanitarian-ecological component includes implementing the interdisciplinary approach and instilling the ecological personal culture.

The component of "soft skills" focuses on the following areas:

Adopting project forms of education enabling students to create their own projects and perform graduate qualification papers in the form of multi-format startups (of social orientation as well);

Expanding opportunities for students to follow their own educational trajectories and choose pools of minors (additional education profiles/cycles), establishing individual practice-oriented and flexible educational trajectories for each student within the curricula framework.

4 Conclusions

An important objective of the educational system as a key institution for social transformations in the current environment is to train new personnel having a good command of new knowledge and competences for sustainable development. An important role in achieving this objective belongs to humanitarian education that promotes a graduate's personality development and implementation of the proactive function of personnel training in the period of the systemic crisis and global challenges. It is only the humanitarian basis that can ensure a student's correspondence to current requirements, develop an ability of overcoming crisis phenomena, think globally and systematically, find his/her own way in the polycultural space and the ever progressive world.

Taking the Berlin conference recommendations into account, we believe that current concepts and technologies of sustainable development should be introduced into curricula of all levels as a universal competence. Disciplines and modules making up the humanitarian component based on the development of the new philosophical and cultural concept must be expanded and presented as a separate educational package (compared with the existing educational standards). A promising course is network interaction between universities and other stakeholders that will promote an effective update of education on the basis of principles and goals of sustainable development. To solve this problem it would be essential to find teachers who will be able to work in the new paradigm and raise their professional skills for this purpose.

The model of proactive education ought, in our opinion, to include the humanitarian component; the authors believe that it should include the following key elements: professional competences, ideological values, spiritual-patriotic and humanitarian-ecological components, digital competences and soft skills. If implemented on the systemic basis, these components will help change society's vision of global problems and promote successful implementation of the sustainable development strategy.

References

1. G. Brundtland, Our Common Future: Report of the 1987. World Commission on Environment and Development, United Nations (1987)
2. World Declaration on Higher Education for the 21st Century: Vision and Action (1998) https://www.conventions.ru
3. S. Ahmad, OIDA International Journal of Sustainable Development, **4(12)**, 79 (2012)
4. N. Abdallah, The Case for Advancing Sustainable Development in Higher Education: An Economic Perspective (2008), http://www.oecd.org
5. N. E. Serebrovskaya, O. I. Pets, Bulletin of the Moscow Region State University: Psychological Sciences series, **2**, 136 (2020)
6. W. Lambrechts, J. Hindson, Environment and School Initiative. Research and innovation in education for sustainable development. Exploring collaborative networks, critical characteristics and evaluation practices, 197, upcommons.upc.edu
7. M. Keiner, Environment, Development and Sustainability, **6(4)**, 379 (2005)
8. H. Singh, Education for sustainable development in the 21st century, Sodh Sanchayan, 1 (2010) https://www.researchgate.net
9. Berlin Declaration on Education for Sustainable Development (2021) https://ru.unesco.org

Youth policy in the human capital management system: Russian practice

Daria Maltseva[1,*], *Olga* Safonova[1], *Anastasia* Dedul[1], *Maria* Petrunina[1], and *Sofia* Tepina[2]

[1]St. Petersburg State University, St. Petersburg, Russia
[2]Secondary general education school № 316 with in-depth study of the English language, St. Petersburg, Russia

Abstract. The study analyzes the modern youth policy of the Russian Federation using the methodological foundations of the theory of human capital. In this context, the criteria for the efficiency of youth policy are defined. The authors argue that youth policy is one of the priority areas of activity that determines the improvement of the country's human capital. Moreover, it is possible to designate the formation and development of well-rounded young people who demonstrate initiative, are responsible and patriotic, have an active civic position, beliefs and a system of values, and take an active part in political and public life.

1 Introduction

Today, in any modern state, youth is a strategically important social group, and the formation and accumulation of human capital and the society as a whole depends on its resources management. As United Nations Secretary General, António Guterres, said: "Peace, economic dynamism, social justice, tolerance, all these and more depend on tapping into the potential of youth" [1,2]. Above-mentioned priorities set the vector for the development of youth policy strategies, which have to become a system for the reproduction of world-class human capital, create added value in the economy, and play a transformative role.

Despite the fact that Russia has already accumulated some experience in strategic youth management, a number of problems still remain unresolved: the lack of methodological recommendations necessary for the development and implementation of strategic documents on state youth policy; gaps in youth policy legislation; insufficient coordination of practices and projects aimed at accumulating the country's human capital, etc.

This makes it necessary to analyze the existing legislation in the field of state youth policy, the experience of its implementation, as well as innovative development projects in assessing the impact on the effectiveness of human capital development.

* Corresponding author: buenafiesta@mail.ru

2 Materials and methods

In recent years, youth policy has become an extremely popular subject of academic research. However, the range of topics covered in this subject area is very limited, which determines the high degree of relevance of this analysis. The main purpose of this study is to examine the existing mechanisms for implementing youth policy, as well as to develop recommendations to public authorities on its harmonization and optimization.

The key feature of the research methodology is the combination of scientific principles and theoretical approaches that allow us to characterize all the variety of processes and effects of the state youth policy implementation in the modern world, primarily in the sphere of human capital management.

The scientific foundation for this article consists of the main methodological developments of the comparative approach, which is used for a comparative analysis of the implemented state programs, innovative projects and practices in the field of state youth policy, as well as statutory instruments. In particular, the study included a thorough analysis of the main regulatory document that defined the key goals and main provisions of the state youth policy in the interests of the state and society at the present stage: the Decree of the Government of the Russian Federation No. 2403-p On Approval of the Fundamentals of the State Youth Policy of the Russian Federation for the Period up to 2025 dated November 29, 2014 [4].

The socio-cultural approach made it possible to study the relationship between the interpretations of youth as the main driver of the accumulation and intensification of human capital and the cultural, historical, political and social specifics of the state [2,3,6]. The competence-based approach was used in this article to determine the main directions of youth development strategies and to assess the regulatory impact of state programs and initiatives in this area.

The methodology of neo-institutionalism used in the study makes it possible to analyze in detail the possibilities of combining the efforts of the state and business in the direction of modernization of youth policy in the context of digital transformations of the management process at all levels [1].

The risk management methodology made it possible to identify both the most effective and potentially hazardous scenarios for the youth policy implementation in the context of digitalization, as well as to assess the effectiveness of state policy in the field of human capital development, to describe the specifics of the interaction process between state and non-state organizations in this area [1,6].

3 Results and discussion

In the course of the analysis conducted, it was proved that the most significant positive effects of the Russian youth policy were shown in the implementation of projects to support talented youth.

Within the framework of the presented research, a fundamental analysis of current projects and initiatives in the area of state youth policy was conducted. During the study, the following strategic directions of program implementation were identified: 1) education; 2) civic and patriotic upbringing; 3) personnel policy; 4) entrepreneurship.

The analysis showed that today in Russia the existing state activities/projects in education, such as "All-Russian Academic Olympics for School Students", "Tavrida", "The Territory of Meanings", "Russia Is a Country of Opportunities", "Sirius", "I'm a Professional", and others are aimed at the identification of the talented youth at the earliest age possible, its further development and professional support by the creation of conditions for the development of their intellectual and personal potential, participation in the education of potential future leaders of the country [7,8]. Most of the projects are implemented throughout Russia, which allows to create equal starting opportunities for people from different regions, as well as to help them to use their innovative potential in the economic and political spheres.

The abundance of youth events, especially in the field of education, reflects the need of the state and society for professional, highly qualified domestic personnel able to lead the management system in various industries in the future.

The next important area of implementation of the state youth policy is patriotic education. One of the most striking manifestations is the creation of the Rospatriot Center (the Russian Center for Civic and Patriotic Education of Children and Youth). Among the main activities supported by the Federal Agency for Youth Affairs are the following: coordination of the implementation of the state program "Patriotic Education of Citizens of the Russian Federation for 2016-2020" [4,9], as well as the program of patriotic education of young people. This activity covers a wide range of spheres of society's life, uniting and coordinating interaction between military and patriotic associations, military units and veteran organizations, and is aimed at creating a value orientation among young people, including a sense of responsibility for the fate of the state and a positive attitude to public service and military service. In practice, this initiative is manifested in an increase in the number and higher activity level of search parties, military history clubs, associations of Cossack youth, etc.

In addition to the support for talent and proactive young people, as well as patriotic education, the presented study recorded an extremely high interest of the Russian state in the future leaders and managers. The development of the most successful personnel requires not only an advanced educational system, but also support for proactive candidates at the first stage of their professional development. One of the most urgent problems of modern youth is its indifferent position and passive attitude to public life. About half of young people do not see the point in exercising their civil and political rights: they do not participate in the elections, they are indifferent to changes in the political situation in the country, and do not demonstrate any civil initiative [10, 11]. In the context of underdeveloped civil society and local self-government, this situation seems logical. But if young people do not take part in the political life of the society, are not interested in governance, then it is extremely difficult to make significant state decisions, develop business and the economy, strengthen the country's position in the international arena. Therefore, the task of the state and socio-political organizations is to create conditions for increasing the political and economic interest and activity of young people, attract them to state development and to the development of the economic sector. One of the most prominent examples of state initiatives in this area is the creation of youth branches of political parties: Molodaya Gvardiya ("the Young Guard", the United Russia), the LDPR Youth Organization, and others. Leaders are in demand not only in the sphere of political governance, but also in the economy and business. That is why modern state policy is aimed at supporting business and micro-entrepreneurship. The economy is the engine of the country's development, bringing funds to the budget, ensuring the influence of the state in the international arena. Russian legislation provides additional guarantees for young entrepreneurs, but in practice these initiatives have been implemented only marginally. My First Business project for participants aged 14 to 23 seems to be effective in introducing young people to economic activity. The successful completion of the project not only gives young people the opportunity to show their potential once, but also lays the foundations for further participation in state youth support programs. The need to support youth entrepreneurship was also established in the above-mentioned Government Decree on Approving the Foundations of the State Youth Policy in the Russian Federation up to 2025.

In the course of the analysis conducted, it was found that effective implementation of support for youth entrepreneurship requires harmonization of three strategic aspects. The first is the legislative aspect: the adoption of a single regulatory act at the federal level to establish clear criteria for identifying young entrepreneurs, as well as guarantees provided to them by the state. The second is educational, which includes the need to develop educational programs for the development of entrepreneurial thinking, planning and business management, strengthening the connection between educational organizations and existing entrepreneurs in order to synthesize practical and theoretical knowledge and skills. The third aspect is economic, which requires the liberalization of economic activity. The latter can be considered as one of the most difficult to implement, since it must take into account the public interest in state control, the degree of state intervention in the economy, as well as the freedom of entrepreneurial activity and the market economy [11,12]. In addition to the economic goal, the support for youth entrepreneurship will ensure the development of centers for youth innovative creativity, as well as the spread of educational programs on technological entrepreneurship.

It is obvious that the modern processes of growing global crises and rapidly developing digital transformations have significantly changed the approaches to the study of youth policy in the human capital management system and created new tasks for the public administration in this area.

The modern history of the formation of youth policy in the Russian Federation began in 1992 with the adoption of the Decree of the President of Russia On priority measures in the field of state youth policy [6, 13, 14] and the Resolution of the Supreme Council of the Russian Federation of 1993 On the main directions of state youth policy in the Russian Federation [15]. The documents became one of the first statutory instruments of the independent Russian Federation aimed at regulating youth policy. The decree provided for the creation of a National Council of Youth Associations, the development of international youth cooperation, and also reestablished state support, including material support. It was supposed to allocate funds not only from the federal budget, but also from the budgets of the corresponding regions to finance a part of youth programs and projects, and there was also an option of issuing long-term loans to young families. At the same time, the provisions of the Decree were aimed at solving the most acute social problems: child neglect and juvenile delinquency. Further, in order to provide socio-economic support for young people in the Russian Federation, in 1996 the Government adopted a resolution On Additional Measures to Support Young People in the Russian Federation [7,8,9].

The fate of the draft law On the Basics of State Youth Policy in the Russian Federation, initiated by deputies of the State Duma of Russia in 1998 is of interest [15]. After a lengthy debate, the draft law On the Basics of State Youth Policy in the Russian Federation adopted in the third reading in July 1999 was rejected by the Federation Council as insufficiently elaborated, declarative, containing general provisions and "expensive" for the country's budget (90 billion rubles for the period of three years). A conciliation commission was established and as a result, the agreed version was approved by both chambers in October 1999. In November 1999, the President of the Russian Federation (B.N. Yeltsin) vetoed the law; this does not happen very often at present, but then, in the period from 1995-1999, President Yeltsin used the right of veto 123 times. In July 2000, the State Duma considered the law again, but after a heated debate and a roll-call vote, it did not overcome the president's veto. In 2004, the law On the Fundamentals of State Youth Policy in the Russian Federation was withdrawn from consideration.

The third (from 1994) Federal Target Program "Youth of Russia (2001-2005)" approved by the decree of the Government of the Russian Federation was aimed at creating and developing legal, economic and organizational conditions for fostering civic awareness among young people in the context of a democratic society, a market economy and the rule of law, personal fulfillment of young people as active participants in the transformations of modern Russian society [16].

The next document which the Government intended to regulate the sphere of youth policy was "The Strategy of State Youth Policy in the Russian Federation" adopted in 2006. It was developed for the period up to 2016 and contained a set of priority areas and tasks related to young people, including the use of the subject-to-subject approach, when young people directly participated in the implementation of national projects [17]. Despite the fact that some lawyers and sociologists noted that the document was the main act regulating relations between the state and the youth, this statement seems controversial, since the proposed measures were of advisory nature. It was this decree of the Government that was replaced in 2014 with the "Fundamentals of the State Youth Policy of the Russian Federation for the Period up to 2025".

An important milestone in the evolution of youth policy was the approval by the Government of the Russian Federation of the "Concept of Long-Term Socio-Economic Development of the Russian Federation for the Period up to 2020". The ninth section of the Concept is devoted to youth policy, which is reflected in its name. The document provides a more detailed description of the goal of the socio-economic aspect of the state youth policy, which is to create, with the help of modern financial and economic tools, "conditions for successful social engagement and effective personal fulfillment of young people, the development of the potential of young people and its use in the interests of innovative development of the country"[4].

The conducted analysis has demonstrated that today the Russian legal system is characterized by the fragmented set of legislative acts of various nature and levels, and without the harmonization of such acts, as well as without the intensification of the most effective scenarios of state youth policy, its regulation seems extremely difficult. The existing statutory instruments regulate a very narrow sphere, namely, the issues of supporting talented youth and military and patriotic education. The study found that in order to solve this problem, it is necessary to consolidate in a single legislative act the powers of federal bodies, such as the Government of the Russian Federation, the Ministry of Science and Higher Education of the Russian Federation, the Ministry of Culture of the Russian Federation, the Ministry of Education, the Ministry of Sports and the Federal Agency for Youth Affairs, as well as to establish the jurisdictions of the regional authorities. The powers of the Government should include the establishment of the general foundations of the state youth policy, as well as providing funding and preparing the budget for this industry. It seems reasonable to entrust the Federal Agency for Youth Affairs the implementation of specific programs and projects in all regions of the country, while the ministries, according to their competence area, after conducting research, will be able to determine standards and regulations on their basis, develop response and interaction measures. At the same time, the authorities of the regions should not be deprived of the opportunity to implement the main goals and objectives. At the regional level, the support will be most effective in adopting laws of a particular region, promotion of local youth initiatives and provision of more guarantees in addition to federal ones.

Thus, it can be concluded that the reform in the sphere of youth policy, which is of particular importance for the nation-wide development and intensification of human capital, requires an integrated and systematic approach with the priority of legal support and regulation.

4 Conclusion

Today, the state is fully aware of the importance of creating conditions for the development of the human capital of the younger generation. The lack of state attention and support over a long period damaged the mechanisms of implementation, and the collapse of the previous political system destroyed the patterns of interaction between the state and the youth. A vast layer of socio-political institutions and organizations disappeared, as well as party-based social elevators that offered promotion for talented and proactive leaders. This led to general aging of the political system simply due to the lack of young talent to replace the current cadre. The risks involved in this are already known from historical experience: these are the stagnation of the political system, which results in passive attitude of young people, their disillusionment with the authorities, and, consequently, a crisis of legitimacy and an increase in protest sentiment.

The current policy of the Russian Federation is aimed at restoring the mechanisms of state support for young people, creating a regulatory framework for their effective implementation and reviving the initiative and civic interest of the younger generation.

For a fruitful solution to the problem of young people's human capital management, it is necessary, first of all, to ensure effective legal regulation of the process, which should not be reduced to casuistics, but also must not leave wide gaps, since this will create lack of legal clarity and lead to potential abuse and discrimination. Having worked out all aspects of youth policy, it is recommended to adopt legislation that clearly distinguishes the powers assigned to the jurisdiction of the Russian Federation and to the jurisdiction of the regions. Competent distribution will promote most effective implementation of the goals and objectives set, allocate the funds of the relevant budget appropriately, without creating an excessive burden on the systems of government authorities. It is also necessary to provide for potential sanctions for specific violations of the established procedure for implementing state policy, since in practice the absence of sanctions creates an impression that it is possible to violate it with impunity. It is also necessary to preserve a certain autonomy of the regions, giving them the opportunity to establish additional guarantees, as well as to implement policies taking into account the specifics of the region (the standard of living, the main ethnic groups).

In addition to the development of legislation, it is necessary to ensure broad access of young people to the provided opportunities. Despite the success of projects to support talented youth, one of the main problems is that certain groups of the population do not have information about the existence of opportunities provided by the state. Therefore, it is often impossible to participate in forums, competitions and projects due to the lack of information on these initiatives and insufficient popularization thereof. This creates the need for marketing youth activity with the involvement of mass media and educational organizations of all levels. By obliging organizations to inform students about the existing opportunities to participate in events will make it possible to attract the largest number of young people.

It should also be noted that for a more effective youth policy, it is necessary to involve public and private organizations in this work. Well-developed social movements, especially those related to history and patriotic education (People's Memorial Books, the Immortal Regiment) will complement state policy and contribute to the development of an active civil policy not only for young people, but also for society as a whole. Private organizations, among other things, will make a great contribution to supporting youth entrepreneurship and scientific and technical projects. For example, during forums and competitions dedicated to new technologies, they will be able to conclude profitable contracts with those who have submitted promising projects, thereby developing their production and contributing to the implementation of state programs.

Thus, the article proved that youth policy is one of the highest priority areas of state activity that determine the improvement of the national human capital. As its ultimate goal, it is possible to designate the formation and development of well-rounded young people who demonstrate initiative, are responsible and patriotic, have an active civic position, beliefs and a system of values, and take an active part in political and public life.

Acknowledgments

The reported study was funded by RFBR and EISR according to the research project № 21-011-33045.

References

1. L. Bennett, S. Livinston, European J. of Communication, **33(2)**, 122 (2018)

2. T. V. Sheludyakova, *Levels of implementation of state youth policy in Russia: problems of delineation of powers*, 8 (2017)

3. T. V. Sheludyakova, *Social support of youth in the constituent entities of the Russian Federation: legal regulation and implementation practice*, 12 (2014)

4. The Decree of the Government of the Russian Federation No. 2403-p On Approval of the Fundamentals of the State Youth Policy of the Russian Federation for the Period up to 2025 dated November 29, 2014, http://www.consultant.ru/

5. Strategy of the State Youth Policy in the Russian Federation" dated 18.12.2006 No. 1760-p (as amended on 16.07.2009), http://www.consultant.ru/

6. S. Guriev, D. Treisman, J. of Economic Perspectives, **33(4)**, 10 (2019)

7. On Priority Measures in the Field of State Youth Policy. Decree of the President of the Russian Federation No. 1075 dated 16.09.1992, http://www.kremlin.ru/

8. S. N. Pershutkin, *The role of the President of Russia in disassembling and implementing the state youth policy: Normative and legal framework, ways and norms of activation*, 31 (2007)

9. N. P. Tsukanova, *Legal regulation of state youth policy in Russia*, 92 (2014)

10. N. A. Samokhvalov, *Implementation of state youth policy as an element of modernization of Russian statehood at the present stage*, 3 (2015)

11. F. Kh. Tsalikova, *Implementation of the state youth policy: on the example of the Republic of North Ossetia-Alania*, 125 (2014)

12. M. V. Vavilin, *Prosecutor's supervision over the implementation of laws in the field of patriotic education of youth as one of the priority directions of prosecutorial activities*, 11 (2017)

13. V. I. Lenin, «Tasks of Youth Unions» III All-Russian Congress of the Russian Communist Youth Union (Moscow, 1920)

14. I. M. Ilyinsky, Art, NRC VKSH at the Central Committee of the Komsomol, 18 (1987)

15. On priority measures in the field of state youth policy: Decree of the President of the Russian Federation dated September 16, 1992 No. 1075 (as amended on April 12, 1999). - SPS "ConsultantPlus"

16. On the main directions of state youth policy in the Russian Federation: Resolution of the Supreme Soviet of the Russian Federation of July 3, 1993 No. 5090-1. - SPS "ConsultantPlus"

17. On state youth policy in the Russian Federation: Draft Federal Law of 08.05.2007 N 428343-4. - SPS "ConsultantPlus

Relationship of the productivity of forest dendrocenoses with the geomorphological profile of soils and their influence on the ecology

V.K. Khlyustov[1,*] , *V.A.* Sedykh[1, 2], *V.D.* Naumov[1], *A.M.* Ganikhin[1], and *N.L.* Kamennykh[1]

[1]Russian State Agrarian University – MSHA named after K. A. Timiryazev», Moscow, Russian Federation
[2]Federal State Research Institute of the Federal Penitentiary Service of the Russian Federation, Moscow, Russian Federation

Abstract. It is well known that there is a relationship between the productivity of forest phytocenoses and soil fertility, which is largely determined by the terrain and soil geomorphology. An important element of assessing the productivity of forests is the substantiation of the methodology for modeling the age dynamics of the growth of forest stands, depending on soil and ground conditions. The solution of this methodological problem is possible only if there is data on forest objects, represented by permanent sample plots and data from a detailed soil survey.

1 Introduction

In our case, the object of the study was the Lesnaya Opytnaya Dacha (Forest experimental site) (FES), which is a scientific station and an educational training ground with an area of 232.3 hectares on the territory of Moscow. In the forest area, long-term periodic (after 10-15 years) observations of the growth, structure and productivity of forest stands are carried out, as well as their impact on the environment is assessed. The organization of the FES territory is associated with the name of the famous scientist-forester Count A.R. Vargas de Bedemar. which in 1863 carried out the first forest inventory [1]. The forest area was divided into 14 quarters with a description of plantings in them and the establishment of permanent sample plots. Later, scientists-foresters of the Petrovskaya and Timiryazevskaya academies V.T. Sobichevsky, M.K. Tursky, N.S. Nesterov, V.P. Timofeev, G.R. Eitingen, A.N. Polyakov and others continued to observe naturally forming forest stands and forest cultures of various mixing and density. In total, 156 test plots were laid in the plantings, on which regular observations have been carried out for more than 150 years. The forest growing conditions of FES are represented by meso-hygrophilic (C3) and hygrophilic (C4) complex subors - relatively rich soils. However, the accepted classification only in the first approximation characterizes the soil-typological conditions found in the forest area.

* Corresponding author: helenr2003@mail.ru

S.V. Zonn [8], B.D. Zaicev [6], N.P. Remezov, P.S. Pogrebnyak [13], L.O. Karpachevsky [9] Yu.P. Demakov et al. [4] indicate that it is the soil that is the leading component of the biogeocenosis, which determines its structural organization, dynamics and productivity.

G.N. Vysotsky is a contemporary and student of K.A. Timiryazev and V.V. Dokuchaeva developed the doctrine of forest-vegetation zoning, developed a diagram of natural zones demonstrating the relationship between the distribution of forest vegetation, depending on climatic, soil and hydrological conditions. The scheme proposed by him has not lost its significance so far. He made a significant contribution to the study of the impact of forests on the environment, to the classification of types of water regime of soils and grounds [12].

Later N.V. Dylis, V. N. Sukachev [5] for the first time formulated the theory of the horizontal structure of forest biogeocenoses (BGC). Starting with the works of L.O. Karpachevsky [9, 10] the idea gradually developed that the soil cover in forest biogeocenoses resembles a honeycomb (parcell) structure. A parcel represents forest areas isolated in space (along a conventional radius), formed by a specific edificator and dominants of the vegetation cover. Within the parcel, along the radius from the edificator, the influx of litter, accumulation of litter, consumption of water and nutrients change.

Yu.P. Demakov, A.V. Isaev [3] indicate that at present, a lot of factual field material has been accumulated, various kinds of measurements for systematization and data processing. This requires only the correct formulation of problems in identifying conformities (relationships), the use of system analysis with elements of mathematical statistics and modeling methods.

In modern economic conditions, large-scale field surveys are very problematic, therefore, the use of field material already accumulated by predecessors, both on soils and on the growth, productivity of forest phytocenoses, should be considered a good basis for successfully solving the problem posed in the study.

V.D. Zelikov [7] it is shown that the granulometric composition of soils and the thickness of the soil profile has an important role in the ontogeny of forest stands. Using statistically reliable data, the author has shown that forest stands with a predominance of pine reach the highest bonitet (I-I ^{a}) in those cases when they grow on deep sandy loamy-sandy deposits, as well as loamy soils up to 120 cm thick. D.A. Danilov and V.P. Tsarenko [2], studying the influence of the granulometric composition of soils on the productivity of naturally formed stands of spruce and pine in the conditions of the Northwest region. They noted that on loamy and sandy loamy soils, the pine element of the forest shows greater productivity and better marketable qualities of wood than spruce. The authors conclude that with age in these soil conditions, pine mixed with spruce covers most of the ecological niche, which affects the overall productivity of forest stands.

O.V. Martynenko [11] provides information on the developed regression model of the productivity of pine forest stands of SCHUOLKh of the Moscow region. At the same time, it is indicated that information on bonitet classes, expressed by the average height of stands at 100 years of age, was taken as an independent variable. And as independent variables in the formation of the model, such variables as the level of groundwater, the content of humus in the root-accessible soil layer and the upper boundary of the appearance of signs of gleying in the mineral part of the profile are involved. At the same time, it is noted that bonitet is a dynamic indicator and not sufficiently reliable for long-term forecasting.

The presence of permanent sample plots of FES opens up great opportunities for conducting research that allows one to assess the role of various elements of dendrocenosis of artificial and natural origin on the soil-forming process and the structure of sod-podzolic soils. At the same time, the data of taxation indicators of forest stands on permanent sample plots make it possible to reveal the influence of soil characteristics on the dynamics of productivity of dendrocenosis elements [16].

The aim of the study was to develop a methodology for assessing the dynamics of growth by average height and productivity of closed stands located at different levels of geomorphological soil surfaces.

Study objectives:

1.Conduct modeling of age-related changes in the average height of the main forest elements based on data from permanent sample plots.

2.Reveal the relationship between the levels of productivity of forest stands located at different elevations of geomorphological surfaces.

2 Research methods

Experimental material is represented by a series of sample plots (Table 1), presented in different parts of the FES geomorphological profile (Fig. 1).

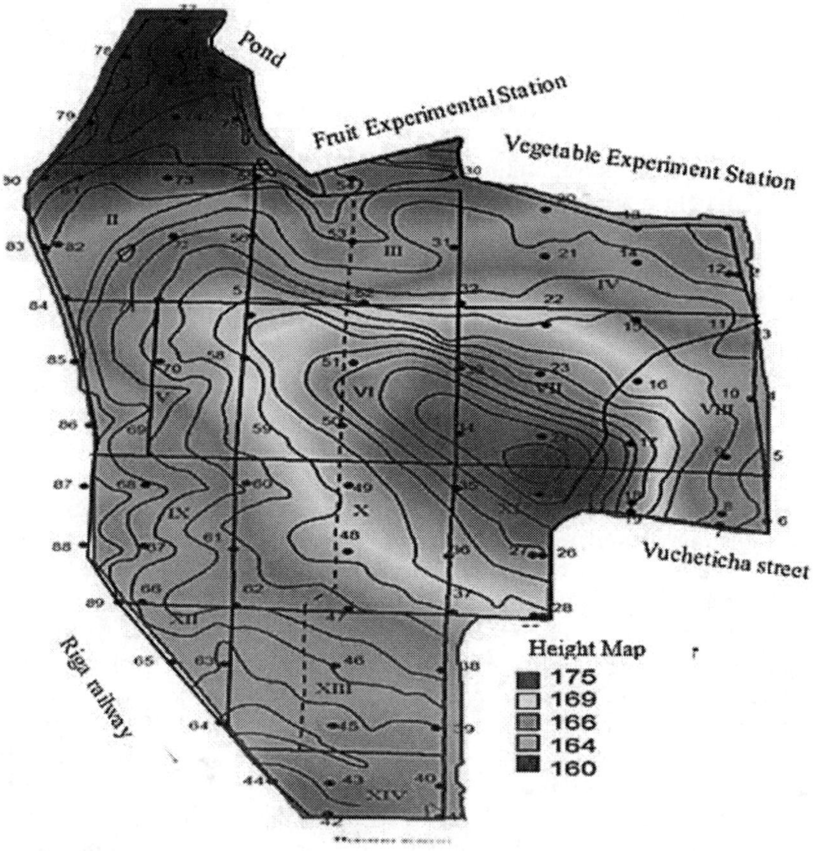

Fig.1. Altitude marks of the FES territory.

Table 1. Soil and forest inventory characteristics of the sample plots.

Quarter no.	Sample plot	Forest stand composition	Origin	Bonitet class	Height mark in the relief, m	Soil classification name
IV	M	9S1Lp	artificial	2	166,1	Sod-podzolic deep-soddy strongly - super-deep podzolic light loamy on moraine medium loam. **Pd** $_{5\text{-}3/5}$ **lsMs**
IV	F	9C1E	artificial	2	165,4	Sod-podzolic deep-soddy strongly - super-deep podzolic light loamy on moraine light loam. **Pd** $_{5\text{-}3/5}$ **lsMls**
IV	L	9S1Lp	artificial	2	166,9	Sod-podzolic medium-sod strongly-deep podzolic deep-gley light loamy on sandy loam moraine deposits. **Pd** $^{GII}{}_{4\text{-}3/4}$ **lsMs**
IV	R	8S1B1E	artificial	2	168,1	Sod-podzolic deep-soddy strongly - super-deep podzolic sandy loam on moraine light loam. **Pd** $_{5\text{-}3/5}$ **sMls**
IV	Z	6S2B1Lp1D	artificial	1	165,4	Sod-podzolic deep-soddy shallow-deep-podzolic deep-gley light loamy on moraine light loam. **PD** $^{GII}{}_{5\text{-}2/4}$ **lsMls**
IV	-	8S2Lp	artificial	2	168,1	Sod-podzolic deep-soddy strongly - super-deep podzolic sandy loam on moraine light loam. **Pd** $_{5\text{-}3/5}$ **sMls**
IV	K	9S1Lp	artificial	2	167,8	Sod-podzolic deep-soddy strongly - super-deep podzolic light loamy on moraine medium loam. **Pd** $_{5\text{-}3/5}$ **lsMs**
IV	V	9S1Lp	artificial	2	165,7	Sod-podzolic deep-soddy strongly - super-deep podzolic deep-gley light loamy on moraine medium loam. **Pd** $^{GL1}{}_{6\text{-}3/5}$ **lsMs**

Table 1. Continued.

Quarter no.	Sample plot	Forest stand composition	Origin	Bonitet class	Height mark in the relief, m	Soil classification name
IV	S	7C2E1B	artificial	2	166,8	Sod-podzolic medium-sod strongly-deep podzolic deep-gley light loamy on sandy loam moraine deposits. **Pd** $^{Gl1}{}_{4\text{-}3/4}$ **lsMs**
IV	U	9C1E	artificial	2	165,8	Sod-podzolic deep-soddy strongly - super-deep podzolic light loamy on moraine medium loam **Pd** $_{5\text{-}3/5}$ **lsMs**
IV	O	10C	artificial	2	165,4	Sod-podzolic deep-soddy strongly - super-deep podzolic light loamy on moraine light loam. **Pd** $_{5\text{-}3/5}$ **lsMls**
IV	U	5S5Lp	artificial	2	165,7	Sod-podzolic deep-soddy strongly-deep podzolic medium loamy on moraine light loam. **P** $_{D5\text{-}3/4}$ **sMls**
VI	C	5O5Lp	artificial	1	170,7	Sod-podzolic deep-soddy strongly - super-deep podzolic light loamy on moraine sand. **P** $_{D5\text{-}3/5}$ **hpMP**
V	D	5S5Lp	artificial	2	168,1	Sod-podzolic deep-soddy strongly - super-deep podzolic profile - gleyic light loamy on moraine sand. **PD** $^{Gl3}{}_{6\text{-}3/5}$ **hpMP**
V	V	6Lp4S	artificial	1	168,6	Sod-podzolic deep-soddy strongly - super-deep podzolic profile - gleyic light loamy on moraine sand. **PD** $^{Gl3}{}_{5\text{-}3/5}$ **hpMP**
VII	P	10L	artificial	1 b	172,8	Sod-podzolic deep-soddy medium-deep podzolic light loamy on moraine light loam. **P** $_{D5\text{-}2/4}$ **hpMls**

Table 1. Continued.

Quarter no.	Sample plot	Forest stand composition	Origin	Bonitet class	Height mark in the re-lief, m	Soil classification name
VII	R	10L	artificial	V	173,4	Sod-podzolic deep-soddy weakly-deep podzolic light loamy on moraine light loam. **P** $_{D5\text{-}1/4}$ **hpMls**
VII	K	8C2Kl	natural	1 b	172,8	Sod-podzolic medium-soddy medium-deep podzolic profile-gley medium loamy, on a sandy loam moraine. **P** $_{D}{}^{Gl3}{}_{5\text{-}2/4}$ **lsMsp**
VII	S	8L2Lp	artificial	1 a	172,8	Sod-podzolic medium sod medium - deep podzolic medium loamy on moraine medium loam. **P** $_{D4\text{-}2/4}$ **lsMls**
VII	E	6L2Kl	artificial	1	173,9	Sod-podzolic deep-soddy medium-deep podzolic deep-gley light loamy, on moraine light loam. **PD** $^{Gl1}{}_{5\text{-}2/4}$ **lsMls**
XI	B2	6Lp4S	artificial	2	175	Sod-podzolic medium-sod medium-deep podzolic sandy loam soil on moraine sand. **P** $_{D4\text{-}2/4}$ **spMP**
XI	M	5Lp4D1B	natural	1	169,6	Sod-podzolic deep-sod medium-deep podzolic deep-gley light loamy on moraine sand. **P** $_{D}{}^{Gl1}{}_{6\text{-}2/4}$ **lsMs**
VIII	N	8D2Lp	natural	2	168,5	Sod-podzolic deep-soddy medium-deep podzolic light loamy on moraine sandy loam. **P** $_{D5\text{-}2/4}$ **lsMsp**
VIII	O	10D	natural	1	165,4	Sod-podzolic deep-soddy medium-super-deep podzolic profile - gleyic light loamy on moraine sand. **PD** $^{Gl3}{}_{5\text{-}2/5}$ **lsMP**

Table 1. Contunued.

Quarter no.	**Sample plot**	**Forest stand composition**	**Origin**	**Bonitet class**	**Height mark in the re-lief, m**	**Soil classification name**
VIII	Z	10Lp	natural	1 b	170,7	Sod-podzolic deep-soddy medium-deep podzolic light loamy on moraine sand. **P** $_{D6-2/4}$ **lsMP**
V	Zh	10L	artificial	1 b	167,5	Sod-podzolic deep-soddy strongly-deep podzolic profile - gleyic light loamy on moraine light loam. **PD** $^{Gl3}{}_{5-2/4}$ **lsMls**
V	U	6L3B1Kl	artificial	1 b	167,5	Sod-podzolic deep-soddy strongly super-deep podzolic profile - gleyic light loamy on moraine sandy sediments. **P** $_{D}{}^{Gl3}{}_{6-3/5}$ **lsMP**
V	M4	6S4B	artificial	2	165,4	Sod-podzolic deep-soddy strongly super-deep podzolic profile - gleyic light loamy on moraine medium loam. **P** $_{D}{}^{Gl3}{}_{5-3/5}$ **lsMs**

Table 1. Continued.

Quarter no.	Sample plot	Forest stand composition	Origin	Bonitet class	Height mark in the relief, m	Soil classification name
IV	NS	7S2D1B	artificial	2	165,4	Sod-podzolic deep-sod medium-super-deep podzolic profile-gley sandy loam on moraine sandy loam sediments.$\mathbf{PD^{Gl3}_{6\text{-}3/5}}$ **lsMs**
IV	Sh	6B4S	natural	1	168,3	Sod-podzolic, extremely shallow, deep-podzolic, profiled, gleyic, light loamy on moraine sand. $\mathbf{P_{D}{}^{Gl3}{}_{1\text{-}3/4}}$ **lsMp.**
IV	B	8S1Lp1D	natural	1	166,4	Sod-podzolic deep-soddy strongly - super-deep podzolic medium loamy on medium moraine loam. $\mathbf{P_{D6\text{-}3/5}}$ **sms**
IV	E	10C	artificial	2	166,6	Sod-podzolic deep-soddy strongly super-deep podzolic profile - gley sandy loam on moraine sandy loam. $\mathbf{PD^{Gl3}_{5\text{-}3/5}}$ **spMsp.**

Table 1. Continued.

Quarter no.	Sample plot	Forest stand composition	Origin	Bonitet class	Height mark in the relief, m	Soil classification name
IV	B	9S1Lp	artificial	1	166,7	Sod-podzolic medium-sod strongly super-deep podzolic deep-gley light loamy on moraine sand. **PD ^{Gl1} $_{4\text{-}3/5}$ lsMP.**
IV	2*	5S3D2Lp	artificial	1	166,9	Sod-podzolic deep-soddy strongly-deep podzolic profile - gley medium loamy on moraine light loam. **PD ^{Gl3} $_{5\text{-}3/4}$ sMls**
VI	P	6S1B1D1Lp1V	artificial	1 a	170,2	Sod-podzolic deep-soddy strongly-super-deep podzolic profile-gleyic light loamy on moraine medium loam. **PD ^{Gl3} $_{5\text{-}3/5}$ lsMs**
VI	B	9B1D	artificial	1 a	171,2	Sod-podzolic deep-soddy medium-deep podzolic surface-gley light loamy on moraine sand. **PD ^{Gl2} $_{5\text{-}2/4}$ lsMP**

Table 1. Continued.

Quarter no.	Sample plot	Forest stand composition	Origin	Bonitet class	Height mark in the relief, m	Soil classification name
XI	E	9D1Lp	natural	1	171,2	Sod-podzolic medium-sod medium-shallow podzolic deep-gley sandy loam on moraine sand. **PD** $^{Gl1}{}_{4\text{-}2/3}$ **spMP**
IV	G	9B1S	natural	1 a	165,4	Sod-podzolic slightly sod medium-super-deep podzolic profile-gley light loamy on moraine sand. **PD** $^{Gl3}{}_{3\text{-}2/5}$ **lsMP**

Judging by the share of tree species in the composition of forest stands, they can be conditionally attributed to pine (57%), larch (14%), linden (14%), oak (8%) and birch (8%).

It should be noted that the generally accepted tool in forestry and forest inventory for assessing the productivity of plantations is the boniteration scales of Professor M.M. Orlov separately for seed and coppice stands, as well as the improved VNIILM quality class scale, which takes into account the growth energy of stands at average height for forest-forming species of slow, moderate and fast growth. Bonitet is mainly a classification characteristic of forest stands by productivity (stock) at the time of taxation. He only states the level of productivity, not linking it with the geomorphology and forest vegetation properties of soils [17, 18].

By the decision of the silvicultural part of the work, it is envisaged to link the geomorphological surfaces of the Lesnaya Opytnaya Dacha with the productivity of forest phytocenoses. It is generally accepted to assess the productivity levels of forest stands either on a bonitet or on a forest typological basis according to the relationship between the average height of the forest stands with age [14]. Observing this condition for each of 38 permanent sample plots, we simulated the age-related change in the average height of the stand according to the growth function of Korsun-Bakman:

$$\mathrm{H} = \exp(k_0 + k_1 lnA + k_2 \ln^2 A)$$

In all cases, the smoothing of the empirical data of the sample plots is characterized by a very high closeness of the relationship, close to the functional one, with a coefficient of determination from 0.995 to 0.999.

Further, for each sample plot, using the indicated regression equations, the average height at 40, 70 and 100 years of age was determined. The obtained values of the average heights were included in the calculation of the numerical regression coefficients. Along with this, a matrix of binary variables was compiled, encoding conventionally given decades with a gradation of 30 years (Table 2) in accordance with the guidelines set out in the monograph [15].

Table 2. Coding of the levels of the average height of stands for three age groups with block fictitious variables (X_i).

Average height at age, years	**Block fictious variables**	
	X₁	***X₂***
40	0	0
70	1	0
100	0	1

Multiple regression analysis made it possible to obtain an equation for the relationship between the average height of forest stands and the height of objects above sea level in accordance with the coordinates of the geolocation of test plots of the type:

$$\mathrm{H}_{sr} = \exp(-25{,}8675 + 5{,}5899 ln\mathrm{H}_{UM} + 0{,}3704\mathrm{X}_1 + 0{,}5292\mathrm{X}_2) \quad (1)$$

$$R^2=0{,}771;\ ES=\pm 13{,}1\%;\ t= |3{,}3;\ 3{,}6;\ 7{,}2;\ 9{,}8|>t_{05}=1{,}96;\ F=39{,}27$$

$$at\ P<0.05;\ \Delta = 0.81\%;\ O_\sigma = \pm\ 12.7\%;\ \lim \mathrm{H_{UM}} = 160 - 175\ \mathrm{m}$$

where:
$\mathrm{H_{cp}}$- average height of the stand, m;
$\mathrm{H_{YM}}$ - height above sea level, m;
$\mathrm{X_1}, \mathrm{X_2}$ - block fictitious variables;
R^2 - coefficient of determination;
ES is the standard error of the regression equation;
t - Student's criterion;
F - Fisher's criterion;
P - the significance of the differences according to the F criterion;
Δ — systematic error of the equation,%;

About $_\sigma$ - random error of the equation,%.

The presented parameters of the model (1) indicate the presence of a close relationship between the average height of the forest stands with the independent variables included in the equation, covering 77% of the variance of the dependent variable. The statistical significance of the parameters of the equation, the absence of a systematic error ($\Delta = 0.81\%$) and the presence of a random error not exceeding ± 13% indicate a sufficiently high accuracy of the model, which takes into account only one physical indicator - the height above sea level.

The degree of correspondence of the theoretical values of heights to the actual ones is clearly shown in Figure 2. The location of the theoretical values of the height along the diagonal of the square of the adequacy assessment confirms the conclusion made.

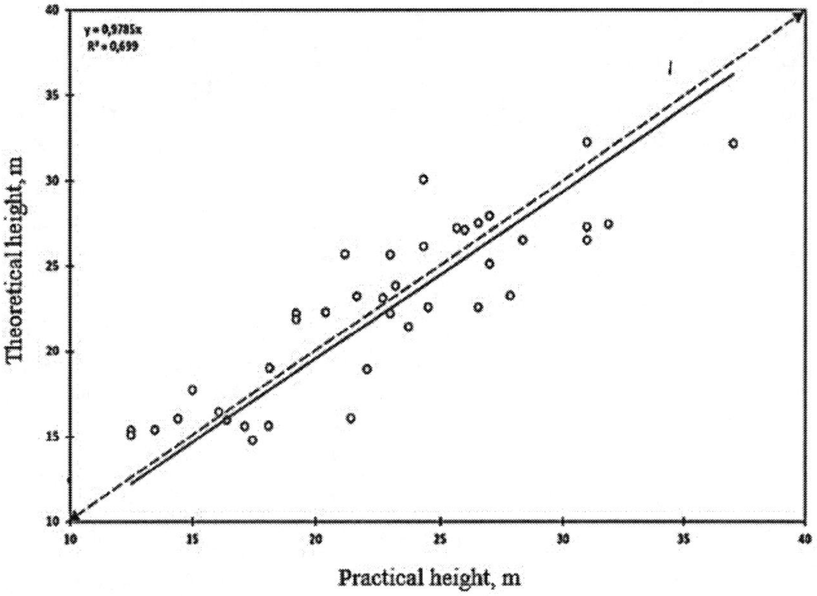

Fig. 2. Visualization of the correspondence of theoretically obtained values of average heights according to model (I) to actual data.

Further, to represent the conformities of change in the average height of forest stands (A, years), taking into account the elevation marks of the geomorphological profile ($H_{\text{УМ}}$) model (1) was transformed into a functional regression equation of the form:

$$\mathrm{H} = \exp(-32{,}01974 + 2{,}54107 ln\mathrm{A} - 0{,}23674 \ln^2 A + 5{,}58992 lnH_{UM}) \qquad (2)$$

$R^2=1{,}0$; $t>t_{05}=1{,}96$

The age-related change in the average heights of forest stands growing at different heights of the geomorphological profile of the FES is shown in Figure 3

A graphical interpretation of the new model is shown in Figure 4. The theoretically obtained tabular data, as well as the three-dimensional graph of changes in the average heights of forest stands in terms of age and height above sea level, clearly indicate a high degree of mutual influence of the variables included in the model.

 https://doi.org/10.1051/e3sconf/202131109001

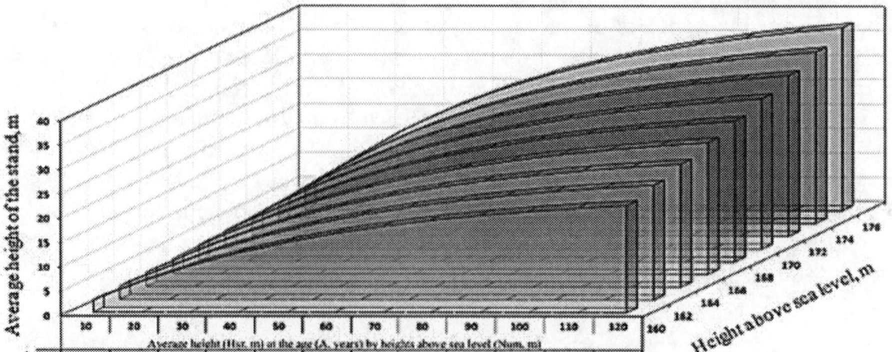

Height above sea level, m	10	20	30	40	50	60	70	80	90	100	110	120
Average height (Hsr, m) at the age (A, years) by heights above sea level (Num, m)												
160	2,6	6,3	9,5	12,2	14,4	16,2	17,7	18,9	19,9	20,7	21,4	22,0
162	2,8	6,7	10,2	13,1	15,4	17,4	19,0	20,3	21,3	22,2	22,9	23,5
164	3,0	7,2	10,9	14,0	16,5	18,6	20,3	21,7	22,8	23,8	24,6	25,2
166	3,2	7,7	11,7	15,0	17,7	19,9	21,7	23,2	24,4	25,5	26,3	27,0
168	3,4	8,3	12,5	16,0	18,9	21,3	23,2	24,8	26,1	27,2	28,1	28,8
170	3,6	8,8	13,4	17,1	20,2	22,7	24,8	26,5	27,9	29,1	30,0	30,8
172	3,9	9,4	14,3	18,3	21,6	24,3	26,5	28,3	29,8	31,1	32,1	32,9
174	4,1	10,0	15,2	19,5	23,0	25,9	28,3	30,2	31,8	33,1	34,2	35,1
176	4,4	10,7	16,2	20,8	24,5	27,6	30,1	32,2	33,9	35,3	36,5	37,4

Fig. 3. Age-related change in the average heights of forest stands growing at different heights of the geomorphological profile of the FES.

The regression surface, supplemented by arrows indicating the value of the systematic elevation excess, allows judging the level of excess of the average heights of forest stands at different heights above sea level from the lowest (H $_{YM}$ = 160 m) to the highest (H $_{YM}$ = 175 m). Thus, the maximum excess is observed in mature stands and is 15 m, which corresponds to four bonitet classes.

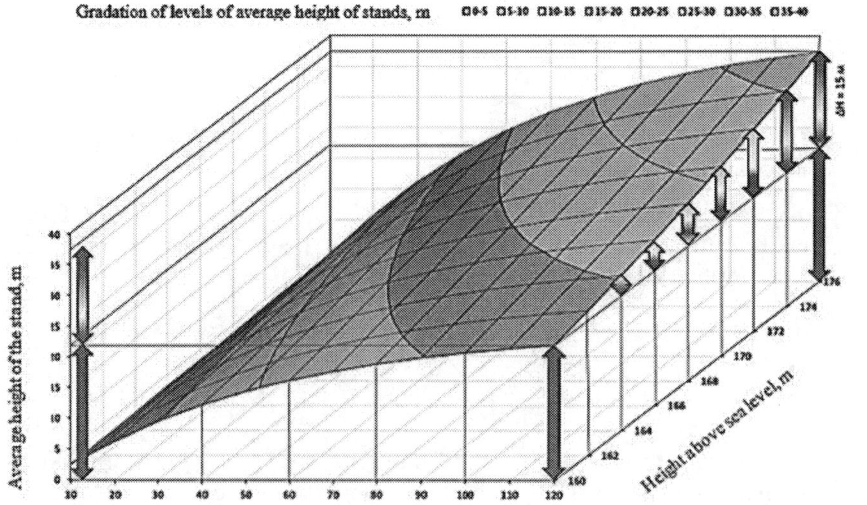

Fig. 4. Graphical representation of the excess of the average height of forest stands at different elevations of the geomorphological profile.

Legend: ⟺ the top of the moraine hill, ⟺ the slopes of the moraine hill, ⟺ the surface of the water-glacial plain, ⟺ the surface of the valley-like depressions

All the features of the revealed conformities are clearly visible in the upper part of the figure, bounded by a green outline. The proposed statistical model of average height (2) is the basis for assessing the tree stock and biological productivity of forest elements both with age and in statics with a known completeness and proportion of the species in the composition of the forest stand. So, in accordance with the normative and reference materials for assessing productivity, it is only necessary to calculate the potential reserve of individual elements of the dendrocenosis (tree species), provided that they have formed pure in composition (10 units) and maximally closed forest stands with the completeness taken as 1.0. To do this, it is enough to use the standard tables compiled for forest-forming species, in which the functional relationship of the stock with the average height of the forest stands is presented. So, the stock in 100-year-old pine stands with H_{UM} = 160 m is from 330 to 400 m ^{3} / ha, with H_{UM} = 175 m, respectively 730-844 m ^{3} / ha; in larch crops with H_{UM} = 160 m - 350-380 m ^{3} / ha, with H_{UM} = 175 m - 680-770 m ^{3} / ha; in birch stands with H_{UM} = 160 m - 220-240 m ^{3} / ha, with H_{UM} = 175 m - 550 m ^{3} / ha; in oak stands with H_{UM} = 160 m - 300 m ^{3} / ha, with H_{UM} = 175 m - 530 m ^{3} / ha. A decrease in the completeness of forest stands and the share of wood species in the composition of the forest stand directly proportionally reduces the productivity of a particular forest element.

Thus, the demonstrated method of collecting and analyzing data located at different elevations with the corresponding structure of soil geomorphology allows us to come close to identifying the conformities of formation of forest phytocenoses in the context of soil types and soil differences.

3 Conclusions

1.The productivity of forest stands is influenced by the location of the stands on the geomorphological profile of the FES territory.
2.In the range of elevations from 160 to 173 m above sea level, the excess of average heights at the age of mature forest stands (120 years) is 15 m, which corresponds to four bonitet classes.
3.The proposed statistical model of average height is the basis for modeling the course of growth of forest stands both in height and in terms of productivity at a given completeness and proportion of tree species in the composition of forest stands.

Reference

1. A. R. Vargas de Bedemar, *Taxation of the forest dacha of the Petrovsk Agricultural Academy*, 281 (1863)
2. D. A. Danilov, V. P. Tsarenko, Izvestia SPbGAU, **38**, 76 (2015)
3. Yu. P. Demakov, A. V. Isaev, Eko-Potential, **4(8)**, 41 (2014)
4. Yu. P. Demakov, A.V. Isaev, N. B. Nureyev, I. I. Mityakova, Bulletin of the Volga State Technological University, **3(39)**, 30 (2018)
5. N. V. Dylis, V. N. Sukachev, *Fundamentals of forest biogeocenology* (Nauka, Moscow, 1964), 574
6. B. D. Zaicev, *Forest and soil* (Publishing house of the forest industry, M., 1964), 159
7. V. D. Zelikov, *Soil and quality of plantings* (Timber Industry, 1971), 120
8. S. V. Zonn, *The influence of the forest on the soil* (Publishing house of the Academy of Sciences, M., 1954), 160
9. L. O. Karpachevsky, *Forest and forest soils* (Lesnaya promyshlennost, M., 1981), 264

10. L. O. Karpachevsky, *The diversity of the soil cover in the forest biogeocenosis* (Moscow State University Publishing House, Moscow, 1977), 312
11. O. V. Martynenko, Influence of soil factors on the growth and productivity of pine plantations on the example of SCHUOLH of the Moscow region (Moscow, 2011), 135
12. M. D. Merzlenko, P. G. Melnik, G. N. Vysotsky, Sustainable forest management, **44** (2015)
13. N. P. Remezov, P. S. Pogrebnyak, *Forest soil science* (Publishing House of the Forest Industry, 1965), 323
14. V. K. Khlyustov, *Comprehensive assessment and management of wood resources: models-standards-technologies <Book I>* (Publishing house of the Russian State Agricultural University named after K.A. Timiryazeva, M., 2015), 389
15. V. K. Khlyustov, *Comprehensive assessment and management of wood resources: models – standards – technologies <Book II>* (Publishing house of the Russian State Agricultural University named after K.A. Timiryazeva, M., 2015), 449
16. A. I. Aristarkhov, *Optimization of plant nutrition and the use of fertilizers in agroecosystems* (TsINAO, M., 2000), 520
17. N. V. Voitovich, *Soil fertility of the Non-Chernozem zone and its modeling* (Kolos, M., 1997), 387
18. A. G. Zamaraev, V. I. Savich, V. G. Sychev, *Energy and mass exchange in the link of field crop rotation, part 2* (VNIIA, M., 2005), 336
19. V. I. Savich, V. A. Sedykh, S. A. Izmailova, Agrochemical Bulletin, **4**, 21 (2012)
20. V. I. Savich, O. D. Sidorenko, L. V. Mosina, J. Norovsuren, D. S. Anikina, Int. agricultural zh-l, **1**, 32 (2019)
21. V. I. Savich, S. P. Torshin, S. L. Belopukhov, V. A. Sedykh, *Agroecological assessment of organo-mineral complex compounds of soils* (OOO Megaprint, Irkutsk, 2017), 298
22. K. M. Tuğrul, *Soil Management in Sustainable Agriculture, Sustainable Crop Production, Mirza Ha-sanuzzaman, Marcelo Carvalho Minhoto Teixeira Filho, Masayuki Fujita and Thiago Assis Rodrigues Nogueira, IntechOpen* (2019)
23. J. D. Phillips, Soil Complexity and Pedogenesis. Soil Science, **182(4)**, 117 (2017)
24. P. Lavelle; A. Spain; M. Blouin; G. Brown; T. Decaëns; M. Grimaldi; J. J. Jiménez; D. McKey; J. Mathieu; E. Velasquez; A. Zangerlé, A Review of Concepts and Future Research Questions. Soil Science, **181(3/4)**, 91 (2016)
25. J. Shengyao, Y. Xianglong, Z. Jianming, L. Guang, Soil Science, **179(4)**, 211 (2014)
26. F. Nicholson, A. Bhogal, M. Taylor and et al., Soil Science, **183(3)**, 89 (2018)
27. W. Dong, X. Zhang, H. Wang and et al., *Effect of Different Fertilizer Application on the Soil Fertility of Paddy Soils in Red Soil Region of Southern China* (2012)
28. W. Jr. Shepherd, P. Liang, X. Jiang, M.P. Doyle, M.C. Erickson, *Microbiological analysis of composts produced on South Carolina poultry farms* (2011)

Comparative effectiveness of preventive medical care when orthopedic pathologies in cows

Vladimir Komarov[1,*], *Boris* Belkin[1], and *Anna* Gutsyna[2]

[1]Oryol State Agrarian University named after N.V. Parakhin, General Rodin st., 69, 302019 Oryol, Russia
[2]Oryol State University named after I.S. Turgenev, Komsomolskaya st., 95, 302026 Oryol, Russia

Abstract. The available bibliographic sources on the study of distal limbs diseases in cows were analyzed. The causes and factors that contribute to limb pathology risk in cows at milk production enterprises are emphasized. Lack of active exercise, inadequate feeding rations, insufficient veterinary hygiene, high contamination of premises and low or insufficient level of preventive medical care have an increase in the incidence of animals up to 40% of the herd. Therefore, the problem of cows' distal limbs pathology treatment and prevention is currently of great importance. In the complex preventive and medical care against the disease of the distal limbs in cows, functional trimming of the hoof horn has an important veterinary hygienic value. The beginning of therapy at an early stage reduces the risk of developing more severe pathologies of the limbs. The use of "EMS-I type A" for the treatment and prevention of distal limb lesions, depending on the degree and size of tissue damage, provides recovery after 2-3 bandages and disease containment in the study groups.

1 Introduction

In modern conditions of dairy farming, cattle limbs diseases cause considerable economic damage to the commercial farm units. The mechanism of mass limbs diseases, including hooves is a complex of various causes: all-year zero pasture of animals, high contamination of premises, manure gutters, pens and places of cow alley, inadequate, insufficient and unbalanced feeding, low level of preventive medical care [1, 2, 3].

Currently, more and more reports are appearing about problems in dairy cattle breeding associated with the occurrence of distal limbs diseases, which, of course, negatively affects the health of animals and their productivity. The increase in the incidence of the distal limbs depends on many factors: all-year zero pasture of cows, high contamination of farms, sections, boxes, pens and places of cow alley, lack of active exercise, inadequate and insufficiently balanced feeding regimes, low level of preventive medical care. The high level of morbidity in cows confirms the complexity of the multifactorial etiology of limb lesions. According to many researchers, it is noted that this problem in our country has

*Corresponding author: komarov.volodya@yandex.ru

begun to manifest itself quite acutely in recent decades with the reduction of domestic breeds of animals and the active import of Holstein cattle, which are predisposed to have a problem with limb diseases. Domestic breeds of cows were created by animal breeders in certain territories with the peculiarities of natural and climatic conditions [4, 5].

Due to the spread of limb pathology in cows, all farms suffer considerable economic damage, which consists in a significant decrease in milk yield and milk quality, an increase in premature cow disposal, a decrease in reproductive capacity and an increase in the cost of treating animals [6, 7].

Based on the above, the aim is to determine the level of incidence of orthopedic pathology in cattle and to conduct a comparative analysis of preventive medical care when distal limb lesions in cows.

2 Methods and methods

The research was carried out on a complex and a dairy farm, located in the Oryol region. The object of the study was cows (healthy and sick with orthopedic pathologies) of a black-and-white Holstein breed. During the clinical examination, the animals' body temperature, pulse rate, and respiration were measured.

According to the principle of analogs, animals were selected in experimental groups. Two series of experiments were carried out. In the first one the therapeutic effectiveness of the drug was studied and in the second, its prophylactic effect.

Spray with an antibiotic, Top-Hooves gel for hooves, oxytetracycline hydrochloride pulvis, copper sulfate pulvis, preparation "EMS-Y type A" were used for the treatment and prevention of distal limbs diseases in cows.

The preparation "EMS-Y type A" is a disinfectant made in the form of a solution for external use, containing free iodine in the form of iodophor (patent No. 2535016) as an active substance and surfactants as auxiliary components that do not have an irritating effect on the skin. The composition includes crystalline iodine, oxyethylidenediphosphonic acid, dimethyl carbinol, neonol, glycerin, water. In appearance, the liquid is dark brown in color with a faint smell of iodine, mixing with water in any ratio. It has significant washable properties and a bactericidal effect against non-sporous microflora. Glycerin softens the effects of iodine on the skin. The drug is developed and mass-produced by the limited liability company “Experimental and Technological Company "Etris"”.

Functional processing and trimming of hooves were carried out with special discs, cutters, forceps and knives.

The results of the studies were subjected to statistical processing.

3 Results and Discussion

Due to the spread of limb pathology in cows, all farms suffer considerable economic damage, which consists of a significant decrease in the quantity and quality of milk, an increase in the loss of cows from the herd, a decrease in reproductive ability and an increased cost of medical and preventive measures [4, 6, 8].

According to scientists, the limbs pathology of highly productive cows is in the second and third place after obstetric and gynecological pathologies and mastitis. In some farms it affects up to 35-40 % of the herd. Studies show that hoof pathology is a polygenic and multifactorial disease, so therapy program should be aimed at the complex prevention of this pathology. The development of safe and effective means and methods of its application for care and prevention of orthopedic pathologies in cows is of great veterinary, zoohygienic and economic importance [2, 7].

At the complex, three groups of dairy cows with 65-70 heads in each were examined. 20-30% of the animals in each study group were identified with obvious signs of limb lesion. These cows visually showed different degrees of lameness when moving.

On the dairy farms 623 milk cows were examined. 210 animals (33.7%) were identified with obvious signs of limb lesion.

Table 1. Analysis of the morbidity of cows with distal limb lesions

	Numbers of cattle	**%**
Healthy animals	413	66,3
Animals with signs of limb lesions	210	33,7
Total amount of animals	**623**	**100**

Changes in the anatomical structure of the hoof, pododermatitis, interdigital phlegmon, necrotic dermatitis, laminitis, interdigital dermatitis, sole exfoliation, plantar ulcers, hoof deformity, calluses (interdigital growths) were recorded during the examination of animals (Figure 1).

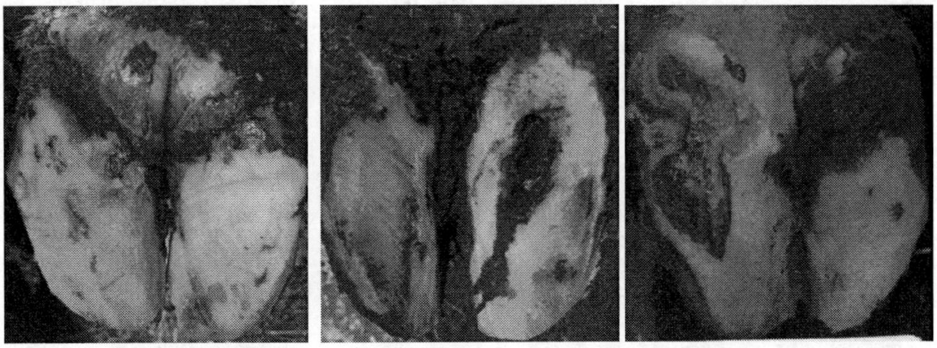

Fig. 1. Clinical signs of distal limbs pathology

Dairy cows, heifers on rearing are periodically passed through hoof baths filled with a 10% solution of copper sulfate, 3-5% solution of formalin and "SKIF-D" for the hoof diseases prevention. In addition, cleaning and individual treatment of hooves are carried out periodically in sick animals, with the use of disinfecting drugs and antibiotics. Despite the measures taken at the dairy complex with a yard housing in different age groups, from 20 to 45% of sick animals with distal limbs pathologies are observed. The significant number of affected hooves on the hind limbs were noted in clinical examination of the herd compared to the forelimbs. Thus, the right hind hooves are affected in 50% of animals, the left hind in 23.7 %, the right forelimbs - 11.3 %, the left forelimbs - 15 %.

On the dairy farms during the period of functional processing and trimming of hooves, the morbidity of limbs in animals that are in winter on tie-up housing was as follows: the right forelimb is 4.8 %; the left forelimb is 2.7 %; the right hind is 48.5 %; the left hind is 44 %.

Causes of the development of limbs pathology in cows at the dairy complex:

1. Lack of active exercise, which would contribute to the natural attrition of the hoof horn.
2. Inadequate feeding rations.
3. Insufficient level of veterinary hygiene.
4. High contamination rate in the premises where the animals are kept.

5. High level of limb injuries.
6. Low or insufficient level of preventive medical care, namely: insufficient examination of animals, untimely functional and preventive trimming of overgrown hoof horn and late treatment of the developed pathology.

Before the experiments, bacteriological and mycological studies of inter-hoof cleft scrapings of sick animals were carried out. In the samples from the affected limbs, the causative agent of necrobacillosis was not isolated, but the pathogenic Staphylococcus aureus (B – hemolytic, plasma-coagulating) was isolated. And during mycological studies, fungi of the genus Aspergillus (A. niger, A. flavus), Penicillium, Mucor were found. Laboratory studies of the microflora growth retardation zone in Petri dishes showed that the isolated microorganisms have a certain sensitivity to antimicrobial drugs and "EMS-I type A" agent.

In the experiment, the agent "EMS-I type A" was tested by individual treatment of sick animals. Before the treatment, the animals' hooves were thoroughly cleaned and trimmed. To clear the hooves of the animals, a set of tools was used: pliers, a hoof knife. The instrument was disinfected before use. A special machine was used to fix the animals. Cleaning hooves of cattle was carried out in the following way. First of all, the hoof was cleaned of mechanical contamination. Then with the help of hoof knife, the old horn with a yellowish color was carefully cut off from the sole and digital torus, and the overgrown hoofed wall was bitten off with pliers, leaving it at the level of the white line. Special milling cutter was used to trim the hoof horn. The damaged areas were cleaned of necrotic tissues and treated with a 3 % solution of hydrogen peroxide. For treatment, a 50% foamy solution of "EMS-I type A" was used on the cleared surface of the hooves and a gauze bandage was applied. The bandage was changed after three days. In the presence of a pathological process, the bandage was applied repeatedly. Figure 2 shows the hooves of cows before and after treatment.

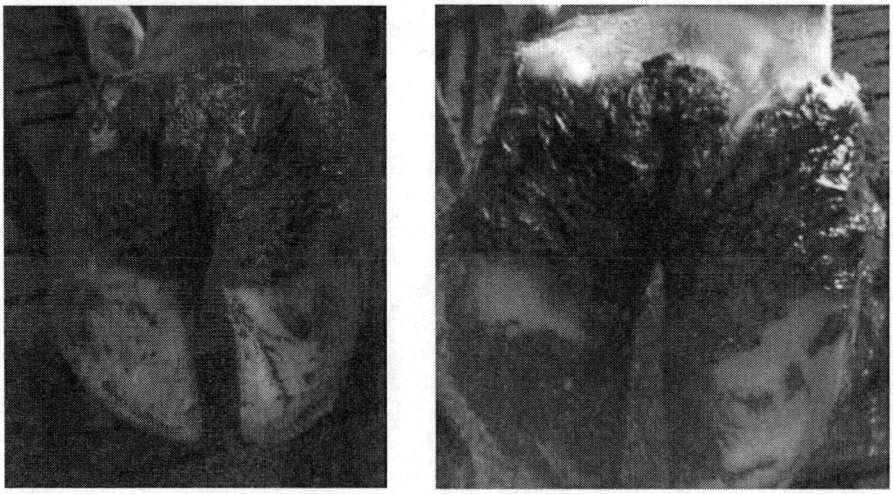

Fig. 2. Images of a cow's limb before and after treatment

Individual clearing and treatment of sick hooves with the use of 50 % foam agent "EMS-I type A" ensured the recovery of animals after 2-3 bandages. The recovery of the animals depended on the degree and area of tissue lesion.

Currently, in veterinary practice, there are a large number of monocomponent disinfectants used as preventive medical care against the lower limbs' pathology of cattle. As preventive measure, farms use formalin baths with a 5% solution of formaldehyde, copper-vitriol baths with a 10% solution of copper sulfate, Hooves solutions and others. However, these disinfectants do not solve this problem. The most effective ones are combined disinfectants; their main advantages are absence of immunosuppressive action, low toxicity to cows, and breadth of biocidal action against potentially pathogenic microorganisms [8, 9, 10].

"EMS-I type A" is an important component in the integrated method of combating widespread pathology when studying the hoof diseases prevention effectiveness in the cows group passing through foot baths filled with a foamy preparation of 10% solution. The control group of animals was passed through hoof baths with a 10 % solution of copper sulfate. Preparation, use and replacement of solutions in hoof baths were carried out according to generally accepted methods (Figure 3).

The cow's hooves were completely immersed in the foam solution, and the animal, coming out of the bath, carried a certain amount of foam on the limbs, which was kept on the hooves for about 10 minutes providing a longer treatment.

Fig. 3. Hoof bath filled with foam and hooves of animals after treatment with foam of 10 % solution of "EMS-I type A

During the period of the experiment, there was a decrease in the number of sick cows in each study group. When treated with a foam solution with a concentration of 10 % of the agent "EMC-I type A", the lameness reduction was 7 %. In the control group of animals, when treated with a 10% solution of copper sulfate, the decrease was also 7 % as in the experimental group.

The cows were examined during the study period. It was found that the preventive effect was shown by all the solutions of drugs used at the initial stage and with a small area of hoof tissue lesion. The studied solutions of drugs did not contribute to wound healing with a stronger lesion of hooves. Timely clearing and trimming with the formation of the correct anatomical shape of the hooves is a significant effect in the prevention of diseases in cows.

4 Conclusions

Orthopedic diseases in cows are a widespread pathology. The level of cows' morbidity with this pathology in farms, depending on the use of preventive medical care is in the range of 20 to 45%, and regardless of veterinary hygiene, the hind limbs are affected. Preventive examination of animals, systematic treatment of hooves is the most essential link in the complex preventive medical measures that are aimed at combating this disease. Individual cleaning and therapy of animals was provided in the distal limbs with the use of "EMS-I type A". This depended on the degree and area of tissue damage, provided recovery of animals after 2-3 bandages. Cattle driving through hoof baths filled with a foam solution of the "EMS-I type A" product in order to prevent pathologies of the hooves has a deterrent effect on the growth of the disease.

References

1. E. Veremey, V. Zhurba, V. Rukol, A. Stekolnikov, B. Semenov, Animal husbandry of Russia, **2** (2017)
2. V. G. Semenov, A. V. Chuchulin, D. A. Baimukanov, V. G. Sofronov, A. Kh. Volkov, N. K. Kirillov, A. V. Aldyakov, IOP Conference Series Earth and Environmental Science, 433 (2020)
3. Samolovov, S. V. Lopatin, Innovation and food safety (Novosibirsk State Agrarian University), **2** (2013)
4. S. Karamaev, N. Cumshewa, K. Valitov, A. Karamaeva, E3S Web of Conferences, **175(5)** (2020)
5. V. G. Semenov, A. V. Chuchulin, Scientific notes of Kazan State Academy of Veterinary Medicine named after N. E. Bauman, **226(2)** (2016)
6. V. B. Andreev, B. L. Belkin, L. S. Gromov, T. E. Ilyin, V. Y. Komarov, Kaluga: Publishing House: OOO Noosfera, 104 (2017)
7. L. E. Green, V. J. Hedges, Y. H. Schukken, R. W. Blowey, A. J. Packington, Journal of Dairy Science, **85** (2002)
8. Ya. Batrakov, Z. K. Zueva, N. N. Teterev, Veterinary Medicine, **5** (2010)
9. E. M. Marin, V. A. Ermolaev, P. M. Lyashenko, A. V. Sapozhnikov, S. N. Hokhlova, A. L. Hokhlov, S. N. Zolotukhin, D. M. Marin, V. I. Ermolaeva, Research Journal of Pharmaceutical, Biological and Chemical Sciences, **8(3)** (2017)
10. R. Uvarov, A. Briukhanov, B. Semenov, A. Nazarova, BIO Web of Conferences, 27 (2020)

Analysis of ecological tourism in Russia and USA

Dennis Dayneko[1,*], *Alevtina* Dayneko[2], and *Victoria* Dayneko[2]

[1]Irkutsk Scientific Centre SB RAS, 664033, Lermontov str., 134-117, Irkutsk, Russia
[2]Irkutsk National Research Technical University, 664074, Lermontov str., 83, Irkutsk, Russia

Abstract. The necessity of development and marketing of ecological tourism is substantiated today by requirements for sustainability of entrenched economic activities in specific regions, as well as by problems of environment preservation. The ecological tourism with active involvement of local residents in particular should become an important element for development of the regions promoting both economic development and preservation of natural and cultural heritage of our countries and the regions. The work presents the modern concept and approaches to ecological tourism, as well as the trends of its development in Russia and USA. The method of sociological survey and comparative approach are suggested as the tools to evaluate the role and effectiveness of ecological tourism, which is the most prospective way of tourism development, and is an important component of the socio-economic development too, capable not only to develop, but to protect natural, historic, and cultural heritage. The preliminary conclusions about the importance of the ecological tourism development for the sustainable development of the regions based on the entrenched economic activities of the local population. There is a need to implement modern effective and efficient eco-tourism practices in specific regions of Russia and USA.

1 Introduction

Today there is a need to support innovative activities in various industries of economy and in tourism in particular. It is the ecological tourism, which is the global activity and an important component to facilitate economic and political relations between our countries and the development of the regions involved and to preserve their natural and cultural values.

Ecological tourism is a popular theme in developed countries but a relatively new concept in many regions in Russia. The distinguishing feature of ecological tourism is that the common tourism in general implies just travelling a distance from home, while ecological tourism is defined specifically as "responsible travel to natural areas that conserves the environment and improves the welfare of local people [1]. Other definitions of the ecological tourism are:

*Corresponding author: dvdayneko@gmail.com

- "tourism involving travel to areas of natural or ecological interest, typically under the guidance of a naturalist, for the purpose of observing wildlife and learning about the environment" [2];
- "tourism that is designed to contribute to the protection of the environment or at least minimize damage to it, often involving travel to areas of natural interest in developing countries or participation in environmental projects" [3];
- tourism to places having unspoiled natural resources [4];
- tourism to exotic or threatened ecosystems to observe wildlife or to help preserve nature [5].

A distinctive feature of ecotourism from conventional tourism is that mass tourism means simply traveling a distance from home, while ecotourism is defined as responsible travel to natural areas that promotes the protection of nature and supports the well-being of the local population and includes interpretation and education. The International Ecotourism Society (TIES) defines ecotourism as a system of travel and tourism that promotes environmental conservation by educating travelers about sustainable lifestyles [1]

2 Development of ecological tourism

The innovative development of the economies requires a deep study of domestic and foreign tourism experience [6]. After studying a variety of modern local, national and international sources about tourism, authors came to the conclusion that eco-tourism should be an important component of the socio-economic development of specific regions that can preserve its natural, historical and cultural values.

According to statistics provided by the World Tourism Organization (UNWTO), ecotourism accounts for about 12-15% of all tourists flows. Moreover, the number of eco-tourists increases by 25-30% annually. It is expected that the share of Russian and American tourists in the total number of ecotourists will significantly increase [7].

Actually, ecotourism received its initial development back in the 1970s in the United States, the country that for many years is the leading tourist region of the North America. The USA is one of the most popular tourist destinations. This is due to the many natural, cultural and historical sights and monuments. The ecological tourism plays an important role, especially now, in the whole United States tourism industry.

According to the UNWTO data, the number of tourists in the United States in 2017 and before the Covid-19 pandemic amounted to more than 75.9 million visitors, which allowed the country to take the third place in the ranking of host countries with tourist spending $210.7 billion [8]. Despite the fact that the United States is one of the largest markets for recreational tourism for residents of European countries [9], it is impossible not to underestimate the role of ecotourism in Russia.

Russia, like the United States, has very wealthy recreational resources that are studied and used further for ecological tourism purposes. Our countries have a huge number of specially protected natural areas (SPNAs) that have become world-famous and are of interest to both tourists and local residents, especially those who are interested in wildlife.

It should be noted that the specifics of American and Russian attractions is not only in their diversity and scale, but also in their geographical location at a considerable distance from each other. And if day trips, thematic voyages, for example, to places known out of westerns, to Indian indigenous territories or to the largest megacities are very popular in the United States, then in Russia, these are, first of all, places of wildlife and natural attractions.

The understanding of the tourism and of ecological tourism in particular in Russia by local residents is usually strongly correlated with backpacking, recreation activities and leisure time. Recreation is considered as a sort of amusement or an activity a person takes part in for pleasure or relaxation rather than as work. Hence local tourist areas are mostly areas for recreational stay, sports, and games, often with children. Considering the climatic conditions of the region, with average annual temperatures below zero degrees Celsius, the outdoor recreation is more closely related to the mass tourism and usually to the sea or to the ocean. For example, sunbathing, ashore and in-water attractions visited are usually both recreational activities and tourist activities, with no regard to the duration and location of the trip. Whereas, a traveler who is going out or backpacking, for example, to the nearby forests, for a day or for a weekend, or a longer trip, if visiting further destinations, can be considered as a tourist participating in an ecological tour.

Every year ecological tours attract many thousands tourist from Russia and abroad. There are thousands of places quite interesting for the visitors not only during the summer but during other seasons too. Though, most of the visits are definitely during the summer months. Some people like to stay in comfortable cottages and bungalows, others prefer to spent nights in tents or aboard the boats. However, there is still no such a developed ecological tourism infrastructure in Russia as we can see in other eco-attractive destinations, and, especially, in the United States.

Today we can name at least seven defining practices to describe ecotourism in international practice, which are to be adopted and implemented in Russia in particular too. The profound studies of ecotourism including University programs use this as the following defining points for the working definition [10]:

a. *It assumes travels to sites of real nature:* These places are usually quite remote areas, whether inhabited or uninhabited, and are officially under the environmental protection at the international, national, regional, municipal or private level.

b. *Minimum impact:* Ecological tourism always strives to minimize the adverse anthropological impact, which can be achieved via, for example, well established and organized tours and use of recycled materials or local construction materials for hotels and other buildings required to provide the tourists infrastructure and environmentally and culturally sensitive architectural design, as well as, alternative energy sources, recycling and/or non-harmful disposal of wastes. Minimization of impact also requires that the behavior of tourists and number of visits are regulated to ensure no damage to the local ecosystem.

c. *It educates environmental awareness:* Ecological tourism always means education, for both tourists and residents of the local communities. It is very important to provide travelers with the relevant information about the country, environment and people, as well as about the formal and informal institutions involved well before the departure to a destination. According to Ecotourism Societies guidelines it is required "to learn about the places and peoples visited" and "to minimize their negative impacts while visiting sensitive environments and cultures". Therefore, good eco-touristic programs should be provided by well-trained, multilingual guides with the knowledge of natural and cultural history, environmental interpretation, ethical principles and effective ways of communication. All ecotourism projects are supposed to educate tourists and local community members, including adults and kids in the host country.

d. It provides direct financial benefits for nature protection and conservation: Ecotourism not only promotes environmental awareness but is directed to raise some funds required for the environmental protection, research and education. This can be achieved via a variety of financial mechanisms, including park entrance fees, eco-tour and transportation companies' contributions, special offers in hotels, eco-taxes and voluntary contributions.

e. It should provide financial benefits and job advantages to local people: It is a common believe that National Parks and other conservation areas will only survive if there are "happy people" around the area. It is getting obvious that local communities must be involved with and receive income and other tangible benefits related to state of life improvement and local infrastructure as a result of the conservation areas promotion and construction of it's tourist facilities. Actually, people from the local communities surrounding an ecotourism attraction should run campsites, lodges, guide services, restaurants and other concessions. Such a partnership is suggested for all sorts of investors planning their activities in parks or other tourist destination. More importantly, since Ecotourism is often viewed as a tool for rural development, it should also allocate most of economic and political control to the local communities. This is the most common mistake of the operators and the foreign partners to neglect the role of the local officials and business people, including those living in villages and nearby settlements.

f. It respects local culture: Local culture is another concern of the ecotourism. Ecological tourism is not only "greener" but also less culturally intrusive and exploitative than any other forms of tourism. Ecotourism always strives to be culturally respectful and have a minimal effect on both the natural environment and the human population of a host country. Sometimes it is not that easy, since ecotourism often involves travel to distant and secluded places where small and isolate communities have had little experience interacting with foreign visitors. Sometimes, even ecotourism involves an unequal relationship of the visitors and the hosts and a commoditization of the relationship through exchange of money. Actually, responsible eco-tourists are learning beforehand about the local customs, dress codes and other social norms and do not intrude into the community unless either invited or as a part of well-organized tour.

g. It should support human rights and local democracy: Traditionally, it is tourism that is loudly hailed as a tool for building international understanding and world peace promotion. However, it is not always the fact or it does not necessarily happen immediately upon a visit. Most often tourism just bolsters the economies of various countries including even those that tend to be or are undemocratic and repressive. While regular tourists pay low attention to the political system of a country they visit, unless a civil unrest brakes out in there and which is harming the tourists too. In fact, rural population living around national parks and close to other ecotourism attractions are in contradictions with the national and regional laws and regulations, and multinational corporations for control of the assets and possible benefits.

3 Marketing in the sphere of ecological tourism

Marketing in the sphere of ecological tourism is the same activity of planning, development, promotion, and selling of touristic products, including goods and services, stimulating their demand and price setting, as for mass tourism. Marketing in Ecological tourism also can be defined as a complex of measures to increase the market share of eco-touristic products and services. Touristic marketing is a system of management and organization of touristic organization for development and provision of new ecological and eco-effective kinds of touristic services and excursions for their further sale to make a better profit resulting from eco-oriented attitude to services and products provided and considering the processes happening in tourism markets. The major task of Ecological tourism marketing is to stimulate eco-oriented demand for recreation activities, to set the prices right, to provide market segmentation and develop a favorable assortment of eco-touristic services and excursions and attractive advertisement.

The specifics of the marketing in the sphere of ecological tourism are that the major product is actually intangible, the source is inseparable from the service offered itself, the great correlation of the service quality with the local culture and level of training for the personnel, and that none of the eco-product can be stored or kept in a storage. The touristic services offered in the ecotourism market are intangible, you cannot taste or try them, really smell or hear before an actual purchase of such a product.

The marketing of any eco-tour should start with marketing analysis of the environment. The overall and detailed perception of the markets and professional feedback is the responsibility of marketing specialists, whose job is not only to stimulate the demand and increase sales but to develop a firm considering employees' potential, financing, management, price formation, customers' feedback. It implies the creation of channels of distributions, marketing programs development and implementation and budgeting.

Marketing is a good tool to stimulate the demand for ecological tourism, the knowledge of which is a responsibility for everyone involved in tourism industry. The major marketing task in ecological tourism is creation of an ecological tour/product/service to motivate and satisfy clients eco-oriented demands. The major task of a professional marketing specialist in the sphere of ecological tourism is to attract and keep aware customers about the advantages of ecological tours and services, to satisfy the existing and create new demand to increase the profitability. Therefore, marketing in the sphere of ecological tourism is to solve such tasks, as, for example, determination and/or formation of needs and demands for ecological services and touristic products in the market, development and offer of high quality products and services at a market price, stimulation of demand and distribution of supply and its availability. Hence, we can name major tasks of the marketing management in the sphere of ecological tourism. These are demand regulation, analysis, planning, ecological tourism products sales and control, and touristic markets monitoring.

The major feature of the concept of ecological tourism marketing is the nature protection compound. Marketing in the sphere of ecological tourism always assumes a high level of responsibility of marketing specialists for the environment and nature. The nature oriented concept of marketing implies environmental friendliness and careful management of nature resources. Touristic companies and ecological tour operators have to consider their potential negative influence on the environment and strive, if not fully eliminate it, than, rather to minimize it via related environmental activities. This is what is called the environmental friendliness. The care for the clients health in places of their stay is the task of managers in eco-oriented touristic companies.

The modern marketing strategy always requires the inclusion of the ecological component. The issue of social responsibility and of ethical behavior are especially important in tourism business today. The social concept of marketing means employees and specialists awareness about the lawful and acceptable business practices and have to know the business dealing principles, which are based on ethics, corporate culture and interests of clients.

The customer oriented approach is the key to success for any touristic company, since, touristic business consists not only in the development of a final touristic product, but, rather, in satisfaction of customers' demands. Herewith, customer demands can be changed and corrected, and specifically due to professional work of marketing specialists. Marketing can manage the demand and promote ecological tourism products at the target markets, can help to evaluate dynamics of sales in every strategic direction and determine prospective incomes. Once the ecological tour aims of a company and possible financing are determined, the marketing can solve the tasks, which have been set, effectively to maximize the profitability.

Marketing planning in ecological tourism is required to forecast levels of ecological tours sales and incomes. Therefore, current and planned sales and their time are analyzed and related strategies are suggested. First of all, marketing specialists have to consider the existing markets shares and further market penetration with ecological tour products and only than the possibilities of new markets development. The following steps are development of new products for existing markets, and offer of new products in new target markets, i.e. diversification.

Once a marketing strategy is developed, the related programs for its implementation are defined. Therefore, the strategy includes frame, resources, and personnel to achieve the strategy aims. The qualified and professional employees are involved, training and educational programs for the personnel are introduced, products and services quality control and improvement programs are implemented, the services and touristic products are promoted, informing about the positive and innovative features.

It is worth to mention the particular importance of the personnel awareness and involvement in touristic companies into the development and implementation of corporate strategies. The firm employees in touristic business are the most important element of successful product and service. Formation of terms and conditions, which encourage good services for clients, is often called *internal marketing* [11]. Today, the most successful tourism enterprises have to offer in the markets not only ecological tourism products, but to develop corporate ethics and mentality to satisfy the customers' specific ecological demands. The employees of travel agencies should understand prestige value of ecological tours and advantages of their promotion. The key aspect of the successful internal marketing is employees' motivation. Therefore, along with various financial advantages, the positive influence plays credibility and prospects of their workplace. Other reinforcements and motivators are, for example, a flexible work schedule, child and kids care programs, and other non-financial benefits. The internal marketing will help tourist companies attract and keep good employees too.

We can specify at least three major current tasks of ecological tourism marketing: 1) maintaining and developing existing markets of ecological tourism; 2) search and development of new markets (new clients and destinations); 3) reduction of seasonality influence on demand.

In order for a touristic company to achieve these tasks there is a need to perform the following at least annually: to correct the marketing plan; to assign group of professionals to develop a program of the company's specific, collaborative, research, and other related activities to improve the marketing plan and attract more customers; to cooperate with other touristic companies to allocate, provide, and serve customers in the market of ecological tourism; to test new ecological tourism products and services; to develop mutual measures to promote ecological tourism products in the target markets; to develop target segments of the market with seasonal character of services; to offer ecological tour packages to multiple destinations according to customers preferences and inquiries.

The major task of the marketing of ecological tourism is to understand the specific ecological interest of customers and offer them the best suiting ecological destinations and services. As a matter of fact, the most successful touristic companies offering ecological tourism products and services, first of all, study from the customers and their demands. It is called "getting closer to the customer". Later they have just to monitor their client and their demands to improve their capabilities and services offered.

Hence, the basic functions of ecological tourism marketing should be the following:

a) creation of ecological tourism products and services to be offered to clients in the region studied;

b) promotion of the ecological touristic products and services in the market, including advertisement, and consequent sale;

c) arrangement of direct and in person sales;

d) setting prices and achieving an allowable level of sales and incomes.

In order to comply with international norms and standards, marketing specialist of ecological tours should follow three main tourism marketing functions proposed by International Tourism Organization:

1. Customer relations and feedback establishment. The major task is to convince clients that ecological destinations offered fully satisfy their expectations and demands;
2. Development, including design of innovations;
3. Control, which is analysis of the results of activities to promote ecological tourism services in the market and verification of results real compatibility to existing touristic capabilities.

Therefore, the especial importance in the marketing of ecological tourism agency plays the relations with the customers. The long lasting relations with clients can be more lucrative and cost less than the marketing expenses, incurred to attract the interest of prospective consumers of ecological tourism services.

If the long lasting relations with ecological tourist are established, it provides a synergy effect, expressed even in adherence and devotion of clients to the touristic agency and their reluctance to change the tour company providing ecological products and services when repeating purchases. It is an interesting fact, that the Pareto effect is clearly observed in the sphere of ecological tourism: 20% of consumers make 80% of a company's income.

Tourism and, especially, ecological tourism is totally different from other consumer industries. Hence, we can indicate its specific characters to be considered by marketing specialists:

1. While planning ecological touristic activities, the central issue to be considered are the preferences and demands of the final consumers;
2. Non-primary essence of ecological tourism services and lack of knowledge about the regions studied. The ecological tourism is not an essential product, though, the importance of the tourism and of ecological tourism, in particular, as the means of recreation and healthy life style, is on the rise;
3. Marketing plays a crucial role in ecological tourism industry. Since, there is no a reference samples of touristic tourism products and the ecological tours sellers have to provide the proof and sound arguments in advantage of their products and services. It is possible only with a well planned marketing program that uses tailored for the target markets instruments of advertising and promotion;
4. There is a need to constantly monitor the quality of ecological tourism products and services in especially due to their unstable provision and seasonality in some regions;
5. Ecological tourism services are unique. The repetitions of all its aspects and provisions are very uncommon. The schedules, routes, terms and conditions of services and prices usually always vary.

Therefore, marketing of ecological tourism for a touristic agency operating in selected regions should become a system of continuous harmonization of services and products provided with services and products demanded, which an ecology oriented company can offer with a profit and more effectively than the competitors in the market. Today, only a firm with creative and ecology oriented marketing concept can succeed in competitive environment, constantly searching new ways to promote their services in changing markets, actively influencing and interacting with the consumers.

4 Results and discussion

After studying the issue of the ecological tourism development in our countries the authors has figured out that the basis of ecological tourism in USA are natural, cultural and historical sights and monuments and in Russia are SPNAs mostly, which include nature reserves, national and natural parks, State reserves, natural monuments, dendrological parks and botanical gardens, medical and recreational areas and resorts, and territories of traditional nature usage.

In Russia there are 105 State nature reserves (*zapovedniks*), covering more than 330,000 km^2 or about 1.5% of the country's territory. There are also nature protected areas where only certain species are protected, and some certain activities are prohibited known as *zakazniks*, 64 national parks that cover approximately 155,672 square kilometers, 57 Federal reserves, 17 natural monuments, and 67 botanical gardens have been created in Russia in total [12]. Detailed information about 102 nature reserves, 39 national parks and 68 Federal reserves of the Russian Federation can be found on the website Specially Protected Natural Territories of Russia. [13] The total area of Federal protected areas is 63.3 million hectares, and together with regional SPNA territories, they comprise the area of 209.5 million hectares.

During the Soviet era only nature reserves (zapovedniks) and preserves (zakasniks) existed in Russia. Only in 1961, Russian geographers visited the United States to study international experience in creating protected areas intended for tourists to relax and teach them to take care of nature. This trip was an introduction to the USA experience in environmental protection. Soviet scientists visited the Yellowstone National Park and the Great Smoky Mountains National Park.

Only after this trip, Innokenty Gerasimov, director of the Institute of Geography, USSR Academy of Sciences, developed further the idea of creating nature parks in Russia. He proposed the creation of a Baikal nature park in 1965. A similar Nature park was designed in the Lake Seliger area on the Valdai Hills too. In 1966 Innokenty Gerasimov and Vladimir Preobrazhensky published an article about the need to create a system of Natue parks in the USSR. Since then, Nature parks had been thought of as recreation places for tourists to relax, but to protect animals and plants too. So, that is why in some territories of Nature parks tourists are not allowed to visit without a guide.[14]

Many of SPNAs are included into the UNESCO world heritage list. Moreover, some of them, such as, for example, lake Baikal in Russia, are under the threat of extinction. In total, there are 11 World natural heritage sites in the territory of the Russian Federation, which include: 13 nature reserves, seven national parks, three Federal reserves, several natural monuments and buffer zones of reserves: "Virgin forests of the Komi Republic "(Pechora-Ilych reserve and the national Park "Yugydva»); "Lake Baikal" (Baikal, Barguzinsky, Baikal-Lensky nature reserves, Pribaikalsky, Zabaikalsky and Tunkinsky national parks (partially), Kabansky and Frolikhinsky nature reserves); "Kamchatka Volcanoes" (Kronotsky reserve and Yuzhno-Kamchatsky wildlife preserve); "Golden mountains of Altai" (Altai and Katunsky reserves); "Western Caucasus" (Caucasian state reserve, Sochi national Park, Bolshoy Thach Natural Park, natural monuments "Buyny ridge", "headwaters of the Tsitsa river", "headwaters of the Psheha and Pshehashha rivers»; as well as part of the buffer zone of the Caucasian reserve, which goes along its Northern and North-Western edge with the width of 1 km); "Central Sikhote-Alin" (Sikhote-Alin reserve and national Park "Bikin"); Wrangel Island reserve; Ubsunurskaya basin reserve; "Putorana Plateau" (Putoransky reserve); national Park "Lena pillars; "landscapes of Dauria" (Daursky reserve).

All the listed above specially protected natural areas are objects of national heritage. The Ministry of natural resources and ecology of the Russian Federation carries out State management in the fields of organization and functioning of specially protected natural territories of Federal significance. However, government agencies do not always cope with their responsibilities, as have been noted by the head of the Greenpeace program for specially protected natural areas. In 2020, the World Heritage Committee considered five the most problematic Russian World heritage sites, namely: "Virgin forests of Komi", "Lake Baikal", "Volcanoes of Kamchatka", "Golden mountains of Altai", and "Western Caucasus" [15].

The protected areas of the United States and their designations are the following: 1) National Park System including National parks, National preserves, National seashores, National lakeshores, National forests, National grasslands; 2) National conservation lands, including National Monuments, National Conservation Areas, Wilderness Areas, Wilderness Study Areas, National Wild and Scenic Rivers, National Scenic Trails, National Historic Trails, Cooperative Management and Protection Areas, Forest Reserves, Outstanding Natural Areas; 3) National Marine Sanctuaries; 4) National Recreation Areas; 5) National Estuarine Research Reserves; 6) National Trails System; 7) National Wild and Scenic Rivers System; 8) National Wilderness Preservation System; 9) National Wildlife Refuge System, and 10) International protected area designations, which are .UNESCO Biosphere Reserves.

In the United States, in turn, there is also an extensive network of national parks and protected areas. Recreational areas in the United States consist of more than 36,000 protected territories with the total area of 1 233 161 km2, which is 14% of the total area of the country. [16] The system of U.S. National parks, consisting of 419 national landmarks, includes 62 major natural monuments, such as Acadia, the Everglades, the Grand Canyon, and Yellowstone, as well as 346 natural, cultural, and 11 other protected areas that are used in the tourism industry. [17]

The SPNAs of the United States are managed by an array of different federal, state, tribal and local level authorities and receive widely varying levels of protection. Some areas are managed as wilderness, while others are operated with acceptable commercial exploitation. The U.S. also had a total of 787 National Marine Protected Areas, covering an additional 3,210,908 km2 or 37 % of the total marine area of the United States. In year 2021, 803 wilderness areas have been designated, totaling 450,691.20 km2, which comprise about 4.5% of the land area of the United States. [18,19]

The National Wilderness Preservation System (NWPS) of the United States protects federally managed wilderness areas designated for preservation in their natural condition. Activity on formally designated wilderness areas is coordinated by the National Wilderness Preservation System. Wilderness areas are managed by four federal land management agencies: the National Park Service, the U.S. Forest Service, the U.S. Fish and Wildlife Service, and the Bureau of Land Management.

More than 400 U.S. national parks are monitored by the U.S. Department of the interior's National Park Service, which works with local communities to help preserve historical monuments and create and maintain recreational areas. [20] Parks are also controlled by regional and local authorities, national and international environmental organizations.

Ecotourism in the United States is one of the fastest growing tourism industries. The goal of ecotourism in USA is to provide a fun, relaxing holiday while protecting the surrounding ecosystem. Ecotourism is also often used to educate and immerse participants in the eco-friendly lifestyle that brings thousands of people to some of the most beautiful regions of the United States, while preserving and protecting these natural wonders. [21]

Along with traditional types of ecological tourism, such as, for example, trips to nature reserves, trips to reservations, adventure tours, scientific tourism, and nature history tours, the United States has also developed such ways of ecotourism as: agricultural tourism, community ecotourism, eco-lodging, eco-tracking, cultural ecotourism, ecotourism associated with endangered species of wildlife, extreme ecotourism, and others. [22,23] There are states that specifically encourage ecotourism. Among those are California, Louisiana, and Alaska. At the same time, environmental tours abroad including to Russia are especially popular in the United States.

Each of these types of ecological tourism has its own features and contributes to the development of national parks, reserves, wildlife preserves, tourist infrastructure, the emergence of new ecological trails and routes, and excursion programs. There are no mandatory certificates to conduct environmental tourism business, but there are several national and international voluntary certificates. The major factors that are taken into account when planning an eco-friendly holiday in the United States are the following: the hotel or resort must adhere to all local and national safety standards, there must be a high level of service quality, and the environment have to be sustainable and all efforts should be directed to maintaining the environmentally responsible protocol.

The detailed analysis of the literature about the development of ecotourism in Russia, as the whole, was conducted for this particular research. The thorough in-depth study of methodological materials on ecotourism, management and interaction of SPNAs with the local population in the United States and in Russia is scheduled too. The further work involves field research, including formal and informal surveys and interviews with scientists, managers, specialists, tour operators and other experts in the field of eco-tourism. The survey is intended to analyze the development and role of ecological tourism in the studied regions. Interviews with stakeholders will also help determine the levels of development and impact of tourism on the preservation of natural complexes and cultural heritage sites. For this purpose, we propose using the comparative approach and has developed the method of sociological survey and regression analysis models to evaluate the effectiveness of ecological tourism. A special questionnaire has been prepared and published for the experts interviewing in Russian and English [24,25,26]. Though, it is quite difficult to conduct expert surveys and formal interviews with representatives of local communities due to the low availability of respondents, administrative and cultural barriers. However, authors have prepared a number of questions in advance, plan to obtain visas and are requesting invitations from universities, institutes and other interested organizations. It is also planned to interview local residents in informal conversations if possible. The visits to some tourist attractions, natural sites, SPNAs have been partially accomplished for full-scale survey, assessment and analysis.

Many regions in Russia and in USA are the most prospective centers of ecological tourism. These days many purely ecological tours are offered. Though, visitors wishing to see the cleanest and most beautiful lakes and the Atlantic ocean can easily reach them and stay at the modern ecological facilities few days or weeks and enjoy the comfort of local small resorts and campings only in USA. Since, the tourism infrastructure and the ecological tours themselves are still underdeveloped in Russia and in its particular regions.

Ecotourism to natural attractions, which is usually accompanied with recreational and educational activities, actually, is a promising means of advancing social, economic, and environmental objectives. It offers new opportunities for small-enterprise investment and employment and increases the national stake in protecting biological resources, as well as improves economic ties and political relations. That is why it is so important for our countries.

However, making ecotourism a positive economic and environmental tool requires policies that foster responsible nature tourism development, broad-based and active local participation in its benefits, and conservation of cultural heritage. It also helps educate travelers and locals and make them environmentally responsible; it even supply funds if not for the conservation but for the minimization of the negative anthropogenic influence; and can directly benefit the economic development and social empowerment of local communities; and it also fosters respect for different cultures and for human rights.

5 Methods and models

We propose the comparative approach to analyze the ecological tourism in our countries, which assumes visits to various places of interest in Russia and in USA to evaluate the role and effectiveness of ecological tourism, and to identify relevant and popular destinations and the ways of its further development, considering the modern environmental requirements. It has been previously revealed, with the help of the comparative approach to ecological tourism, that the recreational, geographical and price advantages of ecological tourism destinations in Russia are significantly attractive for visitors. At the same time, new innovative eco-oriented projects, which are based on the experience of developed countries, are required for implementation in Russia to achieve the goals of socio-economic development of the regions and of the ecosystems sustainability. [24]

When specifying the existing terms and conditions and circumstances of the formation of the competitive tourist environment and of the ecotourism itself as the priority direction of development, and taking into account all institutional factors to address the problems of ecological tourism in popular wild nature, history and culture regions, we have to apply the comparative approach, which helps to reveal and select the best eco-oriented tourism and nature management practices. It is important to train professional personnel in these fields via special educational programs, including mutual ones, both for the local population and for tourists visiting ecological attractions. This will result in positive attitude and promotion of ecological tourism in our countries. The specific feature of ecological tourism is that it stimulates and satisfies the desire to communicate with nature, and at the same time, it prevents or at least minimizes the negative anthropogenic impact on the environment.

In addition to the comparative approach and method of system analysis to evaluate the state of popular attractions in regions, methods of economic, statistical and institutional analysis of the problems of development of ecotourism can be used to determine the advantages and disadvantages of natural, historical and cultural attractions and the institutional problems of their sustainable development.

We suggest the method of sociological survey as the measurement tool for evaluation of the role and effectiveness of ecological tourism. Therefore, a special questionnaire for experts and cross-sectional regression models have been developed and presented to evaluate the role and effectiveness of the ecological tourism. Another wide used analytical tool used in this research is the SWOT analysis. [26]

For the evaluation of the effectiveness of the ecological tourism development we suggest the generalized estimation of the ecological tourism development progress in a specific region in general. The short description of the totals and of regressors presented in the models is given bellow. The following estimation characteristics are selected as the major factors influencing the totals: gross domestic product, government and administrations effectiveness, norms and legislature provisions, the quality of infrastructure and services, availability of educational programs and information and analytical support, informal institutions. As the explaining variables in the models the sights features and their kinds are selected: natural, cultural, historic.

The sociological survey method and regression analysis is substantiated to determine the problems and directions of the further development of ecological tourism, as well as to evaluate its role and effectiveness, when the quality estimates obtained have to be converted into the quantity characteristics, which is required for the interpretation of results and models presented

The model for evaluating the role of ecological tourism in the sustainable development of the specific regions is the following:

$$ET = f(\text{GDPt, GovEfft, Qst, Pat, Iet, Iinft}), \quad (1)$$

where indicator (*ET*) is a generalized assessment of the progress of ecotourism development in the region as a whole (it is an expert assessment of the volume and quality of the institutional structure of ecotourism). The following estimated development characteristics were selected as potential factors influencing the *ET* indicator: GDP: gross domestic product in 2017 prices, representing the share of tourism revenue, GovEfft: efficiency of the government and administrations, including such institutional aspects as regulatory and legislative support and its implementation and infrastructure quality, Qst: quality of service, Pat: popularity of attractions, Iet: implementation of educational programs and information and analytical support, Iinft: informal institutions (traditions and customs) and other factors that determine the attractiveness of the visited place.

To evaluate the effectiveness of ecotourism, the following regression model is proposed:

$$Ef = f\,(An, Ac, Ah, E), \quad (2)$$

where (*Ef*) is a generalized indicator that evaluates the effectiveness of ecotourism development. The explanatory variables of the model are the estimated characteristics of the use of certain types of attractions in the region, namely: An: natural attractions, Ac: cultural attractions, Ah: historical attractions, E: compliance of attractions with expectations.

This model meets the requirements and can be used to explain (analyze) the effect of eco-tourism on the sustainable development of tourism in the region as a whole. It allows us to evaluate the following: 1) how effectively ecotourism programs are implemented in the region; 2) evaluate the role of attractions; 3) determine the importance of natural, cultural and historical attractions in the development of tourism.

This model shows the total impact of attractions due to variations in the values obtained based on expert estimates. From this model, and according to preliminary estimates of experts, it is clear that for the sustainable development of tourism in our countries, special attention should be paid to the development of rooted activities of the local population and provide access to attractions in such a way that it does not cause damage to the environment.

The comparative approach and methods of sociological survey and regression analysis to evaluate the state of natural, historical and cultural attractions and evaluate the effectiveness of ecological tourism in Russia and USA, involves visiting places of natural and cultural heritage and is used along with other methods presented earlier. The scientific novelty of the proposed comparative approach consists in the development of methods of analysis and evaluation of natural, historical and cultural attractions and of ecological tourism as a factor of sustainable socio-economic development on the basis of a questionnaire survey of experts in the field of tourism.

The preliminary questionnaire "Ecological tourism" for the survey of experts has been conducted already. It includes information specific and related to ecological tourism to answer the questions, such as: how attractive the sights are; how important are the factors affecting the sustainable development of eco-tourism; what impedes the development of eco-tourism; what is influenced by the implementation of eco-tourism; how the State is involved; how perfect is the current tourist legislation; what are the main problems of ecological tourism development; what are the major advantages of eco-tourism; is the professional staff in the tourism industry trained well; are visits to ecological attractions comfortable and safe for ecological tourists, etc.

6 Conclusions

The result of activities of the ecological tourism facilities in our countries is to attract funds for the development of tourism and the economy of the regions in which they are located. The importance of ecotourism in the territories selected is currently highly relevant. This type of tourism plays a significant role in preserving the unique natural, cultural and historical heritage used for tourism purposes. It attracts a huge number of tourists and is thus profitable from an economic, social and political attitudes.

The UNWTO experts make forecasts according to which the popularity of ecological tourism in our countries will increase annually and the growth rate of popularity will be very high. This will ensure high efficiency of the activities of those territories that are located around national parks and other SPNAs, as well as create a mentality aimed at taking care of nature.

Wildlife and its habitats in regions under study are becoming increasingly popular attractions for international tourism. The regions specified can provide many of the richest, in terms of unspoiled nature, places with bio diverse inhabitants conserved. The numbers of eco-tourists to our region are growing, despite the crisis and sanctions. The new wave of nature tourists is expected to bring with them the finances to spend. This money will support the local travel industry, which already exists and will create new jobs and incomes for households and local communities in and around places of attractions and in other protected areas. There is a need to establish ecotourism enterprises, special tour agencies and guide services, lodges and private reserves, as well as, such activities *as arts and crafts industries*, already present in Russia and in USA. There is also a need to develop transportation and food services at the local levels in Russia and improve them in USA to foster ecological tourism. Local administrations can use the incomes generated in this way for more efficient operating and protection of natural attractions.

The further research will be tailored to better understand the participation, economics, and marketing of ecological tourism in specific regions of Russia and the United States, and to conduct a comparative analysis of eco-tourism in our countries. In addition, the development of course modules on Ecological tourism development and marketing, and on cultural heritage and its relationship to international tourism development and management is expected, which are part of the University of Central Florida's Rosen College of Hospitality Management Master's program in hospitality management, since one of the author is invited to present his research and provide lectures to students during classes if sponsored by the Fulbright Visiting Scholar program.

References

1. The International Ecotourism Society (2021) http://www.ecotourism.org/
2. Ecological tourism. (n.d.) *American Heritage® Dictionary of the English Language, Fifth Edition* (2011) http://www.thefreedictionary.com/
3. Ecological tourism. (n.d.) *Collins English Dictionary – Complete and Unabridged* (2003). http://www.thefreedictionary.com/
4. Ecological tourism. (n.d.) *Random House Kernerman Webster's College Dictionary* (2010) http://www.thefreedictionary.com/
5. Ecological tourism. (n.d.) *WordNet 3.0, Farlex clipart collection* (2008) http://www.thefreedictionary.com/
6. D.V. Dayneko, V.V. Dayneko, IOP Conf. Ser.: Earth Environ. Sci. **190**, 1 (2018)
7. World Tourism Organization (2021) https://www.unwto.org
8. UNWTO Revealed the final statistics on the international tourist flow in 2017 (2017) https://tourism.interfax.ru/
9. P.O. Grushev, Economica, **1**, 56 (2017)
10. M. Honey, *Ecotourism and Sustainable Development* (2008)
11. L.Y. Prourzin, *Tourist as an economic priority* (2003)
12. Six specially protected natural areas will receive federal status in 2018 https://tass.ru/
13. Specially protected natural territories of Russia, http://oopt.info
14. A.V. Sobisevich, V.A. Snytko IOP Conf. Ser.: Earth Environ. Sci. **579,** 1 (2020)
15. Natural heritage can now be monitored from space (2020) https://greenpeace.ru/
16. Protected Planet (2021) https://www.protectedplanet.net/
17. How Many National Parks are There? (2021) https://www.nationalparks.org/
18. Wilderness Fast Facts (2021) https://wilderness.net/
19. Wilderness Connect for Practitioners (2021) https://www.umt.edu/
20. National Park Service (2021) https://www.usa.gov/
21. C. Rayburn-Trobaug, Ecotourism in the U.S. (2021) https://traveltips.usatoday.com/
22. R. Faith, What are the Types and Advantages of Ecotourism? (2021) https://traveltips.usatoday.com/

23. R. Dawn, Examples of Ecotourism (2018) https://traveltips.usatoday.com/
24. D.V. Dayneko, V.V. Dayneko, *Opportunities for the development of local history and tourist of the Siberian region and adjacent territories,* 14 (2019)
25. D.V. Dayneko, V.V. Dayneko, Geographic foundations and environmental principles of regional environmental policy, 819 (2019)
26. D.V. Dayneko, A.I. Dayneko, V.V. Dayneko, Advances in Economics, Business and Management Research, **90**, 204 (2019)

Magneto-inductive systems as a method to reduce the environmental risks of the existing systems of incoming quality control of metallurgical raw materials

A. A. Fogel*, *S. V.* Kochemirovskaya, *D. A.* Mokhorov, *A. V.* Isayev, and *V. A.* Kochemirovsky

Peter the Great St. Petersburg Polytechnic University (SPbPU), St. Petersburg, Russia

Abstract. The advantages of automated systems for magnetic inductive incoming quality control of raw materials of large metallurgical enterprises are demonstrated. Environmental risks under the existing visual control system were assessed.

1 Introduction

In the "Fundamentals of State Policy in the Field of Environmental Development of Russia for the Period up to 2030", approved by the President of the Russian Federation on April 30, 2012, the strategic goal of state policy in the field of environmental development was defined – the solution of socio-economic problems, providing environmentally-oriented economy growth, conservation of a favorable environment, biodiversity and natural resources to meet the needs of present and future generations, realization of the right of every person to a favorable environment, strengthening the rule of law in the field of environmental protection and environmental safety [1].

The use of secondary mineral resources in industry allows to solve not only set problems, such as resource conservation and environmental protection, but also becomes the reserve for increasing the Russian economy efficiency.

European countries also do not stand aside from the set problems. Thus, the European policy in the Waste Framework Directive (WFD) [2] sets as a priority goal the prevention

* Corresponding author: alena.fogel22@gmail.com

and reduction of waste production and its harmfulness. This can be achieved by optimizing the use of primary resources, introducing technologies that minimize waste generation, developing environmentally friendly products and appropriate technologies for the final processing of hazardous substances contained in waste, and recycling waste, through recycling, reuse, enhancement or other processes that aim to produce secondary raw materials.

A significant part of secondary resources are recycled ferrous metals, which serve as a feedstock for steelmaking, production of ferroalloys, steel and iron castings. As the Government of the Russian Federation emphasizes in the "Strategy for the Development of the Russian Ferrous Metallurgy for 2014-2020 and for the Prospect until 2030", the ferrous scrap industry is strategically important for the socio-economic development of the country [3]. The use of secondary raw materials in metallurgical production will reduce the cost price of final metallurgical products, while simultaneously solving the problem of hazardous waste disposal. Compared to steel production from mineral raw materials, steel production from recycled metals consumes 74% less energy and 40% less water, and reduces pollutants in water by 76% and in air by 86% [4].

As it is indicated by a number of researchers [5 - 7], the recycling of scrap metal and waste in the world economy has recently occupied a significant place. The amount of scrap and waste ferrous metals used as raw materials in global metallurgical production increases every year.

The system for processing scrap and waste metals can be divided into the following stages: collection, sorting, cutting raw materials, decontamination, remelting [8, 9].

The scrap metal supplied by scrap metal recyclers comes from various sources and is heterogeneous in composition and quality. The main quality indicators of scrap as charge material for steelmaking furnaces are: bulk density, chemical homogeneity, the content of non-metallic impurities (earth, sand, oil, organic materials, etc.) and nonferrous metals.

As noted in a number of papers [10, 11], the quality of used scrap is gradually deteriorating, the proportion of lightweight scrap containing impurities of nonferrous metals is increasing, the amount of plastic, glass and other impurities is increasing.

The procedure of reception, storage and methods of testing of scrap and waste ferrous and nonferrous metals are regulated by current standards, respectively, GOST 2787-75 "Ferrous secondary metals. General technical conditions". Scrap metals may contain various inclusions such as sand, paint, ground, rubber, plastic, nonferrous metals, refractories, wood. Scrap contamination is on average 3 - 5% [12]. Although this value seems small, but in the end it leads to the formation of hundreds of thousands of tons of waste, for transportation along with scrap and disposal of which huge funds are spent. The issue of reducing the impurity level in secondary metal legitimately arises.

In practice, many scrap metal suppliers neglect this requirement, using the difficulties encountered by the buyer in unloading of railway gondola cars due to stringent rules to limit downtime of rolling stock regulated by the Federal Law "The Charter of Rail Transport" [13] and the requirements of instructions of the USSR Gosarbitrazh dated June 15, 1965 No. P-6 and dated April 25, 1966 No. P-7 [14, 15] for acceptance of goods in quantity and quality, which are still applied in disputes arising between suppliers and buyers. The main method of falsification is the hidden loading of massive non-metallic cargoes into railway gondola cars, such as stone and concrete blocks, construction waste, earth, scrub, scale, rubber, plastic, and household waste.

Clogging inclusions give the railway gondola car additional technologically useless mass, fixed on the scale car, the report of which serves as the basis for mutual settlements with the Supplier at prices established for conditioned raw materials. As a result, the loss caused by clogging of railroad gondola cars to large metallurgical enterprises is more than 7 billion rubles a year of direct financial losses, not including the cost of unloading, sorting and disposal of contamination.

2 Materials and methods

As methods for assessing the contamination of scrap metal, a visual assessment of car contamination was used by specialists of the Quality Control Department of a large city-forming parent enterprise of a metallurgical holding (hereinafter – "LME"). This method of assessing the scrap metal contamination is used, so far, at all metallurgical plants in Russia. Visual assessment of car contamination is based on the visual presence of earth on the surface of scrap metal and waste on the bottom of gondola car.

As a reference method, the developed specialized means of express control UKNV-2 "Induktomass", based on the magnetic principle of measurements, was used.

Based on the data obtained from measuring system and scale car, speed sensors and gondola car dimensions, the automatic analysis system plotted the distribution of magnetic mass along gondola car length. In the calculations, correction factors or functional dependences of the speed of movement and the gondola car dimensions were used.

In the UKNV-2 modification, instead of spiral-screwed magnetic coil, the blocks of which cover a railway car and an operating mainline, the installation of a system of magnetoresistive and magneto-inductive sensors (magnetic field transducers) in the amount of from 5 to 15 pcs in accordance with the patent [16].

The system of two lidars with specially developed software, installed at a height of 5.0 m from the level of the rail top of the railway track, is used to simultaneously measure the integral bulk density of cargo, the overall characteristics of the cargo and railway gondola car, the car, instant and average speed, the car serial number in rail transport. The magnetic transducer system is used to measure the level of magnetic field fluctuation in the area of railway gondola car passing near the supports of existing railway infrastructure (supports of contact network and bridge crossings).

The fundamentals of method for magnetic detection of nonmagnetic impurities are described in [17].

The mass of waste was determined by studying the contamination of 20 cars with scrap, randomly selected from the trains arriving by rail to the LME's site.

3 Results and discussion

Currently, when accepting a batch of scrap and waste metal, control of the degree of scrap contamination at metallurgical plants is determined visually. Let's make some calculations that characterize the method capabilities.

Bulk density of scrap metal in gondola cars is 550-750 kg\m^3. The bulk density of ground is 1600-2500 kg\m^3. The monolithic metal density is 7850 kg\m^3. The net weight of a fully loaded car is 50-70 tons. The average metal thickness in scrap metal is 3 mm. The gondola car volume is 76-88 m^3. The gondola car area is about 25-35 m^2.

Let's calculate the ground layer on the surface of a railway car, corresponding to 5 – 7% of contamination.

Taking as input data for assessment of the average values of the above we get:

The ground mass with a contamination of 7% in a car with a net mass of 60 tons is 4200 kg. Which corresponds to 4200 : 2000 = 2.1 m^3.

The ground mass with a contamination of 5% in a car with a net mass of 60 tons is 3000 kg. Which corresponds to 3000 : 2000 = 1.5 m^3.

The surface area of 1 m^3 of scrap metal will be 1 : 0.003 x (650: 7850) ≈ 28 m^2.

The scrap metal surface area in the car will be 28 x 80 = 2240 m^2.

Thus, the contamination of 7% corresponds to the visual presence of ground layer on the metal surface, on average, with a thickness of 2.1 : 2240 = 0.0009 m, i.e. less than 1 millimeter (including “waste on car bottom”).

The contamination of 5% corresponds to the visual presence of ground layer on the metal surface, on average, with a thickness of 1.5 : 2240 = 0.00066 m, i.e. less than 0.7 mm (including “waste on car bottom”).

It is quite difficult to distinguish such a contamination visually from the height of car view from the trestle (5-6 m) by QCD specialists of metallurgical enterprises, and, therefore, the contamination percentage will be determined with a large margin of error.

Comparison of carried out results of the UKNV's measurements and QCD's assessments, based on visual observations and sigs of metal, passed shredding showed that, on average, according to the contamination assessment of UKNV, each car contains 5 – 10 % more contamination, determined visually by enterprise's QCD. The monitoring results are illustrated in Table 1.

 https://doi.org/10.1051/e3sconf/202131109004

Table 1. Monitoring results.

Photocontrol	QCD's data, contamination, %	UKNV-2's data, contamination, %
	**5.8 %**	**11.5 %** The ground thickness – at least 0.3 m, the density – at least 1 t\cubic meter (dry soil). Car area 25 sq.m. Total – at least 10 tons. Or at least 11 %.
	**5.1 %**	**14.6 %** The ground thickness – at least 0.5 m, the density – at least 1 t\cubic meter (dry soil). Including metal – 50% of the volume Car area 35 sq.m. Total – at least 8 tons. Or at least 13%.
	**4.3 %**	**7.5 %** The ground thickness – at least 0.1 m, the density – at least 1 t\cubic meter (it is not clear from the photo, the soil may be wet). Car area 35 sq.m. Total – not less than 3.5 tons. Or at least 6% for dry soil and at least 8-9% for wet

Table 2. Continued.

	**4.3 %**	**9.6 %** The ground thickness – at least 0.3 m, the soil is thoroughly mixed with metal (50-60% of the volume), the soil density – at least 1 t\cubic meter (it is not clear from the photo, perhaps the soil is wet, then – up to 2.5 t\cubic meter). Car area 35 sq.m. Total – at least 4 tons. Or at least 7% for dry soil and at least 9-10% for wet soil.
	**8.2 %**	**4.7 %** The car does not contain any visually distinguishable blockages. The main part of contamination according to UKNV is water (the car is heavily watered) 3-4%.

4 Conclusion

As you can see from the table above, the magneto-inductive system gives a more objective result than the visual system. The incoming scrap metal inspection by the magnetic-inductive method will give great savings in LME's funds, which, among other things, can be directed to measures that reduce the damage from storage and disposal of slag dumps and emissions of harmful gases generated from the processing of ballast material that clogs metallurgical raw materials. The introduction of magnetic inductive systems will reduce the amount of environmentally hazardous pollution at the stage of supplying raw materials for large metallurgical enterprises.

References

1. *Fundamentals of state policy in the field of environmental development of the Russian Federation for the period up to 2030* (2012)
2. Directive 2008/98/EC of the European Parliament and of the Council of 19 November 2008 on waste and repealing certain Directives
3. Order of the Ministry of Industry and Trade of the Russian Federation dated May 5, 2014 No. 839. On approval of the Strategy for the Development of the Ferrous Metallurgy of Russia for 2014 - 2020 and for the long term until 2030 and the Strategy for the Development of the Nonferrous Metallurgy of Russia for 2014 - 2020 and for the prospect until 2030
4. *Bureau of International Recycling (BIR). The Industry* (2021)
5. V. V. Smirnov, Auditing, **5(4)**, 30 (2018)
6. X. Hu, C. Wang, M. K. Lim, S. C. L. Koh, J. of Cleaner Production, **243**, 118576 (2019)
7. K. E. Daehn, A. Cabrera Serrenho, J. M. Allwood, Environmental Science & Technology, **51(11)**, 6599 (2017)
8. E.V. Rastyannikova, Innovation and investment, **7**, 81 (2019)
9. A. Damgaard, A.W. Larsen, T. H. Christensen, Waste Management & Research, **27(8)**, 773 (2009)
10. E. A. Popov, Business strategies, **9(2)**, 35 (2021)
11. A. O. Andrianov, O. A. Mednikov, Scientific almanac, **1-1(27)**, 19 (2017)
12. GOST 2787-75, Secondary ferrous metals. General specifications
13. Federal Law "Charter of Railway Transport"
14. Instruction on the procedure for accepting industrial and technical products and consumer goods by quantity (approved by the decree of the State Arbitration of the USSR dated June 15, 1965 N P-6)
15. Instructions on the procedure for accepting products for industrial and technical purposes and consumer goods in terms of quality (approved by the Decree of the State Arbitration of the USSR dated April 25, 1966 N P-7)
16. V. A. Kochemirovsky, A. G. Kuzmin, M. A. Malygin, Patent of the Russian Federation. RU 2717286 C1 System for determining bulk density and non-magnetic contamination of scrap metal in gondola cars in railway vehicles. Bul. 8 (2020)
17. V. A. Kochemirovsky, S. V. Kochemirovskaia, M. A. Malygin, A. G. Kuzmin, M. O. Novomlinsky, A.A. Fogel, L.S. Logunov, Applied Sciences, **9(10)**, 2001 (2019)

 https://doi.org/10.1051/e3sconf/202131109005

Analytical evaluation and systematization of technologies for extraction of rare metals from production waste

L. Ya. Shubov[1], *I. G.* Doronkina[2], *O. N.* Borisova[2], *L. M.* Tyger[2], and *T. R.* Lyubetskaya[3]

[1]Federal State Autonomous Institution Research Institute Center for Environmental Industrial Policy, Moscow, Russia
[2]Russian State University of Tourism and Service, Moscow, Russia
[3]Moscow State University of Food Production, Moscow, Russia

Abstract. The article is a message containing systematized information on resource-saving technologies as the basic for involving metal-containing secondary raw materials in the economic turnover. The set of independent technologies arranged in a certain sequence that solve the problems of resource saving allows us to compare the specific features and effectiveness of the use of each of them, the details and features of the processes. It is shown that the following can be the main sources of rare metals and REE: phosphogypsum (REE, Sr); ash and slag waste from TPPs, ash and slag waste (REE, Ti, Sc); metallurgical slags (Ge, Be, Se, Te, V); sulfuric acid dust (Se, Te); alumina production waste - red mud (Sc, REE); liquid waste from sulphate production of titanium dioxide (Sc, Ti); refinery dust bins (Se, Te); wastes from production of permanent magnets (Nd, Sm). It is noted that the technology uses beneficiation methods, acid leaching and ion-exchange concentration, biotechnology, electrolysis, heat treatment, a combination of different methods. To assess the quality of man-made raw materials and their suitability for complex processing and utilization, the article uses the methodology of analytical research of a set of patented technological solutions (technological samples were taken from real man-made deposits).

1 Introduction

Waste generated in the mining industry and metallurgy contains a large amount of non-ferrous, ferrous, precious, rare metals and various types of dispersed metals. These are man-made raw materials that are stored and accumulated due to the lack of an economical and environmentally friendly technology for processing and disposal.

The rare group of metals used in relatively insignificant quantities (their content on the earth's surface is 0.53 % by weight, including 0.41 % is titanium): Be, Ga, In, Te, Sc, Y, La, Ac, lanthanides and actinides, Ti, Zr, Hf, V, Nb, Ta, Mo, W, Po, Re, Te, elements of group I of the periodic system is conventionally referred to as rare ones.

The analysis shows that man-made mineral raw materials represent a large reserve of the country's mineral resource base. The main sources of production of rare and rare-earth metals based on use of industrial waste can be Phosphogypsum (REE, Sr); ash and slag waste from TPPs, ash and slag waste (REE, Ti, Sc); metallurgical slags (Ge, Be, Se, Te, V); sulfuric acid dust (Se, Te); alumina production waste - red mud (Sc, REE); liquid waste from sulphate production of titanium dioxide (Sc, Ti); refinery dust bins (Se, Te); wastes from the production of permanent magnets (Nd, Sm) [1].

The content of a number of useful components in production wastes allows us to consider them as substitutes for natural raw materials (in this case, production wastes often contain both rare metals and REE) [2].

A necessary condition for involvement of technogenic raw materials in the economic turnover is reliable information about its resources and quality, as well as possible directions of its processing and utilization [3-7].

The results of numerous technological tests in different directions of representative samples of industrial waste, carried out by different researchers, can be used as a method of expert assessments in determining the environmental and economic efficiency and the feasibility of involving technogenic raw materials in processing. In essence, we are talking about the study of technogenic raw materials for washability, as a result of which investment risks are minimized when recyclable materials are involved in the economic turnover.

2 Materials and Methods

To assess the quality of technogenic raw materials as a source of metal extraction and its suitability for complex processing and utilization, the paper uses the methodology of analytical research of a set of patented technological solutions that have novelty and usefulness.

Analysis of patented technologies:

1. The purpose of the patented technology is hydrometallurgical processing of germanium-containing slag of lead production to extract valuable components: germanium and silicon (in the form of dioxides), iron and calcium (in the form of hydroxides), zinc (in the form of salts) [8]. The composition of the slag of lead production (content, %): silicon 11.16, germanium 0.04, zinc 9.5, iron 14.39, calcium 16.52.

2. The purpose of the patented technology is metallurgical processing of beryllium-containing slags from the production of copper-beryllium master alloys and alloys for extraction of beryllium and copper from them [9].

A significant part of the beryllium produced in the world (~ 75 %) is used for production of precipitation-hardening copper-based alloys that are used in high-tech industries (aerospace, electronics, etc.). Copper-beryllium alloys are characterized by high resistance to corrosion, heat and wear, hardness, tear resistance, high electrical and thermal conductivity, and non-magnetic properties. Copper-beryllium alloys are used for the manufacture of parts for radio equipment, bearings, automatic drives, chemical equipment, and automatic weapons main springs.

The widespread use of beryllium is constrained by its high price due to the complexity of the beneficiation and processing of beryllium ores. Hence, it is very important to involve beryllium-containing wastes in processing (they often contain up to 20 % beryllium - many times more than it is contained in ores). Losses of beryllium and copper with slags are associated with slagging of their oxides, their partial dissolution in the slag melt, as well as entanglement of the smallest metal droplets suspended in the slag. This waste has not been processed to date, which creates economic and environmental problems (beryllium and its compounds belong to the first class of hazard).

In the Russian Federation, copper-beryllium master alloy is produced at Basalt Federal State Unitary Enterprise enterprise by direct fusion of metallic copper and beryllium. The slags formed during the smelting process are beads of copper-beryllium alloy (covered from the surface with a film containing beryllium oxide, mono- and divalent copper oxides and impurities that were present in the initial alloyed metals). The total content of Be in the slag is 10-20 %. Slag size of 90 % + 0.1 mm (individual pieces have a size of 20-50 mm).

Waste from metallurgical production in general, and beryllium-containing wastes in particular, it is advisable to process at the source of their formation.

The developed technology is low-waste and provides for:

- melting in an induction furnace at the temperature of 1,250-1,350 ^{0}C beryllium starting slag with addition of the combined fluorine-alkaline flux using magnesium fluoride (exhaust producing metallic beryllium comprises MgF_2 98 %), and soda; the ratio of the components of the flux (based on the mass of the initial slag), %: magnesium fluoride 15-50, soda 5-20; the melt is kept for 15-30 minutes;
- separate discharge: first of the lighter secondary slag, then of the metal phase (with its use as a recycle product in the smelting of copper-beryllium alloy); metal phase yield ~ 82% (alloy of beryllium and copper) - from the initial mass of waste; use of MgF_2 makes it possible to destroy the oxide film covering the beads of metallic beryllium and copper, with the subsequent fusion of the metals;
- secondary slag is melted with beryllium flotation concentrates to obtain beryllium hydroxide, and then metallic beryllium.

The experiments used copper-beryllium master alloy waste of the following composition, %: Be 12.6, Cu 86.7, Mg 0.058, Al 0.024, Fe 0.63, Pb 0.0008.

The patented technology can also be in demand at enterprises producing beryllium-containing alloys and bronzes.

3. The purpose of the patented technology is processing of tungsten-containing ore dressing tailings using separation methods for additional extraction of W [10].

The content of WO_3 in the tailings supplied for processing by gravity and flotation methods is 0.09%, and in the resulting high-grade tungsten concentrate is 62% WO_3 (with about 50% recovery from the operation). The output of the tungsten concentrate from the operation is 0.07 %. The content of WO_3 in the tailings is 0.04 %.

4. The purpose of the patented technology is to extract valuable components (aluminum, scandium, rare earth metals) from red mud (waste of alumina production) by transferring them into solution (leaching with formic HCOOH or acetic acid CH_3COOH) [11].

Composition of red mud, %: SiO_2 9.3, Al_2O_3 12.4, Fe_2O_3 44.3, TiO_2 4.4, P_2O_5 0.75, CaO 13.6, MgO 0, 93, Na_2O 2.9, K_2O less than 0.15, V_2O_5 0.09, Cr_2O_3 0.05, MnO 0.52, SO_3 2.8, REE (sum) 0, 14, Sc $9.0 \cdot 10^{-3}$.

The essence of the technology:

- acid treatment of red mud using formic or acetic acid (or mixtures thereof) as a leaching agent; leaching is carried out with the portionwise addition of red mud to the reagent solution with stirring with pH control (when the pH reaches 2.3-3.8, the addition of the mud is stopped), at the temperature of 80 ^{o}C;
- stirring the slurry for 1 hour after stopping the supply of sludge to the leaching process;
- keeping the suspension for 1 hour after stopping stirring (at the given temperature);
- separation of the solid residue from the solution by filtration (compounds of aluminum, scandium and rare earth metals pass into the solution, they are separated by known methods).

Extraction into solution under optimal conditions, %: aluminum ~ 63, scandium ~ 65, REE ~ 50.

5. The purpose of the patented technology is to extract scandium from red mud (waste of alumina production) using hydrometallurgy methods [12].

Solid phase composition of red mud, %: Fe_2O_3 46.5, Al_2O_3 12.0, CaO 8.0, SiO_2 10.0, TiO_2 4.5, Sc_2O_3 0.0130, ZrO_2 0.1. The ratio of S:L in the pulp is 1: 4; Pulp pH 12.5.

The essence of the technology:

- repeated sequential leaching of red mud with a mixture of solutions of carbonate and sodium bicarbonate when passing through the slurry a gas-air mixture containing CO_2 at the pressure of 3-6 atmg in vibro-cavitation mode; as a gas mixture can use a sintering furnace flue gases containing 8-17% CO_2 (at the pressure of 6 atmg, ^{60}°C and the process duration of 8 hours);
- two-stage holding of the productive solution; in the first step the pH is maintained at 9-9.5 for 3 hours at the temperature of not less than 90$^{\circ C}$ with further filtration (removal of impurity components);
- filtering the carbonized slurry to obtain a productive scandium-containing solution;
- the second stage of exposure at pH ≥12.5 for 3 hours at the temperature of 100-110 °C;
- precipitation of scandium in the form of Sc_2O_3 by adding sodium hydroxide solution.

Extraction of scandium from red mud is low – only about 20 %. The content of Sc_2O_3 in the concentrate is 6.5 %.

A variety of the considered technology has been patented [13]. Its main difference: the extraction of scandium from the scandium-containing solution is carried out by sorption with phosphate ion exchangers, desorption of scandium from the organic phase of the ion exchangers is carried out with mixed carbonate-chloride solutions in a pulsating mode to obtain a scandium-containing eluate, from which stage-by-stage precipitation of poorly soluble scandium compounds is carried out. The recovery of Sc_2O_3 in the eluate is 91-94 %.

6. The purpose of the patented technology is disposal of aluminum slag from secondary processing of aluminum and red mud (waste of alumina production) [14].

Raw material characteristics:

- composition of aluminum slag, %: Al_2O_3 34, KCl 22, NaCl 13, SiO_2 3.5, CaO 1.9, MgO 4.8, Zn 0.6, CuO 0.4, metal aluminum 11, Fe_2O_3 2.6;
- composition of red mud (Bogoslovsky aluminum plant), %: Fe_2O_3 40, $Al_2O_3$15, CaO 14, SiO_2 10, $TiO_2$4, Na_2O 4.

The essence of the technology:

- grinding of aluminum slag;
- separation of metallic aluminum (enrichment methods);
- mixing the residue after the separation of metallic aluminum with red mud and sintering the material at the temperature of 800-900°C (combining fine-grained particles into pieces);
- separation of the oxide and salt components of the cake to isolate the salt component of aluminum oxide (the operation is carried out in an aqueous medium using an upward flow in a pulsating column operating in a closed cycle with a conical settler);
- returning the clarified settler solution to the column to create an upward flow;
- magnetic separation of the solid phase of the lower discharge of the pulsation column (output of the magnetic fraction ~ 75 %).

The mixing ratio of the components aluminum slag: red mud is 1:1.5.

Directions of use of enrichment products:

- magnetic fraction – raw material for ferrous metallurgy;
- mixture enriched with aluminum oxide – return to alumina production;
- the amount of chlorides isolated from the solution is a flux for recycling aluminum scrap.

3 Results

It is necessary to note one more source of REE – wastes from the production of permanent magnets. These wastes are not large-tonnage, but technologies for their processing began to be developed in the United States at the end of the last century; judging by the lack of publications, such works are not carried out in the Russian Federation. In accordance with the US patent, two types of wastes from the production of permanent magnets enriched with REE are identified – sawdust and slag; the technologies for extracting rare-earth elements from them are different, but they use methods of hydrometallurgy [15].

Analysis of the set of patented technological solutions proves the suitability of phosphogypsum (chemical industry), ash and slag waste (fuel and energy industry) and metallurgical waste, which form large accumulations, for reuse and reveals their wide technological capabilities as objects of processing.

It shall be emphasized that the extraction of valuable components (including metals) by itself does not solve the problem of large-tonnage waste – a low yield of useful products; their large-scale utilization can be tied to the construction industry and road construction (after the extraction of metals and, if necessary, enrichment).

The most significant developments are those that have passed large-scale tests, as well as semi-industrial and industrial approbation:

- processing of phosphogypsum with extraction of REE and radioactive elements, obtaining fertilizers and phosphate cement binder [16];
- extraction of rare earth metals and scandium from ash and slag waste [17];
- joint processing of cuprous slag and sludge from copper electrolyte production (extraction of rare and noble metals);
- extraction of vanadium compounds from metallurgical slags [18-20].

Information about these solutions shall be included in the relevant information and technical reference books, at this stage – as promising solutions that meet the modern requirements of the circular economy.

From the point of view of extraction of REE, wastes of mining and metallurgical production are of little importance, but they are serious sources for production of germanium, bismuth, scandium, titanium, beryllium, vanadium [21, 22]. In part, metallurgical slags of lead production, along with wastes from mechanical processing of sediment materials, are the main source of obtaining germanium (as well as bismuth); applied technology – hydrometallurgy. Liquid waste from titanium dioxide production is suitable for extracting not only titanium, but also scandium (extraction technology) – bearing in mind that the main industrial source of scandium production is uranium ore; it is also possible to extract scandium (and REE) from red mud (waste of alumina production). A rich source of beryllium (as well as copper) are slags from the production of copper-beryllium ligatures (they contain up to 20 % beryllium – many times more than it is contained in ores); original thermal technologies are used to extract beryllium. The possibility of recycling dust bins of refining production is shown not only for additional extraction of precious metals, but also for obtaining a concentrate of Se and Te (the content of dust in the dust runways is 4-8 %, Te 6-10 %). Extraction of metals into the final product is about 95 % (the technology used is a combination of hydro- and pyrometallurgy).

4 Conclusions

Therefore, considering the quantitative characteristics of large-tonnage waste, the type and content of valuable components in them, it is possible to identify with a certain accuracy the priority directions of practical actions aimed at arranging the recycling of rare and rare-earth elements:

- extraction of REE from phosphogypsum;
- extraction of Ge and Bi from metallurgical slags;
- extraction of Ti and Sc from liquid titanium production wastes;
- extraction of Be from slags of the production of copper-beryllium alloys;
- extraction of V from metallurgical slags.

The results of assessment of patented technologies indicate that in the 21st century, a wide range of technological solutions have been developed and tested, allowing the extraction of rare metals and REE from technogenic raw materials [23-25]. These solutions can and shall find practical application; one of the approaches may be the inclusion of information on patented technologies in the relevant information and technical reference books, at this stage – as promising solutions that meet the modern requirements of the circular economy, resource and nature management.

References

1. L. Ya. Shubov, K.D. Skobelev, S.I. Ivankov, I.G. Doronkin, Ecology of industrial production, **3(103),** 15 (2018)
2. O. M. Denisyuk, Naukovy Oglyad, **5(58),** 49 (2019)
3. O. N. Borisova, I. G. Doronkina, L. V. Sumzina, A. V. Maximov, M. R. Gozalova, 5th International Multidisciplinary Scientific Conference on social sciences and arts SGEM 2018. Conference proceedings, 545 (2018)
4. O. N. Borisova, I. G. Doronkina, A. V. Demenev, G. Malyutin, 5th International multidisciplinary scientific conference on social sciences and arts sgem 2018. Conference proceedings, 441 (2018)
5. H. I. Abdel-Shafy, M. S. M. Mansour, Egyptian Journal of Petroleum, **27(4),** 1275 (2018)
6. O. Olubajo, O. Osha, U. El-Natafty, H. Adamu, Abubakar Tafawa Balewa University (2017)
7. V. Murko, V. Khyamyalyainen, and M. Baranova, IIIrd International Innovative Mining Symposium, E3S Web of Conferences, **41,** 01042 (2018)
8. L. M. Cheremisinov, A. M. Kuraev, A. I. Demidov, RF patent No. 2423535 C22B7, C22B19, C22B41, Method of hydrometallurgical processing of mineral raw materials (2011)
9. V. E. Matyasova, A. D. Bykov, Z. M. Alekberov, G. V. Goryaev, M. L. Kotsar, RF patent No. 2558588 C22C9, C22C25, C22C1, C22B7, Method of processing beryllium-containing waste (2015)
10. V. A. Kantsel, A. A. Kantsel, S. A. Lapshinov, A. A. Letyushov, M. N. Fomin, V. A. Potapov, RF patent 2424333, C22B3, C22B34, Method of complex processing of tungsten-containing ore dressing tailings (2011)
11. A. V. Bogomazov, A. S. Senyuta, RF patent No. 2544725, C22B7, C22B21, C22B59, C01F7, C22B3, Method of acid processing of red mud (2015)
12. V. S. Anashkin, S. E. Vishnyakov, O. V. Petrakova, S. N. Gorbachev, A. V. Panov, RF patent No. 2562183, C22B59, C22B3, C01F17, Method of obtaining scandium concentrate from red mud (2015)
13. G. N. Klimentenok, V. S. Anashkin, S. E. Vishnyakov, A.V. Panov, RF patent No. 2536714, C22B59, C22B3, Method of obtaining scandium concentrate from red mud (2014)

14. V. A. Nizov, D. F. Rakipov, E. V. Pustynnykh, A. R. Bakirov. RF patent No. 2540317, C22B7, Method of processing aluminum slag (2015)
15. D. O. Skobelev, Ecological systems and devices, **12,** 29 (2018)
16. D. O. Skobelev, Ecological systems and devices, **1,** 12 (2019)
17. I. Kotarska, B. Misera, P. Stefanek, E3S Web of Conferences 41, III International Innovative Mining Symposium, 02013 (2018)
18. M. Samir, F. Alama, P. Buiss, T. van Nylen, O. Ostanin, III International Innovative Mining Symposium. E3S Web of Conferences **41,** 02012 (2018)
19. A. Pinchuk, N. Tkalenko, V. Markhasova, IV-th International Innovative Mining Symposium. E3S Web of Conferences **105,** 04048 (2019)
20. J. Pyssa, E3S Web of Conferences **14,** 02024 (2017)
21. V. Mymrin, K. Alekseev, O. M. Fortini, Y. K. Aibuldinov, C. L. Pedroso, A. Nagalli et al, Journal of Cleaner Production, **161,** 376 (2017)
22. B. Greenberg, US patent No. 5429724, C25C1, Neodymium recovery process (1995)
23. M. A. Konov, R. Kh. Khamizov, RF patent No. 2630072 C22B59, Method of ensuring environmental safety of industrial waste in the form of phosphogypsum with obtaining a two-component target product (2017)
24. T. V. Vlasova, N. S. Velichkina, A. V. Kalashnikov, A. A. Zakharov, P. Yu. Novikov, V. Yu. Koltsov, M. E. Zvonareva, T. B. Yudina, RF No. 2630072, C22B59, 2016 / RF patent No. 2657149 C22B, Method for extracting rare earth metals and scandium from ash and slag waste (2018)
25. N. F. Danilov, V. V. Vdovin, A. A. Karpov, A. A. Kamenskikh, V. P. Kudryashov, RF patent No. 2230128 C22B34, Method of processing vanadium-containing converter slags (2015)

Harm and benefit of Khibiny mining waste

Olga Krinochkina[1], *Dinya* Mamina[1, *], and *Vladislav* Stulov[2]

[1]Moscow State University of Civil Engineerin, Moscow, Russia
[2]LabWare LLC, Moscow, Russia

Abstract. The article is focused on the assessment of the impact of mining waste on the environment. Mining waste (MW) is understood as mine (mining) dumps and dumps of processing plants (tailings dumps). The article also suggests considering MW as technogenic deposits of rare earth and precious metals, the development of which will allow solving many environmental, social and economic problems and using natural resources in a sustainable manner. The objects under study are located on the Kola Peninsula, within the Khibiny mountain range, and are the production assets of the Apatit Mining Complex. This enterprise develops the so-called deposits of the Khibiny Tundra, it also owns apatite-nepheline processing plants.

1 Introduction

The relevance of the research topic is due, on the one hand, to the extremely large number of mineral deposits (MD) being developed in this country, and, on the other hand, is explained by the need for new approaches to protecting the natural environment, which is inevitably transformed during their development. According to the authors, when creating mining enterprises, it is necessary to take into account the specific features of the terrain and the potential response of the natural environment to man-made impacts. This opinion of the authors develops the ideas of M.I. Agoshkov, B.N. Laskorin, N.V. Melnikov, V.V. Rzhevsky, E. M. Sergeev, A.V. Sidorenko, N.F. Khachaturov, S.S. Shvarts, M.I. Glazovskaya, A.I. Perelman, S.N. Kasimov, R.V. Goleva and other domestic and foreign scientists. The current development level of mining production cannot be limited only to environmental actions of a protective nature [1]. It is necessary that such actions should rather be preventive and based on a scientific forecast. The latter should be based on an assessment of the interrelationships of the physical and chemical processes underlying the development of minerals on the one hand and natural conditions on the other.

2 Materials and methods of research

The research is based on the materials obtained during the implementation of the State Contract of the Ministry of Natural Resources of the Russian Federation "Assessment of changes in the quality of groundwater in the areas of mining enterprises as an indicator of

* Corresponding author: dmamina@mail.ru

accumulated environmental damage". The objects of this research are the deposits of the Khibiny Tundra, the existing Kirovsky and Rasvumchorrsky mines, as well as apatite-nepheline processing plants (ANOF-1, ANOF-2 and ANOF-3) and the zones of their impact. All these enterprises are production assets of JSC "Apatit" (Figure 1).

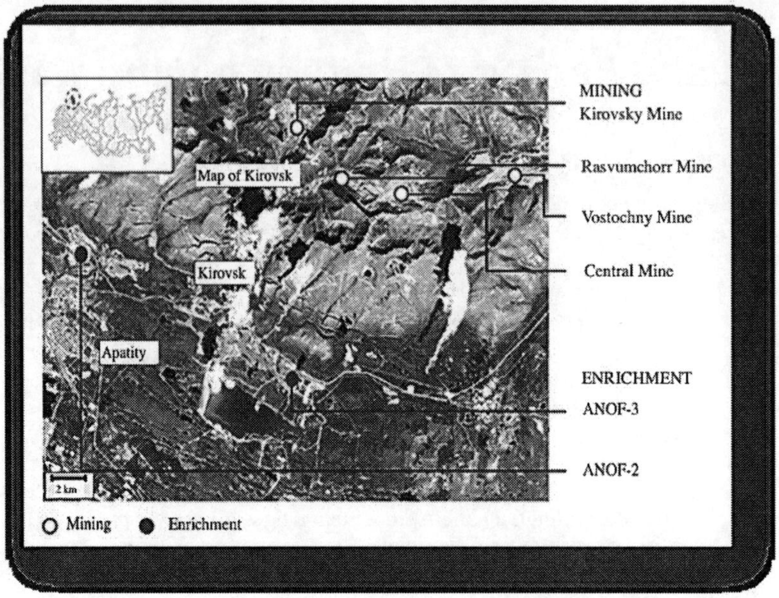

Fig. 1. Location of the objects of research within the area of the Apatit mining complex.

This group of deposits of the Kola Peninsula (Murmansk Region) is characterized by very large reserves of ores, a long period of their mining and a set of dangerous elements polluting the environment, such as rare, rare-earth and radioactive elements, as well as strontium , fluorine and other toxic elements.

The research methods included field observations, sampling (of soils, bottom sediments, surface and underground waters, ores and enclosing rocks, rock dumps by fractions), multi-scale ecological and geochemical mapping, and office processing. The complex of laboratory studies included the following: approximate quantitative atomic emission spectral analysis for 40 elements (for samples of ores, enclosing rocks, mining dumps), atomic absorption for gross forms of mercury (in all samples) and mass spectrometric analysis with inductively coupled plasma for samples of underground and surface waters and mobile forms of elements for soils and bottom sediments (for 46 elements). Office processing of materials was carried out using software in the shell of Argis, Exel, Statistic.

The specific feature of this research methodology was the study of the composition of fractions (according to the particle size) in the mining dumps: mining dumps and tailings dumps (hereinafter referred to as dumps and tailings dumps).

The mine has been existing for more than 85 years. Average mineral composition of apatite-nepheline ore (%) is as follows: apatite (10 – 91.5), nepheline (3.16 – 73.5), pyroxene (3.21 – 17.57), sphene (signs – 36.12). The content of the main components in the ore (%) is as follows: P_2O_5 – 14.48; SrO – 1.18; Al_2O_3 – 14.23; Ga – 0.0015; Rb_2O – 0.0095; SeO – 0.0000025; TiO_2 –1.84; Nb_2O_5 – 0.16 [2]. Such a long-term development of deposits associated with the above-mentioned material composition creates serious negative environmental changes. This is primarily due to the fact that due to the predominance of alkaline rocks in the region, the migration abilities of most toxic elements are greatly increased. Naturally, this is explained by sharp predominance of the ligand-ion of OH. In addition, the alkaline dumps of ore deposits also contribute to the accumulation of such strong ligands-complexing agents, such as fluorine and organic substances, for example. All this activates the migration of many toxic elements into underground and surface waters.

Existing measures for the protection and recovery of the environment mainly include the purification of drain waters from suspended particles, petroleum products and nitrogen compounds. According to G.B. Melentyev [3], the environmental monitoring system installed within the mines and the area of operation of apatite-nepheline processing plants (ANOF-1,2,3) does not track the behavior of many toxic components, and therefore additional ecological and geochemical studies were conducted.

To assess the impact of tailings dumps and mining dumps on the environment, the authors carried out a complex of works in the studied area, including:

1) selection of geochemical samples from dumps, tailings dumps and from watercourses associated with dumps and tailings dumps;

2) separation of the debris material of dump samples into gain sizes – 5-2 mm, 2-1 mm, 1-0.5 mm, 0.5-0.25 mm, 0.25-0.1 mm, 0.1-0.044 mm, 0.044-0.02 mm and less than 0.02 mm and tailings – 0.5-0.25 mm, 0.25-0.1 mm, 0.1-0.044 mm, 0.044-0.02 mm and less than 0.02 mm (separation of the debris material of dump samples into particle fractions was carried out using sieve and slurry techniques);

3) determination of the trace element composition of the samples by a set of methods, namely: the material of gain sizes in the dumps– by spectral semi-quantitative analysis (SSQA), and individual lithochemical and surface and underground water samples – by mass spectrometry (ICP MS and ICP OES).

Further processing of the obtained materials showed the proximity of the trace element composition of the conjugated size classes. Therefore, to simplify the procedure of the composition analysis, the classes were enlarged, i.e. combined in pairs (5.0 -1.0 mm; 1.0-0.25 mm; 1.0-0.25 mm; 0.044-0.02 mm and less than 0.02 mm). But within the enlarged classes, the chemical composition showed significant differences, both in mine dumps and in tailings dumps.

3 Research results

Upon completion of the field observations, sampling and appropriate laboratory studies, as well as office processing of materials, the results were summarized in Table 1.

Table 1. Summarized results.

Mine dumps	
Particle size (mm)	**Element associations**
5-1	B, Sc, Ti, Ba
1-0.25	B, Sc, Ti, Ba, (Cr)
0.25-0.044	P, Ga, Cr, Sr
<0.044	Cu, Pb, Zn, Be, Y, Yb, Sc, Zr, Nb
Concentration not expressed	Ce, La, Su, As, Ag, Co, Ni, V, Mn
Tailings dumps	
Particle size (mm)	Element associations
1-0.25	B, Sc, P, Mo
0.25-0.044	Nb, Ga, Zn, V, Mn, Ti

Table 1. Continued.

<0.044	Pb, Zn, Cu, Co, Mn, Y, Yb, Ce, La, P, Ti, V, Cr, Be, Sn, As, Sr Nb
Concentration not expressed	Zr, Ag, Ni, Ba

As follows from the Table, one can observe significant variations in the distribution of element concentrations across the combined groups of fractions of detrital material [4] of mining dumps and tailings dumps.

When comparing the trace element compositions of particle sizes in the mine dumps, attention is drawn to the identity of the associations of accumulation elements (B, Sc, Ti, Ba) in the two largest sizes-5.0 – 1.0 mm and 1.0 – 0.25 mm. The next largest class (1.0 – 0.25 mm) is also characterized by the concentration of a relatively small number of elements – P, Ga, Cr and Sr, while in the thinnest class (0.044 – 0.02 mm and less than 0.02 mm) their number increases significantly.

Tailing dumps as a whole are also characterized by growth of the accumulation of the number of chemical elements in fine grain sizes. Cu, Pb, Zn, Co, Mn, Y, Yb, Ce, La, P, Ti, V, Cr, Be, Sn, As, Sr, Nb are concentrated here, so the number of accumulation elements is greatly higher than in a similar association of the mining dumps [5].

The noted similarities and differences in the trace element compositions of fine grains in the tailings dumps and rock dumps can be associated both with the primary natural enrichment of rocks and ores with certain elements, and with the features of the enrichment technology of processed ores [6].

Intensive enrichment of fine grains of dumps and tailings with various chemicals, including toxic elements (TE), determines interest in them as the most likely and active sources of pollution of natural waters and the environment as a whole on the one hand, and as technogenic deposits of these elements on the other hand. At the same time, it is natural that the pollution occurs more intensively during the migration of elements from tailings dumps, which is due, firstly, to higher concentrations of TE here, and secondly, due to disintegration of the material here, its active contact surface with the host medium is increased. The high intensity of the TE migration from them is attributed to active interaction of water and debris due to its intensive grinding. Thus, when comparing the stale “tails” from ANOF-1 enrichment (recultivation was carried out only by seeding plants) and the modern ”tails” of the ANOF-2 processing plant, it was discovered found that the content of toxic elements such as beryllium, plumbum, arsenic (Be, Pb, As) decreases several times over the years (Figure 2). It is clear that these elements mainly penetrate into natural waters, i.e. they are potential pollutants of surface and underground waters in the area of activity of JSC Apatit.

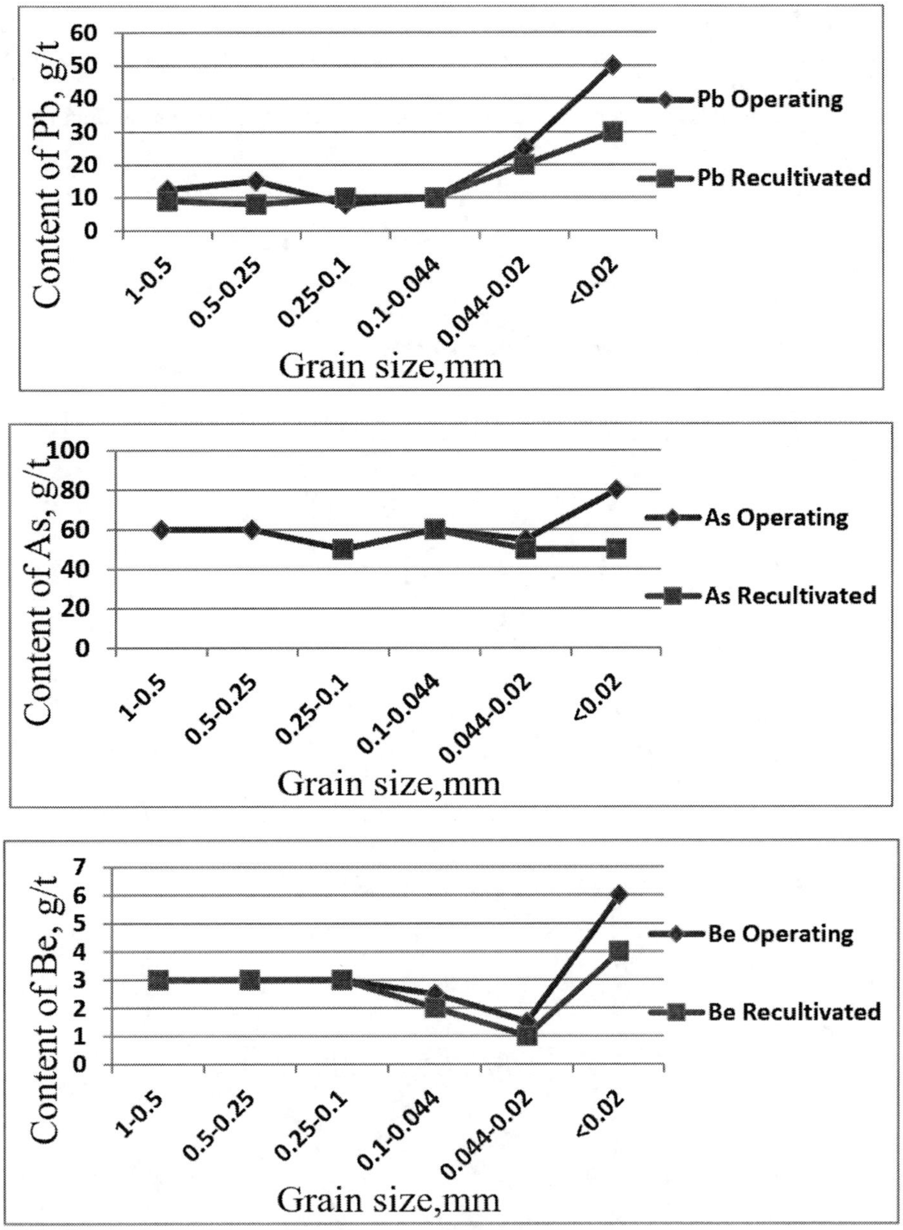

Fig. 2. Grain distribution of several toxic elements in the existing and recultivated tailings dumps [7].

According to monitoring data in the of ANOF-1 impact zone (recultivated tailings storage) in the underground waters of the peat-marsh and technogenic aquifers, arsenic and plumbum concentrations exceeding the background ones by 1.5-2 times are observed in individual samples. Within the ANOF-1 impact zone (the operating tailings storage facility), background excess for the same components is more often found in surface water reservoirs, which is most likely due to dusting of loose dumps.

Let us now consider the fine particles of mining waste from the deposits of the Khibiny Tundra as a concentrate of rare, rare-earth (REM) and precious metals. The first in the list is gold. It was first found by the authors in individual samples of MW in concentrations up to 3-4 g/t according to the SSQA data, and, according to mass spectrometric analysis (ICP-MS), it was 0.1 – 0.2 g/t. It should be noted that gold is characterized by an extremely uneven distribution.

An idea of the content (concentrations) of the main REM gains can be obtained from Figure 3.

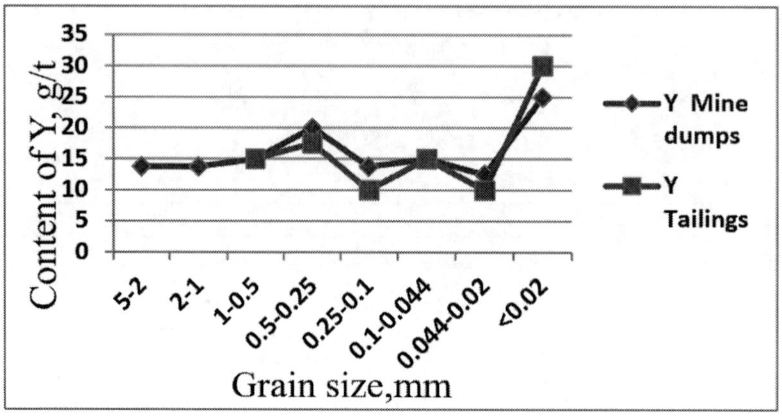

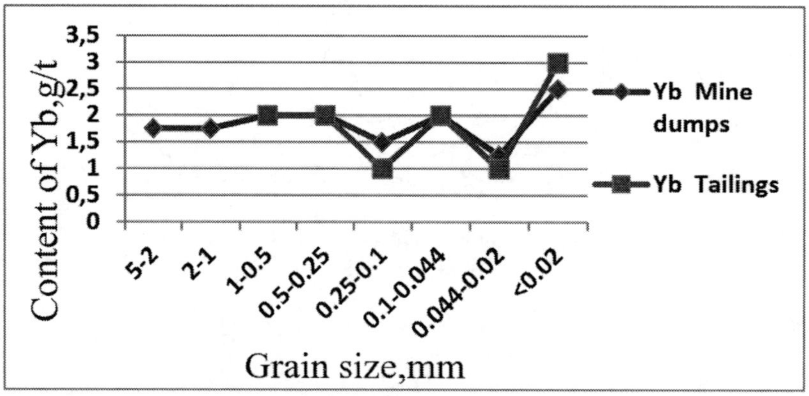

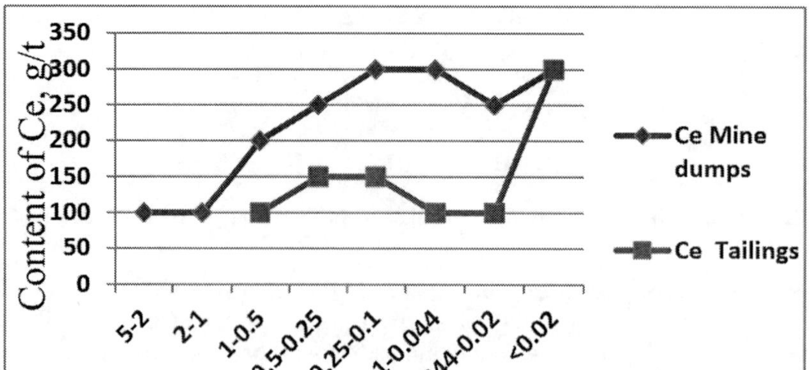

 https://doi.org/10.1051/e3sconf/202131109006

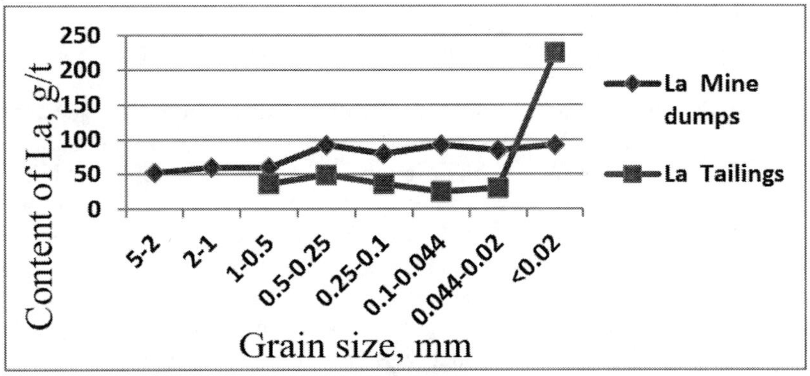

Fig. 3. Average concentrations of some rare-earth elements in the isolated fractions of the Khibiny mining dumps and tailings dumps (g/t).

As can be seen from the accumulation series (see Table 1 – The associations of elements are arranged in rows in descending order of their concentration coefficients) and the graphs above, cerium is found in the largest quantities of REM (according to the authors' data, Ce content is up to 300 g/t), then goes lanthanum (up to 220 g/t), and together they total 520 g/t. Next comes yttrium and elements of the yttrium group, and these metals are two or even three orders of magnitude more expensive than the more common metals of the cerium group [8].

Below is some data on the presence of REM in various large deposits of Kazakhstan:

— the Zhanatas deposit has an average ∑REM content of 614 g/t, including yttrium – 300 g/t, ytterbium – 4.2 g/t, cerium – 110 g/t, lanthanum – 200 g/t;

— the Aksai deposit has an average ∑REM content of 938 g/t, including: yttrium – 440 g/t, ytterbium – 440 g/t, cerium – 210 g/t, lanthanum – 280 g/t;

— in the field Chulaktau the average content of ∑REM is 1088 g/t, including: yttrium – 520 g/t, ytterbium and 8.1 g/t, cerium – 250 g/t, lanthanum – 310 g/t;

— the Kokjon field has an average content of ∑REM 63 g/t, including: yttrium – 312 g/t, ytterbium – 5.5 g/t, cerium – 138 g/t, lanthanum – 175 g/t;

— the Koksu deposit has an average ∑REM content of 700 g/t, including: yttrium – 376 g/t, ytterbium – 6.3 g/t, cerium – 166 g/t, lanthanum – 152 g/t.

When comparing the above data, the conclusion arises that it is necessary to consider the Khibinsky MW as promising technogenic deposits of REM and gold. Using this resource, several environmental, social and national economic tasks are addressed at once. And namely: minerals are extracted from the end concentrate, public employment rises due to the creation of new jobs, and mining waste that have polluted the environment for many years is disposed of. Of course, a lot of work is needed to develop a technology for extracting valuable metals, studying their mineralogical and metallogenic features, but given an extremely high demand in them, this will significantly increase the role of MW as a valuable natural resource, including this resource in complex technological chains of processing apatite-nepheline ore. The disposal of MW will improve the environmental situation in the region under consideration

4 Conclusions

Toxic elements concentrated in the MW, and especially in the material of small fractions (less than 0.044 mm) of mine dumps and tailings dumps can become a source contamination source of the environment.

1. Geochemical assessment of the grain size fractions of the MW material can be used to predict the potential threat of environmental contamination from the operation of mining enterprises.

2. The results obtained indicate the possibility of identifying a unique technogenic deposit of rare earth metals (REM) with gold within the Khibiny mining area, as well as the need for revaluation of MW with their involvement in processing. As is known, about 82% of the balance reserves of REM are apatite ores, and more than 70% of the reserves are associated with apatite-nepheline ores of the Khibiny Tundra deposits. However, they are still not extracted from the ores mined there. The only exception is the Lovozerskoye deposit field. In the remaining deposits of the Khibinsky group, REM remains in the tailings of enrichment. Although it is estimated that their extraction is quite cost-effective. Complex processing of all apatite raw materials would allow producing about 40 thousand tons of rare earth per year. According to the authors' calculations, the reserves of cerium (Ce) alone in the tailings of ANOF-2 enrichment plant are worth about 150 thousand tons (it should be noted that there are much more Ce in mine dumps). In addition, these reserves are in the form of finely dispersed fractions, i.e. it is a ready-made "concentrate" of gold and REM. All this is especially relevant now, when REM is extremely demanded by the industry.

References

1. M. Pevsner, V. Kostovetsky, *Ecology of mining production* (1990)
2. S. Ikorsky, V. Nivin, V. Pripachkin, *Geochemistry of gases of endogenous formations* (1992)
3. G. Melentyev, O. Vdovina, E. Malinina, I. Karimova, A. Popova, Proceedings of the collection of scientific articles of the regional scientific and practical jubilee conference on November 13-15, 93 (2013)
3. G. Melentyev, O. Vdovina, E. Malinina, I. Karimova, A. Popova, Collection of scientific articles of the regional scientific and practical jubilee conference on November 13-15, 93 (2013)
4. O. Vdovina, K. Naumov, N. Stulova, Collection of scientific articles of the regional scientific and practical jubilee conference on November 13-15, 123 (2013)
5. I. Spiridonov, O. Vdovina, K. Naumov, R. Vysokinskaya. Proceedings of the of the XIII International Conference, 25 (2014)
6. P. J. Beckett, S. Pappin-Willanen, G. M. Courtin, Proceedings of the ISGE (GEOENV`97) Istanbul, Turkey, 252 (1997)
7. O. Vdovina, A. Lavrusevich, G. Melentyev, I. Evgrafova, K. Naumov, D. Yelchin, K. Polyakova., E. Shubina, MGSU Bulleten, **12**, 152 (2014)
8. O. Krinochkina, V. Ivanov, Proceedingof the All-Russian Conference, 107 (2016)

Recirculating carbon dioxide refrigeration unit for freezing food

Evgeniy Neverov[1], *Igor* Korotkiy[1], and *Ilya* Prib[1,*]

[1]Kemerovo State University, 6 Krasnaya Street, 650000 Kemerovo, Russia

Abstract. The article describes the development process of a refrigeration unit using carbon dioxide as a refrigerant. A feature of the plant is the possibility of recirculating carbon dioxide. The authors substantiate the use of the natural substance CO2 as a refrigerant and consider the prospects for using the refrigerant R744. The efficiency of the cycle of a one-stage subcritical refrigeration unit is analyzed. The authors have designed a recirculating carbon dioxide refrigeration unit based on the traditional subcritical cycle. The essential and auxiliary components of the refrigeration machine were presented and described. The developed scheme allows freezing food products by direct contact of the product with snow-like carbon dioxide. After that, the sublimation process takes place in the freezing chamber and the gas-liquid mixture is sent through the liquid separator for recirculation. The article describes the operation of the system diagram, the principle of operation, and the purpose of the elements of automation and regulation.

1 Introduction

Natural refrigerants are substances that occur naturally in nature. Unnatural or synthetic refrigerants are man-made chemicals that do not occur naturally. The correctness of the term "natural refrigerants" is sometimes controversial. People today distinguish between substances whose chemical properties and safety have been fully investigated, and these chlorine and fluorinated gases, the negative impact of which on the ozone layer leads to global warming and a threat to environmental safety. As a result, the scientific community is constantly discussing the problem of using these gases. [1] The most common natural refrigerants include ammonia (NH3, R717), carbon dioxide (CO2, R744), and hydrocarbons (HC) such as propane (R290), isobutane (R600a), and propylene (R1270) (known as propane).

In addition, mixtures of ammonia and dimethyl ether (R723) and various mixtures of hydrocarbons characterized by optimized operational and safety properties (isobutane and propane R441, etc.) are created.

Using CO2 as a refrigerant has expanded significantly in recent years. Using traditional refrigerants is restricted by various regulations and is more stringent around the world. Natural refrigerants are increasingly used [2].

* Corresponding author: prib_ilya123@mail.ru

Carbon dioxide is used as an alternative refrigerant. Carbon dioxide is a type of natural gas that is environmentally friendly in terms of efficiency and safety as it is present in the Earth's atmosphere and biosphere. The advantage of CO2 is that it is harmless to the environment, neutral to the ozone layer, low cost, non-flammable, non-explosive, and compatible with mineral oil. To use CO2 as a refrigerant, you need to use water-cooled condensers and pipes with thicker walls than other refrigerants. Carbon dioxide is most commonly used in two-stage low temperature applications. The use of CO2 has been well shown in small refrigerators, domestic refrigerators, and heat pumps. CO2 is found in natural refrigerants like propane, butane, ammonia, water, each of which has own disadvantages. Propane and butane are flammable and explosive, ammonia is poisonous, water has a limited range of applications. Most of the essential and auxiliary equipment works with refrigerants such as ammonia and freon. Ammonia is used in large and medium-sized installations, while freon is more commonly used in medium-sized, small refrigerators, as it is less energy efficient. Synthetic freon has other disadvantages: ozone depletion, low efficiency and is not compatible with many oils. The transition to natural refrigerants is promising. Carbon dioxide has the following advantages: high refrigeration capacity, non-toxicity, and safety, not inertness regarding the material, low cost, and availability. Using CO2 is widely used in low temperature technology for its high efficiency and high refrigeration capacity. Figure 1 shows a comparison of the cooling capacity of CO2, R22, and NH3 where the advantage of carbon dioxide is clearly seen. [2]

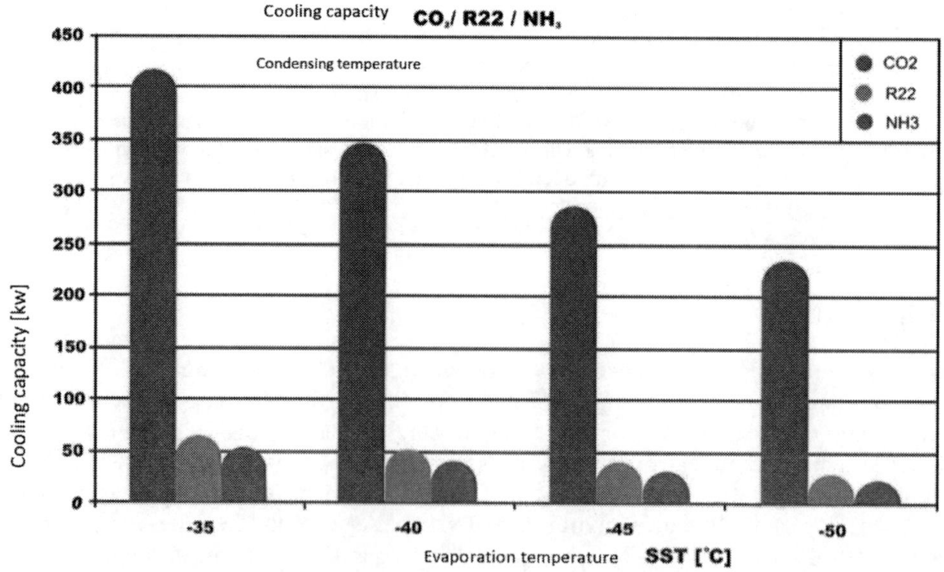

Fig. 1. Comparative graph of refrigerating capacity of refrigerants

The major disadvantages are high operating pressure up to (10 MPa) and low critical temperature (31°C). Carbon dioxide mixed with other refrigerants, propane and butane must eliminate these disadvantages. Because of its properties, carbon dioxide is becoming more and more popular as a refrigerant.

CO2 at atmospheric pressure is in the solid or gas phase. In the aggregate state of dry ice at a temperature (-78.4°C). The triple point is reached at pressure (0.52 MPa) and temperature (-56.6°C), at this moment all three phases are in equal proportions. The critical point is reached when the temperature (+ 31.1°C) is reached. At this moment, the densities of the liquid and gas phases are equal. It can also be used as a subcritical and supercritical refrigerant. [2]

2 Materials and methods

This work develops a recirculating carbon dioxide refrigeration unit for freezing food products that allow freezing food products in an environment of gaseous carbon dioxide and recuperating it. In the first stage, we built the operation cycle of the installation in the P-h diagram (figure 2) and identified key points. Below, we will consider the cycle and the essential and auxiliary units, shut-off and control valves, and automation elements of the refrigeration unit.

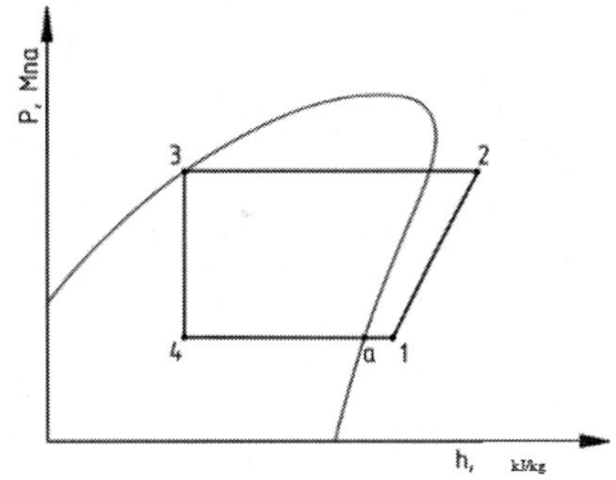

Fig. 2. Single stage subcritical refrigeration cycle.

Process a – 1 overheating of a substance in a liquid separator
Process 1-2 gas compression in a compressor
Process 2-3 gas condensation in a shell-and-tube condenser
Process 3-4 throttling gas into the chamber

Figure 3 shows a scheme of a designed refrigeration unit. The recirculating carbon dioxide refrigeration unit for freezing food is composed of essential and auxiliary elements.

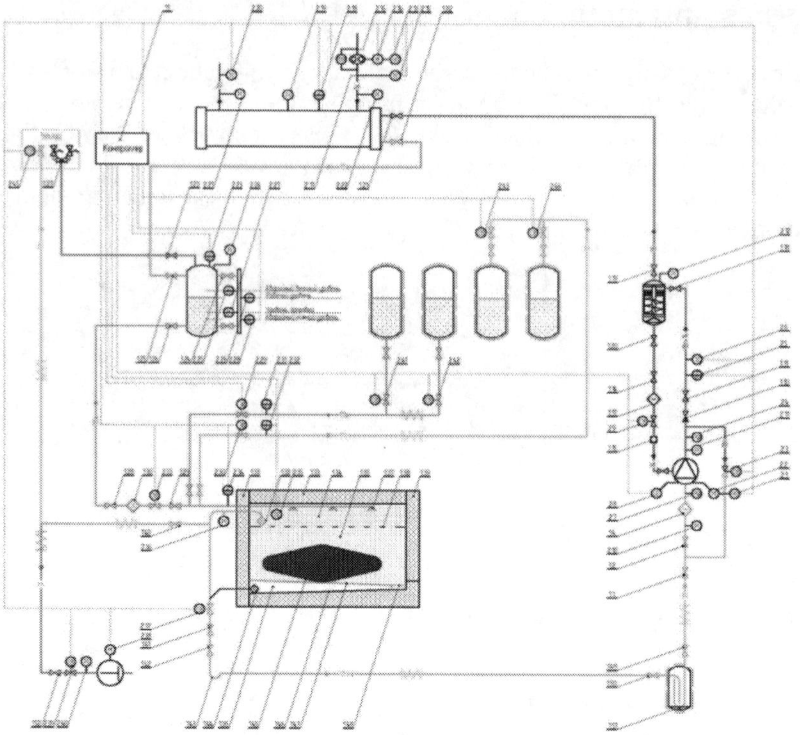

Fig. 3. Scheme of the designed refrigeration unit.

3 Results and discussion

The main units of the refrigeration machine include a compressor unit (1), an oil separator (2) of a cyclone type, a water-cooled condenser (3) of shell-and-tube design, a linear receiver (4), a cooling chamber (5), and a liquid separator (6).

The auxiliary elements of the installation are a cold water supply pump, a condenser (7), a vacuum pump (8), cylinders for supplying liquid carbon dioxide (9), and cylinders for supplying gaseous carbon dioxide (10), a controller (11).

Each of the essential and auxiliary components of the refrigeration machine includes auxiliary automation elements and valves.

The control and automation elements of the compressor unit (1) include a shut-off valve on the suction line (1.1), a check valve (1.2), a low pressure gauge (2.10) with a shut-off valve, a fine filter (1.4), a low pressure switch (2.7) , compressor motor speed converter (2.1), solenoid valve (2.3), high pressure gauge (2.11) with shut-off valve, high pressure switch (2.4), check valve (1.10), discharge shut-off valve (1.11), sensor pressure (2.5), temperature sensor (2.6), oil shut-off valves (1.14), sight glass (1.15), solenoid valve (2.9), oil filter (1.17).

The unit for the functioning of the oil separator (2) includes a shut-off valve at the inlet to the apparatus (1.18), a valve at the outlet of the vessel (1.19), a high-pressure manometer (2.12), and a shut-off valve on the oil outlet line (1.20).

The water condenser automation kit (3) includes a temperature sensor (2.20) for a warm water circuit, a high pressure manometer (2.19) with a shut-off valve, a refrigerant pressure sensor (2.18), a water pump (7), a water pump differential pressure switch (2.16-2.17), a water pump motor shaft speed converter (2.13), a supply water temperature sensor (2.15), a gas line shut-off valve (1.21), a liquid line shut-off valve (1.22), pressure gauges for measuring incoming (2.22) and heated (2.21) water.

The operability of the linear receiver is maintained by a valve for connecting emergency pressure relief valves (1.23), a pressure sensor (2.23), a pressure gauge with a shut-off valve (2.24), valves for connecting a level and automation column (1.24), a filling level switch (2.27), an emergency lower liquid level (2.28), liquid refrigerant inlet valve from the condenser (1.25), liquid supply valve to consumers (1.26), upper alarm liquid level switch (2.25), operating level switch (2.26) and emergency pressure relief valves (1.27).

The elements of regulation, automation, and design of the refrigerated chamber (5) are valves on the liquid refrigerant supply line (1.28) and (1.29), filter drier (1.30), electronic expansion and control valve with a stepper motor (2.33), pressure sensor (2.34), removable chamber cover (1.31), coarse filter (1.32), temperature sensor (2.35), chamber body (1.33).

To describe the principle of operation of the refrigerating machine, the inner cavity of the refrigerated chamber is conventionally divided into three volumes: the upper volume (1.34), the middle volume (1.35), and the lower volume (1.36).

The refrigerated chamber assembly (5) includes nozzles (1.37), an upper partition (1.38), a chamber door (1.39), a valve for connecting a vacuum pump and a solenoid relief valve (1.40), a pressure gauge with a shut-off valve (2.36), a solenoid valve (2.37), non-return valve (1.41), suction shut-off valve (1.42), oil lifting loop (1.43), coarse filter (1.44), refrigerated product (1.45), lower floor surface of the chamber (1.46), lower partition (1.47), mesh opening (1.48), and the pressure relief solenoid valve (2.45).

The liquid separator (6) operates with the shut-off valves (1.49) and (1.50). An oil return hole (1.51) is provided at the bottom of the compressor suction pipe loop inside the vessel.

The vacuum pump (7) is connected to the refrigeration circuit through a shut-off valve (1.52), a solenoid valve (2.39), a pressure gauge with a shut-off valve (2.40).

The operation of the liquid carbon dioxide supply cylinders (9) is carried out using solenoid valves (2.41) and (2.42), as well as using an electronic expansion and control valve with a stepper motor (2.29) and a pressure sensor (2.31).

The carbon dioxide gas cylinders (10) work in tandem with the solenoid valves (2.43) and (2.44), the electronic expansion valve (2.30), and the pressure sensor (2.32). [8]

4 Conclusion

The technological development of the refrigeration industry in recent years allows the use of CO2 in both subcritical and transcritical refrigeration systems.

The article designed a recirculating carbon dioxide refrigeration plant based on a traditional subcritical cycle for freezing food. The circuit makes it possible to freeze food by direct contact with the food with snow-like carbon dioxide. After that, the sublimation process takes place in the freezing chamber and the gas-liquid mixture is sent through the liquid separator for recirculation. With the help of such a scheme and method of freezing, the refrigerating capacity increases, therefore, the heat transfer coefficient increases. All this allows to increase the shelf life of the product and reduce the time for its freezing, which means that the risk of bacteria formation is reduced.

The essential and auxiliary components of the refrigeration machine have been presented and described in a typical diagram. We have presented a description of the operation of the system diagram, the principle of operation, and the purpose of the elements of automation and regulation.

References

1. On natural refrigerants, Survey information, Ozon Program, https://www.ozoneprogram.ru/
2. Refrigerant R744. A new breakthrough in air conditioning systems, Survey information, Laduga, Automotive engineering, http://laduga.ru/blog/r744/, free.
3. E. N. Neverov, I. A. Korotkiy, A. N. Rasshchepkin, M. I. Ibragimov, A. N. Grinyuk, IOP Conference Series: Earth and Environmental Science. International Conference on Production and Processing of Agricultural Raw Materials, 072022 (2021)
4. E. N. Neverov, I. A. Korotkiy, P. S. Korotkih, R. U. Skhaplok, I. A. Prib, IOP Conference Series: Earth and Environmental Science. Cep. "International Science and Technology Conference "Earth Science"" , 012042 (2021)
5. E. N. Neverov, I. A. Korotkiy, L. V. Lifentseva, A. N. Rasshchepkin, Proceedings of Nizhnevolzskiy Agrouniversity complex: Science and Higher vocational education, **1(61)**, 326-337 (2021)
6. E. N. Neverov, I. A. Korotkiy, E. V. Korotkaya, A. N. Rasshchepkin, Periodico Tche Quimica, **18(37)**, 1-12 (2021)
7. E. N. Neverov, I. A. Korotkiy, P. S. Korotkih, L. V. Lifenceva, IOP Conference Series: Earth and Environmental Science. Current Problems and Solutions, 012039 (2019)
8. I. A. Prib, I National Conference -Kuzbass: Refrigeration Engineering and Biotechnology – 2019, 641-664, 182-185 (2019)

Self-attitude and coping-strategies of children with eating disorders

Elena Yurievna Bakaeva, and *Tatiana Sergeevna* Pilishvili

Peoples' Friendship University of Russia

Abstract. Nowadays, the topic of eating behavior is becoming more and more popular. People with deviant forms of eating are described as mysterious "THEY" who deceive a lot in order to hide the true form of eating and deep feelings about this, are tyrants in their families, strive for perfection and are extremely difficult to treat. This paper focuses on the attitudes of patients with eating disorders and their habitual strategies for coping with stress behavior. The study is based on the question: are there differences in self-attitudes and coping strategies among adolescents with different eating behaviors?

1 Introduction

Disorder of the eating behaviour is a group of disorders characterized by anxiety and pathological belief in the presence of a significant physical disability, sometimes of a hypochondriacal nature, which the patient seeks to compensate through manipulation of his body, with particular attention to the issues of nutrition and calorie expenditure.

Anorexia nervosa is characterized by a conscious refusal to eat in order to correct, even at the cost of one's own life, one's "vice", which is often false. The disease can manifest itself both within an independent nosological unit and as an endogenous disease syndrome.

There are 4 stages of the course of the disease:

1. Initial, which is characterized by the so-called Korkina triad.

As a reason for discovering a hated defect in themselves, patients cite a painfully accepted reproach, a mockery of appearance or nutrition.

This stage is characterized by persistent dysmorphophobic and dysmorphomanic experiences - a feeling of self-rejection and actions aimed at correcting a "defect in appearance" through excessive physical exertion and strict diets.

2. Anorectic - it is characterized by a significant loss of body weight (up to 50 percent), the addition of secondary somato-endocrine disorders.

Depending on the premorbid personality traits, food deprivation, and the comorbidity of an eating disorder with affective pathologies, patients are either satisfied with their appearance or are convinced that it is necessary to continue to lose weight. Various pathological obsessions and fears regarding nutrition are relevant - fear of losing control over food intake, fear of gaining body weight, fear of painful sensations in the epigastric region after eating, etc. Behind them are rigid thoughts about titanic efforts over oneself, suffering in the pursuit of compensation for one's shortcomings, about an unbearable struggle against hunger and refusal to devalue them.

3. Cachectic - at this stage, a lethal outcome is possible due to suicide, cardiac disorders, pneumonia, pulmonary edema, gastric perforations, internal bleeding and other intercurrent diseases. (L.I. Zakharova, 2019), (E.M. Solodovnik, L.A. Nepovinnykh, 2020).

Asthenic syndrome comes to the fore, patients are adynamic and emaciated.

4. Reduction of anorexia nervosa - with the slightest weight gain, dysmorphomanic experiences are again actualized.

Most authors studying eating disorders argue that patients with anorexia nervosa are characterized by the following emotional and personal characteristics: perfectionism, punctuality, scrupulousness, rigidity, high intelligence are expressed. The authors note that patients are emotionally immature and dependent individuals. (V.V. Marilov, M.V. Korkina, A.M. Tsivilko, 1986), (M. Hirsch, 2018), (M. Woodman, 2011), (F. Alexander, 2017).

Bulimia nervosa is Greek for bovine hunger and is characterized by hyperphagic eating behavior. This disorder can manifest itself both within the framework of an independent borderline mental pathology, and within the framework of anorexia nervosa.

E.K. Kislova classifies bulimic behavior in anorexia nervosa as follows:

- Bulimia as a symptom of anorexia nervosa: patients experience intolerable hunger, but the fear of gaining body weight stops them from overeating, as a breakdown, one can see that patients chew food and spit it out;
- Bulimia as a stage of anorexia nervosa: occurs against the background of prolonged fasting, patients experience uncontrolled bouts of hunger, as a result of which there is a strong overeating followed by cleansing behavior. Obsessive fears of weight gain and body dysmorphic experiences are relevant;
- Bulimia as a variation of anorexia nervosa is formed in a short time after refusal to eat, as a response to unbearable fasting (E.K.Kislova, 2004), (G.F.Muzychenko, p.135).

Most modern authors consider eating disorder in terms of a biopsychosocial model. (Malkina-Pykh, 2007) It is believed that it is the combination of all three factors that forms this disease. It is about the following phenomena:

The biological factor considers eating behavior from the point of view of the need for nutrition and satisfying hunger, as a means of saturating the energy necessary to maintain the life and health of the individual. In addition, geneticists have found that all patients suffering from an eating disorder have similarities in the genetic code - the presence of a certain gene (HTR2A gene for serotonin receptor 5-HT2A, as well as gene for brain neutrophic factor (BDNF)).

Physiologists, for their part, argue that metabolic disturbances can also lead to the formation of deviant eating behavior. (A.O. Kibitov, G.E. Mazo, 2016)

The psychological factor opens up new facets of the possibility of considering the eating process. When eating, people enjoy themselves, reward or punish themselves, accept or hate themselves, build interpersonal contacts or avoid them. By developing different diets and diets in order to shape their body image, a person seeks recognition, acceptance, aesthetics and health. And it is amazing how self-deception is hidden behind the apparent self-care - having developed a nutrition program and achieving real results in the form of a slimming body and approval from others, a person receives positive reinforcement of his actions and does not notice how his hobby turns into addiction. Obsessive fears of losing control over hunger and food intake, and gaining body weight become relevant. As a result, weight continues to decline to the degree of emanciation (cachexia). Obsessive thoughts about food, fear of eating and feelings of guilt for the eaten food are expressed. There are also obsessive actions in the form of constant measurement of the volume of one's body, daily weighing, purification behavior, which manifests itself in inducing vomiting after eating, and sometimes until the stomach is completely washed (vomiting till the transparent water).

An interesting phenomenon of "confusion" in understanding one's physical and psychological state at the moment when the patient experiences negative emotions. We noticed that people with anorexia nervosa describe the state of satiety at the moment of experiencing frustration based on physicality - "an unpleasant feeling of satiety", "an obsessive desire to vomit", "abdominal fullness", "squeezing in the throat." It can be assumed that for patients with anorexia nervosa, the physiological feeling of satiety is identical to the emotional overflow of negative emotions.

It is important to note that patients with eating disorders are characterized by high rigidity, which manifests itself in stubbornness, obsession, difficulty in switching attention, difficulties in adapting to new conditions. We are inclined to assume that it is with this personality trait that the difficulty of correcting deviant eating behavior is associated.

Social factor.

In this context, it is relevant to characterize the influence of society on the mentality of the individual.

Fashion, as a social factor, imposes its ideals on us. "If I do not match, then I will not be accepted." An adolescent with a "neurotic" form of perfectionism, in the terminology of Hamachek (N.G. Garanyan, 2006), begins to do everything to become accepted by society, to become perfect. He strives for model forms, for the maximum achievements of his activities, as a result, he becomes a prisoner of his own high requirements. In his pursuit of perfection, an adolescent can no longer independently break out of the rigid framework. Thus, a person turns into a robot, exhausted and has no room for error. Consequently, food addiction is formed on the basis of low adaptation in society, emotional immaturity and pathological rigidity of the patient.

It is impossible not to note religious views and their possible influence on the formation of eating behavior. According to the works of A. Freud, self-restraint, refusal to eat is symbolic of fasting of repentance, cleansing from sins, asceticism. If you try to analyze the relationship of this protective mechanism with religion, then the following points draw your attention: during the monthly period, a woman is prohibited from entering the church, she is considered “dirty”. It can be assumed that growing up, healthy work of the reproductive system and, as a consequence, the appearance of sexual impulses are the “vices” of women, which the spiritual world denies. The child is considered by church canons as a pure being, for whom the road to Paradise is open. The relationship is seen that the child is accepted and clean, the adult woman is judged and "dirty". Therefore, the way to maintain acceptance of the church is to refuse to grow up or to “wash away sins” through self-sacrifice. For a deeper understanding, let us quote the words of a priest of a Moscow church: "the martyrs suffered and ended up in Paradise, so we must humble ourselves and accept our fate."

Based on the above, it is clear that patients with an eating disorder have a low degree of adaptation and displace, and perhaps forbid themselves to experience and express negative emotions.

The aim of this work is to study ways of coping with stress behavior and characteristics of self-attitude in patients of pubertal age with different eating behaviors.

We hypothesized that there are significant differences in the characteristics of self-attitude and coping strategies that adolescent girls with different eating behaviors resort to in anxious situations.

The results obtained can be used in the development of group psychotherapeutic programs aimed at correcting self-attitude in patients with eating disorders, as well as developing constructive coping strategies under stress conditions.

In the presented work, we conducted a study in which 3 groups of respondents aged 16-17 years took part:

- 21 people - girls with restrictive eating behavior (diagnosed with anorexia nervosa), BMI$\leq$17.5;

- 20 people - girls with an external and emotiogenic type of eating behavior, 20.5≤ BMI≤32;
- 20 people - girls with a harmonious type of eating behavior, 18.5≤ BMI≤25.

The study was carried out on the basis of the children's gynecological department of the Federal State Budgetary Institution National Medical Research Center for Obstetrics, Gynecology and Perinatology named after V.I. Academician V.I. Kulakov of the Ministry of Health of the Russian Federation.

The following psychodiagnostic techniques were used for the study:

1. Questionnaire of coping strategies (coping strategies) by R. Lazarus, S. Folkman (adaptation by T. L. Kryukova, E. V. Kuftyak, M.S. Zamyshlyaeva;
2. The scale of self-attitude of V. Stolin-S. Panteleev.

With the subsequent application of methods of mathematical and statistical data processing (comparative analysis - Mann Whitney U-test).

2 Results of the conducted empirical research

The results of coping behavior diagnostics are shown in Figure 1 (mean values).

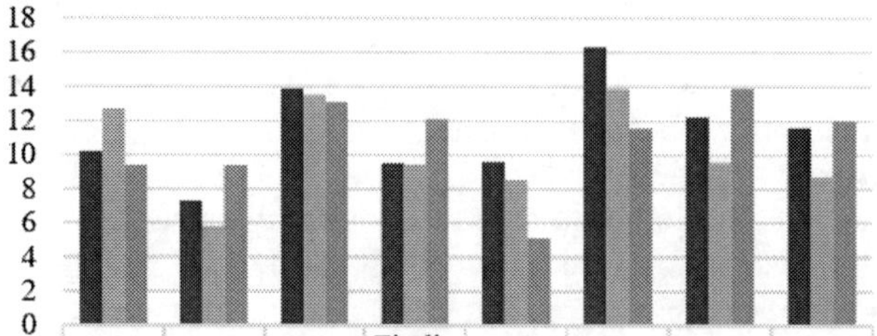

	Confrontation	Distancing	Self-control	Finding social support	Taking responsibility	Escape	Planning a solution	Positive revaluation
■ Restrictive EB	10,2	7,3	13,9	9,5	9,6	16,3	12,2	11,6
■ Emocyogenic and external EB	12,7	5,8	13,5	9,45	8,55	13,85	9,55	8,7
■ Harmonious EB	9,4	9,4	13,1	12,1	5,1	11,55	13,9	12

Fig. 1. Results of diagnostics of coping strategies in girls with different eating behaviors.

Figure 1 shows that in the group of girls with restrictive behavior strategies of escape and self-control prevail, while distancing is expressed in them to the least degree. Hence, when faced with difficulties, girls with anorexia nervosa tend to avoid solving the problem, control their emotions and behavior, while attempts to overcome negative experiences in connection with the problem due to a subjective decrease in the significance of the problem are not typical for them.

In the group of girls with hyperphagic behavior, a high level of severity of escape and self-control strategies is observed, while distancing is the least characteristic of them. This indicates that in difficult situations, they tend to avoid solving problems and over-control their emotions and behavior, while overcoming the problem by reducing its importance is not typical for them ...

In the group of girls with harmonious eating behavior, the strategies of planning problem solving and self-control are most pronounced, while the acceptance of responsibility is expressed to the least degree. This suggests that in difficult situations, they seek to find a real way out of the situation and control their emotions, while underestimating their contribution to the occurrence of the problem may be observed.

Thus, girls with anorexia nervosa, as well as girls with bulimic eating behavior, are distinguished by the predominance of escape and self-control strategies, while the strategy of distancing is expressed in them to the least degree. For girls with harmonious eating behavior, the predominance of strategies for planning problem solving and self-control is characteristic, and the strategy of taking responsibility is expressed in them to the least degree.

Let's analyze the features of self-attitude in groups. Figure 2 shows the distribution of respondents by the level of the integral indicator of self-attitude.

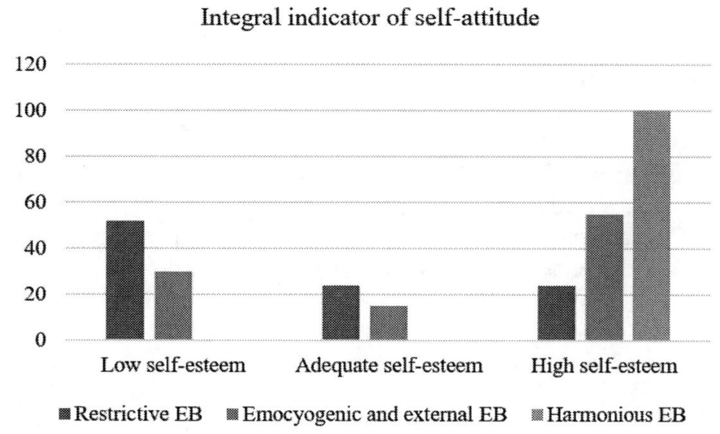

Fig. 2. Distribution of girls with different eating behavior according to the level of the integral indicator of self-attitude.

In the group of girls with a restrictive type of eating behavior, an underestimated level of self-esteem prevails (52%), while an equal number of respondents (24%) have adequate self-esteem and an overestimated level. This indicates that most girls have problems with self-esteem, manifested in inadequate self-esteem.

In the group of girls with hyperphagic eating behavior, 55% of the surveyed have an overestimated level of self-esteem, 30% are characterized by an underestimated level, and 15% have an adequate level. This suggests that this group is also dominated by girls with negative features of self-attitude.

In the group of girls with a harmonious type of eating behavior, 100% have an overestimated self-esteem. This suggests that they are characterized by an inadequate attitude towards themselves, which consists in an inadequately overestimated assessment of their qualities and abilities.

Thus, in the group of girls with anorexia nervosa, an underestimated level of self-esteem prevails, for girls with bulimic form of eating behavior overestimated self-esteem is most characteristic, and girls with harmonious eating behavior also have an overestimated self-esteem.

In the course of processing the research results, we carried out a comparative analysis of coping strategies and self-esteem features in groups of adolescents with different eating behaviors using the Mann-Whitney U-criterion.

The results obtained demonstrate statistically significant differences both in the methods of coping with stress and in the characteristics of self-esteem in groups of adolescents with various forms of eating behavior.

2.1 Description of the results of a comparative analysis of coping strategies in adolescents with different types of eating behavior

2.1.1 In a comparative analysis of coping strategies in groups of adolescents with restrictive and hyperphagic eating behavior, the following results were obtained:

- The CONFRONTATION indicator is significantly higher in the group of girls with emotiogenic and external types of eating behavior (U = 77.5, p = 000.1, cf. rank in group 1 = 14.69, cf. rank in group 2 = 27.63). Consequently, girls with hyperphagic eating behavior in a difficult situation more often respond with high behavioral activity, including those aimed at changing the situation, than girls with anorexia nervosa. It may be connected with the fact that for adolescents who are prone to excessive food restriction, active activity in stress situations is extremely energy-intensive, because aggravates the depletion of the body, which makes this strategy a little typical for them.
- The indicator ACCEPTANCE OF RESPONSIBILITY is significantly higher in the group of girls with a restrictive type of eating behavior (U = 130.5, p = 0.035, compare rank in group 1 = 24.79, compare rank in group 2 = 17.04). Consequently, girls with anorexia nervosa are more likely to respond with excessive self-criticism in difficult situations than girls in the other group. This is probably due to the fact that girls with anorexia use excessive self-accusation as punishment, with which they justify inadequate food restriction.
- The "Escape" indicator is significantly higher in the group of girls with a restrictive type of eating behavior (U = 84.5, p = 0.001, compare rank in group 1 = 26.98, compare rank in group 2 = 14.7). This suggests that girls with a restrictive type of behavior tend to avoid solving problems more often than girls with hyperphagic eating behavior. This is probably due to the fact that girls of the first group are distinguished by a large subjective feeling of helplessness, formed against the background of prolonged exhaustion.
- The indicator PLANNING A SOLUTION TO THE PROBLEM is significantly higher in the group of girls with a restrictive type of eating behavior (U = 58, p = 0.001, compare rank in group 1 = 28.24, compare rank in group 2 = 13.4). This suggests that girls in the first group in a stressful situation are more often directed at planning actions to solve the problem than girls in the second group. This can be explained by the fact that in girls with bulimic type of eating behavior, emotional response prevails over rational understanding of the situation, and therefore coping strategies that imply a rational approach are less typical for them.
- The "Positive overestimation" indicator is significantly higher in the group of girls with a restrictive type of eating behavior (U = 58.5, p = 0.001, compare rank in group 1 = 28.21, compare rank in group 2 = 13.3). This suggests that girls with anorexia are more likely to positively rethink a stressful situation than girls in the second group. This is probably due to the fact that girls with a restrictive type of eating behavior justify their food restrictions by striving for self-development, due to which they perceive stressors as an opportunity to approach their ideal image of a strong-willed, strong person.

2.1.2 In a comparative analysis of the peculiarities of coping strategies in girls with restrictive and harmonious eating behavior, the following results were obtained:

- The "Distancing" indicator is significantly higher in the group of girls with a harmonious type of eating behavior (U = 119, p = 0.016, M1 = 16.6, M2 = 25.5). It turns out that girls with a harmonious type of eating behavior in a difficult situation tend to reduce its importance more often than adolescents with a restrictive type of eating behavior. This is probably due to the fact that girls with anorexia are characterized by pronounced rigidity, which makes it difficult for them to switch from a stressful situation to other stimuli.

- The indicator “Search for social support” is significantly higher in the group of girls with a harmonious type of eating behavior (U = 76, p = 0.001, M1 = 14.6, M2 = 27.7). Consequently, girls with harmonious eating behavior more often in a difficult situation resort to the search for social resources to solve it than adolescents of the first group. This is probably due to the fact that girls with anorexia have a lower level of social adaptation: due to a negative attitude towards themselves, they expect the same attitude from others, which causes them to strive for isolation.
- The indicator “Acceptance of responsibility” is significantly higher in the group of adolescents with a restrictive type of eating behavior (U = 5.5, p = 0.001, M1 = 30.7, M2 = 10.7). It turns out that girls with anorexia are more likely to react with excessive self-criticism in difficult situations than respondents from the other group. This is probably due to the fact that adolescents with a restrictive type of eating behavior, due to low self-esteem, are characterized by high self-accusation, i.e. they blame themselves rather than other people or circumstances for any problems that arise.
- The Runaway indicator is significantly higher in the group of girls with a restrictive type of eating behavior (U = 20, p = 0.001, M1 = 30.05, M2 = 11.5). Consequently, adolescents with anorexia tend to avoid problem solving more often than respondents with harmonious eating behavior. This is probably due to the fact that girls of the first group are distinguished by a great subjective feeling of helplessness, and any stressful situations cause severe manifestations of mental exhaustion in them, which makes them avoid them.
- The indicator "Planning a solution to the problem" is significantly higher in the group of adolescents with a harmonious type of eating behavior (U = 78.5 p = 0.001, M1 = 14.7, M2 = 27.5). This suggests that girls of the second group in a stressful situation are more often directed at planning actions to solve the problem than adolescents of the first group. This is probably due to the fact that girls with a harmonious type of eating behavior are distinguished by a more realistic vision of the situation and a positive approach to it, while patients with anorexia see the problem as clearly hopeless.

2.1.3 As a result of a comparative analysis of the peculiarities of coping strategies in adolescents with hyperphagic and harmonious eating behavior, the following results were obtained:

- The “Confrontation” indicator is significantly higher in the group of girls with bulimic symptoms (U = 36, p = 0.001, M1 = 28.7, M2 = 12.3). It turns out that girls of the first group are more inclined to respond with non-targeted behavioral activity in stressful situations. This is probably due to the fact that hyperphagous eating behavior is formed on the basis of impulsivity, which is also true for the strategy of confrontation.
- The indicator "Distancing" is significantly higher in the group of girls with a harmonious type of eating behavior (U = 34, p = 0.001, M1 = 12.2, M2 = 28.8). Consequently, adolescents with a harmonious type of eating behavior in a difficult situation tend to reduce its importance more often than respondents with an emotiogenic and external type of eating behavior. This is probably due to the fact that girls of the first group are characterized by increased emotional reactivity, as a result of which the strategies for reducing the significance of the problem due to intellectual efforts turn out to be inoperative for them.
- The indicator "Search for social support" is significantly higher in the group of adolescents with a harmonious type of eating behavior (U = 43, p = 0.001, M1 = 12.6, M2 = 28.3). It turns out that respondents with a harmonious type of eating behavior more often in a difficult situation resort to looking for social resources to solve it than girls of the first group. This is probably due to the fact that adolescents with hyperphagic eating behavior

have a lower level of social adaptation associated with the expectation of a negative assessment from others.

- The indicator “Acceptance of responsibility” is significantly higher in the group of girls with bulimic type of eating behavior (U = 6.5, p = 0.001, M1 = 30.1, M2 = 10.8). This suggests that girls with hyperphagic eating behavior are more likely to react with excessive self-criticism in difficult situations than adolescents from the other group. This is probably due to the fact that the respondents of the first group are more prone to self-accusation reactions, formed as a response to compulsive overeating.
- The "Escape" indicator is significantly higher in the group of girls with emotiogenic and external types of eating behavior (U = 75, p = 0.001, M1 = 26.7, M2 = 14.2). Consequently, adolescents of the first group tend to avoid solving problems more often than girls with harmonious eating behavior. This is probably due to the fact that the respondents of the first group are more likely to have a negative perception of themselves and the world around them, as a result of which they do not believe in their ability to resolve life difficulties.
- The indicator "Planning a solution to the problem" is significantly higher in the group of girls with a harmonious type of eating behavior (U = 9 p = 0.001, M1 = 10.9, M2 = 30.05). Consequently, the adolescents of the second group in a stressful situation are more often directed at planning actions to solve the problem than the respondents of the first group. This is probably due to the fact that girls with a harmonious type of eating behavior are distinguished by a more realistic vision of the situation and a positive approach to it, while patients with eating disorders are characterized by a negative outlook on the problem and a belief in its insolubility.
- The "Positive overestimation" indicator is significantly higher in the group of adolescents with a harmonious type of eating behavior (U = 48 p = 0.001, M 1 = 12.9, M 2 = 28.5). This means that girls with harmonious eating behavior are more likely to perceive difficulties as stimuli for personal growth than girls from the other group. This is probably due to the fact that girls with emotional and external eating behavior as a result of the predominance of a negative emotional background more often see the problem as a hindrance to their activity, ignoring the possible experience that it carries.

2.2 Description of the results of a comparative analysis of the characteristics of self-attitude in adolescents with different types of eating behavior

2.2.1 In a comparative analysis of the characteristics of self-attitude in adolescents with restrictive, emotional and external eating behavior, the following data were obtained:

- in terms of self-attitude indicators "Self-esteem" (U = 117.5 p = 0.015, average rank in group 1 = 16.6, average rank in group 2 = 25.63), "Self-interest" (U = 131 p = 0.034, average rank in group 1 = 17.2, average rank in group 2 = 24.9) and "Self-understanding" (U = 107.5, p = 0.006, compare rank in group 1 = 16.1, cf. rank in group 2 = 26.13), the level of which is higher in girls with a hyperphagic type of eating behavior. Consequently, respondents with emotional and external eating behavior have a more positive attitude towards themselves than adolescents of the second group. This is probably due to the fact that patients with anorexia have more pronounced high demands on themselves and excessive striving to meet certain ideals, which does not give the opportunity to enjoy their own activities, causes constant dissatisfaction with themselves and contributes to a decrease in self-esteem. Presumably, girls with anorexia nervosa find it more difficult to overcome imposed ideals due to their high rigidity.

2.2.2 In a comparative analysis of the characteristics of self-attitude in adolescents with restrictive and harmonious eating behavior, significant differences are observed in all indicators of self-attitude:

- "Integral self-attitude" (U = 15.5, p = 0.001, M1 = 11.7, M2 = 30.7);
- "Self-esteem" (U = 19, p = 0.001, M1 = 11.9, M2 = 30.5);
- "Selfsympathy" (U = 12.5 p = 0.001, M1 = 11.6, M2 = 30.88);
- "General expected attitude towards oneself" (U = 52, p = 0.001, M1 = 13.4, M2 = 28.9);
- "General self-interest" (U = 46, p = 0.001, M1 = 13.1, M2 = 29.2);
- "Self-confidence" (U = 66.5, p = 0.001, M1 = 14.1, M2 = 28.1);
- "Expected attitude" (U = 83.5, p = 0.001, M1 = 14.9, M2 = 27.33);
- "Self-acceptance" (U = 2.5, p = 0.001, M1 = 11.1, M2 = 31.3);
- "Self-guidance" (U = 102, p = 0.003, M1 = 15.8, M2 = 26.4);
- "self-accusation" (U = 27.5, p = 0.001, M1 = 29.6, M2 = 11.88);
- "Self-interest" (U = 33, p = 0.001, M1 = 12.5, M2 = 29.8);
- "Self-understanding" (U = 25, p = 0.001, M1 = 12.1, M2 = 30.2).

The differences indicate that adolescents with harmonious eating behavior have a more positive self-attitude than patients with anorexia. This is probably due to the fact that negative self-attitude is a risk factor for the formation of eating disorders: against the background of an inadequate attitude towards oneself, the need for acceptance grows, which is achieved through the desire to achieve ideals.

2.2.3 In a comparative analysis of the characteristics of self-attitude in adolescents with hyperphagic and harmonious eating behavior, significant differences were found in the following indicators of self-attitude:

- "Integral self-attitude" (U = 12, p = 0.001, M1 = 11.1, M2 = 29.9);
- "Self-esteem" (U = 38 p = 0.001, M1 = 12.4, M2 = 28.6);
- "Self-sympathy" (U = 12 p = 0.001, M1 = 11.1, M2 = 29.9);
- "General expected attitude towards oneself" (U = 14.5, p = 0.001, M1 = 11.2, M2 = 29.7);
- "General self-interest" (U = 43.5, p = 0.001, M1 = 12.6, M2 = 28.33);
- "Self-confidence" (U = 57, p = 0.001, M1 = 13.3, M2 = 27.65);
- "Expected attitude" (U = 74.5, p = 0.001, M1 = 14.2, M2 = 26.7);
- "Self-acceptance" (U = 2, p = 0.001, M1 = 10.6, M2 = 30.4);
- "Self-accusation" (U = 32.5, p = 0.001, M1 = 28.8, M2 = 12.1);
- "Self-interest" (U = 16, p = 0.001, M1 = 11.3, M2 = 29.7);
- "Self-understanding" (U = 82.5, p = 0.001, M1 = 14.6, M2 = 26.3).

The revealed differences indicate that the parameters of self-attitude are more favorable in the group of adolescents with a harmonious type of eating behavior. This is probably due to the fact that girls with a hyperphagic type of eating behavior, due to bouts of overeating and the inability to cope with negative emotions in a different way, feel their inferiority, which is aggravated by the inability to positively interact with the environment due to fear of evaluation.

3 Conclusion

As a result of the study, we came to the conclusion that there are statistically significant differences in the characteristics of self-esteem and coping strategies in puberty children with different eating behaviors.

It was found that adolescents with an eating disorder are more likely to resort to non-constructive coping strategies in a difficult situation for them - escape and self-control. It is worth noting the following features:

- the group of patients with anorexia nervosa is most characterized by coping strategies of taking responsibility and escape.
- the group of adolescents with hyperphagic eating behavior uses the coping strategy of confrontation more often than others.
- the group of respondents with harmonious eating behavior more often than others resort to the strategy of distancing, planning a solution to the problem and positive overestimation.

With regard to self-esteem, we came to the conclusion that all three groups of adolescents have inadequate forms of it, but there are significant differences:

- adolescents with anorexia nervosa have low self-esteem. These patients are dominated by exaggerated demands on themselves, striving for ideality, and each discrepancy deeply wounds and strikes a blow to the image of the "I", which probably causes the need to punish oneself with hunger;
- girls with hyperphagic eating behavior have a more positive attitude towards oneself than patients with restrictive eating behavior. It should be noted that this group of respondents hides vulnerability and emotional immaturity behind the external overestimated self-esteem. It can be assumed that these adolescents have a desire to seize their experiences and against the background of overeating, the inability to break the pathological circle, they experience a feeling of inferiority, which entails a new attack of filling themselves with food.
- respondents with harmonious eating behavior also showed a positive attitude towards themselves and overestimated self-esteem, manifested in an excessively high and not always justified assessment of their qualities and capabilities.

References

1. V. V. Marilov, M. V. Korkina, A. M. Tsivilko, *Anorexia nervosa,* 176 (1986)
2. G. F. Muzychenko, *Projective method "Non-existent animal". Guidance and results of psychodiagnostic research of adult patients with various disorders of the emotional and personal sphere, 556* (2013)
3. A. O. Kibitov, G. E. Mazo, Social and Clinical Psychology (2016)
4. I. G. Malkina-Pykh, *Eating behavior therapy,* 1040 (2007)
5. N. G. Garanyan, *Difficult Patient Journal for Physicians* (2006)
6. M. Hirsch, *"This is my body ... and I can do with it whatever I want": A psychoanalytic view of dissociation and body dramatization, 381* (2018)
7. F. Alexander, *Psychosomatic medicine. Principles and application*, 250 (2017)
8. E. K. Kislova, *Bulimic disorders in anorexia nervosa* (2004)
9. E. Wufong, P. Rhodes, J. Conti, Journal of Eating Disorders, **7,** 5 (2019)

Legal formation of the national systems of organic production in Russia and the EAEU

Denis Matytsin, *Iolanta* Baltutite*, and *Agnessa* Inshakova

Volgograd State University, 400062, prosp. Universitetsky 100, Volgograd, Russia

Abstract. The chapter substantiates the conclusion that an organic product is a food product produced on basis of natural agriculture and animal husbandry, without the use of pesticides and other plant protection products, chemical fertilizers, growth stimulants and animal fattening. Having studied the agricultural legislation of Russia and other EAEU countries, the authors note that these legal acts contain a large number of common approaches and methods of state regulation of the production and turnover of organic products, although it differs in terms of individual production conditions, prohibitions and restrictions. Despite the fact that the adoption of laws on organic products in Russia and the EAEU countries was a great victory for agricultural producers and consumers, there is still a lot of work to done. It is necessary to bring the EAEU standards in line with international requirements, to develop specific measures to support producers of organic agricultural products, to fine-tune control mechanisms and to establish sanctions for violations in this area. In particular, it is advisable to establish legal liability for the use of organic labeling without proper grounds. As economic measures to support producers of organic products, it is necessary to establish a regulatory obligation of the state to purchase only organic products for military personnel, hospitals, orphanages, as well as other similar social needs.

1 Introduction

The current state of the market for environmentally friendly products characterized by rapid development, since where these technologies introduced, there is an increase in yields and an improvement in soil fertility [1] .According to experts, in the developed countries of the world, about 80% of consumers prefer environmentally friendly (organic) products, and the incentive for the development of its market is the fact that programs to support organic agriculture belong to the measures of the" green basket " of the WTO, and the costs are not limited to them [2] .But organic farming is not just a "green" project, but also a real tool for the development of modern competitive agriculture in the global food market [3].The environmental aspect of consumers' choice of organic technologies is also due to the fact that chemicals in the form of fertilizers and pesticides can pollute groundwater and streams through runoff cause algae blooms and oxygen depletion in water bodies, contribute to soil

*Corresponding author: u938om@yandex.ru

acidification, kill beneficial insects and potentially poison wildlife and its reproductive system [4]. Along with this, consumers may have a number of additional considerations.

Until recently, Russia was among them, since it did not implement legal regulation of the production of environmentally friendly products. The situation changed only after the adoption of Federal Law No. 280-FZ of August 3, 2018 "On Organic Products and on Amendments to Certain Legislative Acts of the Russian Federation» [5].The Law is important for the sustainable development of agriculture and ensuring food security. The unique territorial and climatic conditions of our country create an optimal springboard for the development of ecological agriculture and the accompanying market for organic (environmentally friendly) products.

2 Materials and Methods

The legislative base of the study made up of the laws of the Russian Federation and other EAEU countries regulating various aspects of the activities of agricultural producers for the cultivation of organic products. The authors analyzed the provisions of the Russian federal laws "On Organic products and on Amendments to Certain Legislative Acts of the Russian Federation", "On the Quality and Safety of Food Products", "On Technical Regulation» [6]. The authors analyzed the laws of other EAEU countries. The Law of the Republic of Kazakhstan "On the Production of Organic Products" of November 27, 2015 [7].The Law of the Republic of Belarus of November 9, 2018 "On the production and circulation of Organic Products" [8].The Law of the Republic of Armenia of April 8, 2008 "On Organic Agriculture" [9].The Law of the Kyrgyz Republic of May 18, 2019 "On Organic Agricultural production in the Kyrgyz Republic» [10]. These laws regulate the conditions and requirements for producers of organic products, prohibitions and restrictions aimed at ensuring the proper quality of organic products, procedures for voluntary certification, maintaining a single register and labeling organic products.

The problems that arise in the field of organic production have studied in the works of scientists from Russia and the CIS countries, as well as the works of representatives of European countries and the United States. Thus, Z. Yu. Belyakova, Yu. N. Budanova, B. A. Voronin, and S. D. Belyakova devoted their works to the problems in the sphere of production and turnover of organic products. Isvaliev, T. V. Kurman, A. I. Meniv, E. S. Navasardova, K. A. Selivanova, R. S. Yarandaykin, and also J. J. Czarnezki, K. L. Harrison, J. Hass, E. Morgera, K. Ong, P. Robért, V. J. Watnick and many others. The opinions and conclusions expressed by them used by the authors of this chapter to formulate their own doctrinal position.

In the process of research, general scientific methods are used, such: formal-logical, dialectical, system-structural, critical cognition. Methods of synthesis, classification, and generalization used to interpret the results of the study. The paper also uses private scientific methods: formal-legal, the principle of assessing legal processes, the method of comparative analysis, etc.

3 The concept of organic (environmentally friendly) products and the legal regulation of their production in the Russian Federation

Until recently, the Russian law did not have the term "environmentally friendly product" or any other concept that denotes it. Federal Law No. 280-FZ of August 3, 2018 "On

Organic Products and on Amendments to Certain Legislative Acts of the Russian Federation" established that organic products are environmentally friendly agricultural products, raw materials and food, the production of which meets the requirements established by this law. In addition, Article 1 of Federal Law No. 29-FZ of January 2, 2000 (as amended on 13.07.2020)"On the quality and safety of food products"contains the concept of"food quality". This is a set of characteristics of safe food products that meet the requirements established in accordance with the legislation of the Russian Federation, the terms of the contract, the sample, standardization documents, technical documentation that determine their consumer properties, nutritional value, authenticity, grade (caliber, category, etc.), and meet the physiological needs of a person [11]. According to Article 15 of Law No.29-FZ, food products intended for sale must meet the physiological needs of a person in the necessary substances and energy. They must meet the mandatory requirements established by the legislation of the Russian Federation for the permissible content of chemical (including radioactive), biological substances and their compounds, microorganisms and other biological organisms that pose a danger to the health of current and future generations.Thus, Russian legislation suggests that there are two approaches to the assessment of agricultural products: it can be simply safe, not have negative consequences for health and meet the physiological needs of a person, or it can be useful, and grown in compliance with special environmental technologies [12].

This means that environmentally friendly production and environmentally safe production are two systems of agricultural activity that do not completely coincide, requiring a different approach in legal regulation. It seems that products that do not have a harmful effect on nature and human health and meet the safety requirements (sanitary standards, environmental regulations or technical regulations) can considered environmentally safe. Eco-friendly products are a more neutral term than organic products. In the moment Article 4 of Federal Law No. 280-FZ of August 3, 2018 "On Organic Products and on Amendments to Certain Legislative Acts of the Russian Federation" provides for a number of requirements for the production of organic products.

The requirements of the Russian legislation regarding the concept and conditions of organic production determined by the content of scientific discussions that have been going on in Russian and world science for a long time.

Thus, the first group of authors believes that organic agriculture is a system of growing plants, livestock and fish farming with respect for the principles of environmental protection, using "natural" farming methods aimed at ensuring the long-term health of the population, the productivity of ecosystems, improving the quality of products and protecting nature [13].

Representatives of the second group of authors believe that in the legal sense environmentally friendly products are not so much the degree of their compliance with the level of requirements imposed in regulatory acts. It is proposed to consider environmentally friendly products that have increased nutritional value, strengthen health, do not contain toxic substances do not have carcinogenic, mutagenic or other adverse effects on the human body [14].

Summarizing it seems that an organic product is a food product produced on basis of natural agriculture and animal husbandry, without the use of pesticides and other plant protection products, chemical fertilizers. Without growth stimulants and fattening of animals, antibiotics, hormonal and veterinary drugs, artificial food additives and GMOs, without refining, mineralization, addition of artificial flavors, dyes and other additives that change the natural properties of such a product, which is enshrined in national legislation.

4 Problems of labeling and certification of organic products in Russia and other EAEU countries

The Russian legislation provides for several tools of state management in the field of organic production, including a number of technological prohibitions (for example, on the use of food additives, ionizing radiation, etc.), as well as means of state regulation: certification, labeling, a single register of manufacturers.This means that the ecological entrepreneur has more responsibilities than the producer of traditional agricultural' products. However, at the same time, it receives certain advantages in the consumer market, allowing it significantly increase profits.

The main question at the same time is how to guarantee the consumer high-quality organic products, with the help of which information means he should receive information about such a product, and how the state should control this process. For example, the U.S. Government controls the production and circulation of organic products through three main methods: certification, authentication, and labeling [15].

In the EU, environmental certification (state-owned and produced on a voluntary basis by non-profit organizations) is the most common way of environmental information, which allows you to maintain consumer confidence in organic products. However not environmental certification is given to the competence of private organizations. In the United States, Sweden and a number of other countries, the state is responsible for controlling environmentally friendly production. The American National Organic Standard developed by the US Department of Agriculture.

In Russia, after the normative consolidation of the concept of "organic product", there is no mandatory certification of organic production or products, that is, it made on a voluntary basis. According to Article 21 of Federal Law No. 184-FZ of 27.12.2002 "On Technical Regulation", voluntary confirmation of compliance carried out at the initiative of the applicant on the terms of the contract between the applicant and the certification organ.

Voluntary conformity assessment can carried out to establish compliance with standardization documents, voluntary certification systems, and contract' terms.The certification organ confirms the conformity of objects to these standards; issues certificates of conformity for objects; grants applicants the right to apply the mark of conformity; suspends or terminates the validity of the certificates of conformity issued to them. The system of voluntary certification can created by a legal entity or an individual entrepreneur (their associations)[16].The conformity mark informs the buyer of the product that the product certified and meets the established quality standards. Currently, there are four national certification bodies in Russia: "Organic-Expert", "Roskachestvo", "Rosselkhoznadzor" and "Organic-Certification».

In Russia created a special system for labeling organic products, which includes legal norms that establish mandatory requirements within the framework of state regulation, as well as a system of organs, that monitor the implementation of these norms. According to Article 7 of Federal Law No. 280-FZ of August 3, 2018 "On Organic Products and on Amendments to Certain Legislative Acts of the Russian Federation", producers of organic products, after confirming the conformity of production of organic products, have the right to place the marking that is a distinctive feature of organic products. Marking is made in the form of a combination of inscriptions and a graphic image (sign) of organic products of a single sample on the packaging, consumer and (or) transport packaging of organic products or on other information carriers attached to it or placed in it.

No less important than certification and labeling, an element of public administration in the Russian Federation was the introduction of the Unified State Register of Organic

Producers. This register allows you to distinguish environmental entrepreneurs engaged in the production of organic products from the total mass of traditional industries.

In other EAEU countries, we see the following situation.

a) The Republic of Kazakhstan has the Law of the Republic of Kazakhstan "On the production of organic products" dated November 27, 2015 No. 423-V (as amended on 28.10.2019). The Law contains rules and requirements comparable to the Russian legislation in the field of production and turnover of organic products. In particular, it provides (Article 11) conditions for the production of organic products (for example, including a ban on the use of synthetic substances, pesticides, hormones, antibiotics and food additives). As in the Russian Federation, in the Republic of Kazakhstan, the confirmation of conformity of organic production is voluntary (Article 12). The result of the confirmation of conformity of organic production is a certificate of conformity.

b) The Law of the Republic of Belarus of November 9, 2018 No. 144-Z "On the production and circulation of organic products", in contrast to the laws of other EAEU countries, regulates in sufficient detail the powers of various government organs (agriculture, health, etc.). A special feature of the law is also the regulation of seed production and beekeeping. In other respects, the law contains comparable requirements for the production of organic products, as well as other EAEU countries, as well as methods of public administration – labeling, registration,voluntary certification of products.

c) The Law of the Republic of Armenia of April 8, 2008 No. 23-N "On Organic Agriculture" aimed at the development of organic production, promotion of innovations, introduction of modern technologies, export of organic products within the framework of interstate economic cooperation. An independent certification system for organic products provided.

d) Law of the Kyrgyz Republic No. 65 of May 18, 2019 "On Organic Agricultural Production in the Kyrgyz Republic". As in the rest of the EAEU countries, this law contains a standard set of management methods, including maintaining a register, voluntary certification and labeling of organic products. However, unlike the Russian Federation, in the Republic of Kyrgyzstan, the law directly prescribes measures of state support for producers of organic products.

Thus, the legislation of Russia and other EAEU countries contains a large number of common approaches and methods of state regulation of the production and turnover of organic products, although it differs in terms of individual production conditions, prohibitions and restrictions [17].In addition, the duration of the transition period differs in the EAEU countries, and their laws do not mention the quality of natural objects other than land as a condition that can negatively affect the quality of organic products grown by them [18]. Further development of the legislation of the EAEU countries on organic products will influenced by international standards, including in force in the EU countries.

5 Results and Discussion

Within the framework of the regulatory regulation of the production of organic (environmentally friendly) agricultural products in various countries of the EAEU, a number of issues remained unresolved that require the attention of representatives of agricultural and legal science. Among them, we note the following:

1) since there are three basic models of organizing agricultural production (traditional model; using GMOs; organic farming), the national authority must approve a special Provision regulating the coexistence of these models. The fact is that if a field where organic products are grown is located next to a field where GMO crops grow, bee pollination of both fields may occur, which will violate the purity of organic products. In

Russia, the cultivation of agricultural products using GMOs currently prohibited, although in other countries the issue of the distance between organic fields and fields with GM crops, the difference in crop rotation, information issues, etc. is being discussed [19].

2) the question follows from the previous one: can there be GMO impurities in the composition of organic products, and if so, in what amount. For the EAEU countries, the experience of the United States is of particular interest, where the labeling of organic products has four levels, setting the degree of "organic" products in the range from less than 70 % to 100 %. The general principle underlying such labeling rules is that the degree of labeling increases as the organic content of the product increases. Thus, the higher the organic content of the product, the more clearly this organic content shown to the consumer [20]. In the EU, there are three categories of "organic" goods: if the product contains more than 95 % organic ingredients, it called organic; if there are 70-94% organic components, the word "organic" used in the list of ingredients; less than 70 % organic ingredients-the word "organic" used on the package [21,22]

3) the agricultural clusters can created for the cultivation of organic products. The category "cluster" is widely used in Russian legislation (for example, in relation to the creation of industrial clusters) [26].

Agricultural clusters in the Russian Federation and other EAEU countries should focus on strengthening their innovation orientation in the near future. This will require the addition of the Federal Law of 29.12.2006 (as amended on 15.10.2020) "On the development of Agriculture" with the special term "agricultural cluster", indicating the legal regime, incentives and restrictions for agricultural producers engaged in the cultivation of organic products within its borders.

4) the management of agricultural production, including the cultivation of organic products, is closely linked to climate problems. On the one hand, climate change strongly affects the efficiency of agricultural production (for example, due to droughts). On the other hand, the question arises about the impact of such production on climate change

5) today in many countries of the world (including Russia), the market of organic products has been captured by large producers. Obviously, it already makes sense to think about a system of legal incentives and restrictions that will allow small farms that produce organic products to remain on the market. The change in legislation could allow local authorities to purchase organic products for schools and other social facilities. The promotion of organic products could also serve as an adjustment of school educational programs.

6) On August 28, 2011, the state of Vermont (USA) experienced one of the worst natural disasters - tropical storm Irene (Irene). It is therefore important to ensure that public authorities properly regulate this aspect of organic production cultivation. Meanwhile, regulations in many countries do not adequately protect future organic crops from growing on previously flooded soils, and this raises important consumer protection issues for buyers of organic products [27].

6 Conclusion

An organic product is a food product produced on basis of natural agriculture and animal husbandry, without the use of pesticides and other plant protection products, chemical fertilizers, growth stimulants and fattening of animals, antibiotics, hormonal and veterinary drugs, artificial food additives and GMOs. And without refining, mineralization, the addition of artificial flavors, dyes and other additives that change the natural properties of such a product, which is enshrined in national legislation. The legislation of Russia and other EAEU countries contains a large number of common approaches and methods of state regulation of the production and turnover of organic products, although it

https://doi.org/10.1051/e3sconf/202131110002

differs in terms of individual production conditions, prohibitions and restrictions. In addition, the EAEU countries differ in the duration of the transition period, and their laws do not mention the quality of other (other than land) natural objects as a condition that can negatively affect the quality of organic products grown by them.

Despite the fact that the adoption of laws on organic products in Russia and the EAEU countries was a great victory for agricultural producers and consumers, there is still a lot of work to done to bring the EAEU standards in line with international requirements. It is necessary to develop specific measures to support producers of organic agricultural products, to fine-tune control mechanisms and to establish sanctions for violations in this area. In particular, it is advisable to establish legal liability for the use of organic labeling without proper grounds. As economic measures to support producers of organic products, it is necessary to establish a regulatory obligation of the state to purchase only organic products for military personnel, hospitals, orphanages, as well as other similar social needs. Producers engaged in environmentally friendly production should granted land tax benefits, as well as other taxes in Russia and the EAEU countries. The proposed directions of the state agrarian policy due to fact, that production and economic activities for the cultivation of organic products have certain features, carried out with minimal human intervention using the natural laws of the development of agricultural plants and animals. Given the relatively recent appearance of laws on the regulation of organic agricultural production in Russia and other EAEU countries, the number of topical issues will only increase.

Acknowledgments

The reported study was funded by RFBR according to the research project No. 18-29-16132.

References

1. A.O. Inshakova, E.E. Frolova, M.V. Galkina, E.P. Rusakova, International Journal of Sociology and Social Policy, **ahead-of-print**, ahead-of-print (2020)
2. B. A. Voronin, Agrarian and Land Law, **9**, 72 (2013)
3. S. Zankovsky, V. Bezbakh, A. Inshakova, E. Rusakova, *Social consequences of economic globalization: experience of developed and developing countries and perspectives of optimization* (2020)
4. J.J. Czarnezki, Stanford Environmental Law Journal, **30**, 3 (2011)
5. Federal Law, On Organic Products and On Amendments to Certain Legislative Acts of the Russian Federation (2018)
6. Federal Law, On Technical Regulation (2002)
7. Law of the Republic of Kazakhstan, On the Production of Organic Products (2015)
8. Law of the Republic of Belarus, On the Production and Circulation of Organic Products (2018)
9. Law of the Republic of Armenia, On Organic Agriculture (2008)
10. Law of the Kyrgyz Republic, On Organic Agricultural Production in the Kyrgyz Republic (2019)
11. E. Morgera, C.B. Caro, G.M. Duran, Organic agriculture and the law. Food and agriculture organization of the United Nations (2012)
12. K. Ong, Environmental Law Journal, **17**, 883 (2008)

13. D. E. Matytsin, E.P. Rusakova Strategy of quality management in industry 4.0 and formation of cognitive economy based on industrial and manufacturing engineering in the Russian Federation and countries of the EU (2021)
14. K.L Harrison, Environmental Law Review, **25,** 211 (2008)
15. Y.N. Bydanova, Management of innovative development of agri-food systems at the national and regional levels: proceedings of the II International Scientific and Practical conference. Voronezh: Voronezh State Agrarian University (2020)
16. D.E. Matytsin, Internet-investing as a remote algorithm of the retail investment financing (2021)
17. E. S. Navasardova, A. N. Zakharin, Humanities and Legal Studies, **2**, 141 (2019)
18. A.I. Meniv, Legal support for the use of GMOs in the cultivation of agricultural products of plant origin in Ukraine and the EU (2016)
19. J.J. Czarnezki, Utah Environmental Law Review, **31-2**, 263 (2011)
20. J. Hass, William & Mary Environmental Law and Policy Review, **34**, 589 (2010)
21. R.S. Yarandaikin, Organizational and legal problems of production and sale of environmentally friendly agricultural products (1999)
22. P. Robért, Journal of Environmental Law, **14**, 303 (2012)

Modern biotechnology: search for an optimal model of legal regulation

Denis Matytsin, *Olesya* Kazachenok, and *Agnessa* Inshakova*

Volgograd State University, Volgograd, Russia

Abstract. This chapter justifies the conclusion that GMO technologies as a form of biotechnology are the greatest technological breakthrough of our time. With the help of GMOs, it is possible to increase the yield, which allows not only to solve the problems of world food security, but also to preserve untouched lands from human impact (including through the creation of national parks or other types and forms of specially protected natural areas there). The use of GMO technologies reduces the use of pesticides and agrochemicals in agriculture. However, in conditions where the harm from GMO technologies and products not convincingly proven, the search for an optimal model for their use should continue. The authors note that the concept of sustainable development implies a balance of environmental, economic and social interests. Deviation from this balance in any direction entails a number of negative consequences for the entire society, both in terms of problems with the realization of the right to food, as well as in terms of environmental protection. In this situation, it seems appropriate not to ban GMOs absolutely (as in Russia), which means the uselessness of the law of the relevant state as a regulator of public relations. Considering the measures of state regulation of the use of GMO technologies and products, the authors focus on the prospects for the use of certification and labeling of GMO products, which in a number of BRICS and EAEU countries has already had a positive effect.

1 Introduction

Other studies of the potential dangers of GMOs are more objective, while the discussion between supporters and opponents of GMOs, who differently assess the degree of safety of products produced with their help, has no end. Despite the high public demand, there is still no accurate information about the presence of danger (or its absence) in the use of agricultural GMO products. It has not yet been possible to prove harm to human health from GMO technologies, although some types of products produced in violation of technologies have caused such harm, which has already proven several times in the courts of a number of countries.

In addition to possible harm to health, a separate problem of handling GMOs is the presence of potential or real harm to the environment. The fact is that GMO technologies

* Corresponding author: ainshakova@list.ru

can not only improve the resistance to climate factors or weeds of cultivated agricultural plants, but also lead to increased survival of the weeds themselves.

Taking in to account above circumstances it is necessary to study the problem of the use of GMO technologies in agriculture from the standpoint of legal science. Of course, such a study will not be able to give recommendations on the use of a certain type of GMO products (this question is rather from the field of biological or medical science). However, it is through legal means that the use of GMO technologies in individual countries or their regions can recommended or prohibited, and a list of measures is defined to assess the quality of the product and give the results of such an expert assessment legal significance. Finally, it is the legal means, that allow us to guarantee the information rights of consumers, to regulate the procedural features of the consideration of cases for compensation for harm caused by the use of GM products.

2 Materials and methods

In the course of the research, general scientific methods used, including formal-logical, dialectical, system-structural, and critical cognition. Methods of synthesis, classification, and generalization used to interpret the results of the study. Private scientific methods also used in the work: formal-legal, the principle of assessing legal processes, the method of comparative analysis, etc.

3 Results and discussion

Today, more than 100 lines of transgenic plants grown on a production scale in the world, the undisputed leader among which is soy, which accounts for 98% of the total number of genetically modified plants. Two-thirds of all genetically modified crops in the world are cultivated in the United States. The primary objectives of creating genetic modifications were to obtain high yields, achieve resistance to drought and cold, and a number of others. Now we are talking about the creation of vitamin-rich organisms, with a regulated fat content. The use of GMOs in the production of agricultural products and food products is of great importance in terms of food security. Despite this, the prevailing opinion about the use of GMOs is that there is a potential threat to human health and the gene pool of the nation as a whole, and there is no single definition of GMOs in legal science.

According to some authors, a genetically modified organism is an organism whose genetic material has been modified using genetic engineering methods [1]. Others believe that a genetically modified organism used to grow agricultural products of plant origin is a plant organism (its parts, including seeds). In which the genetic material has received such changes that cannot obtained in natural conditions in the process of natural recombination, and which is used to increase crop yields, change the quality indicators of agricultural products [2]. Agreeing with the latter definition it noted that GMO technologies should considered in a broader context – as a kind of biotechnology that has long used by humanity.

The term "biotechnology" first proposed in 1917 by the Hungarian engineer K. Jereki [3]. Trying to eliminate the gaps in national law, in the second half of the XX century, A. Sasson (1987) [4] defined biotechnology as a technological process implemented using biological systems - living organisms and components of a living cell. At moment, there is no unified scientific opinion in the understanding of biotechnologies. They interpreted as a set of methods and techniques for obtaining useful products and phenomena for humans using biological agents [5]. As a special social activity aimed at the practical transformation of the surrounding world and oneself by a person through use of biological processes and agents [6]. As a science of ways to obtain target products using biosynthesis controlled by environmental parameters or genetic-engineering manipulations or a combination of these effects [7]. The latter position seems to us the most successful.

So biotechnology is a field of applied biology, which involves the use of living organisms and biological processes in engineering, technology, medicine and other areas in which biological products needed. In the legal sense, "biotechnology" means any type of technology related to the use of biological systems, living organisms or their derivatives for the manufacture or modification of products (processes) for the purpose of their direct use. The Russian Federation has joined the Convention on Biological Diversity in 1992, but it has not signed the Protocol on Biosafety (the Cartagena Protocol) to this convention.

Summarizing all of the above, we note that a GMO is an organism or several organisms, any non-cellular, unicellular or multicellular formation capable of reproducing or transmitting hereditary genetic material, other than natural organisms, obtained using genetic engineering methods and containing genetically engineered material, including genes, their fragments or combinations of genes. Modification of the genome of an organism is one of the methods of biotechnology (although not the only one), and modern agricultural biotechnologies are almost identical to the use of genetic engineering and the use of GMOs. Taking into account the polarity of opinions on the usefulness / danger of GMO technologies and products for human health and the environment, we will try to systematize the positions of different sides of this discussion.

In turn, opponents of GMOs report the first death of a person who ate food obtained using GMO technologies in a Spanish hospital in 2015. Medical examiners then determined the cause of death as result of consumption of tomatoes containing fish-related genes and antibiotic-resistant genes that prevented the development of white blood cells. There are also examples showing that concerns about the safety of GM foods have been justified in relation to potential exposure and risk in a number of other countries. For example, a study conducted by the Department of Food Science and Technology at the University of Nebraska found that a gene produced from Brazil nuts when introduced into GM soy causes allergies in humans [8].

The danger of GMO technologies is much more reasonable from the point of view of its impact on the environment. Opponents of GMOs are undoubtedly right in their concerns in the sense that ignoring the laws of evolution and population genetics and invading the genetic program of an organism is very dangerous, especially when it caused by the short-term financial benefits of individuals, companies or financial groups [9]. Such risks are associated with the emergence of super weeds, the formation of new, poison-resistant insect populations, genetic contamination, and the permanent loss of traditional crop varieties [10].

On the other hand, all officially registered cases of harm to the health of citizens or nature occurred not because of the use of the achievements of genetic engineering as such, but because of the fact that specific companies-producers of agricultural GMO products incorrectly applied the appropriate technology. Accordingly, genetic engineering is a manifestation of scientific and technological progress, which cannot prohibited [11]. Therefore, in our opinion, further discussions on the applicability of GMO technologies in agriculture conducted not from the standpoint of a complete ban or permission of GMOs, but in the direction of finding a balance of private and public interests, a compromise between the authorities, the public and agricultural producers. At the same time, it is necessary to look for new criteria of objectivity when evaluating the results of the use of genetic engineering methods, since this innovative industry has not yet received its development and acceptance by all countries of the world.

It seems that over time the results of the use of GMO technologies in agriculture in those countries where they allowed, will allow us to conduct convincing studies that can make an unambiguous conclusion about the danger or harmlessness of GMO products. At moment, the task of legal science is to offer an intermediate compromise, with the guarantee of the rights and legitimate interests of citizens (the development of labeling, expertise, judicial guarantees, etc.).

From the point of view of international law, there are two main approaches to the permissibility of the use of GMOs in agriculture:

1) the principle of equivalence of products obtained using GMO technologies and "natural" products. This principle assumes the safety of GMO-containing products, which confirmed by the regulatory requirements for the quality and safety of food products.

This approach used in Canada, the United States, and some other countries. Accordingly, the main object of control is not the manufacturing process, but the product itself, which is subject to appropriate labeling.

2) the precautionary principle of handling GMO products. This principle is a significant restriction (or complete ban) on the use and import of GM organisms in a particular country. It based on the thesis that at moment the world science is not sufficiently aware of the results of the impact of GMOs on the environment and human health. In particular, this principle implemented in the Russian Federation.

The latter principle closely related to one of the basic principles of environmental law, provided for by both international and national law. For example in Russia and other EAEU countries, the principle of prohibiting economic and other activities whose consequences are unpredictable for the environment is established (Article 3 of the Federal Law "On Environmental Protection" of January 7, 2002). The implementation of this principle is one of the additional guarantees of protecting the right of everyone to a favorable environment. In order for this guarantee to be able to work in full force, we consider it necessary to amend the civil legislation, recognizing GMO technologies as a type of sources of increased danger with the resulting consequences.

It seems that even in countries where there is a complete ban on the use of GMO technologies in agriculture, it is necessary to develop research activities in this area. Since the appropriate scientific support for genetic engineering will allow in the future to more accurately determine the nature and extent of the impact of GMO technologies and products on the environment and human health. At the same time, the results of such studies should be open for general access – this will ensure the implementation of the constitutional principles of transparency, public participation in solving problems of environmental protection, and the formation and improvement of the level of environmental and legal culture of citizens.

It is necessary to give a positive assessment of the experience of the European Union, which provides for strict requirements for the labeling of GMO products. This expressed in the fact that the notification of the placement of GMOs on the market must contain a proposal for labeling. The imperative of this rule is extremely positive. Since the norm informs the buyer about the content of the product and is regarded as an effective trend in the legislation on the labeling of GMO products. Along with the legal regulation of the production of GMO products, EU legal acts pay much attention to such a property of GMOs as "traceability" (traceability). This means the ability to detect GMOs and products made from GMOs at all stages of their placement on the market through, the use of production and sales chains [12].

Among other measures for the legal regulation of relations in the field of the use of GMOs, it is necessary to highlight the existing bans in a number of European countries on the cultivation of genetically modified varieties of agricultural plants. Bans on lands that intended for the production of organic products and raw materials, as well as those that have granted the status of a special zone for the production of raw materials used for the production of baby food and dietary food. Therefore, it seems appropriate to adopt special rules for the coexistence of genetically modified varieties of agricultural plants with traditional and organic agricultural crops. In addition, the issue of a single list of GMOs that do not pose a threat to human health or the environment also needs to discuss. And also on measures of state support for genetic engineering research, scientific and practical developments in the field of biological and genetic safety in the creation and practical use of GMOs for economic purposes.

It is also necessary to fix in national legislation the obligation to create buffer zones for GMO producers when growing these agricultural products, separating the crops of GMO crops from traditional varieties of agricultural plants. At the same time, such zones established on land plots. At the expense of their area of the commodity producer, that uses GMOs in the cultivation of agricultural products of plant origin [5].

4 Conclusion

Currently, GMO crops cover more than 200 million hectares. This is 1.5% of the US territory or more than 10% of the world's crop. Since 1996, when GMOs began to used commercially for the first time, their plantings have grown more than 100 times. This is the fastest growing technology. The reason this is happening is because it the greatest technological breakthrough of our time. This is a revolution comparable to the smelting of the first metals or to the Neolithic revolution, with the domestication of domestic animals and plants. With the help of GMOs, it is possible to increase the yield, which allows not only to solve the problems of world food security, but also to preserve untouched lands from human impact (including through the creation of national parks or other types and forms of specially protected natural areas there).

The use of GMO technologies reduces the use of pesticides and agrochemicals in agriculture. Undoubtedly, the introduction of such a complex technology into production inevitably creates certain risks for nature and human health, the study and discussion of which continued. However, in conditions where the harm from GMO technologies and products not convincingly proven, it is necessary to find a compromise between private and public interests. As you know, the concept of sustainable development involves the search for a balance of environmental, economic and social interests. Deviation from this balance in any direction entails negative consequences for the entire society, both in terms of problems with the realization of the right to food, as well as in terms of the state of the environment. In this situation, it seems appropriate not to absolutely ban GMOs (as in Russia), which means that the law of the relevant state as a regulator of public relations is useless. It is necessary to build a complex legal compromise, for example, with the prohibition of not all GMO products, but only the part of it that intended for children or pregnant women. Otherwise, the whole society loses.

Instead of banning GMOs, it is necessary to develop and improve the legal mechanism for the production of agricultural products using GMOs. In particular, it is necessary to regulate the issues of state control and examination of product quality, its labeling, certification, licensing, and improve the mechanism of legal responsibility.

Acknowledgments

The reported study was funded by RFBR according to the research project No. 18-29-16132.

References

1. N. V. Bogatyreva, Bulletin of the Saratov State Law Academy, **2**, 135 (2015)
2. O. I. Meniv, *Legal support for the use of GMOs in the cultivation of agricultural products of plant origin in Ukraine and the EU* (2016)
3. M. Bhardwaj, Asian Biotechnology and Development Review, **6(1)**, 39 (2003)
4. A. Sasson, *Biotechnology: achievements and hopes* (1987)
5. E. I. Karavaeva, R. V. Kravtsov, Siberian Legal Bulletin, **3**, 7 (2005)
6. E. V. Vorontsova, Izvestiya Kursk State Technical University, **3**, 113 (2010)
7. T. G. Volova, *Biotechnology* (1999)
8. S. K. Balashanmugam, BRICS Law J., **5(1)**, 27 (2018)
9. D. E. Matytsin, *Internet-investing as a remote algorithm of the retail investment financing. Modern Global Economic System: Evolutional Development Vs. Revolutionary Leap* (2021)
10. A. H. Scott, Villanova Environmental Law J., **26**, 145 (2015)
11. V. V. Tarakanov, A. O. Inshakova, A. I. Goncharov, Proceedings of the International Scientific Conference in Memory of Oleg Inshakov (1952-2018) (2020)
12. O. P. Kazachenok, A. V. Lavrentyeva, Studies in Computational Intelligence, **826**, 333 (2019)

Trichloroethylene adsorption extraction under kinetic conditions by carbon materials in the processes of water treatment of food production

Alena Gorelkina[1], *Evgeniy* Neverov[1,*], *Irina* Timoshchuk[1], *Olga* Altshuler[1] and *Elena* Ostapova[1]

[1]Kemerovo State University, Krasnaya Street, 6, 650000 Kemerovo, Russia

Abstract. The high level of environmental pollution currently represent a great danger to ecosystems and human health, since a significant part of the generated industrial wastewater is discharged into reservoirs either insufficiently treated or without treatment. In order to reduce the organic substances concentration in the discharged effluents, the use of carbon sorbents is promising. The sorption process requires a comprehensive study, including the process statics, kinetics and dynamics which allows us to establish the features and extraction mechanism. In this paper a model of the studied sorbents granules structure is determined, the mass transfer mechanism in the adaptive control systems is a trichloroethylene (trilene) aqua solution is established, the external mass transfer coefficients which is necessary for engineering calculations are determined. The research results scientific significance lies in the development of the organic substances adsorption theory on the example of trilen by carbon sorbents that differ in the surface structure and properties. The practical significance lies in obtaining kinetic data which is necessary for calculating the wastewater treatment process technological parameters.

1 Introduction

Degradation of water bodies as a result of anthropogenic load on them and the main factor is the discharge of effluents into water bodies with a significant excess of polluting components. Decrease in ecological functions is also the result of long-term pollution by sewage, which affects not only the state of the reservoir, but also human health, since water bodies are often used as sources of household drinking water supply.

Trilene is dangerous for aquatic organisms and can cause significant changes in the ecosystem. The maximum permissible concentration of trichloroethylene in the water bodies water for economic and drinking and cultural and household water use is 0.005 mg/dm3 in accordance with Sanitary Regulations and Norms 1.2.3685-21 "Hygienic

[*] Corresponding author: neverov42@mail.ru

standards and requirements for ensuring the safety and (or) harmlessness of environmental factors for humans".

The compounds negative influence related to trihalomethanes dictates the need to extract trichloroethylene from wastewater for this purpose the adsorption methods using carbon materials use is promising. The sorption equipment development, the technologically effective sorbing material selection requires a comprehensive adsorption study including the process statics, kinetics and dynamics which allows us to establish the extraction features and mechanism.

The trilene sorption under static conditions from aqueous solutions by various sorption materials is presented in paper 6.

The process of extracting trichloroethylene from the "active carbon-aqueous solution" system is determined by the chemical properties of the extracted component and is difficult to determine the adsorption mechanism. To obtain reliable information about the extraction mechanism under kinetic conditions, a conceptual approach is needed that will allow us to assess the adequacy of existing models and, when comparing theoretical and experimentally obtained data, determine the possibility of using a particular model to characterize the adsorption process under kinetic conditions. This stage is necessary in the comprehensive study of the adsorption process and, subsequently, the rational choice of purification modes based on adsorption extraction.

The study aim was to establish with the help of various models which are describing the sorption extraction kinetics, the mass transfer mechanism and the calculation of mass transfer coefficients necessary for performing engineering calculations.

2 Materials and methods

The research objects in this paper are sorbents of the following brands AG-OV-1, KAU, KsAU, Purolat-Standard, their main characteristics are presented in Table 1.

The extraction of trilene was carried out in accordance with the limited volume method. The determination of the content of trilene was carried out according to the method in accordance with GOST R31951-2012. [7]. Pseudo-first-order and pseudo-second-order models were used to model the kinetics of trichloroethylene adsorption on AU.

Table 1. Carbon sorbent specifications

Brand AU	**AG-OB-1**	**KAU**	**KsAU**	**Purolat-Standart**
Manufacturer	JSC "Sorbent", Perm	JSC "Sorbent", Perm	"Eurocarb", Great Britain	JSC "Synthesis", Rostov-on-Don
Raw material	Coal of rank KS, wood chemical and coke chemical resin	Apricot seed shells	Coconut shells	anthracitic coal
Graining (shape)	Granular cylinder	granulated, unequal	granulated, unequal	granulated, unequal
Carbonation and activation	Double-stage, gas-vapor	Double-stage, gas-vapor	Double-stage, gas-vapor	Single-stage, gas-vapor
Particle size, mm	1-5		1,5 – 3	0.1 – 3
Bulk density, g/cm³	0.52	0.417	0.49	0.68
toughness, %	70	90	97	70 – 80

Table 1. Continued

Specific pore surface area (S_{BET}), m^2/g	682	1419	1512	311
Total pore volume by water, cm^3/g	1.03	0.97	0,951	0,500
Pore volume, cm^3/g: micro-	0.22	0.31	0,622	0,070
meso-	0.24	0.66	0,110	0,000
macro-	0.57	-	0,219	0,430

The pseudo-first order equation is widely used to describe the pollutants adsorption rate from aqueous solutions [8]. In linear form this equation looks like this:

$$ln(a_e - a) = lna_e - k_1 t$$

where k_1 – pseudo-first-order velocity constant, min^{-1} .

This equation describes the film diffusion cases that controls the rate of adsorption during the first few minutes in mixing experiments [9]. The pseudo-second-order equation [10] is widely used to describe the kinetic laws of adsorption. In the integrated form this equation can be represented as follows [10]:

$$t/a = 1/(k_2 \cdot a_e{}^2 + t/a_e)$$

where k_2 – the pseudo-second order adsorption rate constant, g mg^{-1} $min^{-1/2}$.

In practice, the applicability of a particular mathematical model is possible using a graphical representation by the method of data linearization, and the overlap of experimental data and those obtained using a particular model shows compatibility. Obtaining the determination coefficients tending to unity demonstrates the applicability of the selected model. When choosing a model for describing the process of extracting contaminants under kinetic conditions, it is necessary to have information on the grain structure, for carbon sorbents there are two models of quasi-homogeneous or biporous.

"If an adsorbent grain is considered as a system formed by a disorderly interweaving of micro-and meso-pores in which it is impossible to distinguish between areas with only or predominantly meso-pores and areas containing only micro-pores then such a grain can be attributed to a quasi-homogeneous model." [11; 12]

The difference between the biporous model and the quasi-homogeneous one is in the delimitation of the space inside the grain, which determines the order of movement of the extracted components through the porous structure.

Matusovsky R. M. [11] developed a theory of intra-diffusion adsorption kinetics, to assess the possibility of its use in describing the process of trilene extraction under kinetic conditions, a theoretical kinetic equation was obtained that reflects the dependence of the indicator γ (defined as the ratio of the adsorption value at a time to the value of equilibrium adsorption (at/ap=γ)) from the indicator of the proportional time of the process T.

$$T = A\beta_n t, \quad A = (V_з/V_p + k_1);$$

where β_n - external mass transfer coefficient;

V_3 – the adsorbent mass total volume, cm3;

Vp – the solution volume in contact with the adsorbent;

k_1 – constant $k_1 = 1/k_r$ (k_r – the Henry coefficient $k_r = a_p/C_p$).

The dependence of the value T on τ is expressed by a straight line coming out of the origin.

To determine the limiting phase of the adsorption process under kinetic conditions, we compare the graphical dependencies for various functions, i.e. If the adsorption kinetics is limited by external mass transfer, the dependence deviating from the direct one reflects the smooth transition of the process from external diffusion control to internal control, when the control over the extraction of the component from the aqueous medium passes to internal mass transfer. This is explained by the fact that when approaching equilibrium, the gradient of specific adsorption decreases, and, consequently, the rate of internal mass transfer, proportional to the gradient, also decreases.

3 Results and their considerations

During the experiment the primary information about the adsorption rate, i.e. the granule with adsorbate saturation rate, is represented by kinetic curves (Fig.1) characterizing the change in the adsorption value over time.

Studies have shown that a high adsorption rate observed for the studied activated carbons (AG-OV, CAU, CsAu, Purolate-Standard) at the initial stage. speed fires in the time interval. And the curve deviates from the classical form, which is due to the nonlinearity of the achievement of adsorption in the equilibrium period.

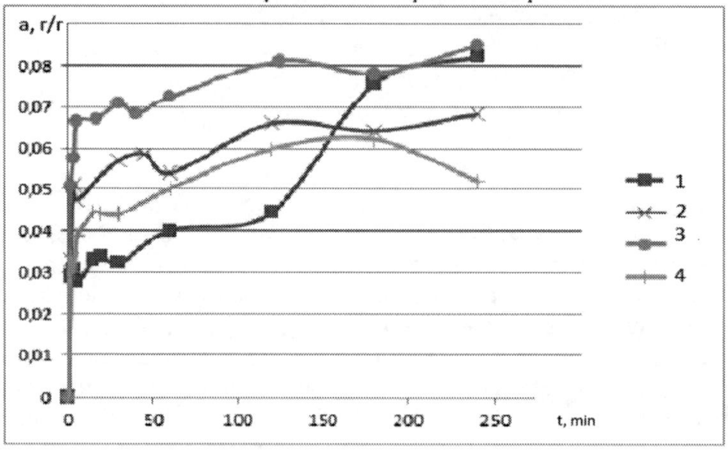

Fig. 1. Kinetic curves of trichloroethylene adsorption on active coals 1 AG-OV, 2 Purolate-Standard, 3 CAU, 4 CsAu

The experimental data satisfactorily approximate the used pseudo-first-order and pseudo-second-order kinetic models (Fig. 2).

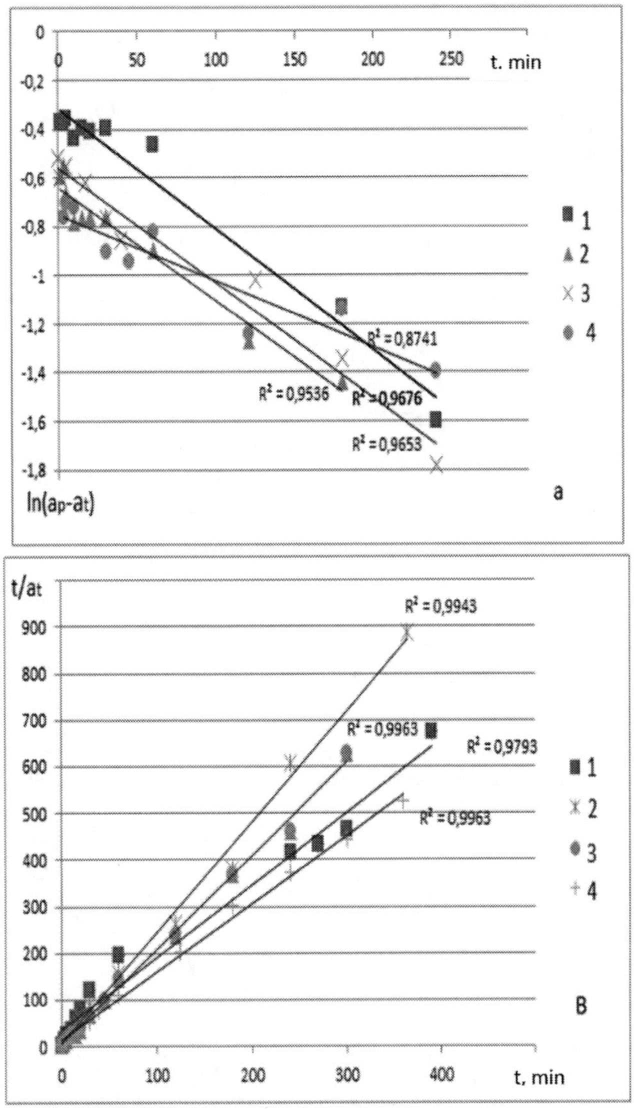

Fig. 2. The adsorption kinetics of trichloroethylene AU grades 1 AG-OV, 2 CAU, 3 Purolate-Standard, 4 CsAu description in the coordinates of pseudo-first-order (a) and pseudo-second-order (b)models

The kinetic curves are also analyzed in the coordinates γ from τ (Fig. 3) and the dimensionless coefficient T from τ (Fig.4) used in the R. M. Marutovsky method. The external mass transfer coefficients ($\beta_п$) are calculated by the tangent of the slope angle of the direct dimensionless coefficient T onτ dependence: β CsAu = 0.0308 sec^{-1}, β AG-OV-1 = 0.0282 sec^{-1}, β KAU = 0.0411 sec^{-1}, β Purolate-Standard = 0.2517 sec^{-1}.

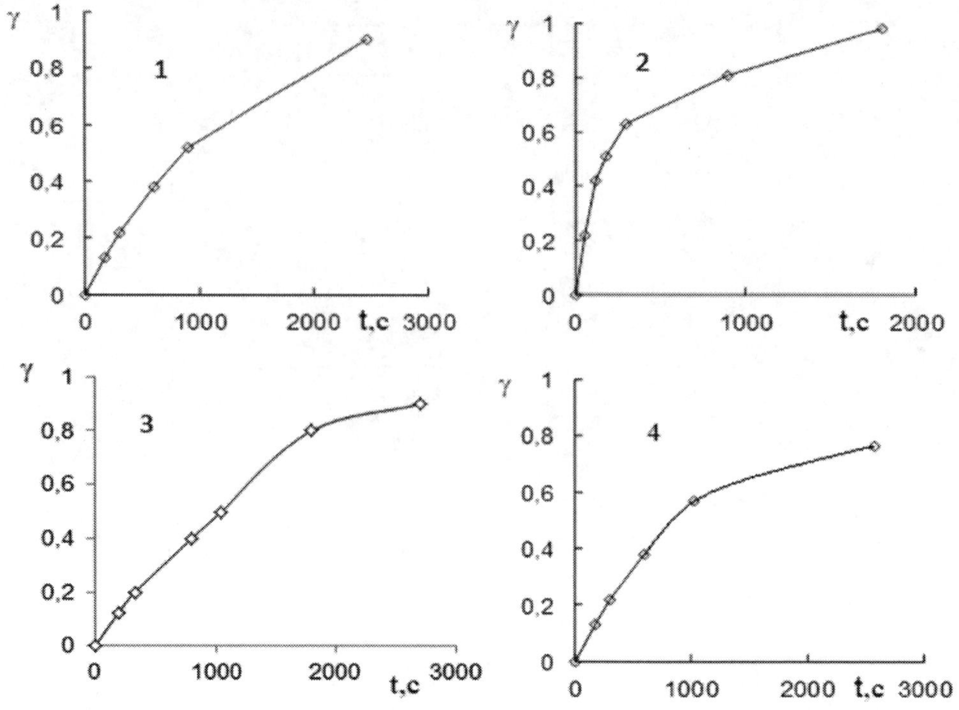

Fig. 3. Dependences of the adsorption equilibrium γ achievement degree on the adsorption time τ: 1 AG-OV, 2 CAU, 3 Purolate-Standard, 4 CsAu

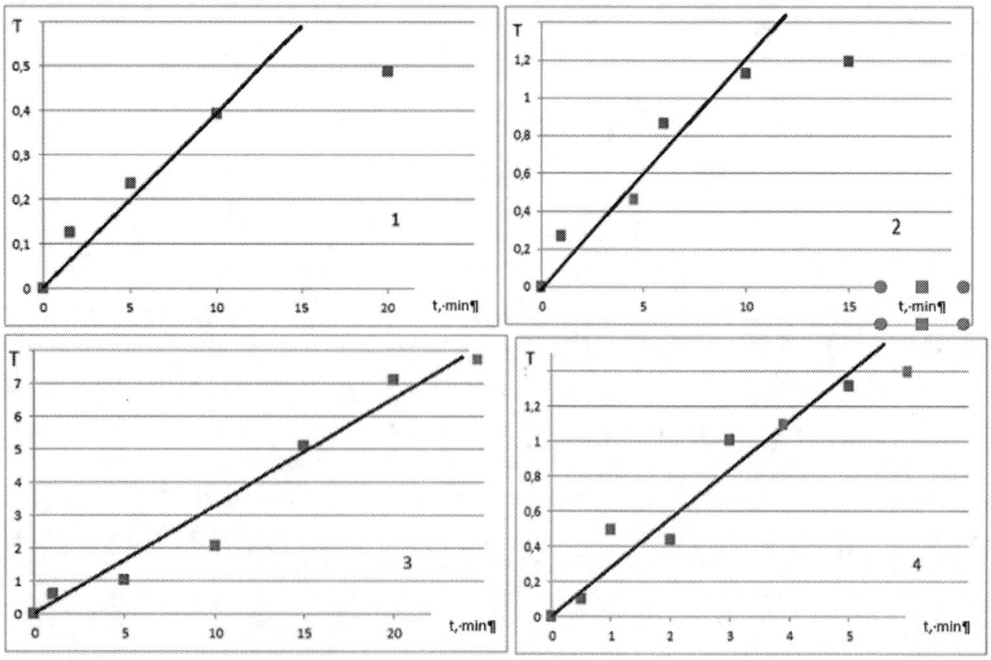

Fig. 4. Adsorption theoretical kinetic curves in the system of activated carbon are a trichloroethylene aqueous solution: 1 AG-OV, 2 CAU, 3 Purolate-Standard, 4 CsAu.

The pseudo-first-order equation satisfactorily (R^2=0.87-0.96) describes the adsorption laws at the initial stages of the adsorption process (Fig. 2a) when the phenomenon of external diffusion has a significant influence on the process. The dissolved substance diffuses under the influence of the concentration gradient that occurs in the liquid filling the adsorption space.

The pseudo-second order equation describes the experiment with significant accuracy (R2 = 0.97-0.99). The process is influenced by diffusion and is determined by the possibility of participating in the process adsorption of functional groups of the sorbent. And allows to use all the considered kinetic theories to describe the kinetics of trichlorethylene adsorption by the studied AC brands.

The kinetic dependences presented in Figure 3 allow us to find out the model of the active carbon granules structure. The initial section of the dependence curve of the achievement degree of adsorption equilibrium on time is rectilinear up to γ=0.6-0.9. High values of γ on a straight section suggest that the used sorbents granules correspond to a quasi-homogeneous model so the sorbent is considered as a system formed by a disorderly interweaving of micro-and meso-pores in which it is impossible to distinguish between areas with only or predominantly meso-pores and areas containing only micro-pores.

The direct interval on the dependence T=f(τ) (Fig.4), based on the aspects of the theory of R. M. Marutovsky, gives reason to believe that the first 5-25 minutes of the extraction of trilene occurs due to the external movement of components along the grain sorbent, for a short period of contact the component occupies the pores available in size and the adsorption process approaches equilibrium, the rate at which adsorption proceeds inside the grain indicates that the external diffusion process determines the control stage of extraction [12], the deviation from the direct dependence demonstrates the predominance of internal diffusion, and the subsequent transfer or redistribution of trilene inside the activated carbon grain has little effect on the amount of adsorption, but is simply associated with the reorientation of molecules and the possible formation of a cluster structure.

For the coal-water solution of trilene system, the extraction process under kinetic conditions was studied and as a result, the external mass transfer coefficient (βn) was obtained for various grades, it has its own value, for coal of the KsAU brand β=0.0308 sec-1, AG-OV-1 β= 0.0282 sec-1, KAU β=0.0411 sec-1, Purolate-Standard β=0.2517 sec-1. The coefficients of external mass transfer for active coals of the grades KsAU, AG-OV-1, KAU are close and have fairly high values for coal of the Purolat-Standard brand the value β is an order of magnitude higher which is probably due to the most developed system of transport pores. The obtained data allow us to expect a high component extraction rate from the treated water when filtered through a fixed sorbent layer.

4 Conclusion

The conceptual approach chosen in the study of trichloroethylene adsorption of their "active carbon-aqueous solution component" system allowed us to analyze a number of existing theories that allow us to describe the kinetics under different extraction mechanisms; the diffusion model showed the presence of an external diffusion mass transfer that determines adsorption in a short period of time, but at the same time sufficient to saturate the sorbent grain with the extracted component; the "pseudo-first" and "pseudo-second" order theories confirmed the diffusion theory, and allowed us to assume a mixed-diffusion character adsorption under kinetic conditions. The values of the calculated mass transfer coefficient (βn) are applicable for modeling and optimizing the extraction process under dynamic conditions, and the values allow us to expect a high recovery rate of the component during cleaning in stationary adsorbers.

References

1. N. A. Egorova, A. A. Bukshuk, G. N. Krasovsky, Hygiene and Sanitation, **2**, 18-23 (2013)
2. M. J. Nieuwenhuijsen, J. Grellier, R. Smith, N. Iszatt, J. Bennett, N. Best, Philosophical Transaction of The Royal Society A: Physical, Mathematical and Engineering Sciences, 367(1904), 4043-4076 (2009)
3. J. M. Wright, J. Schwartz, D. W. Dockery, Environmental Health Perspectives, **112(8)**, 920-925 (2004)
4. D. Baytak, A. Sofuoglu, F. Inal, S. C. Sofuoglu, Science of The Total Environment, **407(1)**, 286-296 (2008)
5. E. Y. Shkatova, O. G. Masaltseva, A.V. Oksuzyan, *Toxic technical liquids*, 32 (2013)
6. T. A. Krasnova, I. V. Timoshchuk, A. K. Gorelkina, Yu. S. Shulzhenko, Nauka i studia, **2**, 77-79 (2018)
7. GOST R31951-2012 DRINKING WATER, Determination of the content of volatile organohalogen compounds by gas-liquid chromatography
8. L. K. Neuchudina, A. Ya. Golub, Yu. G. Yatluk, Butler's Messages, **27(14)**, 55-68 (2011)
9. S. Azizian, J. Coll. Inter. Sci., **276**, 47-62 (2014)
10. S. A. Anurova, Water: Chemistry and Ecology, **6**, 70-75 (2012)
11. D. P. Timofeev, *Kinetics of adsorption*, 252 (1962)
12. I. M. Fedotkin, A. M. Kohanovsky, I. G. Roda, R. M. Muratovski, Physical chemistry, **48(2)**, 473-475 (1994)

The use of antibiotics in food technology: the case study of products from Moscow stores

Viktoriya Erofeeva[1,2,*], *Yulia* Zakirova[2], *Sergey* Yablochnikov[1,3], *Evgeny* Prys[4], and *Irina* Prys [4]

[1]Moscow Technical University of Communications and Informatics, 111024 Moscow, Russia
[2]RUDN University, 117198 Moscow, Russia
[3]Moscow State University of Food Production, 125080 Moscow, Russia
[4]Academy of the FPS of Russia, 390000 Ryazan, Russia

Abstract. The global threat of the 21st century is the uncontrolled use of antibiotics. Many bacteria have developed resistance to antibacterial drugs, resulting in drug-resistant superbugs. The article presents the results of the case study of Russian food products from the stores in the city of Moscow and analyzes them for the presence of antibiotics. The object of the study is the Russian market of animal products. The subject of the study is detecting the presence of antibiotics in food products on the Russian market. As a result of studying of the range of food products of animal origin, it has been revealed that the amount of harmful substances (such as antibiotics) meets the MPC standards, according to the Russian GOST standards. The research results obtained in laboratory conditions are of interest for assessing the prevalence of antibiotic residues in meat and dairy products. Based on the monitoring studies, it has been found out that most of the residual antibiotics were present in poultry meat. In all three samples, the antibiotic enrofloxacin was found but the MPC of the antibiotic was not exceeded (the norm is not more than 100 μg / kg).

1 Introduction

The uncontrolled use of antibiotics has become a global threat of the 21st century. Most bacteria have developed resistance to antibacterial drugs, resulting in drug-resistant superbugs. Animal products are the most important link in this chain being one of the basic components of a healthy diet. In the modern world, the struggle for a healthy lifestyle continues in the literal sense of this word. The effect of substances contained in food on a human body is being studied by scientists in different countries of the world. Modern scientific research methods make it possible to assess the health effects of various components of food. An increasing number of people understand that the diet, the consumed food have a major impact on human health and performance. Besides, due to publications on the significant effect of microbial resistance in animals on the treatment of infectious diseases in humans and medicine in general, the use of antibiotics in animal husbandry has become a crucial issue. The greatest danger in terms of prevalence and toxicity is posed by antibiotics because of their massive use in animal husbandry, poultry farms, beekeeping and in the process of industrial fish farming. Antibiotics are substances of microbial, animal or plant origin that can inhibit the growth of microorganisms or cause their death.

In the 21st century, 70% of all manufacturers add antibiotics to healthy animals' feed to prevent, rather than cure, disease [1-4]. Antibiotics remain in meat, milk and eggs even after heat treatment and contribute to the development of allergies, gastrointestinal disorders in humans, as well as drug-resistant superbugs [5, 6]. Along with resistance, the ineffectiveness of antibiotic therapy in the treatment of people, the latest generations of antibiotics included, is also increasing. Since 2004, scientists have not discovered any new group of antibiotics. Scientists believe that resistance to the groups of fluoroquinolones, macrolides, penicillins has been transmitted to people from animal husbandry, since the use of antibiotics by humans cannot be compared with their use in the agro-food industry to stimulate growth and prevent diseases: 1000 times more antibiotics are introduced into the body of animals and fish by weight, than used by humans [7-9]. The World Health Organization (WHO) and the World Bureau of Epizootics (OIE) have compiled a list of the most essential antibiotics used in veterinary medicine and medicine. A rational approach to the use of antibiotics in humans and animals is important for maintaining the effectiveness of the drug and its use in medicine. Maintaining animal health is critical to providing the population with safe food that meets the needs of a growing global population for more available animal protein in their diets. Overall, livestock accounts for 50% of all antibiotics produced in the world. On the one hand, during the life of an animal, the owner is interested in the use of antibiotics, since they can significantly increase the gain in live weight by increasing the use of nutrients in the diet, which in turn makes it possible to reduce feed costs and shorten the feeding period. On the other hand, interest in antimicrobial drugs in the production, storage and sale processes is due to the possibility of extending the shelf life of products. It also prevents premature death of animals. 70% of all manufacturers add antibiotics to healthy animals for the purpose of preventing, not treating, disease. It has also been found out that antibiotics remain in meat, milk and eggs even after being cooked. Antimicrobial drugs in meat and milk contribute to the development of allergies, gastrointestinal disorders in humans, as well as drug-resistant superbugs. The situation is aggravated by the fact that over the past 20 years, no new classes of antibiotics have been discovered, and the few developments that are underway at the moment will not be effective in the treatment of super-resistant strains [10-11].

The main type of milk used in the dairy industry is cow's. Antibiotics are used to treat inflammatory conditions, the most common of which is inflammation of the mammary glands (mastitis). Depending on the season, ten to eighty percent of dairy cows are susceptible to mastitis. When treating this disease, antimicrobial drugs are detected in milk. According to sanitary standards, milk from treated cows cannot be sold for three to fourteen days. Due to the lack of extended incoming control of incoming products, milk containing antibiotics is not disposed of, but mixed with milk from healthy animals. The content of antimicrobial drugs in milk decreases, but the residual amount of harmful substances continues to be detected. The Russian technical regulation for milk and dairy products, adopted in 2008, introduced a ban on the presence of antimicrobial drugs. However, in 2015, the specifications changed, which led to changes in the availability of antibiotics, in components for baby food included.

The detection of antibiotics in raw milk is important for dairy processing plants. The presence of these substances in milk reduces the quality of the product.

In 2017, there was a tendency towards a decrease in the production of feed antibiotics, which can be seen in Figure 1, however, the problem is still relevant, since in some of the studied samples of chicken meat, minced meat and milk in 2016-2019, residues of antimicrobial drugs were detected.

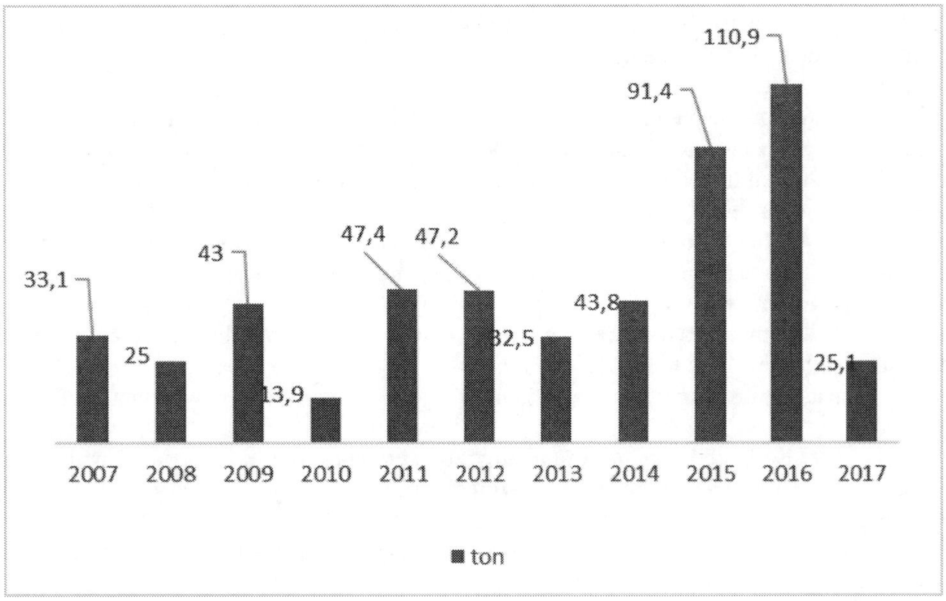

Fig. 1. Production of feed antibiotics in Russia in 2007-2017

The consequences of the irrational use of antibiotics in animal husbandry and their pollution with the remains of environmental objects (farm wastewater, food products, soil in places where manure was applied) resulted in a colossal acceleration of microbial variability, the formation of resistance and the emergence of superbugs that are resistant to most antimicrobial drugs. The problem is that since 2004, scientists have not discovered any new group of antibiotics (Figure 2).

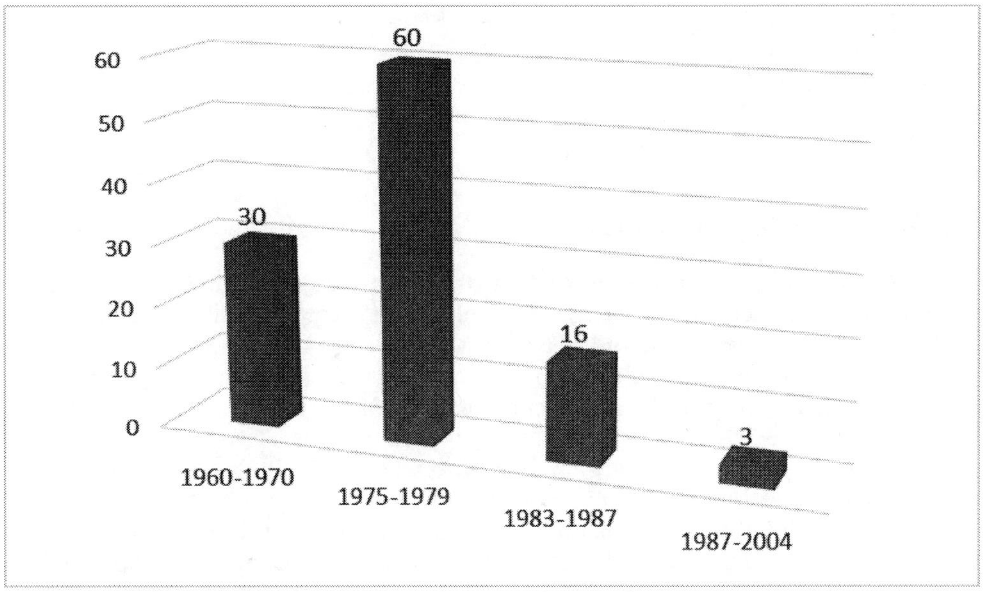

Fig. 2. Discovery of antibiotics

Along with resistance, the ineffectiveness of antibiotic therapy in the treatment of people, including the latest generations of antibiotics, is also growing.

In 2019, a national program was developed in Russia to solve the problem of the use of antibiotics in animal husbandry. The program is aimed at introducing restrictions on the use of antimicrobial drugs, restricting their sales. In the long term, abandoning of the use of antibiotics in animal husbandry is supposed. The program also provides for the creation of a register of drugs prohibited and restricted for use in veterinary medicine. Responsibility for the use of antibiotics in livestock production lies primarily with the producers. However, even if there is a ban on the use of antibiotics, in order to increase the profitability of production, farmers will continue to use antimicrobial drugs, which cannot but satisfy pharmaceutical companies that are looking for new markets for antibiotics in countries with an insufficiently developed legal culture.

To date, the following serious problems of food contamination with antibiotics have emerged.

1. The increase in and spread of non-infectious human diseases associated with changes in the normal intestinal microflora: allergies, dysbiosis, atopy, obesity, irritable bowel syndrome;
2. When using organic fertilizers from animals that have been given antimicrobial drugs, the frequency of contamination with resistant bacteria of plants as well as the spread of micro doses of antibiotics increase and the chain reaction of resistance occurs.
3. Induction of horizontal transfer of resistance genes from foreign microorganisms to biotechnological microflora in its fermentation products.
4. The presence of multi-resistant strains in ready-to-eat dairy and meat products leads to their entry into the digestive tract in a live form, which is a serious health risk factor.

Despite the fact that since 2017 there has been a tendency towards a decrease in the production of feed antibiotics, the problem is still relevant, since in most of the samples studied – chicken meat, minced meat and milk (in 2018-2019) – residues of antimicrobial drugs were detected.

All of the above determines the relevance of this work.

The aim of our study is to determine the effect of antibiotics used in animal husbandry on human health. In accordance with this goal, in this work, we studied the range of food products and identified the amount and prevalence of harmful substances (for example, antibiotics) in milk, minced meat and poultry.

2 Materials and methods

The studies were carried out in 2018-2020 in the veterinary laboratory of the city of Serpukhov near Moscow. In order to determine the presence of antibacterial drugs in milk, Delvotest SP was used in accordance with MUK 4.1.3534-18. The material for the study was the store-bought milk of cattle of various manufacturers: "Sample No. 1" 3.2% fat, purchased in the store "Auchan", "Sample No. 2" 3.2% fat from the store "Perekrestok" and "Sample No. 3" 3.2% fat, purchased in the hypermarket "Lenta".

To determine the residual amount of antibacterial drugs in minced meat, the "Antibiotest" was used to control meat and animal slaughter products for the presence of antibiotics and other antimicrobial chemotherapeutic substances in accordance with GOST R 55481-2013. The material for the study was minced meat from beef and pork of various manufacturers: homemade minced meat "Sample No. 1", bought in the store "Auchan", minced meat "Sample No. 2" from the store "Perekrestok" and chilled minced meat "Sample No. 3" purchased in the hypermarket "Lenta".

To determine the residual amount of antibacterial drugs in the carcasses of broiler chickens, the "Antibiotest" was used to control meat and animal slaughter products for the presence of antibiotics and other antimicrobial chemotherapeutic substances in accordance with GOST R 55481-2013.

The material for the study was the carcasses of broiler chickens from various manufacturers: a chicken carcass "Sample No. 1" purchased at the "Auchan" store, a chicken carcass "Sample No. 2" from the store "Perekrestok" and a chicken carcass "Sample No. 3" purchased in the hypermarket "Lenta".

3 Results

In accordance with the methodological guidelines MUK 4.1.3534-18, we analyzed the following product groups for the presence of antibiotics:

1) dairy products;
2) mixed mince (pork, beef);
3) poultry.

According to the results of the study of milk of the trade marks "Sample No. 1", "Sample No. 2" and "Sample No. 3", no antibiotics were found in the milk "Sample No. 1". The samples of milk "Sample No. 2" and "Sample No. 3" contained the antibiotic streptomycin, at a concentration of 0.01 per 1 liter of milk. This concentration of antibiotics in the studied milk samples is lower than the MPC value (Table 1).

Table 1. Detecting antibiotics in milk

Product / Store	Presence of antibiotics in the product	Exceeding of the MPC (yes/no)
milk "Sample No. 1", 3.2% / "Auchan"	no	no
milk "Sample No. 2", 3.2% / "Perekrestok"	streptomycin 0.01	no
milk "Sample No. 3", 3.2% / "Lenta"	streptomycin 0.01	no

Thus, as the results of our research show, all milk samples are fit for consumption causing no harm to human health according to the permissible MPC values (Table 2).

Table 2. Permissible levels, mg / kg / l, of antibiotics in milk and dairy products in accordance with TR CU 033/2013 "On the safety of milk and dairy products"

Antibiotic groups	MPC, mg / l
streptomycin	less than 0,2
tetracyclines	less than 0,01
penicillins	less than 0,004
chloramphenicol	less than 0,0003

According to the results of the study of homemade minced meat of various brands, "no antibiotics were found in samples "No. 1"and "No. 2". The antibiotic sulfamethazine at a dose of 17.33 μg / kg was found in the minced meat of the manufacturer "Sample No. 3", bought in the hypermarket "Lenta", the concentration of which in the sample of minced meat does not exceed the MPC value. In the Russian Federation and the countries of the Customs Union, the content of sulfonamides in food is limited by the "Unified Sanitary-Epidemiological and Hygienic Requirements for Goods Subject to Sanitary-

Epidemiological Surveillance". According to this document, the total content of residues of all sulfanilamide preparations in meat, fat, liver and kidneys of all types of slaughtered animals and poultry should not exceed 0.1 mg / kg. Our research has shown that samples of minced meat "Sample No. 1", "Sample No. 2" and "Sample No. 3" are fit for consumption inflicting no harm to human health (Table 3).

Table 3. Detecting antibiotics in minced meat

Product/Store	The presence of antibiotics in the product µg / kg	Exceeding of the MPC (yes / no)
minced meat "Sample No. 1"	not detected	no
minced meat "Sample No. 2"	not detected	no
minced meat "Sample No. 3"	sulfamethazine 17.33	no

According to the results of the study of carcasses of broiler chickens "Sample No. 1", "Sample No. 2" and "Sample No. 3" the antibiotic enrofloxacin was detected in all three samples. In "Sample No. 1" the concentration of enrofloxacin was 3.72 µg / kg, in "Sample No. 2" the concentration was 2.3 µg / kg, in the carcass "Sample No. 3" the concentration of enrofloxacin was 8.31 µg / kg which is a high value (Table 4).

Table 4. Detecting antibiotics in poultry meat

Product/Store	The presence of antibiotics in the product µg / kg	Exceeding of the MPC (yes / no)
the carcass of broiler chickens "Sample No. 1"	enrofloxacin 3,72	no
the carcass of broiler chickens "Sample No. 2"	enrofloxacin 2,3	no
the carcass of broiler chickens "Sample No. 3"	enrofloxacin 8,31	no

Thus, the concentration of the antibiotic enrofloxacin does not exceed the MPC. Thus, all three samples are usable not posing any threat to human health.

4 Conclusions

1. As a result of studying the range of food products of animal origin, it has been revealed that the amount of harmful substances (for example, antibiotics) meets the MPC standards, according to GOST R 55481-2013 GOST R 54904-2012, GOST R 52480-2005, GOST R ISO 17604, GOST R 51447, GOST 19496-2013, GOST 9225-81.

2. The study of Russian products from Moscow stores (such as meat, milk and poultry) for the possible presence of unacceptable components, has shown the following.

All milk samples of the trade marks "Sample No. 1", "Sample No. 2" and "Sample No. 3" meet the standard not exceeding the MPC value and are fit for consumption posing no threat to human health.

Samples of minced meat "No. 1", "No. 2" and "No. 3" are fit for consumption causing no harm to human health.

The samples of broiler chicken carcasses from various manufacturers "Sample No. 1", "Sample No. 2" and "Sample No. 3" are suitable for consumption inflicting no harm to human health. The concentration of detected antibiotics is below the MPC value.

3. In all three samples, the antibiotic of the fluoroquinolone series, enrofloxacin, was found, the maximum permissible concentration of which was not exceeded.

Streptomycin was detected in two out of three test samples of milk, MPCs were not exceeded. In the studied samples of minced meat, the residual amount of the antibacterial substance was detected only in one sample "No. 3", however, the MPC of the substance sulfamethazine was not exceeded, which is not dangerous for humans.

References

1. N. G. Phillips, R. Ackley, E. R. Crosson et al., Environmental Pollution, **173**, 1-4 (2013)
2. J. P. Kaye, I. C. Burke, A. R. Mosier, Ecological applications, **14(4)**, 975-981 (2004)
3. Y. Mayorova, V. Glebov, V. Erofeeva, S. Yablochnikov, B. Laver, Physiological Evaluation of the Cardiology System of Nonresident Students at Different Training Periods, E3S Web of Conferences **169**, 5 (2020)
4. O. Maslennikova, V. Erofeeva, Biological contamination of soils in urbanized ecosystems by Toxocara sp. Eggs, E3S Web of Conferences 169, 6 (2020)
5. V. V. Erofeeva, O. V. Maslennikova, Hygiene and sanitation, **8**, 897-902 (2019)
6. V. V. Erofeeva, V. V. Vasenev, Influence of Environmental Factors on the Development and Survival of *Toxocara Sp*. Eggs in Various Soil Substrates, Springer Geography, 52–57 (2020)
7. D. V. Ivanov, Izvestiya vysshikh uchebnykh zavedeniy, Geology and exploration, **4**, 63-67 (2009)
8. G. A. Alexandrov, M. A. Sokolov, A. L. Stepanov, Soil science, **10**, 1192-1194 (1996)
9. V. M. Semenov, I. K. Kravchenko, T. V. Kuznetsova, et. al., Microbiologiya. **73**, 423-429 (2004)
10. O. Yu. Goncharova, O. V. Semenyuk, G. V. Matyshak, A. A. Bobryk, Vestnik Mosk. un-ta, Soil Science, **17(2)**, 3-10 (2016)
11. Interaction of soil and atmospheric air, MSU Publishing House, 107 (1985)

AUTHOR INDEX

AUTHOR INDEX

AUTHOR INDEX